GLUCAGON: Its Role in Physiology and Clinical Medicine

GLUCAGON: Its Role in Physiology and Clinical Medicine

Edited by
Piero P. Foà • Jasbir S. Bajaj • Naomi L. Foà

Editorial Assistant
Lillie Bivens

Springer-Verlag New York Heidelberg Berlin

Piero P. Foà
Department of Physiology
Wayne State University School of Medicine
Detroit, Michigan 48201/USA

Naomi L. Foà
Department of Research
Sinai Hospital of Detroit
Detroit, Michigan 48235/USA

Jasbir S. Bajaj
Department of Medicine
All-India Institute of Medical Sciences
New Delhi 110016/India

Library of Congress Cataloging in Publication Data

Main entry under title:

Glucagon • its role in physiology and clinical medicine.

Includes index.
1. Glucagon—Congresses. I. Foà, Piero P. II. Bajaj, Jasbir S.
III. Foà, Naomi L.
QP572.G5G59 616.4′62′07 77-13657

Softcover reprint of the hardcover 1st edition 1977

9 8 7 6 5 4 3 2 1

ISBN-13: 978-1-4612-6368-5 e-ISBN-13: 978-1-4612-6366-1
DOI: 10.1007/978-1-4612-6366-1

PREFACE

The growing interest in glucagon, almost universal among diabetologists, made the decision to hold a satellite symposium immediately after the IX Congress of the International Diabetes Federation easy, indeed almost unavoidable. The climate, the beauty of its uniquely picturesque canals and houseboats, of its lakes and its mountains and above all, the friendliness of its people made the choice of Srinagar equally easy. Problems of transportation and housing which appeared of Himalayan proportions from thousands of miles away were resolved with deceptive ease: as if the late autumn sun of New Delhi and Srinagar had melted the snow that already covered many areas of the United States. For this, we thank the Executive Council and the Scientific Program Advisory Committee of the Congress, the Chairman and Co-Chairman of the local committee, Drs. Ali Mohammed Jan and S. N. Ahmed Shah; the Organizing Secretary, Dr. Syed Zahoor Ahmed; the joint Secretary, Dr. M. Y. Alvi; the Secretary of the Scientific Session, S. N. Dhar and the other committee members, Drs. G. Q. Allaqband, Wm. Riberio, Girja Dhar, J. A. Naqashbandi and Messrs. D. P. Zutshi, K. Amla and Ajit Singh. We are deeply grateful to His Excellency Sheikh Abdullah, Chief Minister of Jammu and Kashmir for his interest in the symposium and for the unforgettable hospitality offered in the name of his people. The suggestions, criticism and understanding of many colleagues helped us select topics and speakers for a representative rather than a comprehensive program. Yet, contributions from authors who could not attend the symposium or who could attend only the glucagon workshop held during the main congress in New Delhi, provided additional material which converted this book from a volume of proceedings to a monograph.

Our financial burdens were lightened by the National Science Foundation, Washington; the Secretariat of the Congress, New Delhi; the Sinai Hospital Guild, Detroit; the U.S.V. Pharmaceutical Corporation, Tuckahoe, New York and Wyeth Laboratories, Philadelphia.

The Editors
Detroit and New Delhi

CONTENTS

III. SECRETION OF PANCREATIC GLUCAGON

IV. MODE OF ACTION OF GLUCAGON

CONTRIBUTORS

ALFORD, F. P., Melbourne, Australia

ANDERSSON, A., Uppsala, Sweden

AOKI, T. T., Boston, Massachusetts, U.S.A.

ASHBY, J. P., Edinburgh, U.K.

ASSAN, R., Paris, France

ASSIMACOPOULOS-JEANNET, F. D., Nashville, Tennessee, U.S.A.

AYUSO-PARRILLA, M. S., Madrid, Spain

BAETENS, D., Copenhagen, Denmark

BAJAJ, J. S., New Delhi, India

BATAILLE, D., Paris, France

BAXTER-GRILLO, D. L., Ile-Ife, Nigeria

BIRNBAUMER, L., Houston, Texas, U.S.A.

BLÁZQUEZ, E., Madrid, Spain

BLOOM, S. R., London, U.K.

BRICKER, L. A., Miami, Florida, U.S.A.

BROLIN, S. E., Uppsala, Sweden

BUCHANAN, K. D., Belfast, U.K.

BUCHER, N. L. R., Boston, Massachusetts, U.S.A.

BUHLER, E., Basel, Switzerland

CAHILL, G. J., Jr., Boston, Massachusetts, U.S.A.

CAMPBELL, L., Sydney, Australia

CANTERBURY, J. M., Miami, Florida, U.S.A.

CARPENTIER, J.-L., Genève, Switzerland

CASEY, J. H., Sydney, Australia

CHERRINGTON, A. D., Nashville, Tennessee, U.S.A.

CHISHOLM, D. J., Melbourne, Australia

CONLON, J. M., Belfast, U.K.

CREPALDI, G., Padova, Italy

CROWTHER, R. L., Ann Arbor, Michigan, U.S.A.

DAY, J. L., Ipswich, U.K.

DOBBS, R. E., Dallas, Texas, U.S.A.

DOI, K., Toronto, Ontario, Canada

DUNBAR, J. C., Detroit, Michigan, U.S.A.

EATON, R. P., Albuquerque, New Mexico, U.S.A.

EXTON, J. H., Nashville, Tennessee, U.S.A.

FAJANS, S. S., Ann Arbor, Michigan, U.S.A.

FANSKA, R., San Francisco, California, U.S.A.

FEDELE, D., Padova, Italy

FELIG, P., New Haven, Connecticut, U.S.A.

FERRE, P., Paris, France

FLOYD, J. D., Jr., Ann Arbor, Michigan, U.S.A.

FOÀ, P. P., Detroit, Michigan, U.S.A.

FRANDSEN, E. K., Copenhagen, Denmark

GAROTTI, M. C., Padova, Italy

GEORGI, H., Basel, Switzerland

GERICH, J. E., San Francisco, California, U.S.A.

GESPACH, C., Paris, France

GEY, F., Basel, Switzerland

GIRARD, J. R., Paris, France

GONEN, B., Chicago, Illinois, U.S.A.

GRILLO, T. A. I., Ile-Ife, Nigeria

GRODSKY, G. M., San Francisco, California, U.S.A.

HARRISON, J., Louisville, Kentucky, U.S.A.

HELLERSTRÖM, C., Uppsala, Sweden

HOJVAT, S., Hines, Illinois, U.S.A.

HOLST, J. J., Copenhagen, Denmark

HUEN, H-J., Chicago, Illinois, U.S.A.

HUTSON, N. J., Nashville, Tennessee, U.S.A.

JACOBSEN, H., Copenhagen, Denmark

JARROUSSE, C., Paris, France

JASPAN, J. B., Chicago, Illinois, U.S.A.

KANETO, A., Tokyo, Japan

KERVRAN, A., Paris, France

KIRMAN, E., Detroit, Michigan, U.S.A.

KIRSTEINS, L., Hines, Illinois, U.S.A.

KISLA, J., Maywood, Illinois, U.S.A.

KRAEGEN, E. W., Sydney, Australia

LAWRENCE, A. M., Hines, Illinois, U.S.A.

LAZARUS, L., Sydney, Australia

LECLERCQ-MEYER, V., Bruxelles, Belgium

LEFÈBVRE, P. J., Liège, Belgium

LEFFERT, H. L., San Diego, California, U.S.A.

LEHOTAY, D. C., Miami, Florida, U.S.A.

LEVEY, G. S., Miami, Florida, U.S.A.

LICKLEY, L., Toronto, Ontario, Canada

LUNDQUIST, I., San Francisco, California, U.S.A.

LUYCKX, A. S., Liège, Belgium

MALAISSE, W. J., Bruxelles, Belgium

MALAISSE-LAGAE, F., Genève, Switzerland

MARASA, J. C., St. Louis, Missouri, U.S.A.

MARCHAND, J., Bruxelles, Belgium

MARTIN-REQUERO, A., Madrid, Spain

MATSUYAMA, T., Osaka, Japan

MELER, H., Sydney, Australia

MOODY, A. J., Copenhagen, Denmark

MORITA, S., Toronto, Ontario, Canada

MUGGEO, M., Padova, Italy

MÜLLER, W., Geneve, Switzerland

MURPHY, R. F., Belfast, U.K.

NAKABAYASHI, H., Dallas, Texas, U.S.A.

NISHIKAWA, M., Osaka, Japan

NOE, B. D., Atlanta, Georgia, U.S.A.

NONAKA, K., Osaka, Japan

NOSADINI, R., Padova, Italy

OKUNO, G., Itami, Hyogo, Japan

ORCI, L., Geneve, Switzerland

ÖSTENSON, C.-G., Uppsala, Sweden

PALMER, W. K., Davis, California, U.S.A.

PALOYAN, V., Hines, Illinois, U.S.A.

PARRILLA, R., Madrid, Spain

PEGORIER, J. P., Paris, France

PEK, S., Ann Arbor, Michigan, U.S.A.

PEREZ-DIAZ, J., Madrid, Spain

PETERSSON, B., Uppsala, Sweden

PICTET, R., San Francisco, California, U.S.A.

PODOLSKY, S., Boston, Massachusetts, U.S.A.

RASKIN, P., Dallas, Texas, U.S.A.

ROSS, G., Toronto, Ontario, Canada

ROSSELIN, G., Paris, France

RUBENSTEIN, A. H., Chicago, Illinois, U.S.A.

RUIZ, E., Miami, Florida, U.S.A.

SAMOLS, E., Louisville, Kentucky, U.S.A.

SHERWIN, R., New Haven, Connecticut, U.S.A.

SHIMA, K., Osaka, Japan

SILVERMAN, H., Detroit, Michigan, U.S.A.

SPEAKE, R. N., Macclesfield, U.K.

SRIKANT, C. B., Dallas, Texas, U.S.A.

SUNDBY, F., Copenhagen, Denmark

SWARTZ, T. L., Houston, Texas, U.S.A.

TAI, T.-Y., Ann Arbor, Michigan, U.S.A.

TAKENAKA, H., Itami, Hyogo, Japan

TANAKA, R., Osaka, Japan

TARUI, S., Osaka, Japan

TAYLOR, C. I., Ann Arbor, Michigan, U.S.A.

TELNER, A. H., Ann Arbor, Michigan, U.S.A.

TIENGO, A., Padova, Italy

TOYOSHIMA, H., Osaka, Japan

TRIMBLE, E. R., Belfast, U.K.

TUNG, A. K., Toronto, Ontario, Canada

UNGER, R. H., Dallas, Texas, U.S.A.

VALVERDE, I., Madrid, Spain

VAUCLIN, N., Paris, France

VOLPE, J. J., St. Louis, Missouri, U.S.A.

VRANIC, M., Toronto, Ontario, Canada

WALSH, D. A., Davis, California, U.S.A.

WEIR, G. C., Boston, Massachusetts, U.S.A.

YIP, C., Toronto, Ontario, Canada

YOSHIDA, T., Osaka, Japan

ZACHARIN, M., Syndey, Australia

INTRODUCTION

GLUCAGON 1977

Piero P. Foà

This introductory foreword reviews the biologic properties of "glucagon", the chemically defined pancreatic hormone and of a group of related peptides, secreted by the mucosa of the gastrointestinal tract and by the salivary glands. These peptides have been called collectively "glucagon-like immunoreactivity" (GLI) or, more simply, "immunoreactive glucagon" (IRG) because they react with antisera raised against pancreatic glucagon. Emphasis is placed on information published during the last two years and on other matters mentioned only in passing or not at all in the ensuing text, such as some pharmacologic effects of glucagon which can be exploited for diagnostic and therapeutic purposes. The number of papers cited represents a relatively small fraction of those published during this period, but a wealth of bibliographic information can be found in this book and in other reviews and monographs (38-40, 45, 65, 72, 123, 125).

Chemistry of Glucagon and of IRG (see pp. 31-176 and ref. 125)

A. *Glucagon* is a 29 amino acid peptide with a molecular weight of 3485 daltons*, secreted by the A cells of the pancreas and of the oxyntic mucosa of the stomach (64, 86). Its primary structure appears to be the same in all mammals, except the guinea pig, and to vary somewhat in birds and fishes. In dilute aqueous solutions, glucagon shows little ordered structure, but at high concentrations or in the presence of lipid micelles an interaction between the hydrophobic C-terminal portions of adjacent molecules results in helix formation. This is the form in which glucagon is stored in the α granules and which appears necessary for specific binding to the membrane receptors of the target cells. The C-terminal sequence of the molecule is believed to contain also the antigenic

*An essentially homogeneous glucagon preparation, with minimal amounts of impurities and of aggregated molecules has been offered as the first international glucagon standard (16).

determinant for the production of antibodies "specific" for pancreatic glucagon, that is, of antibodies that do not crossreact with extracts of intestinal mucosa. On the other hand, the N-terminal portion of the molecule is believed to contain the antigenic determinant of "non-specific" antibodies, that is, antibodies which react with glucagon and all other IRG fractions. Perhaps not surprisingly, the entire molecule appears to be necessary for the full expression of its biologic activity, at least in the hepatocytes.

B. *Immunoreactive glucagon or IRG,* as defined by its property to react with non-specific antisera, is a mixture of substances with a molecular weight of 2000 Δ or less (possibly degradation products of glucagon), of 7000 to 12,000 Δ (possibly biosynthetic precursors of glucagon) and of larger molecules such as the "interference factor" and the "big plasma glucagon" which weigh as much as 160,000 Δ. These may represent the ancestral gene which gave rise to all other IRG peptides, including those of the invertebrates (118).

The possibility that IRG and, in particular, the relatively abundant peptides of intermediate size may have a biologic role other than as "proglucagons" is suggested by numerous phenomena, such as their release following the ingestion of food or glucose (43, 54, 81), their suppressibility by somatostatin and insulin and their hypersecretion in patients and animals with insulin deficient diabetes, with gastrectomy or pyloroplasty, with glucagon producing tumors or with a specific genetic defect (11, 24, 80, 82, 132, 134). Under certain circumstances, IRG binds competitively with the glucagon receptors and stimulates adenyl cyclase, even if not as effectively as glucagon itself. Thus, depending upon the experimental conditions or the experimenter's point of view, IRG may act as a competitive moderator of glucagon or as an "auxiliary" hormone. Either hypothesis may explain why IRG appears to play a relatively minor role in the pathogenesis of diabetic hyperglycemia, even though it may be present in very large amounts (82, 134).

<u>The Secretion and Metabolic Clearance of Glucagon</u> (see pp. 1, 23, 31, 51, 177-286, 551-594, 699-734; refs. 4, 8, 34, 38, 42, 125).

The islets of Langerhans are a complex organ composed of different endocrinologic cell types which secrete glucagon, insulin, somatostatin, human pancreatic polypeptide and probably gastrin (91). The orderly arrangement, the presence of gap junctions between cells of the same and of different type and their contact with autonomic nerve terminals provide the structural basis for the regulatory function that some of these cells exert on each other (105, 123, 124) and for their responses to autonomic stimuli (115; Fig. 1). The evolutionary process which produced the pancreatic islets, the embryologic development of this well integrated organ and their relationship to the nutrient intake and to the metabolic requirements of the individual have been the object of a recent symposium (49).

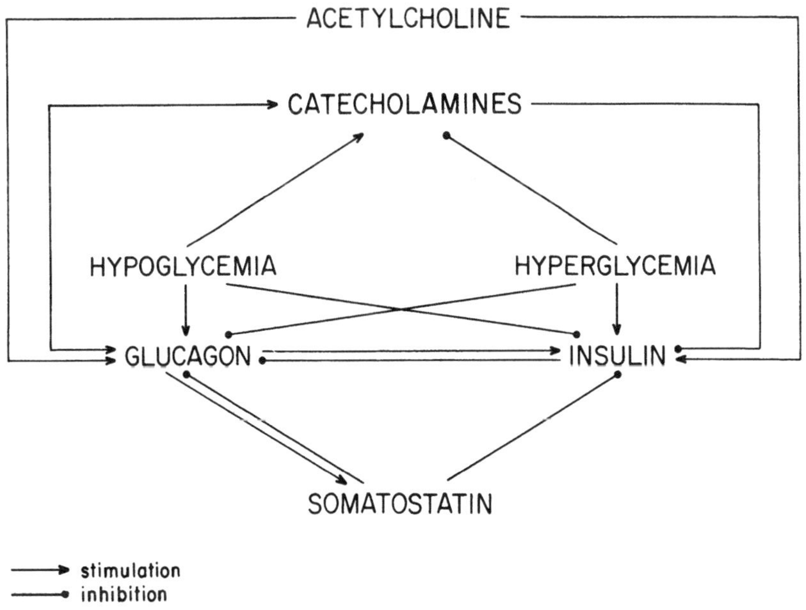

Fig. 1. The secretion of glucagon by the A cells is inextricably related to that of insulin, somatostatin and autonomic transmitters produced by other structures in the pancreatic islets. ⟶ = stimulation; —• = inhibition.

The secretion of glucagon starts with the synthesis of proglucagon, followed by its conversion to glucagon and storage in the α granules. These are composed of glucagon trimers, often have a crystalline structure and, under proper stimulation, are released by exocytosis. The A cells contain several dehydrogenases, respire actively and metabolize a variety of substances. However, the secretion of glucagon continues, indeed is accelerated when these processes are inhibited by lack of substrate or by metabolic blockers. The meaning of this seemingly paradoxical relationship between secretion and expenditure of energy is obscure. Perhaps energy is required not for the release of glucagon, but to prevent its "downhill" transport from a compartment of high concentration, the A cell, to a compartment of low concentration, the extracellular fluid.

The secretion of glucagon is regulated by nervous, endocrine, metabolic and pharmacologic factors. Thus, glucagon is released upon stimulation of the ventromedial nuclei of the hypothalamus, the vagus, the splanchnic and the pancreatic nerves or following the administration of α-adrenergic agonists and β-adrenergic blockers (115), a mechanism which may participate in the pathogenesis of stress-induced hyperglucagonemia *(vide infra)*. Several hormones contribute to the regulation of glucagon secretion. Some of them, such as insulin (38, 129), somatostatin (6, 29, 50, 110), serotonin (95) and glucagon itself (28) are inhibitory and may exert their regulatory function within the islets.

Other hormones and pharmacologic agents have a stimulatory effect. Among them are the vasoactive intestinal peptide, the gastric inhibitory peptide, growth hormone, the catecholamines, histamine, prostaglandin E_2, neurotensin and xenopsin, l-dopa, diesterase inhibitors, diazoxide, the sulfonylureas, scorpion toxin, guanidine, ergocriptine, aspirin and cyclic AMP (57, 62, 75, 78, 83, 87, 96, 98, 106, 119, 136). Undoubtedly other substances have escaped my attention, much as the discovery of a logical pattern for these different agents has escaped my search: perhaps their common final effector is a change in the concentration of Ca^{++} (63, 71, 136). On the other hand, the metabolic factors which stimulate glucagon secretion, i.e. a drop in the extracellular fluid concentration of utilizable sugars and sugar derivatives (48, 58, 109) or fatty acids and a rise in the concentration of amino acids, appear to reflect two major physiologic situations: increased demand for endogenous nutrients, as during prolonged fasting, physical exercise or stress, and impending hypoglycemia, such as that due to amino acid-induced insulin secretion, following a protein meal. Under normal conditions, glucagon tends to elevate the serum levels of glucose and fatty acids and to lower that of amino acids *(vide infra)*. Thus, it sets in motion a regulatory feedback of its own secretion.

In contrast to the inhibitory effect of plasma glucose and fatty acids on the secretion of pancreatic glucagon, the ingestion of carbohydrates and fats stimulates the release of IRG from the intestinal mucosa, suggesting that IRG may have a role in the regulation of food intake.

The circulating levels of pancreatic glucagon vary significantly in the animal kingdom, in particular between mammals, birds and lizards, possibly reflecting their different nutritional habits and metabolic requirements (99). In man, the average fasting plasma glucagon level is 50-85 pg/ml (34, 42) and, of course, represents a balance between secretion, about 0.10-0.15 mg/24 h (37) and metabolic clearance. The latter is a function of the liver, the kidneys and the skeletal muscles (27, 32, 72) and may be sufficiently decreased in hepatic and renal disease (108, 109) and in diabetes mellitus to account for the hyperglucagonemia often observed in patients with these diseases (1, 61).

The Mode of Action of Pancreatic Glucagon (see pp. 287-402; refs. 7, 14, 20, 23, 53, 67, 84, 103, 117, 121).

Pancreatic glucagon is released into the portal vein and carried first to the liver where it binds to specific receptors. This interaction leads to the activation of adenyl cyclase by separation of its regulatory from its catalytic moiety and, hence, to the production of cyclic 3',5'-adenosine monophosphate (cAMP) from ATP (Figs. 2-4). Optimal receptor binding and cyclase activation require guanosine triphosphate and bivalent cations (69) and may be inhibited by guanosine diphosphate (GDP), by adenosine (a product of cAMP degradation) or by a specific peptide produced in the liver itself (p. 373). According to the classic hypo-

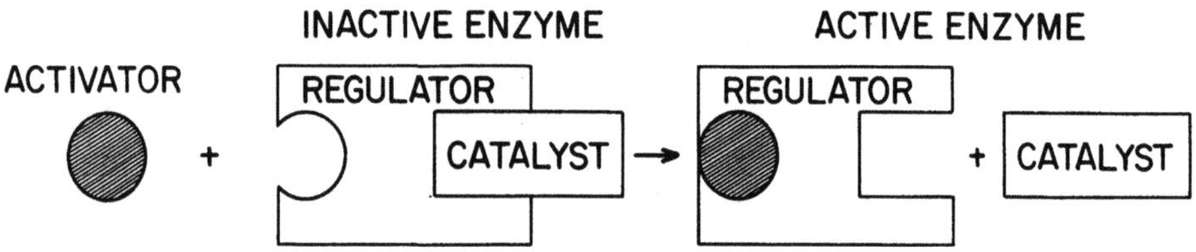

Fig. 2. Enzyme activation by separation of the catalytic moiety from its regulator.

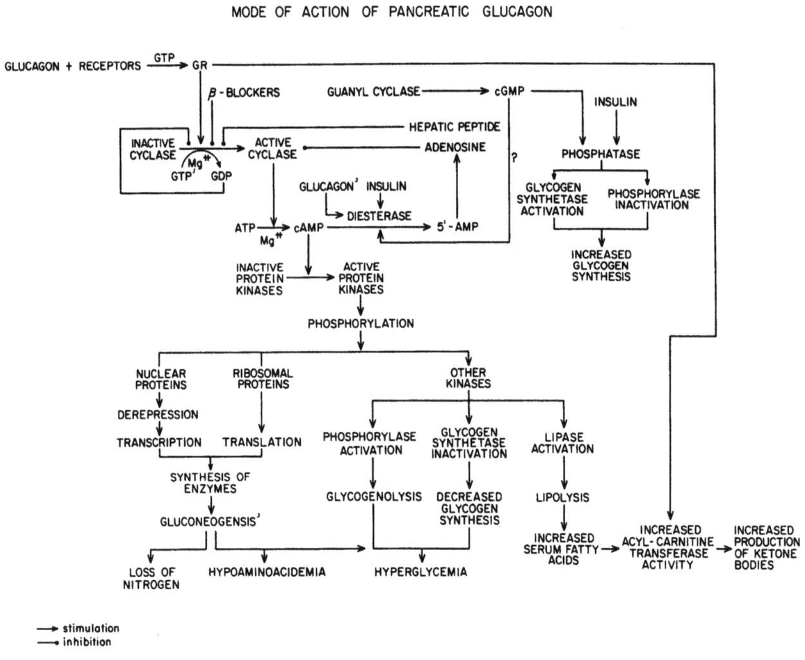

Fig. 3. [1]Guanosine triphosphate (GTP) is required for optimal cyclase activation and, at the same time, may accelerate the dissociation of glucagon from its receptors. Guanosine diphosphate (GDP) may inhibit cyclase activation. This arrangement results in a rapid "on-off" mode of action. In some tissues, the activity of guanyl cyclase and, hence, the formation of cyclic guanosine-3,5-monophosphate may be stimulated by insulin. [2]The stimulation of DNA synthesis in hepatocytes requires the synergistic action of insulin and is aided by the addition of epidermal growth factor. [3]See Fig. 4.

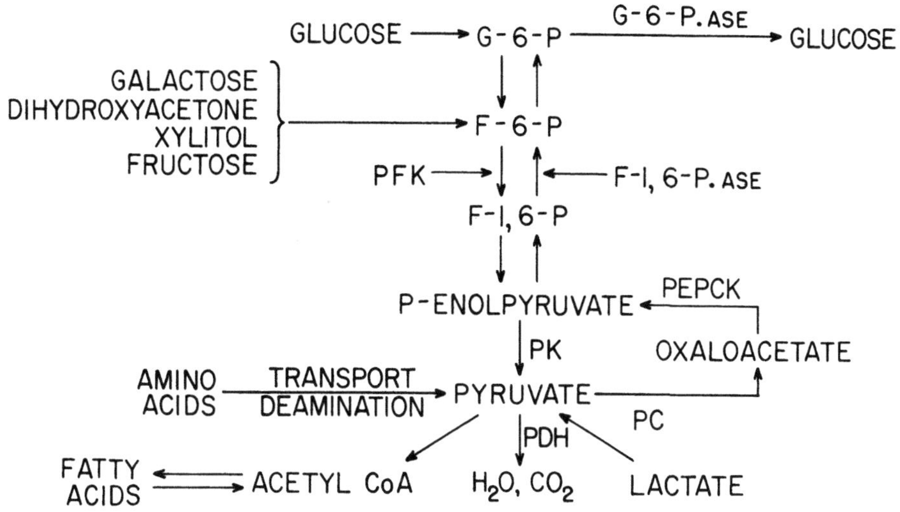

Fig. 4. Gluconeogenesis from substrates that enter the metabolic pathway at the level of pyruvate or below requires bypassing the irreversible phosphokinase (PK) step. This is the role of pyruvate carboxylase (PC) and of phosphoenolpyruvate carboxykinase (PEPCK). Gluconeogenesis is accelerated by the activation of fructose diphosphatase (F-1, 6-P.ase) and of glucose 6-phosphatase (G-6-P.ase), which also promotes glucose production from substrates that enter the cycle at the level of fructose 6-phosphate. Glycolysis is inhibited by a decrease in the activities of phosphofructokinase (PFK), pyruvate kinase (PK) and pyruvate dehydrogenase (PDH).

thesis of Sutherland, cAMP, acting as an endocellular messenger, conveys the stimulus delivered by glucagon on the surface of the cell to the effector enzymes in the cytoplasm. Here, much like glucagon had done for adenyl cyclase, cAMP combines with the regulatory moiety of protein kinases and activates them by liberating the catalytic subunits. This leads to the phosphorylation of nuclear and ribosomal proteins (103), of other kinases and still other enzymes. Thus, the rates of transcription and of translation of the genetic code increase, new enzyme molecules are synthesized, while the activity of existing enzyme molecules is modified. Among the enzymes whose synthesis is accelerated by glucagon are pyruvate carboxylase and phosphoenolpyruvate carboxykinase. This effect, combined with the glucagon-stimulated transfer of glucogenic amino acids from muscle to liver and into the hepatocyte (88), with an increase of transaminase and ureogenetic enzymes activity (137) and with an inhibition of hepatic protein synthesis (25), increases the rate of gluconeogenesis from amino acids and other precursors (56, 92, 93). The abundance of glucose is further increased by the glucagon-induced stimulation of fructose diphosphatase and glucose-6-phosphatase, by the inhibition of phosphofructokinase and pyruvate kinase (10, 21, 126) and by a suppression of glycolysis and lipogenesis. The latter is due to a decreased activity of muscle pyruvate dehydrogenase and hepatic acetyl-CoA carboxylase. In addition, glucagon stimulates the phosphorylation of glycogen phosphorylase and synthetase, activating the former, inactivating the latter and thus causing net glycogen breakdown. Finally, glucagon stimulates adipocyte lipase and hepatic acyl-carnitine transferase and decreases the concentration of malonyl CoA (22), a repressor of fatty acid oxidation (78), increasing the availability of fatty acids and their conversion to ketones (77). Predictably, the results of this cascade of enzymatic activity are increased serum levels of glucose, fatty acids and ketone bodies, decreased levels of amino acids and increased urinary excretion of nitrogen.

Thus, glucagon has a general catabolic action brought about, in part, by the *de novo* synthesis of gluconeogenetic enzymes. Whether this limited anabolic effect is related to the synergistic effect of glucagon on insulin-stimulated liver regeneration after injury (15, 35, 133) in a matter of speculation. The observations that glucagon increases ornithine decarboxylase activity and the synthesis of spermine (75, 117) and inhibits the diurnal rise in mevalonate and lowers serum cholesterol (7), are hereby duly recorded, although they cannot be fitted easily into the overall metabolic scheme.

Many effects of glucagon in the hepatocyte are accompanied by an efflux of Ca^{++}, K^+, Na^+ and H^+ and it is possible that this redistribution of ions may be essential for the activation of protein kinase. On the other hand, a glucagon-induced release of mitochondrial Ca^{++} into the cytoplasm, accompanied by an increase in the concentration of cyclic guanosine monophosphate (cGMP) may lead to a decreased abundance of cAMP and consequently, inhibit glucose production (33). This may be one explanation for the vanishing effect of glucagon during sustained administration (13, 110). Other possible reasons for this "down regulation" are a decrease in receptor binding or in glucagon-responsive adenyl cyclase.

Several actions of glucagon are antagonized directly by insulin. Thus, while insulin may or may not alter the concentration of hepatic cAMP (23), it appears to decrease its effect on protein kinases, decreasing their activity relative to that of the corresponding phosphatases. This results in dephosphorylation of glycogen synthetase and phosphorylase and consequently in net glycogen deposition (53). In addition, insulin stimulates glycolysis and lipogenesis, inhibits lipolysis, ketogenesis and gluconeogenesis and, at least in diabetic animals, increases the production of mevalonate and the synthesis of cholesterol.

The Physiologic Role of Pancreatic Glucagon (see pp. 483-594).

Since the injection of glucagon brings about the secretion of insulin, the metabolic effects of glucagon can best be demonstrated in insulin-deprived diabetic animals and patients or when the secretion of insulin is suppressed by somatostatin. Using these models, it has been shown that hepatic glucose production, lipolysis and ketogenesis are stimulated in the presence of excessive amounts of circulating glucagon (2, 18, 45) and depressed in glucagon deficiency (68). Although, under normal conditions, the action of glucagon may be of short duration (13, 36) and limited by the secretion of insulin, its release appears adequate to mobilize hepatic glucose and to provide alternate fuels (fatty acids and ketone bodies) during carbohydrate or total starvation, impending hypoglycemia or whenever the metabolic demand exceeds exogenous supply (Fig. 5). Indeed, plasma glucagon is elevated following exercise, trauma, pain, sepsis, burns, hemorrhage, cardiogenic shock and other forms of stress (65, 123, 125).

Other Effects of Pancreatic Glucagon

Glucagon has a variety of effects other than those on liver and adipose tissue (Fig. 6). Among the targets of glucagon is the myocardium where glucagon has an inotropic and a chronotropic action associated with an increase in adenyl cyclase and protein kinase activities, Ca^{++} transfer, oxygen uptake and lactate

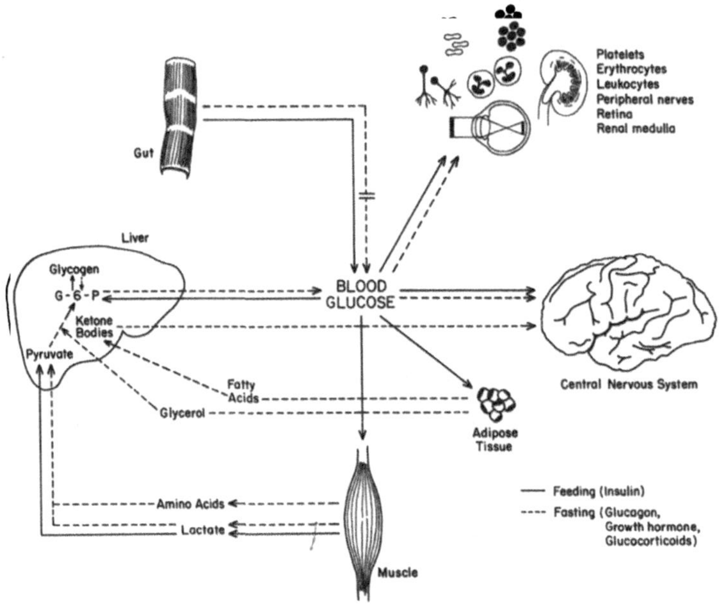

Fig. 5. Role of glucagon and insulin in the regulation of fuel economy.

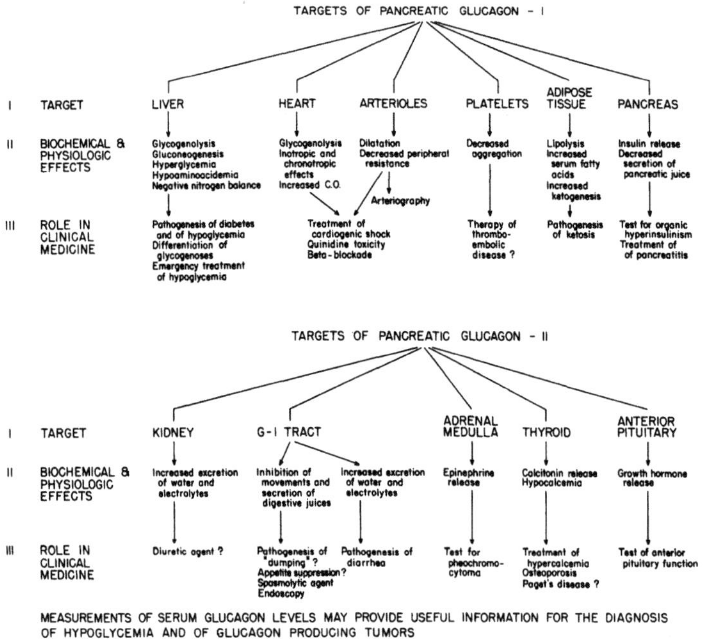

Fig. 6

production (3, 112, 135). This action may be mediated or enhanced by catecholamines released under glucagon stimulation. Glucagon causes marked vascular responses *in vivo*. For example, it prevents the constrictor effect of norepinephrine, angiotensin and vasopressin on the hepatic arterial bed, increases coronary, hepatic, renal, pulmonary, intestinal and peripheral blood flow, and decreases blood flow to the gastric mucosa (9, 70, 100, 101, 116, 122, 130). Whether or not these pharmacologic effects reflect a role of glucagon in cardiovascular regulation, the glucagon-induced changes in myocardial function and the associated decrease in coronary and peripheral resistance have been exploited for the treatment of quinidine toxicity (97), β-adrenergic blockade (128) and cardiogenic shock, even though in chronic congestive heart failure, myocardial cyclase appears to be glucagon resistant (3, 67).

The marked and generally inhibitory effects of glucagon on gastrointestinal movements and secretions also may or may not reflect a physiologic role of the hormone, but in any case, the parenteral administration of glucagon decreases the contractions of the lower esophageal sphincter, of the stomach, the intestine, the sphincter of Oddi and the intestinal villi, inhibits the secretion of hydrocholoric acid, digestive enzymes and pancreatic juice and increases the flow of bile (5, 19, 55, 59). Thus, glucagon has been recommended for the relief of visceral cramps, for the treatment of pancreatitis, as an aid to the expulsion of gallstones, as an adjuvant in the treatment of gastric bleeding, as an appetite suppressant, and as a smooth muscle relaxant in endoscopy and arteriography (47, 89, 90, 130). On the other hand, glucagon increases the intestinal loss of water and electrolytes (78) and may cause watery diarrhea, especially when accompanied by hypersecretion or administration of gastrin and vasoactive intestinal peptide.

Glucagon increases the excretion of most urinary components, producing osmotic diuresis. This effect may be mediated by a glucagon-sensitive adenyl cyclase (85) and appears to be due in part to an anti-mineralocorticoid action in the renal tubules (115), in part to mild increases of renal plasma flow and glomerular filtration rate.

Glucagon stimulates the release of other hormones, such as the catecholamines, growth hormone, insulin and calcitonin. A property that has been exploited for the exploration of endocrine function (39, 65, 102) and for the treatment of hypercalcemia, osteoporosis and Paget's disease. Finally, it has been reported that glucagon suppresses the secretion of gastric inhibitory peptide (30), decreases platelet aggregation, and stimulates the synthesis of prostaglandins E and F in the liver (94), that it has an anti-inflammatory effect (66), possibly due to a release of adrenal corticosteroids and a calorigenic effect, possibly due to catecholamine release and to enhanced fatty acid oxidation.

The Role of Glucagon in the Pathogenesis of Disease (see pp. 595-781).

Glucagon has been implicated in the pathogenesis of several clinical entities. Prominent among them is *diabetes mellitus*. Indeed, in patients and animals with genetic, autoimmune or experimentally induced diabetes, the plasma levels of glucagon are often normal or elevated in spite of the coexisting hyperglycemia, and rise paradoxically to even higher levels following the ingestion of glucose (26, 124). Conversely, the administration of glucagon can aggravate the metabolic derangement of diabetes, while glucagon-producing tumors are often associated with glucose intolerance *(vide infra)*. On the other hand, in non-diabetic animals or patients, glucagon does not cause significant or lasting metabolic impairment, suggesting that the diabetogenic potential of hyperglucagonemia can be readily suppressed by insulin (29, 37).

Two explanations have been offered for the failure of hyperglycemia to inhibit glucagon secretion in diabetes. One is that, in some way, insulin is necessary for the proper function of the A cells; the other is that, in diabetes A and B cells malfunction independently, possibly due to a genetically determined defect of the glucose receptors (46). The choice between these alternatives is of more than academic interest, for it may help determine if the restoration of A cell function and the suppression of glucagon are to be expected as probable benefits of insulin replacement (17, 129, 132) or if they should be sought independently, e.g. by means of somatostatin (36, 41, 44, 45, 124). The latter course of action has given encouraging results (44, 50), but has potential drawbacks. Among them are our almost total ignorance about the physiologic role of this ubiquitous peptide; its short biologic half-life; its possible, even if unconfirmed, harmful effects on coagulation (50, 52); the simultaneous suppression of any residual B cell activity; evidence that the somatostatin-induced "improvement" of oral glucose tolerance may not be due to glucagon suppression, but to decreased absorption of glucose (127) and, above all, the increasing awareness that the role of glucagon excess in the pathogenesis of diabetic hyperglycemia is secondary to that of insulin insufficiency (29, 36, 109, 110). These considerations do not diminish the possible importance of A cell malfunction as a sign of prediabetes and gestational diabetes (60) or of glucagon suppression in the hypoglycemic effect of the sulfonylureas (120).

Glucagon producing tumors. The major cell types of the pancreatic islets can give rise to hormone secreting tumors. Among them are tumors which produce glucagon and proglucagon in different amounts and, sometimes in variable proportions (24, 132). In most cases, these glucagonomata are associated with carbohydrate intolerance, whose severity may depend upon the relative abundance of neoplastic A and B cells (12, 51). Dermatitis, anemia and weight loss are characteristic of this syndrome, but their pathogenetic relationships with glucagon are not understood. Glucagonomata may bear some endocrinologic resemblance with a transplantable islet cell tumor of the hamster which secretes both insulin and glucagon (28, 73).

Glucagon deficiency syndromes. The pancreatic islets of certain species of teleost fish, urodele amphibians and carnivorous birds secrete insulin, but little or no glucagon. These animals have relatively low fasting blood sugar levels and following pancreatectomy, become diabetic. In contrast, lizards, ducks and chickens, whose pancreas secretes both insulin and glucagon, have a relatively high fasting blood sugar, tolerate insulin well and, when depancreatized, show signs of a double hormonal insufficiency: post-prandial hyperglycemia and glucose intolerance and profound, often fatal fasting hypoglycemia (113). These animal models and knowledge of the mode of action of glucagon, led to the suggestion that glucagon insufficiency may cause *hypoglycemia* also in man. This possibility is supported by morphologic evidence, by preliminary measurements of serum glucagon levels in a few patients with idiopathic hypoglycemia (105), by the hypoglycemia of a patient with a mesothelioma, apparently secreting an A cell inhibitor (111), and by the hypoglycemic action of antiglucagon sera (29). Glucagon insufficiency or glucagon resistance may also cause *hyperlipedemia* (31), since glucagon inhibits the synthesis of VLDL apoprotein and triglycerides and may reduce serum triglyceride levels. Finally, although glucagon does not appear to play a role in the pathogenesis of the *glycogenoses* (53), the blood sugar response to a test dose of glucagon may help identify the underlying enzymatic defect.

REFERENCES

1. Alford, F. P., Bloom, S. R. and Nabarro, J. D. N. (1976): Glucagon metabolism in man. Studies on the metabolic clearance rate and the plasma acute disappearance time of glucagon in normal and diabetic subjects. J. Clin. Endocrinol. Metab. 42, 830-838.

2. Altszuler, N., Gottlieb, B. and Hampshire, J. (1976): Interaction of somatostatin, glucagon, and insulin on hepatic glucose output in the normal dog. Diabetes 25, 116-121.

3. Antonaccio, M. J. (1977): Cardiovascular Pharmacology. Raven Press, New York, pp. ix-534.

4. Assan, R., Attali, J. R., Ballerio, G., Boillot, J. and Girard, J. R. (1977): Glucagon secretion induced by natural and artificial amino acids in the perfused rat pancreas. Diabetes 26, 300-307.

5. Balabaud, C., Peron-Marc, D., Noel, M. and Dangoumau, J. (1976): Effect of glucagon on bile production in the rat. J. Pharmacol. 7, 265-274.

6. Barden, N., Lavoie, M., Dupont, A., Côté, J. and Côté, J.-P. (1977): Stimulation of glucagon release by addition of antisomatostatin serum to islets of Langerhans in vitro. Endocrinology 101, 635-638.

7. Beytıa, E. D. and Porter, J. W. (1976): Biochemistry of polyisoprenoid biosynthesis. Ann. Rev. Biochem. 45, 113-142.

8. Bhathena, S. J., Smith, S. S., Voyles, N. R., Penhos, J. C. and Recant, L. (1977): Studies on submaxillary gland immunoreactive glucagon. Biochem. Biophys. Res. Commun. 74, 1574-1581.

9. Bisgard, G. E. and Will, J. A. (1977): Glucagon and aminophylline as pulmonary vasodilators in the calf with hypoxic pulmonary hypertension. Chest 71, 263-265.

10. Blair, J. B., Cimbalan, M. A., Foster, J. L. and Morgan, R. A. (1976): Hepatic pyruvate kinase. Regulation by glucagon, cyclic adenosine 3',5'-monophosphate, and insulin in the perfused rat liver. J. Biol. Chem. 251, 3756-3762.

11. Boden, G. and Owen, O. E. (1977): Familial hyperglucagonemia. An autosomal dominant disorder. New Engl. J. Med. 296, 534-538.

12. Boden, G., Owen, O. E., Rezvani, I., Elfenbein, B. I. and Quickel, K. E. (1977): An islet cell carcinoma containing glucagon and insulin. Chronic glucagon excess and glucose homeostasis. Diabetes 26, 128-137.

13. Bomboy, J. D. Jr., Lewis, S. B., Lacy, W. W., Sinclair-Smith, B. C. and Liljenquist, J. E. (1977): Transient stimulatory effect of sustained hyperglucagonemia on splanchnic glucose production in normal and diabetic man. Diabetes 26, 177-184.

14. Bonfils, S., Fromageot, P. and Rosselin, G., Eds. (1977): Hormonal Receptors in Digestive Tract Physiology. North-Holland Publ. Co., Amsterdam, pp. xxvii-514.

15. Bucher, N. L. R. and Swaffield, M. N. (1975): Regulation of hepatic regeneration in rats by synergistic action of insulin and glucagon. Proc. Natl. Acad. Sci. USA 72, 1157-1160.

16. Calam, D. H. and Storring, P. L. (1975): The heterogeneity and degradation of glucagon studied by polyacrylamide gel electrophoresis. J. Biol. Stand. 3, 263-265.

17. Charles, M. A., Imagawa, W., Forsham, P. H. and Grodsky, G. M. (1976): Islet transplantation into rat liver: *In vitro* secretion of insulin from the isolated perfused liver and *in vivo* glucagon suppression. Endocrinology 98, 738-742.

18. Christensen, S. E., Hansen, AA.-P., Iversen, J., Lundbaek, K., Ørskov, H. and Seyer-Hansen, K. (1974): Somatostatin as a tool in studies of basal carbohydrate and lipid metabolism in man: modifications of glucagon and insulin release. Scand. J. Clin. Lab. Invest. 34, 321-325.

19. Christiansen, J., Holst, J. J. and Kalaja, E. (1976): Inhibition of gastric acid secretion in man by exogenous and endogenous pancreatic glucagon. Gastroenterology 70, 688-692.

20. Clark, J. H., Kleen, W., Levitzki, A. and Wolff, J., Eds. (1976): Hormone and Antihormone Action at the Target Cell. Abakon Verlagsgesellschaft, Berlin, pp. 226.

21. Clark, M. G. (1976): Glucagon regulation of pyruvate kinase in relation to phosphoenol pyruvate substrate cycling in isolated rat hepatocytes. Biochem. Biophys. Res. Commun. 68, 120-126.

22. Cook, G. A., Lakshmanan, M. R. and Veech, R. L. (1977): The effect of glucagon on hepatic malonyl-coenzyme A concentration and on lipid synthesis. Feder. Proc. 36, 672.

23. Czech, M. P. (1977): Molecular basis of insulin action. Ann. Rev. Biochem. 46, 359-384.

24. Danforth, D. N. Jr., Triche, T., Doppman, J. L., Beazley, R. M., Perrino, P. V. and Recant, L. (1976): Elevated plasma proglucagon-like component with a glucagon-secreting tumor. Effect of streptozotocin. New Engl. J. Med. 295, 242-245.

25. Dich, J. and Glund, C. N. (1976): Effect of glucagon on cyclic AMP, albumin metabolism and incorporation of ^{14}C-leucine into proteins in isolated parenchymal rat liver cells. Acta Physiol. Scand. 97, 457-469.

26. Dobbs, R., Sakurai, H., Sasaki, H., Faloona, G., Valverde, I., Baetens, D., Orci, L. and Unger, R. (1975): Glucagon: role in the hyperglycemia of diabetes mellitus. Science 187, 544-547.

27. Duckworth, W. C. (1976): Insulin and glucagon degradation by the kidney. I. Subcellular distribution under different assay conditions. Biochim. Biophys. Acta 437, 518-530.

28. Dunbar, J. C., Walsh, M. F. and Foà, P. P. (1976): Secretion of immunoreactive insulin and glucagon in hamsters bearing a transplantable insuloma. Diabète et Métab. 2, 165-169.

29. Dunbar, J. C., Walsh, M. F. and Foà, P. P. (in press): The serum glucose response to glucagon suppression with somatostatin, insulin or antiglucagon serum. Diabetologia.

30. Ebert, R., Arnold, R. and Creutzfeldt, W. (1977): Lowering of fasting and food stimulated serum immunoreactive gastric inhibitory polypeptide (GIP) by glucagon. Gut 18, 121-127.

31. Elkeles, R. S. and Hambley, J. (1976): Glucagon resistance as a cause of hypertriglyceridaemia. Lancet 2, 18-21.

32. Emmanouel, D. S., Jaspan, J. B., Kuku, S. F., Rubenstein, A. H., Katz, A. I. and Huen, A. H. J. (1976): Pathogenesis and characterization of hyperglucagonemia in the uremic rat. J. Clin. Invest. 58, 1266-1272.

33. Epand, R. M. and Prosser, C. (1976): Regulation of glucagon-stimulated production of glucose in rat liver by guanosine 3',5'-cyclic phosphate. Can. J. Physiol. Pharmacol. 54, 834-837.

34. Faloona, G. R. and Unger, R. H. (1974): Glucagon. In: Methods of Hormone Radioimmunoassay. B. M. Jaffe and H. R. Behrman, Eds., Academic Press, New York, pp. 317-330.

35. Farivar, M., Wands, J. R., Isselbacher, K. J. and Bucher, N. L. R. (1976): Effect of insulin and glucagon on fulminant murine hepatitis. New Engl. J. Med. 295, 1517-1519.

36. Felig, P., Wahren, J., Sherwin, R. and Hendler, R. (1976): Insulin, glucagon, and somatostatin in normal physiology and diabetes mellitus. Diabetes 25, 1091-1099.

37. Felig, P., Wahren, J., Sherwin, R. and Palaiologos, G. (1977): Amino acid and protein metabolism in diabetes mellitus. Arch. Intern. Med. 137, 507-513.

38. Foà, P. P. (1972): The secretion of glucagon. In: Endocrine Pancreas. D. F. Steiner and N. Freinkel, Eds., Handbook of Physiology, Sec. 7, Vol. 1, pp. 261-277.

39. Foà, P. P. (1973): Glucagon: an incomplete and biased review with selected references. Amer. Zool. 13, 613-623.

40. Foà, P. P. (1976): The metabolic role of pancreatic glucagon (IRG_p), of enteroglucagon (IRG_e) and of other glucagon-like immunoreactive materials (GLI) in health and disease. In: Diabetes Mellitus. Diagnosis and Treatment, Vol. 4. K. E. Sussman and R. J. S. Metz, Eds., American Diabetes Association, New York, pp. 23-27.

41. Foà, P. P., Matsuyama, T., Dunbar, J. C., Hoffman, W. H. and Wider, M. D. (1976): The possible role of glucagon suppression in the therapy of diabetes mellitus. In: Diabetes Mellitus in Asia. S. Baba, Y. Goto and I. Fukui, Eds. Excerpta Med. I.C.S. 390, pp. 296-307.

42. Foà, P. P., Matsuyama, T. and Foà, N. L. (1977): Radioimmunoassay of glucagon. In: Handbook of Radioimmunoassay. G. E. Abraham, Ed., Marcel Dekker, Inc., New York, pp. 299-318.

43. Frame, C. M. (1976): The contribution of the distal gastrointestinal tract to glucagon-like immunoreactivity secretion in the rat. Proc. Soc. Exp. Biol. Med. 152, 667-670.

44. Gerich, J. (1977): Somatostatin: its possible role in carbohydrate homeostasis and the treatment of diabetes mellitus. Arch. Int. Med. 137, 656-666.

45. Gerich, J. E., Charles, M. and Grodsky, G. (1976): Regulation of pancreatic insulin and glucagon secretion. Ann. Rev. Physiol. 38, 353-388.

46. Gerich, J. E., Langlois, M., Noacco, C., Lorenzi, M., Karam, J. H. and Forsham, P. H. (1976): Comparison of the suppressive effects of elevated plasma glucose and free fatty acid levels on glucagon secretion in normal and insulin-dependent diabetic subjects. Evidence for selective alpha-cell insensitivity to glucose in diabetes mellitus. J. Clin. Invest. 58, 320-325.

47. Gerlock, A. J. and Hooser, C. W. (1976): Oviduct response to glucagon during hysterosalpingography. Radiology 119, 727-728.

48. Goto, Y., Seino, Y., Taminato, T., Matsukura, S. and Imura, H. (1977): Inhibition by mannose of arginine-induced glucagon secretion in the isolated perfused rat pancreas. Horm. Metab. Res. 9, 243-244.

49. Grillo, T. A. I., Leibson, L. and Epple, A. (1976): The Evolution of Pancreatic Islets. Pergamon Press, Oxford, pp. xii-377.

50. Hansen, A. P. and Lundbaek, K. (1976): Somatostatin: a review of its effects, especially in human beings. Diabète et Métab. 2, 203-218.

51. Hayashi, M., Floyd, J. C. Jr., Pek, S. and Fajans, S. S. (1977): Insulin, proinsulin, glucagon and gastrin in pancreatic tumors and in plasma of patients with organic hyperinsulinism. J. Clin. Endocrinol. Metab. 44, 681-694.

52. Henry, R. L., Long, M. W., Micossi, P., Pontiroli, E., Dunbar, J. C. and Foà, P. P. (1977): Effects of somatostatin on hemostasis and platelet function in the rat. Intern. Symp. on Somatostatin, Freiburg, Sept. 25-27.

53. Hers, H. G. (1976): The control of glycogen metabolism in the liver. Ann. Rev. Biochem. 45, 167-189.

54. Holst, J. J., Christiansen, J. C. and Kuhl, C. (1976): The enteroglucagon response to intrajejunal infusion of glucose, triglycerides, and sodium chloride, and its relation to jejunal inhibition of gastric acid secretion in man. Scand. J. Gastroenterol. 11, 297-304.

55. Ihász, H., Koiss, I., Németh, E. P., Folly, G. and Papp, M. (1976): Action of caerulin, glucagon or prostaglandin E_1 on the motility of intestinal villi. Pflüger's Arch. Eur. J. Physiol. 364, 301-304.

56. Jakob, A. (1976): Interactions of glucagon and fructose in the control of glycogenolysis in perfused rat liver. Mol. Cell. Endocrinol. 6, 47-58.

57. Johnson, D. G. and Ensinck, J. W. (1976): Stimulation of glucagon secretion by scorpion toxin in the perfused rat pancreas. Diabetes 25, 645-649.

58. Kajinuma, H., Kuzuya, T., Ide, T. and Tyler, J. M. (1975): Effects of glucosamine on insulin and glucagon secretion in dogs and ducks. Endocrinol. Jpn. 22, 517-523.

59. Khettery, J., Effmann, E., Grand, R. J. and Treves, S. (1976): Effect of pentagastrin, histalog, glucagon, secretin, and perchlorate on the gastric handling of ^{99m}Tc pertechnetate in mice. Radiology 120, 629-631.

60. Kuhl, C. and Holst, J. J. (1976): Plasma glucagon and the insulin:glucagon ratio in gestational diabetes. Diabetes 25, 16-23.

61. Kuku, S. F., Jaspan, J. B., Emmanouel, D. S., Zeidler, A., Katz, A. I. and Rubenstein, A. H. (1976): Heterogeneity of plasma glucagon. Circulating components in normal subjects and patients with chronic renal failure. J. Clin. Invest. 58, 742-750.

62. Leblanc, H., Lachelin, G. C. L., Abu-Fadil, S. and Yen, S. S. C. (1977): The effect of dopamine infusion on insulin and glucagon secretion in man. J. Clin. Endocrinol. Metab. 44, 196-198.

63. Leclercq-Meyer, V., Marchand, J. and Malaisse, W. J. (1976): The role of calcium in glucagon release, interactions between glucose and calcium. Diabetologia 12, 531-538.

64. Lefèbvre, P. and Luyckx, A. (1977): Factors controlling gastric-glucagon release. J. Clin. Invest. 59, 716-722.

65. Lefèbvre, P. J. and Unger, R. H. (1972): Glucagon. Molecular Physiology, Clinical and Therapeutic Implications. Pergamon Press, Oxford, pp. xiii-370.

66. Leme, J. G., Morato, M. and Souza, M. Z. A. (1975): Antiinflammatory action of glucagon in rats. Br. J. Pharmacol. 55, 65-68.

67. Levey, G. S. (1976): Hormone-Receptor Interaction. Molecular Aspects. Marcel Dekker, Inc., New York, pp. xii-474.

68. Liljenquist, J. E., Mueller, G. L., Cherrington, A. D., Keller, U., Chiasson, J. L., Perry, J. M., Lacy, W. W. and Rabinowitz, D. (1977): Evidence for an important role of glucagon in the regulation of hepatic glucose production in normal man. J. Clin. Invest. 59, 369-374.

69. Lin, M. C., Nicosia, S., Lad, P. M. and Rodbell, M. (1977): Effects of on binding of [^{3}H]-glucagon to receptors in rat hepatic plasma membranes. J. Biol. Chem. 252, 2790-2792.

70. Lindberg, B. and Darle, N. (1976): Effect of glucagon on hepatic circulation in the pig. Arch. Surg. 111, 1379-1383.

71. Lundquist, I., Fanska, R. and Grodsky, G. M. (1976): Interaction of calcium and glucose on glucagon secretion. Endocrinology 99, 1304-1312.

72. Luyckx, A. (1974): Sécrétion de l'insuline et du glucagon. Etude clinique et expérimentale. Masson et C.ie, Paris, pp. xxiv-338.

73. Luyckx, A. S. and Lefèbvre, P. J. (1976): Glucagon secretion by the transplantable islet-cell tumor of the syrian hamster. In: Hypoglycemia. D. Andreani, P. J. Lefèbvre and V. Marks, Eds., G. Thieme, Stuttgart, pp. 27-33.

74. Luyckx, A. S. and Lefèbvre, P. J. (1976): Pharmacological compounds affecting plasma glucagon levels in rats. Biochem. Pharmacol. 25, 2703-2708.

75. Lyons, R. T., Pitot, H. C. (1976): The regulation of ornithine aminotransferase synthesis by glucagon in the rat. Arch. Biochem. Biophys. 174, 262-272.

76. McGarry, J. D. and Foster, D. W. (1977): Hormonal control of ketogenesis. Biochemical considerations. Arch. Intern. Med. 137, 495-501.

77. McGarry, J. D., Mannaerts, G. P. and Foster, D. W. (1977): A possible role for malonyl-CoA in the regulation of hepatic fatty acid oxidation and ketogenesis. J. Clin. Invest. 60, 265-270.

78. MacFerran, S. N., Mailman, D. (1977): Effects of glucagon on canine intestinal sodium and water fluxes and regional blood flow. J. Physiol. (London) 266, 1-12.

79. Marco, J., Calle, C., Hedo, J. A. and Villanueva, M. L. (1976): Glucagon-releasing activity of guanidine compounds in mouse pancreatic islets. FEBS Lett. 64, 52-54.

80. Marco, J., Hedo, J. A. and Villanueva, M. L. (1977): Inhibition of intestinal glucagon-like immunoreactivity (GLI) secretion by somatostatin in man. J. Clin. Endocrinol. Metab. 44, 695-698.

81. Marco, J., Hedo, J. A., Villanueva, M. L., Calle, C., Corujedo, A. and Segovia, J. M. (1977): Effect of food ingestion on intestinal glucagon-like immunoreactivity (GLI) secretion in normal and gastrectomized subjects. Diabetologia 13, 131-135.

82. Matsuyama, T., Wider, M. D., Tanaka, R., Shima, K., Tarui, S., Nishikawa, M. and Foà, P. P. (in press): Pancreatic and gut glucagon in normal and depancreatized dogs. Inhibition by somatostatin and insulin.

83. Micossi, P., Pontiroli, A. E., Baron, S. H., Tamayo, R., Lengel, F. Bevilacqua, M., Raggi, U., Norbiato, G. and Foà, P. P. (in press): Aspirin stimulates the secretion of insulin and glucagon and increases glucose tolerance in normal and diabetic subjects.

84. Müller, P., Singh, A., Orci, L. and Jeanrenaud, B. (1976): Secretory processes, carbohydrate and lipid metabolism in isolated mouse hepatocytes. Aspects of regulation by glucagon and insulin. Biochim. Biophys. Acta. 428, 480-494.

85. Mulvehill, J. B., Hui, Y. E., Barnes, L. D., Palumbo, P. J. and Dousa, T. P. (1976): Glucagon-sensitive adenylate cyclase in human renal medulla. J. Clin. Endocrinol. Metab. 42, 380-384.

86. Muñoz-Barragan, L., Rufener, C., Srikant, C. B., Dobbs, R. E., Shannon, W..A. Jr., Baetens, D. and Unger, R. H. (1977): Immunocytochemical evidence for glucagon-containing cells in the human stomach. Horm. Metab. Res..9, 37-39.

87. Nagai, K. and Frohman, L. A. (1976): Hyperglycemia and hyperglucagonemia following neurotensin administration. Life Sci. 19, 273-279.

88. Pariza, M. W., Butcher, F. R., Kletzien, R. F., Becker, J. E. and Potter, V. R. (1976): Induction and decay of glucagon-induced amino acid transport in primary cultures of adult rat liver cells: paradoxical effects of cycloheximide and puromycin. Proc. Natl. Acad. Sci. USA 73, 4511-4515.

89. Paul, F. (1977): Intravenöse Langzeitinfusion von Glukagon zur Abtreibung von Gallenwegskonkrementen. Fortschr. Endoskopie 8, 161-163.

90. Paul, F. und Freyschmidt, J. (1976): Anwendung von Glukagon bei endoskopischen und röntgenologischen Untersuchungen des Gastrointestinaltrakts. Fortschr. Röntgenstr. 125, 31-37.

91. Pelletier, G. (1977): Identification of four cell types in the human endocrine pancreas by immunoelectron microscopy. Diabetes 26, 749-756.

92. Pilkis, S. J., Riou, J. P. and Claus, T. H. (1976): Hormonal control of [^{14}C]-glucose synthesis from [U-^{14}C]-dihydroxyacetone and glycerol in isolated rat hepatocytes. J. Biol. Chem. 251, 7841-7852.

93. Pi-Sunyer, F. X., Conway, J.M., Lavau, M., Campbell, R. G. and Eisenstein, A. B. (1976): Glucagon, insulin, and gluconeogenesis in fasted odd carbon fatty acid-enriched rats. Am. J. Physiol. 231, 366-369.

94. Polonovski, J., Bard, D. and Bereziat, G. (1974): Modifications de la synthèse des prostaglandines dans le foie de rat sous l'influence de l'insuline et du glucagon. C. R. Soc. Biol. (Paris) 168, 1208-1210.

95. Pontiroli, A. E., Micossi, P. and Foà, P. P. (in press): Effects of serotonin, of its biosynthetic precursors and of the antiserotonin agent metergoline on the release of glucagon and insulin from rat pancreas. Horm. Metab. Res.

96. Pontiroli, A. E., Micossi, P. and Foà, P. P. (in press): Effects of histamine, histidine and antihistaminic agents on the release of glucagon and insulin from the rat pancreas.

97. Prasad, K. (1977): Use of glucagon in the treatment of quinidine toxicity in the heart. Cardiovasc. Res. 11, 55-63.

98. Raptis, S., Escobar-Jimenez, F., Maier, V., Müller, R. and Rosenthal, J. (1977): Somatostatin and pentoxifylline modulation of insulin, glucagon and growth hormone following stimulation by arginine. Acta Endocrinol. Suppl. 208, 51-52.

99. Rhoten, W. B. (1976): Glucagon levels in pancreatic extracts and plasma of the lizard. Am. J. Anat. 147, 131-137.

100. Richardson, P. D. I. and Withrington, P. G. (1976): The inhibition by glucagon of the vasoconstrictor actions of noradrenaline, angiotensin and vasopressin on the hepatic arterial vascular bed of the dog. Br. J. Pharmacol. 57, 93-102.

101. Richardson, P. D. I. and Withrington, P. G. (1976): The vasodilator actions isoprenaline, histamine, prostaglandin E_2, glucagon and secretin on the hepatic arterial vascular bed of the dog. Br. J. Pharmacol. 57, 581-588.

102. Rochiccioli, P., Enjeaume, P., Dutau, G., Ribot, C. and Auger, D. (1976): Stimulation par le test propranolol-glucagon de la sécretion somatotrope chez 71 enfants. Arch. Franç. Péd. 33, 453-465.

103. Rubin, C. S., Rosen, O. M. (1975): Protein phosphorylation. Ann. Rev. Biochem. 44, 725-774.

104. Saccà, L., Perez, G., Carteni, G. , Trimarco, B. and Rengo, F. (1977): Role of glucagon in the glucoregulatory response to insulin-induced hypoglycemia in the rat. Horm. Metab. Res. 9, 209-212.

105. Samols, E. and Harrison, J. (1976): Intraislet negative insulin-glucagon feedback. Metab. Clin. Exp. 25, 1443-1447.

106. Samols, E., Weir, G. C., Patel, Y. C., Arimura, A., Loo, S. W. and Gabbay, K. H. (1977): Autonomic and pharmacologic control of canine A, B, D and PP cell secretion. Proc. 27th Congr. Intern. Union Physiol. Sci. 12, 517.

107. Seino, Y., Taminato, T., Inoue, Y., Goto, Y., Ikeda, M. and Imura, H. (1976): Xylitol: stimulation of insulin and inhibition of glucagon responses to arginine in man. J. Clin. Endocrinol. Metab. 42, 736-743.

108. Sherwin, R. S., Bastl, C., Finkelstein, F. O., Fisher, M., Black, H., Hendler, R. and Felig, P. (1976): Influence of uremia and hemodialysis on the turnover and metabolic effects of glucagon. J. Clin. Invest. 57, 722-731.

109. Sherwin, R. S., Fisher, M., Hendler, R. and Felig, P. (1976): Hyperglucagonemia and blood glucose regulation in normal, obese and diabetic subjects. New Engl. J. Med. 294, 455-461.

110. Sherwin, R. S., Hendler, R., DeFronzo, R., Wahren, J. and Felig, P. (1977): Glucose homeostasis during prolonged suppression of glucagon and insulin secretion by somatostatin. Proc. Natl. Acad. Sci. USA 74, 348-352.

111. Silbert, C. K., Rossini, A. A., Ghazvinian, S., Widrich, W. C., Marks, L. J. and Sawin, C. T. (1976): Tumor hypoglycemia: deficient splanchnic glucose output and deficient glucagon secretion. Diabetes 25, 202-206.

112. Simaan, J. and Fawaz, G. (1976): The cardiodynamic and metabolic effects of glucagon. Naunyn-Schmiedeberg Arch. Pharmacol. 294, 277-283.

113. Sitbon, G. and Miahle, P. (in press): Pancreatic hormones and plasma glucose: regulation mechanisms in the goose under physiological conditions. I. Pancreatectomy and replacement therapy. Horm. Metab. Res.

114. Smith, P. H. and Porte, D. Jr. (1976): Neuropharmacology of the pancreatic islets. Ann. Rev. Pharmacol. Toxicol. 16, 269-285.

115. Spark, R. F., Arky, R. A., Boulter, P. R., Saudek, C. D. and O'Brian, J. T. (1975): Renin, aldosterone and glucagon in the natriuresis of fasting. New Engl. J. Med. 292, 1335-1340.

116. Stuehlinger, W., Turnheim, K. and Gmeiner, R. (1974): Effects of glucagon on the pulmonary circulation in the dog. Eur. J. Pharmacol. 28, 241-243.

117. Tabor, C. W. and Tabor, H. (1976): 1,4 diaminobutane (putrescine), spermidine and spermine. Ann. Rev. Biochem. 45, 285-306.

118. Tager, H. S., Markese, J., Kramer, K. J., Speirs, R. D. and Childs, C. N. (1976): Glucagon-like and insulin-like hormones of the insect neurosecretory system. Biochem. J. 156, 515-520.

119. Taminato, T., Seino, Y., Goto, Y., Inoue, Y., Kadowaki, S., Mori, K., Nozawa, M., Yajima, H. and Imura, H. (1977): Synthetic gastric inhibitory polypeptide. Stimulatory effect on insulin and glucagon secretion in the rat. Diabetes 26, 480-484.

120. Tsalikian, E., Dunphy, T. W., Bohannon, N. V., Lorenzi, M., Gerich, J. E., Forsham, P. H., Kane, J. P. and Karam, J. H. (1977): The effect of chronic oral antidiabetic therapy on insulin and glucagon responses to a meal. Diabetes 26, 314-321.

121. Tung, A. K., Rosenzweig, S. A. and Foà, P. P. (1977): Glucagon from avian pancreatic islets: radioreceptor studies. Can. J. Biochem. 55, 915-918.

122. Ueda, J., Nakanishi, H., Miyazaki, M. and Youichi, A. (1977): Effects of glucagon on the renal hemodynamics of dogs. Eur. J. Pharmacol. 41, 209-212.

123. Unger, R. H. and Orci, L. (1976): Physiology and physiopathology of glucagon. Physiol. Rev. 56, 778-825.

124. Unger, R. H. and Orci, L. (1977): Role of glucagon in diabetes. Arch. Intern. Med. 137, 482-491.

125. Unger, R. H., Srere, P., Bromer, W. W. and Felig, P. (1976): Glucagon Symposium. Metabolism 25, (Suppl. 1), 1303-1533.

126. Vandenheede, J. R., Keppens, S. and De Wulf, H. (1975): The activation of liver phosphorylase *b* kinase by glucagon. FEBS Lett. 61, 213-217.

127. Wahren, J. and Felig, P. (1976): Influence of somatostatin on cabohydrate disposal and absorption in diabetes mellitus. Lancet 2, 1213-1216.

128. Ward, D. E. and Jones, B. (1972): Glucagon and beta-blocker toxicity. Brit. Med. J. 2, 151.

129. Warne, G. L., Alford, F. P., Chisholm, D. J. and Court, J. (1977): Glucagon and diabetes. II. Complete suppression of glucagon by insulin in human diabetes. Clin. Endocrinol. 6, 277-284.

130. Warrick, M. W. and Lin, T.-M. (1977): Action of glucagon and aspirin on ionic flux, mucosal blood flow and bleeding in the fundic pouch of dogs. Res. Commun. Chem. Pathol. Pharmacol. 16, 325-335.

131. Weber, C. J., Lerner, R. L., Felig, P., Hardy, M. A. and Reemtsma, K. B. (1975): Effect of pancreatic islet transplantation on insulin and glucagon levels in diabetic rats. Surg. Forum 26, 192-194.

132. Weir, G. C., Horton, E. S., Aoki, T. T., Slovik, D., Jaspan, J. and Rubenstein, A. H. (1977): Secretion by glucagonomas of a possible glucagon precursor. J. Clin. Invest. 59, 325-330.

133. Whittemore, A. D., Kasuya, M., Voorhees, A. B., Jr. and Price, J. B., Jr. (1975): Hepatic regeneration in the absence of portal viscera. Surgery 77, 419-426.

134. Wider, M. D. and Foà, P. P. (1977): Plasma immunoreactive glucagon fractions in normal and depancreatized dogs. The Physiologist 20.

135. Winokur, S., Nobel-Allen, N. L. and Lucchesi, B. R. (1975): Positive inotropic effect of glucagon in the chronically failed right ventricle as demonstrated in the isolated cat heart. Eur. J. Pharmacol. 32, 349-356.

136. Wollheim, C. B., Blondel, B., Renold, A. E. and Sharp, G. W. G. (1976): Stimulatory and inhibitory effects of cyclic AMP on pancreatic glucagon release from monolayer cultures and the controlling role of glucagon. Diabetologia 12, 269-277.

137. Yamazaki, R. K. and Graetz, G. S. (1977): Glucagon stimulation of citrulline formation in isolated hepatic mitochondria. Arch. Biochem. Biophys. 178, 19-25.

I. ONTOGENY OF THE A CELL AND THE MORPHOLOGY OF GLUCAGON RELEASE

ONTOGENY OF THE ENTERO-PANCREATIC A CELLS AND OF GLUCAGON PRODUCTION

T. A. I. Grillo and D. L. Baxter-Grillo

Division of Human Biology and Behaviour, Faculty of Health Sciences,
University of Ife
Ile-Ife, Nigeria, W. Africa

A review of the literature, histochemical and electron microscopic observations, radioimmunoassay and phosphorylase activation measurements indicate that the appearance of phosphorylase activity correlates with that of pancreatic glucagon secretion and is a prerequisite for the response of hepatic and cardiac glycogen to the hormone. The early appearance of glucagon in the embryonic gut suggests that this hormone may play a role in development. It is important to point out that although the gut continues to produce glucagon-like materials after birth, its ability to produce insulin is restricted to the embryonic and neonatal periods.

INTRODUCTION

The study of the ontogeny of the production and the secretion of glucagon has presented us with a number of problems. It has been necessary to answer the following questions: Where in the embryo do the cells which produce the hormone originate? Is there a common stem cell for the endocrine and exocrine cells of the pancreas and the gut? If there is a common stem cell, is it possible to stimulate that cell to develop after birth? Are the endocrine, acinar and gut cells interchangeable, i.e. can they be induced to transform into one another? What evidence is there that glucagon is effective in embryonic life? Is there an embryologic origin of A cell dysfunction?

Embryonic Origin of A cells

While there has been general agreement that the pancreatic acinar cells migrate into the pancreas from the embryonic foregut, there has been some doubt that the islet cells have a similar origin. In 1969, Pearse (35) observed that a number of endocrine cells which secrete polypeptide hormones have certain common chemical characteristics, such as the production and secretion of biogenic amines. Because these amines are fluorogenic, they can be identified histochemically by the Falck-Hillarp technique. Other chemical characteristics of the cells are their ability to take up amine precursor and to decarboxylate them. Because of these characteristics, Pearse called them amine precursors uptake and decarboxylation or APUD cells. Pearse (35) remarked that the APUD cells had other cytochemical characteristics, i.e. a high non-specific esterase or cholinesterase activity; a high side chain carboxyl or carboxyamide content; a high content of α-glycerophosphate menadione reductase and a specific immunofluorescence. In addition, Pearse observed that all the cells have common ultrastructural characteristics, i.e. membrane bound secretion vesicles; low levels of rough endoplasmic reticulum; high levels of smooth endoplasmic reticulum; electron-dense, fixation-labile mitochondria; high content of free ribosomes; prominent microtubules, including centrosomes and the tendency to produce fine protein microfibrils.

Pearse suggested that the APUD cells may share these common characteristics: they have evolved similar biochemical mechanisms for the production of similar polypeptide hormones although they were of diverse embryologic origin; or because they have evolved a similar set of biochemical mechanisms in response to a common specific secretory stimulus, e.g. cholinergic or aminergic; or because, having differentiated from a common embryologic source, the neural crest (36), they have retained a common and distinct set of ancestral functions.

Recently, Dieterlen-Lièvre and Beaupain (13) reported the result of their experiments with chick embryos to test the third of Pearse's propositions. They carefully separated the splanchnopleura from the somatopleura of the chick embryos, then grafted fragments of the splanchnopleura into the coeloma of 3 day-old quail embryos.

If Pearse's theory of a neural crest origin of pancreatic endocrine cells had been correct, the cells of the splanchnopleura graft should have differentiated into islet cells in the gut of the host quail embryos. The results of their experiments clearly showed that islet cells did develop from the graft.

They also demonstrated using the immunohistochemical methods of Mason et al (30) that both glucagon and insulin were produced by the islet cells which had differentiated. Thus, their results conclusively showed that the islet cells derive from the endoderm and do not develop from the neural crest, although they may share all the chemical characteristics of the APUD cells, some of which may indeed originate from the neural crest.

These important observations made it easier for us to concentrate on the ontogeny and migration of these hormone producing cells from their site of origin in the embryonic gut to their position in the pancreas and to reflect on the fact that some of these cells do not migrate at all.

Differentiation of islet cells in chick and human embryos

Villamil (43) reported that two types of islet cells, the dark and the light, could be recognized in the pancreas of the chick embryo on the 8th day of incubation. On that day, A granules could also be recognized while B granules appeared on the 12th day and D granules by the 14th day. Dieterlen-Lièvre (11) confirmed that A cells could be identified on the 8th day although they did not assume their definitive character until the 13th day of incubation, on which day they became full of A granules. In a more recent study, using immunohistochemical techniques, Beaupain and Dieterlen-Lièvre (5) showed that on the 3rd day of incubation the dorsal pancreatic rudiment contained glucagon. However, Tanizawa and Fujita (39) could not find A cells with granules in chick embryonic pancreas before the 6th day.

The most comprehensive study of the histogenesis of the human embryonic pancreas was made by Conklin (9) although it is important to pay tribute to the earlier contributions of Pearse (34) and of Weichselbaum and Kyrl (44). Conklin (9) stated that between the 30 and the 40 mm stage of development, i.e. at about the 8th week of gestation, the pancreas shows the growth of many epithelial buds from the interlobular and intercalated ducts. These epithelial buds are located in the center of the lobule. In the 34 mm embryo some of the buds contain a few cells in which argyrophilic granules can be seen. There is a slight increase in the number of these argyrophilic cells at the 40 mm stage, when ganglionic cells also appear in the stroma of the pancreas. At this stage of development, these ganglionic cells are innervated by preganglionic fibers but no post-ganglionic fibers are found. At the 50 to 60 mm stage, basophilic cells can be demonstrated with Mason A. Conklin concludes that these cells are A cells. These basophilic A cells increase in number rapidly between 60 and 90 mm and are found to contain tryptophan when treated with dimethylamino-benzaldehyde, while the argyrophilic cells do not contain the amino acid.

At the 65 mm stage, D cells can be seen, but B cells differentiate only between the 90 mm and the 110 mm stage of development. While some of these B cells assume a central position in the islets and are surrounded by other cells, i.e. a "mantle-islet" pattern is formed, other B cells remain within the walls of the ducts. Between the 110 and the 150 mm stage, the vascular pattern becomes better established within the islets. By the 160 mm stage, the A cells become degranulated while the argentaffin cells increase in numbers. At about 180 mm, the argentaffin cells decrease in number. Periodic degranulation of A and B cells were observed by Conklin who suggested that the degranulation may be considered evidence of hormone secretion during fetal life.

Robb (37) divided the differentiation of the islets of man into four stages: the stage of budding islets, during the 3rd gestational month when the islet cells bud off from the ducts and when A and B cells are present; the bipolar stage, when the islets grow and the B cells are polarized to the tip furthest from the base of the islet while the A cells are at the base; the "mantle-islet" stage, where the A cells migrate and surround the B cells, occurring during the 5th month of gestation; and the stage of mature islets when, during the 8th month, the islets have the irregular appearance of mature adult islets and the A cells intermingle with the B cells.

Ontogeny of the A Cells and of Glucagon Production

Lin and Potter (29) claim that two generations of islets appear during development. During the 8th week of gestation, primary islets arise from the ducts. These primary islets reach maturity during the 5th month, after which most of them degenerate. The secondary islets develop from the cells of the terminal ducts from the 3rd month of gestation and constitute the permanent islets of adult life. It is the opinion of Lin and Potter that the A and B cells develop directly from the cells of the ducts without passing through an intermediate stage and that they never are transformed from one to the other. In this regard it is important to note that, although some Russian investigators (15) have reported the existence of acinar-islet cells in the developing pancreas, the occurrence of these mixed cells is not as common as it should be if the transformation of acinar to islet cell were the normal process of islet cell differentiation. Acinar-islet cells are known to be widespread among vertebrates such as the toad (6, 28), reptiles (24), the duck (7), the rat (25) and man (45). Thus, it is important to bear in mind the possibility of a transformation of acinar to islet cells not only in the adult, but in the embryo. Dieterlen-Lièvre (12) has reported observing acinar-islet cells with A granules in the 21 day-old chick embryo and finding mixed cells containing A, B and zymogen granules.

Histochemistry of the Developing Pancreatic Islets

While there have been great efforts to study the histochemistry of the pancreatic B cells, relatively little is known about the histochemistry of the A cells. In studies of B cell ontogeny, the histochemistry of SH groups was used as an index for the onset of insulin synthesis, for Barrnett et al (2) had demonstrated the presence of SH groups in the B cells with their dihydroxy-dinaphthyl-disulphide (DDD) method. Thommes (40) obtained positive DDD reaction in B cells of chick embryos from 7 days onwards, with a marked increase at day 11 to 13, while Grillo (22) could only be certain of definite SH groups in B cells of chick embryos from the 14th day onwards. Using the pseudoisocyanin reaction of Schiebler and Schiessler (38), believed to be specific for insulin, positive results were obtained in fetal rat from the 16th day of gestation onwards.

Since tryptophan is one of two amino acids present in glucagon and absent in insulin, attempts have been made to use the histochemistry of tryptophan as an index for the functional differentiation of A cells (19). Indeed, the post coupled p-dimethylaminobenzylidene reaction for tryptophan was used by its inventors to demonstrate the amino acid in the A cells of the adult pancreas. Attempts to use this reaction for the demonstration of A cells in embryonic pancreas have not been very successful because the acinar cells also gave a positive reaction for the amino acid. Nevertheless, it has been shown that the islet cells of the chick embryo pancreas contain tryptophan (Fig. 1). However, it is important to note that somatostatin also contains tryptophan and that the somatostatin secreting cells will also give a positive reaction for this amino acid.

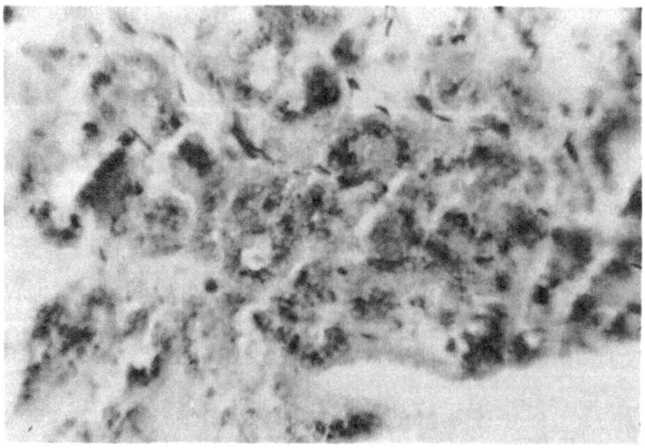

Fig. 1. A cells in the pancreatic islets of a 14-day-old chick embryo. Tryptophan stain.

Ontogeny of the A Cells and of Glucagon Production

Enzyme Histochemistry

The enzyme histochemistry of the B cells has been extensively studied and correlated with the function of B cells in embryonic and in adult vertebrates. Unfortunately, very little is known of the enzyme histochemistry of A cells, apart from the fact that the APUD series of polypeptide hormone-producing cells contain non-specific esterases or cholinesterase and α-glycerophosphate menadione reductase.

Glucagon in the Entero-Pancreatic Complex of Embryos

Since it is recognized that the islet cells originate in the embryonic gut and since it has been shown that functioning glucagon-secreting cells exist in the adult gastrointestinal tract (42), it becomes important to study the production of glucagon in the gut and pancreas of the embryo.

For the purposes of this investigation, we used Rhode Island red and brown Leghorn chick and human embryos. After incubation for various length of time the pancreata of the chick embryos were carefully dissected free of the duodenum and the two organs were collected in separate pools. The pancreata of human embryos were collected at the University College Hospital, Ibadan, Nigeria. The extraction of glucagon was carried out by the procedure of Kenny (27). The assay for glucagon was based on the activation of phosphorylase in liver homogenates (26), as described by Okuno et al. (33). This method of bioassay was selected because of doubts about the reliability of antibodies used in the radioimmunochemical methods. Glucagon was demonstrated in frozen sections of pancreata by the immunofluorescent method of Grodsky et al (17). Control sections were preincubated with cysteine, as described by Nonaka et al (32), to eliminate the possibility of insulin immunofluorescence. The antiglucagon sera used for conjugation with fluorescence were donated by Dr. Foà. No attempt was made to eliminate the possibility of crossreaction with somatostatin. Insulin was extracted from the tissues as described by Grillo (22) and estimated by radioimmunoassay.

For electron microscopy the duodena of chick embryos of various ages were fixed in 2% gluteraldehyde at pH 7.2 for 2 hours, washed in phosphate buffered saline for 12 hours and postfixed in 1% OsO_4 in distilled water for 1 1/2 hours. The tissues were embedded in epon and, after sectioning, were stained with 1% uranyl acetate.

RESULTS

Glucagon was found in the pancreata and duodena of chick embryos of all ages examined (4). In the pancreas, the level of glucagon rose to the highest level on the 16th day of incubation, fell on the 18th and 20th day and rose again after hatching. By contrast, in the duodenum, the highest level of glucagon was seen in the youngest embryos examined, i.e. on the 12th day. The level of glucagon was higher in the duodenum than in the pancreas at almost all ages and in particular at hatching (Table 1). In contrast, the level of insulin was higher in the pancreas than in the duodenum at all embryonic ages and at hatching (Table 2).

TABLE 1

GLUCAGON CONTENT OF DUODENUM AND PANCREAS OF THE CHICK EMBRYO
mgP/100 mg Fresh Tissue

Age in days	Organ	
	Pancreas	Duodenum
12	0.0330 ± 0.0052	0.43 ± 0.05
14	0.0353 ± 0.0035	0.22 ± 0.03
16	0.0710 ± 0.0071	0.10 ± 0.02
18	0.0213 ± 0.0021	0.05 ± 0.01
20	0.0328 ± 0.0041	0.08 ± 0.005
Newly hatched	0.0676 ± 0.0153	0.08 ± 0.009

TABLE 2

INSULIN CONTENT OF DUODENUM AND PANCREAS OF THE CHICK EMBRYO
Bovine insulin equivalents. μg/g Wet Tissue

Age in days	Organ	
	Pancreas	Duodenum
12	2.33 ± 0.10	1.97 ± 0.03
14	2.13 ± 0.08	2.11 ± 0.05
16	4.00 ± 0.10	0.79 ± 0.02
18	3.02 ± 0.08	1.76 ± 0.03
20	2.97 ± 0.08	0.31 ± 0.01
Newly hatched	1.27 ± 0.03	0.29 ± 0.01

The assay of extracts of human foetal pancreata showed that the level of glucagon was lowest in the youngest pancreas studied, i.e. at a fetal length of 158 mm, highest at 190 mm and was still high at 212 mm, the longest fetus studied (Table 3).

TABLE 3

GLUCAGON CONTENT OF THE HUMAN FETAL PANCREAS
mgP/100 mg Fresh Tissue

Fetal CR mm	
158	0.0018 ± 0.0005
183	0.0044 ± 0.0004
190	0.0084 ± 0.0010
212	0.0072 ± 0.0013

Immunohistochemical Localization of Glucagon

Tentative confirmation that the biologic activity forming the basis of the assay was indeed indicative of the presence of glucagon, was obtained by the immunohistochemical localization of the hormone in the epithelial cells of the duodenum of chick embryos from 12 days of incubation onwards (3). Similarly, glucagon was localized by immunohistochemical methods in the pancreas of the 174 and 190 mm human fetuses (Figs. 2-4). In both types of embryos, insulin immunofluorescence was eliminated by prior treatment of the sections with cysteine. However, the possibility of somatostatin fluorescence was not ruled out.

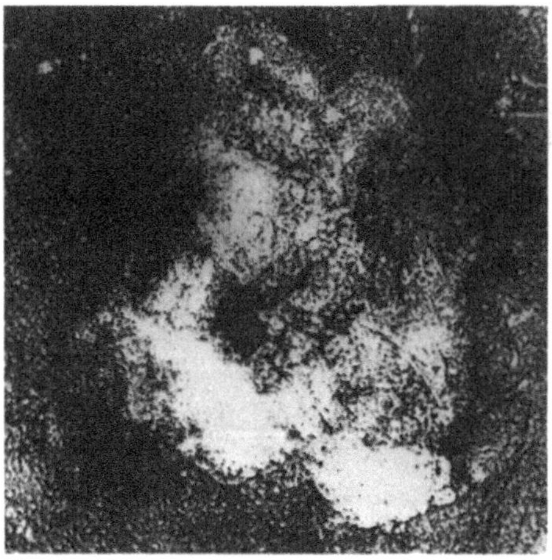

Fig. 2. Glucagon immunofluorescence in the islet of a 174 mm human fetal pancreas.

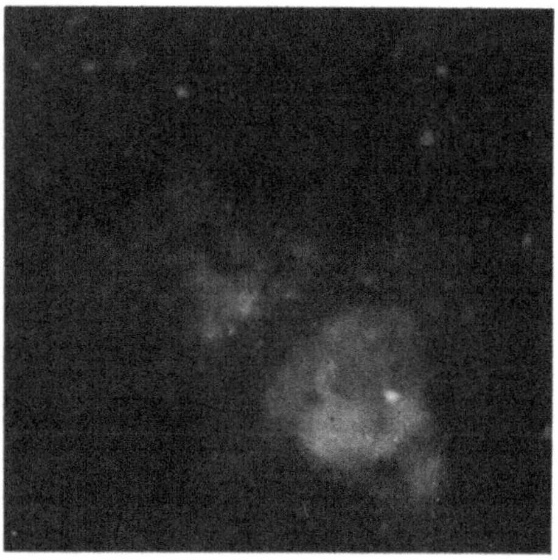

Fig. 3. Glucagon immunofluorescence in the islet of a 190 mm human fetal pancreas.

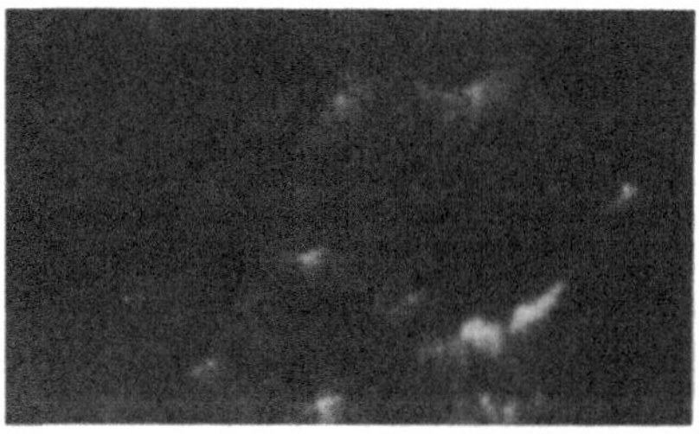

Fig. 4. Glucagon immunofluorescence in the epithelial cells of the duodenum of a 14-day-old chick embryo.

Ontogeny of the A Cells and of Glucagon Production

Differentiated endocrine pancreatic cells have been observed as early as the 6th day in the chick embryo (39). Although these cells were considered immature, they were presumed to be A cells. Beaupain and Dieterlen-Lièvre (5) have obtained conclusive evidence that A cells are present in the dorsal pancreatic rudiment as early as the third day of development. This early differentiation of A cells in the chick embryo is in contrast to the late differentiation of the same type of cells in the mouse, in which Munger (31) observed that while B cells have differentiated by the 13th day of gestation, the A cells are not present until birth. In the rat, Angervall et al. (1) have demonstrated that only A_1 cells (i.e. D cells) differentiate before birth. In spite of this lack of a histochemical demonstration of A_2 cells (i.e. A cells) in the embryonic rat pancreas, Okuno et al. (33) have been able to demonstrate the presence of a glucagon-like substance from the 14th day of gestation onwards. Both types of A cells have been observed in the developing pancreas of the guinea pig before birth. However, A_2 cells were observed to differentiate earlier (as early as the 26th day) than A_1 cells. It may be suggested that the apparent conflict in the results of electron microscopic studies of mammalian embryonic pancreas may be due to the fact that the chemo-differentiation of the A cells in mammals precedes the histodifferentiation, i.e. the production of glucagon precedes the development of the A granules.

In any case, it may be concluded that, in most mammalian embryos as well as in the chick embryo, glucagon secreting cells have differentiated in the pancreas by the end of gestation or incubation.

The electronmicroscopic examination of the developing gastrointestinal tract of the chick embryo has revealed the presence of cells containing granules typical of the pancreatic A cells (Fig. 5) and of cells containing granules similar to those of the B cells (4). It is therefore not surprising that both insulin-like and glucagon-like substances have been observed in the gut of the chick embryo (4; Fig. 6).

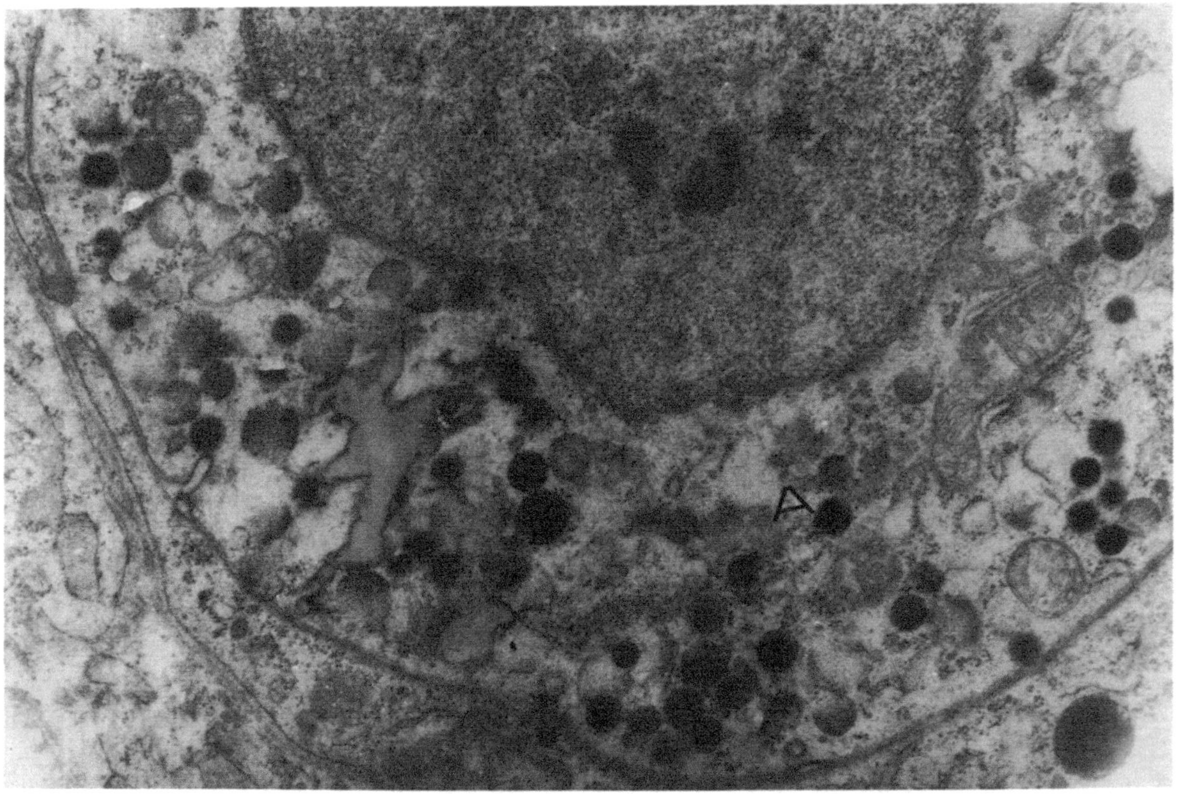

Fig. 5. Electronmicrograph of the duodenum of a 19-day-old chick embryo showing the presence of A granules in an epithelial cell. X 30,000.

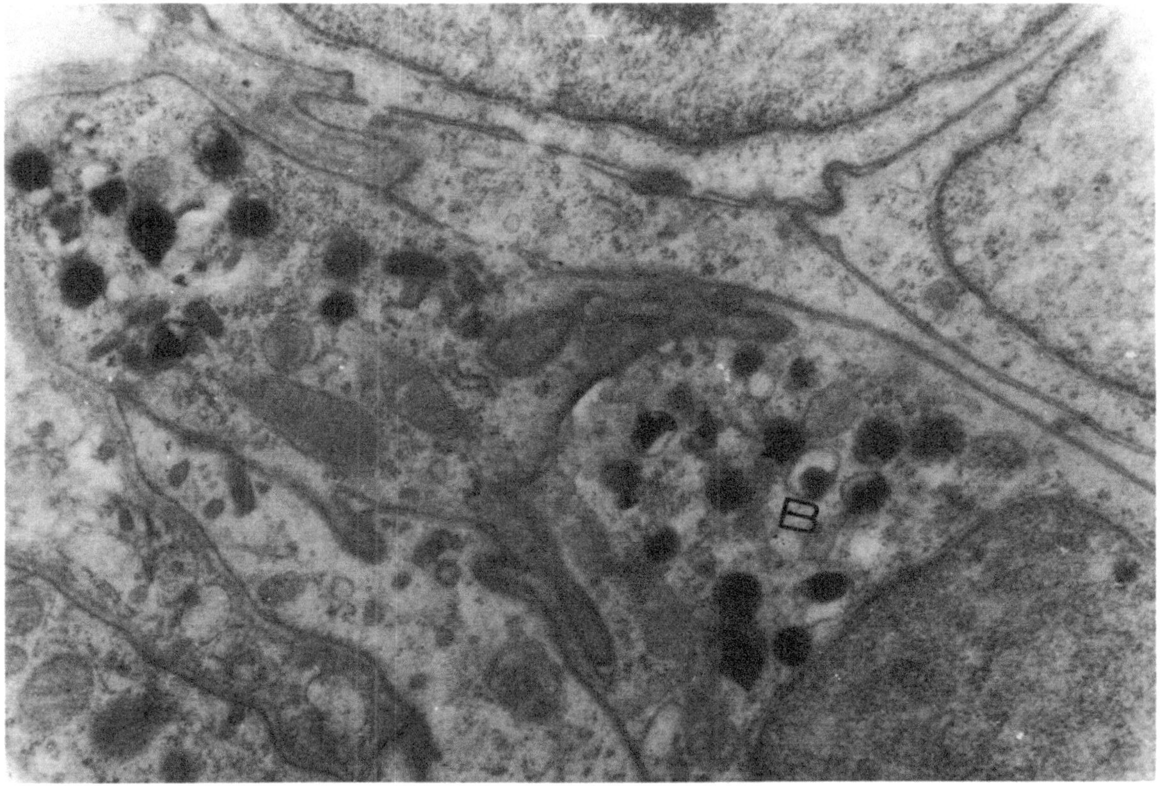

Fig. 6. Electronmicrograph of the duodenum of a 19-day-old chick embryo showing the presence of B-like granules in epithelial cells. X 30,000.

Correlation of A Cell Function with Glycogen Metabolism in the Embryo

Glycogen metabolism begins in the liver of the chick embryo as early as on the 6th day of incubation (10, 18). Shortly after this, by the 7th day, the chick embryonic liver shows the activity of phosphorylase (22). Exogenous glucagon has also been shown to induce hepatic glycogenolysis in the chick embryo as early as the 7th day (20, 23). While the resynthesis of glycogen following the effect of exogenous glucagon was rapid in the 16-day-old chick embryo, the reappearance of glycogen in the liver of the 9-day-old chick embryo was slow (Fig. 7). The resynthesis of glycogen in the liver of the 9-day-old chick embryo was more rapid if insulin was injected simultaneously with glucagon. From this result it could be deduced that the greater ability of the 16-day-old embryo to resynthesize glycogen is an index of the secretion of insulin in the older embryo in response to the effect of the exogenous glucagon.

Skeletal muscle phosphorylase in the chick embryo was found to be active from the 14th day of development which explains why, in this tissue, unlike in the liver and heart, adrenaline has no effect on glycogen metabolism until that day (16; Figs 8-12). It is relevant to note that the chick embryonic adrenals contain histochemically demonstrable adrenaline and nonadrenaline as early as the 6th day of incubation (18; Figs. 13, 14).

A possible correlation of A cell function with glycogen metabolism in mammals is indicated also by the fact that the fetal rat liver shows an increasing activity of phosphorylase from the 14th day when the fetal rat pancreas has been found to produce significant amounts of glucagon (33).

Repeated injection of glucagon into pregnant rats has been shown to result in the development of fetuses of lower weight (41). These results are similar to those obtained by Elrick et al. (14) in the chick embryo, but contrary to those of Cavallero (8) who earlier reported an increase in the weight of glucagon-treated embryos. No explanation has yet been given for these results. It is relevant to recall that glucagon has been found to inhibit the growth of tumours and to

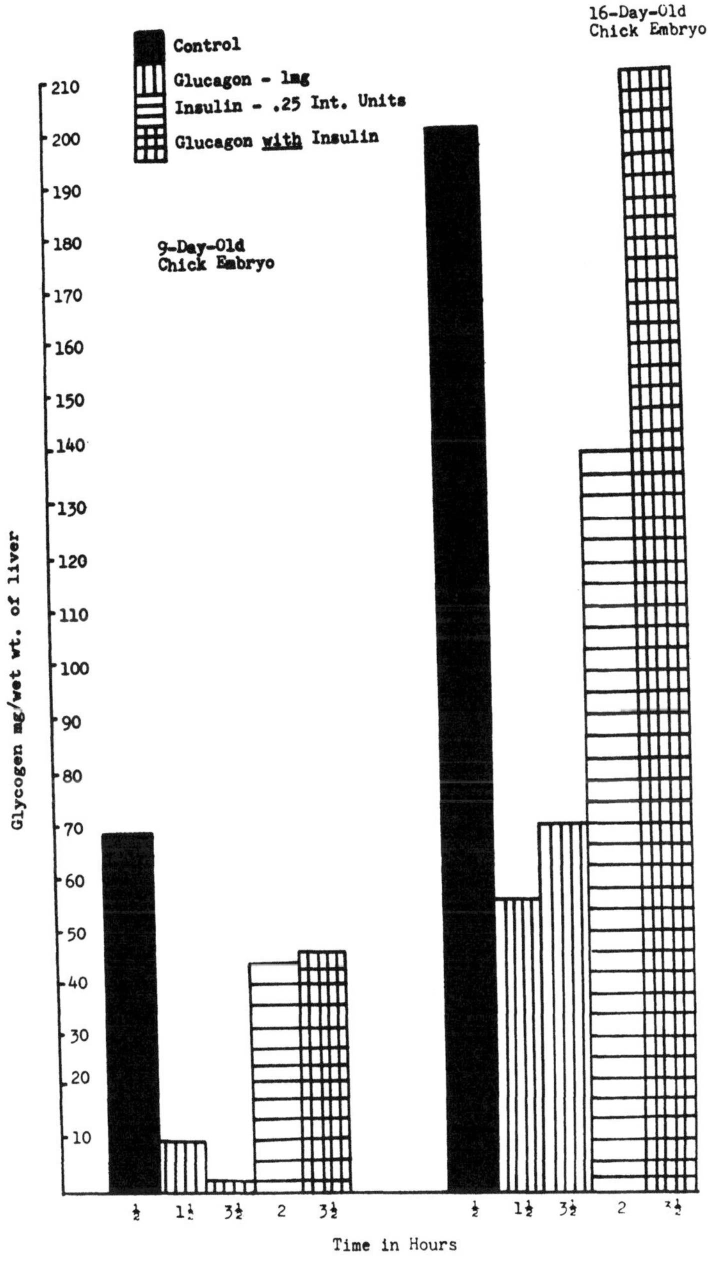

Fig. 7. Histogram showing the effect of glucagon, insulin and glucagon plus insulin on the hepatic glycogen content in 9 and 16-day-old chick embryos.

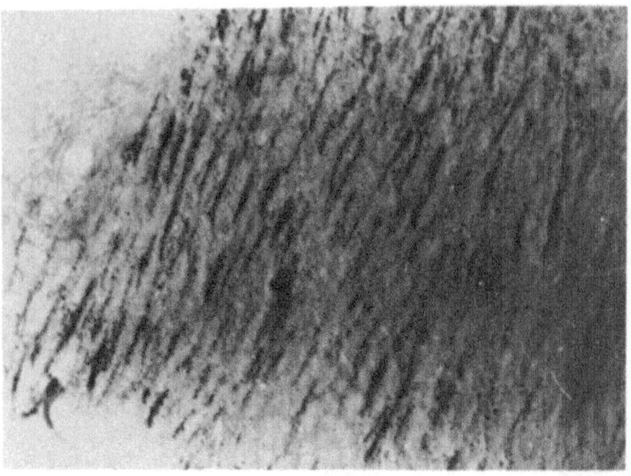

Fig. 8. Glycogen in the skeletal muscle of a 6-day-old chick embryo demonstrated histochemically by the periodic acid Schiff method.

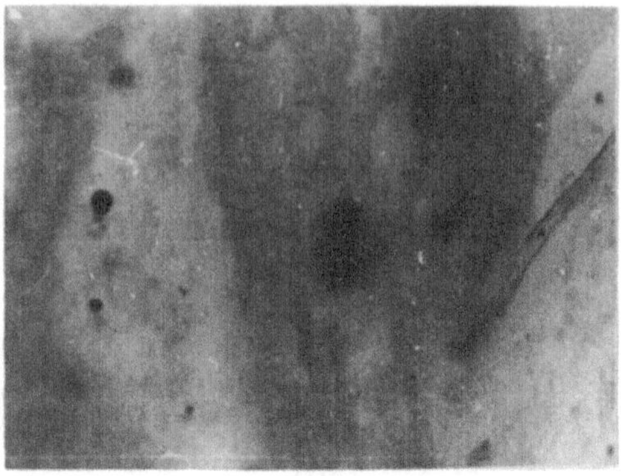

Fig. 9. No phosphorylase could be demonstrated histochemically in the skeletal muscle of a 10-day-old chick embryo.

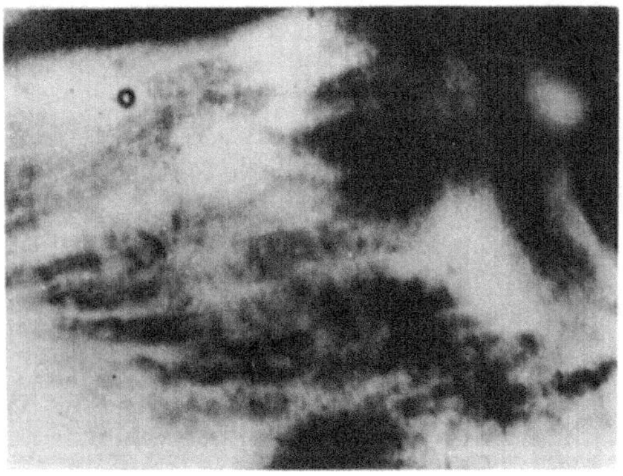

Fig. 10. Phosphorylase was demonstrated histochemically in the skeletal muscle of an 18-day-old chick embryo.

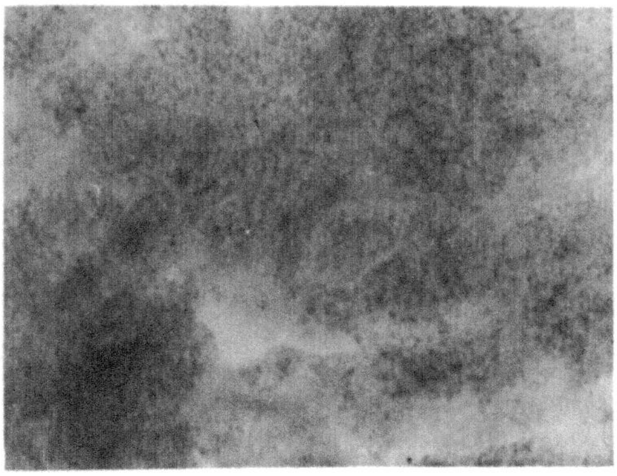

Fig. 11. The hepatic phosphorylase activity in an 18-day-old chick embryo was low when demonstrated histochemically in the absence of cyclic adenosine monophosphate.

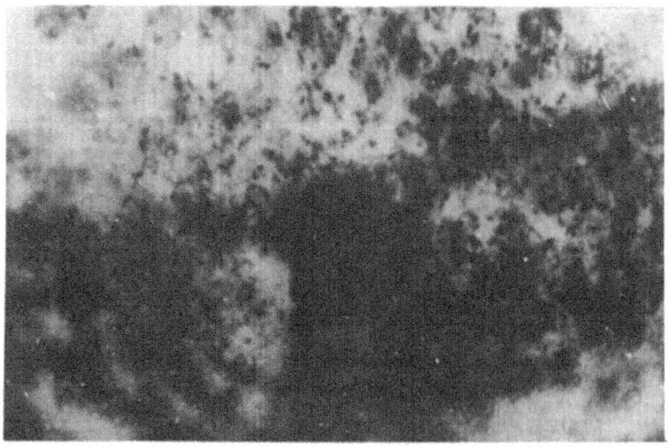

Fig. 12. The hepatic phosphorylase activity of an 18-day-old chick embryo was relatively high when demonstrated histochemically in the presence of cyclic adenosine monophosphate. The results shown in Figs. 11 and 12 demonstrate the importance of cyclic adenosine monophosphate in the interrelation of phosphorylase and glucagon at this early stage of life.

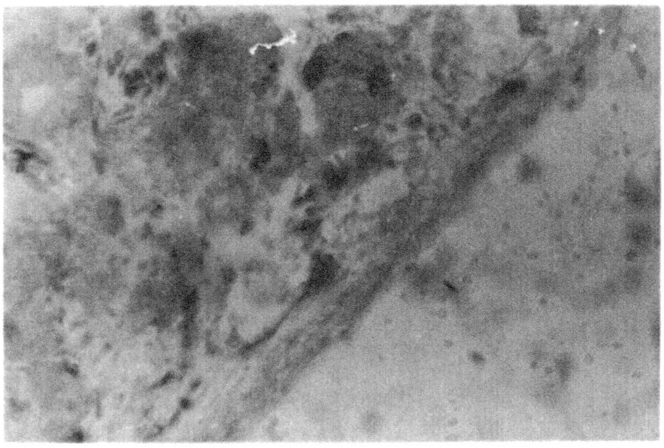

Fig. 13. Section of the adrenal gland of a 6-day-old chick embryo showing the presence, at the periphery, of adrenaline- and noradrenaline-containing chromaffin cells.

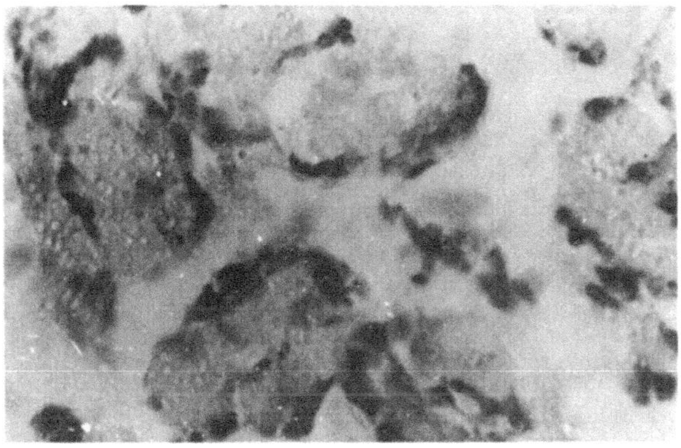

Fig. 14. Section of the adrenal gland of a 13-day-old chick embryo showing chromaffin cells scattered within the gland. This distribution is characteristic of birds and reptiles.

enhance the release of lysosomal enzymes which, in turn, may cause the death of cells. It will be essential to investigate more closely the effects of glucagon on embryonic lysosomal enzymes. The third explanation may be that the substances injected into the embryos might have contained somatostatin as well as glucagon and that the somatostatin could have inhibited the growth of the embryos.

CONCLUSIONS

It has been established that the A cells of the pancreatic islets (indeed like the B cells) arise from the embryonic foregut. These A cells produce and secrete glucagon in the embryo. There is a correlation between the secretion glucagon and the appearance of phosphorylase activity in the embryonic liver and heart. There is, however, a delay in the occurrence of phosphorylase in the embryonic skeletal muscle. The response of liver and cardiac glycogen to glucagon depends upon the presence of phosphorylase in the two organs. In contrast, the synthesis of glycogen in most tissues of the embryo occurs before the development of phosphorylase. This means that the anabolism of glycogen precedes its catabolism, perhaps because the formation of insulin in embryonic tissues occurs early in development. The formation of glucagon occurs much earlier, particularly in the embryonic gut, a finding leading to the speculation that glucagon may have a role in development, separate from its action on glycogen metabolism.

It is important to stress that while glucagon producing cells continue to be functional in both the gastrointestinal tract and the pancreas after embryonic life, the production of insulin in the gut is restricted only to the embryonic and neonatal periods and that the presence of functional B cells in the adult gut is considered abnormal.

A hormone of great metabolic importance such as glucagon has been found to appear very early in embryologic development, probably before any other hormone. What role does it play in differentiation? How does the dysfunction of the cells which produce it affect the individual in later life? These questions remain to be answered.

Acknowledgement

The authors wish to express their gratitude to Messrs. J. Oyarekna, S. Faoye and F. Mekoma for their technical assistance and to Miss Subuola Akinosho for preparing the manuscript for publication. We are also grateful to the International Atomic Energy Agency for the grants which made part of this work possible.

REFERENCES

1. Angervall, L., Hellström, C. and Hellman, B. (1962): Development of the endocrine pancreas of rats as manifested by the appearance of argyrophilia and granulation. Path. Microbiol. (Basel) 25, 389-399.

2. Barrnett, R. J., Marshall, R. B. and Seligman, A. M. (1955): Histochemical demonstration of insulin in the islets of Langerhans. Endocrinology 57, 419-437.

3. Baxter-Grillo, D. L. (1970): Enzyme histochemistry and hormones of the developing gastrointestinal tract of the chick embryo. III. Enterochromaffin cells - their possible products, glucagon, 5-hydroxytryptamine and the relation of monoamine oxidase. Histochemie 21, 129-135.

4. Baxter-Grillo, D. L. (1973): Enzyme histochemistry and hormones of the developing gastrointestinal tract of the chick embryo. V. The ultrastructure and biochemical evidence for the presence of an insulin like substance, enteroglucagon and serotonin in the developing chick duodenum. Histochemie 33, 281-286.

5. Beaupain, D. and Dieterlen-Lièvre, F. (1974): Etude immunocytologique de la differenciation du pancréas endocrine chez l'embryon de poulet. II. Glucagon. Gen. Comp. Endocrinol. 23, 421-431.

6. Bencosme, S. A., Bergman, B. J. and Martinez-Palomo, A. (1965): The principal islet of bullhead fish (Ictalurus nebulosus). A correlative light and electron microscopic study of islet cells and of their secretory granules isolated by centrifugation. Rev. Can. Biol. 24, 141-154.

7. Bjorkman, N. and Hellman, B. (1964): Ultrastructure of the islets of Langerhans in the duck. Acta Anat. 56, 348-367.

8. Cavallero, C. (1956): Pancreatic islets and growth. In: Internal Secretions of Pancreas. G. E. W. Wolstenholme and C. O'Connor, Eds. Little Brown and Co., Boston, pp. 266-284.

9. Conklin, J. L. (1961): The development of the pancreas of the human foetus. University Microfilm Inc., Ann Arbor, Mich., pp. 61-90.

10. Dalton, A. J. (1973): The functional differentiation of the hepatic cells of the chick embryo. Anat. Rec. 68, 393-409.

11. Dieterlen-Lièvre, F. (1957): Contribution à l'histogenèse du pancréas endocrine chez l'embryon de poulet. Arch. Anat. Micr. Morphol. 46, 61-80.

12. Dieterlen-Lièvre, F. (1965): Etude morphologique et expérimentale de la differentiation du pancreas chez l'embryon de poulet. Bull. Biol. Fr. Belg. 49, 3-116.

13. Dieterlen-Lièvre, F. and Beaupain, D. (1976): Immunocytological study of endocrine pancreas ontogeny in the chick embryo: normal development and pancreatic potentialities in the early splanchnopleura. In: Evolution of Pancreatic Islets. T. A. I. Grillo, L. Leibson and A. Epple, Eds., Pergamon Press, Oxford, pp. 37-50.

14. Elrick, H., Konigsberg, I. R. and Arai, Y. (1958): Effect of glucagon on growth of chick embryo. Proc. Soc. Exp. Biol. Med. 97, 542-544.

15. Gerlovin, E. Sh. (1976): Some principles of cytodifferentiation of pancreatic islets in vertebrata during onto- and phylogeny from the standpoint of molecular biology and genetics. In: Evolution of Pancreatic Islets. T. A. I. Grillo, L. Leibson and A. Epple, Eds., Pergamon Press, Oxford, pp. 97-112.

16. Gill, P. M. (1938): The effect of adrenaline on embryonic glycogen in vitro as compared with its effect in vivo. Biochem. J. 32, 1792-1799.

17. Grodsky, G. M., Hyashida, T., Peng, C. T. and Geshwind, I. I. (1961): Production of glucagon antibodies and their role in metabolism and immunoassay of glucagon. Proc. Soc. Exp. Biol. Med. 107, 491-494.

18. Grillo, T. A. I. (1960): A histochemical study of carbohydrate metabolism in the developing chick embryo. Thesis. Cambridge University.

19. Grillo, T. A. I. (1961): The ontogeny of function of A cells of the pancreas of the chick embryo. J. Anat. 95, 284.

20. Grillo, T. A. I. (1961): The response of embryonic tissues to glucagon. Am. J. Dis. Child. 102, 568-569.

21. Grillo, T. A. I. (1961): The histochemical study of phosphorylase in the tissue of the chick embryo. J. Histochem. Cytochem. 9, 389-391.

22. Grillo, T. A. I. (1961): The ontogeny of insulin secretion in the chick embryo. J. Endocrinol. 22, 285-292.

23. Grillo, T. A. I. (1965): The ontogenesis of the endocrine pancreas and carbohydrate metabolism. In: Organogenesis. R. L. Dettaan and H. Ursprung, Eds. Holt, Rinehart and Winston, New York, pp. 513-538.

24. Grillo, T. A. I., Ito, R. and Watanabe, K. (1976): The endocrine pancreas of some West African reptiles. Electron microscopy, histochemistry and hormone content. In: Evolution of Pancreatic Islets. T. A. I. Grillo, L. Leibson and A. Epple, Eds. Pergamon Press, Oxford, pp. 189-249.

25. Herman, L, Sato, T. and Fitzgerald, P. J. (1963): Electron microscopy of "acinar-islet" cells in the rat pancreas. Fed. Proc. 22 603.

26. Hers, H. G. (1959): Etude enzymatiques sur fragments hépatiques: application à la classification des glycogènoses. Rev. Intern. Hépat. 9, 35-55.

27. Kenney, A. J. (1955): Extractable glucagon of the human pancreas. J. Clin. Endocr. 15, 1089-1105.

28. Kobayashi, K. (1966): Electron microscopic studies of the Langerhans islets in the toad pancreas. Arch. Histol. Jap. 26, 439-482.

29. Lin-Hsiang, M. and Potter, E. (1962): Development of the human pancreas. Arch. Path. 74, 439-452.

30. Mason, T. E., Phifer, R. F., Spicer, S. S., Swallow, R. A. and Dreskin, R. B. (1969): An immunoglobulin-enzyme bridge method for localizing tissue antigens. J. Histochem. Cytochem. 17, 563-569.

31. Munger, B. (1958): A light and electron microscopic study of cellular differentiation in the pancreatic islets of the mouse. Am. J. Anat. 103, 275-312.

32. Nonaka, K., Grillo, T. A. I. and Foà, P. P. (1970): Glucagon secretion in normal rats and hamsters and in hamsters with a transplantable islet cell tumor. In: The Structure and Metabolism of the Pancreatic Islets. S. Falkmer, B. Hellman and I. B. Taljedal, Eds. Pergamon Press, Oxford, pp. 149-155.

33. Okuno, G., Price, S., Grillo, T. A. I. and Foà, P. P. (1964): Development of phosphorylase and phosphorylase-activating (glucagon-like) substances in the rat embryo. Gen. Comp. Endocrinol. 4, 446-451.

34. Pearce, R. M. (1903): The development of the islets of Langerhans in the human embryo. Am. J. Anat. 2, 445-455.

35. Pearse, A. G. E. (1969): The cytochemistry and ultrastructure of polypeptide hormone-producing cells of the APUD series and the embryologic, physiologic and pathologic implications of the concept. J. Histochem. Cytochem. 17, 303-313.

36. Pearse, A. G. E., Polak, J. M. and Heath, C. M. (1973): Development, differentiation and derivation of the endocrine polypeptide cells of the mouse pancreas. Immunofluorescence, cytological and ultrastructural studies. Diabetologia 9, 120-129.

37. Robb, P. M. (1961): Development of the islets of Langerhans in man. Nature 190, 1018.

38. Schiebler, T. H. and Schiessler, S. (1959): Über den Nachweis von Insulin mit den metachromatisch reagierenden Pseudoisocyaninen. Histochemie 1, 445-465.

39. Tanizawa, Y. and Fujita, H. (1966): Some observations on the fine structure of the development of the pancreas in the chick embryo. Arch. Hist. Japon. 26, 535-546.

40. Thommes, R. C. (1960); A histochemical study of insulin in the chick embryo pancreas. Growth 24, 69-80.

41. Tyson, J. E., Kinch, R. A. H. and Stevenson, J. A. F. (1968): Glucagon and fetal growth in the rat. Am. J. Obst. Gynec. 101, 834-838.

42. Unger, R. H., Ketterer, H. and Eisentraut, A. M. (1966): Distribution of immunoassayable glucagon in gastrointestinal tissues. Metabolism 15, 865-867.

43. Villamil, N. F. (1942): Citogenesis del pancreas exo-y endocrino en embriones de pollo. Rev. Soc. Argent. Biol. 18, 416-424.

44. Weichselbaum, A. and Kyrl, J. (1909): Über das Verhalten der Langerhansschen Inseln des menschlichen Pankreas in fötalen und postfötalen Leben. Arch. Mickroskop. Anat. 74, 223-258.

45. Zagury, J., DeBrux, J., Adgla, J. and Leger, L. (1961): Etude des îlot de Langerhans du pancréas humain au microscope éléctronique. Presse méd. 69, 887-890.

MORPHOLOGIC STUDY OF GLUCAGON RELEASE

J.-L. Carpentier, F. Malaisse-Lagae, W. Müller and L. Orci

Institute of Histology and Embryology, and
Institute of Clinical Biochemistry
Geneva, Switzerland

The authors studied the release of glucagon from isolated pancreatic islets using thin section preparations, freeze-fracture techniques and radioimmunoassay. The evidence strongly suggests that exocytosis is a significant mechanism for glucagon release during arginine stimulation, but not for the release induced by calcium deprivation.

While it is generally accepted that specific membrane-bound cytoplasmic granules of the pancreatic A cells (A_2 cells) represent the main storage form of glucagon within the islet of Langerhans (1, 2), doubts remain about the structural basis of glucagon release. Several authors have suggested that this release occurs through intracytoplasmic dissolution of the secretory granule (11, 15, 16) while others have proposed that the granule's content is extruded by exocytosis (3, 5, 7, 10, 13, 17). We have attempted therefore to clarify this issue by studying simultaneously the functional and morphologic events involved in glucagon secretion. The morphologic approach was accomplished by morphometric evaluation of thin section preparations and freeze-fracture replicas of isolated islets, while the functional approach consisted of biochemical measurements of glucagon release.

Batches of islets were isolated from the pancreata of fed rats by collagenase digestion (6). Twenty islets were incubated at 37°C for 30 minutes in 1 ml of medium. Two different stimuli of glucagon secretion were studied: arginine (40 mM) in the absence of glucose (4, 14) and deprivation of calcium in the presence of 16.5 mM glucose (8, 9). At the end of the incubation period, the media were kept for the assay of glucagon and the islets fixed for further electron-microscopic processing (thin section or freeze-fracturing). Since in normal rat islets, the different cell types cannot be identified in freeze-fracture replicas, we used an indirect method to reveal the A cells: it consisted of degranulating the B cells by glibenclamide, thus leaving A cells as the most numerous granulated islet cell type (12). Preliminary experiments indicated that control or glibenclamide-treated islets released equivalent amounts of glucagon in basal as well as in stimulated conditions. As shown in Figure 1, the two stimuli of glucagon secretion induced a significant increase of glucagon release by isolated islets. Such data allowed us to conclude that the model used was satisfactory to perform the morphologic analysis of A cell secretion.

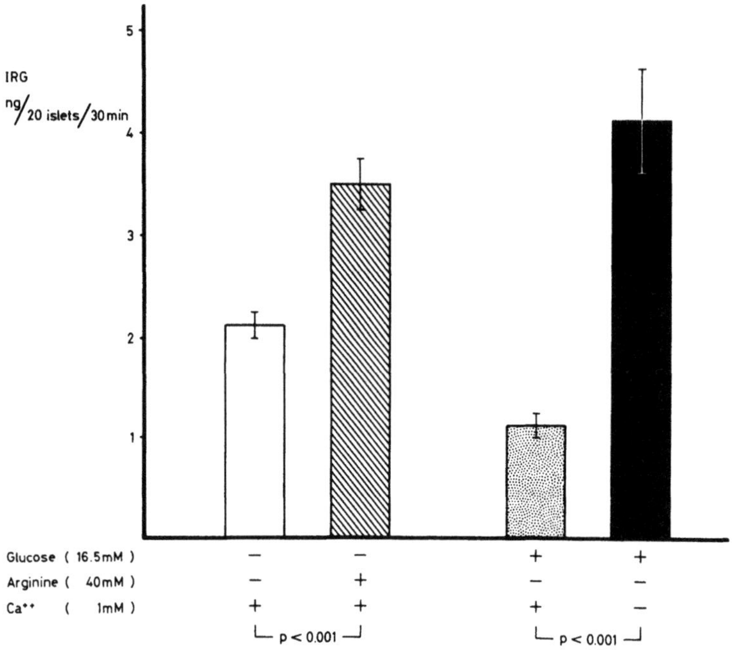

Fig. 1. Glucagon (IRG) release into the medium at the end of a 30 min incubation period in different experimental conditions. In a glucose-free medium, arginine (40 mM) increased significantly the release of glucagon, while in the presence of glucose (16.5 mM), glucagon release was significantly enhanced by the deprivation of calcium. The suppressive effect of glucose upon glucagon release in a medium of normal calcium composition is also apparent from comparison of the open column and stippled column, referring to unpaired experiments.

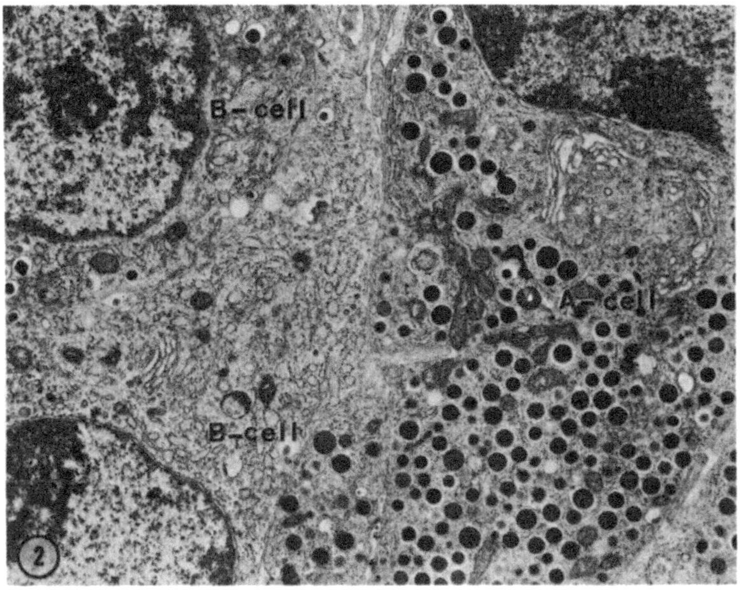

Fig. 2. Portion of two B and one A cells from a glibenclamide treated rat. The B cells contain only a few secretory granules while the A cell is well granulated. X 10.500.

On thin section preparations of isolated islets from glibenclamide-treated rats, B cells appeared conspicuously deprived of their secretory granules (Fig. 2), while A cells did not seem to be affected by the drug. After 30 minutes of incubation in the presence of arginine, secretory granules in A cells were often seen in close contact with the plasma membrane and some of them were found undergoing exocytosis (Fig. 3). In a medium deprived of calcium (Fig. 4), the extracellular space was enlarged, but the A cells appeared unaffected and secretory granules were rarely seen in contact with the cell membrane. When the number and margination index* of secretory granules of A cells were assessed quantitatively by morphometry in the two experimental conditions, it was found that under the influence of arginine, the margination index of α granules was doubled while the

*The margination index evaluates the number of granules in contact with the plasma membrane (margination).

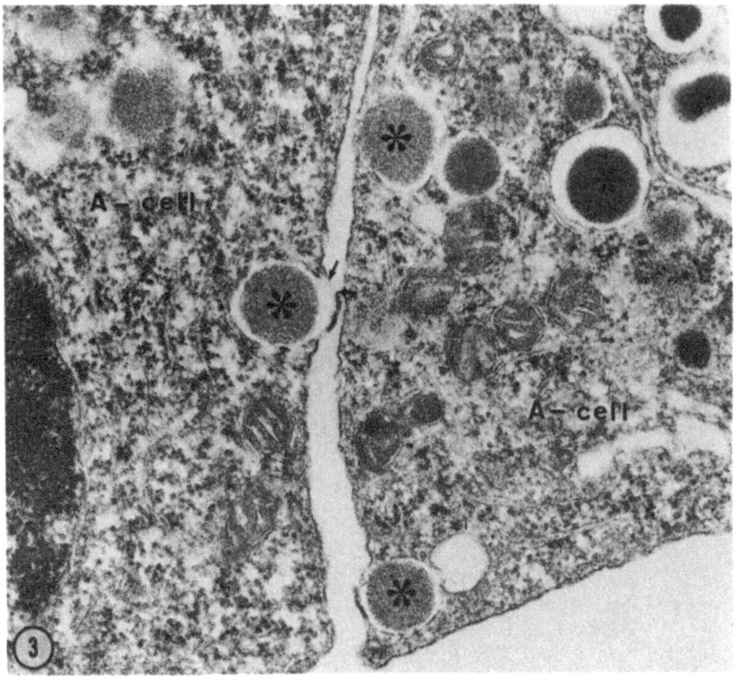

Fig. 3. Portion of two A cells from isolated islets incubated for 30 min in a medium containing arginine (40 mM). Secretory granules are seen in close contact with the plasma membrane (*); the arrows indicate the opening created by the coalescence of the granule-limiting membrane with the plasma membrane through which the granule core is being discharged into the extracellular space. X 38.500.

number of granules within the cell was not significantly affected. On the contrary, calcium deprivation significantly decreased the quantity of stored granules but did not influence the margination index. A direct evaluation of exocytosis, i.e. counting the number of exocytotic stomata at the surface of A cells in freeze-fracture replicas (Fig. 5), confirmed the indirect evidence given by calculating the margination index. As shown in Figure 6, the number of exocytotic stomata during stimulation of glucagon release by arginine was significantly increased while no increase was observed when glucagon secretion was stimulated by calcium deprivation.

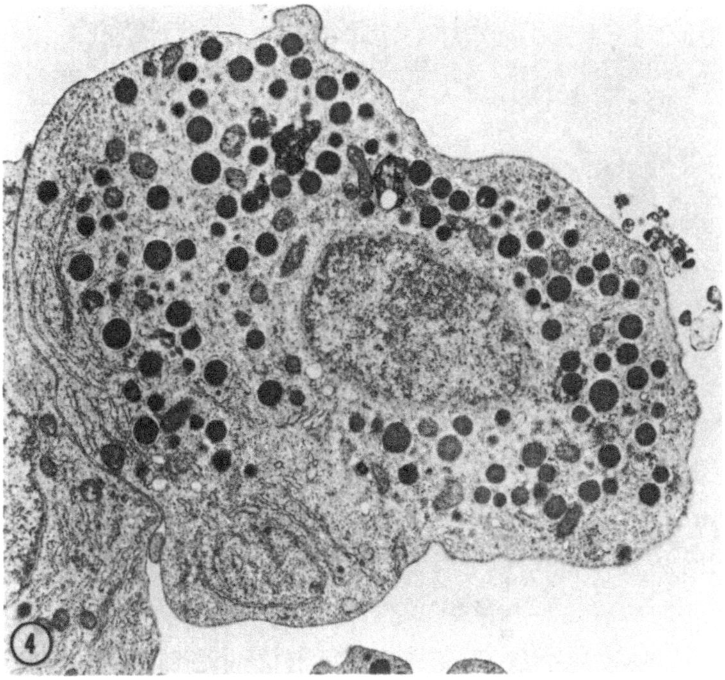

Fig. 4. A cell from an isolated islet incubated for 30 min in a calcium-free medium. Note that, except in a limited area, most of the A cell surface has lost the normal close relationship with neighbouring cells. X 11.500.

Our results thus demonstrate that exocytosis is a significant mechanism of glucagon release during arginine stimulation. On the other hand, the sizable glucagon release induced by calcium deprivation, although associated with a significant reduction of α granule stores, did not seem to involve exocytosis, at least as defined by our morphologic criteria and at the end of the stimulatory incubation. Thus, our data would confirm that at this time, exocytosis appears to be an important mode of glucagon release by the A cell, as suggested by previous non-quantitative data (3, 5, 7, 10, 13, 17), although the possibility of other mechanisms of release (11, 15, 16) cannot be excluded. Some degree of caution is therefore needed in assuming that different secretagogues of hormones such as glucagon act through a similar process.

Fig. 5. Freeze-fracture replica of an isolated islet incubated in the presence of 40 mM arginine. Several exocytotic stomata are seen (solid arrows) on the exposed face of the plasma membrane of the granulated cell (presumably A cell). The dotted arrow indicates a granule core undergoing extrusion. X 14.000.

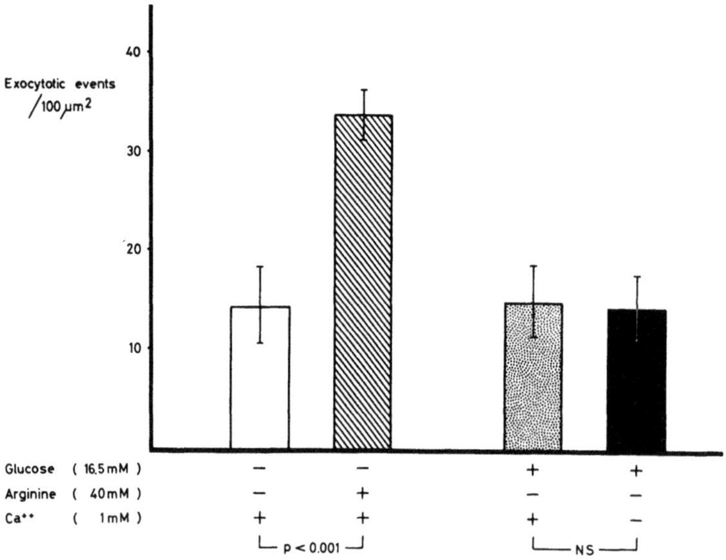

Fig. 6. Quantification of the number of exocytotic events per 100 µm2 of exposed face of plasma membrane from granulated cells in different experimental conditions.

Acknowledgement

Supported by Swiss National Science Foundation.

REFERENCES

1. Bussolati, G., Capella, C., Vassalo, G. and Solcia, E. (1971): Histochemical and ultrastructural studies on pancreatic A cells. Evidence for glucagon and non-glucagon components of the alpha granule. Diabetologia 7, 181-188.

2. Erlandsen, S. L., Parsons, J. A., Burke, J. P., Redick, J. A., Van Orden, D. E. and Van Orden, L. S. (1975): A modification of the unlabeled antibody enzyme method using heterologous antisera for the light microscopic and ultrastructural localization of insulin, glucagon and growth hormone. J. Histochem. Cytochem. 23, 666-677.

3. Esterhuizen, A. C. and Howell, S. L. (1970): Ultrastructure of the A cells of cat islets of Langerhans following sympathetic stimulation of glucagon secretion. J. Cell Biol. 47, 593-598.

4. Gerich, J. E., Charles, M. A. and Grodsky, G. M. (1974): Characterization of the effects of arginine and glucose on glucagon and insulin release from the perfused rat pancreas. J. Clin. Invest. 54, 833-841.

5. Gomez-Acebo, J., Parrilla, R. and Candela, J. L. R. (1968): Fine structure of the A and D cells of the rabbit endocrine pancreas in vivo and incubated in vitro. I. Mechanism of secretion of the A cells. J. Cell Biol. 36, 33-44.

6. Lacy, P. E. and Kostianovsky, M. (1967): Method for the isolation of intact islets of Langerhans from the rat pancreas. Diabetes 16, 35-39.

7. Lazarus, S. S., Shapiro, S. and Volk B. W. (1968): Secretory granule formation and release in rabbit pancreatic A cells. Diabetes 17, 152-160.

8. Leclercq-Meyer, V., Marchand, J. and Malaisse, W. J. (1976): The role of calcium in glucagon release. Interactions between glucose and calcium. Diabetologia 12, 531-538.

9. Leclercq-Meyer, V., Rebolledo, O., Marchand, J. and Malaisse, W. J. (1975): Glucagon release: paradoxical stimulation during calcium deprivation. Science 189, 897-899.

10. Machino, M., Onoe, T. and Sakuma, H. (1966): Electron microscopic observations on the islet alpha cells of the domestic fowl pancreas. Electron Micr. 15, 249-256.

11. Munger, B. L. (1962): The secretory cycle of the pancreatic islet α-cell An electron microscopic study of normal and synthalin-treated rabbits. Lab. Invest. 11, 885-901.

12. Orci, L., Malaisse-Lagae, F., Ravazzola, M., Rouiller, D., Renold, A. E., Perrelet, A. and Unger, R. H. (1975): A morphological basis for intercellular communication between alpha and beta cells in the endocrine pancreas. J. Clin. Invest. 56, 1066-1070.

13. Orci, L., Stauffacher, W., Renold, A. E. and Rouiller, Ch. (1971): The ultrastructural aspect of A-cells of non-ketotic and ketotic diabetic animals: indications for stimulation and inhibition of glucagon production. In: Current Topics on Glucagon. M. Austoni, C. Scandarelli, G. Federspil and A. Trisotto, Eds. Cedam, Padova, pp. 15-30.

14. Pagliara, A. S., Stillings, S. N., Hover, B., Martin, D. M. and Matchinsky, F. M. (1974): Glucose modulation of amino acid-induced glucagon and insulin release in the isolated perfused rat pancreas. J. Clin. Invest. 54, 819-832.

15. Rhoten, W. B. (1973): The cellular response of saurian pancreatic islets to chronic insulin administration. Gen. Comp. End. 20, 474-489.

16. Smith, P. H. (1975): Effects of selective partial pancreatectomy on pancreatic islet cell morphology in the Japanese quail. Gen. Comp. Endo. 26, 310-320.

17. Unger, R. H. (1976): Diabetes and the alpha-cell. Diabetes 25, 136-151.

II. BIOSYNTHESIS, NATURE AND SITES OF ORIGIN OF PANCREATIC GLUCAGON AND OF GLUCAGON-LIKE IMMUNOREACTIVE MATERIALS

GLUCAGON BIOSYNTHESIS IN ANGLERFISH ISLETS: IDENTIFICATION OF A 4,900 DALTON METABOLIC CONVERSION INTERMEDIATE

B. D. Noe

Department of Anatomy, Emory University
Atlanta, Georgia USA

^{3}H-tryptophan- and ^{14}C-isoleucine-labeled anglerfish islet peptides have been separated and analyzed by gel filtration, polyacrylamide gel electrophoresis (PAGE) and ion exchange chromatography and tested for immunologic activity by the Unger 30K antiserum. The results indicate that anglerfish proglucagon (~12,100 daltons) undergoes sequential metabolic cleavage through intermediates of ~9,100 (I) and ~4,900 daltons (II) to yield 3,500 dalton glucagon. Proglucagon and intermediate I proved to be more electronegative than glucagon during high pH PAGE, whereas intermediate II was found to be more electropositive than glucagon. Conversion of intermediate II to glucagon occurs at a very slow rate. Limited treatment with trypsin rapidly degrades proglucagon and both intermediates. Therefore, trypsin or an enzyme having a similar specificity may participate in the metabolic conversion of anglerfish proglucagon to glucagon. Preliminary results from experiments in which anglerfish islet mRNA was utilized to direct the synthesis of cell-free translation products in a wheat germ system indicate that the maximum possible size for an anglerfish pre-proglucagon is less than 18,000 daltons.

The biosynthesis of glucagon has been studied in a number of experimental systems. These include isolated pigeon islets (28, 33-35), guinea pig islets grown in culture (10-14), isolated perfused rat pancreas (26) and isolated human pancreatic islets (24). In all of these investigations, it has been possible to incorporate radioactively labeled tryptophan (a component of glucagon but not present in any proinsulin characterized to date) into glucagon immunoreactive peptides of 9,000 daltons or larger. However, in most of these investigations, attempts to demonstrate metabolic transfer of radioactivity from these larger synthetic products to glucagon or glucagon-related peptides have been unsuccessful. As a result of one recent study using microdissected pigeon pancreatic islets, O'Connor and Lazarus (28) have concluded that there is no precursor-product relationship between an 8,500 dalton "big glucagon" and 3,500 dalton "little glucagon."

These results differ from those obtained in studies of glucagon biosynthesis using the isolated islet of anglerfish. The anglerfish is unique in that it provides a source of large quantities of pancreatic islet tissue (50-250 mg per animal) essentially free of any exocrine contamination. Results from biosynthetic studies using this tissue indicate that ^{3}H-tryptophan, incorporated during pulse incubations into large glucagon immunoreactive peptides (m.w. 9,000-12,000), is transferred to a peptide having the approximate molecular size of glucagon during chase incubations. This transfer was observed to occur both in the presence of cycloheximide (22) or of excess unlabeled amino acid in isotope free medium (23). Further analysis of this transfer indicates that it occurs in a step-wise fashion suggesting that one or more intermediates are involved in the metabolic processing of anglerfish proglucagon (23). The data reported herein provide evidence for the participation of a 4,900 ± 400 dalton peptide as a cleavage intermediate in the conversion process.

METHODS

The techniques for preparation and incubation of anglerfish islets, tissue extraction, gel filtration, glucagon radioimmunoassay, and polyacrylamide gel electrophoresis (PAGE) have been described previously (22,23). However, in the present study, the separatory gels for PAGE were polymerized with N, N'-methylenebisacrylamide rather than with diallyltartardiamide and the gels were hand-sliced at 1 mm per slice after freezing with powdered dry ice. DEAE Bio-Gel A was obtained from Bio-Rad Laboratories, Rockville Center, N. Y. Amino acid analyses were performed on a JEOL JLC-6AH amino acid analyzer. Further details of experimental conditions are given in the figure legends.

RESULTS

Gel filtration of tissue extracts in propionic acid. In previous studies using anglerfish islet, gel filtration of tissue extracts was performed in 1 M acetic acid. The data shown in Fig. 1 were obtained by gel filtration of extracts of tissue incubated continuously for 2 hr (A), 4 hr (B) and 5 hr (C) in 1 M propionic acid. Filtration in propionic acid provides better resolution of the various peptides in the extract than does filtration in acetic acid. Isoleucine is present neither in anglerfish glucagon nor in a ~9,000 dalton glucagon containing peptide isolated from anglerfish islets (32). Tryptophan is a component of anglerfish glucagon (32) but is not found in anglerfish proinsulin (21, 40). Therefore, as shown in Fig. 1, proinsulin (peak 1) and insulin (peak 2) become selectively labeled with ^{14}C-isoleucine and glucagon-related peptides are selectively labeled with ^{3}H-tryptophan (peaks a-d).

The peptide forming peak a in Fig. 1 has been identified tentatively as anglerfish proglucagon (m.w. ~12,000). The peptide forming peak b has been identified as a glucagon containing peptide (m.w. $\leq$ 9,000) which may serve as a metabolic conversion intermediate in cleavage of proglucagon to glucagon (23). Peaks a and b are not resolved into separate peaks by filtration in 1 M acetic acid. Note that with increasing duration of incubation there is a shift in relative predominance from peak 1 to peak 2 and from peak a to peak b. Peaks c and d are poorly resolved by filtration in 1 M propionic acid, but are resolved even less well during filtration in 1 M acetic acid.

When used to calibrate the column used for the filtration shown in Fig. 1, the elution differential between bovine insulin and porcine glucagon markers is 10-12 1.5 ml fractions. This is approximately the resolution observed between peak 2 and peak d in Fig. 1. However, the ^{3}H-tryptophan radioactive peptide responsible for forming peak c has been previously identified as labeled anglerfish glucagon (22, 23). There are several reasons why peak c was thought to represent labeled glucagon. First, although the falling phases of all ^{3}H-tryptophan-labeled peaks, observed to elute more slowly than anglerfish insulin in previous gel filtration runs using acetic acid, were rarely as symmetrical as those observed for proinsulin and insulin, a distinctly biphasic nature, such as that

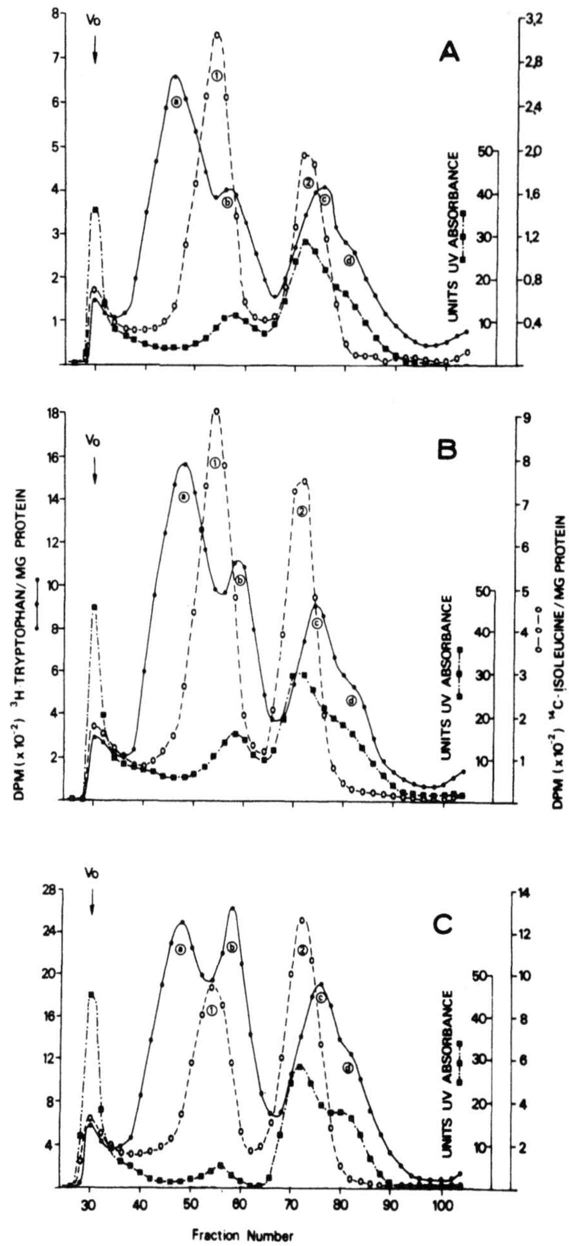

Fig. 1. Continuous incubation of 370 mg of anglerfish islet for 2 hr (A), 232 mg for 4 hr (B), and 300 mg for 5 hr (C). Incubations proceeded in the presence of 10 μCi L-^{14}C -isoleucine (313 mCi/mmol) and 100 μCi L-^{3}H -tryptophan (1.15 Ci/mmol). After acid ethanol-methylene chloride-acetic acid extraction, each extract was lyophilized and then resuspended in 1M propionic acid. Gel filtration was performed on a 1.6 x 90 cm. column of Bio-Gel P-10 (100-200 mesh) in 1M propionic acid. Fractions of 1.5 ml were collected. Peaks 1 and 2 represent ^{14}C-labelled proinsulin and insulin respectively. Peaks a-d are glucagon-related peptides labeled with ^{3}H-tryptophan.

shown in the regions for peaks c and d in Fig. 1, was not discerned. Second, high pH PAGE of the peptides contained in peaks 2, c and d resulted in the localization of two ^{3}H-tryptophan labeled bands which possess glucagon immunoreactivity, as well as the electrophoretic mobility of anglerfish glucagon and monodesamidoglucagon (22, 23; Fig. 3A). It was thus thought that this region of the gel filtration eluate contained ^{3}H-labeled glucagon and monodesamidoglucagon only. In the light of more recent results, this conclusion now seems to have been in error.

Comparison of radioactivity and immunoreactivity profiles by gel filtration. To more carefully investigate the relationship between ^{3}H-radioactivity and glucagon immunoreactivity eluting in the insulin and glucagon containing region of gel filtration eluates, refiltration of protein concentrates obtained from portions of numerous eluates was performed. The result of one such refiltration is shown in Fig. 2 (upper frame). Small amounts of proinsulin (peak 1) and tryptophan labeled peptides larger than insulin (peaks a and b) contaminate the insulin and glucagon containing pools due to incomplete separation by the initial gel filtration. The predominant peak of glucagon immunoreactivity elutes more closely to ^{3}H-labeled peak d than to peak c. Peak d is therefore more likely to represent ^{3}H-labeled glucagon. The lack of complete coincidence of the IRG peak and peak d suggests that much of the glucagon present in the cells at the time of homogenization and extraction was unlabeled.

Several of the insulin-glucagon pools were subjected to limited treatment with trypsin at an enzyme to protein ratio of 1:100 prior to refiltration. The result after refiltration of one such trypsin treated pool is shown in the lower frame of Fig. 2. Proinsulin and the ^{3}H-tryptophan labeled peptides larger than insulin were removed by trypsin treatment. In accordance with the results of Yamaji et al (40), insulin (peak 2) appears to be affected relatively little by limited trypsin treatment. The ^{3}H-labeled peak c has been removed by trypsin treatment whereas a portion of peak d remains, as well as smaller ^{3}H-labeled tryptic products. If peak d is anglerfish glucagon, then these results correlate well with those of Trakatellis et al. (32). They found that anglerfish glucagon has a COOH-terminal lysine which is quite susceptible to tryptic action, but glucagon itself is more resistant to tryptic action than larger glucagon-containing peptides.

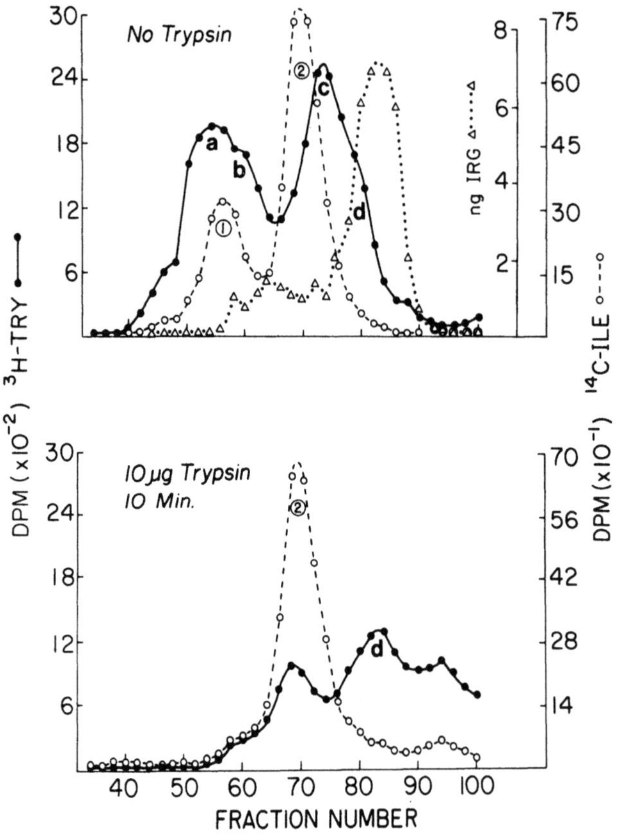

Fig. 2. Analysis of the distribution of ^{3}H-tryptophan radioactivity and glucagon immunoreactivity in the insulin and glucagon containing portions of gel filtration eluates. One mg samples of ^{3}H-try/^{14}C-ile labeled protein obtained from lyophilized insulin and glucagon containing gel filtration pools were suspended in 1 ml of 0.1 M Tris-HCl buffer pH 8.5 and subjected to incubation at 37° in the presence or absence of trypsin (Sigma Type XI) for varying periods of time. After incubation, the samples were immediately acidified, diluted and lyophilized. The distribution of radioactivity and immunoreactive glucagon (IRG) after filtration on a 1.6 x 96 cm column of Bio-Gel P-10 is shown for a control (upper frame) and a sample incubated with trypsin (lower frame). Peaks are labeled in Fig. 1. IRG-immunoreactive glucagon.

Since ^3H-labeled peaks a and b are thought to be proglucagon and an initial conversion intermediate, respectively, the question as to the nature of peak c then arises. Does peak c represent a peptide which may serve as a second conversion intermediate between proglucagon and glucagon, or is this peptide possibly unrelated to glucagon? A further consideration which made the determination of the nature of peak c more difficult is that nearly all the ^3H-tryptophan radioactivity migrating toward the anode during high pH PAGE of peptides in peaks 2, c and d had been shown to be present in anglerfish glucagon and monodesamidoglucagon (22, 23); Fig. 3A). Glucagon and desamidoglucagon presumably elute as components of peak d on gel filtration. What, then, is the disposition of the peak c peptide during high pH electrophoresis?

Bidirectional electrophoresis at high pH. In 1973, Tager and Steiner (30) reported the complete amino acid sequence of a 4,500 dalton peptide isolated from crystalline glucagon preparations which proved to be glucagon with 8 additional amino acids at the COOH-terminus. This peptide was found to be highly basic. The similarity in molecular size between this probable fragment of mammalian proglucagon and the anglerfish peak c peptide is striking. This similarity prompted us to attempt to determine if the peptide in peak c is also basic. Figure 3 shows the results of a high pH PAGE run in which portions of the same insulin-glucagon pool used for the refiltrations in Fig. 2 were run toward the anode (A) and equivalent portions were run in tubes inverted so that the separatory gel was oriented toward the cathode (B). The pattern of distribution of radioactivity and immunoreactive glucagon in the anode gel was the same as has been found in previous studies. Stained bands 1 and 2 in Fig. 3A are anglerfish glucagon and monodesamidoglucagon respectively. Band 3 is anglerfish insulin. Nearly all the ^3H-radioactivity and glucagon immunoreactivity are coincident in glucagon and monodesamidoglucagon. The predominant portion of the ^{14}C-radioactivity coincides with the migration position of the stained insulin band.

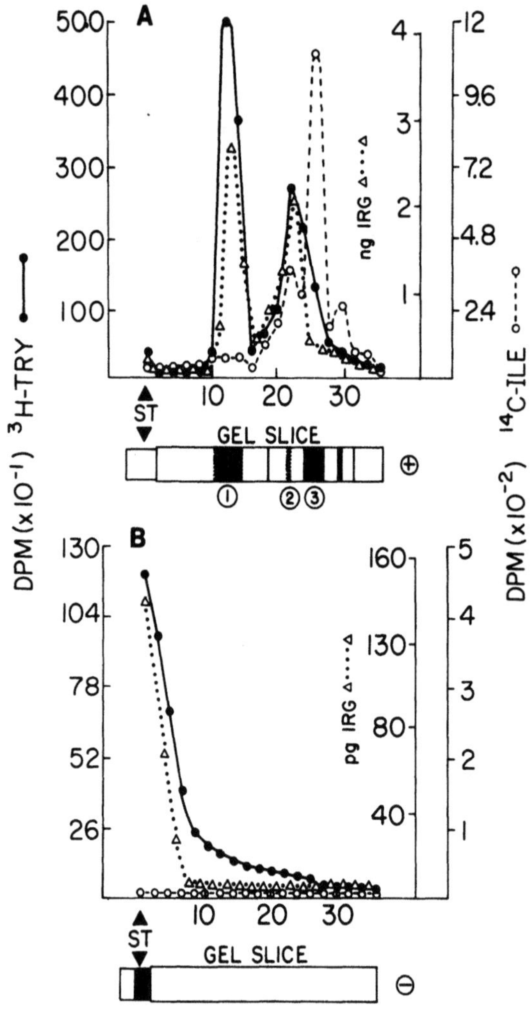

Fig. 3. Gel electrophoresis at pH 9.5 of protein samples from the same preparation used for the experiments shown in Fig. 2. Two mg samples were suspended in 2 ml 0.1 M Tris-HCl buffer pH 8.5. Aliquots were distributed to 4 electrophoresis tubes in which separatory and stacking gels had been polymerized. Two gels were run toward the anode and two were inverted to run toward the cathode. After electrophoresis, one gel in each pair was stained with Coomassie Blue and the other sliced for determination of the distribution of radioactivity and immunoreactive glucagon (IRG) in alternate slices. A-anode gel, B-cathode gel, ST-stacking gel. Bands 1 and 2 in A are anglerfish glucagon and desamidoglucagon. Band 3 is anglerfish insulin.

The results shown in Fig. 3B indicate that there is a peptide in the insulin-glucagon pool which migrates slowly toward the cathode during high pH PAGE. This peptide possesses ^{3}H-tryptophan radioactivity, as well as glucagon immunoreactivity and was localized in the stacking gel by Coomassie blue staining. Since the electrophoretic system used stacks at pH 8.9 and runs at pH 9.5, the estimated isoelectric point of this peptide is slightly above pH 8.9. It is noteworthy that no ^{14}C-radioactivity in excess of background levels migrated toward the cathode. Limited trypsin treatment prior to electrophoresis abolished migration of the glucagon immunoreactive component toward the cathode (data not shown). These results are interpreted as an indication of the disposition of the peptide forming gel filtration peak c during high pH PAGE. The peptide in peak C is apparently quite basic and is thus not observed after conventional high pH PAGE.

DEAE ion exchange chromatography of insulin-glucagon pool peptides obtained by gel filtration. To further investigate the nature of this glucagon immunoreactive basic peptide, insulin-glucagon pool protein from gel filtration was subjected to ion exchange chromatography on DEAE Bio-Gel A, at pH 8.8. The results are shown in Fig. 4. Three peaks which possess both ^{3}H-labeling and glucagon immunoreactivity were observed (peaks 1-3; Fig. 4). Peak 1 eluted soon after sample application. This suggests the presence of a peptide with little or no affinity for the ion exchanger under these conditions. Peak 2 also eluted before any indication of an increase in conductivity due to initiation of the salt gradient was noted in the eluate. This suggests the presence of a peptide which was slightly retarded in its passage through the ion exchanger but not completely bound. Peak 3 eluted as the eluate conductivity reached approximately 2 mmho. Peak 4 which eluted at a conductivity of ~4 mmho, possessed essentially no glucagon immunoreactivity and was labeled primarily with ^{14}C-isoleucine.

To determine the identity of the peptides eluting in each peak, appropriate fractions were pooled and lyophilized. The resulting residue was then desalted on a Bio-Gel P-2 column in 1 M acetic acid. Portions of the resulting lyophilized

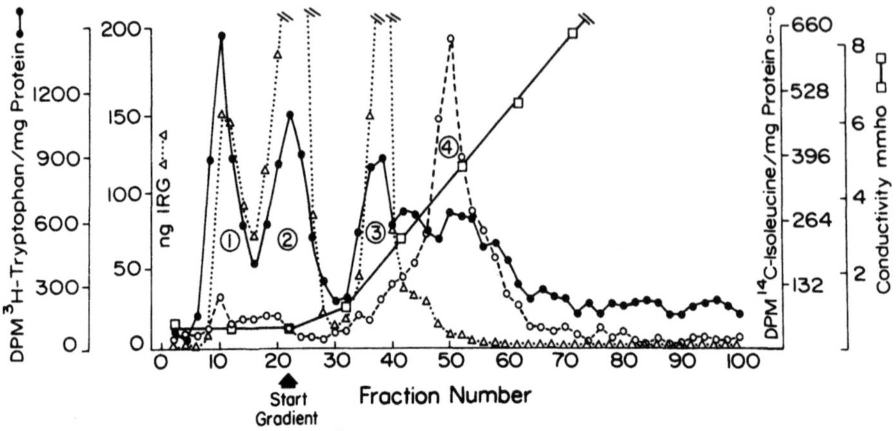

Fig. 4. Ion exchange chromatography of protein from the same preparation used for the experiments shown in Figs. 2 and 3. Two mg of protein were suspended in 5 ml of starting buffer (0.01 M Tris-HCl, pH 8.8, 0.02% NaN_3) and siphoned onto a 0.9 x 20 cm column of DEAE Bio-Gel A previously equilibrated with starting buffer. The flow rate was 25 ml/hr and fractions of 2 ml were collected. After elution of proteins not attracted to the ion exchanger under these conditions, a salt gradient was begun by placing 250 ml starting buffer in the mixing chamber and 250 ml of the same solution 0.5 M in NaCl in the feeder chamber of a Pharmacia GM-1 gradient mixer. Aliquots of appropriate size were taken for assay of radioactivity and glucagon immunoreactivity (IRG). The conductivity of the column eluate was monitored with a Serfass conductivity bridge. Starting conductivity was 0.45 mmho.

P-2 void volume protein were subjected to high pH PAGE. One half of each sample was run on a gel oriented toward the anode and one-half on a gel oriented toward the cathode. The distribution of the Coomassie blue stained bands obtained is shown in Fig. 5. Only DEAE peak 1 contained material which migrated toward the cathode. Therefore, cathode gels for peaks 2, 3 and 4 are not shown.

The major component of DEAE peak 1 proved to be a basic peptide which migrated toward the cathode. This band appears to represent the same glucagon immunoreactive peptide which was shown to be ^{3}H-tryptophan labeled in Fig. 3B. Minor contaminants of DEAE peak 1 include anglerfish glucagon and two unidentified peptides which migrated more slowly toward the anode on high pH PAGE (Fig. 5, second bar from top).

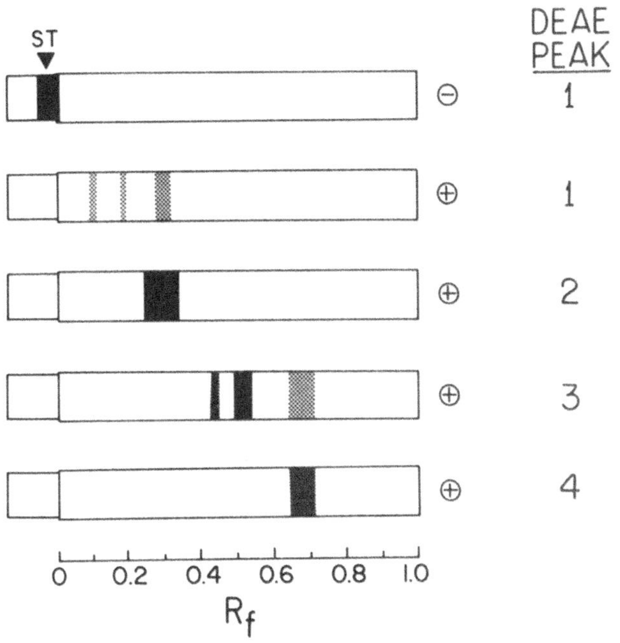

Fig. 5. Analysis of contents of DEAE ion exchange eluate pools by polyacrylamide gel electrophoresis. The eluate fractions encompassing peaks 1-4 from ion exchange chromatography (Fig. 4) were separately pooled, lyophilized and desalted in 1 M acetic acid on a 2.5 x 20 cm column of Bio-Gel P-2. Portions of the protein found in the resulting lyophilized P-2 void volumes were suspended in 0.1 M Tris-HCl buffer pH 9.0 for electrophoresis. Sample pairs were run with one gel oriented toward the anode and the other toward the cathode. The bars in the figure show the distribution and relative intensity of all protein bands which could be demonstrated by Coomassie Blue staining. Anode gel + , Cathode gel - .

DEAE peak 2 is comprised of anglerfish glucagon (R_f ~0.3). The primary component of DEAE peak 3 was monodesamidoglucagon (R_f ~0.5) with small amounts of contaminating insulin (R_f ~0.68) and a more slowly migrating unidentified peptide. Finally, DEAE peak 4 consisted primarily of anglerfish insulin.

The elution pattern of peaks 1, 2 and 3 from the DEAE ion exchanger thus suggests the presence of a glucagon-containing peptide which is more basic and possesses higher specific radioactivity (DPM/ng IRG) than either glucagon or monodesamidoglucagon. This peptide also possesses much lower immunospecificity (ng IRG/ng protein) than glucagon or monodesamidoglucagon (data not shown). These data, taken in conjunction with the finding that this basic peptide probably elutes as peak c during gel filtration (Figs. 1 and 2), suggests that it may serve as a cleavage intermediate in the metabolic processing of proglucagon to glucagon.

DISCUSSION

The anglerfish pancreatic islet is an excellent experimental system in which to study islet hormone biosynthesis. Large amounts of islet tissue free of exocrine contamination can be obtained with very little mechanical and no chemical perturbation. Recent studies in which insulin-, glucagon- and somatostatin-containing cells were localized in anglerfish islet by immunohistochemical means, revealed an approximate ratio of 9:6:4 of B:D:A cells (insulin:somatostatin: glucagon). Camera lucida drawings prepared from adjacent serial sections showed a close association between, but very little overlap in the distribution of cells containing each of the three hormones (15).

In other very recent studies, it has been determined that in addition to insulin and glucagon, somatostatin is synthesized in anglerfish islets (25). During the course of this work, it was found that a large proportion of anglerfish somatostatin is lost during the acid ethanolmethylene chloride-acetic acid extraction procedure employed in our previous studies. Although prosomatostatin may be retained in acid ethanol extracts, such extracts are largely free of somatostatin. Furthermore, when present in islet extracts (anglerfish somatostatin is quite soluble in 2 M acetic acid), somatostatin was found to elute in a position near the salt volume of Bio-Gel P-10 columns. This region is completely separate from the insulin- and glucagon-containing portion of the eluate. Since the acid ethanol procedure was employed to obtain the islet protein for the present study, and since gel filtration was performed using Bio-Gel P-10, the labeled peptides observed are primarily insulin and glucagon and their respective precursors.

The data in Figs. 1-5 provide further evidence that metabolic cleavage of anglerfish proglucagon occurs in a step-wise fashion. For the reasons given in the results section, consideration of the data in Figs. 1 and 2 indicates that ^{3}H-tryptophan labeled peak d, rather than peak c, represents anglerfish glucagon. The data in Fig. 3A confirm that one of the ^{3}H-labeled peptides in the gel filtration region encompassing peaks c and d is indeed anglerfish glucagon.

The results shown in Figs. 3B, 4 and 5 suggest that the ^{3}H-tryptophan-labeled peptide which is intermediate in size between insulin and glucagon (peaks c; Figs. 1 and 2) is a basic peptide having glucagon immunoreactivity. Estimates of the molecular size of this peptide, based on plots of log m.w. versus K_{av} from the elution positions shown in Fig. 1, resulted in a molecular size of 4,900 ± 400 daltons. The size of this peptide and the fact that, after radioisotope incorporation, it possesses higher specific radioactivity (DPM/ngIRG) than glucagon or monodesamidoglucagon (Fig. 4), suggest that it may serve as an intermediate in the metabolic processing of proglucagon to glucagon. In addition to acting as a highly electropositive peptide during both high pH PAGE and DEAE ion exchange chromatography, the basic nature of this peptide has been confirmed by results of preliminary amino acid analyses of the contents of peaks 1 and 2 from DEAE chromatography. The data indicate that peak 1 peptides have a higher arginine and lysine content than those in peak 2 (unpublished data). In many respects, this basic peptide appears to be quite similar to the fragment of mammalian proglucagon isolated and characterized by Tager and Steiner (30).

The diagram in Fig. 6 shows our current working hypothesis with regard to the relative size of anglerfish proglucagon and its metabolic conversion intermediates. Estimates of molecular size are based on calculations of the elution differentials observed in Fig. 1. The accuracy of these estimations is ± 8 per cent. The depiction of an NH_2-terminal histidine and a C-terminal lysine on the anglerfish glucagon molecule and an NH_2-terminal histidine on intermediates I and II is based on the results of Trakatellis et al (32). A histidine residue was found at the amino-terminus of both anglerfish glucagon and a glucagon containing peptide having a size similar to intermediate I. The placement of a tryptophan residue near position 25 is by analogy with mammalian glucagon (3-5, 31). Trakatellis et al. (32) found one tryptophan residue in anglerfish glucagon. The relationship between anglerfish proglucagon and intermediate I has been described previously (23). The identification of intermediate II is described in the present communication.

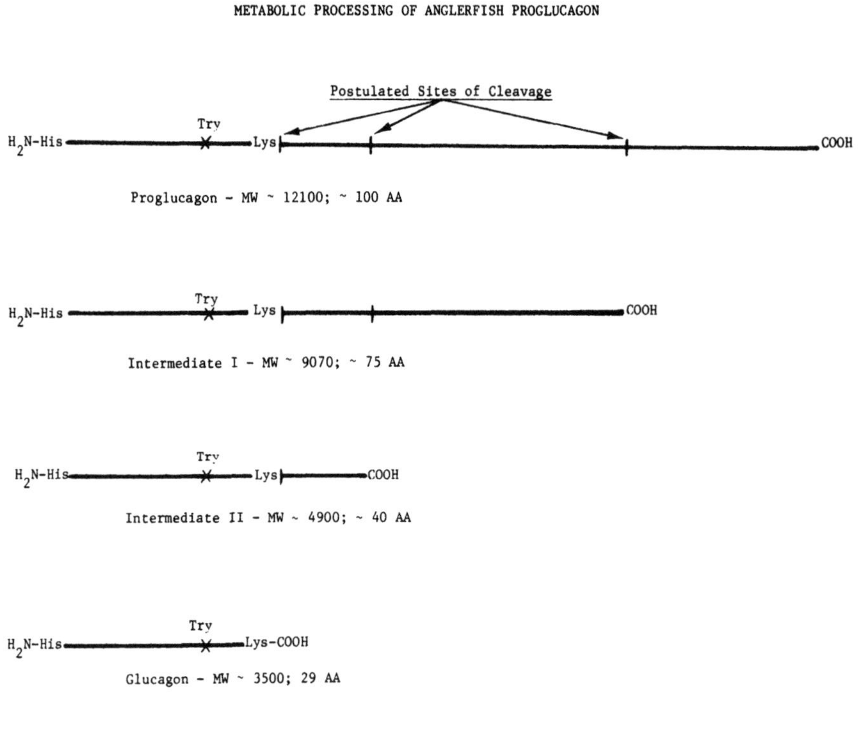

Fig. 6. A working hypothesis with regard to the relative size of various peptides involved in anglerfish glucagon biosynthesis. Confirmation awaits isolation and determination of the primary structures of all these peptides.

There is no evidence to support the placement of glucagon at the N-terminus of proglucagon. The extension of approximately 25 residues, which when added to intermediate I form proglucagon, could be positioned at the amino-terminus of proglucagon. If this were the case, then proglucagon would consist of peptide extensions at both the N- and C-terminal ends of glucagon. This arrangement would be similar to that found in the procollagen pro-α chains (9).

In contrast to intermediate II, both proglucagon and intermediate I act as acidic peptides during high pH PAGE and ion exchange chromatography. Both are more electronegative than glucagon (22, and unpublished results). Results from recent isoelectric focusing studies place the pI of these peptides near 5.5 and 5.9 (B. D. Noe and A. J. Moody, unpublished results). These results agree well with amino acid composition determinations on anglerfish "proglucagon" (m.w. ~9,000) by Trakatellis (32). O'Connor and Lazarus (27) recently reported the isolation and determination of several physico-chemical and biologic properties of a bovine "big pancreatic glucagon" (m.w. ~8,200). This peptide was also found to be more electronegative than glucagon.

The data available at present do not preclude the possibility that a peptide larger than 12,000 daltons which is not acid ethanol-acetic acid soluble may serve as a precursor or intermediate in anglerfish glucagon biosynthesis. Further, the 12,000 m.w. peptide, which does appear in acid ethanol-acetic acid extracts of anglerfish islets, is probably not the primary gene product in synthesis. Results from recent work in which (pro)insulin and (pro)PTH synthesis was studied using isolated mRNA from islets or parathyroids translated in heterologous cell-free systems, indicate that proinsulin and pro-PTH which can be extracted from the tissue of origin are not the primary gene products in synthesis (6, 16, 18, 41). Similar results have been obtained using isolated message for prolactin (8, 20). In all instances, the primary gene product is somewhat larger than the "precursor" peptides found in tissue extracts.

These discoveries correlate very well with the results of other recent studies in which mRNA isolated from exocrine pancreatic cells and murine myeloma was translated in cell-free systems (1, 2, 7). It has been found without exception that peptides destined for export from the cell have an amino terminal extension attached to the "precursor". The extension peptides from the various pro-enzymes (7) and proinsulin (6) show remarkable sequence homology. Dr. Blobel and his co-workers (1, 2) first suggested that these extension peptides may serve as a "signal sequence" which functions in the transfer of the newly synthesized peptide across the microsomal membrane. If the "signal" peptide is cleaved immediately upon entry into the endoplasmic reticulum cisternae, the "pre-proenzymes" or "pre-prohormones" would not be expected to be present in significant quantities in tissue extracts.

Are pre-prohormones utilized in synthesizing insulin, somatostatin and glucagon in anglerfish islets? D. Shields at Rockefeller University has recently been successful in translating anglerfish islet mRNA in a wheat germ system. High yields or radioactively labeled peptides were obtained. Size determinations of these cell-free translation products using SDS-PAGE and autoradiography revealed 3 major bands which had migration distances correlating with molecular sizes near 17,000, 15,000 and 11,500 daltons. One minor band having a molecular size near 13,000 daltons was also discerned. The 11,500 dalton peptide has eeen identified as anglerfish pre-proinsulin and amino acid sequence determination is in progress (D. Shields, unpublished results). Although attempts to identify any of the other translation products as being glucagon- or somatostatin-related have thus far been unsuccessful, the implications of the available data could be considered quite important.

Many reports appearing recently in the literature describe the identification of glucagon-like immunoreactivity associated with large peptides found in the plasma of normal humans and patients diagnosed as having glucagonomas. These peptides have also been found in plasma and pancreatic extracts of various experimental animals. Gel filtration of plasma samples indicates that the immunoreactive peptides range in size from ~9,000 to ~160,000 daltons (19, 29, 36-39). An important question which arises from these observations is whether any or all of these components represent normal A cell products. If one assumes that one of the three unidentified cell-free translation products in Shield's work represents anglerfish pre-proglucagon, then its maximum size would be near 17,000 daltons (the size of the largest translation product observed). This would mean that the maximum expected size for an A cell secretory product would be somewhat less than 17,000 daltons. Therefore, the glucagon-immunoreactive components found in plasma which exceed this size should not be considered "proglucagons". The physicochemical characteristics, biologic significance and origin of "big plasma glucagons" thus remain to be more clearly elucidated.

Acknowledgements

Appreciation is expressed for the technical assistance of Ms Ann Hunter.

This work was supported in part by Grant GB 43456 from The National Science Foundation and by Grant AM 16921 from the National Institute of Arthritis, Metabolism and Digestive Diseases, U. S. Public Health Service.

REFERENCES

1. Blobel, G. and Dobberstein, B. (1975): Transfer of proteins across membranes. I. Presence of proteolytically processed and unprocessed nascent immunoglobulin light chains on membrane-bound ribosomes of murine myeloma. J. Cell Biol. 67, 835-851.

2. Blobel, G. and Dobberstein, B. (1975): Transfer of proteins across membranes. II. Reconstitution of functional rough microsomes from heterologous components. J. Cell Biol. 67, 852-862.

3. Bromer, W. W., Boucher, M. E. and Koffenberger, J. E. Jr. (1971): Amino acid sequence of bovine glucagon. J. Biol. Chem. 246, 2822-2827.

4. Bromer, W. W., Boucher, M. E., Patterson, J. M., Pekar, A. H. and Frank, B. H. (1972). Glucagon structure and function. I. Purification and properties of bovine glucagon and monodesamidoglucagon. J. Biol. Chem. 247, 2581-2585.

5. Bromer, W. W., Sinn, L. G. and Behrens, O. K. (1975): Amino acid sequence of porcine glucagon. V. Location of amide groups, acid degradation studies and sequential evidence. J. Amer. Chem. Soc. 79, 2807-2810.

6. Chan, S. J., Keim, P. and Steiner, D. F. (1976): Cell-free synthesis of rat preproinsulin: characterization and partial amino acid sequence determination. Proc. Nat. Acad. Sci. 73, 1964-1968.

7. Devillers-Thiery, A., Kindt, T., Scheele, G. and Blobel, G. (1975): Homology in amino-terminal sequence of precursors to pancreatic secretory proteins. Proc. Nat. Acad. Sci. 72, 5016-5020.

8. Evans, G. A. and Rosenfeld, M. G. (1976): Cell-free synthesis of a prolactin precursor directed by mRNA from cultured rat pituitary cells. J. Biol. Chem. 251, 2842-2847.

9. Fessler, L. I., Morris, N. P. and Fessler, J. H. (1975): Procollagen: biological scission of amino and carboxyl extension peptides. Proc. Nat. Acad. Sci. 72, 4905-4909.

10. Hellerström, C., Howell, S. L., Edwards, J. C. and Andersson, A. (1972): An investigation of glucagon biosynthesis in isolated pancreatic islets of guinea pigs. FEBS Letters 27, 97-101.

11. Hellerström, C., Howell, S. L., Edwards, J. C. and Andersson, A. (1973): Aspects of the biosynthesis of glucagon in mammals. Postgrad. Med. J. 49, 601-603.

12. Hellerström, C., Howell, S. L., Edwards, J. C., Andersson, A. and Östenson, C. G. (1974): Biosynthesis of glucagon in isolated pancreatic islets of guinea pigs. Biochem. J. 140, 13-23.

13. Howell, S. L., Hellerström, C. and Tyhurst, M. (1974): Intracellular transport and storage of newly synthesized proteins in the guinea pig pancreatic A cell. Horm. Metab. Res. 6, 267-271.

14. Howell, S. L., Hellerström, C. and Whitfield, M. (1974): Radioautographic localization of labelled proteins after incubation of guinea-pig islets of Langerhans with (^{3}H)-tryptophan. Biochem. J. 140, 22-23.

15. Johnson, D. E., Torrence, J. L., Elde, R. P., Bauer, G. E., Noe, B. D. and Fletcher, D. J. (1976): Immunohistochemical localization of somatostatin, insulin, and glucagon in the principal islets of the anglerfish (Lophius americanus) and the channel catfish (Ictalurus punctata.) Amer. J. Anat. 147, 119-124.

16. Kemper, B., Habener, J. F., Ernst, M. D., Potts, J. T. Jr. and Rich, A. (1976): Preproparathyroid hormone: analysis of radioactive tryptic peptides and amino acid sequence. Biochem. 15, 15-19.

17. Kemper, B., Habener, J. R., Mulligan, R. C., Potts, J. T. Jr. and Rich, A. (1974): Preproparathyroid hormone: a direct translation product of parathyroid messenger RNA. Proc. Nat. Acad. Sci. 71, 3731-3735.

18. Kemper, B., Habener, J. F., Potts, J. T. Jr. and Rich, A. (1976): Preproparathyroid hormone: fidelity of the translation of parathyroid messenger RNA by extracts of wheat germ. Biochem. 15, 20-25.

19. Kuku, S. F., Zeidler, A., Emmanouel, D. S., Katz, A. I. and Rubenstein, A. H., Levin, N. W. and Tello, A. (1976): Heterogeneity of plasma glucagon: patterns in patients with chronic renal failure and diabetes. J. Clin. Endocr. Metab. 42, 173-176.

20. Maurer, R. A., Stone, R. and Gorski, J. (1976): Cell-free synthesis of a large translation product of prolactin messenger RNA. J. Biol. Chem. 251, 2801-2807.

21. Neumann, P. A., Koldenhof, M. and Humbel, R. E. (1969): Amino acid sequence of insulin from anglerfish (Lophius piscatorius). Hoppe-Seyler's Z. Physiol.

Chem. 350, 1286-1288.

22. Noe, B. D. and Bauer, G. E. (1971): Evidence for glucagon biosynthesis involving a protein intermediate in islets of the anglerfish (Lophius americanus). Endocrinology 89, 642-651.

23. Noe, B. D. and Bauer, G. E. (1975): Evidence for sequential metabolic cleavage of proglucagon to glucagon in glucagon biosynthesis. Endocrinology 97, 868-877.

24. Noe, B. D., Bauer, G. E., Steffes, M. W., Sutherland, D. E. R. and Najarian, J. S. (1975): Glucagon biosynthesis in human pancreatic islets: Preliminary evidence for a biosynthetic intermediate. Horm. Metab. Res. 7, 314-322.

25. Noe, B. D., Weir, G. C. and Bauer, G. E. (1976): Somatostatin biosynthesis in anglerfish islets. Biol. Bull. (in press).

26. O'Connor, K. J., Gay, A. and Lazarus, N. R. (1973): The biosynthesis of glucagon in perfused rat pancreas. Biochem. J. 134, 473-480.

27. O'Connor, K. J. and Lazarus, N. R. (1976): The purification and biological properties of pancreatic big glucagon. Biochem. J. 156, 265-277.

28. O'Connor, K. J. and Lazarus, N. R. (1976): Studies on the biosynthesis of pancreatic glucagon in the pigeon (Columba livia). Biochem. J. 156, 279-288.

29. Recant, L., Perrino, P. V., Bhathena, S. J., Danforth, D. N. and Lavine, R. L. (1976): Plasma immunoreactive glucagon fractions in four cases of glucagonoma: increased "large glucagon immunoreactivity". Diabetologia 12, 319-326.

30. Tager, H. S. and Steiner, D. F. (1973): Isolation of a glucagon-containing peptide: primary structure of a possible fragment of proglucagon. Proc. Nat. Acad. Sci. 70, 2321-2325.

31. Thomsen, J., Kristiansen, K., Brunfeldt, K. and Sundby, F. (1972): The amino acid sequence of human glucagon. FEBS Letters 21, 315-319.

32. Trakatellis, A. C., Tada, K., Yamaji, K. and Gardiki-Kouidou, P. (1975): Isolation and partial characterization of anglerfish proglucagon. Biochem. 14, 1508-1512.

33. Tung, A. K. (1973): Biosynthesis of avian glucagon: Evidence for a possible high molecular weight biosynthetic intermediate. Horm. Metab. Res. 5, 416-424.

34. Tung, A. K. (1974): Glucagon biosynthesis in avian pancreatic islets: evidence for medium-sized biosynthetic intermediates. Can. J. Biochem. 52, 1081-1088.

35. Tung, A. K. and Zerega, F. (1971): Biosynthesis of glucagon in isolated pigeon islets. Biochem. Biophys. Res. Commun. 45, 387-395.

36. Valverde, I., Dobbs, R. and Unger, R. H. (1975): Heterogeneity of plasma glucagon immunoreactivity in normal, depancreatized, and alloxan-diabetic dogs. Metabolism 24, 1021-1028.

37. Valverde, I., Lemon, H. M., Kessinger, A. and Unger, R. H. (1976): Distribution of plasma glucagon immunoreactivity in a patient with suspected glucagonoma. J. Clin. Endocr. Metab. 42, 804-808.

38. Valverde, I., Villanueva, M. L., Lozano, I. and Marco, J. (1974): Presence of glucagon immunoreactivity in the globulin fraction of human plasma ("big plasma glucagon"). J. Clin. Endocr. Metab. 39, 1090-1098.

39. Weir, G. C., Knowlton, S. D. and Martin, D. B. (1975): High molecular weight glucagon-like immunoreactivity in plasma. J. Clin. Endocr. Metab. 40, 296-302.

40. Yamaji, K., Tada, K. and Trakatellis, A. C. (1972): On the biosynthesis of insulin in anglerfish islets. J. Biol. Chem. 247, 4080-4088.

41. Yip, C. C., Hew, C. L. and Hsu, H. (1975): Translation of messenger ribonucleic acid from isolated pancreatic islets and human insulinomas. Proc. Nat. Acad. Sci. 72, 4777-4779.

HIGH MOLECULAR WEIGHT GLUCAGONS FROM AVIAN ISLETS AND FETAL BOVINE PANCREAS

A. K. Tung

Department of Medicine, University of Toronto
Toronto, Canada

Extracts of pigeon pancreatic islets and fetal bovine pancreas contain a 9000 dalton and a 69,000 dalton molecule with glucagon-like immunoreactivity. The avian 9000 dalton molecule, purified by gel filtration and DEAE-cellulose ion-exchange chromatography is composed of 76 amino acids, including all those of turkey glucagon.

In the past several years, glucagon biosynthesis has been the subject of intensive research in several laboratories. The available data suggest the involvement of precursor molecules with molecular weights ranging from 69,000 daltons to 4500 daltons (3, 4, 6, 7, 9). Our own studies have suggested the synthesis of two large glucagon-related molecules (69,000 Δ, and 9000 Δ) in the isolated pigeon islet system (4, 9). The present communication is concerned with additional studies on the isolation and purification of these large size proteins possessing glucagon immunoreactivity (IRG). See Fig. 1.

Studies Using Pigeon Islets

Islets obtained from collagenase digested pigeon pancreata were treated with TCA and extracted with ethanol-hydrochloric acid mixture containing 10% 0.25 M benzamidine hydrochloride (7). The acid ethanol soluble proteins were filtered on a Sephadex G-50 column with 1 M acetic acid. The gel filtration profile of acid ethanol soluble islet proteins shows 3 IRG peaks (Fig. 2): a peak corresponding to the void volume (peak I), a peak eluting in the 9000 Δ region (peak II) and glucagon (Peaks III, IV).

Partial Purification of a 9000 Dalton IRG Protein from Avian Islets

Fractions corresponding to peak II were pooled, concentrated by lyophilization and rechromatographed on Sephadex G-50 (Fig. 3). The twice gel-filtered

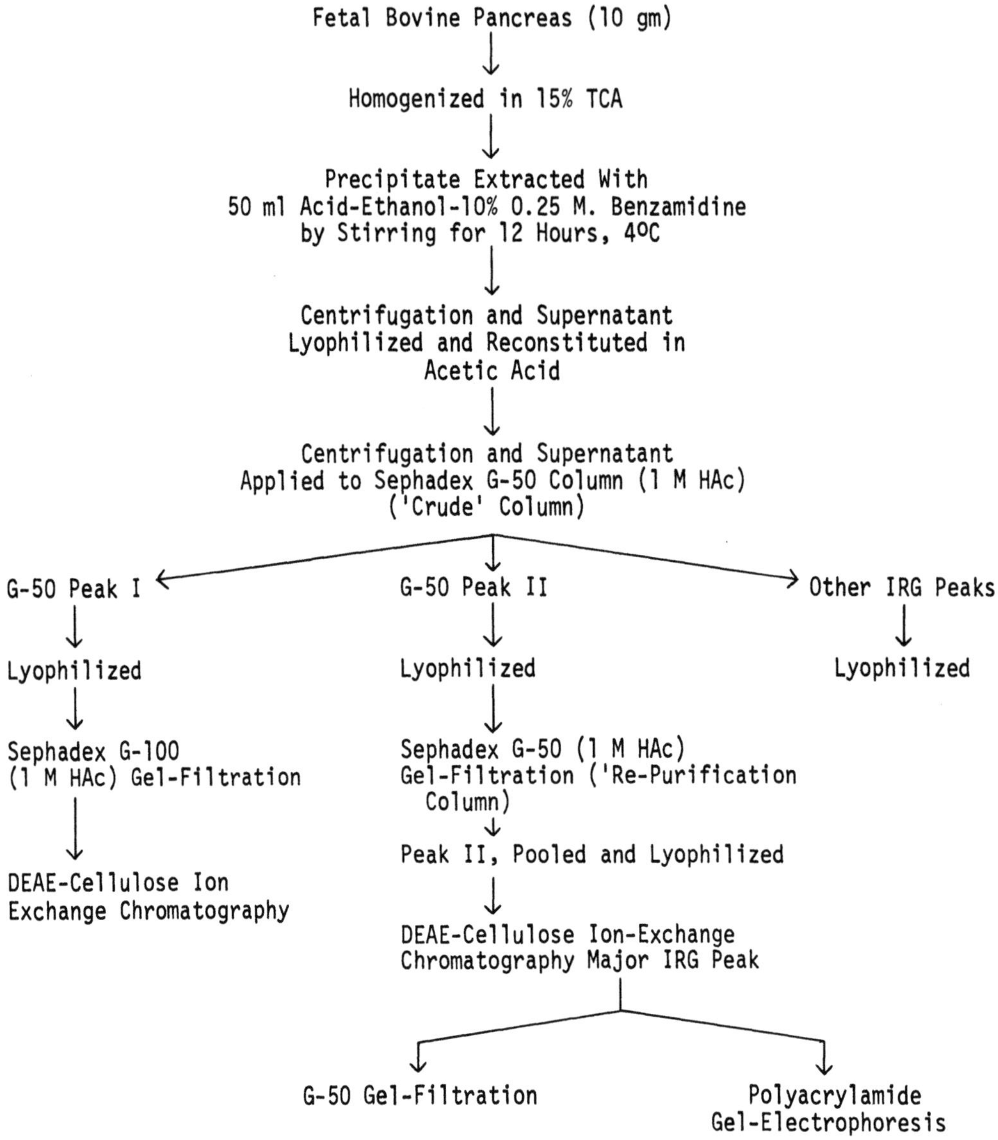
Fetal Bovine Pancreas (10 gm)
Homogenized in 15% TCA
Precipitate Extracted With
50 ml Acid-Ethanol-10% 0.25 M. Benzamidine
by Stirring for 12 Hours, 4°C
Centrifugation and Supernatant
Lyophilized and Reconstituted in
Acetic Acid
Centrifugation and Supernatant
Applied to Sephadex G-50 Column (1 M HAc)
('Crude' Column)
G-50 Peak I
Lyophilized
Sephadex G-100
(1 M HAc) Gel-Filtration
DEAE-Cellulose Ion
Exchange Chromatography
G-50 Peak II
Lyophilized
Sephadex G-50 (1 M HAc)
Gel-Filtration ('Re-Purification
Column)
Peak II, Pooled and Lyophilized
DEAE-Cellulose Ion-Exchange
Chromatography Major IRG Peak
G-50 Gel-Filtration
Polyacrylamide
Gel-Electrophoresis
Other IRG Peaks
Lyophilized

Fig. 1

peak II material was dissolved in Tris-HCl-urea buffer and submitted to DEAE-cellulose ion exchange column chromatography (tris-HCl, urea). When the column was eluted with 0.01 M NaCl, a small amount of immunoreactivity (DEAE IIa) appeared. When eluted with 0.1 M NaCl, DEAE IIb fraction containing more than 90% of the immunoreactivity was obtained (Fig. 4). Electrophoresis of the peak IRG material on polyacrylamide gel electrophoresis showed one main band, appearing more cationic than glucagon (Fig. 5). When DEAE IIb was iodinated and applied to a Sephadex G-50 column, a single radioactive peak of 9000 daltons was obtained (Fig. 6). Fig. 3-6 amino acid analysis showed that the DEAE IIb peak is composed of 76 amino acids including all those of turkey glucagon (2). Its amino acid composition is similar, but not identical to that of angler fish "proglucagon" (Table 1).

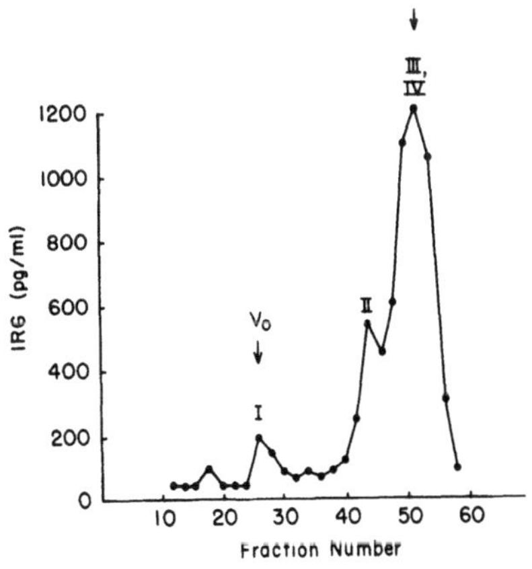

Fig. 2. Sephadex G-50 gel filtration of acid-ethanol islet proteins.

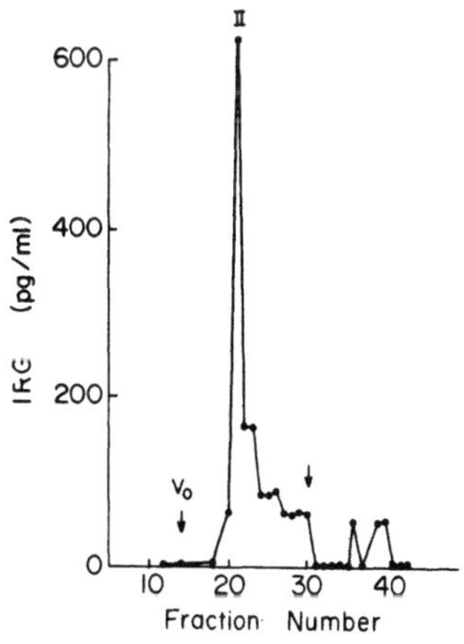

Fig. 3. Sephadex G-50 rechromatography of peak II on a sephadex G-50 column.

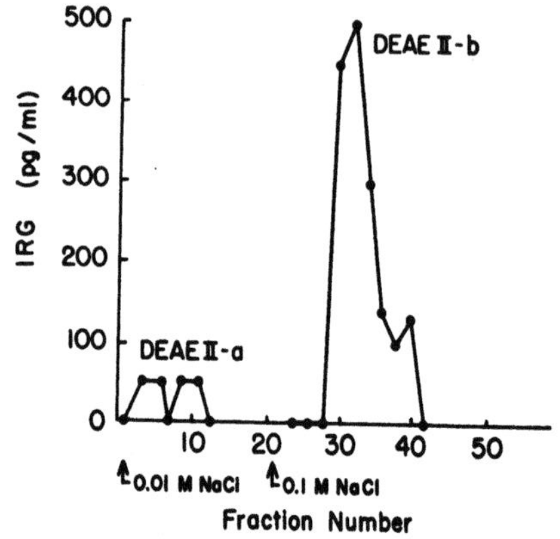

Fig. 4. DEAE-cellulose ion-exchange chromatography of peak II (Fig. 3) after refractionation on a Sephadex G-50 column. Proteins, extracted from islets in the presence of benzamidine-HCl, were chromatographed. Aliquots of 0.1 ml in 0.1 ml of aprotinin solution (1000 KIU) were dialyzed against distilled water. The material was lyophilized and dissolved in 0.5 ml of sodium barbital-albumin buffer for radioimmunoassay.

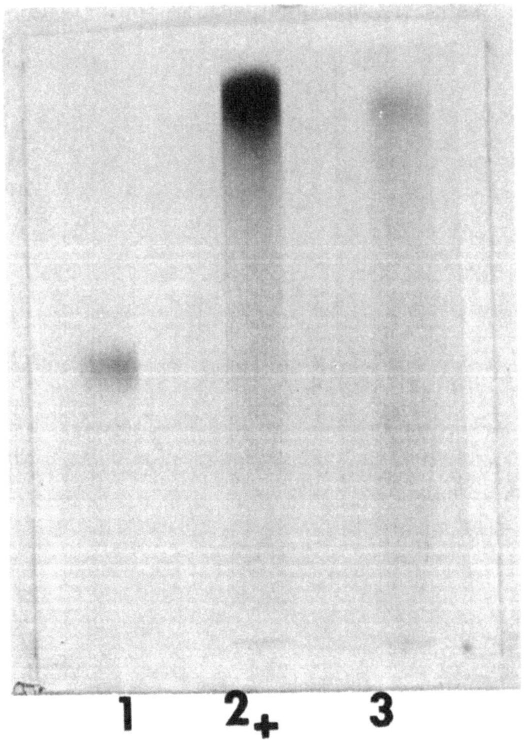

Fig. 5. Polyacrylamide gel-electrophoresis of DEAE-Peak-IIb. Fractions corresponding to DEAE-Peak-IIb (Fig. 4) were pooled and dialyzed against distilled water. The protein recovered by lyophilization was dissolved in 100 µl of 0.01 M HCl and subjected to electrophoresis in urea-containing gels, pH 3.5. (1) Pancreatic glucagon; (2) DEAE-Peak-IIa; (3) DEAE-Peak-IIb. +, anode; - cathode.

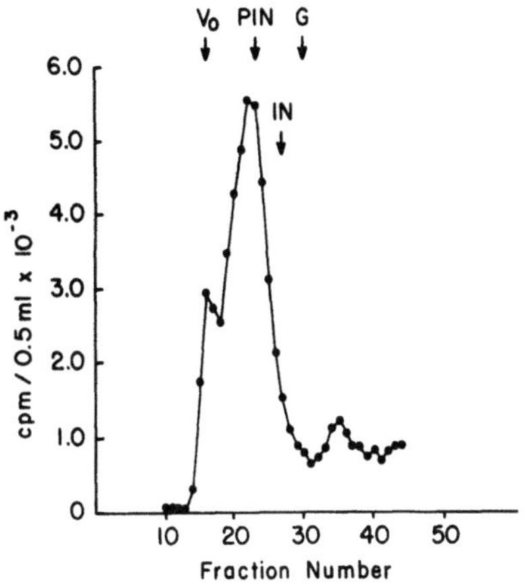

Fig. 6. Sephadex G-50 (1 M HAc) gel filtration of ^{125}I-DEAE-IIb

TABLE 1

AMINO ACID COMPOSITIONS OF THE 9000 DALTON GLUCAGON IMMUNOREACTIVE ISLET PROTEIN (DEAE-PEAK IIb), AVIAN GLUCAGON, AND ANGLERFISH "PROGLUCAGON"

	Amino Acid Residues Per Molecule		
Amino Acid	DEAE-peak* II-b	Avian** Glucagon (10)	Fish Proglucagon* (18)
Ala	7	1	8
Arg	3	2	4
Asp	7	3	9
Cys 1/2	0	0	0
Glu	10	3	14
Gly	13	1	6
His	1	1	2
Ileu	2	0	0
Leu	3	2	16
Lys	4	1	3
Met	1	1	0
Phe	2	2	2
Pro	4	0	1
Ser***	10	5	6
Thr	4	3	2
Trp	N.D.	1	1
Tyr	2	2	2
Val	3	1	2

*Based on a molecular weight of 9000 daltons. Numbers in parentheses represent the nearest integers. **Based on a molecular weight of 3500 daltons. ***Serine content was increased by 10% to account for destruction during hydrolysis.

Studies on Peak I

Peak I obtained from Sephadex G-50 gel filtration was fractionated on Sephadex G-100 (1 M HAc). Fig. 7 shows that it appears as a single IRG peak eluting like bovine serum albumin. When G-100-fractionated peak I was submitted to a radioimmunoassay using two different antiglucagon sera, almost identical dose-response curves were obtained (Fig. 8).

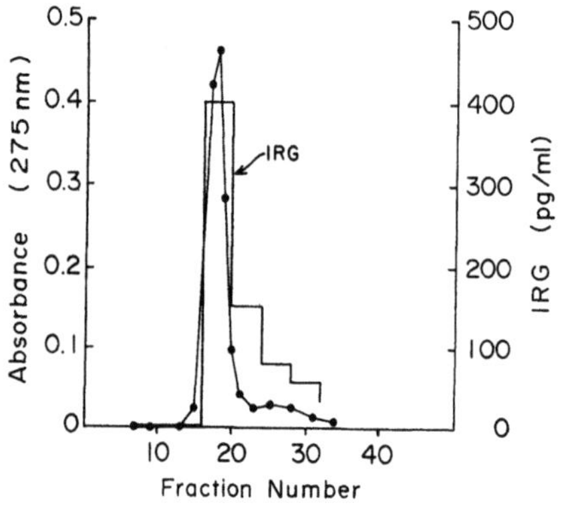

Fig. 7. G-100, Sephadex (1 M acetic acid) gel-filtration of G-50-peak I. AGS-30K was used for RIA.

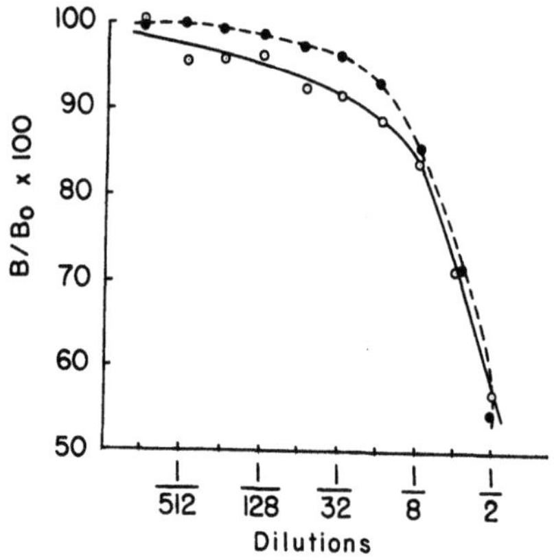

Fig. 8. Radioimmunoassay of Sephadex G-100-peak I (69,000 dalton) employing AGS 10, AGS 18.

Studies Using Fetal Bovine Pancreas

In the past several years, extensive use has been made of the fetal bovine pancreas for the study of insulin biosynthesis (8) since the fetal pancreas contains a greater proportion of endocrine cells than the adult pancreas (1). In addition, proteolytic enzyme activities in the fetal pancreas are much lower than in that of the adult. Because of these considerations and the fact that glucagon and glucagon-related proteins are extremely susceptible to proteolytic degradation (5), we have decided to use fetal bovine pancreas for the large-scale isolation and purification of glucagon related proteins.

Extraction of Glucagon-Related Proteins

Acid ethanol extraction has been the standard procedure for the preparation of glucagon from pancreatic tissue for many years. Recently, we have employed an extraction procedure involving TCA treatment of islets prior to extraction with acid ethanol containing benzamidine. The procedure was thought to improve the recovery of glucagon and large glucagon. We have used 3 different methods to extract fetal bovine pancreas: urea-acetic acid, ethanol-hydrochloric acid, and ethanol-hydrochloric with TCA precipitation, benzamidine hydrochloride (10%-0.25 M) being present in all procedures. The soluble proteins were chromatographed on Sephadex G-50 columns (1 M acetic acid; Fig. 9). Results show that approximately equal amounts of peak I IRG were obtainable by all 3 methods. However, significantly greater amounts of the 3500 Δ glucagon was extracted by the TCA-ethanol-HCL method than by either of the other 2 methods (data not shown). When Sephadex G-50-peak I was rechromatographed on Sephadex G-100 (1 M acetic acid), peak I (69,000 daltons) and varying amounts of glucagon, depending on the extraction method, were obtained (Fig. 10). Peak I prepared by extraction with TCA-ethanol-HCl is far more stable than components prepared by either of the other two methods.

Partial Purification of a 9000 Dalton IRG Protein From Fetal Bovine Pancreas

Fractions corresponding to peak II from the Sephadex G-50 gel filtration of acid ethanol soluble proteins (after TCA precipitation) were pooled, lyophilized, and resubmitted to G-50 gel filtration. Figure 11 shows peak II eluting as a 9000 dalton peak. In addition, IRG appeared in the insulin (indicated by the marker) and glucagon region. In order to further purify peak II, it was fractionated on a DEAE cellulose column, equilibrated in Tris-HCl, 3 M urea, 0.01 M NaCl. Elution was done using a stepwise NaCl gradient: 0.05 M, 0.075 M and 0.1 M. A major glucagon immunoreactive peak was found to be eluted by 0.05 M NaCl (Fig. 12). When the DEAE-peak material was dialyzed against distilled water

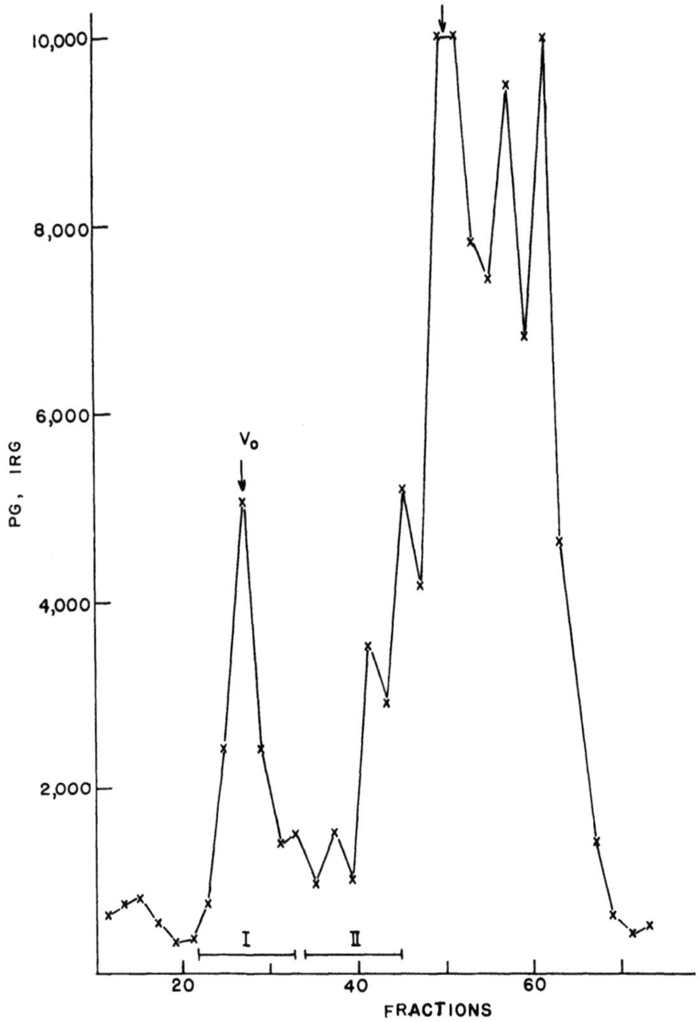

Fig. 9. Sephadex G-50 (1 M acetic acid) gel-filtration of acid ethanol soluble proteins obtained from fetal bovine pancreas. Arrow: position of insulin marker. This refers to an experiment in which TCA was used.

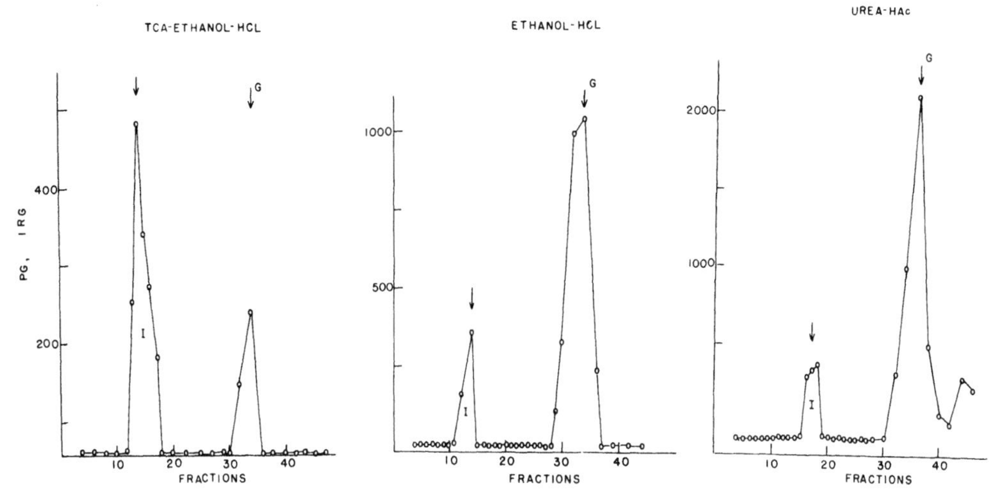

Fig. 10. Sephadex G-100 gel-filtration (1 M acetic acid) of G-50-peak I. Fetal bovine pancreas was (a) homogenized in 10% TCA, and the TCA precipitate extracted with ethanol-HCl containing benzamidine, (b) extracted with ethanol-HCl-benzamidine, (c) extracted with urea-acetic acid. ↓ elution position of serum albumin. (AGS 18 was employed in all RIAs).

and rechromatographed on Sephadex G-50, it appeared as a single IRG component corresponding to a m. w. of approximately 9000 daltons (Fig. 32). Polyacrylamide gel electrophoresis, in alkaline urea-containing gels, showed a single peak slightly more anionic (negative) than glucagon, suggesting that the final product is immunometrically homogeneous (Fig. 14). The G-100 peak I has been purified via DEAE-cellulose showing one main peak. Sodium dodecosulfate (SDS) gel electrophoresis of the material showed most of the IRG remaining close to the origin further suggesting its large size.

CONCLUSIONS

Our studies support the existence of a 9000 dalton and a 69,000 dalton high molecular weight molecule with glucagon immunoreactivity. Both molecules are clearly demonstrable in isolated avian islets and in the fetal bovine pancreas.

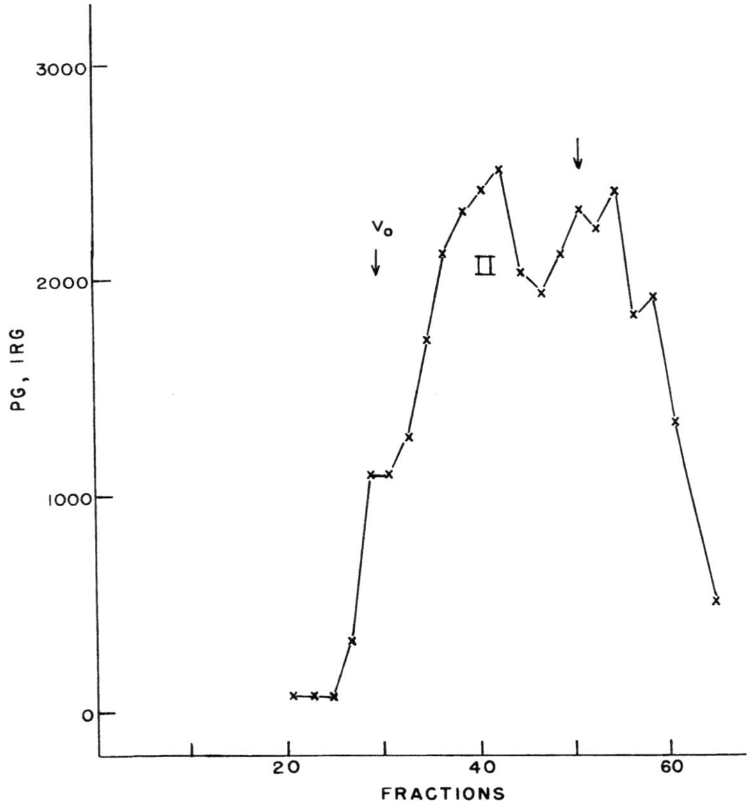

Fig. 11. Sephadex G-50 (1 M acetic acid) gel-filtration of G-50-peak II. Arrow: position of insulin.

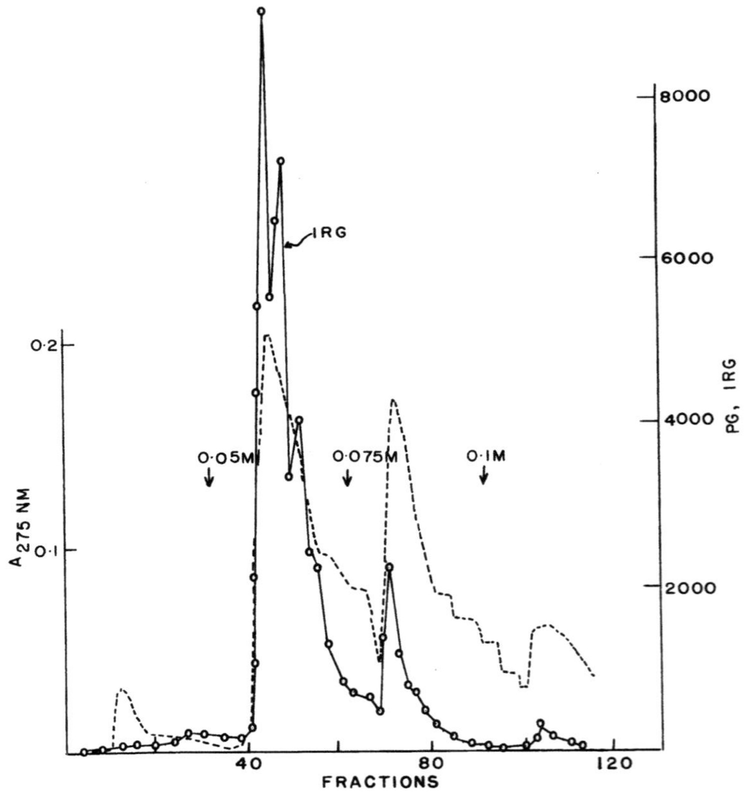

Fig. 12. DEAE-cellulose ion-exchange column chromatography (0.01 M Tris-HCL, 3 M urea, pH 8.6) of G-50-peak II, recovered by lyophilization, and reconstituted in 0.01 M Tris-HCl, 3 M urea, 0.01 M NaCl, pH 8.6. A stepwise NaCl gradient (0.01, 0.05, 0.075, 0.1 M) was used to develop the chromatogram. Continuous line, IRG; discontinuous, A275 nm.

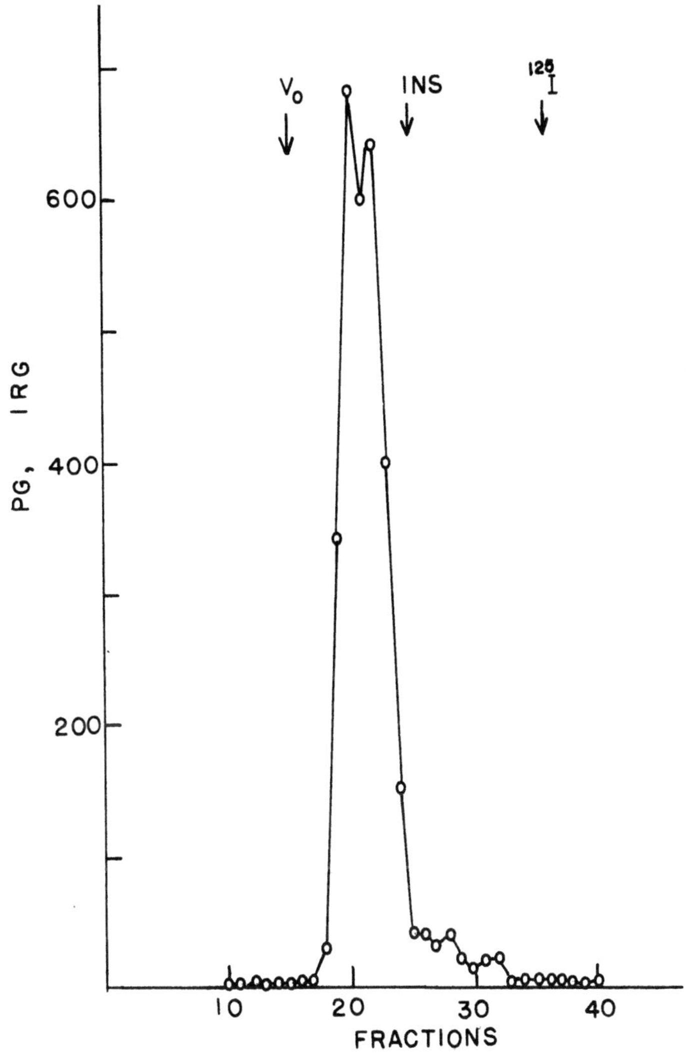

Fig. 13. Sephadex G-50 gel-filtration of the major IRG-peak. Obtained from DEAE-cellulose ion-exchange column, showing the elution of peak II as a molecule of 9000 Δ. Proinsulin, elutes at fraction 22.

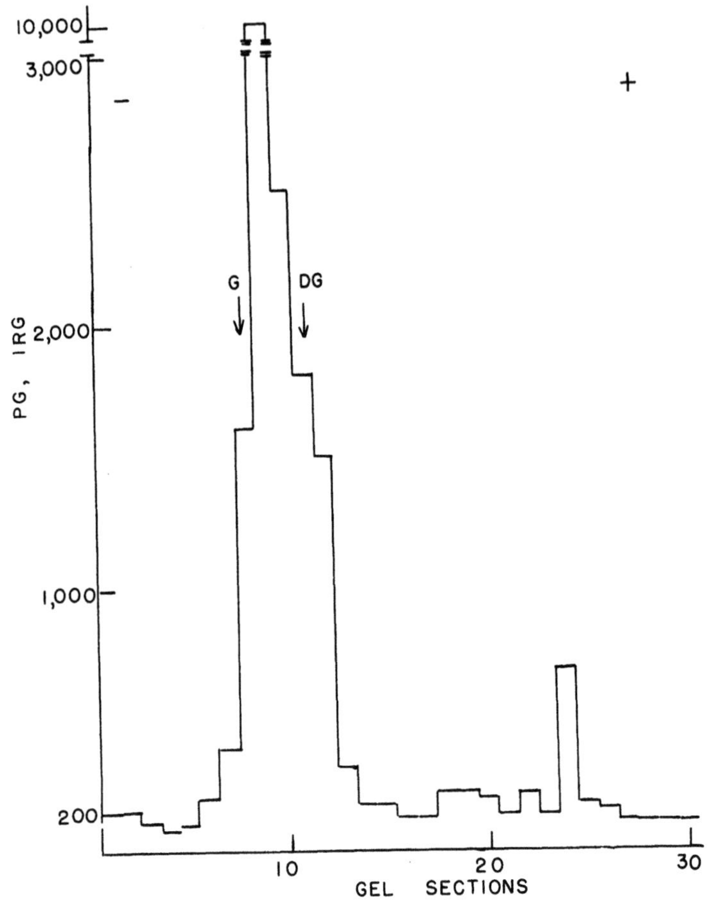

Fig. 14. Polyacrylamide gel-disc electrophoresis (7.5% gel, pH 8.9, urea containing gels) of the DEAE-cellulose purified peak II, showing that it is immunometrically homogenous. Gel slices were extracted in RIA diluent and then assayed for glucagon.

The 9000 dalton species has been partially purified from both avian islets and the fetal bovine pancreas by a combination of gel filtration and DEAE-cellulose ion exchange chromatography. The partially purified protein appears stable upon repeated gel filtration and polyacryamide gel disc electrophoresis. Very high molecular weight glucagon forms (approximately 69,000 daltons) are present in both the avian islets and the bovine fetal pancreas. These forms, like the 9000 dalton species, are very sensitive to proteolytic degradation. Thus, the trichloroacetic acid precipitation step of the acid ethanol extraction procedure is essential in preparing components which do not degrade into glucagon on gel refiltration.

REFERENCES

1. Lazarus, S. S. and Volk, B. W., Eds. (1962): The Pancreas in Human and Experimental Diabetes. Grune and Stratton, New York.

2. Markussen, J., Frandsen, E., Heding, L. G. and Sundby, F. (1972): Turkey glucagon: crystallization, amino acid composition and immunology. Horm. Metab. Res. 4, 360-363.

3. Noe, B. D. and Bauer, G. E. (1971): Evidence for glucagon biosynthesis involving a protein intermediate in islets of the anglerfish (Lophius americanus). Endocrinology 89, 642-651.

4. Noe, B. D. and Bauer, G. E. (1973): Further characterization of a glucagon precursor from anglerfish islet tissue. Proc. Soc. Exp. Biol. Med. 142, 210-213.

5. Traketellis, A. C., Tada, K., Yamaji, K. and Gardiki-Kouidou, P. (1975): Isolation and partial characterization of anglerfish proglucagon. Biochem. 14, 1508-1512.

6. Tung, A. K. (1973): Biosynthesis of avian glucagon: Evidence for a possible high molecular weight biosynthetic intermediate. Horm. Metab. Res. 5, 416-424.

7. Tung, A. K., Rosenzweig, S. A. and Foà, P. P. (1976): Glucagon from avian pancreatic islets: Purification and partial characterization of a 9000 dalton species with glucagon immunoreactivity. Proc. Soc. Exper. Biol. Med. 153, 344-349.

8. Tung, A. K. and Yip, C. C. (1969): Biosynthesis of insulin in bovine fetal pancreatic slices: the incorporation of titrated leucine into a single-chain proinsulin, a double-chain intermediate, and insulin in subcellular fractions. Proc. Nat. Acad. Sci. (Wash.) 63, 442-449.

9. Tung, A. K. and Zerega, F. (1971): Biosynthesis of glucagon in isolated pigeon islets. Biochem. Biophys. Res. Commun. 45, 387-395.

ASSESSMENT OF GLUCAGON IMMUNOREACTIVITY IN PLASMA

G. C. Weir

Diabetes Unit and Department of Medicine
Massachusetts General Hospital and Harvard Medical School
Boston, Massachusetts USA

The measurement of glucagon in plasma by radioimmunoassay, even with so-called pancreatic specific antisera, is known to be fraught with complexities. A problem not widely appreciated is that significant numbers of diabetic subjects with a history of insulin therapy appear to have endogenous glucagon antibodies. Of particular importance is the finding that materials of varying size with glucagon immunoreactivity (160,000; 10-20,000; 9000; 3500; and less than 2000 daltons) can be found in plasma in various circumstances. The 3500 molecular weight form is felt to represent pancreatic A cell true glucagon. The other forms, however, have not been fully characterized and their contribution to metabolic regulation in health and disease remains to be determined.

Shortly after the basic principles of the insulin radioimmunoassay were developed by Yalow and Berson (38), Unger *et al*. (27) applied the same methods to the measurement of glucagon. There were, however, a number of unexpected obstacles which made the accurate measurement of glucagon in plasma difficult. Degradation of labelled glucagon by plasma led to artifactually high assay results, a problem which was remedied initially with the proteolytic enzyme inhibitor Trasylol (4) and later benzamidine (5). There were, in addition, peptides of gastrointestinal origin with glucagon-like immunoreactivity which reacted with antisera used in the early assays. Finally antisera were developed which were relatively specific for pancreatic glucagon and reacted only weakly with the gastrointestinal peptides It then became possible to define many of the basic characteristics of A cell secretion in both normal and pathological states. A recent complicating discovery has been that the gastrointestinal tract of dogs, primarily the stomach, contains and secretes a peptide, indistinguishable from pancreatic glucagon, which under certain circumstances contributes significantly to the amount of glucagon immunoreactivity in plasma (12, 13, 22, 33). It seems likely, however, that in humans this "gastric glucagon" is present in physiologically insignificant amounts (1). These assay complexities, which have been described only superficially here, can be read about in detail in this monograph and elsewhere (6).

In addition to the above problems, it has gradually become clear that 3500 molecular weight pancreatic glucagon is not the only substance in plasma with glucagon immunoreactivity. For instance, gamma globulin was reported to react in the assay (26). Then a series of investigations evolved following attempts to make glucagon-free plasma by adsorption with charcoal (37). Charcoal could very easily adsorb ^{125}I-glucagon or crystalline porcine glucagon which was added to plasma, leading to the expectation that charcoal-treated plasma would be devoid of glucagon immunoreactivity. It was found instead, that large amounts of glucagon immunoreactivity (as determined with pancreatic specific antiserum 30K) remained and this material was termed the "interference factor". The amount of interference factor varied strikingly between individuals, being undetectable in some, and accounting for over 90% of the total glucagon immunoreactivity in others. An estimate of the amount of pancreatic glucagon could be obtained by subtracting the amount of interference factor from the total glucagon immunoreactivity. Subsequent studies with chromatographic techniques, which will be described below, suggest that the residual immunoreactivity following this subtraction is almost entirely pancreatic glucagon. Following the infusion of arginine, marked increases of pancreatic glucagon occurred whereas there was no change in the amount of interference factor. When fasting plasma from a large population of diabetics and non-diabetics was studied, the interference factor accounted for approximately 65% of the total glucagon immunoreactivity (36). Furthermore, the amount of interference factor, pancreatic glucagon, and total glucagon immunoreactivity did not significantly differ between diabetic and nondiabetic subjects.

The use of the subtraction technique is far from ideal. The charcoal treatment procedure is cumbersome and two assays must be done for a single sample. With the error inherent in radioimmunoassay techniques, the subtraction of one value from another clearly leads to unwanted variability, particularly in view of the fact that the plasma pancreatic glucagon concentration is so low, in the range of 20-50 pg/ml. Another approach has been to remove pancreatic glucagon, and other small peptides (which are probably present in only negligible amounts in normal subjects) with an acetone extraction technique (34). These extracts can then be dried, reconstituted in buffer and assayed, with the result being an estimate of the amount of glucagon in plasma.

With this background, column chromatography was then used to estimate the size of the various substances in plasma with glucagon immunoreactivity. Using G-200 Sephadex, the interference factor was estimated to have a molecular weight of approximately 160,000 (36). Furthermore it was found that charcoal treatment of plasma did not influence either the amount or the chromatography characteristics of the interference factor. The existence of this high molecular weight material in plasma was confirmed by Valverde et al. (31), and termed "big plasma glucagon" (BPG) even though it would seem more accurate to call it "big plasma glucagon immunoreactivity" (BPGI).

The identity of BPGI has remained a mystery. Conditions designed to dissociate non-covalent bonds (8 M urea) did not change the chromatographic pattern, but incubation of BPGI with trypsin led to the appearance of a moiety with a size comparable to glucagon. Despite the presence of large molecular weight material in extracts of pancreas and islets, there is no convincing evidence that BPGI is secreted by the A cell. Recent provocative information indicates that BPGI has hepatic glycogenolytic activity which is at least as potent as that of crystalline porcine glucagon (25). The utilization of cumbersome charcoal and acetone techniques to permit estimation of the amount of pancreatic glucagon in plasma was based partially on the assumption that only glucagon itself had significant biologic activity and that BPGI was inactive. Now, however, with the likelihood that BPGI is biologically active, it may be useful to have measurements of total plasma glucagon immunoreactivity if one wishes to equate immunoreactivity with biologic activity. On the other hand, the use of total plasma glucagon immunoreactivity can be misleading if one is assessing A cell function because concentrations of plasma BPGI do not appear to be influenced by acute changes of A cell secretion. This has become important recently because of the increased interest in A cell suppressibility. Most laboratories measure total glucagon immunoreactivity and following glucose and somatostatin infusions significant amounts of glucagon immunoreactivity persist. It is possible that the A cell is fully suppressed, that pancreatic glucagon levels are close to zero, and that the residual immunoreactivity is accounted for by BPGI.

In addition to the glucagon immunoreactivity in plasma contributed by BPGI and glucagon itself, it is apparent that other immunoreactive forms are present which include a 9000 dalton species (30) as well as material with a molecular weight of less than 2000 daltons (28, 31). Also, a 10-20,000 molecular weight form has recently been described in a family with increased levels of total glucagon immunoreactivity (19). Most normal subjects do not have detectable amounts of any immunoreactivity other than BPGI and glucagon (8, 9, 31). The identity of the 10-20,000 dalton form is unknown; the small form of less than 2000 daltons is also a mystery, but may be a glucagon fragment. Current concepts about the 9000 molecular weight species will be discussed below.

There has recently been growing appreciation of the clinical features of the glucagonoma syndrome which include a dermatitis, termed necrolytic migratory erythema, weight loss, anemia, stomatitis, impaired glucose tolerance, hypoaminoacidemia, and elevated plasma glucagon levels (11, 14). This has led to the identification of increasing numbers of patients with these glucagon secreting pancreatic A cell tumors. We recently had the opportunity to characterize the distribution of plasma glucagon immunoreactivity in five patients with glucagonoma (35). Using column chromatography with Biogel P30, three peaks of immunoreactivity could be identified. The void volume immunoreactivity was designated peak A and probably represents BPGI, although accurate size estimations with other chromatography techniques were not undertaken. Peak B contained material of approximately 9000 daltons and peak C material contained what is presumably true glucagon. The amount of total glucagon immunoreactivity in the fasting state was clearly elevated in all five subjects ranging from 760 to 2700 pg/ml. The striking finding in these subjects was both an absolute and a relative increase in the amount of 9000 dalton peak B immunoreactivity as compared with normal subjects. Peak B accounted for between 24.0 and 97.5 percent of the total plasma glucagon immunoreactivity. There were also large increases in the amount of the true glucagon peak C species, but the void volume peak A immunoreactivity remained within normal limits. The chromatographic distribution of plasma glucagon immunoreactivity has been found to be similar in other patients with glucagonoma (20, 29). In addition, total plasma glucagon immunoreactivity has been known to be increased in chronic renal failure (2) and now it is apparent that 9000 dalton peak B material accounts for the largest amount of this immunoreactivity (8, 9).

Studies were then carried out in these glucagonoma patients to determine whether these forms were actually A cell secretory products (35). Increases of total plasma glucagon immunoreactivity were seen after intravenous arginine and oral glucose, this latter response being a paradoxical phenomenon found in some patients with glucagonoma (10). A decrease of total immunoreactivity was seen after intravenous glucose (Fig. 1). The major contribution to these changes was from the true glucagon peak C species, but there were also clear changes of peak B material which presumably reflect either increased or decreased A cell secretion. There were no convincing changes in the amount of the peak A moiety which probably contains BPGI.

These questions were also approached by studying the venous effluent from tumors of two of the subjects at the time of surgery. There was a marked "step-up" of total glucagon immunoreactivity and peak C material in the effluent (Fig. 2). Particularly noteworthy was the finding of increased amounts of the 9000 dalton species in the venous effluent indicating the secretion of this material by the neoplastic A cells. There is also evidence for secretion of 9000 dalton material from the canine pancreas (28, 30) as well as from isolated

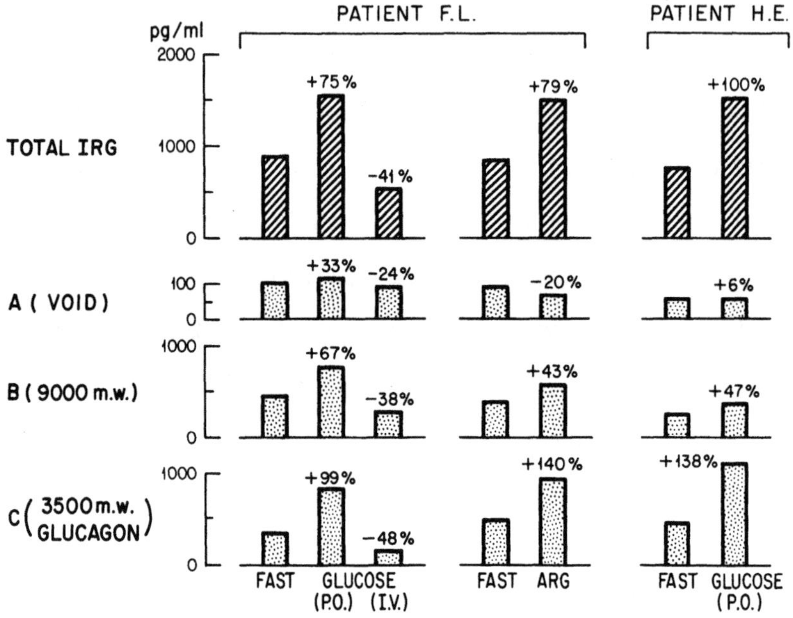

Fig. 1. Distribution of glucagon immunoreactivity in plasma of two patients with glucagonomas as determined by column chromatography. The percent changes refer to the fluctuations found after oral and intravenous glucose and intravenous arginine. The data used for these graphs has been published (35).

pigeon islets (18). In the glucagonoma patients there appears to be increased amounts of peak A immunoreactivity in the venous effluent, but this may be artifactual because of overlap between the very small peak A and the large peak B. The secretory characteristics of neoplastic A cells may therefore be similar to those of neoplastic B cells. It is now well accepted that insulinomas often secrete inordinate amounts of proinsulin and these data with glucagonoma patients suggest the secretion of inordinate amounts of a glucagon biosynthetic precursor. These characteristics may eventually have value for diagnostic purposes, as well as an aid in monitoring the progress of therapy.

There is only fragmentary information about the properties and identification of the 9000 molecular weight peptide with glucagon immunoreactivity found in plasma. It may be identical to a similar sized immunoreactive material which has

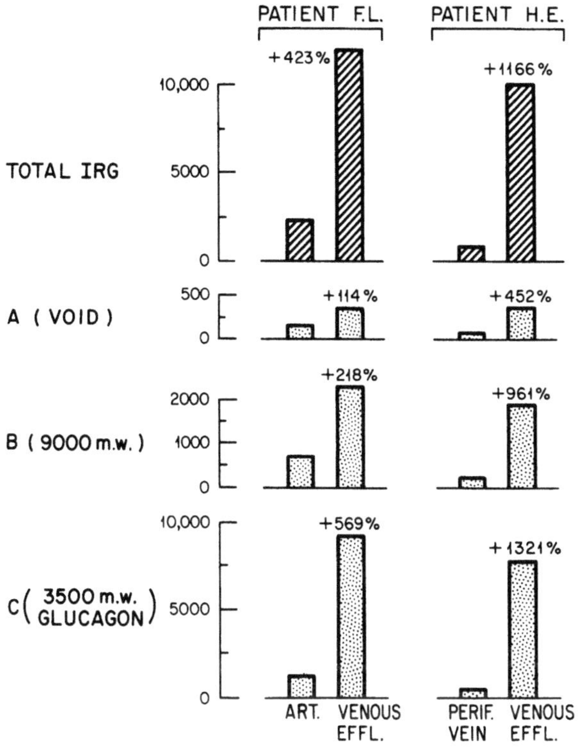

Fig. 2. Distribution of glucagon immunoreactivity in plasma of the venous effluent from two glucagonomas, obtained at surgery. The arterial and peripheral venous samples were obtained simultaneously. These data have been published previously (35).

been found in extracts of pancreatic tissue, and has been termed large glucagon immunoreactivity (LGI) (21) or "big glucagon" (17). There is much circumstantial evidence to suggest that this material in both tissue and plasma is a biosynthetic precursor of glucagon. Studies in islet tissue from a number of species, including humans (7, 15, 16, 18) indicate that ^{3}H-tryptophan is incorporated into a peptide of about 9000 daltons. Furthermore studies of immunologic displacement characteristics suggest that the 9000 dalton form and glucagon have common antigenic determinants (20, 35). Rigopoulou et al. (21), studying canine LGI were unable to demonstrate any glycogenolytic activity in the perfused rat liver or hyperglycemia in response to an in vivo injection of the material into dogs.

O'Connor and Lazarus (17) found that big glucagon from bovine pancreatic extracts, estimated to have a molecular weight of 8200, had an interesting spectrum of biologic effects. Big glucagon, which is probably the same as LGI, bound poorly to hepatic membrane glucagon receptors, actually diminished basal hepatic adenylate cyclase activity, and elicited no glycemic response following injection into rats. On the other hand, it was able to stimulate insulin release from islets and lipolysis in isolated avian fat cells. Recant (20) studied the biologic activity of the 9000 and 3500 dalton fractions isolated from the plasma of two patients with glucagonoma. Measuring cyclic AMP generation in isolated rat hepatocytes, the 3500 dalton fraction was as active as crystalline porcine glucagon, but the 9000 dalton species had reduced activity, and that which was found may have resulted from contamination by small amounts of the 3500 dalton form. Very little is known about the metabolic clearance rate of the 9000 molecular weight material. O'Connor and Lazarus (17) did, however, find that the half-life for the disappearance of tissue big glucagon was 6.9 minutes as compared with 3.2 for glucagon.

Another factor which may complicate the assessment of glucagon immunoreactivity in plasma is the presence of endogenous glucagon antibodies in diabetics receiving insulin therapy. In our studies of the plasma of 48 diabetic subjects with a history of insulin treatment, eight (17%) were found to have increased binding of ^{125}I glucagon by plasma, as assessed with a charcoal separation technique (36). On the other hand, we have never found increased binding in nondiabetic subjects or diabetics with no history of insulin injections. There are both positive (3, 23) and negative (24) reports of glucagon antibodies in plasma and Villalpando and Drash (32) have recently reported increased binding in twenty-four (12%) of 205 insulin-dependent diabetic children. They provided further evidence that this binding was caused by antibodies because the bound labelled glucagon was found in the gamma globulin fraction of plasma. It is not known what led to the generation of these antibodies, but perhaps they are the result of the use of insulin preparations contaminated with small amounts of glucagon. It is also possible that glucagon injections given for hypoglycemic reactions led to antibody development. This finding of increased glucagon binding by the plasma of significant numbers of diabetics is important because of the possible introduction of artifacts into the radioimmunoassay. Artifactually high values may be seen in a double antibody system whereas low values may be found with adsorption separation techniques, such as charcoal or talc.

Acknowledgements

This work was supported by research grants from the National Institutes of Health (AM 18230), the Juvenile Diabetes Foundation, and the L. A. Parker Foundation.

Present address for Dr. Weir: Department of Medicine, Medical College of Virginia, MCV Station, Richmond, Virginia 23298.

REFERENCES

1. Barnes, A. J. and Bloom, S. R. (1976): Pancreatectomized man: A model for diabetes without glucagon. Lancet 1, 219-220.

2. Bilbrey, G. L., Faloona, G. R., White, M. C. and Knochel, J. P. (1974): Hyperglucagonemia of renal failure. J. Clin. Invest. 53, 841-847.

3. Cresto, J. C., Lavine, R. L., Perrino, P., Recant, L., August, G., Hung, W. (1974): Glucagon antibodies in diabetic patients. Lancet 1, 1165.

4. Eisentraut, A. M., Whissen, N. and Unger, R. H. (1968): Incubation damage in the radioimmunoassay for human plasma glucagon and its prevention with "Trasylol". Amer. J. Med. Sci. 255, 137-142.

5. Ensinck, J. W., Shepard, C., Dudl, R. J. and Williams, R. H. (1972): Use of benzamidine as a proteolytic inhibitor in the radioimmunoassay of glucagon in plasma. J. Clin. Endocrinol. Metab. 35, 463-467.

6. Faloona, G. R. and Unger, R. H. (1974): Glucagon. In: Methods of Hormone Radioimmunoassay. B. M. Jaffe and H. R. Behrman, Eds. Academic Press, Inc., New York, pp. 317-330.

7. Hellerström, C., Howell, S. L., Edwards, J. C., Anderson, A. and Östenson, G. C. (1974): Biosynthesis of glucagon in isolated pancreatic islets of guinea pigs. Biochem. J. 140, 13-23.

8. Kuku, S. F., Jaspan, J. B., Emmanouel, D. S., Zeidler, A., Katz, A. I. and Rubenstein, A. H. (1976): Heterogeneity of plasma glucagon: Circulating components in normal subjects and patients with chronic renal failure. J.

Clin. Invest. 58, 742-750.

9. Kuku, S. F., Zeidler, A., Emmanouel, D. S., Katz, A. I., Rubenstein, A. H., Levin, N. W. and Tello, A. (1976): Heterogeneity of plasma glucagon: Patterns in patients with chronic renal failure and diabetes. J. Clin. Endocr. Metab. 42, 173-176.

10. Leichter, S. B., Pagliara, A. S., Greider, M. H., Pohl, S., Rosai, J. and Kipnis, D. M. (1975): Uncontrolled diabetes mellitus and hyperglucagonemia associated with an islet cell carcinoma. Am. J. Med. 58, 285-293.

11. Mallinson, C. N., Bloom, S. R., Warin, A. P., Somomom, P. R. and Cox, B. (1974): A glucagonoma syndrome. Lancet 2, 1-5.

12. Mashiter, K., Harding, P. E., Chou, M., Mashiter, G. D., Stout, J., Diamond, D. and Field, J. B. (1975): Persistent pancreatic glucagon, but not insulin response to arginine in pancreatectomized dogs. Endocrinology 96, 678-693.

13. Matsuyama, T. and Foà, P. P. (1974): Plasma glucose, insulin, pancreatic, and enteroglucagon levels in normal and depancreatized dogs. Proc. Soc. Exp. Biol. Med. 147, 97-102.

14. McGavran, M. H., Unger, R. H., Recant, L., Polk, H. C., Kilo, C. and Levin, M. E. (1966): A glucagon-secreting alpha cell carcinoma of the pancreas. N. Eng. J. Med. 274, 1408-1413.

15. Noe, B. G. and Bauer, G. E. (1975): Evidence for sequential metabolic cleavage of proglucagon to glucagon in glucagon biosynthesis. Endocrinology 97, 868-877.

16. Noe, B. D., Bauer, G. E., Steffes, M. W., Sutherland, D. E. R. and Najarian, J. S. (1975): Glucagon biosynthesis in human pancreatic islets: Preliminary evidence for a biosynthetic intermediate. Horm. Metab. Res. 7, 314-322.

17. O'Connor, K. J. and Lazarus, N. R. (1976): The purification and biological properties of pancreatic big glucagon. Biochem. J. 156, 265-277.

18. O'Connor, K. J. and Lazarus, N. R. (1976): Studies on the biosynthesis of pancreatic glucagon in the pigeon (Columba livia). Biochem. J. 156, 265-288.

19. Palmer, J. P., Werner, L., Benson, J. W. and Ensinck, J. W. (1976): Dominant inheritance of large molecular weight immunoreactive glucagon (IRG). Diabetes 25 (Suppl. 1) 326.

20. Recant, L., Perrino, P. V., Bhathena, S. J., Danforth, D. N. Jr. and Lavine, R. L. (1976): Plasma immunoreactive glucagon fractions in four cases of glucagonoma: increased "large glucagon" immunoreactivity. Diabetologia 12, 319-326.

21. Rigopoulou, D., Valverde, I., Marco, J., Faloona, G. and Unger, R. H. (1973): Large glucagon immunoreactivity in extracts of pancreas. J. Biol. Chem. 245, 496-501.

22. Sasaki, H., Rubalcava, B., Baetens, D., Blázquez, E., Srikant, C. B., Orci, L. and Unger, R. H. (1975): Identification of glucagon in the gastrointestinal tract. J. Clin. Invest. 56, 135-145.

23. Stahl, M., Nars, P. W., Herz, G., Baumann, J. B. and Girard, J. (1972): Glucagon antibodies after long-term insulin treatment in a diabetic child. Horm. Metab. Res. 4, 224-225.

24. Thomsen, H. G. (1971); Insulin treatment does not induce glucagon antibodies. Horm. Metab. Res. 3, 57-58.

25. Unger, R. H. (1976): Personal communication.

26. Unger, R. H. and Eisentraut, A. M. (1976): Etudes recentes sur la physiologie du glucagon. J. Ann. Diabetol. Hôtel Dieu 7, 7-18.

27. Unger, R. H., Eisentraut, A. M., McCall, M. S., Keller, S., Lanz, H. C. and Madison, L. L. (1959): Glucagon antibodies and their use for immunoassay of glucagon. Proc. Soc. Exp. Biol. Med. 102, 621-623.

28. Valverde, I., Dobbs, R. and Unger, R. H. (1975): Heterogeneity of plasma glucagon immunoreactivity in normal, depancreatized, and alloxan-diabetic dogs. Metabolism 24, 1021-1028.

29. Valverde, I., Lemon, H. M., Kessinger, A., Unger, R. H. (1976): Distribution of plasma glucagon immunoreactivity in a patient with suspected glucagonoma. J. Clin. Endocr. Metab. 42, 804-808.

30. Valverde, I., Rigopoulou, D., Marco, J., Faloona, G. R., Unger, R. H. (1970): Molecular size of extractable glucagon and glucagon-like immunoreactivity (GLI) in plasma. Diabetes 19, 624-629.

31. Valverde, I., Villaneuva, M. L., Lozano, I. and Marco, J. (1974): Presence of glucagon immunoreactivity in the glucagon factor of human plasma (big plasma glucagon). J. Clin. Endocr. Metab. 39, 1090-1098.

32. Villalpando, S. and Drash, A. (1976): Circulating glucagon antibodies in childhood diabetes, clinical significance and characterization. Diabetes 25 (Suppl. 1), 334.

33. Vranic, M. S., Pek, S. and Kawamori, R. (1974): Increased "glucagon immunoreactivity" in plasma of totally depancreatized dogs. Diabetes 23, 905-912.

34. Walter, R. M., Dudl, R. J., Palmer, J. P., Ensinck, J. W. (1974): The effect of adrenergic blockade on glucagon responses to starvation and hypoglycemia in man. J. Clin. Invest. 54, 1214-1220.

35. Weir, G. C., Horton, E. S., Aoki, T. T., Slovik, D., Jaspan, J. and Rubenstein, A. H. (1977): Secretion by glucagonomas of a possible glucagon precrusor. J. Clin. Invest., in press.

36. Weir, G. C., Knowlton, S. D. and Martin, D. B. (1975): High molecular weight glucagon-like immunoreactivity in plasma. J. Clin. Endocr. Metab. 40, 296-302.

37. Weir, G. C., Turner, R. C. and Martin, D. B. (1973): Glucagon radioimmunoassay using antiserum 30K: Interference by plasma. Horm. Metab. Res. 5, 241-244.

38. Yalow, R. S. and Berson, S. A. (1960): Immunoassay of endogenous plasma insulin in man. J. Clin. invest. 39, 1157-1175.

QUANTIFICATION OF PLASMA GLUCAGON IMMUNOREACTIVE COMPONENTS IN NORMAL AND HYPERGLUCAGONEMIC STATES

I. Valverde

Fundacion Jimenez Diaz, Universidad Autonoma de Madrid
Madrid, Spain

Chromatography of human and dog plasma on Bio Gel P-30 columns revealed the presence of as many as four glucagon fractions reacting with a "pancreatic glucagon-specific" antiserum (30K). These fractions have approximate molecular weights of 2000, 3500, 9000 and 160,000 and respond differently to stimulation with arginine and suppression with glucose. The 3500 and 9000 m. w. fractions account for most of the rise in plasma glucagon observed in chronic hyperglucagonemic states, such as cirrhosis of the liver, diabetes, phlorizin-induced hypoglycemia, renal failure, pancreatectomy, and glucagonoma. It is possible that the immunoreactive forms larger than pancreatic glucagon (3500 m. w.) may represent biosynthetic precursors and that those of smaller size may represent their metabolic products.

The report (3) of an antiserum that crossreacted very weakly with gut extracts and showed a decline in plasma immunoreactive glucagon values after oral glucose, whereas many other antisera indicated a rise and crossreacted highly with gut extracts (Glucagon-like immunoreactivity of intestinal origin, gut-GLI), led to the concept of "pancreatic glucagon-specific antiserum".

The first information concerning the molecular size of plasma glucagon was reported by Valverde _et al_ (28). Using a "glucagon specific antiserum" (G58), two glucagon immunoreactive peaks were found by gel filtration of acid-alcohol extracts of canine plasma collected during amino acid infusion. The major peak eluted with the labeled glucagon and the smaller one, less than 10% of the total immunoreactivity, eluted immediately before the labeled insulin. The immunoreactivity present in the 3500 molecular weight zone appeared to have the same glycogenolytic activity in the perfused rat liver of an equivalent amount of glucagon extracted from dog pancreas. The small peak may be related to the equally sized "large glucagon immunoreactivity" (LGI) described by Rigopoulou _et al_ (20) in pancreas extracts. The same report investigated the molecular size of gut GLI during its stimulation by intraduodenal glucose in dogs whose A-cell bearing portion of the pancreas had been resected. Two immunoreactive peaks of ~9000 and 3500 m.w. similar to those found after glucagon stimulation, were detected with a gut-GLI crossreacting antiserum (G-128) but they were barely detected by a specific antiserum" (G58).

From these data it could be assumed that using a "specific antiserum", the only detectable immunoreactivity in total plasma (provided that Trasylol was used to prevent glucagon degradation; 4) was the intact biologically active hormone. Nevertheless, the discrepancies of total plasma glucagon values obtained by various "glucagon specific antisera" led to the suspicion that other immunoreactive forms may exist. Weir et al (37) reported a new source of plasma glucagon immunoreactivity that was not removed by charcoal treatment of the plasma; they called it "interference factor".

The improvement of the sensitivity of the glucagon radioimmunoassay (5) gave the possibility of performing gel filtrations of unextracted plasma and assaying the eluates without prior or ulterior concentration of the samples.

By column chromatography of total plasma on Bio Gel P-30 (Fig. 1), as many as four glucagon immunoreactive fractions could be detected in plasma from humans and from several animal species with Unger's 30K (a "specific glucagon antiserum" which does not react with more than 2% of plasma gut-GLI; 5). The first eluting immunoreactive fraction appears in the void volume and has been named big plasma glucagon or BPG. On G-100 columns BPG elutes with the globulins and runs electrophoretically with the α_2 globulin. Treatment with 8 M urea- 1 M acetic acid does not alter its chromatographic pattern. Trypsin treatment of BPG produces glucagon immunoreactive molecules of smaller size down to one which approximates the 3500 m.w. zone. A similar fraction was detected in pancreatic extracts (31). The "interference factor" was later identified as BPG, and shown to have an approximate m.w. of 160,000 on G-200 and Sepharose A-15 columns (32, 36). The second fraction with an approximate m.w. of 9000 may well correspond to the "large glucagon immunoreactivity" (LGI) of pancreatic extracts, considered a candidate for the role of proglucagon (17). The third fraction with a m. w. of 3500 is considered to be the real hormone or "true glucagon". The final eluting fraction has a m.w. of approximately 2000.

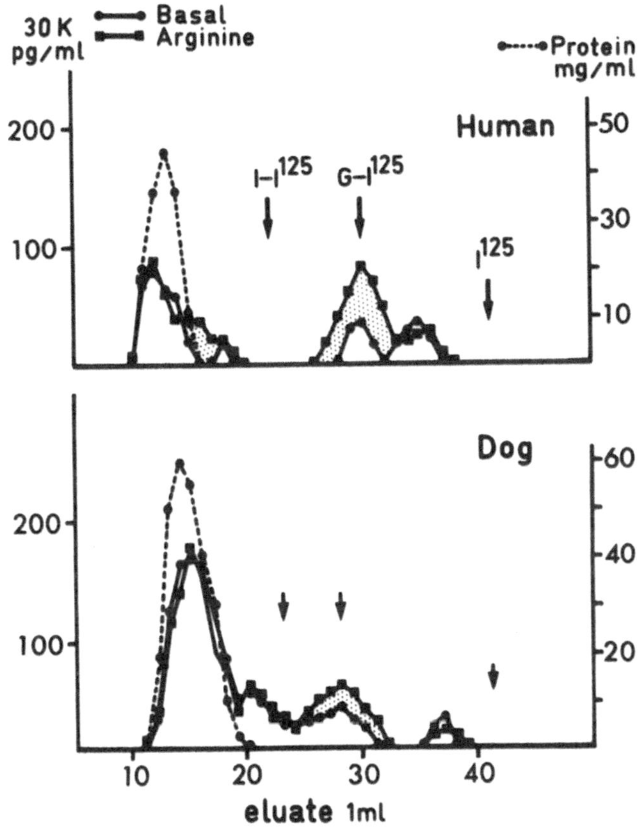

Fig. 1. Distribution of plasma 30K glucagon immunoreactivity in basal and post-arginine samples (3 ml loaded). Bio Gel P-30. The arrows indicate the elution volume of the radioactive markers (insulin-^{125}I; I-I^{125}: glucagon-^{125}I:G-I^{125}; ^{125}I).

Plasma Glucagon Immunoreactive (GIR) Components in Normal Basal State

The mean values of the four GIR components obtained in plasma samples from normal humans and dogs after an overnight fast are shown in Table 1.

TABLE 1

BASAL PLASMA 30K IMMUNOREACTIVE COMPONENTS. BIO GEL P-30

Humans N = 20	Total Plasma pg/ml	BPG pg/ml	9000 pg/ml	3500 pg/ml	2000 M. W. pg/ml
Mean	155	91	19	27	19
S. D.	96	54	17	25	16
% of total (range)		28-100	0-31	0-39	0-33
Dogs N = 10					
Mean	70	39	11	12	9
S. D.	32	23	10	11	5
% of total (range)		28-86	0-29	0-31	0-35

For each group, the mean BPG value represented the highest percentage of the total immunoreactivity recovered from the column (58% and 55%, respectively). The 9000 m.w. component was detectable in 75% of the humans and in 70% of the dogs studied; the mean value represented 12% and 15% of the total immunoreactivity, respectively. The 3500 m.w. component was detectable in 85% of the humans and in 80% of the dogs; the mean value represented 17% of the total in both species. The 2000 m.w. component was detectable in 85% of the human cases and in all but one dog.

Effect of Glucagon Stimulation on Plasma GIR Components

An arginine infusion in ten normal subjects revealed a significant increase in the 3500 m. w. component as compared to the respective basal samples (225 ± 165 vs 31 ± 29 pg/ml, mean ± SD, p <.005). The BPG rose slightly in 70% of the cases, but the mean value did not show a statistically significant difference (128 ± 80 vs 109 ± 71 pg/ml). No significant change was observed in the other two fractions (Fig. 2). In normal dogs the elevation of the glucagonemia during arginine or alanine infusion was also accounted for mainly by an increase of the 3500 m. w. component, although a slight rise of the 9000 m. w. component was observed in two out of five dogs studied. Weir et al. (37) found no change of immunoreactivity in charcoal treated plasma after arginine stimulation.

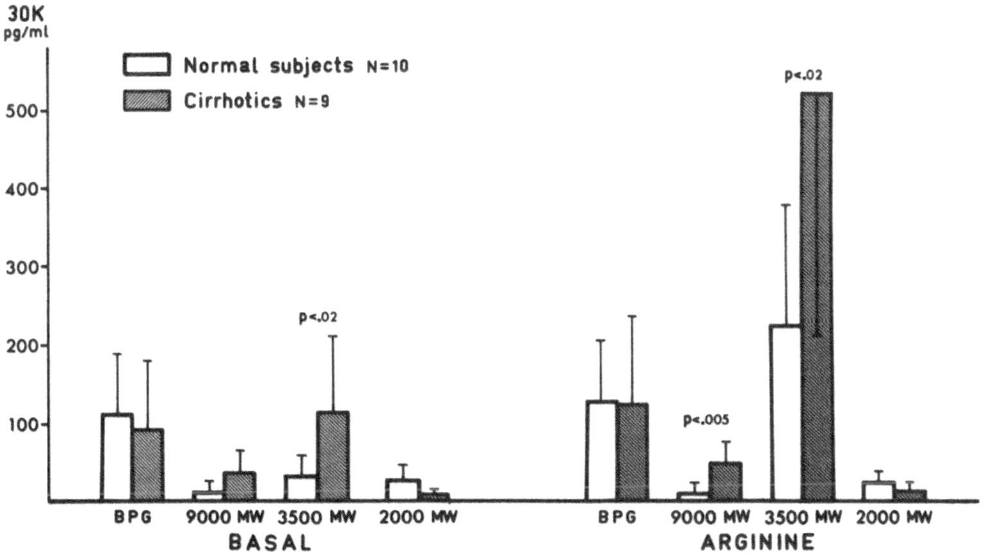

Fig. 2. Basal and post-arginine 30K plasma GIR components in cirrhotic patients compared to normals. Bio-Gel P-30. Mean ± SD. Student's t test.

Effect of Glucagon Suppression on Plasma GIR Components

Intravenous glucose in three normal subjects for 30 minutes (1.7 g/minute) produced a decline of total plasma glucagon from 104 ± 21 to 59 ± 12 pg/ml (mean ± SD). The small values of the 9000 and 3500 m. w. components of the basal samples (14 ± 12 and 9 ± 4 pg/ml, respectively) became undetectable during the hyperglycemic state, whereas the means of BPG and of the 2000 m. w. component were slightly reduced (18 and 5 pg/ml, respectively) but not significantly different from the basal values.

Oral glucose load (1.7 g/Kg of body weight) in seven normal subjects resulted in a drop of total plasma glucagon which was accounted for by undetectable values in the 3500 m. w. zone and a slight reduction of the 9000 m. w. component.

Plasma GIR Components in Chronic Hyperglucagonemic States

The possible participation of each of these components has been studied in the following chronic pathologic and experimental hyperglucagonemic states:

Cirrhosis of the liver. Nine patients were selected from the group studied by Marco *et al.* (11). Those with portocaval shunts were not included. In basal samples (Fig. 2), the 3500 m. w. fraction was the only one showing a statistically significant elevation when compared to ten normal individuals (115 ± 98 vs. 31 ± 29 pg/ml, mean ± SD, $p < .02$). The mean value of the 9000 m. w. component was more than two fold that of normals (35 ± 32 vs 12 ± 16 pg/ml) but the difference was not significant.

After arginine infusion, a significant rise from the basal value was observed in the 3500 m. w. fraction (115 ± 98 vs. 521 ± 309 pg/ml, $p < .005$). The 9000 m. w. component increased from 35 ± 32 to 50 ± 29 pg/ml, and the BPG changed from 97 ± 90 to 127 ± 111 pg/ml, but these differences were not statistically significant.

When the mean values of the GIR components of the post-arginine samples were compared in cirrhotics and normals, both 9000 and 3500 m. w. components showed a significant elevation (50 ± 29 vs. 10 ± 16 pg/ml, $p < .002$, and 521 ± 309 vs. 225 ± 165 pg/ml, $p < .02$, respectively).

Diabetes. Gel filtration of hyperglucagonemic plasma from three diabetic subjects after an overnight fast showed an increase in the mean value of BPG, 9000, and 3500 m. w. components when compared to normal human plasma, the last two being significant. In basal plasma from alloxan-diabetic dogs the same findings were observed when compared to normal dog plasma (Table 2).

TABLE 2

FASTING PLASMA 30 K IMMUNOREACTIVE COMPONENTS IN TWO TYPES OF DIABETES
BIO GEL P-30

Diabetic Humans	N = 3	Total Plasma pg/ml	BPG pg/ml	9000 pg/ml	3500 pg/ml	2000 M. W. pg/ml
	Mean	307	124	54	91	19
	S. D.	167	8	43	109	23
	* P vs. normals N = 20	n.s.	n.s.	<.02	<.02	n.s.
Alloxan-Diabetic Dogs	N = 4					
	Mean	159	63	38	55	15
	S. D.	59	84	17	35	17
	* P vs. normals N = 10	.01	n.s.	<.005	<.005	n.s.

*Student's t test. The group of normals are those of Table 1.

In one of the diabetic patients, arginine infusion resulted in a clear increase of the 3500 m. w. fraction and a slight rise of the 9000 m. w. component (Fig. 3).

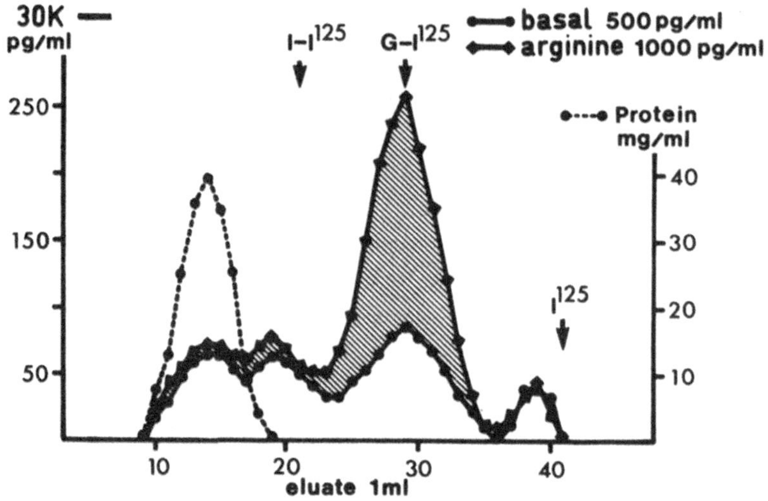

Fig. 3. Distribution of plasma 30K glucagon immunoreactivity in basal and post-arginine samples from a diabetic patient (3 ml loaded). Bio Gel P-30.

Another one of these patients was studied before and one week after starting insulin treatment (Table 3). The reduction of the total plasma glucagon levels achieved after treatment was due to a decrease of the 9000 and 3500 m. w. components to one-half and one-third of the original values, respectively.

Weir et al (36) did not find statistically significant differences in the charcoal treated plasma from 60 diabetic subjects when compared to 30 nondiabetics.

Phloridzin-induced hypoglycemia. Sustained phloridzin-induced hypoglycemia in dogs produced a marked increase in plasma glucagon (492 ± 245 pg/ml, mean ± SD, n = 5, range 240-800 pg/ml). A highly significant elevation was found in the 9000 m. w. component as well as in the 3500 m. w. fraction when compared

TABLE 3

PLASMA 30 K IMMUNOREACTIVE COMPONENTS IN A DIABETIC PATIENT BEFORE AND ONE WEEK AFTER INSULIN TREATMENT. BIO GEL P-30

	Total Plasma pg/ml	BPG pg/ml	9000 pg/ml	3500 pg/ml	2000 M. W. pg/ml
Before Treatment					
3 - 5 a.m.*	210	119	42	61	11
1 - 3 p.m.	360	140	65	127	12
After Treatment					
3 - 5 a.m.	148	108	19	24	19
1 - 3 p.m.	188	129	37	54	11

*The values correspond to a pool of samples taken at the time indicated.

to ten normal dogs (83 ± 43 pg/ml, p <.001 and 257 ± 131 pg/ml, p <.002). During a 30 minute somatostatin infusion there was a reduction of both components with a disappearance rate of 16 and 3 minutes, respectively (29). In one of the dogs a slight decline of the BPG was demonstrated during the somatostatin infusion.

Renal failure. The hyperglucagonemia present in fasting plasma from patients undergoing long term hemodialysis (427 ± 215 pg/ml, n = 9) was found to be due to a highly significant increase in the levels of the 9000 and 3500 m. w. components when compared to ten normal subjects (242 ± 165 vs. 27 ± 14 pg/ml, p <.001 and 84 ± 65 vs. 23 ± 22 pg/ml, p <.02, respectively). The mean of BPG, although higher than in normals, was not statistically elevated (119 ± 85 vs. 73 ± 22 pg/ml. The finding of much higher levels of the 9000 m. w. component than those of the 3500 m. w. component, previously reported by Kuku et al (9), was confirmed.

In five one-day anephric dogs, plasma glucagon rose dramatically to values ranging from 720 to 2080 pg/ml. This total immunoreactivity was proportionally and equally distributed as in dialysed patients. In one dog, somatostatin reduced both fractions, 9000 and 3500 m. w., with half-disappearance times approximately two fold those in phloridzinized dogs (27). These findings are in good agreement with the results of Lefèbvre et al. (10) and Forrest et al. (6), demonstrating the importance of the kidney in glucagon degradation.

Glucagonoma. Recent reports (8, 19, 26, 35) agree in finding very high levels of the 9000 m. w. component in the plasma of patients with glucagon-secreting tumors. In some of the patients the amount of the 9000 m. w. component exceeded that of the 3500 m. w. component, which was also increased. Our patient had a total glucagon level of 2600 pg/ml; 63% of it was the 3500 m. w. component and 27% corresponded to the 9000 m. w. component. The other two fractions, BPG and 2000 m. w. component, were not different from normal values.

Pancreatectomy. An original form of hyperglucagonemia was observed in insulin-deprived, totally depancreatized dogs (2, 12, 13, 34). This immunoreactivity is thought to represent glucagon secreted by A cells located in the gastric mucosa. Plasma gel filtration from seven pancreatectomized dogs yield the same four glucagon immunoreactive fractions as those from intact animals with a significant increase in the 3500 and the 9000 m. w. components compared to ten normal dogs (110 ± 92 pg/ml, $p < .02$ and 92 ± 77 pg/ml, $p < .02$). Arginine infusion in two depancreatized dogs produced an increase in the 3500 m. w. component of 102 and 110 pg/ml, and a modest rise in the 9000 m. w. component (32 and 34 pg/ml, respectively). Somatostatin or insulin infusion suppressed the 3500 m. w. and markedly reduced the 9000 m. w. component (25).

Although A cells seem to be present in the gastric mucosa of humans (16), no evidence of high circulating levels of glucagon in pancreatectomized subjects has been shown so far. Plasma from a pancreatectomized patient was studied by Villanueva et al (33), showing that the total immunoreactivity of the basal sample (≃100 pg/ml) was distributed between the BPG and the 2000 m. w. fraction. As in the reports from Müller et al (15) and Barnes and Bloom (1), no arginine response was found, and the elution pattern remained unchanged. The lack of glucagon response to arginine in these patients could be explained by the fact that they were on insulin treatment and that all of them had some kind of gastric resection. The remaining immunoreactivity could be related to A cells, possibly still present but not stimulated. In 15% of the normal subjects studied in basal state neither 9000 nor 3500 m. w. components were detectable.

The heterogeneity of plasma glucagon is another example of how most, if not all, peptide hormones are found in circulating blood as well as in their tissue of origin (38). The immunoreactive forms larger than the known biologically active hormones may represent their precursors and those of smaller size their metabolic products.

There has been accumulating evidence of the presence of large glucagon immunoreactive molecules in the pancreas of several species from that of 9000 m. w. up to the size of BPG (7, 17, 21, 24). Biosynthetic studies provide evidence that the 9000 m. w. molecule may be a candidate for proglucagon but they do not exclude the possibility that a larger molecule or molecules be the original precursor(s) (17). A presumed fragment of bovine or porcine proglucagon containing glucagon as part of its molecule, has been chemically characterized by Tager and Steiner (22). From anglerfish pancreas, Trakatellis et al. (23) have isolated and partially characterized a 78-amino acid molecule that yielded anglerfish glucagon by tryptic digestion.

The identity of the 3500 m. w. plasma GIR component, measured by a "specific antiserum", leaves no doubt that it is the fully active biologic hormone. Its levels readily and appropriately change with any given state of stimulation or suppression of glucagon secretion or impaired degradation. It has the same half-disappearance time as crystalline glucagon and its glycogenolytic activity has been demonstrated.

Concerning the 9000 m. w. plasma GIR component, the clues are that it represents a proglucagon or an intermediate proglucagon moiety. It is highly augmented in plasma from patients with glucagon secreting tumors as occurs with precursor moieties in other functioning endocrine tumors (14). It can also be increased in situations of sustained hypersecretion of A cells or of impairment of its degradation. It has a slower turnover than glucagon. Proglucagon isolated from dog pancreas is devoid of glycogenolytic activity (20, 21). The biologic activity of this plasma component from a glucagonoma patient, assessed by Recant et al. (19) on rat hepatocytes, was shown to have one-third of the capacity of glucagon in generating cAMP, but no data are available in plasma of non-tumoral hyperglucagonemic states.

The nature of BPG is still uncertain. The speculation that it is related to a secretory product of the A cells still holds up, at least for part of it. The lack of significant changes in BPG in any of the situations here reported, may be due to the large individual variations. When the same individual served as his own control, arginine elevated the BPG in 70% of the normal or cirrhotic subjects studied. Somatostatin infusion in one dog produced a slight decline of BPG. Presumably this moiety has a very slow turnover rate and significant data cannot be obtained unless glucagon suppression is maintained for several hours.

Sporadic cases of very great amounts of BPG in humans have been reported (18, 30). An attempt to relate these very high levels of BPG to the IgG content of the plasma failed (30).

The fact that BPG is diversely measurable by various "glucagon specific antiserum" does not exclude that this moiety is a substance related to glucagon biosynthesis. The different affinity of proinsulin for various insulin antisera is well known. Nevertheless the possibility that BPG is a pool of different substances must be considered.

As for the 2000 m. w. plasma GIR component it could be plausible to think that it corresponds to a metabolic product of glucagon or of its precursors, but its identification awaits further studies.

Awareness of these different forms in glucagon immunoreactivity detectable with "specific glucagon antisera" should be taken into account in evaluating total plasma glucagon values. Further studies are needed to clarify their relationship to biosynthetic glucagon products and their importance as biologically active molecules.

Acknowledgements

I wish to express my deep gratitude to Dr. R. H. Unger for initiating me in the glucagon field, for his generous supply of glucagon antisera and for his direct collaboration in part of these studies; to Dr. F. Vivanco for his constant interest and support; to Dr. M. L. Villanueva for her dedicated work and continuous help, and to all the collaborators that contributed to the realization of these studies.

The valuable assistance of Ms. G. Calvo, T. Aragón, I. García-Muñoz and A. Ramirez is gratefully acknowledged.

I wish to thank Dr. R. Guillemin (Salk Institute) for the supply of somatostatin and Dr. R. Chance (Eli Lilly) for providing the crystalline glucagon used for iodination and standards.

This work was supported in part by grants from the following foundations: Conchita Rábago, Pedro Barrié de la Maza, Eugenio Rodriguez Pascual, from the Insituto Nacional de Previsión (12-894-74) and by NIH grant AN02700-15.

REFERENCES

1. Barnes, A. J. and Bloom, S. R. (1976): Pancreatectomized man: a model for diabetes without glucagon. Lancet 1, 219-220.

2. Dobbs, R., Sakurai, H., Sasaki, H., Faloona, G., Valverde, I., Baetens, D., Orci, L. and Unger, R. (1975): Glucagon: role in the hyperglycemia of diabetes mellitus. Science 187, 544-546.

3. Eisentraut, A. M., Ohneda, A. and Parada, E. (1968): Immunologic distinction between pancreatic glucagon and enteric glucagon-like immunoreactivity (GLI) in tissue and plasma. Diabetes 17, 321-322.

4. Eisentraut, A., Whissen, N. and Unger, R. H. (1968): Incubation damage in the radioimmunoassay of human plasma glucagon and its prevention with "Trasylol". Am. J. Med. Sci. 255, 137-142.

5. Faloona, G. R. and Unger, R. H. (1974): Glucagon. In: Methods of Hormone Radioimmunoassay. B. M. Jaffee and H. R. Behrman, Eds. Academic Press, New York, pp. 317-330.

6. Forrest, J. N., Fisher, M., Hendler, R., Soman, V., Sherwin, R. and Felig, P. (1976): Contrasting roles of the kidney in the disposal and hormonal action

of physiological concentrations of glucagon. Clin. Res. 24, 400A.

7. Hellerström, C., Howell, S. L., Edwards, J. C., Andersson, A. and Östenson, G. C. (1974): Biosynthesis of glucagon in isolated pancreatic islets of guinea pigs. Biochem. J. 140, 13-23.

8. Holst, J. J. and Bang Pedersen, N. (1975): Glucagon producing tumors of the pancreas. Acta Endocr. (Kbh) 199, 379.

9. Kuku, S. F., Zeidler, A., Emmanouel, D. S., Katz, A. I., Rubenstein, A. H., Levine, N. W. and Tello, A. (1976): Heterogeneity of plasma glucagon: patterns in patients with chronic renal failure and diabetes. J. Clin. Endocrinol. Metab. 42, 173-176.

10. Lefèbvre, P. J. and Luyckx, A. S. (1975): Effect of acute kidney exclusion by ligation of renal arteries on peripheral plasma glucagon levels and pancreatic glucagon production in the anesthetized dog. Metabolism 24, 1169-1176.

11. Marco, J., Diego, J., Villanueva, M. L., Diaz-Fierros, M., Valverde, I. and Segovia, J. M. (1973): Elevated glucagon levels in cirrhosis of the liver. New Engl. J. Med. 289, 1107-1111.

12. Mashiter, K., Harding, P. E., Chou, M., Mashiter, G. D., Stout, J., Diamond, D. and Field, J. B. (1975): Persistent pancreatic glucagon but not insulin response to arginine in pancreatectomized dogs. Endocrinology 96, 678-693.

13. Matsuyama, T. and Foà, P. P. (1974): Plasma glucose, insulin, pancreatic and enteroglucagon levels in normal and depancreatized dogs. Proc. Soc. Exp. Biol. Med. 147, 97-102.

14. Melani, F. (1974): Pro-hormones in tissues and in circulation. Horm. Metab. Res. 6, 1-8.

15. Müller, W. A., Brennan, M. F., Tan, M. H. and Aoki, T. T. (1974): Studies of glucagon secretion in pancreatectomized patients. Diabetes 23, 512-516.

16. Muñoz Barragan, L., Rufener, C., Srikant, C. B., Dobbs, R. E., Shannon, W. A. Jr., Baetens, D. and Unger, R. H. (1977): Immunocytochemical evidence for glucagon-containing cells in the human stomach. Horm. Metab. Res. 9, 37-39.

17. Noe, B. D., Bauer, G. E., Steffes, M. W., Sutherland, D. E. R. and Najarian, J. S. (1975): Preliminary evidence for a biosynthetic intermediate. Horm. Metab. Res. 7, 314-322.

18. Palmer, J. P., Werner, P. L., Benson, J. W. and Ensinck, J. W. (1976): Dominant inheritance of large molecular weight immunoreactive glucagon (IRG). Diabetes 25, 326.

19. Recant, L., Perrino, P. V., Bhathena, S. J., Danforth, D. N., Jr. and Lavine, R. L. (1976): Plasma immunoreactive glucagon fractions in four cases of glucagonoma. Diabetologia 12, 319-326.

20. Rigopoulou, D., Valverde, I., Marco, J., Faloona, G. and Unger, R. H. (1970): Large glucagon immunoreactivity in extracts of pancreas. J. Biol. Chem. 245, 496-501.

21. Srikant, C. B., McCorkel, K. and Dobbs, R. E. (1976): A biologically active "macro 'glucagon' (IRG)"; glucagon secreting tissues of the dog. Diabetes 25, 326.

22. Tager, H. S. and Steiner, D. F. (1973): Isolation of a glucagon-containing peptide. Primary structure of a possible fragment of proglucagon. Proc. Natl. Acad. Sci. USA 70, 2321-2324.

23. Trakatellis, A. C., Tada, K., Yamaji, K. and Gardiki-Kouidou, P. (1975): Isolation and characterization of anglerfish proglucagon. Biochemistry 14, 1508-1512.

24. Tung, A. K. (1973): Biosynthesis of avian glucagon: evidence for a possible high molecular weight biosynthetic intermediate. Horm. Metab. Res. 5, 416-424.

25. Valverde, I., Dobbs, R. and Unger, R. H. (1975): Heterogeneity of plasma glucagon immunoreactivity in normal, depancreatized and alloxan-diabetic dogs. Metabolism 24, 1021-1028.

26. Valverde, I., Lemon, H. M., Kessinger, A. and Unger, R. H. (1976): Distribution of plasma glucagon immunoreactivity in a patient with suspected glucagonoma. J. Clin. Endocrinol. Metab. 42, 804-808.

27. Valverde, I., Matesanz, R., Lozano, I., Plaza, J. J. and Casado, S. (1976): Evaluation of plasma glucagon immunoreactive fraction in renal failure. Diabetologia 12, 423.

28. Valverde, I., Rigopoulou, D., Marco, J., Faloona, G. R. and Unger, R. H. (1970): Molecular size of extractable glucagon and glucagon-like immunoreactivity (GLI) in plasma. Diabetes 19, 624-629.

29. Valverde, I., Roman, D. and Dobbs, R. (1975): Heterogeneity of plasma glucagon in depancreatized, alloxanized and phloridzinized dogs. Diabetes 24, 412.

30. Valverde, I. and Villanueva, M. L. (1976): Heterogeneity of plasma immunoreactive glucagon. Metabolism 25, 1393-1395.

31. Valverde, I., Villanueva, M. L., Lozano, I., Marco, J. (1974): Presence of glucagon immunoreactivity in the globulin fraction of human plasma (big plasma glucagon). J. Clin. Endocrinol. Metab. 39, 1020-1028.

32. Villanueva, M. L. (1976): Ph.D. Thesis, Universidad Complutense de Madrid.

33. Villanueva, M. L., Hedo, J. and Marco, J. (1976): Plasma glucagon immunoreactivity in a totally pancreatectomized patient. Diabetologia 12, 613-616.

34. Vranic, M., Pek, S. and Kawamori, R. (1974): Increased "glucagon immunoreactivity" in plasma of totally depancreatized dogs. Diabetes 23, 905-912.

35. Weir, G. C., Horton, E. S. and Aoki, T. T. and Slovik, D. M. (1976): Increased circulating large glucagon immunoreactivity in the glucagonoma syndrome. Diabetes 25, 326.

36. Weir, G. C., Knowlton, S. D. and Martin, D. B. (1975): High molecular weight glucagon-like immunoreactivity in plasma. J. Clin. Endocrinol. Metab. 40,

296-302.

37. Weir, G. C., Turner, R. C. and Martin, D. B. (1973): Glucagon radioimmunoassay using antiserum 30K: interference by plasma. Horm. Metab. Res. 5, 241-244.

38. Yalow, R. S. (1974): Heterogeneity of peptide hormone. Recent Progr. Horm. Res. 30, 597-633.

CIRCULATING GLUCAGON COMPONENTS: SIGNIFICANCE IN HEALTH AND DISEASE

J. B. Jaspan, A. H-J. Huen, B. Gonen and A. H. Rubenstein

Department of Medicine, University of Chicago
Chicago, Illinois USA

Plasma immunoreactive glucagon (IRG) represents a family of compounds with different molecular weights. Although the physiologic role of these fractions is largely unknown, their relative abundance may vary under different physiologic and pathologic conditions. This paper describes the pattern of IRG components in normal subjects and in patients with renal failure, diabetes mellitus, glucagonoma and other conditions. The role of the kidneys and of the liver in the metabolism of glucagon is reviewed and discussed.

It has recently been recognized that immunoreactive glucagon (IRG) circulates in a number of different forms in both normal subjects and patients with a variety of disorders. The components of IRG differ in their contribution to total circulating glucagon immunoreactivity in various conditions and in different animal species. At present the significance of these fractions is uncertain. Thus, their relationship to native pancreatic glucagon, biologic potency, response to stimuli, molecular size and even nomenclature are still debated. The overall picture is confused further by the use of different glucagon antisera, which react variably with each circulating immunoreactive glucagon fraction.

Patterns in Normal Subjects

In the initial years after the glucagon radioimmunoassay was established, it was believed that the hormone circulated as a single component with a molecular weight of 3500 daltons. Subsequent studies by Weir *et al.* (47) and Valverde *et al.* (44) showed that IRG in the plasma of normal and diabetic subjects was heterogeneous. The IRG eluted in two fractions, namely a high molecular weight component in the globulin region of the column and 3500 dalton glucagon. As the relationship of this high molecular weight material to the 3500 dalton glucagon was uncertain, it was named big-plasma glucagon (B.P.G) or "interference factor", suggesting that it might be an immunologically crossreacting peptide sequence in an unrelated protein, possibly a gamma-globulin. However, the possibility remained that it might be closely related to glucagon, either in the form of a prohormone or as a result of binding of glucagon to an unidentified large molecular weight protein. Valverde, et al. (42) identified four glucagon immunoreactive fractions in healthy and diabetic dogs, with molecular weights of <20,000, approximately 9000, 3500 and <2000 daltons. Although the concentrations of the void volume and 2000 molecular weight fractions were relatively constant, both the 3500 and 9000 dalton components increased markedly in response to phloridzin-induced hypoglycemia. These four components were also found in normal, diabetic and pancreatectomized dogs, in which the source of IRG measured with the 30-K antiserum is the A cells present in the proximal gastrointestinal tract.

Fig. 1. Elution patterns of plasma immunoreactive glucagon (IRG) on 1 x 50 cm Bio-Gel P-30 columns in 1 control subject and 6 diabetic patients. Vo = void volume (m. w. <40,000), P = proinsulin (m. w. 9000), G = pancreatic glucagon (m. w. 3500). All columns were calibrated with ^{125}I-gamma globulin, ^{131}I-proinsulin and ^{125}I-glucagon. In panel 6, 1500 pg/ml represents a spurious value due to the presence of glucagon antibodies. O----O represents the profile obtained after incubating the plasma with ^{125}I-glucagon. Additional clinical details in these patients are as follows: Panel 1: Healthy subject (20 yrs). Panel 2: Diabetic (19 years old; diabetes 7 yrs). Plasma glucose: fasting, 125 mg/dl; two hour post prandial, 245 mg/dl; urinary glucose excretion, 15.6 gm/24 h. Patient was in good control, with normal renal function. Panel 3: Diabetic (58 years old; diabetes 8 yrs). Normal weight, treated with acetohexamide. Plasma glucose: fasting, 148 mg/dl; two hour post prandial, 340 mg/dl. Urinary glucose excretion, 88 gm/24 h. Diabetic control poor. Normal renal function. Panel 4: Seventy-one year old obese female, previously unrecognized to be diabetic. Presented in coma; plasma glucose, 1080 mg/dl; plasma ketones, trace. Panel 5: Twenty-two year old female, previously well, with no history of diabetes. Presented in diabetic ketoacidosis; plasma glucose, 720 mg/dl. Panel 6: Twenty-two year old male, diabetic since the age of five, control generally poor, retinopathy with blindness, nephropathy with chronic renal failure (creatinine 2.7 mg/dl). Panel 7: Fifty-eight year old female. Presented with one month history of necrolytic skin rash, stomatitis and 10 pound weight loss: multiple hepatic tumor nodules identified as A cells on biopsy. Probable tumor blush in the tail of pancreas on angiography. Blood glucose: fasting, 180; one hour, 412; two hours, 342 mg/dl.

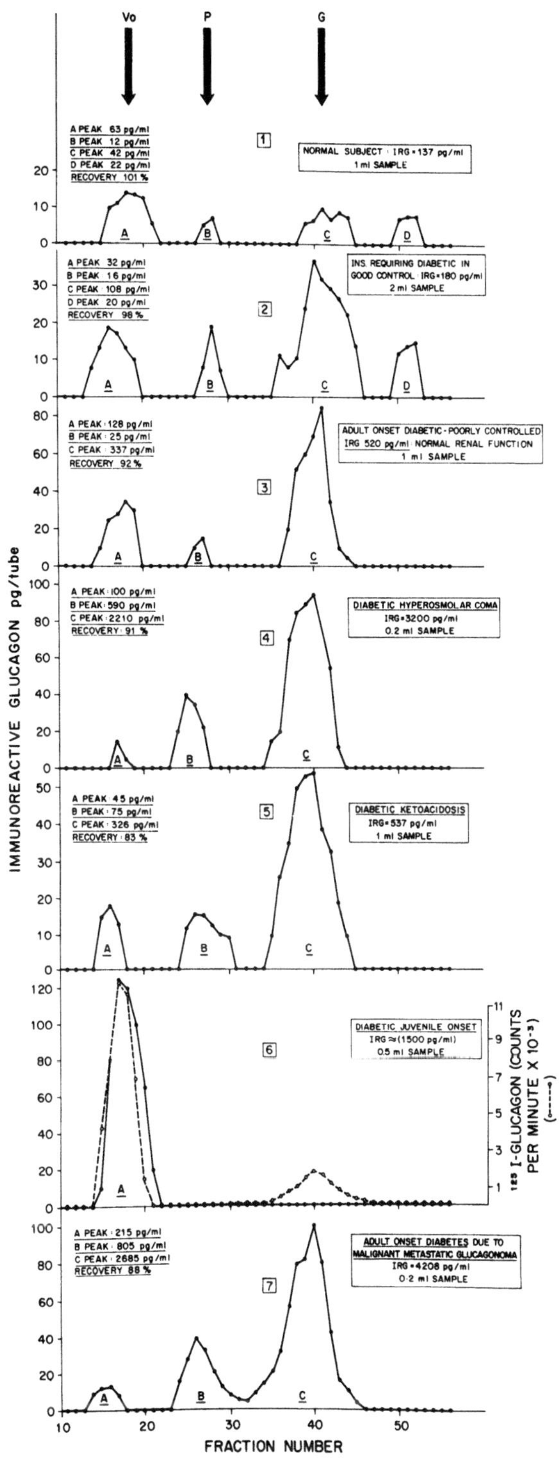
Vo
P
G
A PEAK 63 pg/ml
B PEAK 12 pg/ml
C PEAK 42 pg/ml
D PEAK 22 pg/ml
RECOVERY 101 %
1
NORMAL SUBJECT : IRG = 137 pg/ml
1 ml SAMPLE
A PEAK 32 pg/ml
B PEAK 16 pg/ml
C PEAK 108 pg/ml
D PEAK 20 pg/ml
RECOVERY 98 %
2
INS. REQUIRING DIABETIC IN GOOD CONTROL IRG = 180 pg/ml
2 ml SAMPLE
A PEAK : 128 pg/ml
B PEAK : 25 pg/ml
C PEAK : 337 pg/ml
RECOVERY 92 %
3
ADULT ONSET DIABETIC - POORLY CONTROLLED
IRG 520 pg/ml NORMAL RENAL FUNCTION
1 ml SAMPLE
A PEAK : 100 pg/ml
B PEAK : 590 pg/ml
C PEAK : 2210 pg/ml
RECOVERY : 91 %
4
DIABETIC HYPEROSMOLAR COMA
IRG = 3200 pg/ml
0.2 ml SAMPLE
A PEAK 45 pg/ml
B PEAK : 75 pg/ml
C PEAK : 326 pg/ml
RECOVERY : 83 %
5
DIABETIC KETOACIDOSIS
IRG = 537 pg/ml
1 ml SAMPLE
6
DIABETIC JUVENILE ONSET
IRG ≈ (1500 pg/ml)
0.5 ml SAMPLE
A PEAK : 215 pg/ml
B PEAK : 805 pg/ml
C PEAK : 2685 pg/ml
RECOVERY 88 %
7
ADULT ONSET DIABETES DUE TO MALIGNANT METASTATIC GLUCAGONOMA
IRG = 4208 pg/ml
0.2 ml SAMPLE
A
B
C
D
IMMUNOREACTIVE GLUCAGON pg/tube
125 I-GLUCAGON (COUNTS PER MINUTE X 10⁻³)
FRACTION NUMBER

Fig. 1

Kuku et al (23) confirmed the molecular heterogeneity of glucagon in man. They observed that the large molecular weight component comprised 37 to 87% of plasma immunoreactive glucagon in normal subjects. Although the 9000 dalton glucagon was not initially detected in these plasma samples, it has subsequently been shown to be present in low concentrations (approximately 5-15 pg/ml) in many normal individuals (Fig. 1, panel 1). It is probable that this component is usually present in normal plasma, but may not be detected by column chromatography, where it is distributed into four or five tubes, falling below the limits of detection of the immunoassay which is usually about 4-6 pg/tube. It is possible, however, to amplify this fraction by adsorbing the IRG from larger volumes of plasma and rechromatographing the immunoadsorbed material. The circulating level of this component in dogs is generally higher than in man. Very small amounts of glucagon immunoreactivity with a molecular weight of less than 2000 have also been found in plasma taken from control subjects. Preliminary evidence suggests that its concentration may increase in plasma in which proteolytic activity has not been fully inhibited, in which case it would represent an in-vitro degradation product. However, it is possible that this component represents a circulating fragment derived from the glucagon molecule during its in-vivo catabolism and which retains its immunoreactivity with certain glucagon antisera.

Patterns in Renal Failure

Bilbrey et al (7) reported that renal failure was characterized by high levels of IRG which were normalized by renal transplantation but not by hemodialysis (6). They also demonstrated that in uremic patients IRG levels did not suppress to the same degree as in controls after administering glucose. The hyperglucagonemia of renal failure was confirmed by Kuku et al (22) who reported IRG levels of 534 ± 32 pg/ml in 36 stable uremic subjects as compared to 113 ± 9 pg/ml in 32 controls. The reason for the impaired suppressibility of IRG in response to glucose in the renal failure patients became clear when it was shown that the greatest proportion of their IRG (56.5 ± 3.4%) was present in the 9000 dalton fractions (designated the "B" peak). This component did not decline after glucose administration, while the 3500 dalton component (designated the "C" peak), which comprised 27 ± 4% of IRG was almost completely suppressible. The failure of glucose to suppress the void volume component (designated the "A" peak), which comprised 53.6 ± 10.4% of IRG in normal subjects explained why IRG concentrations seldom declined to values less than 50% of the fasting level. This observation may also explain the results of experiments in which the administration of somatostatin failed to suppress plasma IRG levels much below 40-50% pg/ml. The "B" peak, which is considered to be proglucagon, rose minimally after arginine compared to the increase in "true" glucagon, thus accounting for the smaller percentage increase in IRG in chronic renal failure patients when compared to normals. However, the proglucagon component does respond to a small extent to agents which stimulate or suppress A cell function (Fig. 2).

Patterns in Diabetes

Abnormalities in circulating glucagon levels have been considered to play a role in the pathogenesis of the metabolic disturbances of diabetes and the degree of elevation of plasma glucagon may reflect the quality of diabetic control. However, there are two situations in which the plasma IRG level may not be related to the severity of the diabetic state: The presence of glucagon antibodies and glucagonoma.

A. Antiglucagon antibodies. The presence of glucagon as an impurity in commercial insulin preparations may induce antiglucagon antibodies in some insulin requiring diabetics. Stahl *et al* (39) found antiglucagon antibodies in the plasma of a 17 year old diabetic treated with insulin from 13 months of age, but not in 71 other insulin treated juvenile onset diabetics. Similarly, Thomsen (40) did not find glucagon antibodies in 100 diabetics treated with insulin for more than 10 years. On the other hand, Weir *et al.* (47) identified glucagon antibodies in

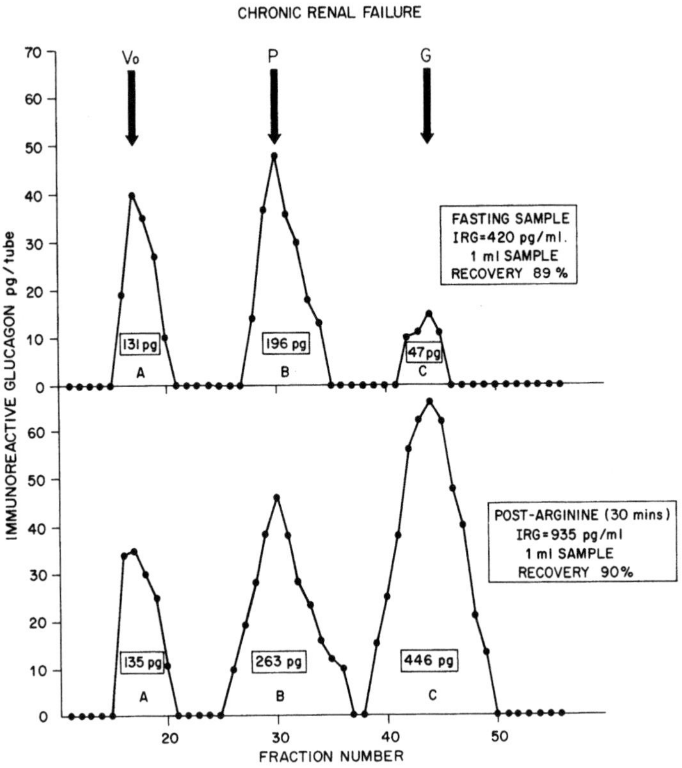

Fig. 2. Elution patterns of immunoreactive glucagon on 1 x 50 cm Bio-Gel P-30 columns in a patient with chronic renal failure (glomerular filtration rate <5 ml/min). Samples were taken fasting and after intravenous arginine hydrochloride (250 mg/kg). Column calibrations as in Figure 1.

8 of 60 (12%) insulin requiring diabetics, Villalpando and Drash (45) in 24 of 205 diabetic children (12%) and Cresto et al (11) in 3 of 85 insulin treated patients. These investigators (11) pointed out that a prolonged period of insulin therapy is not invariably necessary for the development of antibodies, as one of their patients had been treated with insulin for less than two years. Jaspan et al. (20) pointed out that these glucagon antibodies compete with the glucagon antiserum used in the assay for the labeled glucagon resulting in a falsely high value in the double antibody radioimmunoassay method. If charcoal is used to

separate bound from free hormone, the glucagon value will be spuriously low in the presence of glucagon antibodies. Recently, Baba et al (4) reported the development of antibodies to glucagon in a 66 year old diabetic patient who was only treated with tolbutamide. They concluded that the antibodies arose as a result of autoimmunization to glucagon, but few details of the condition were given. Further information regarding this syndrome will undoubtedly be of great interest.

B. Glucagonoma. This tumor is associated with a diabetic syndrome which is usually mild. The plasma glucagon levels, however, may be markedly elevated, ranging from 900 to 7800 pg/ml (see below).

The various components of circulating IRG differ markedly in their response to stimuli (22) and in their biologic potency (34, 38). It is therefore important to define the plasma IRG profiles when evaluating the glucagon disturbances in diabetes and their contribution to the metabolic derangements. Figure 1 illustrates the elution profiles of plasma IRG in a variety of diabetic patients.

Patterns in Glucagonoma

The syndrome associated with glucagonoma is now well recognized. A number of reports have delineated a clinical picture with distinctive features (29, 30) which are discussed in more detail elsewhere in this symposium.

We have had the opportunity to study plasma taken from six patients with proven glucagon secreting tumors. The results of these studies are shown in Table 1, Fig. 1 (panel 7) and Fig. 3. All patients had diabetes which varied from mild glucose intolerance to severe hyperglycemia. In one of the patients (panel 4 of Fig. 1) the diabetes was a major problem which necessitated frequent hospital admissions and ultimately led to her death.

Plasma IRG levels and its distribution are summarized in Table 1. Ten to ninety four percent of IRG eluted in the 9000 dalton region (representing 185-1100 pg/ml), although the 94% value may be an overestimate, as discussed in the legend of Table 1. These results are in close agreement with those of Valverde et al.

(43) who reported a patient with 8% (190 pg/ml) IRG in the "A" peak, 27% (625 pg/ml) in the "B" peak, 63% (1435 pg/ml) in the "C" peak and 2% (35 pg/ml) in the "D" peak. Recant et al. (34) showed qualitatively similar patterns in four cases with glucagonoma, but the percentage of 9000 molecular weight material was 2 to 3 fold higher.

In two patients (Fig. 3, panels 2 and 3) dilution studies were carried out on the three IRG components. In both instances, the "C" peak and "B" peak material diluted proportionately, while the A peak material was not parallel to the glucagon standard. This result suggests that the region of this latter molecule ("A" peak) which reacts with glucagon antisera is not identical to a native glucagon or that the glucagon immunoactive component is complexed to a larger protein in a manner which partially interferes with its accessibility to the glucagon antibodies. On the other hand, a portion of the proglucagon ("B" peak) molecule is presumably closely related in its structure to native glucagon.

Fig. 3. Elution patterns of plasma immunoreactive glucagon on 1 x 50 cm Bio-Gel P-30 columns in five patients with glucagonoma. Column calibrations as in Fig. 1. Additional clinical details in these patients are as follows: Panel 1: Fifty-three year old female, ten year history of diabetes, initially treated with tolbutamide and subsequently with up to 35 units of insulin per day with poor control. Large malignant pancreatic glucagonoma with extensive hepatic and peritoneal metastases. Panel 2: Thirty-nine year old male, four year history of necrolytic skin rash and weight loss; hypoaminoacidemia; diabetic glucose tolerance curve; 4 x 4 cm glucagonoma in the tail of the pancreas with invasive characteristics, but no metastases. Panel 3: Forty year old male, two year history of weight loss, seven month history of necrolytic skin rash; stomatitis, anemia, hypoaminoacidemia and diabetic glucose tolerance curve, glucagonoma in the body of the pancreas and multiple hepatic metastases. Panel 4: Sixty-two year old female, five year history of diabetes, poorly controlled on insulin with recurrent episodes of non-ketotic hyperglycemia requiring hospitalization, and eventually leading to death. Postmortem: 0.4 cm glucagonoma in the body of pancreas with invasive characteristics, but no metastases. Panel 5: Thirty-seven year old male, one and one-half year history of skin rash, weight loss and diabetes. Malignant glucagonoma with hepatic metastases.

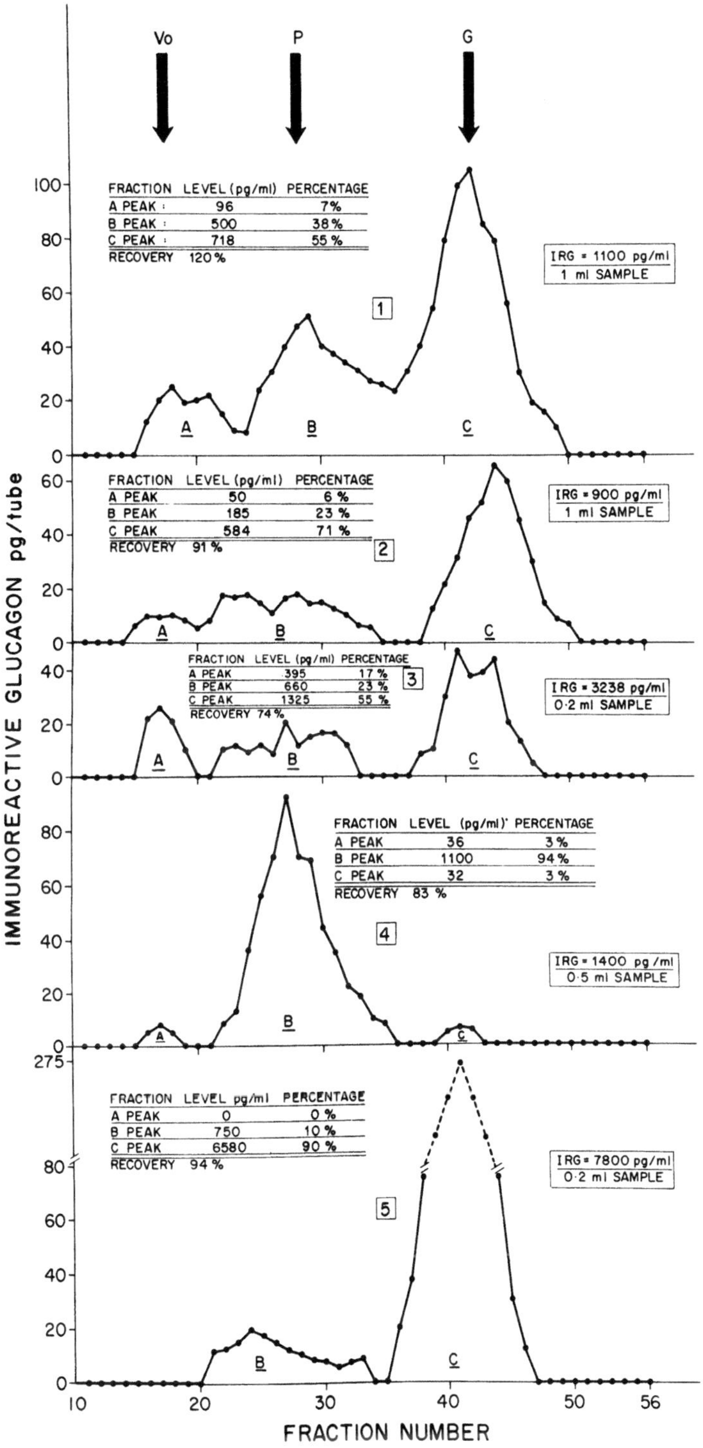
Vo
P
G
FRACTION LEVEL (pg/ml) PERCENTAGE
A PEAK : 96 7%
B PEAK : 500 38%
C PEAK : 718 55%
RECOVERY 120%
IRG = 1100 pg/ml
1 ml SAMPLE
1
A
B
C
FRACTION LEVEL (pg/ml) PERCENTAGE
A PEAK 50 6%
B PEAK 185 23%
C PEAK 584 71%
RECOVERY 91%
IRG = 900 pg/ml
1 ml SAMPLE
2
FRACTION LEVEL (pg/ml) PERCENTAGE
A PEAK 395 17%
B PEAK 660 23%
C PEAK 1325 55%
RECOVERY 74%
3
IRG = 3238 pg/ml
0·2 ml SAMPLE
FRACTION LEVEL (pg/ml) PERCENTAGE
A PEAK 36 3%
B PEAK 1100 94%
C PEAK 32 3%
RECOVERY 83%
4
IRG = 1400 pg/ml
0·5 ml SAMPLE
FRACTION LEVEL pg/ml PERCENTAGE
A PEAK 0 0%
B PEAK 750 10%
C PEAK 6580 90%
RECOVERY 94%
IRG = 7800 pg/ml
0·2 ml SAMPLE
5
IMMUNOREACTIVE GLUCAGON pg/tube
FRACTION NUMBER

Fig. 3

TABLE 1

Patient Number	IRG	A Peak		B Peak		C Peak		Recovery
	pg/ml	pg/ml	%	pg/ml	%	pg/ml	%	%
1	1100	96	7	500	38	718	55	120
2	900	50	6	185	23	584	71	91
3	3238	395	17	660	28	1325	55	74
4*	1400	36	3	1100	94	32	3	83
5	7800	0	0	750	10	6580	90	94
6	4208	215	6	805	22	2685	72	88
MEAN ± SEM	3449 ± 1124	151 ± 71 pg/ml	7 ± 3%	580 ± 111 pg/ml	24 ± 4%	2378 ± 1114 pg/ml	69 ± 6%	92 ± 6%

Plasma IRG levels and distribution of plasma glucagon immunoreactivity in six patients with glucagonoma. The percentage of immunoreactive glucagon recovered in each fraction is shown. Elution profiles of patients 1-5 are shown in corresponding panels of Fig. 3. The profile of patient 6 is illustrated in panel 7 of Fig. 1.

*This sample was taken postmortem. Because it seems probable that degradation of 3500 molecular weight glucagon had taken place, we omitted the results from the calculations.

The finding of immunologic heterogeneity in circulating IRG is similar to that seen in patients with insulinoma and some other endocrine tumors. It appears that higher plasma glucagon levels are associated with malignant tumors, particularly those with metastases. In this regard it should be noted that the glucagon value in cases with very high circulating IRG levels may be underestimated unless measured in suitable dilution. The course of these patients' disease suggests that pathologic overproduction of glucagon may induce diabetes in the absence of overt insulin deficiency. That this does not simply represent the development of a glucagonoma in a patient with diabetes mellitus is indicated by the relief of diabetes which follows the successful resection of the tumor (30). In this context, it is of interest that the liver can down-regulate its response to continued high levels of glucagon (28). Nevertheless, there may be a limit to which this control mechanism can compensate for the excessively high levels of IRG.

Yoshinaga et al. (49) described a patient with glucagonoma who had severe diabetes and a number of other reports have confirmed that diabetes may be a significant aspect of the illness (27, 46). The finding of markedly elevated IRG levels in a diabetic patient should therefore alert one of the possibility of an underlying glucagonoma. Gel filtration of plasma will facilitate the diagnosis. Chronic renal failure which is the only condition which may mimic the elution profile can be separated on clinical grounds. This is of importance particularly as these patients, unlike patients with insulinoma, who frequently declare themselves in a rather dramatic clinical fashion, may otherwise remain with the diagnosis of "idiopathic" diabetes. In fact a number of glucagonomas have been recognized in patients who have been diabetic for long periods of time or who had necrolytic migratory erythema for periods exceeding 10 years (30). Although this pathognomonic skin rash serves as an important clinical sign of glucagonoma, it is often absent as was the case in three of four patients reported by Recant et al (34). Resection of a benign tumor is followed by clinical cure and reversal of the diabetes, while malignant tumors have responded dramatically to streptozotocin (12).

Patterns in Other Conditions

Elution profiles of plasma IRG have been studied in patients with hepatic and pancreatic disease as well as other disorders. Representative patterns in a number of these conditions are illustrated in Fig. 4. The patterns of IRG in pregnancy, severe obesity, insulinoma, disseminated carcinomatosis and septicemia were similar to those found in normal subjects (22, 23). However, significant levels of B peak material were present in many patients with hepatocellular damage or carcinoma of the pancreas. Palmer et al. (33) have recently described a family with hyperglucagonemia with an autosomal dominant inheritance. The IRG in these healthy individuals is comprised predominantly of large amounts of high molecular weight material.

Metabolism of Glucagon

Fischer et al. (18) reported that the metabolic clearance rate of glucagon was 537 ± 27 ml/m^2/min, a value comparable to that of insulin. They calculated the basal systemic delivery rate of glucagon (which excluded that proportion of portal vein glucagon removed by the liver) to be 51 ± 5 ng/m^2/min, equivalent to a 24 hour post-hepatic delivery rate of 139 ± 14 μg/day. The removal of glucagon

Fig. 4. Elution patterns of plasma immunoreactive glucagon on 1 x 50 cm Bio-Gel P-30 columns in patients with hepatic and pancreatic disorders. Column calibration as in Fig. 1. Additional clinical details in these patients are as follows: Panel 1: Twenty-four year old female, acute viral hepatitis progressing to subacute hepatic necrosis and liver decompensation. Panel 2: thirty-nine year old male, with history of alcoholism, presented with severe acute alcoholic hepatitis and liver failure with underlying cirrhosis. Panel 3: Twenty-six year old male, first attack of acute alcoholic pancreatitis. Panel 4: Sixty-one year old male, carcinoma of the pancreas, normal glucose tolerance curve. Panel 5: Eighty-year old male, carcinoma of the pancreas, mild diabetes.

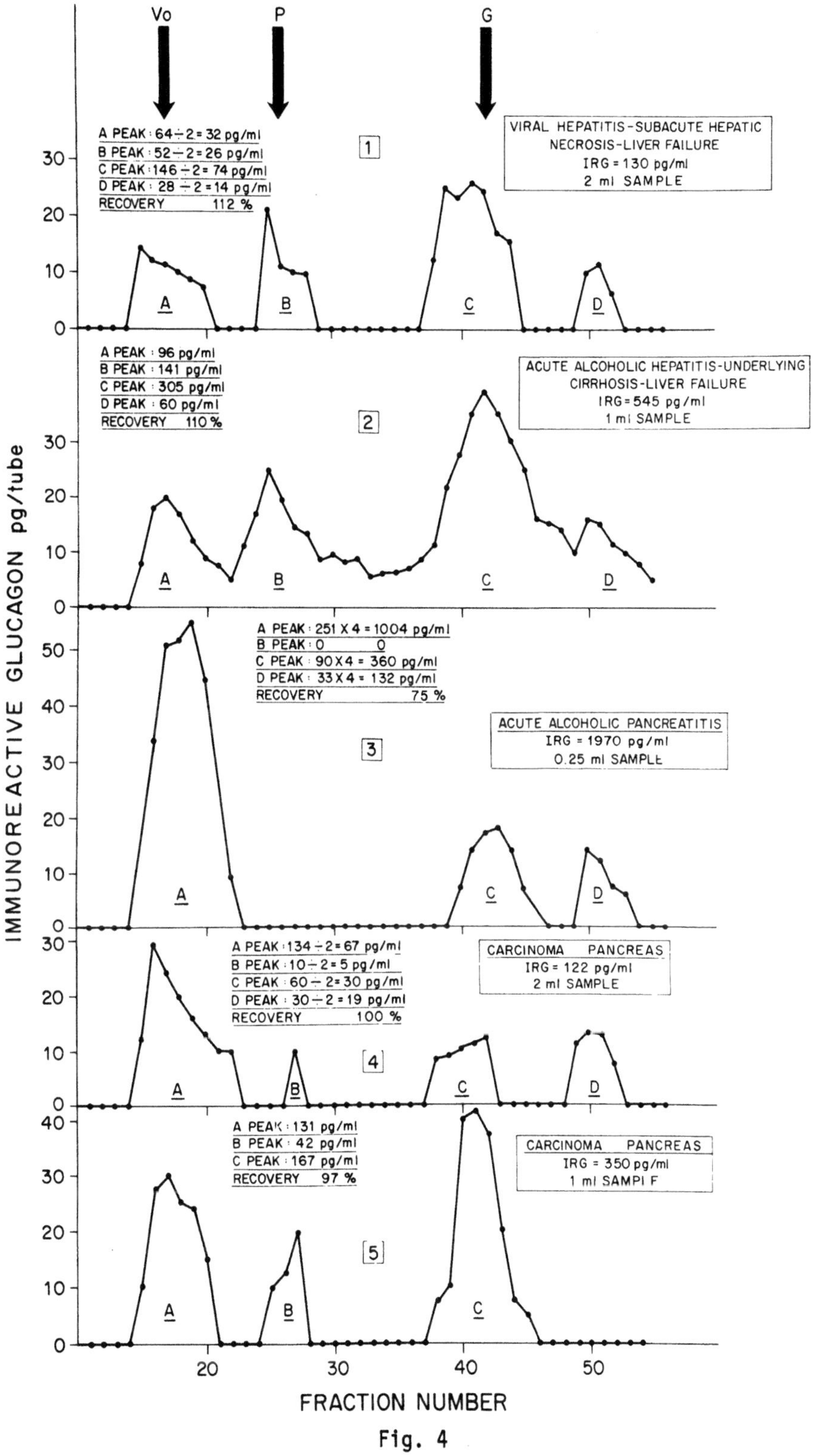
Vo
P
G
A PEAK: 64÷2=32 pg/ml
B PEAK: 52÷2=26 pg/ml
C PEAK: 146÷2=74 pg/ml
D PEAK: 28÷2=14 pg/ml
RECOVERY 112 %
1
VIRAL HEPATITIS-SUBACUTE HEPATIC NECROSIS-LIVER FAILURE
IRG=130 pg/ml
2 ml SAMPLE
A PEAK: 96 pg/ml
B PEAK: 141 pg/ml
C PEAK: 305 pg/ml
D PEAK: 60 pg/ml
RECOVERY 110 %
2
ACUTE ALCOHOLIC HEPATITIS-UNDERLYING CIRRHOSIS-LIVER FAILURE
IRG=545 pg/ml
1 ml SAMPLE
A PEAK: 251 X 4 = 1004 pg/ml
B PEAK: 0 0
C PEAK: 90 X 4 = 360 pg/ml
D PEAK: 33 X 4 = 132 pg/ml
RECOVERY 75 %
3
ACUTE ALCOHOLIC PANCREATITIS
IRG = 1970 pg/ml
0.25 ml SAMPLE
A PEAK: 134÷2=67 pg/ml
B PEAK: 10÷2=5 pg/ml
C PEAK: 60÷2=30 pg/ml
D PEAK: 30÷2=19 pg/ml
RECOVERY 100 %
4
CARCINOMA PANCREAS
IRG= 122 pg/ml
2 ml SAMPLE
A PEAK: 131 pg/ml
B PEAK: 42 pg/ml
C PEAK: 167 pg/ml
RECOVERY 97 %
5
CARCINOMA PANCREAS
IRG = 350 pg/ml
1 ml SAMPLE
A
B
C
D
IMMUNOREACTIVE GLUCAGON pg/tube
FRACTION NUMBER
20
30
40
50

Fig. 4

from the circulation was linear over a range of plasma concentrations from 200-600 pg/ml. These authors (18) suggested that in early starvation hyperglucagonemia results primarily from decreased glucagon removal rather than hypersecretion. Studies by Alford et al. (1) in controls and diabetics showed similar MCR values of 9.0 ± 0.6 and 11.4 ± 1 ml/kg/min, respectively. Despite these similar metabolic clearance rates, the acute disappearance time for glucagon was considerably delayed in diabetics, suggesting that the kinetics of the in vivo metabolism of pancreatic glucagon are not identical in these two groups of subjects.

Role of the kidney in glucagon metabolism. In 1958 Narahara et al. (32), used autoradiographic techniques to show that kidney tubules took up ^{131}I-glucagon. The finding of hyperglucagonemia in renal failure (7) stimulated investigations into the role of the kidney in glucagon metabolism. Glucagon was shown to be metabolized by renal tissue as evidenced by the high extraction ratio and relatively low urinary clearance rates of the hormones (5, 26) and by the effects on plasma glucagon levels of bilateral kidney exclusion in dogs (25). Sherwin et al. (35) reported that the MCR of exogenous glucagon was markedly prolonged in uremic patients, while the calculated basal systemic delivery rate (an index of glucagon secretion) remained unchanged. Lefèbvre and Luyckx (24) demonstrated that acute renal pedicle ligation in dogs led to a prompt and substantial elevation in circulating IRG, while simultaneously measured pancreatic secretion rates remained unchanged.

With the demonstration of molecular heterogeneity of circulating IRG in man (23), it became important to assess the role of the kidney in determining the patterns of circulating IRG. Kuku et al (22) observed that the hyperglucagonemia of renal failure was due largely to a striking elevation in the "B" peak (9000 dalton) fraction, which comprised 56.5 ± 3.4% of the total IRG (534 ± 32 pg/ml). A moderate elevation in the "C" peak fraction (27 ± 4%) also occurred. The mechanism underlying the hyperglucagonemia in renal failure was studied by

Emmanouel et al. (15), who showed that the 3500 dalton IRG component was handled by both glomerular filtration and peritubular uptake, while the metabolism of the 9000 dalton fraction was mainly dependent upon uptake from post-glomerular peritubular blood rather than on glomerular filtration. This observation was qualitatively similar to the renal metabolism of insulin, proinsulin and C-peptide (21) as well as a variety of other peptide hormones and other peptides. In this regard it is interesting that Duckworth et al (14) have suggested that insulin and glucagon are degraded by closely related renal enzyme systems residing in both the brush border and cytosol of proximal tubular cells.

Role of the liver in glucagon metabolism. Portal vein levels of glucagon are higher than in the peripheral circulation. This finding has been confirmed in dogs by Unger et al. (41) and Buchanan et al (10) and in man by Assan et al. (3). Although it is known that plasma IRG is elevated in many patients with compensated cirrhosis, liver failure and porto-systemic shunting (31, 36, 37) and an absence of a portal-peripheral gradient in patients with large porto-caval shunts has been reported (2), there is still uncertainty regarding the importance of the liver in the overall metabolism of glucagon. Small portal-peripheral differences have been reported by some investigators (16-18) leading to the conclusion that the liver plays a negligible role in the degradation of glucagon. However, other studies have suggested (9, 19, 48) that the liver is a major site of glucagon extraction and this is supported by the studies of Blackard et al (8) and Dencker et al. (13) who found portal-peripheral ratios of 1.7 ± 0.5 in normal man.

Recent studies in our laboratory (unpublished data) have revealed a portal-peripheral ratio of 2.80 ± 0.25 in 26 healthy rats. Gel filtration analysis of paired portal and peripheral samples showed that extraction of the 3500 dalton component accounted for virtually all the difference in total IRG. There were only minimal changes in the other immunoreactive fractions. Nephrectomized rats, with markedly elevated levels of the 9000 dalton component, had a portal-peripheral ratio of 1.66 ± 0.18. No differences between portal and peripheral levels of "B" peak and "A" peak components were noted. In contrast, there was a large gradient in "C" peak material. Similar studies in normal dogs, in which species a significant "B" peak component is usually present, confirmed the selective hepatic extraction of 3500 dalton glucagon. These data indicate that as the contribution of the "A" and "B" peak material to the total plasma IRG increases, the lower will be the portal-peripheral ratio as calculated from the total plasma IRG concentration. This observation may account for the variable results which have been reported in the literature for portal-peripheral ratios.

CONCLUSION

In view of the heterogeneity of circulating immunoreactive glucagon and of the differences in the biologic activity and metabolism of different immunoreactive components, it is of importance to analyze the contributions of these components to total plasma IRG in order to interpret glucagon dynamics in health and disease.

Acknowledgement

Supported by Grants AM 13941 and AM 17046, Diabetes Endocrinology Center from the National Institutes of Health.

REFERENCES

1. Alford, F. P., Bloom, S. R., Nabarro, J. D. N. (1976): Glucagon metabolism in man. Studies on the metabolic clearance rate and plasma acute disappearance time of glucagon in normal and diabetic subjects. J. Clin. Endocr. Metab. 42, 830-838.

2. Assan, R. (1972): In vivo metabolism of glucagon. In: Glucagon: Molecular Physiology, Clinical and Therapeutic Implications. P. J. Lefèbvre and R. H. Unger, Eds. Pergamon Press Ltd., Oxford, pp. 47-49.

3. Assan, R., Tchobroutsky, G. and Gross, G. (1971): Abstract, Meet. Europ. Assoc. Study of Diabetes, Southampton, September 1971.

4. Baba, S., Morita, S., Mizuno, N. and Okada, K. (1976): Autoimmunity to glucagon in a diabetic not on insulin. Lancet 2, 585.

5. Bastl, C., Finkelstein, F. O., Sherwin, R. S., Felig, P. and Hayslett, J. (1975): Mechanism of hyperglucagonemia in renal failure. Abstract, 8th Meet. Amer. Soc. Nephrol., p. 59.

6. Bilbrey, G. L., Faloona, G. R., White, M. G., Atkins, C., Hull, A. R. and Knochel, I. P. (1975): Hyperglucagonemia in uremia: reversal by renal transplantation. Ann. Int. Med. 82, 525-528.

7. Bilbrey, G. L., Faloona, G. R., White, M. G. and Knochel, J. P. (1974): Hyperglucagonemia of renal failure. J. Clin. Invest. 53, 841-847.

8. Blackard, W. G., Nelson, N. C. and Andrews, S. S. (1974): Portal and peripheral vein immunoreactive glucagon concentrations after arginine or glucose infusions. Diabetes 23, 199-202.

9. Buchanan, K. D., Solomon, S. S., Vance, J. E., Porter, H. and Williams, R. M. (1968): Glucagon clearance by the isolated perfused rat liver. Proc. Soc. Exp. Biol. Med. 128, 620-623.

10. Buchanan, K. D., Vance, J. E., Dinstl, K. and Williams, R. H. (1969): Effect of blood glucose on glucagon secretion in anaesthetized dogs. Diabetes 18, 11-18.

11. Cresto, J., Lavine, R., Perrino, P. and Recant, L. (1974): Glucagon antibodies in diabetic patients. Lancet 1, 1165.

12. Danforth, D. N., Triche, T., Doppman, J. C., Beazley, R. M., Perrino, P. V. and Recant, L. (1976): Elevated plasma proglucagon-like component with a glucagon secreting tumor: effects of streptozotocin. New Engl. J. Med. 295, 242-245.

13. Dencker, M., Hedner, P., Holst, J. and Tranberg, K.-G. (1975): Pancreatic glucagon response to an ordinary meal. Scand. J. Gastroenterol. 10, 471-474.

14. Duckworth, W. C., Heinemann, M. and Goessling, M. (1976): Enzymatic mechanisms for insulin and glucagon degradation by kidney. Clin. Res. 24, 359A.

15. Emmanouel, D. S., Jaspan, J. B., Kuku, S. F., Rubenstein, A. H. and Katz, A. I. (1976): Pathogenesis and characterization of hyperglucagonemia in the uremic rat. J. Clin. Invest. 58, 1266-1272.

16. Felig, P., Gusberg, R., Hendler, R., Gump, F. and Kinney, T. (1974): Concentrations of glucagon and the insulin:glucagon ratio in the portal and peripheral circulation. Proc. Soc. Exp. Biol. Med. 147, 88-90.

17. Field, J. B., Bloom, G., Petruska, M., Brosnin, K. and Chou, C. (1976): Effects of stimulation of insulin and glucagon secretion on hepatic glucose production, insulin/glucagon ratio and hepatic extraction of insulin and glucagon. Clin. Res. 24, 485A.

18. Fischer, M., Sherwin, R. S., Hendler, R. and Felig, P. (1976): Kinetics of glucagon in man: effects of starvation. Proc. Natl. Acad. Sci. USA 73, 1735-1739.

19. Goldner, M. G., Jauregui, R. M. and Weisenfeld, S. (1954): Disappearance of HGF from insulin after liver perfusion. Am. J. Physiol. 179, 25-28.

20. Jaspan, J. B. Kuku, S. F., Lucker, J. D., Huen, A. H-J, Emmanouel, D. S., Katz, A. I. and Rubenstein, A. H. (1976): Heterogeneity of plasma glucagon in man. Metabolism 25 (Suppl. 1), 1397-1401.

21. Katz, A. I. and Rubenstein, A. H. (1973): Metabolism of proinsulin, insulin, and C-peptide in the rat. J. Clin. Invest. 52, 1113-1121.

22. Kuku, S. F., Jaspan, J. B., Emmanouel, D. S., Zeidler, A., Katz, A. I. and Rubenstein, A. H. (1976): Heterogeneity of plasma glucagon: circulating components in normal subjects and patients with chronic renal failure. J. Clin. Invest. 58, 742-750.

23. Kuku, S. F., Zeidler, A., Emmanouel, D. S., Katz, A. I., Rubenstein, A. H., Levin, N. W. and Tello, A. (1976): Heterogeneity of plasma glucagon: patterns in patients with chronic renal failure and diabetes. J. Clin. Endocrinol. Metab. 42, 173-176.

24. Lefèbvre, P. J. and Luyckx, A. S. (1975): Effect of acute kidney exclusion by ligation of renal arteries on peripheral plasma glucagon levels and pancreatic glucagon production in the anesthetized dog. Metabolism 24, 1169-1176.

25. Lefèbvre, P. J. and Luyckx, A. S. (1976): Plasma glucagon after kidney exclusion. Experiments in somatostatin infused and in eviscerated dogs. Metabolism 25, 761-768.

26. Lefèbvre, P. J., Luyckx, A. S. and Nizet, A. H. (1974): Renal handling of endogenous glucagon in the dog: comparison with insulin. Metab. Clin. Exp. 23, 753-761.

27. Leichter, S. B., Pagliara, A. S., Greider, M. M., Pohl, S., Rosai, J. and Kipnis, D. M. (1975): Uncontrolled diabetes mellitus and hyperglucagonemia associated with an islet cell carcinoma. Am. J. Med. 58, 285-293.

28. Liljenquist, J. E., Chiasson, J. L., Cherrington, A. D., Keller, U., Jennings, A. S., Bomboy, J. D. and Lacy, W. W. (1976): An important role for glucagon in the regulation of glucose production in vivo. Metabolism 25 (Suppl. 1), 1371.

29. McGavran, M. H., Unger, R. H., Recant, L., Polk, H. C., Kilo, C. and Levin, M. E. (1966): A glucagon secreting alpha cell carcinoma of the pancreas. New Engl. J. Med. 274, 1408-1413.

30. Mallinson, C. N., Bloom, S. R., Warin, A. P., Salmon, P. R., Cox, B. (1974): A glucagonoma syndrome. Lancet 2, 1-5.

31. Marco, J., Diego, J., Villaneuva, M., Dizz-Ferros, M., Valverde, I. and Segovia, J. (1973): Elevated glucagon levels in cirrhosis of the liver. New Engl. J. Med. 289, 1107-1111.

32. Narahara, H. T., Everett, N. B., Simmons, B. S. and Williams, R. H. (1958): Metabolism of insulin-I^{131} and glucagon-I^{131} in the kidney of the rat. Am. J. Physiol. 192, 227-231.

33. Palmer, J. P., Werner, L., Benson, J. W. and Ensinck, J. W. (1976): Dominant inheritance of large molecular weight immunoreactive glucagon. Diabetes 25 (Suppl. 1), 326.

34. Recant, L., Perrino, P. V., Bhathena, S. J., Danforth, D. N. and Lavine, R. L. (1976): Plasma immunoreactive glucagon fractions in four cases of glucagonoma. Increased "large glucagon-immunoreactivity". Diabetologia 12, 319-326.

35. Sherwin, R. S., Bastl, C., Finkelstein, F. O., Fisher, M., Black, H., Hendler, R. and Felig, P. (1976): Influence of uremia and hemodialysis on the turnover and metabolic effects of glucagon. J. Clin. Invest. 57, 722-731.

36. Sherwin, R. S., Joshi, P., Hendler, R., Felig, P. and Conn. H. (1974): Hyperglucagonemia in Laennec's cirrhosis. The role of portal systemic shunting. New Engl. J. Med. 290, 239-242.

37. Soeters, P., Weir, G., Ebeid, A. M., James, J. H. and Fischer, J. E. (1975): Insulin and glucagon following porto-caval shunt. Gastroenterology 69, A-67, 867.

38. Srikant, C. B., McCorkle, K. and Dobbs, R. E. (1976): A biologically active "macro-'glucagon' (IRG)" in glucagon secreting tissues of the dog. Diabetes 25 (Suppl. 1), 326.

39. Stahl, M., Nars, P. W., Herz, G., Baumann, J. B. and Girard, J. (1972): Glucagon antibodies after long term insulin treatment in a diabetic child. Horm. Metab. Res. 4, 224-225.

40. Thomsen, H. G. (1971): Insulin treatment does not induce glucagon antibodies. Horm. Metab. Res. 3, 57.

41. Unger, R. H., Ohneda, A., Valverde, I., Eisentraut, A. M. and Exton, J. H. (1968): Characterization of the responses of circulating glucagon-like immunoreactivity to intraduodenal administration of glucose. J. Clin. Invest. 47, 48-65.

42. Valverde, I., Dobbs, R. and Unger, R. H. (1975): Heterogeneity of plasma glucagon immunoreactivity in normal, depancreatized and alloxan-diabetic dogs. Metabolism 24, 1021-1028.

43. Valverde, I., Lemon, H. M., Kessinger, A. and Unger, R. H. (1976): Distribution of plasma glucagon immunoreactivity in a patient with suspected glucagonoma. J. Clin. Endocrinol. Metab. 42, 804-808.

44. Valverde, I., Villaneuva, M. L., Lozano, I. and Marco, J. (1974): Presence of glucagon immunoreactivity in the globulin fraction of human plasma ("Big Plasma Glucagon"). J. Clin.Endocrinol. Metab. 39, 1090-1098.

45. Villalpando, S. and Drash, A. (1976): Circulating glucagon antibodies in childhood diabetes, clinical significance and characterization. Diabetes 25 (Suppl. 1), 334.

46. Weir, G. C., Horton, E. S., Aoki, T., Slovik, D. M., Jaspan, J. B. and Rubenstein, A. H. Secretion by glucagonoma of a possible glucagon precursor. J. Clin. Invest. In press.

47. Weir, G. C., Knowlton, S. D. and Martin, D. B. (1975): High molecular weight glucagon-like immunoreactivity in plasma. J. Clin. Endocrinol. Metab. 40, 296-302.

48. Williams, R. M., Hay, J. S. and Tjaden, M. (1959): Degradation of insulin-^{131}I and glucagon-^{131}I and factors influencing it. Ann. N. Y. Acad. Sci. 74, 513-529.

49. Yoshinanga, T., Okuno, G., Shinji, Y., Tsujii, T., Nishikawa, M. (1966): Pancreatic alpha cell tumor associated with severe diabetes mellitus. Diabetes 15, 709-713.

PLASMA IMMUNOREACTIVE GLUCAGON IN DEPANCREATIZED ANIMALS*

T. Matsuyama, R. Tanaka, K. Shima, K. Nonaka, S. Tarui, M. Nishikawa and P. P. Foà

The Second Department of Internal Medicine and Department of Medicine and Geriatrics, Osaka University Medical School, Osaka, Japan
and
Department of Research, Sinai Hospital of Detroit
Detroit, Michigan USA

The basal level of immunoreactive A cell glucagon (GI, as measured with a pancreatic glucagon specific antiserum) and of gut glucagon-like immunoreactive material (gut GLI: calculated as the difference between total GLI measured with a crossreacting antiglucagon serum and GI) were elevated in the systemic blood of depancreatized dogs and rats. Glucose stimulated gut GLI secretion more effectively when introduced into the ileum than when introduced into the jejunum of normal dogs and this response was enhanced following pancreatectomy. On the other hand, the high levels of gut GI and GLI in depancreatized dogs, were reduced by insulin therapy and by somatostatin although the latter did not cause a decline of blood glucose levels in depancreatized dogs. The secretion of gut GI was not stimulated by insulin-induced hypoglycemia but increased following arginine infusion. These glucagon abnormalities are similar to those observed in insulin deficient diabetics. Thus, high levels of gastrointestinal glucagon fractions may contribute to the overall glucagon abnormality characteristic of insulin deficient diabetes.

INTRODUCTION

Extracts of gastric and intestinal mucosa with hyperglycemic and glycogenolytic properties were prepared many years ago (2, 14) and, following the development of the radioimmunoassay for glucagon (15) were shown to contain glucagon-like immunoreactive materials (GLI; 16). Little is known about the physiologic role of GLI and about its contribution to the glucagon abnormalities associated with diabetes mellitus (1). In this study, the physiologic and pathophysiologic significance of GLI was examined in depancreatized dogs.

ANALYTICAL METHODS

Glucagon was measured by means of a previously described radioimmunoassay (8), or using polyethylene glycol (6) to separate bound from free radioactivity.

---

*Portions of this paper have been published elsewhere (8, 10).

Two rabbit antiglucagon sera (AGS) were used: AGS 10 which binds pancreatic glucagon and gut extract equally well and measures total glucagon-like immunoreactive material (total GLI) and AGS 18, which binds only 5% of the GLI present in our gut extracts (Fig. 1). Thus, the crossreactivity of AGS 18 in our immunoassay system was comparable to that of other pancreatic glucagon specific AGS and was used to measure immunoreactive glucagon of pancreatic or A-cell origin (GI).

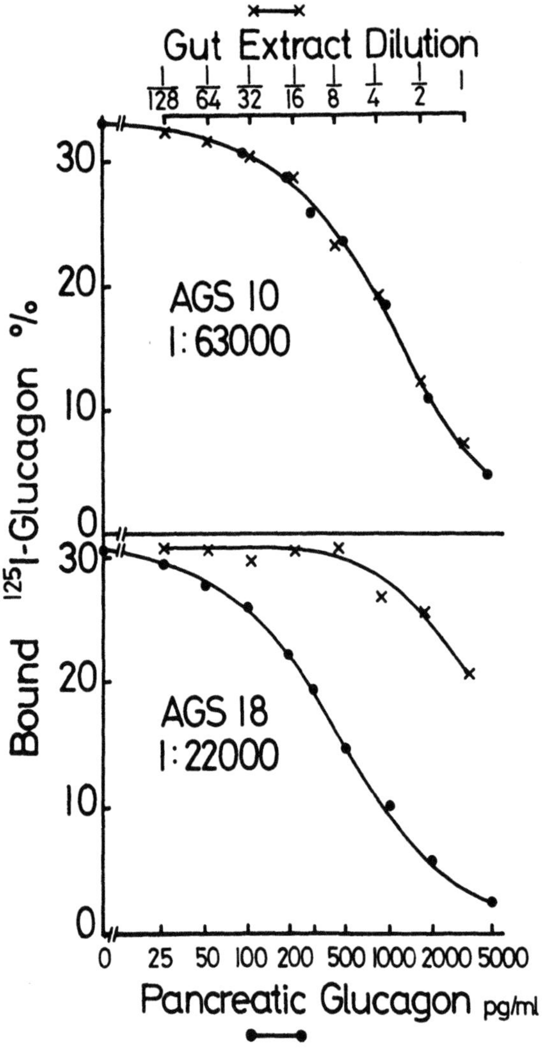

Fig. 1. Standard curves obtained with AGS 10 and AGS 18. Gut extract dilution 1 contains 0.1 mg/ml of crude acid-alcohol extract of dog small intestine.

Gut GLI was calculated as the difference between total GLI and GI. Plasma glucose was measured enzymatically using the AutoAnalyzer* or with an o-toluidine method**. Plasma immunoreactive insulin (IRI) was measured with the method of Hales and Randle (5) or using a solid phase method***.

Pancreatectomy in the Dog and the Rat

Healthy mongrel dogs were depancreatized and received no insulin. Blood samples were obtained from the antecubital vein before and after pancreatectomy. As shown in Fig. 2, pancreatectomy was followed not only by the expected decrease of plasma IRI and increase in plasma glucose, but also by a progressive increase in the concentration of total GLI, GI and gut GLI. In the two dogs that survived 9 days, these values reached levels almost ten times higher than normal.

High level of circulating gut GLI and persistent GI were also observed in rats, 3 days after pancreatectomy (Fig. 3). Increased GI has been observed in insulin deprived depancreatized dogs using another pancreatic glucagon specific AGS (20) and glucagon undistinguishable from the pancreatic hormone has been identified in the duodenal extract of the pig by Sasaki et al (13). We also confirmed the presence of small but significant amounts of immunoreactive glucagon measured by means of pancreatic glucagon specific AGS 18 in the extract of gastrointestinal mucosa of man, rat, dog and pig (Fig. 4). Thus, at least one of the sources of circulating GI after pancreatectomy is the upper gastrointestinal tract.

Intestinal Loop Experiments

To examine the secretion of glucagon-like materials by the intestine, we prepared intestinal loops in normal dogs by tying both ends of a 30 cm intestinal

*Auto-Test; Boehringer Mannheim Corp., New York, New York 10017, USA
**Glucose kit N; Nippon Shoji Kaisha, Ltd., Osaka, Japan
***Phadebas kit; Pharmacia Laboratories, Inc., Uppsala, Sweden

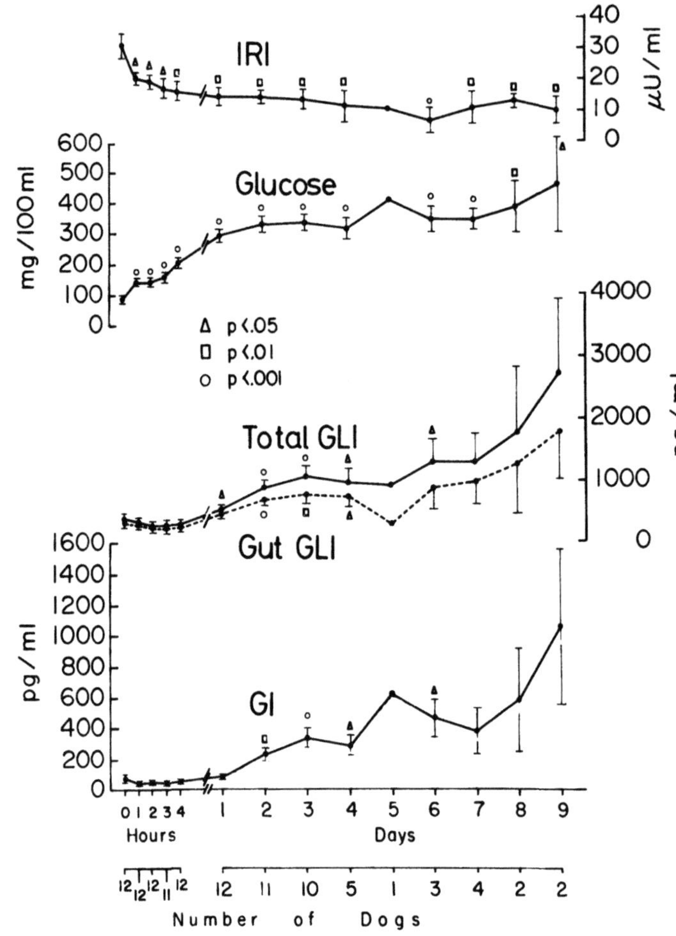

Fig. 2. Plasma levels of IRI, glucose, total GLI, gut GLI and GI after pancreatectomy in the dog. Mean ± S.E.

segment, introduced 50 ml of 5% glucose solution into the intestinal lumen and collected blood samples from the regional mesenteric vein. When glucose was placed into a jejunal loop very small changes were noted except for a significant increase in blood glucose (Fig. 5). On the other hand, when glucose was placed into a loop of ileum, a marked GLI response was observed although the concentration of blood glucose did not change significantly (Fig. 5). No significant

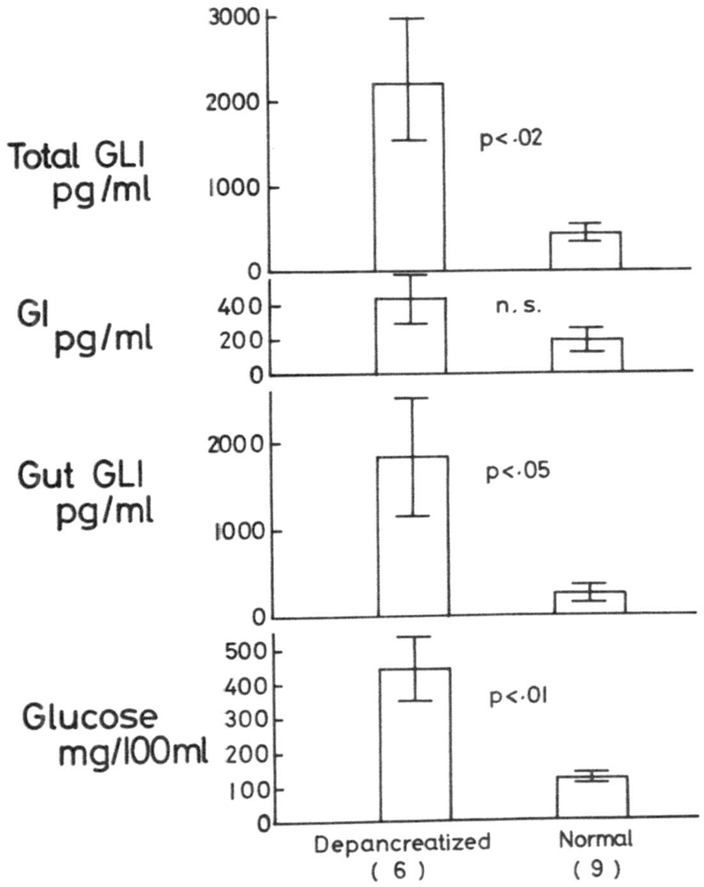

Fig. 3. Plasma levels of total GLI, gut GLI, GI and glucose in normal and depancreatized rats. Mean ± S.E. Number of rats in parentheses.

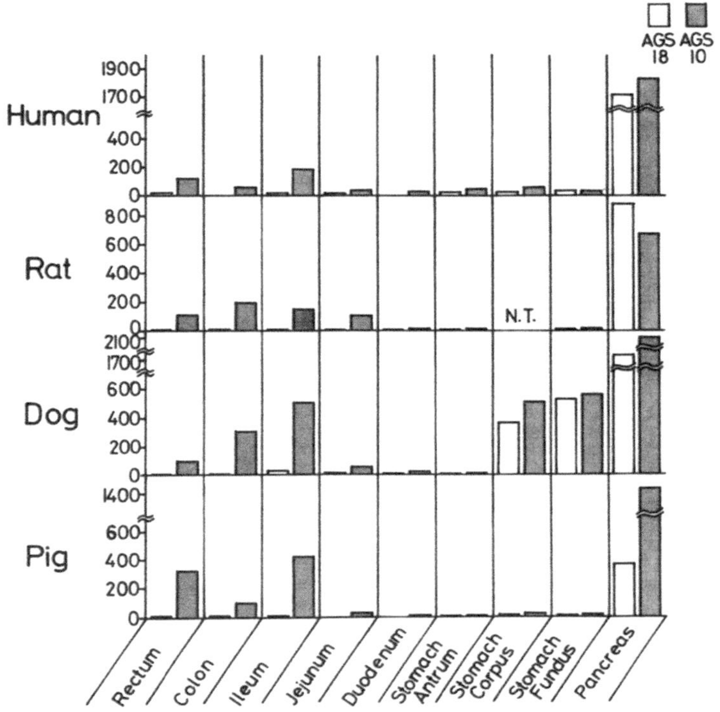

Fig. 4. GI (AGS 18) and total GLI (AGS 10) content of the mucosa of the gastrointestinal tract (ng Eq/g tissue).

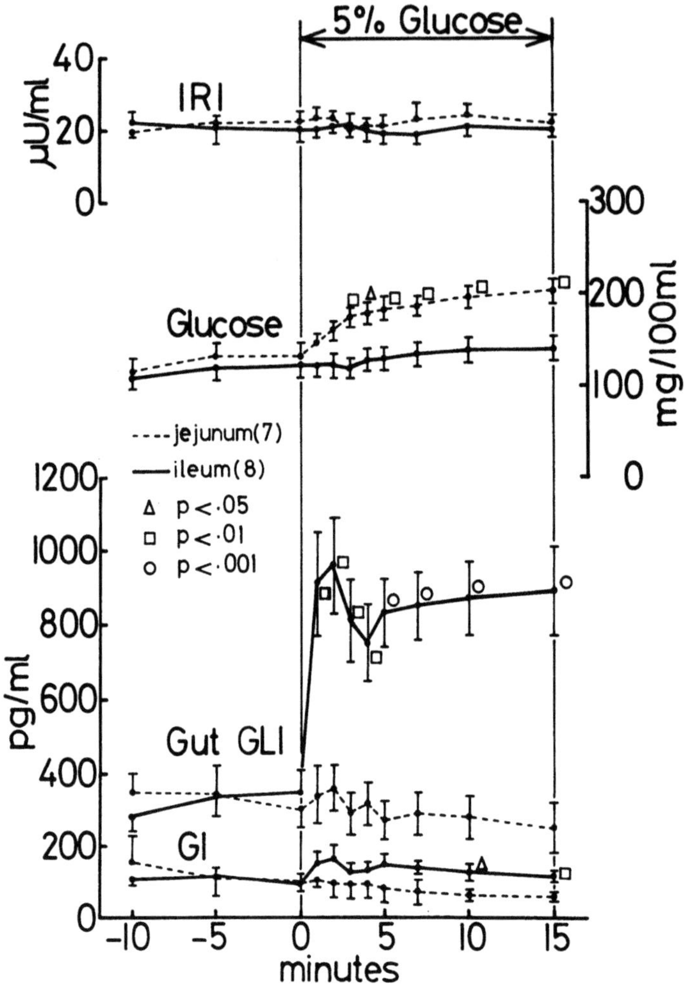

Fig. 5. Plasma IRI, glucose, gut GLI and GI in the regional mesenteric blood following the injection of 50 ml of a 5% glucose solution into a loop of small intestine of normal dogs. Mean ± S.E. P values compare the responses of jejunal and ileal loops. Number of experiments in parentheses.

changes in GI were noted. In view of the meager response of the jejunum in normal dogs, loops of ileum were used in depancreatized dogs 4 to 10 days after surgery. In these animals, the GLI response to glucose was significantly greater than that seen in normal dogs (Fig. 6). Thus, not only did the basal level of GLI increase progressively after pancreatectomy, but so did its response to glucose,

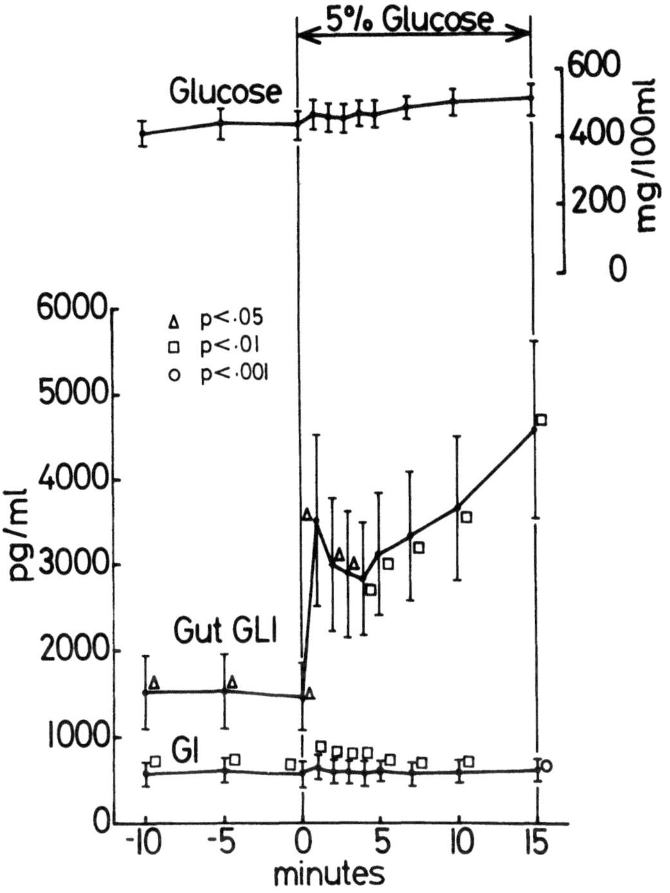

Fig. 6. Plasma glucose, gut GLI and GI in the regional mesenteric blood following the injection of 50 ml of a 5% glucose injection into a loop of ileum in 8 depancreatized dogs. Mean ± S.E. P values compare the responses of depancreatized and normal dogs.

as indicated by the incremental area observed during 15 minutes after glucose introduction into the ileum (Fig. 7).

Effect of Insulin

When a single intramuscular injection of regular insulin (0.4 U/kg) was given to depancreatized dogs 3 days after surgery, a significant elevation of plasma IRI level but only a small decrease in glucose were noted. However, there

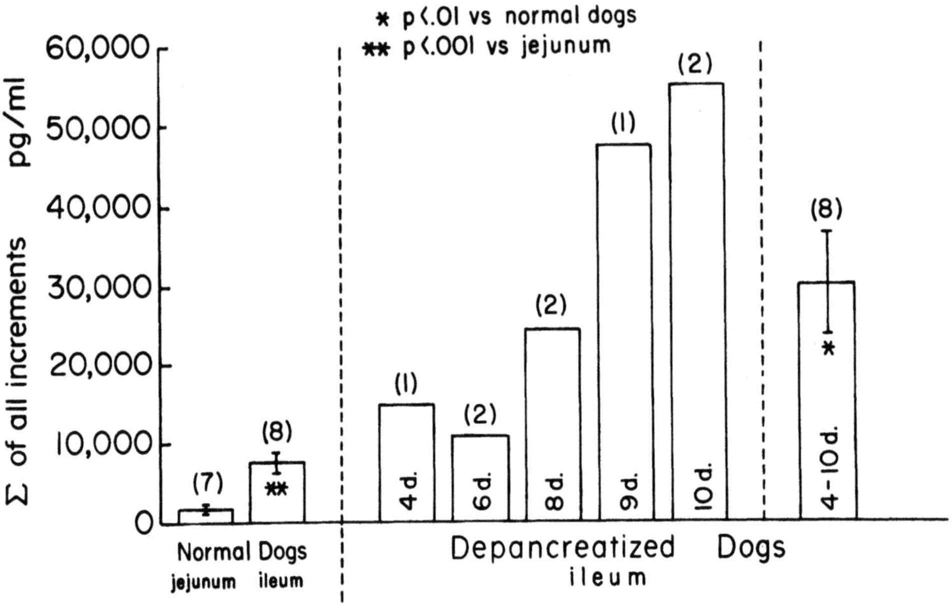

Fig. 7. Incremental area measuring the gut GLI response to the injection of 50 ml of a 5% glucose solution into a loop of small intestine of normal and depancreatized dogs. Number of dogs in parentheses. Day after pancreatectomy indicated.

was a significant decrease in all glucagon fractions, reaching nadir 2 hours after insulin injection (Fig. 8). These results are consistent with the notion that insulin suppresses the secretion of GLI and GI by the gut.

Effect of Somatostatin

In 5 depancreatized dogs, somatostin (SRIF) was infused intravenously 7 days after pancreatectomy, at the dose of 3 μg/min for 30 min. The GI concentration in systemic blood decreased rapidly by as much as 97% within 30 minutes while the decrease of gut GLI was not significant. The concentration of blood glucose did not change significantly. On the other hand, the response of gut GLI to the introduction of 5% glucose into a loop of ileum in depancreatized dogs was also reversed by SRIF. In 5 other depancreatized dogs, SRIF was infused for 90 minutes at the

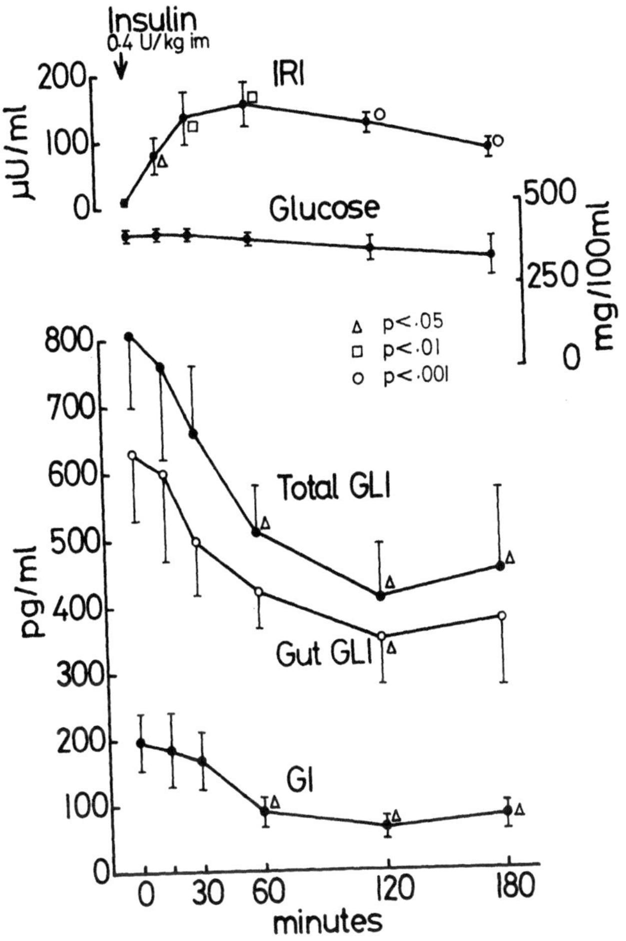

Fig. 8. Plasma IRI, glucose, total GLI, gut GLI and GI in the peripheral blood of depancreatized dogs following a single injection of regular insulin. Mean of 7 experiments ± S.E.

rate of 10 μg/min. A rapid decrease in GI concentration and a slow decrease in gut GLI concentration were noted (Fig. 9). Again, no significant changes in plasma glucose were observed during the experiment. These results suggest that the metabolic clearance rate of gut GLI is slower than that of GI, as indicated also by the fact that the liver removes more GI than gut GLI from the perfusing fluid (4). GI of gastrointestinal origin is believed to have hyperglycemic activity (13) and the suppression of GI by SRIF causes a decline in blood glucose

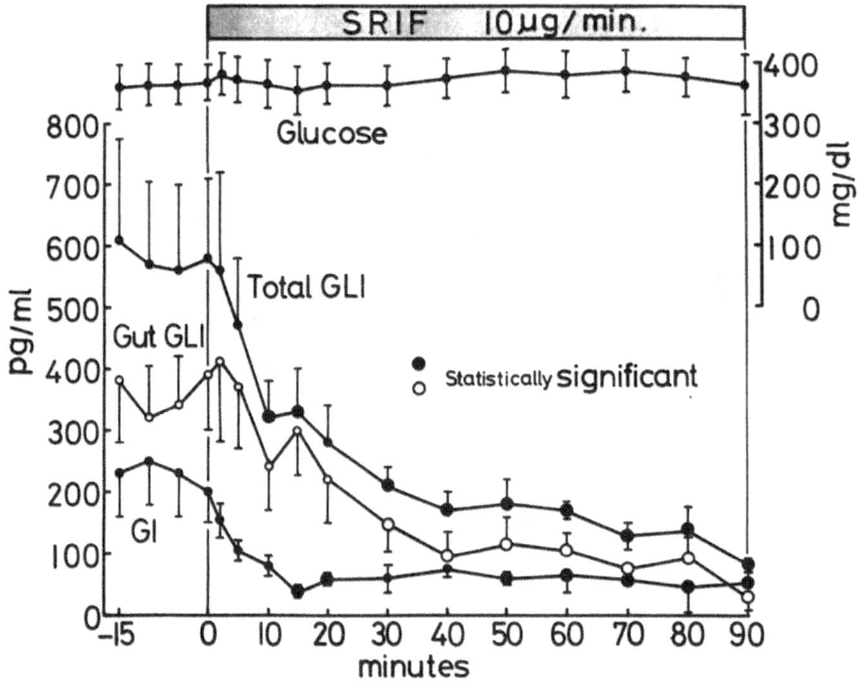

Fig. 9. Plasma glucose, total GLI, gut GLI and GI in the peripheral blood of 5 depancreatized dogs during somatostatin infusion. Mean ± S.E.

concentration of alloxan diabetic dogs (11). SRIF infusion during and after pancreatectomy prevents the increase of GI and the hyperglycemia (12). Yet, in our experiments, the suppression of GI did not cause a lowering of blood glucose, perhaps because our depancreatized dogs were never treated with insulin.

Insulin-induced Hypoglycemia

Another group of depancreatized dogs was treated with 12 units of monocomponent lente insulin daily for 4 to 7 days after surgery. After a 24 hour fast the dogs were anesthetized with pentobarbital sodium and monocomponent actrapid insulin (0.5 U/kg) was injected into the antecubital vein. Control normal dogs received insulin at the dose of 0.2 U/kg. As shown in Fig. 10, the same decline

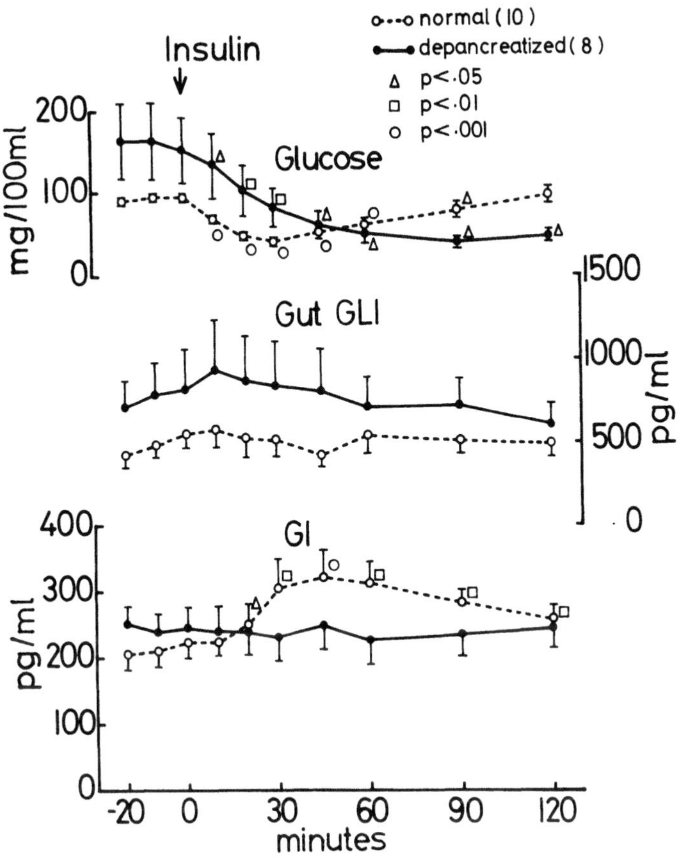

Fig. 10. Plasma glucose, gut GLI and GI in the peripheral blood of normal and insulin-treated depancreatized dogs during insulin tolerance test (ITT). Mean ± S.E. Number of dogs in parentheses.

in blood glucose was achieved in both groups of animals. The concentration of GI did not rise in the depancreatized dogs, while it rose significantly in the normal animals. These findings suggest that the GI response to hypoglycemia requires the presence of the pancreas and that the gastrointestinal A cells do not respond to it. The concentration of gut GLI did not change significantly after insulin injection. The basal levels of GI and gut GLI were not abnormally high in these insulin-treated depancreatized dogs, although the averages were slightly higher than those of normal dogs. These results support the hypothesis that insulin

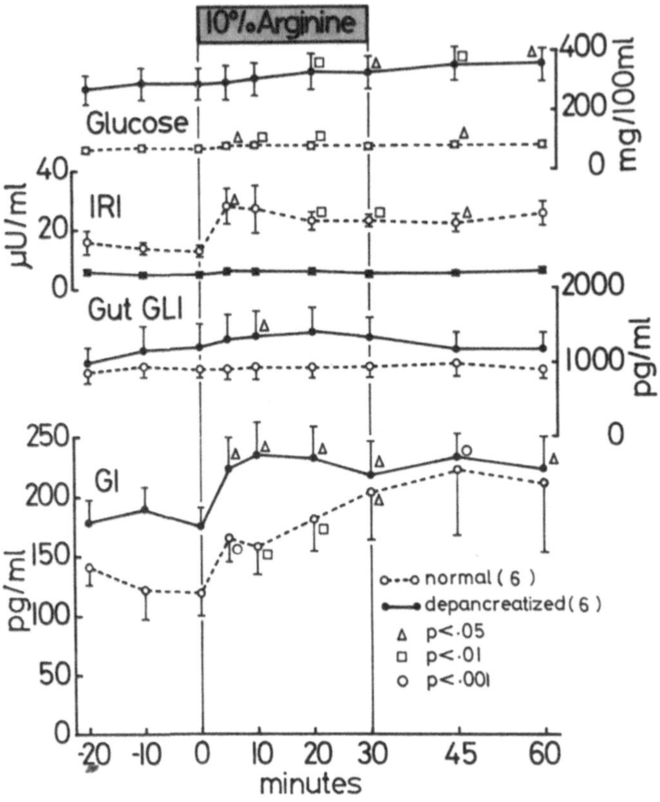

Fig. 11. Plasma glucose, IRI, gut GLI and GI in the peripheral blood of normal and depancreatized dogs during arginine tolerance test (ATT). Mean ± S.E. Number of dogs in parentheses.

insufficiency is required for the hypersecretion of GI and GLI by the gut.

Arginine Infusion

The intravenous infusion of 7 g of arginine in a 10% solution stimulated GI secretion in depancreatized as well as in normal dogs (Fig. 11). No IRI response was seen in the depancreatized animals. These findings confirm those of Mashiter et al (7) and suggest that the gastrointestinal A cells respond to intravenous arginine infusion as do their pancreatic counterparts. The GI, gut GLI, IRI and

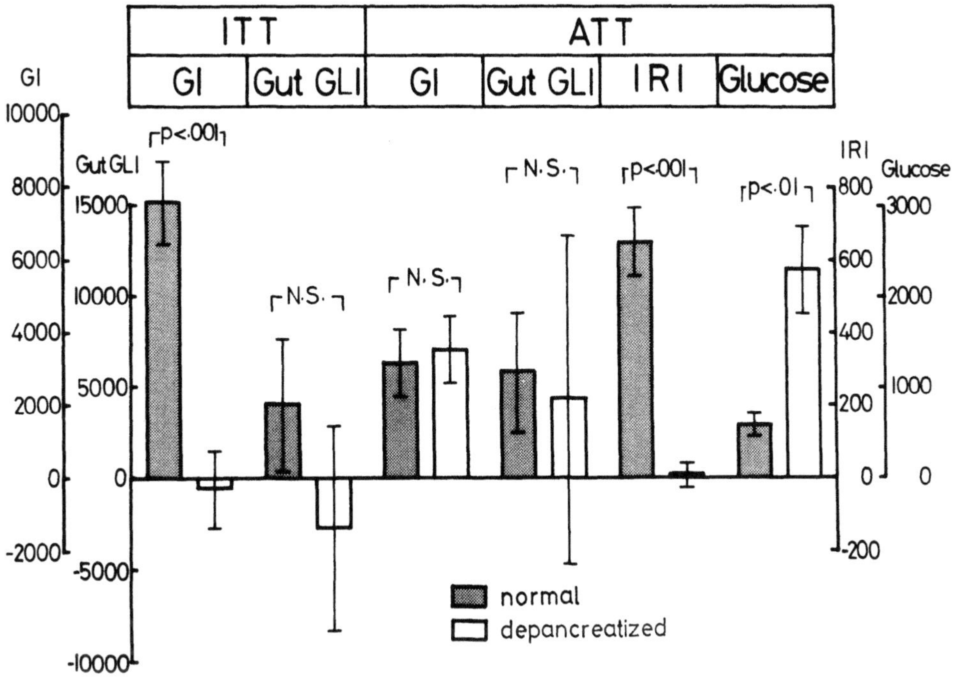

Fig. 12. Incremental area of the GI, gut GLI, IRI and glucose responses during ITT and ATT in normal and depancreatized dogs. Mean ± S.E.

glucose responses to insulin and arginine are shown in Figure 12. In depancreatized dogs, the GI response to insulin-induced hypoglycemia was absent whereas its response to arginine infusion was comparable to that of normal dogs. Again, arginine infusion caused no IRI response, but did result in a greater rise of blood glucose in depancreatized dogs.

Gerich et al (3) reported that in patients with juvenile onset diabetes, insulin-induced hypoglycemia did not stimulate glucagon secretion but that a supernormal glucagon response occurred following intravenous arginine infusion. These similarities between depancreatized dogs and diabetic patients as well as our previous findings that insulin reduced hyperglucagonemia in newly diagnosed diabetic children (9) and in depancreatized dogs suggest that gastrointestinal glucagon may play a role in the metabolic alterations of insulin deficient diabetes similar to that played by pancreatic glucagon (17).

REFERENCES

1. Aguilar-Parada, E., Eisentraut, A. M. and Unger, R. H. (1969): Pancreatic glucagon secretion in normal and diabetic subjects. Amer. J. Med. Sci. 257, 415-419.

2. Foà, P. P., Berger, S., Santamaria, L., Smith, J. A. and Weinstein, H. R. (1953): Extraction of the hyperglycemic factor (HGF) of the pancreas with liquid ammonia. Science 117, 82-84.

3. Gerich, J. E., Langlois, M., Noacco, C., Karam, J. H. and Forsham, P. H. (1973): Lack of glucagon response to hypoglycemia in diabetes: evidence for an intrinsic pancreatic alpha cell defect. Science 182, 171-173.

4. Gutman, R. A., Fink, G., Voyles, N., Selawry, H., Penhos, J. C., Lepp, A. and Recant, L. (1973): Specific biologic effects of intestinal glucagon-like materials. J. Clin. Invest. 52, 1165-1175.

5. Hales, C. N. and Randle, P. J. (1963): Immunoassay of insulin with insulin-antibody precipitate. Biochem. J. 88, 137-146.

6. Henquin, J. C., Malvaux, P. and Lambert, A. E. (1974): Glucagon immunoassay using polyethylene glycol to precipitate antibody-bound hormone. Diabetologia 10, 61-68.

7. Mashiter, K., Harding, P. E., Chou, M., Mashiter, G. D., Stout, J., Diamond, D. and Field, J. B. (1975): Persistent pancreatic glucagon but not insulin response to arginine in pancreatectomized dogs. Endocrinology 96, 678-693.

8. Matsuyama, T. and Foà, P. P. (1974): Plasma glucose, insulin, pancreatic, and enteroglucagon levels in normal and depancreatized dogs. Proc. Soc. Exp. Biol. Med. 147, 97-102.

9. Matsuyama, T., Hoffman, W. H., Dunbar, J. C., Foà, N. L. and Foà, P. P. (1975): Glucose, insulin, pancreatic glucagon and glucagon-like immunoreactive materials in the plasma of normal and diabetic children. Effect of the initial insulin treatment. Horm. Metab. Res. 7, 452-456.

10. Matsuyama, T., Tanaka, R., Shima, K., Nonaka, K. and Tarui, S. (1976): Secretory mechanism of gastrointestinal glucagon. Abstracts, V Intern. Congr. Endocrinol., Hamburg 1976, p. 340.

11. Sakurai, H., Dobbs, R. and Unger, R. H. (1974): Somatostatin-induced changes in insulin and glucagon secretion in normal and diabetic dogs. J. Clin. Invest. 54, 1395-1402.

12. Sakurai, H., Dobbs, R. E. and Unger, R. H. (1975): The role of glucagon in the pathogenesis of the endogenous hyperglycemia of diabetes mellitus. Metabolism 24, 1287-1297.

13. Sasaki, H., Rubalcava, B., Baetens, D., Blázquez, E., Srikant, C. B., Orci, L. and Unger, R. H. (1975): Identification of glucagon in the gastrointestinal tract. J. Clin. Invest. 56, 135-145.

14. Sutherland, E. W. and DeDuve, C. (1948): Origin and distribution of the hyperglycemic-glycogenolytic factor of the pancreas. J. Biol. Chem. 175, 663-674.

15. Unger, R. H., Eisentraut, A. M., McCall, M. S., Keller, S., Lanz, H. C. and Madison, L. L. (1959): Glucagon antibodies and their use for immunoassay for glucagon. Proc. Soc. Exp. Biol. Med. 102, 621-623.

16. Unger, R. H., Ketterer, H. and Eisentraut, A. M. (1966): Distribution of immunoassayable glucagon in gastrointestinal tissues. Metabolism 15, 865-867.

17. Unger, R. H. and Orci, L. (1975): Hypothesis: the essential role of glucagon in the pathogenesis of diabetes mellitus. Lancet 1, 14-16.

18. Vranic, M., Pek, S. and Kawamori, R. (1974): Increased "glucagon immunoreactivity" in plasma of totally depancreatized dogs. Diabetes 23, 905-912.

HETEROGENEITY OF GUT GLUCAGON-LIKE IMMUNOREACTIVITY (GLI)

A. J. Moody, H. Jacobsen, F. Sundby, E. K. Frandsen, D. Baetens
and L. Orci

Novo Research Institute, Copenhagen, Denmark
and
Institut d'Histologie et d'Embryologie, Université de Genève
Genève, Switzerland

One component of gut glucagon-like immunoreactivity (GLI-1) has been purified and characterized. It is proposed that gut GLIs of a given species contain the full homologous glucagon sequence and are immunologically related to the glucagon precursors found in pancreatic A cells. The essential difference between the gastrointestinal cells which produce GLIs and the pancreatic A cells is the degree to which these cells shorten a common primary gene product before storage and secretion.

Gut glucagon-like immunoreactive (GLI) materials are proteins of intestinal origin which react with certain anti-glucagon sera via an immunodeterminant similar to that found in the (2-23) region of glucagon (1,3,4), but which do not bind to other anti-glucagon sera known to react with the C-terminal portion of glucagon.

This communication summarizes our present knowledge concerning the similarities between gut GLIs and glucagon and proposes a relationship between them.

MATERIALS

Porcine gut GLI-1 was prepared according to Sundby *et al*. (7). Anti-glucagon serum K 4023, which reacts fully with gut GLI-1, was prepared and donated by L. G. Heding (4).

Immunochemistry. Porcine gut GLI-1 (7), a highly purified protein (MW 11625) having full molar reactivity with glucagon (2-23)-specific antisera, was coupled to human serum albumin with glutaraldehyde and antisera were raised against the complex in rabbits. The antisera were of varying quality but one (R 64) proved useful. This antiserum has been used in a radioimmunoassay for gut GLI-1 using standards of gut GLI-1 and ^{125}I-labelled gut GLI-1. The assay detects gut GLI in extracts of porcine intestine in amounts slightly smaller than those found with a glucagon (2-23)-specific antiserum (K 4023). The antiserum did not bind ^{125}I-labelled glucagon and glucagon did not displace ^{125}I-labelled gut GLI-1 from the antiserum.

The specificity of R 64 was further investigated by measuring the distribution of K 4023 and of R 64 immunoreactivities during an ion exchange fractionation of a crude extract of porcine jejunum + ileum (Fig. 1). Both antisera detected similar amounts of immunoreactivity in fractions 50 - 80 and these fractions represent gut GLI-1 and closely related proteins. A small amount of R 64 immunoreactivity, which was not associated with K 4023 immunoreactivity, eluted in front of these fractions. A small peak of K 4023 immunoreactivity, eluting behind the main gut GLI peak, was not associated with R 64 immunoreactivity. This peak of gut GLI (temporarily called gut GLI-3) has been partially purified, and is smaller than gut GLI-1. It is concluded that porcine gut GLI-1 contains at least two immunodeterminants: one which reacts with glucagon (2-23)-specific antisera such as K 4023 and the other which reacts with antisera such as R 64. Furthermore the R 64 immunodeterminant is absent from smaller forms of gut GLI such as gut GLI-3.

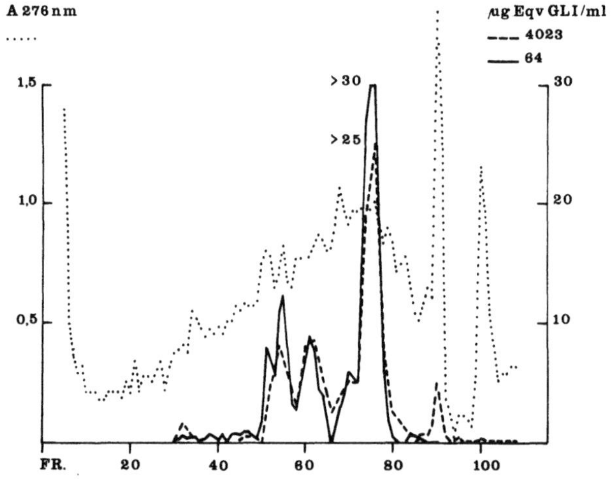

Fig. 1. Ion-exchange chromatography of the gut GLIs in crude gut extract. A crude extract of porcine intestine was fractionated on a column of carboxymethyl cellulose. The sample was applied in 0.02 M sodium acetate, pH 4.5, and eluted with a linear gradient of 0.02 M-0.3 M sodium acetate at pH 4.5.

R 64 has been used for indirect immunofluorescence, and preliminary data suggest that it binds to "gut GLI" cells in the gastro-intestinal tract. This antiserum was used to detect the possible presence of "gut GLI" in the rat pancreas (Fig. 2). Serial sections of a rat islet were positively stained after incubation with a specific anti-glucagon serum (30 K). The anti-gut GLI-1 serum (R 64) bound to the same cells. The binding of the anti-glucagon serum was completely inhibited by adsorption with glucagon but not with gut GLI-1, and the binding of R 64 was eliminated by adsorption with gut GLI-1 but not with glucagon. These findings suggest that the same cells in the rat islet contain "immuno-glucagon" and "immuno-gut GLI-1".

Chemistry. The highly purified gut GLI-1 contains the amino acids of glucagon but does not share the known N-terminal sequence of glucagon (7). The C-terminal decapeptide has been cleaved from gut GLI-1 by cyanogen bromide and sequenced (5). The proposed sequence is:

(Met)-Asn-Thr-Lys-Arg-Asn-Lys-Asn-Asn-Ile-Ala
90 91 92 93 94 95 96 97 98 99 100

The decapeptide was devoid of homoserine and was therefore cleaved from the C-terminal side of a methionine. It is therefore proposed that the 11th residue from the C-terminus is Met. Of interest is that the Met-Asn-Thr sequence from the N-terminus is identical to the last tripeptide of glucagon.

The gut GLI-1 (90-100) sequence is very similar to that of the C-terminal portion of the proposed proglucagon fragment of Tager and Steiner (8) and these two sequences are shown in Fig. 3. The only difference in the sequences is that the Lys-Asn of gut GLI-1 (96-97) are inverted in the proglucagon fragment.

Gut GLI-1 and glucagon have been radio-iodinated and the labelled material digested with chymotrypsin (5). Autoradiography of peptide maps of the digests showed that gut GLI-1 contained two tyrosine-containing peptides, Thr-Ser-Asp-Tyr and Ser-Lys-Tyr, known to be formed by chymotryptic digestion of glucagon (2) and which form the glucagon (7-13) sequence.

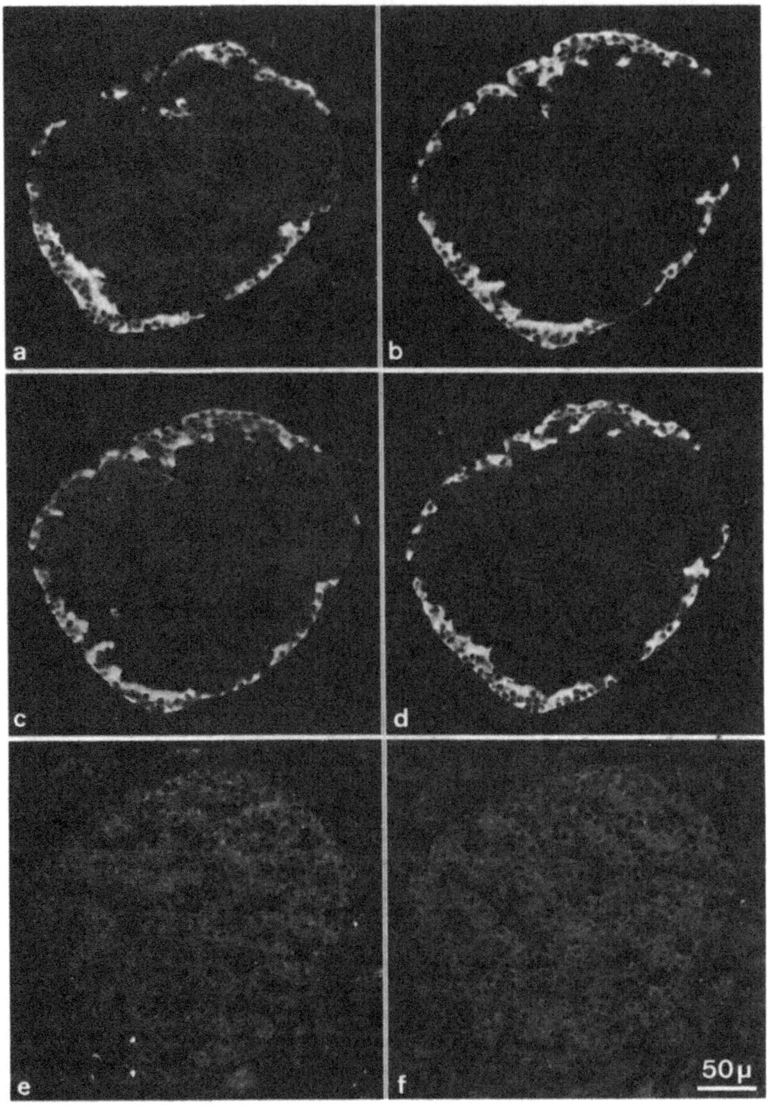

Fig. 2. Indirect immunofluorescence of rat islets. Serial sections of a single islet from the splenic portion of rat pancreas. a) anti-glucagon (30K); b) anti-glucagon (30K) + GLI-1; c) anti-GLI (R64); d) anti-GLI (R64) + glucagon; e) anti-GLI (R64) + GLI-1; f) anti-glucagon (30K) + glucagon.

Glucagon	29 -Met-Asn-Thr
Proglucagon Fragment	29 ... 37 -Met-Asn-Thr-Lys-Arg-Asn-Asn-Lys-Asn-Ile-Ala
Gut GLI-1	92 ... 100 -Met-Asn-Thr-Lys-Arg-Asn-Lys-Asn-Asn-Ile-Ala

Fig. 3. The C-terminal sequences of glucagon, of a proposed pro-glucagon fragment and of gut GLI-1. The inversion of Lys and Asn in the pro-glucagon and in gut GLI-1 is underlined.

The proposed partial sequence of gut GLI-1 is aligned with those of glucagon and of the proglucagon fragment in Fig. 4. Since gut GLI-1 reacts with glucagon 2-23 specific antisera, it is probable that both molecules have a similar sequence in this region and the tyrosine-containing peptides of gut GLI-1 have been placed in a position homologous to glucagon 7-13.

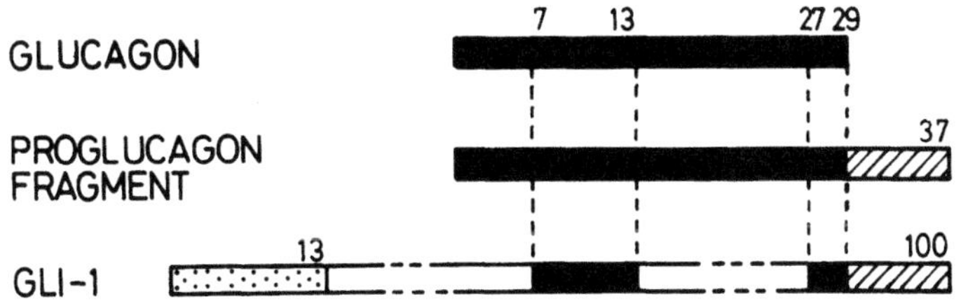

Fig. 4. A schematic representation of the proposed homologous sequences of glucagon, a proglucagon fragment and porcine gut GLI-1.

DISCUSSION AND CONCLUSION

It is considered that enough evidence has been obtained so far by a study of the chemistry and immunochemistry of glucagon and gut GLI-1 to justify the hypothesis that the gut GLIs of a given species contain the full homologous glucagon sequence and that the glucagon precursors found in pancreatic A cells are immunologically related to gut GLIs. This hypothesis is shown schematically in Fig. 5. The essential difference between the gastro-intestinal cells which synthesize and release gut GLIs and the pancreatic A cells (which synthesize glucagon via "gut GLIs") is thus the degree to which the cells shorten their common primary gene product before storage and secretion.

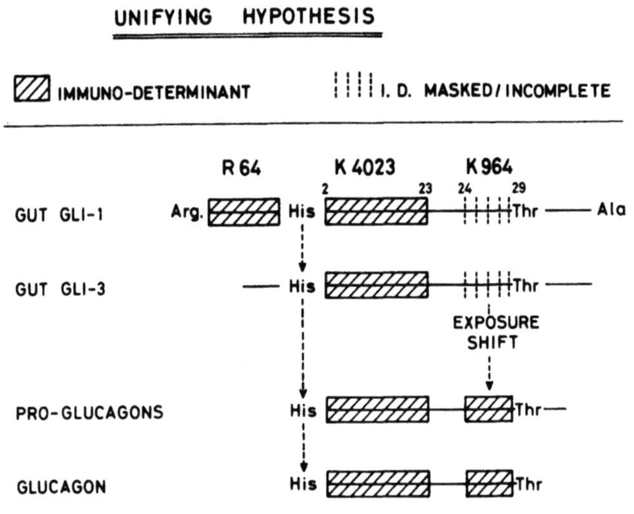

Fig. 5. A schematic representation of the proposed relationship between gut GLI-1 and glucagon.

The plethora of gut GLIs found in extracts of intestine (6, 9) is probably the result of the progressive shortening of a precursor gut GLI. It is not yet known whether all the gut GLIs found in extracts are native forms of the gut GLIs or whether they are formed during extraction.

Acknowledgements

The authors gratefully acknowledge the technical assistance of Ms. I. Bryder-Nielsen, K. Christensen, A. Demandt, G. Duetoft, D. Fisher, E. Gammelgaard and W. Jørgensen.

Supported by grant Nr 3.553.75 from Fonds National Suisse de la Recherche Scientifique.

REFERENCES

1. Assan, R. and Slusher, N. (1972): Structure-function and structure-immunoreactivity relationships of the glucagon molecule and related synthetic peptides. Diabetes 21, 843-855.

2. Bromer, W. W., Boucher, M. E. and Koffenberger, J. E. Jr. (1971): Amino acid sequence of bovine glucagon. J. Biol. Chem. 246, 2822-2827.

3. Buchanan, K. D. (1976): The role of enteric hormones in glucose homeostasis. Horm. Metab. Res. 6 (Suppl.) 80-84.

4. Heding, L. G., Frandsen, E. K. and Jacobsen, H. (1976): Structure-function relationship: immunologic. Metabolism 25 (Suppl. 1) 1327-1329.

5. Jacobsen, H., Demandt, A., Moody, A. J. and Sundby, F. (1976). Europ. J. Biochem. Submitted for publication.

6. Moody, A. J., Frandsen, E. K., and Sundby, F. (1975): Fractionation of gut glucagon-like activities by isoelectric focusing in polyacrylamide gel. In: Progress in Isoelectric Focusing and Isotachophoresis. P. G. Righetti, Ed. North-Holland Publishing Co., pp. 179-191.

7. Sundby, F., Jacobsen, H. and Moody, A. J. (1976): Purification and characterization of a protein from porcine gut with glucagon-like immunoreactivity. Horm. Metab. Res. 8, 366-371.

8. Tager, H. S. and Steiner, D. F. (1973): Isolation of a glucagon-containing peptide: primary structure of a possible fragment of proglucagon. Proc. Nat. Acad. Sci. USA 70, 2321-2325.

9. Valverde, I., Rigopoulou, D., Exton, J., Ohneda, A., Eisentraut, A. M. and Unger, R. H. (1968): Demonstration and characterization of a second fraction of glucagon-like immunoreactivity in jejunal extracts. Am. J. Med. Sci. 255, 415-420.

GLUCAGON-LIKE MATERIALS IN PORCINE AND RAT INTESTINAL MUCOSA

J. M. Conlon, K. D. Buchanan and R. F. Murphy

Departments of Biochemistry and Medicine
Queen's University of Belfast
Belfast, Northern Ireland

Extracts of gastrointestinal tissue contain glucagon-like immunoreactive (GLI) materials with molecular weights varying between 2000 and 30,000. Some of them react with antisera specific for the C-terminal, others with antisera specific for the N-terminal portion of the glucagon molecule. In response to perfusion with glucose the serosal surface of rat small intestine releases increased amounts of some GLI fractions.

Glucagon-like immunoreactive (GLI) material is defined as a mixture of polypeptides which compete with glucagon for binding sites on antibodies raised against glucagon. GLI exists in multiple molecular forms and is distributed throughout the gastrointestinal tract of a wide variety of vertebrate and invertebrate species (for reviews see 4, 8). The method of immunoaffinity chromatography (5) has been used to isolate GLI from extracts of porcine pancreas and gastrointestinal tissue and to study the *in vitro* release of GLI from isolated rat small intestine in response to a glucose load.

The glucagon molecule contains at least two antigenic sites: one associated with the hydrophobic C-terminal region and another associated with the N-terminal and central portion (1). Glucagon antibodies binding the C-terminal fragments (19-29, C-terminal specific) and antibodies binding the N-terminal fragments (1-18, N-terminal specific) were coupled to Sepharose-4B, using either cyanogen bromide or 1,4-bis-(2,3-epoxypropoxy) butane as immobilising agent. Extracts of porcine tissue were chromatographed on columns of these conjugates connected in series (C-terminal specific column first). After removal of non-biospecifically adsorbed material, GLI was eluted from the columns by irrigation with aqueous formic acid, pH 2.0 and lyophilised. Immunoreactive polypeptides were isolated from porcine pancreas, stomach, small intestine and colon and the molecular weights of the GLI species (Table 1) were determined by comparative elution from

Sephadex gels. The pylorus and duodenum contained only trace amounts of immunoreactive material.

TABLE 1

APPROXIMATE MOLECULAR WEIGHT OF GLI SPECIES ISOLATED FROM PORCINE TISSUES

Tissue	Antibody-Sepharose Conjugate Used for Purification	
	C-terminal specific	N-terminal specific
Pancreas	3,000 - 4,000*	8,000 - 10,000
Stomach (Fundus and Body)	3,000 - 4,000	10,000 - 12,000
Jejunum and Ileum		3,000 - 4,000 and
		10,000 - 12,000*
Colon	(before SDS** treatment)	(after SDS** treatment)
	1,000 - 2,000; >30,000*	3,000 - 4,000
		6,000 - 8,000
		10,000 - 12,000*

* Predominant species of GLI in tissue

**Sodium dodecyl sulphate

The immunoreactive material from pancreatic and stomach extracts bound by the C-terminal specific column was immunometrically and electrophoretically indistinguishable from glucagon, but higher molecular weight species of GLI (M. W. 8-10,000 ex pancreas; M. W. 10-12,000 ex stomach) were bound by the N-terminal specific column. These polypeptides reacted at least 10 times more strongly with an antibody which bound N-terminal fragments of glucagon (YY 57) than with an antibody which bound C-terminal fragments (YY 89), under radioimmunoassay conditions. Immunoreactive polypeptides could be isolated from jejunal and ileal extracts by the N-terminal specific column only and were eluted from Sephadex gels in the positions corresponding to GLI peak I and GLI peak II (3), but both species migrated as a heterogeneous mixture on polyacrylamide gel electrophoresis (PAGE). Although the intestinal tract of the pig may be associated with several GLI species of closely related structure, this heterogeneity may have arisen from random proteolysis occurring within the tissue during the collection and extraction procedure. Consequently, attention was directed to a less proteolytically active area of the gut: the colon. When colonic extracts were chromatographed on the antibody-Sepharose conjugates, immunoreactive material was bound by both the C-terminal specific and N-terminal specific columns and could be eluted from Sephadex G-50 gels (exclusion limit, M. W. 30,000) as two peaks. A major component (>90%) emerged at the void volume and a minor component in the region of 1-2,000 M. W. After incubation of the high molecular weight species with 0.1% sodium dodecyl sulphate for several hours at room temperature and rechromatography, immunoreactive polypeptides were eluted as three peaks in the regions of 10-12,000 (major component), 6-8,000 and 3-4,000 M. W. Non-immunoreactive material emerged at the void volume and was heterogeneous on PAGE. Colonic GLI, therefore, is bound in the native state to carrier proteins by non-covalent bonds (2). The 1-2,000 M. W. component may represent a mixture of degradation products of larger immunoreactive species. The purified GLI species from the colon reacted at least 50 times more strongly with the N-terminal specific antibody YY 57 than with the C-terminal reactive antibody, YY 89, under radioimmunoassay conditions.

Administration of glucose, both intraduodenally and intravenously, leads to an increase in concentration of GLI in plasma (10). Isolated rat jejunum and ileum were perfused with a 14 mM solution of glucose under the experimental conditions of O'Connor (7). The GLI released from the serosal surface of the gut was collected directly into acid-alcohol to minimize proteolysis. The immunoreactive polypeptides were isolated by affinity chromatography using the C-terminal specific and N-terminal specific antibody-Sepharose conjugates as described previously. Molecular sizes of the secreted GLI species were determined by gel filtration. Immunoreactive material bound by the N-terminal specific column showed no reactivity towards the C-terminal specific antibody, YY 89, in radioimmunoassays and was eluted from Sephadex G-50 gels as two-peaks in the regions of 3-4,000 (major component) and 10-12,000 M. W. In contrast, GLI bound by the C-terminal specific column was eluted as a single peak in the region of 4,500-5,500 M. W. and was immunometrically similar to glucagon, reacting equally well with both the C- and N-terminal specific antibodies in radioimmunoassays (Table 2). Thus, this material resembles the glucagon-containing peptide of M. W. 4,500 isolated by Tager and Steiner (9) which has been proposed as a possible fragment of proglucagon.

TABLE 2

GLUCAGON-LIKE MATERIALS SECRETED FROM THE SEROSAL SURFACE OF RAT SMALL INTESTINE IN RESPONSE TO PERFUSION WITH GLUCOSE

Approximate M. W.	Reaction With Glucagon Antibodies	
	N-terminal Reactive	C-terminal Reactive
3,000 - 4,000	+	-
4,500 - 5,000	+	+
10,000 - 12,000	+	-

The C- and N-terminal reactive GLI species may be derived from the N-terminal reactive GLI species of 10-12,000 M. W. by specific proteolysis and so the experiment provides evidence that the large molecular species of GLI contain both the C-terminal and N-terminal antigenic sites of glucagon in the same region of the molecule. In the native conformation, however, the C-terminal antigenic site of the large GLI is sterically unavailable to antibodies specific for the C-terminal region of glucagon.

Acknowledgement

This work was carried out with the support of the Medical Research Council, U. K.

REFERENCES

1. Assan, R. and Slusher, N. (1972): Structure/function and structure/immunoreactivity relationships of the glucagon molecule and related synthetic peptides. Diabetes 21, 843-855.

2. Conlon, J. M., Flanagan, R. W. J., Buchanan, K. D. and Murphy, R. F. (1976): Characterisation of complexes of glucagon-like immunoreactivity (GLI) with proteins from plasma and tissue extracts. Current Topics in Diabetes Research. Abstracts IX Congress of the International Diabetes Federation. New Delhi, October 31-November 5, 1976.

3. Markussen, J. and Sundby, F. (1970): Protides of the Biological Fluids (17th Colloquium, Bruges, 1969) ed. H. Peeters, Pergamon Press, Oxford, p. 471.

4. Moody, A. J. (1972): Gastrointestinal glucagon-like immunoreactivity. In: Glucagon, Molecular Physiology, Clinical and Therapeutic Implications, 1st ed., P. J. Lefèbvre and R. H. Unger, eds., Pergamon Press, Oxford, p. 319.

5. Murphy, R. F. (1974): Immunoaffinity chromatography. Biochem. Soc. Trans. 2, 1298.

6. Murphy, R. F., Buchanan, K. D. and Elmore, D. T. (1973): Isolation of glucagon-like immunoreactivity of gut by affinity chromatography on anti-glucagon antibodies coupled to Sepharose-4B. Biochem. Biophys. Acta 303, 118-127.

7. O'Connor, F. A., Gardner, M. L. G. and Buchanan, K. D. (1976): Method for studying the release of GLI from the isolated perfused rat small intestine. Submitted for publication.

8. Sasaki, H., Faloona, G. R. and Unger, R. H. (1974): Candidate hormones of the gut. XIV. Enteroglucagon. Gastroenterology 67, 746-748.

9. Tager, H. S. and Steiner, D. F. (1973): Isolation of a glucagon-containing peptide: Primary structure of a possible fragment of proglucagon. Proc. Nat. Acad. Sci. U. S. A. 70, 2321-2325.

10. Unger, R. H., Ohneda, A., Valverde, I., Eisentraut, A. M. and Exton, J. (1968): Characterisation of the response to circulating glucagon-like immunoreactivity to intraduodenal and intravenous administration of glucose. J. Clin. Invest. 47, 48-65.

IMMUNOREACTIVE GLUCAGON IN THE SALIVARY GLANDS OF MAN AND ANIMAL

S. Hojvat, L. Kirsteins, J. Kisla, V. Paloyan and A. M. Lawrence

Medical and Research Services, Veterans Administration Hospital, Hines, Illinois, and the Departments of Medicine and Biochemistry, Loyola University Stritch School of Medicine, Maywood, Illinois

Recent observations that immunoreactive glucagon levels rise following pancreatectomy and even after total evisceration suggested that extrapancreatic sources of glucagon other than the gastrointestinal tract probably did exist. Acid-saline extractions of many tissues failed to reveal glucagon-like material except in the salivary glands of rats, mice, rabbits, dogs, and man. This material appeared to be in greatest quantity in the submaxillary glands of rodents. Sephadex column fractionation and gel electrophoresis indicate that this material has a molecular weight of approximately 70,000. Refractionation or treatment with urea and acetic acid failed to alter the chromatographic pattern of this material. Injection of rat submaxillary gland glucagon-like immunoreactive material produced hyperglycemia equivalent to that obtained with immunoequivalent amounts of pure porcine pancreatic glucagon. Low glucose concentrations stimulated and high glucose concentrations suppressed the release of this material into tissue culture medium and arginine was a potent stimulus to the release of salivary gland glucagon. Despite a chromatographic profile quite different from that of native pancreatic glucagon, this material cross-reacted identically with the 30K glucagon antiserum thought to be specific for A cell glucagon. Electron microscopy of the submaxillary glands revealed marked accumulation of electron dense secretory granules in the basilar portion of ductular cells and occasional acinus cells with similar appearing granules. Immunoperoxidase studies revealed dense precipitation of the antiserum peroxidase complex in the basilar portion of large ductular cells primarily, although an occasional acinus contained such staining characteristics. The significance of these findings, especially as they may relate to carbohydrate homeostasis and diabetes, remains to be established.

Although extrapancreatic sources of immunoreactive glucagon-like materials have been recognized almost from the advent of the radioimmunoassay for glucagon, relatively little attention was paid to these materials as they were initially thought not to be of significant importance to carbohydrate homeostasis in man and animal (13). Soon after the more general introduction of the immunoassay technique for measuring glucagon, many laboratories discovered and reported the presence of materials cross-reacting with glucagon antisera extractable from various sites in the stomach, small and large bowel (8).

However, when it was discovered that, using what was believed a pancreatic glucagon specific antiserum, glucagon levels persist and even rise following total removal of the pancreas in dogs (1, 6, 7, 14), interest in extrapancreatic sources of "true" glucagon was rekindled. Although initially thought to be only of gastrointestinal origin, it was shown by Penhos and associates that glucagon levels could be measured and even rise in animals deprived of their pancreas and entire gastrointestinal tract (9). The source of glucagon in these eviscerated pancreatectomized animals was and still is a mystery. Nevertheless, these findings prompted us to examine other tissues for the presence of glucagon-like immunoreactive material. In the course of such an exploration, it was discovered that a large, approximately 70,000 molecular weight, immunoreactive glucagon-like material can be extracted from the salivary glands and primarily from the submaxillary glands of animals and man (3-5). To date we have been unable to demonstrate the presence of glucagon immunoreactive material in saliva of man.

MATERIALS AND METHODS

Results to be described were drawn from studies in which submaxillary glands were excised from adult Sprague-Dawley rats, placed immediately in chilled 0.9% saline acidified with 0.1 normal HCl to pH 2.82. Or, glands were placed in Krebs Ringer bicarbonate buffer at 37°C and incubated on a Dubnoff water bath shaker. Glands for extraction were thoroughly homogenized, spun at 2000 rev/min for 15 minutes and the supernatant removed and frozen. Subsequently such fractions were re-spun and the supernatant applied to a 50 x 1.5 cm, G100 Sephadex column, and fractions eluted with 0.2 M glycine buffer, pH 7.5, or with 0.09% NaCl. Two ml fractions were generally collected, frozen and subsequently measured in a single glucagon immunoassay, using a double antibody procedure developed in this laboratory. Trasylol or Aprotinin, 2000 KIU, was added to the incubation systems. Traditional Kenny acid-alcohol extractions of submaxillary glands failed to reveal significant amounts of immunoreactive glucagon.

Incubations of paired submaxillary glands were carried out in buffer with potential glucagon secretagogues added to one gland of each pair. Glucose, 30 mg% and 300 mg%, and arginine, 0.16 m, were studied in these incubations. In addition to measuring release of glucagon into the incubation medium, this material was pooled, lyophilized, reconstituted in distilled water, and fractionated on Sephadex columns.

In addition to the approaches described above, submaxillary glands were also subjected to electron microscopic and immunoperoxidase light microscopic examination.

RESULTS

Salivary gland immunoreactive glucagon: Immunoassay of acid saline extracts of salivary glands from mice, rats, rabbits, dogs and man showed the presence of some immunoassayable glucagon in each of these exocrine glands but content was clearly greatest in the submaxillary glands of the mouse and rat (Table 1).

TABLE 1

SUBMAXILLARY GLAND IMMUNOREACTIVE GLUCAGON

		ng/g wet wt
Mouse	--	7,000
Rat	--	5,000
Man	--	20
Rabbit	--	10
Dog	--	Trace

Human material was obtained either from the surgical pathology laboratory or up to 12 hours postmortem, and the influence of medications on the quantity of GLI in the human organs was not assessed. It should be noted that in these studies, glucagon was measured with the widely used Unger K-30 antiserum. When submaxillary gland glucagon was serially diluted, measured values fell along the pancreatic glucagon standard curve thus suggesting immunologic identity (Figure 1).

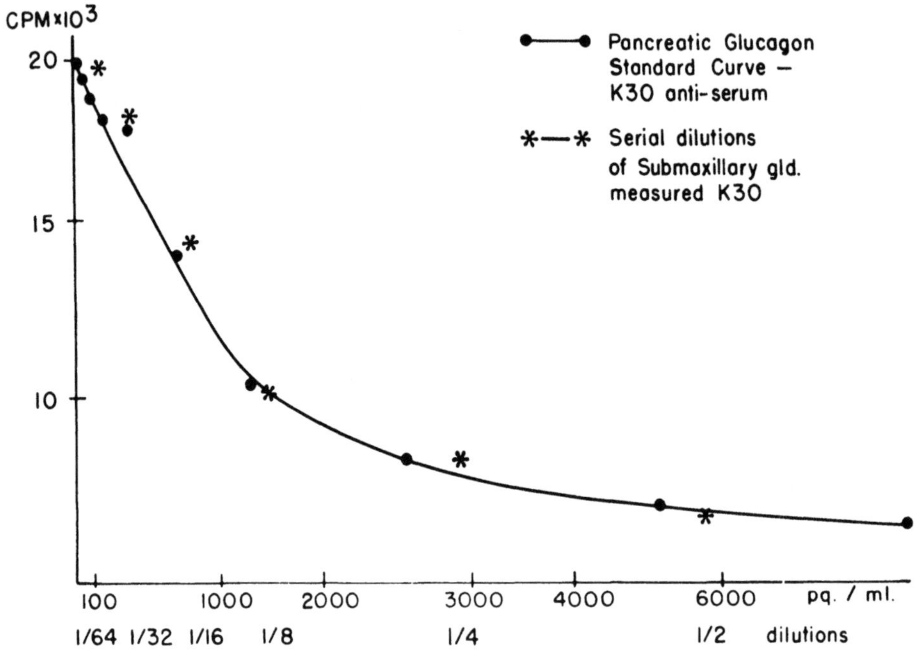

Fig. 1. Demonstration that serial dilutions of rat submaxillary gland glucagon cross-reacts with the K-30 pancreatic glucagon antiserum in a manner identical to that of pancreatic glucagon (5, with permission).

Effect of intravenous injection of immunoequivalent amounts of submaxillary gland glucagon: When an amount of rat submaxillary gland glucagon immunoequivalent to 0.65 µg of pancreatic glucagon was injected into the tail vein of lightly anesthetized male rats, comparable hyperglycemia was obtained and this was significantly greater ($p < .001$) than that obtained with a control saline injection (Figure 2).

Column fractionation profile: Despite the apparent immunoidentity of rat submaxillary gland glucagon, it was noted that Sephadex column fractionation of

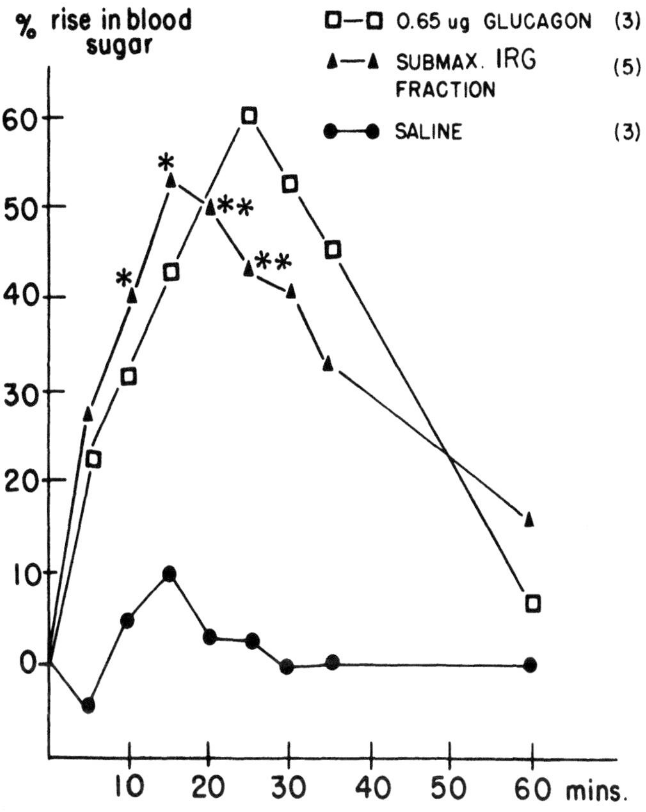

Fig. 2. Comparable hyperglycemia elicited with immunoequivalent amounts of rat submaxillary gland glucagon as with I. V. injection of crystalline porcine pancreatic glucagon. A single asterisk denotes a significant difference from the saline controlled at $p < .05$; a double asterisk, $p < .001$ (4, with permission).

this material revealed a glucagon immunoreactive peak appearing shortly after a radiolabeled gammaglobulin marker and well before radiolabeled insulin and glucagon markers applied simultaneously (Figure 3). It is unlikely that this material, with an estimated molecular weight of 70,000, represents covalent bonding to a larger protein or simple dimerization of glucagon as this peak retained its chromatographic profile after repeated freezing and thawing, repeated chromatography, and even after incubation with 8 M urea and 1 M acetic acid for 16 hours.

In vitro incubation release: When paired glands were incubated *in vitro* in buffer at 37°C on a Dubnoff shaker, a gradual increase in media content of gluca-

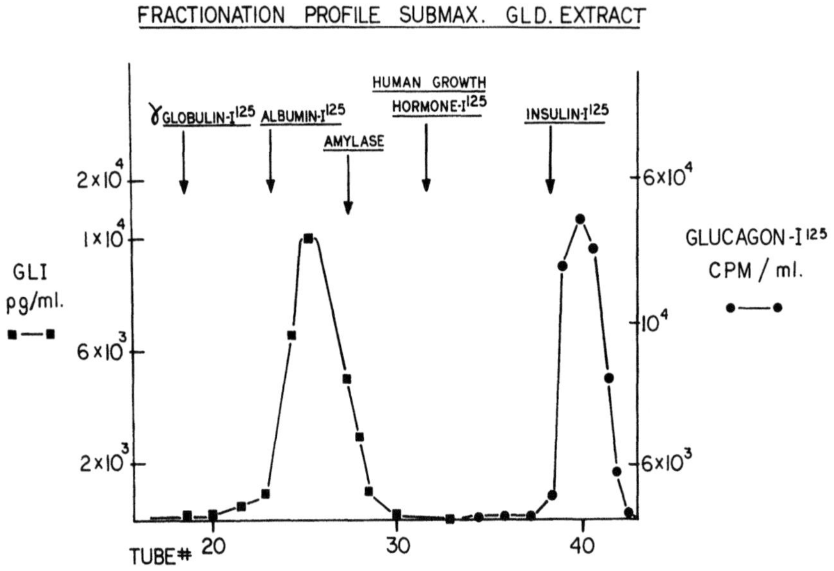

Fig. 3. Fractionation profile from a Sephadex G100 column of immunoassayed glucagon-like material from acid-aqueous extracts of rat submaxillary gland (■-■). This peak is compared to peak activity derived from column fractionation of other radiolabeled peptide markers and ^{125}I pancreatic glucagon (● - ●). (5, with permission).

gon was noted with sampling over a 120 minute period. When glucose was added to the medium, release of glucagon into the medium was greatly enhanced although high glucose was seen to suppress and low glucose to stimulate release of this material (Figures 4 and 5). Arginine added to the medium proved to be a potent in vitro stimulus to submaxillary gland glucagon release (Figure 6) and in preliminary studies it has not been possible to inhibit this phenomenon by the addition of somatostatin.

When media were harvested from the arginine in vitro studies, lyophilized, reconstituted and column fractionated, it could be shown that the material released into the medium consisted of two immunoreactive glucagon peaks, one coinciding with the material extracted from the gland and the other coinciding with the radioactive pancreatic glucagon marker in a ratio of approximately 2:1.

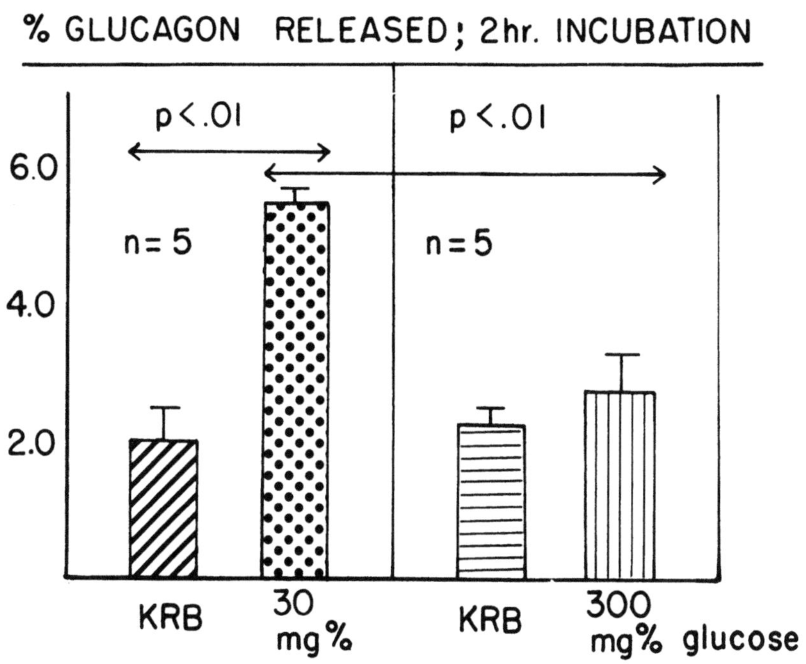

Fig. 4. The augmentive effect of glucose added to the incubation medium on release of glucagon-like immunoreactive material from paired rat submaxillary glands.

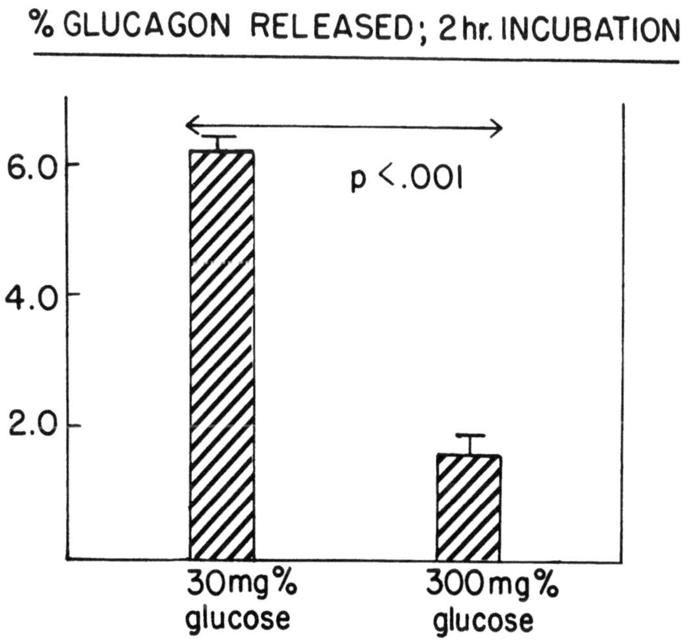

Fig. 5. Demonstration that low glucose stimulates and high glucose suppresses release of rat submaxillary gland glucagon-like immunoreactive material.

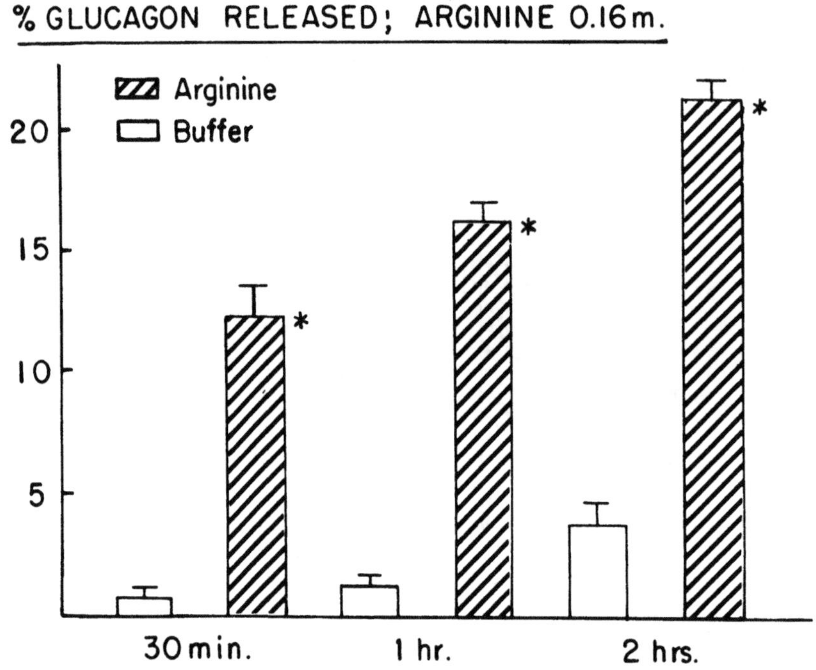

Fig. 6. The stimulatory effect of arginine (0.16 mM) on release of immunoreactive glucagon from rat submaxillary glands incubated in either buffer alone (▭) or in buffer with added arginine (▨). A single asterisk denotes a significant difference between glucagon released in buffer alone compared to the presence of arginine at each time period noted, $p < .001$. The vertical lines represent the standard errors of the mean for each group of experiments (4, with permission).

Microscopic examinations: Although electron microscopy of the submaxillary glands failed to reveal cells containing granules identical to those described in pancreatic A cells, there was marked accumulation of electron dense secretory granules in the basilar portion of ductular cells (Figure 7). These granules were distinctly different from granules contained in easily identified acinus serous and mucoid secreting cells. This observation was all the more interesting as immunoperoxidase examination of rat submaxillary glands also revealed the peroxidase-glucagon antiserum complex to be conspicuously located in the basilar portion of larger ductular cells although an occasional acinus stained as well (Figure 8).

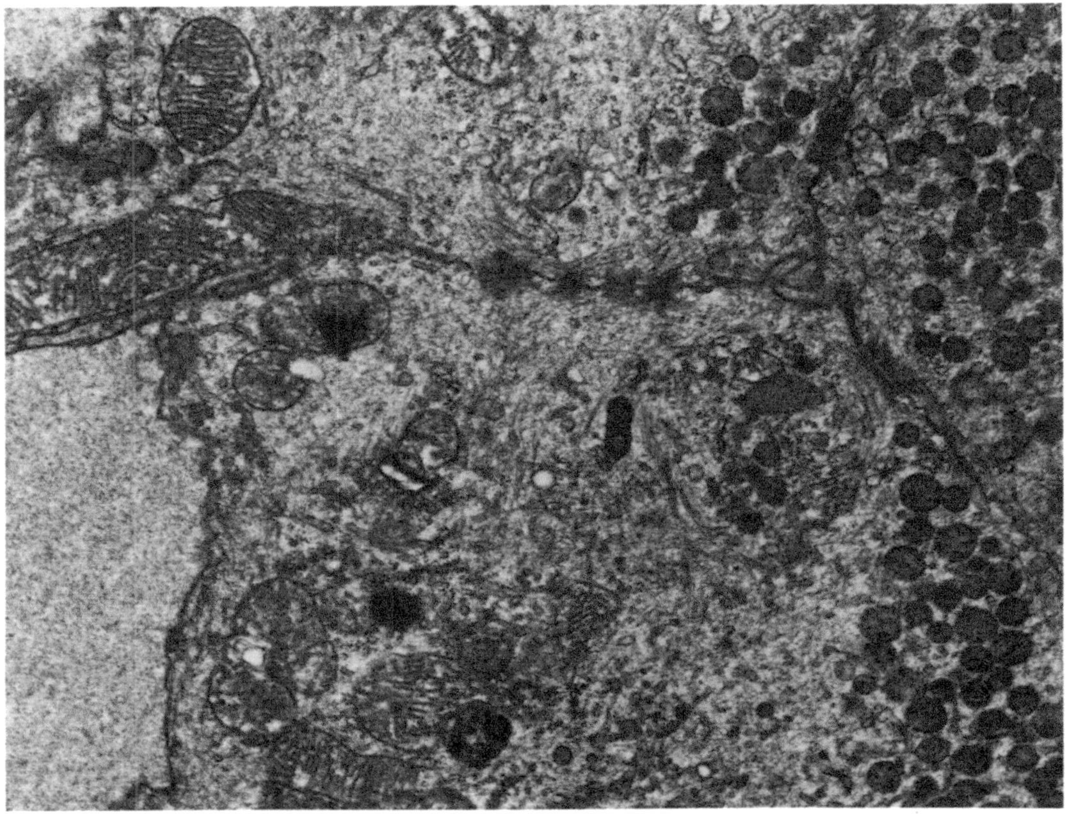

Fig. 7. Electron photomicrograph of araldite embedded rat submaxillary gland (21,800x). Note electron dense granules distributed in the basilar portion of ductular cells.

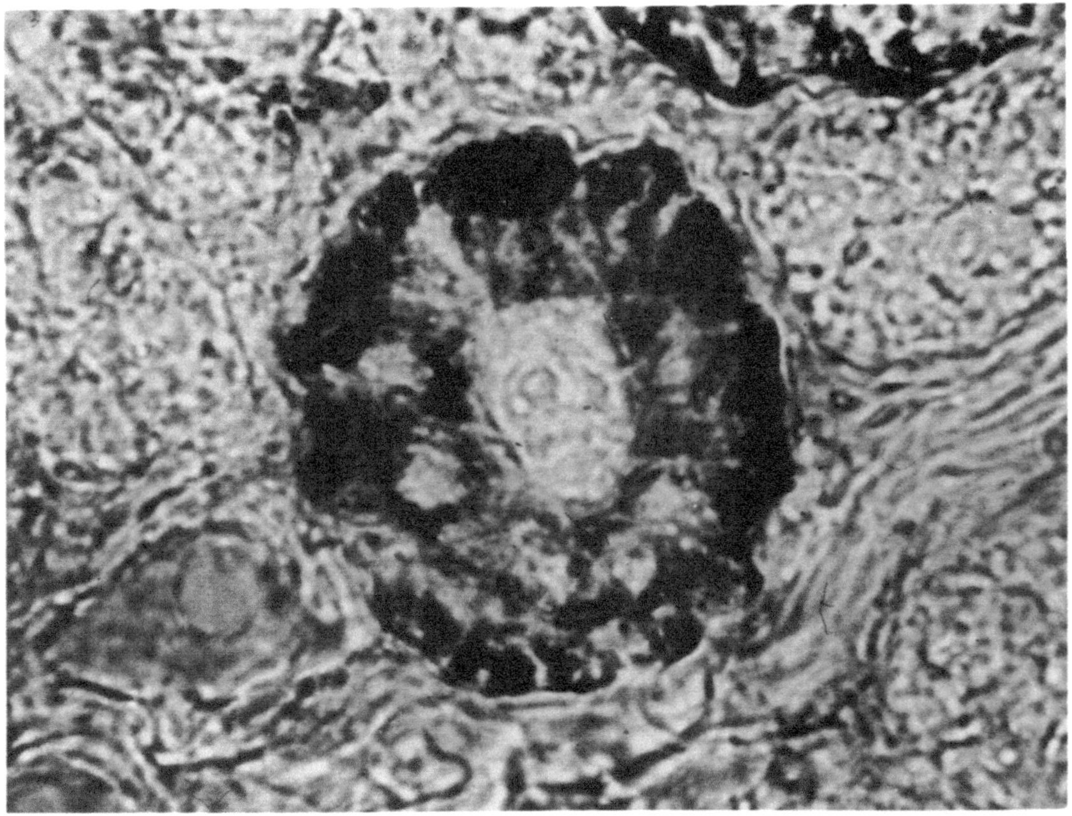

Fig. 8. Immunoperoxidase staining of rat submaxillary gland. Note distribution of the glucagon antibody peroxidase complex in the basilar portion of larger ductular cells.

DISCUSSION

In 1974, Silverman and Dunbar first reported the presence of small amounts of a glucagon immunoreactive type material in the salivary glands of rats (10). In their studies these workers cite the application of the traditional Kenny acid-alcohol method for extraction of glucagon from the pancreas. As we were unable to measure significant quantities of glucagon by this method, we assume this slight modification of the Kenny technique did allow the detection of some of the substantially larger molecular weight glucagon substance in salivary glands. Assuming a molecular weight in the vicinity of 70,000, it seems unlikely that the acid-alcohol method would yield submaxillary gland glucagon by that procedure. Using a simple acid saline extraction, however, it has been possible to show that there are substantial quantities of a large glucagon-like material, possibly even a proglucagon, existing in the submaxillary glands of rodents and man. Whether the substantially lesser amounts found in human material reflects on preanesthetic medication or delay in processing the tissues remains to be determined.

In these studies, submaxillary gland glucagon, although several-fold larger than pancreatic glucagon, nevertheless appeared to cross-react identically with antisera raised to pure porcine glucagon (K-30). When immunoequivalent amounts are injected into rats, a blood sugar rise comparable to that obtained with pancreatic glucagon was noted. In vitro incubation studies showed that release appeared dependent on glucose in the medium but that high glucose suppressed and that low glucose concentrations stimulated "secretion". Similar to what obtains for pancreatic islet glucagon, it could be shown that arginine was a potent stimulus to release of submaxillary gland glucagon into the incubation medium. When this material was fractionated, however, it was observed that as much as 25% of the released material eluted with a radioactive pancreatic glucagon marker. This observation seems to add additional weight to the possibility that submaxillary gland glucagon exists in a precursor state and that it can be converted by parenchymal cells to a material closely resembling, if not identical, to pancreatic glucagon. Furthermore, it may well be that the glucagon rise in response to arginine in pancreatectomized animals reflects secretion of a larger molecular weight material as it has been shown that the major component of plasma glucagon in response to an arginine stimulus in pancreatectomized animals is substantially larger than pancreatic glucagon (14).

The real significance of these findings remains to be determined. Clearly the role of this material in carbohydrate homeostasis and in diabetes, in particular, will need to be explored. The likelihood that the exocrine salivary glands serve an endocrine function as well seems to be further strengthened by these studies, and it would appear that immunoreactive glucagon can be added to the list of other hormones known to exist in rodent salivary glands such as nerve growth factor (2), epidermal growth factor (11), and possibly even gastrin (12).

Acknowledgement

Supported, in part, by the Medical Research Service of the Veterans Administration, and by a grant-in-aid from the Kroc Foundation.

REFERENCES

1. Dobbs, R., Sakurai, H., Sasaki, H., Faloona, G., Valverde, I., Baetens, D., Orci, L. and Unger, R. (1975): Glucagon: Role in the hyperglycemia of diabetes mellitus. Science 187, 544-546.

2. Frazier, W. A., Angeletti, R. H. and Bradshaw, R. A. (1972): Nerve growth factor and insulin. Science 176, 482-488.

3. Lawrence, A. M., Kirsteins, L., Hojvat, S., Rubin, L. and Paloyan, V. (1975): Salivary gland glucagon: A potent extrapancreatic hyperglycemic factor. Clin. Res. 23, 563A.

4. Lawrence, A. M., Tan, S., Hojvat, S. and Kirsteins, L. (1977): Salivary gland hyperglycemic factor: An extrapancreatic source of glucagon-like material. Science 195, 70-72.

5. Lawrence, A. M., Tan, S., Hojvat, S., Kirsteins, L. and Mitton, J. (1976): Salivary gland glucagon in man and animals. Metabolism 25, 1405-1408.

6. Mashiter, K., Harding, P. E., Chou, M., Mashiter, G. D., Stout, J., Diamond, D. and Field, J. B. (1975): Persistent pancreatic glucagon but not insulin response to arginine in pancreatectomized dogs. Endocrinology 96, 678-693.

7. Matsuyama, T. and Foà, P. P. (1974): Plasma glucose, insulin, pancreatic and enteroglucagon levels in normal and depancreatized dogs. Proc. Soc. Exp. Biol. Med. 147, 97-102.

8. Moody, A. J. (1972): Gastrointestinal glucagon-like immunoreactivity. In: Glucagon, Molecular Physiology, Clinical and Therapeutic Implications. P. J. Lefèbvre and R. H. Unger, Eds. Pergamon Press, New York, p. 319.

9. Penhos, J. C., Ezequiel, M., Lepp, A and Ramey, E. R. (1975): Plasma immunoreactive insulin (IRI) and immunoreactive glucagon (IRG) after evisceration with and without a functional liver. Diabetes 24, 637-640.

10. Silverman, H. and Dunbar, J. C. (1974): The submaxillary gland as a possible source of glucagon. Bull. Sinai Hosp. Detroit 22, 192-193.

11. Starkey, R. H., Cohen, S. and Orth, D. N. (1975): Epidermal growth factor: Identification of a new hormone in human urine. Science 189, 800-802.

12. Takeuchi, T., Takemoto, T., Tani, T. and Miwa, T. (1973): Gastrin-like immunoreactivity in salivary gland and saliva. Lancet 2, 920.

13. Valverde, I., Rigopolou, D., Marco, H. J., Faloona, G. R. and Unger, R. H. (1970): Characterization of glucagon-like immunoreactivity (GLI). Diabetes 19, 614-623.

14. Vranic, M., Pek, S. and Kawamori, R. (1974): Increased "glucagon immunoreactivity" in plasma of totally depancreatized dogs. Diabetes 23, 905-912.

ROLE OF THE SUBMAXILLARY GLAND AND OF THE KIDNEY IN THE HYPERGLUCAGONEMIA OF EVISCERATED RATS

J. C. Dunbar, H. Silverman, E. Kirman and P. P. Foà

Department of Physiology, Wayne State University School of Medicine
and Department of Research, Sinai Hospital of Detroit
Detroit, Michigan USA

Significant quantities of A cell and total glucagon were found in the systemic blood of eviscerated rats, especially following arginine infusion. These findings and the hyperglucagonemia previously observed in depancreatized and in eviscerated animals, prompted us to look for glucagon sources other than the pancreas and the gastrointestinal tract. Significant amounts of immunoreactive glucagon (IRG) were extracted from rat submaxillary and parotid gland tissue using a modification of the method of Kenny. This IRG could be measured with an antiglucagon serum specific for pancreatic or A cell glucagon and with an antiserum that crossreacts with extracts of intestinal mucosa. Intravenous injections of submaxillary gland extracts caused a significant increase in blood glucose in normal, but not in eviscerated rats. A-V differences in glucagon concentration between aorta and facial vein suggest that the salivary glands release glucagon into the circulation. Similarly, A-V differences between the blood of the aorta and of the renal vein suggest that the kidneys remove glucagon from the circulation in normal, but not in eviscerated rats. Indeed, the kidneys of eviscerated rats may release previously stored glucagon into the blood. No glucagon was found in the saliva. Glucagon release by the salivary glands and the kidneys may contribute to the hyperglucagonemia characteristic of depancreatized and eviscerated animals.

INTRODUCTION

It has been known for a number of years that extracts of the gastrointestinal tract contain materials that cause glycogenolysis and hyperglycemia (6, 27) and crossreact with pancreatic glucagon antisera (19, 23). Recent attempts to define the physiologic role of pancreatic and extrapancreatic glucagon have yielded the surprising result that their concentration in the general circulation did not decline, but increased markedly, following total removal of the pancreas (1, 18, 19, 28). Equally surprising was the observation that, in the rat, the removal of the entire gastrointestinal tract, the pancreas and the liver, also resulted in an increase in the level of circulating glucagon-like materials (20, 21). These findings led us to investigate other likely sources of glucagon, such as the salivary glands and the kidney. The salivary glands were selected because there is evidence that they may play a role in carbohydrate metabolism. Thus, extracts of beef salivary gland were found to increase the blood sugar of rats (3), while an increased tolerance for glucose was noted following the removal of the parotid glands (2) or the ligature of the parotid ducts (29). Indeed, on the assumption that the salivary glands may produce substances with hyperglycemic properties or capable of inhibiting the release of insulin (10), the submaxillary glands were removed from 86 diabetic patients resistant to conventional forms of treatment (11). It should be noted that other investigators (8, 9, 26) have not confirmed these findings. The kidneys were selected because they are not only organs of glucose production and a target for pancreatic glucagon (5), but because there is evidence that they play an important role in its metabolic clearance (15-17).

MATERIALS AND METHODS

Sprague-Dawley rats, weighing 300 to 350 g, were kept in individual cages with free access to rat chow and water. Evisceration, performed under amytal sodium anesthesia (70 mg/kg, intraperitoneally) according to the procedure of Penhos et al (21), resulted in the removal of the entire gastrointestinal tract, the pancreas and the spleen. The liver remained in situ, but was not functioning because it was deprived of its blood supply. The experiments were performed under amytal anesthesia, four to five hours after evisceration or, in the normal rats, after an 18-hour fast.

Glucagon was extracted as follows: the tissues to be analyzed were removed from the animal, cut into small pieces and extracted using a modification of the method of Kenny (13). For this purpose the material was homogenized with acid alcohol (3 ml of concentrated HCl in 100 ml of 70% ethanol) in a ground glass homogenizer, allowed to stand overnight at 4°C, and centrifuged at 2000 rpm. The supernatant fluid was collected and, after adjustment of its pH to 7.0 with NH_4OH, was allowed to stand in the refrigerator for approximately 2 hr. The resultant precipitate was removed by centrifugation and discarded. After the addition of of a 1:3 alcohol-ether mixture, the extract was allowed to stand overnight, centrifuged and the precipitate was collected, dried at room temperature and dissolved in veronal buffer, pH 8.6, for glucagon assay.

The effect of submaxillary gland extracts on blood glucose was studied in normal and eviscerated rats. For this purpose, the jugular vein of the test animal was cannulated under amytal anesthesia and an amount of submaxillary gland extract containing approximately 10 ng of radioimmunoassayable glucagon was injected into it. Blood samples were collected through the same catheter, at the indicated time intervals. Glucose was measured using the GOD Perod enzymatic method (Boehringer Mannheim Corp., Mannheim, West Germany).

The release of salivary glucagon was studied by analyzing blood samples collected simultaneously from the jugular vein and the abdominal aorta, the facial vein and the abdominal aorta, the renal vein and the carotid artery or from the cannulated jugular vein before and after a 20 min arginine infusion.

Glucagon was measured according to Shima and Foa (24), using antiserum AGS 18, "specific" for A cell glucagon (IRG_a, 7) and, whenever the size of the sample allowed it, with antiserum AGS 10, which crossreacts with glucagon-like materials present in extracts of intestinal mucosa and measures "total" immunoreactive glucagon (IRG_t).

RESULTS

Three hours after evisceration and exclusion of the liver, we noted a significant decrease in serum glucose and a significant increase in serum IRG_a (Table 1). Different amounts of IRG could be extracted not only from pancreas and intestines, but from the salivary glands as well. In the case of the salivary glands, IRG_a represented approximately 60 to 80% of extractable IRG_t (Table 2).

The intravenous injection of extracts obtained from one submaxillary gland caused a significant hyperglycemia in normal rats, while a similarly prepared liver extract had no significant effect (Fig. 1).

TABLE 1

SERUM GLUCOSE, IMMUNOREACTIVE INSULIN, AND IMMUNOREACTIVE GLUCAGON IN NORMAL AND EVISCERATED RATS

Ave. ± S. E. Number of Animals in Parentheses

	Glucose	A Cell Glucagon (IRG_a) pg/ml	Total Glucagon (IRG_t) pg/ml	Insulin (IRI) U/ml
Normal Rats	173 ± 19 (12)	161 ± 38 (7)	446 ± 66 (11)	42 ± 4 (7)
Eviscerated Rats	82 ± 7 (18)**	281 ± 34 (6)*	526 ± 71 (20)	23 ± 2 (10)**

* P <0.05, ** P <0.01, vs normal rats

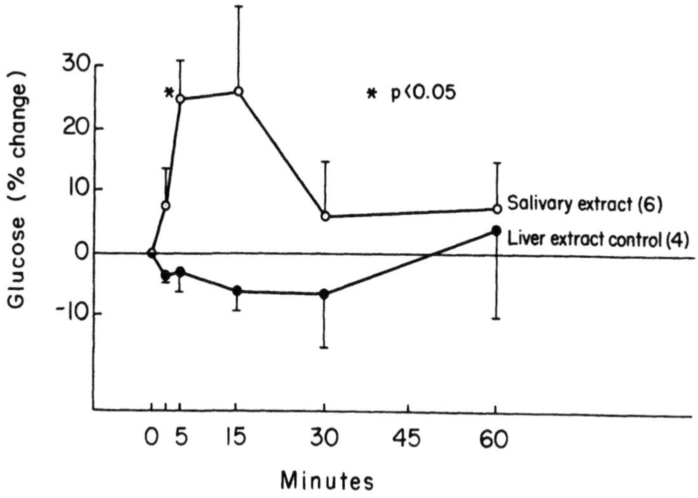

Fig. 1. Effect of an intravenous injection of submaxillary gland extract on blood glucose in normal rats. Ave. ± S.E. Number of experiments in parentheses.

TABLE 2

IMMUNOREACTIVE GLUCAGON EXTRACTABLE FROM THE TISSUES OF NORMAL RATS

Pg/ml. Ave. ± S. E. No. of Experiments in Parentheses

Tissue	IRG_a	pg/mg	IRG_t
Pancreas (2)	2835 ± 639		---
Small Intestines (4)	851 ± 94		---
Submaxillary Glands (11)	35.9 ± 11		41.4 ± 7
Parotid Glands (7)	11.3 ± 11		18.3 ± 3
Kidney (2)	2.8 ± 1		---
Liver (4)	0		---

In Table 3, we show that the blood of the aorta contained more IRG than that of the jugular vein in the normal, but not in the eviscerated rats. In the latter, the A-V difference became negative, although there were large individual variations. This change became statistically significant when, instead of analyzing the blood of the jugular vein, we analyzed that of the facial vein, which drains the salivary glands more directly (Table 4). No IRG was found in the saliva.

The intravenous infusion of arginine caused a significant increase in the serum glucagon (IRG_a) levels of eviscerated rats, whereas saline had no effect (Fig. 2).

The analysis of carotid artery and renal vein blood demonstrated a significant uptake of glucose by the kidney of normal rats and a significant release by that of eviscerated rats (Fig. 3). Similarly, there was a significant uptake of glucagon in the normal ($p <.025$), but not in the eviscerated rats (Fig. 4). Thus, evisceration eliminated the negative A-V difference and apparently decreased or stopped the glucagon uptake by the kidney.

TABLE 3

SERUM GLUCAGON (IRG_a) IN THE BLOOD OF THE AORTA AND OF THE JUGULAR VEIN OF NORMAL AND EVISCERATED RATS

Pg/ml. Ave. ± S. E. No. of Experiments in Parentheses

	Aorta	Jugular Vein		A-V
Normal Rats	226 ± 73 (8)	175 ± 83 (8)	N.S.	51 N.S.
Eviscerated Rats	227 ± 44 (13)	244 ± 56 (13)		17 N.S.

TABLE 4

DIFFERENCE BETWEEN THE CONCENTRATION OF SERUM GLUCAGON IN THE BLOOD OF THE AORTA AND OF THE FACIAL VEIN OF NORMAL AND EVISCERATED RATS

No. of Experiments in Parentheses

	$\frac{A - V}{A} \times 100$
Normal Rats	-1.25 ± 10 (8)
Eviscerated Rats	-35.1 ± 12 (14)*

* P <0.1 vs normal rats

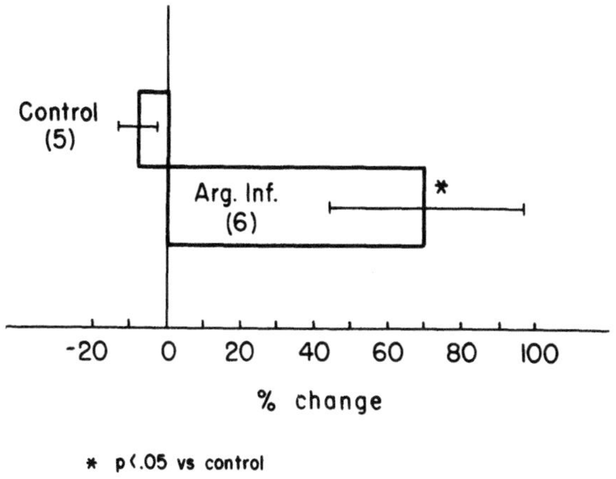

Fig. 2. Effect of arginine infusion (200 mg/kg) on serum glucagon (IRG_a) in eviscerated rats. Ave. ± S. E. Number of experiments in parentheses.

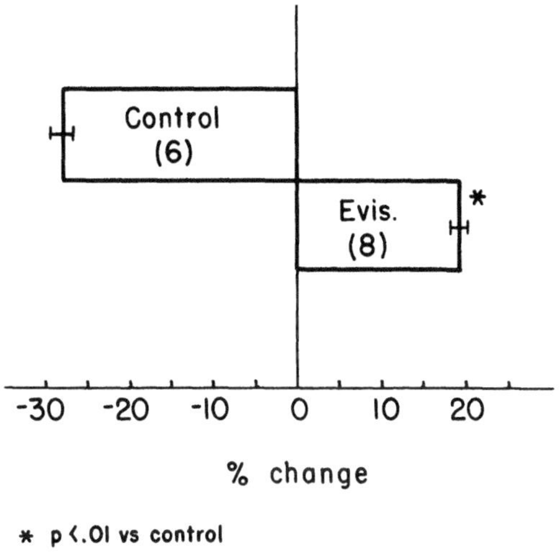

Fig. 3. Difference between the concentration of glucose in the blood of the aorta and of the renal vein of normal and eviscerated rats. Ave. ± S. E. Number of experiments in parentheses.

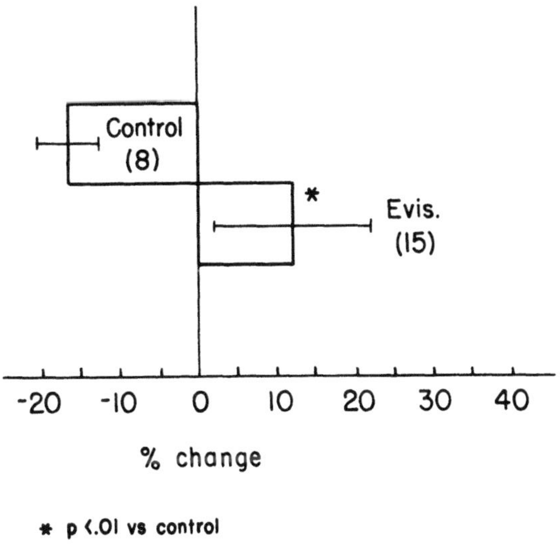

Fig. 4. Difference between the concentration of glucagon in the blood of the aorta and of the renal vein of normal and eviscerated rats. Ave. ± S. E. Number of experiments in parentheses.

DISCUSSION

Our results confirm the observation (16) that evisceration in the rat is followed by an increase in that fraction of circulating glucagon that can be measured by an antiserum specific for pancreatic glucagon. In addition we have observed a persistent elevation of other IRG components. The finding of IRG and in particular of IRG_a in extracts of rat salivary glands, first reported from our laboratory (25), has been confirmed by other investigators (14) and by the results described in this paper. These glucagon fractions are biologically active, as indicated by their release following the intravenous administration of arginine *in vivo* and *in vitro*, and by their hyperglycemic effect in normal rats. The latter, probably, is the result of an action on the liver since it cannot be demonstrated in eviscerated rats.

Whether or not the salivary glucagon plays a physiologic role in the normal animal is not known. One possibility is that it may be a part of the entero-insular axis and participate in the hormonal responses to food intake. Thus, there might be an oral phase of glucagon secretion comparable to that observed for insulin (12). The salivary glands may also contribute to the hyperglucagonemia of pancreatic and other forms of diabetes, while the removal of a significant source of glucagon may explain the reported amelioration of diabetes noted following ablation of the salivary glands (11) as well as the diabetogenic effect of salivary gland extract and of salivary gland implants (22).

An additional source of glucagon in eviscerated animals appears to be the kidney which clears glucagon in normal rats, but fails to do so in the eviscerated animals. Indeed, after evisceration, the kidney may return to the circulation some of the glucagon which it had previously taken up and stored.

Acknowledgements

This work was aided by Grant AM 06034 from the National Institute of Arthritis, Metabolism and Digestive Diseases and by the American Diabetes Association-Michigan Affiliate.

Present address for Dr. Silverman: Embry Hills Medical Center, Chamblee, Georgia.

REFERENCES

1. Blázquez, E., Muñoz-Barragan, L., Patton, G. S., Orci, L., Dobbs, R. E. and Unger, R. H. (1976): Gastric A-cell function in insulin-deprived depancreatized dogs. Endocrinology 99, 1182-1188.

2. Brinkrant, W. B. (1941): The influence of the parotid gland on blood sugar. J. Lab. Clin. Med. 26, 1009-1011.

3. Brinkrant, W. B. (1942): The influence of a parotid extract on blood sugar and structure of the pancreas in the rat. J. Lab. Clin. Med. 27, 510-518.

4. Chisholm, D. J. and Alford, F. P. (1977): The nature of the A cell abnormality in human diabetes mellitus. In: Glucagon, Its Role in Physiology and Clinical Medicine. P. P. Foà, J. S. Bajaj and N. L. Foà, Eds. Excerpta Medica, Amsterdam. This volume, p. 651-662.

5. Foà, P. P. (1973): Glucagon: An incomplete and biased review with selected references. Amer. Zool. 13, 616-623.

6. Foà, P. P., Berger, S., Santamaria, L., Smith, J. A. and Weinstein, H. R. (1953): Extraction of the hyperglycemic factor (HGF) of the pancreas with lipid ammonia. Science 117, 82-84.

7. Foà, P. P., Matsuyama, T. and Foà, N. L. (1977): Radioimmunoassay of glucagon. In: Handbook of Radioimmunoassay. G. E. Abraham, Ed. Marcel Dekker, Inc., New York and Basel, pp. 299-318.

8. Garcia, A. R., Blackard, W. G., Trail, M. L. (1971): Submaxillary gland removal effects on glucose and insulin homeostasis. Arch. Otolar. 93, 597-598.

9. Gault, S. D. (1954): The effect of parotidectomy on blood sugar levels in rat and mouse. J. Lab. Clin. Med. 43, 119-121.

10. Godlowski, Z. Z., Colandra, J. C. (1961): Argentaffine cells in the submaxillary glands of dogs. Anat. Rec. 140, 45-47.

11. Godlowski, Z. Z., Gazda, M., Withers, B. T. (1971): Ablation of salivary glands as an initial step in the management of selected forms of diabetes mellitus. The Laryngoscope 81, 1337-1358.

12. Hommel, H., Fischer, U., Retzloff, K., Knofler, H. (1972): The mechanism of insulin secretion after oral glucose administration. Diabetologia 8, 111-116.

13. Kenny, A. J. (1955): Extractable glucagon of the human pancreas. J. Clin.

Endocr. Metab. 15, 1089-1105.

14. Lawrence, A. M., Tan, S., Hojvat, T. S. and Kirsteins, L. (1977): Salivary gland hyperglycemic factor: An extrapancreatic source of glucagon-like material. Science 195, 70-72.

15. Lefèbvre, P. J. and Luyckx, A. S. (1977): Glucagon and the kidney. In: This volume, p. 167-176.

16. Lefèbvre, P. J., Luyckx, A. S., Nizet, A. H. (1974): Renal handling of endogenous glucagon in dogs: comparison with insulin. Metabolism 23, 753-761.

17. Lefèbvre, P. J., Luyckx, A. S. and Nizet, A. H. (1976): Independence of glucagon and insulin handling by the isolated perfused dog kidney. Diabetologia 12, 359-365.

18. Mashiter, K., Harding, P. E., Chou, M., Mashiter, R. D., Stout, J., Diamond, D. and Field, J. B. (1975): Persistent pancreatic glucagon but not insulin response to arginine in depancreatized dogs. Endocrinology 96, 678-693.

19. Matsuyama, T. and Foà, P. P. (1974): Plasma glucose, insulin, pancreatic and enteroglucagon levels in normal and depancreatized dogs. Proc. Soc. Exper. Biol. Med. 147, 97-102.

20. Penhos, J. C., Ezequiel, M., Lepp, A., Ramey, E. R. (1975): Plasma immunoreactive insulin (IRI) and immunoreactive glucagon (IRG) after evisceration with and without a functional liver. Diabetes 24, 637-640.

21. Penhos, J. C., Woodbury, C., Tizabi, Y., Ramey, E. R. (1975): Metabolic studies in eviscerated rats with functional livers. Proc. Soc. Exp. Biol. Med. 148, 1159-1163.

22. Pisanty, J., Dieck, M. N., Garza, M. L., Garza, J. M., Gomez, M. F. (1975): Diabetogenic effect of submaxillary gland implantation and submaxillary gland extract injections in dogs and mice. IRCS Medical Sci. 3, 521

23. Samols, E., Tyler, J., Megyesi, C. and Marks, V. (1966): Immunochemical glucagon in human pancreas, gut, and plasma. Lancet 2, 727-729.

24. Shima, K. and Foà, P. P. (1968): Double antibody assay for glucagon. Clin. Chem. Acta 22, 511-520.

25. Silverman, H. and Dunbar, J. C. (1974): The submaxillary gland as a possible source of glucagon. Bulletin, Sinai Hospital of Detroit 22, 192-193.

GLUCAGON AND THE KIDNEY

P. J. Lefèbvre and A. S. Luyckx

Division de Diabetologie, Institut de Médicine, Université de Liège
Liège, Belgium

The uptake of glucagon by the kidney is quantitatively important while urinary glucagon excretion is very low. Insulin and glucagon are apparently handled independently by the kidney. Bilateral kidney exclusion in the dog results in a rise in circulating glucagon caused by cessation of kidney glucagon uptake rather than by an increase in the secretion of this hormone. The high plasma glucagon values reported in severe renal failure probably result from decreased catabolism and, together with increased sensitivity to the hyperglycemic action of the hormone, may contribute to the glucose intolerance frequently observed in this condition.

For the past few years, several groups of investigators have been interested in studying the renal handling of glucagon and the effect of renal failure on the circulating levels of this hormone. In the present paper, we will summarize our own contribution to this question with only limited references to the work of others. Therefore this paper is by no means an exhaustive review of the problem.

Renal Handling of Glucagon

We investigated the renal handling of glucagon using two experimental models: (a) the dog kidneys acutely transplanted to the neck vessels of a perfusing anesthetized dog and (b) the isolated dog kidney perfused with whole blood. The experiments using dog kidneys acutely transplanted to the neck vessels of a perfusing anesthetized dog provided the first precise and quantitative data on the renal handling of glucagon (10). As shown in Table 1, glucagon uptake by dog kidney under basal conditions averaged 89 pg/min/g of kidney and was significantly reduced by about 50% when arterial plasma glucagon was decreased by massive glucose infusion. The urinary clearance of glucagon represented only 4.1% of the total kidney glucagon clearance. In these experiments, no correlation was found between arterial glucagon and kidney glucagon uptake; this was attributed to the fact that the range of arterial concentrations of glucagon reached was relatively narrow.

These observations were confirmed using the isolated dog kidney perfused with whole blood (11). In these experiments, exogenous glucagon was perfused in a system permitting the investigation of kidney glucagon uptake at arterial plasma concentrations ranging from 26 to 1184 pg/ml. As shown in Fig. 1, kidney glucagon uptake increased as a function of arterial plasma glucagon ($r = 0.940$; $p < 0.01$). The renal clearance rate of glucagon by the kidney was similar at low, medium or high arterial plasma glucagon concentrations and averaged about 0.5 ml/min/g of kidney.

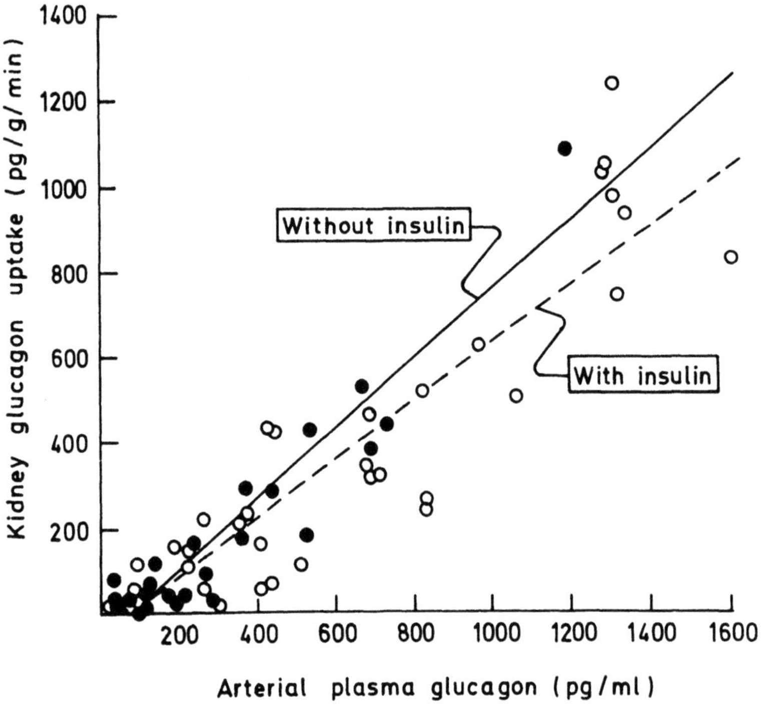

Fig. 1. Correlation between arterial plasma glucagon (A) and kidney glucagon uptake (U) in the absence (•) and in the presence (o) of insulin. U = 0.686 ± 0.046 A in the absence, and 0.622 ± 0.032 A in the presence of insulin. At a 95% threshold, the confidence limits were 0.592 - 0.780 in the absence, and 0.557 - 0.687 in the presence of insulin. Data from Lefèbvre et al. (11, with permission).

TABLE 1

HANDLING OF GLUCAGON BY THE TRANSPLANTED KIDNEY UNDER BASAL CONDITIONS AND DURING MASSIVE GLUCOSE INFUSION (10)

	Basal Conditions	During Glucose Infusion	Statistical Comparison
Plasma Glucose (mg/100 ml)	136 ± 12 (n = 12)	682 ± 36 (n = 10)	$p < 0.001$
Plasma Glucagon (pg/ml)			
Arterial	193 ± 14 (n = 14)	136 ± 21 (n = 10)	$p < 0.005$*
Venous	135 ± 16 (n = 14)	112 ± 22 (n = 10)	N.S.
Renal plasma flow (ml/min/g of kidney)	1.53 ± 0.07 (n = 15)	1.74 ± 0.13 (n = 10)	N.S.
Glucagon uptake (pg/min/g of kidney)	89 ± 14 (n = 14)	42 ± 5 (n = 10)	$p < 0.02$
Q = quantity of glucagon entering the kidneys (pg/min/g of kidney)	288 ± 37 (n = 14)	242 ± 48 (n = 10)	N.S.
Glucagon uptake in percent of Q	37.1 ± 5.8 (n = 14)	23.5 ± 5.7 (n = 10)	N.S.
Glucagon metabolic clearance (ml of plasma/min/g of kidney)	0.482 ± 0.063 (n = 14)	0.315 ± 0.059 (n = 9)	N.S.

*Paired comparison of arterial plasma glucagon before and after glucose infusion in 8 dogs revealed a highly significant reduction in the glucagon values: - 46 ± 14 pg/ml ($p < 0.01$). Number of periods in parentheses. Mean ± S.E. The mean (± S.E.) weight of both kidneys in this series of ten experiments was 51 ± 6.1 g (range 24.5 - 85).

Effect of Insulin on the Renal Handling of Glucagon

Recent investigations have demonstrated the presence in the kidney and muscle of a glucagon degrading enzyme and it was suggested that the same enzyme might be involved in the degradation of both insulin and glucagon (4-6). After isolation and purification of this enzyme from rat skeletal muscle, it was demonstrated that insulin served as a competitive inhibitor of glucagon degradation and that glucagon itself was a competitive inhibitor of insulin degradation (6). In view of the potential physiologic significance of this finding, we investigated the effect of insulin on the magnitude of glucagon uptake by the isolated perfused dog kidney at various glucagon concentrations (11). As indicated in Table 2, the clearance rate of glucagon by the kidney was similar at low, medium or high arterial insulin concentrations and, as shown in Fig. 1, the calculated regression lines plotting kidney glucagon uptake versus arterial plasma glucagon were similar in the presence and in the absence of insulin. Thus, over a wide range of arterial concentrations, the glucagon uptake by the kidney was not significantly affected by the presence of insulin nor was the insulin uptake by the kidney affected by the presence of glucagon (11). These findings suggest that insulin and glucagon are handled independently by the kidney.

Effect of Kidney Exclusion on Circulating Glucagon Levels

As illustrated in Fig. 2, bilateral kidney exclusion by ligation of the renal arteries in the anesthetized dog resulted in an immediate increase in arterial plasma glucagon (8). This rise results from interruption of kidney glucagon uptake rather than from an increase in glucagon production, for the following reasons: 1) it is not inhibited by glucose infusion (Fig. 3; 8); 2) it is present when endogenous glucagon production is inhibited by somatostatin and replaced by a constant infusion of exogenous glucagon (Fig. 4; 9); 3) it is still present when endogenous glucagon production is completely suppressed by abdominal evisceration and replaced by a constant infusion of exogenous glucagon (Fig. 5;

9); 4) it is not accompanied by any increase in the rate of production of glucagon by the pancreas (8).

Therefore, we consider that the abrupt cessation of kidney glucagon uptake is the major factor responsible for the rise in peripheral plasma glucagon observed after ligation of renal arteries. Such an explanation may also apply to the rise in plasma glucagon observed after bilateral ureteral ligation in dogs (2). A similar conclusion that hyperglucagonemia in uremic patients is primarily a result of decreased catabolism rather than hypersecretion of this hormone was most recently drawn by Sherwin et al. (12) and is in agreement with data of Assan (1). In addition, Kuku et al (7) recently suggested that the various plasma immunoreactive glucagon fractions might be handled differently by the kidney.

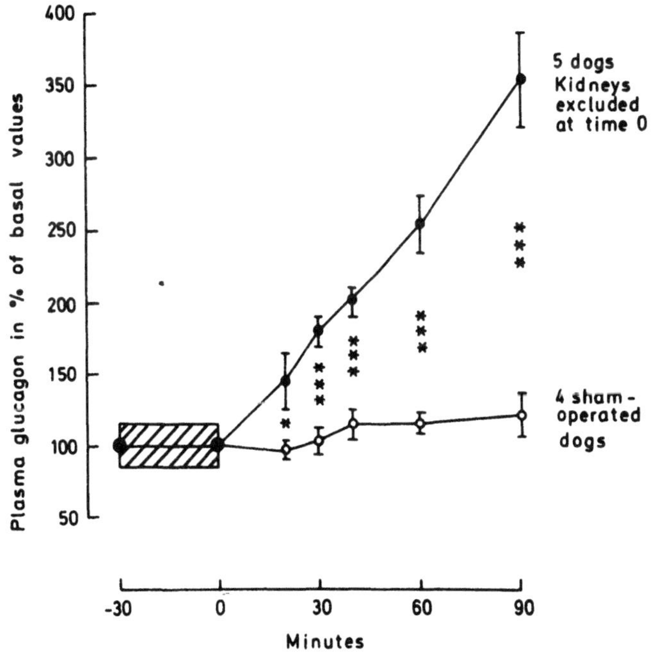

Fig. 2. Changes in arterial plasma glucagon after acute kidney exclusion in dogs (•——•) and kidney manipulation in sham-operated controls (o——o). Results are expressed in per cent of basal values. Mean ± S.E. indicated by the bars, and for basal values, by the hatched area. * and *** correspond to p <0.05 and <0.01, respectively. Data from Lefèbvre and Luÿckx (8, with permission).

TABLE 2

CLEARANCE RATE OF GLUCAGON BY THE KIDNEY AT VARIOUS INSULIN CONCENTRATIONS (11)

Arterial plasma insulin	Low	Medium	High
(μU/ml)	(0 -10)	(150 - 300)	(400 - 500)
Glucagon clearance*	0.45 ± 0.06	0.50 ± 0.06	0.43 ± 0.12
(ml/min/g of kidney)	(n = 26)	(n = 14)	(n = 80

Low vs Medium: N.S.; Medium vs High: N.S.; Low vs High: N.S.

*Mean ± S.E.; N.S. = not statistically significant

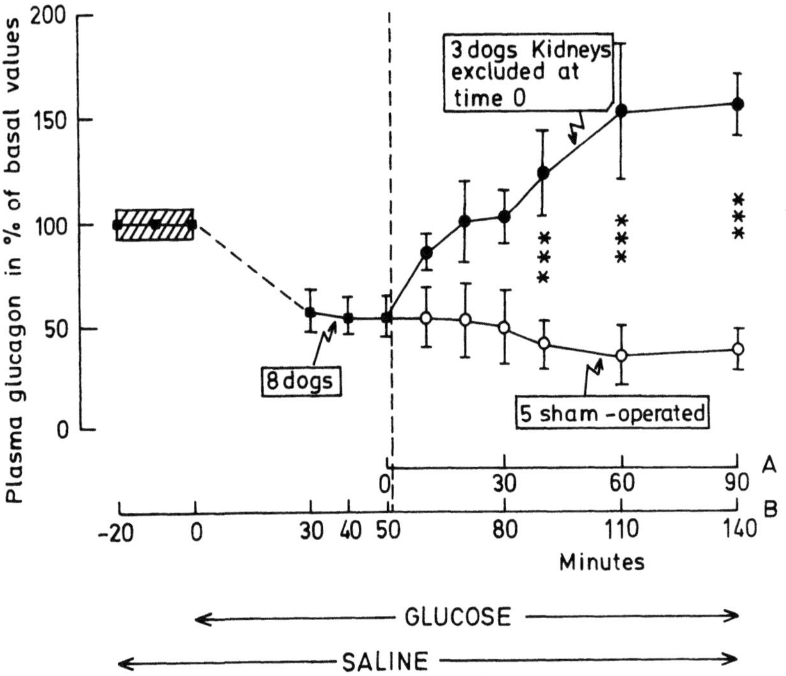

Fig. 3. Arterial plasma glucagon levels before and after glucose infusion (see text) in three dogs with kidney exclusion and in five sham-operated control dogs. Time scale (A) indicates the time of kidney exclusion or manipulation (indicated by the dotted line); time scale (B) concerns the whole experiment. Results are expressed as mean ± S. E. ***corresponds to $p < 0.01$. Data from Lefèbvre and Luyckx (8, with permission).

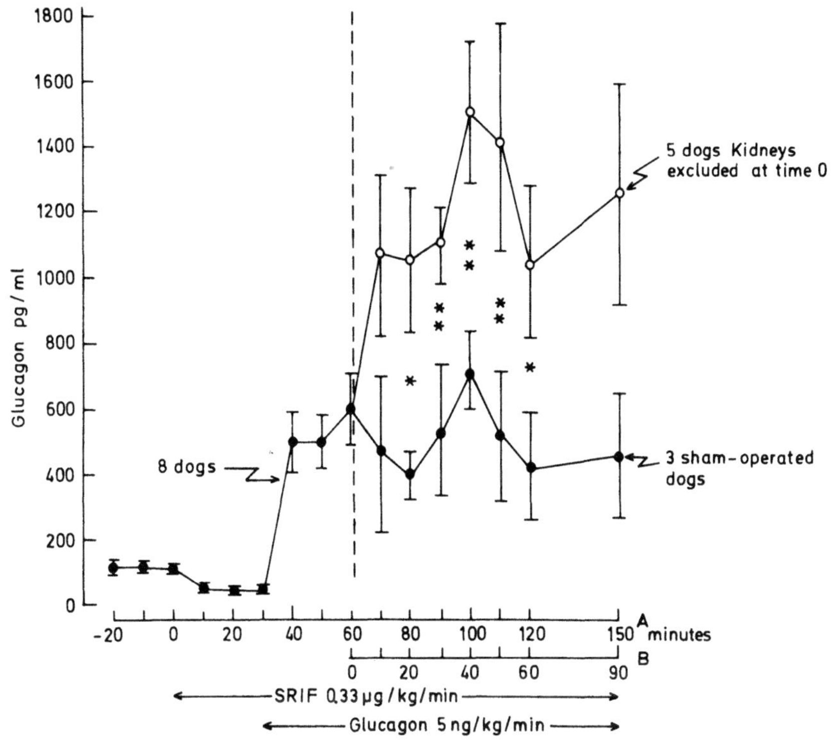

Fig. 4. Changes in arterial plasma glucagon after somatostatin (0.33 µg/kg/min) and exogenous glucagon infusion in 5 dogs with kidney exclusion and 3 sham-operated dogs. Results are expressed as mean ± S.E. * and ** correspond to $p < 0.05$ and < 0.02, respectively. Data from Lefèbvre and Luyckx (9, with permission).

Plasma Immunoreactive Glucagon in Uremic Patients

Like others (1, 2, 7, 12), we have observed high plasma immunoreactive glucagon (IRG) levels in patients with severe renal failure (3). As illustrated in Fig. 6, all plasma IRG values in 18 uremic patients, but two, were out of the range found in normal subjects. As emphasized by Kuku et al (7), the relative amounts of the various plasma IRG fractions is usually modified in uremia with a significant portion of the glucagon immunoreactivity being present in a fraction with a molecular weight of approximately 9000, consistent with the characteristics of proglucagon. In addition, Sherwin et al. (12) elegantly showed that the hyper-

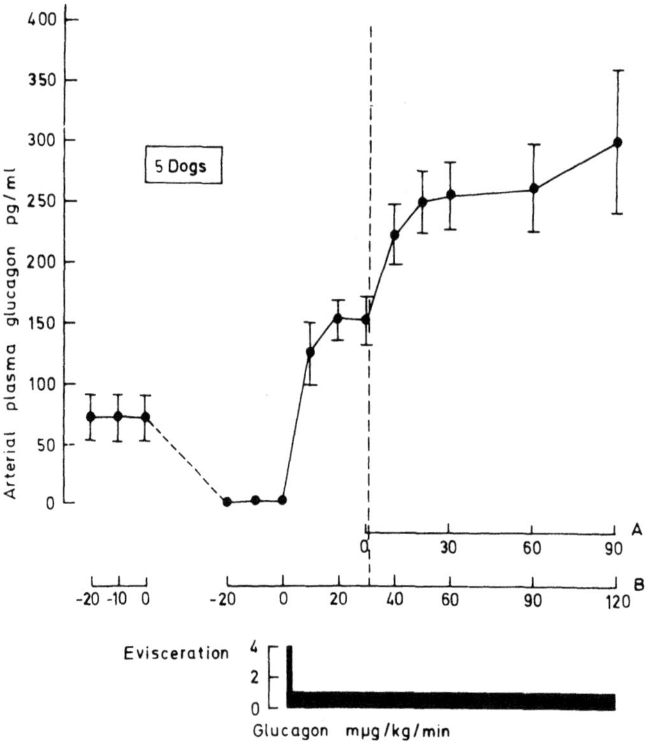

Fig. 5. Changes in arterial plasma glucagon after evisceration and exogenous glucagon infusion in 5 dogs. Results are expressed as mean ± SEM. Time scale A refers to the time of kidney exclusion (indicated by the vertical interrupted line). Time scale B concerns the whole experiment. Data from Lefèbvre and Luyckx (9, with permission).

glycemic effect of physiologic increments in glucagon was increased in nondialyzed uremic patients, suggesting that, in addition to decreased hormonal catabolism, altered tissue sensitivity might contribute to the pathophysiologic action of glucagon in uremia.

Acknowledgement

A. S. Luyckx is an Established Investigator of the Fonds National de la Recherche Scientifique.

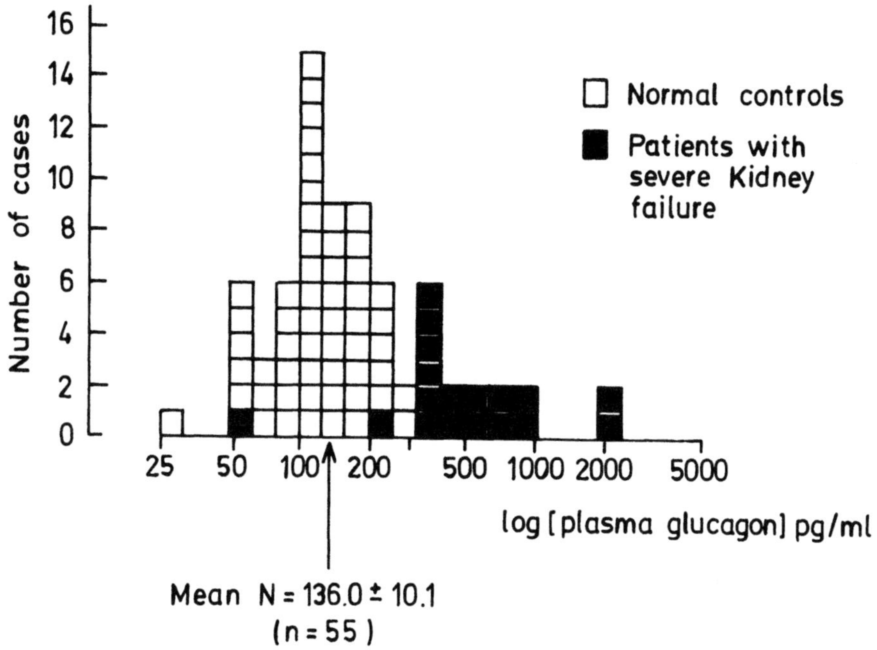

Fig. 6. Plasma glucagon values (30K antiserum) in 18 patients with severe chronic renal failure (▬) in comparison with the values obtained in 55 healthy control subjects (▭). Data from Daubresse et al (3, with permission).

REFERENCES

1. Assan, R. (1972): In vivo metabolism of glucagon. In: Glucagon: Molecular Physiology, Clinical and Therapeutic Implications, Chapter 4. P. J. Lèfebvre and R. H. Unger, Eds. Pergamon Press, Oxford, pp. 47-59.

2. Bilbrey, G. L., Faloona, G. R., White, M. G. and Knochel, J. P. (1974): Hyperglucagonemia of renal failure. J. Clin. Invest. 53, 841-847.

3. Daubresse, J. C., Lerson, G., Plomteux, G., Rorive, G., Luyckx, A. S. and Lefèbvre, P. J. (1976): Lipids and lipoproteins in chronic uraemia. A study of the influence of regular haemodialysis. Europ. J. Clin. Invest. 6, 159-166.

4. Duckworth, W. C., Heinemann, M. and Kemp, K. (1975): Insulin (I) and glucagon (G) degradation by kidney. Clin. Res. 23, 11A.

5. Duckworth, W. C., Heinemann, M. and Kitabchi, A. E. (1975): Proteolytic degradation of insulin and glucagon. Biochem. Biophys. Acta 377, 421-430.

6. Duckworth, W. C. and Kitabchi, A. E. (1974): Insulin and glucagon degradation by the same enzyme. Diabetes 23, 536-543.

7. Kuku, S. F., Zeidler, A., Emmanouel, D. S., Katz, A. I., Rubenstein, A. N., Levin, N. W. and Tello, A. (1976): Heterogeneity of plasma glucagon: Patterns in patients with chronic renal failure and diabetes. J. Clin. Endocrinol. Metab. 42, 173-176.

8. Lefèbvre, P. J. and Luyckx, A. S. (1975): Effect of acute kidney exclusion by ligation of renal arteries on peripheral plasma glucagon levels and pancreatic glucagon production in the anesthetized dog. Metabolism 24, 1169-1176.

9. Lefèbvre, P. J. and Luyckx, A. S. (1976): Plasma glucagon after kidney exclusion: Experiments in somatostatin-infused and in eviscerated dogs. Metabolism 25, 761-768.

10. Lefèbvre, P. J., Luyckx, A. S. and Nizet, A. H. (1974): Renal handling of endogenous glucagon in the dog. Comparison with insulin. Metabolism 23, 753-781.

11. Lefèbvre, P. J., Luyckx, A. S. and Nizet, A. H. (1976): Independance of glucagon and insulin handling by the isolated perfused dog kidney. Diabetologia 12 359-365.

12. Sherwin, R. S., Bastl, C., Finkelstein, F. O., Fisher, M., Black, H., Hendler, R. and Felig, P. (1976): Influence of uremia and hemodialysis on the turnover and metabolic effects of glucagon. J. Clin. Invest. 57, 722-731.

III. SECRETION OF PANCREATIC GLUCAGON

THE ROLE OF CALCIUM IN GLUCAGON SECRETION

J. P. Ashby and R. N. Speake

The Metabolic Unit, Department of Medicine, Western General Hospital Edinburgh and I. C. I. Pharmaceuticals Division, Alderley Park Macclesfield, Cheshire U. K.

A review of experiments using perfused rat pancreas or perifused rat islets suggests that Ca^{++} plays an active role in the secretion coupling mechanism of the A cell.

Calcium plays an important role in the secretory processes of a variety of endocrine and exocrine tissues (13). Its effects on insulin release from the pancreatic B cell have been extensively studied and it is now clear that insulin secretion is absolutely dependent on the presence of extracellular calcium (3). Since the secretory response to glucose and other agents is associated with a proportional uptake of calcium into isolated islets, it has been proposed that an increase in the cytoplasmic concentration of calcium within the B cell triggers the release of insulin (9). It is now becoming clear that calcium is also involved in the regulation of glucagon secretion from the A cell, although its precise function is still a matter of debate.

The role of calcium in glucagon secretion has been assessed in several ways. Gerich et al. (2) studying hormone release from a perfused rat pancreas preparation found that arginine-stimulated glucagon secretion was markedly inhibited when calcium was omitted from the perfusion medium. The reintroduction of calcium into the system promptly restored secretion towards control values (Fig. 1). Similarly Leclercq-Meyer et al. (7) found that the A cell lost its positive secretory response to decreasing glucose concentrations when the calcium level in the medium perfusing a rat pancreas was reduced to approximately 0.2 mM. These results therefore indicate that the presence of extracellular calcium is necessary if the A cell is to show its normal secretory response to amino acids or hypoglycemia. Since verapamil, a blocker of calcium transport has been shown to inhibit the secretory response to arginine (5) it would appear that calcium must also be transported into the A cell in order for glucagon secretion to occur in response to this stimulus. We have investigated the possibility that such an influx of calcium directly triggers the release of glucagon by studying the effects of two bivalent cation ionophores on glucagon release from isolated perifused rat islets (1). These compounds have the ability to transport bivalent cations across biological membranes and if used with a favorable concentration gradient permit the introduction of calcium into cells in a concentration dependent manner.

Ionophore A23187 which specifically binds alkaline earth cations (12) caused a monophasic increase in the rate of glucagon secretion which was absolutely dependent on the presence of calcium in the perifusion medium (Fig. 2a). Ionophore X537A which not only binds bivalent cations, but also transports monovalent cations (11) caused a rapid transient increase in glucagon secretion which was not reduced when calcium was omitted from the system (Figs. 2b, 2c). Experiments using ^{45}Ca as a tracer suggested that the secretory response to both ionophores was associated with an increased calcium influx into the islet cells together with a simultaneous mobilization of calcium from intracellular pools. The results of these studies therefore indicate that calcium may play an active role in the secretion coupling mechanism of the A cell, and they are consistent with the findings of Wollheim et al. (14) who have studied the effects of A23187 on glucagon secretion from pancreatic monolayer cultures. In this system the requirement for calcium was fairly specific and substitution with either magnesium or strontium abolished the secretory response to ionophore A32187. However, it was found that barium could substitute for calcium and that this cation consistently increased glucagon secretion in the absence of any other stimulus.

The ability of ionophore X537A to stimulate glucagon release in the absence of extracellular calcium in our own experiments may be partly explained by a mobilization of intracellular calcium stores, but is also likely to be related to its capacity for transporting substances other than cations. For example, since X537A has a high affinity for various amines (11) this response may be associated

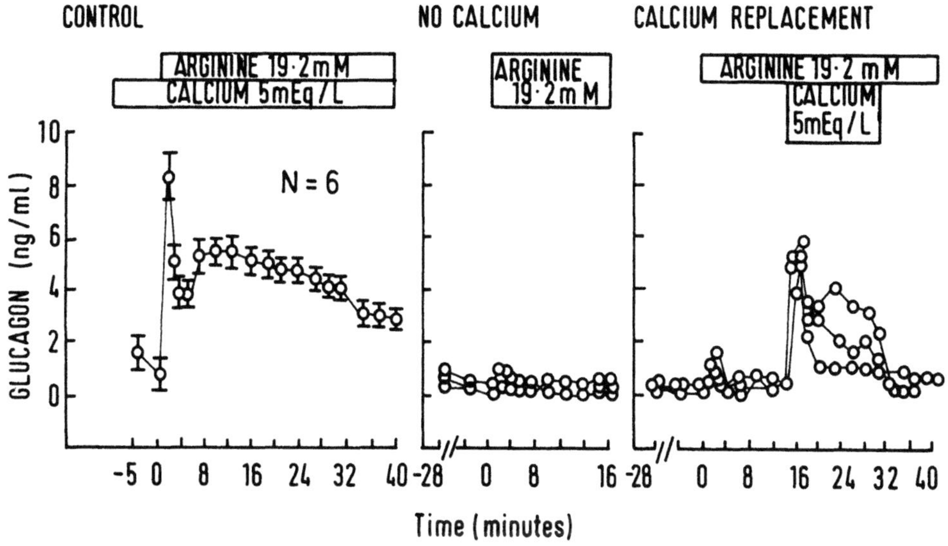

Fig. 1. The effect of calcium deprivation and restoration on arginine-stimulated glucagon secretion from the perfused rat pancreas (2, with permission).

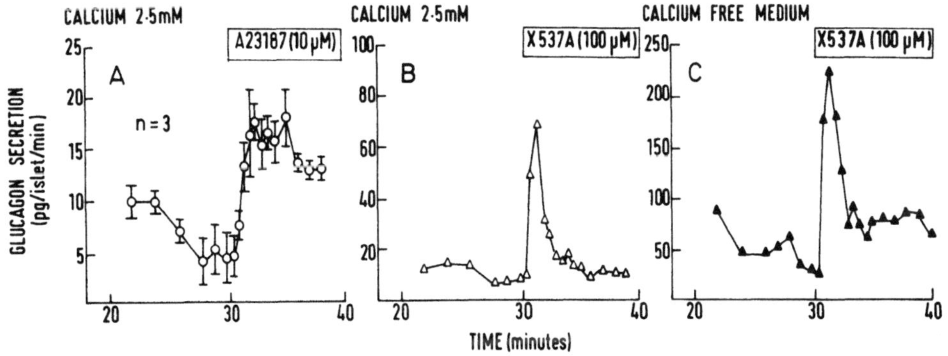

Fig. 2. The effect of ionophores A23187 and X537A on glucagon secretion from isolated perifused islets of Langerhans (1, with permission).

with the release of biogenic amines from sympathetic nerve terminals within islet tissue. This hypothesis is supported by the findings of Lundquist et al. (8) who have reported that chemical sympathectomy abolishes the spike of glucagon released when a perfused rat pancreas preparation is stimulated with a low concentration of calcium. Thus, the response to calcium may be partially dependent on a catecholaminergic stimulus.

Several investigators have attempted to stimulate glucagon secretion by exposing the A cell to supraphysiological concentrations of calcium, but conflicting results have emerged. For example, Kuzuya et al. (4) found that when calcium is administered directly into the pancreatic artery of the dog the secretion of insulin was increased, while the amount of glucagon reaching the circulation remained unchanged. Similar findings have been made using pancreas pieces from duct ligated rats (6), but Ohneda et al. (10) found that the intrapancreatic administration of calcium to the dog actually reduced glucagon secretion. On the other hand, Lundquist et al. (8) have observed a "spike" of glucagon secretion following the addition of a supraphysiological concentration of calcium to a perfused rat pancreas preparation and it has been demonstrated that glucagon release from pancreatic monolayer cultures is also enhanced by excess calcium (14). While these discrepancies may partly reflect the different experimental conditions used, the failure of some investigators to observe glucagon release in response to calcium does not suggest that the cation plays no part in the glucagon secretory process. If for example, physiological concentrations of calcium are sufficient to maintain the activity of the A cell, additional calcium may be ineffective in provoking further secretion.

Finally, although the weight of experimental evidence would suggest that glucagon release is a calcium-dependent process, Leclercq-Meyer and colleagues (7) have demonstrated a paradoxical increase in the secretion of glucagon when the calcium concentration in the medium perfusing a rat pancreas preparation is abruptly reduced from 1.77 mM to 0.14 mM (Fig. 3). Calcium deprivation also results in an elevation of the basal rate of glucagon release from pancreas pieces (6), isolated perifused rat islets (1) and from pancreatic monolayer cultures (14). These findings therefore suggest that calcium has an inhibitory influence on glucagon secretion, but it is not known at present whether they reflect changes in the metabolism of the A cell or are caused by an alteration in the biophysical properties and stability of the cell membranes.

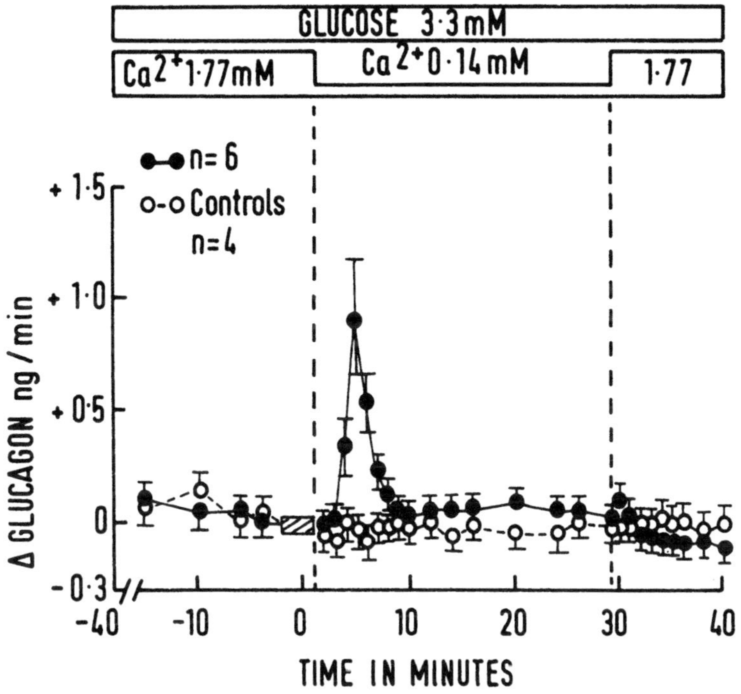

Fig. 3. The effect of calcium deprivation on glucagon release from the perfused rat pancreas in the presence of 3.3 mM glucose (7, with permission).

SUMMARY

The role of calcium in glucagon secretion is complex. While calcium deprivation undoubtedly increases the amount of glucagon released in vitro, the physiological relevance and the mechanism of this effect is as yet unknown. It is clear, however, that the positive secretory response of the A cell to arginine or a hypoglycemic stimulus is dependent on the presence of extracellular calcium and it would therefore appear that under physiological conditions, the secretion of glucagon, like that of insulin, is a calcium-dependent phenomenon. Furthermore, since experiments with verapamil have suggested that the secretory response to arginine is associated with an inward movement of calcium, it is possible that secretion may be mediated by an increase in the cytoplasmic concentration of calcium within the A cell. Experiments with calcium ionophores have confirmed this possibility and it is therefore reasonable to conclude that calcium may play an important role in the secretion coupling mechanism of the A cell.

REFERENCES

1. Ashby, J. P. and Speake, R. N. (1975): Insulin and glucagon secretion from isolated islets of Langerhans. The effects of calcium ionophores. Biochem. J. 150, 89-96.

2. Gerich, J. E., Frankel, B. J., Fanska, R., West, L., Forsham, P. H. and Grodsky, G. M. (1974): Calcium dependency of glucagon secretion from the in vitro perfused rat pancreas. Endocrinology 94, 1381-1385.

3. Grodsky, G. M. and Bennett, L. L. (1966): Cation requirements for insulin secretion in the isolated perfused pancreas. Diabetes 15, 910-913.

4. Kuzuya, T., Kajinuma, H. and Ide, T. (1974): The effect of intrapancreatic injection of potassium and calcium on insulin and glucagon secretion in dogs. Diabetes 23, 55-60.

5. Leclercq-Meyer, V. and Marchand, J. (1976): Differential need for an inward calcium transport in the inhibition by glucose and stimulation by arginine of glucagon release. Studies with verapamil. Diabetologia 12, 405 (abst.).

6. Leclercq-Meyer, V. Marchand, J. and Malaisse, W. J. (1973): The effect of calcium and magnesium on glucagon secretion. Endocrinology 93, 1360-1370.

7. Leclercq-Meyer, V. Marchand, J. and Malaisse, W. J. (1976): The role of calcium in glucagon release. Interactions between glucose and calcium. Diabetologia 12, 531-538.

8. Lundquist, I., Fanska, R. and Grodsky, G. M. (1976): Direct calcium stimulated release of glucagon from the isolated perfused rat pancreas and the effect of chemical sympathectomy. Endocrinology 98, 815-818.

9. Malaisse-Lagae, F. and Malaisse, W. J. (1971): The stimulus secretion coupling of glucose induced insulin release. III. Uptake of ^{45}Calcium by isolated islets of Langerhans. Endocrinology 88, 72-80.

10. Ohneda, A., Matsuda, K., Horigome, K., Ishii, S. and Yamagata, S. (1974): Effect of intrapancreatic administration of calcium on glucagon secretion in dogs. Tohoku J. Exp. Med. 113, 301-311.

11. Pressman, B. C. (1972): Carboxylic ionophores as mobile carriers for the divalent ions, In: Role of Membranes in Metabolic Regulation. Eds. M. A. Mehlman and R. W. Hanson, Academic Press, New York, pp. 149-164.

12. Reed, P. W. and Lardy, H. A. (1972): Antibiotic A23187 as a probe for the study of calcium and magnesium function in biological systems, In: Role of Membranes in Metabolic Regulation. Eds. M. A. Mehlman and R. W. Hanson, Academic Press, New York, pp. 111-131.

13. Rubin, R. P. (1970): The role of calcium in the release of neurotransmitter substances and hormones. Pharmacol. Rev. 22, 389-428.

14. Wollheim, C. B., Blondel, B. and Sharp, G. W. G. (1976): Calcium induced glucagon release in monolayer culture of the endocrine pancreas. Studies with ionophore A23187. Diabetologia 12, 287-294.

THE VERSATILE ROLE OF CALCIUM IN GLUCAGON RELEASE

V. Leclercq-Meyer, J. Marchand and W. J. Malaisse

Laboratory of Experimental Medicine, Free University of Brussels
Brussels, Belgium

Data on the interrelationships between calcium and the secretory activity of the A cell are reviewed and discussed. It is postulated that calcium has a multiple role in the secretory activity of the pancreatic A cell. First, a cellular accumulation of calcium may mediate the glucagonotropic action of arginine and of a mixture of pyruvate, fumarate and glutamate. Second, the presence of a sufficient amount of extracellular calcium seems to be a prerequisite for the identification of certain environmental factors, especially glucose in high concentration, by the A cell. Third, under appropriate experimental conditions, e. g. during prolonged exposure to a low glucose level, the accumulation of intracellular calcium may exert a feedback inhibitory effect upon the secretory process.

For a number of years, we have investigated the role of calcium in glucagon release. While progressing in this study, it became more and more apparent that the interrelationships between calcium and the secretory activity of the A cells are complex, a fact corroborated by other observations reported in the literature. The following presentation deals with the most salient findings made in our laboratory. We shall also attempt to integrate these data with those reported by other investigators, and finally, we shall try to put forward a working hypothesis as to the role, or roles, of calcium in the A cell.

The Enhancing Effect of Calcium Deprivation Upon Glucagon Release

We have consistently observed an enhancement of glucagon release upon calcium deprivation (12-14). In the *in vitro* perfused rat pancreas, such an enhancement varies in both shape and amplitude as a function of the environmental concentration of glucose. Thus, at low glucose levels (3.3 mM), a sharp early peak of glucagon release was observed, the secretion resuming thereafter at a rate close to that seen in control experiments not entailing periods of calcium deprivation. With increasing glucose concentrations (5.5, 8.3 and 16.6 mM), this early peak progressively decreased in amplitude, whereas a secondary phase of release became apparent and progressively more pronounced. The enhancement of glucagon release was readily reversed upon restoration of normal calcium levels (13, 14).

These observations were confirmed in more recent experiments, although two small differences are noteworthy when comparing these results (Figs. 1 and 2) to our earlier work. First, at high glucose (16.6 mM), there was a better dissociation between the first and the second phase of glucagon release (Fig. 2). Second, at low as well as at high glucose concentration (3.3 and 16.6 mM), a sharp and pronounced "off-response" occurred upon restoration of normal calcium levels. This outburst of glucagon was followed by a return of the secretory rates to their respective control values. The foregoing differences are likely to be accounted for by the type of colloid added to the perfusate (4% dextran and 0.5% albumin (11).

It should be duly emphasized that, in the absence of any substrate other than glucose, the secretion of glucagon was never reduced by the lack of extracellular calcium (see also Figs. 6 and 7). Other authors have reported similar findings (3, 4, 27), and the fact that a significant secretion of glucagon can occur in the presence of a calcium-free medium has recently been confirmed (16).

Altogether, these results suggest that calcium normally can exert an inhibitory role upon the A cell, a conclusion supported by the observation that the intrapancreatic administration of calcium in the dog inhibits the release of glucagon (18). Our results, however, seem to be at variance with the recent report of a direct stimulation of glucagon release by calcium (16). These apparently conflicting findings may be reconciled. Indeed, as illustrated in Figs 1 and 2, outbursts of glucagon similar to those observed by Lundquist et al (16) were noted upon restoration of normal calcium levels, when the perfusate contained dextran rather than albumin alone. Whether such outbursts represent a direct stimulatory effect or an "off-response" may be considered a simple problem of terminology rather than a true difference in experimental results.

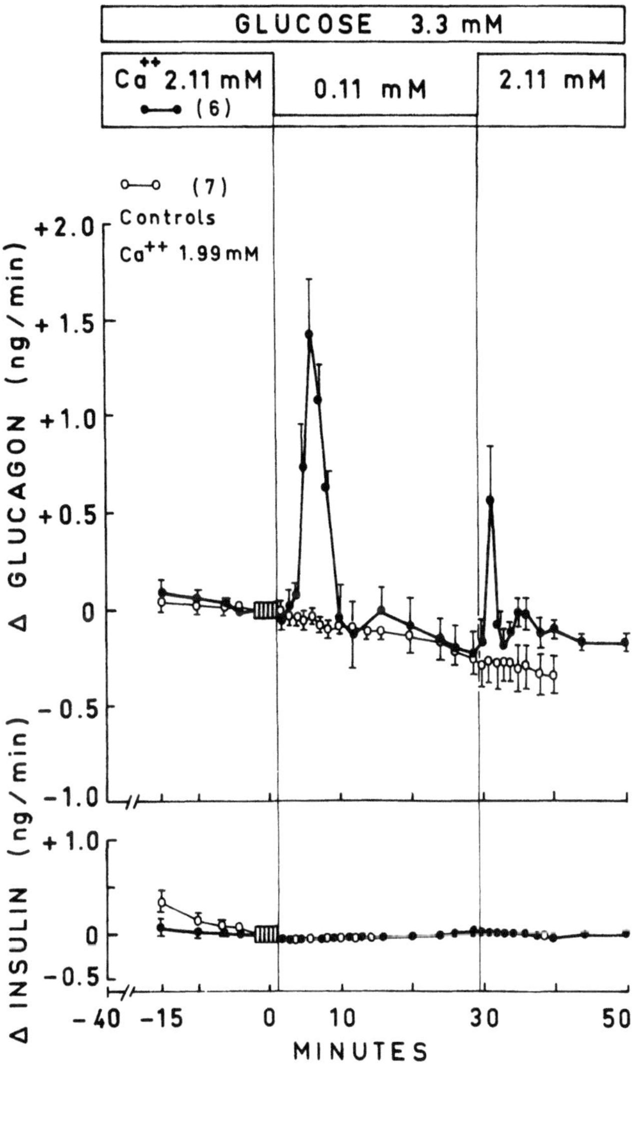

Fig. 1. The effects of calcium deprivation and restoration upon glucagon and insulin release in the presence of a 3.3 mM concentration of glucose (closed circles). Mean changes in secretion (± SEM), relative to the mean hormonal output recorded at the end of the equilibration period (hatched areas), are indicated, with the number of individual experiments in each group (n). In the control experiments (open circles), the normal calcium levels were maintained throughout the perfusions and amounted to 1.99 ± 0.12 mM (n = 7).

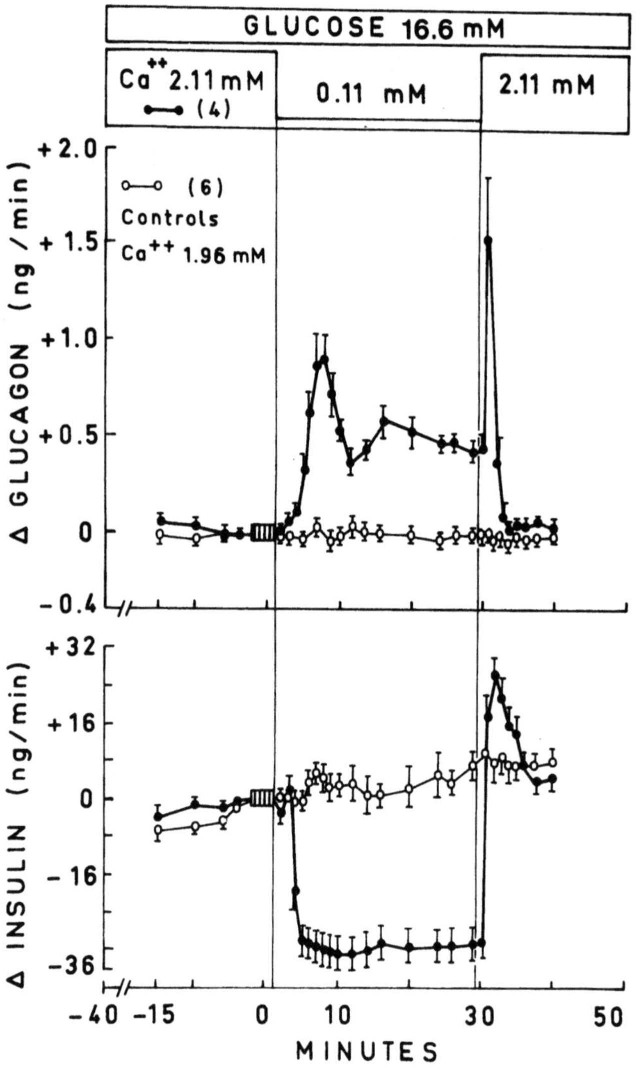

Fig. 2. The effects of calcium deprivation and restoration upon glucagon and insulin release in the presence of a 16.6 mM concentration of glucose (closed circles). Presentation as in Fig. 1. In the control experiments (open circles), the normal calcium levels amounted to 1.96 ± 0.04 mM (n = 6).

It might be asked whether the enhancement of glucagon release, seen upon calcium deprivation in the presence of 16.6 mM glucose, represents a true stimulation or a relief from the inhibitory effect of glucose upon the A cell. Indeed, when the results are expressed in absolute values (pg/min), it becomes obvious that the secretory rates recorded during the late phase of calcium deprivation at high glucose concentration are very similar to those seen in experiments performed in the presence of a low glucose concentration and at a normal calcium level (Fig. 3).

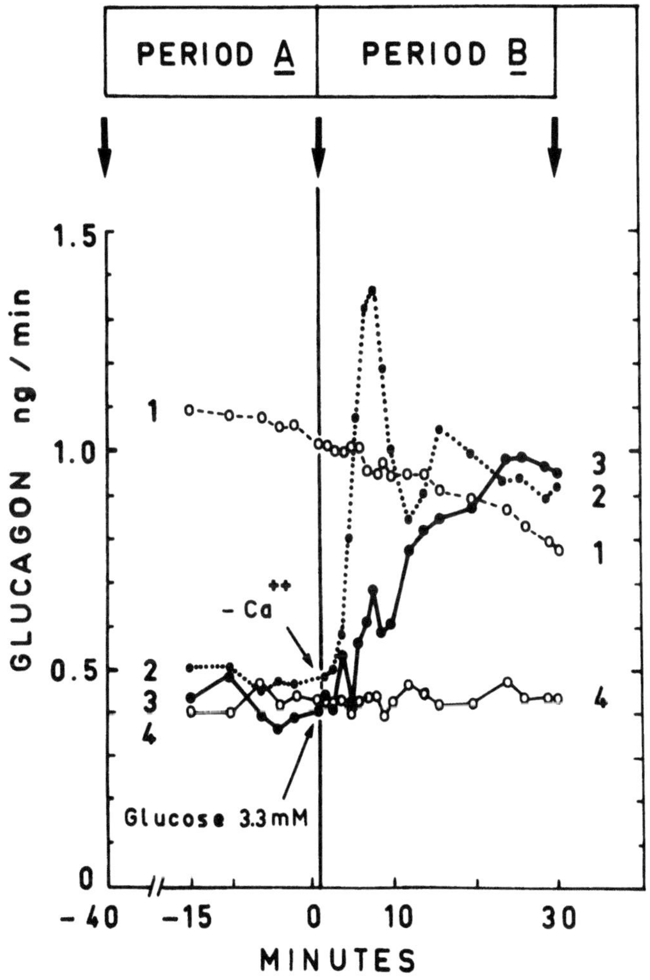

Fig. 3. Mean glucagon output from the perfused rat pancreas in various conditions. Curve no. 1: Glucose 3.3 mM throughout, normal calcium concentrations throughout (2 mM) (n = 7); Curve no. 2: Glucose 16.6 mM throughout, normal calcium during period A, 0.11 mM calcium during period B (n = 4); Curve no. 3: Glucose 3.3 mM during period A; glucose 16.6 mM during period B, normal calcium throughout (n = 5); Curve no. 4: Glucose 16.6 mM throughout, normal calcium throughout (n = 5).

The mechanisms involved in the enhancement of glucagon release are largely unknown. We have investigated the possibility that a reduced inward transport of calcium may participate in the phenomenon, by using verapamil, a drug thought to inhibit the inward transport of calcium in many tissues (5). The infusion of verapamil at a 10 μM concentration induced a small increase in glucagon release at low (3.3 mM) glucose concentration (Fig. 4), but virtually failed to affect the secretory rates recorded in the presence of a high (16.6 mM) glucose concentration (Fig. 5). Thus, it seems that the enhancement of glucagon release seen upon extracellular calcium deprivation is only partially (low glucose) or not at all (high glucose) due to a reduced inward transport of calcium. Incidentally, it should be stressed that verapamil never inhibited glucagon release, although it was quite effective in reducing glucose induced insulin release (Fig. 5). The latter finding is in agreement with the postulated calcium-antagonistic properties of verapamil in the B cell (2, 23). Therefore, our results indicate that the role of inward transported calcium, like that of extracellular calcium, is vastly different in the A and B cells, at least when glucose is the sole exogenous energy substrate.

The Requirement of Extracellular Calcium for the Effectiveness of Newly Applied Stimuli

The glucose stimulus. After a prolonged period of calcium deprivation (40 min), raising or decreasing the concentration of glucose did not affect the secretion of glucagon to any marked extent (13, Figs. 6 and 7). Such reduced responsiveness of the A cell to glucose could also be seen very early (10 min) after the onset of the period of calcium deprivation (13). Thus, the presence of a normal extracellular calcium concentration seems to be a prerequisite for allowing glucose to exert its classical influence upon the secretory activity of the A cell.

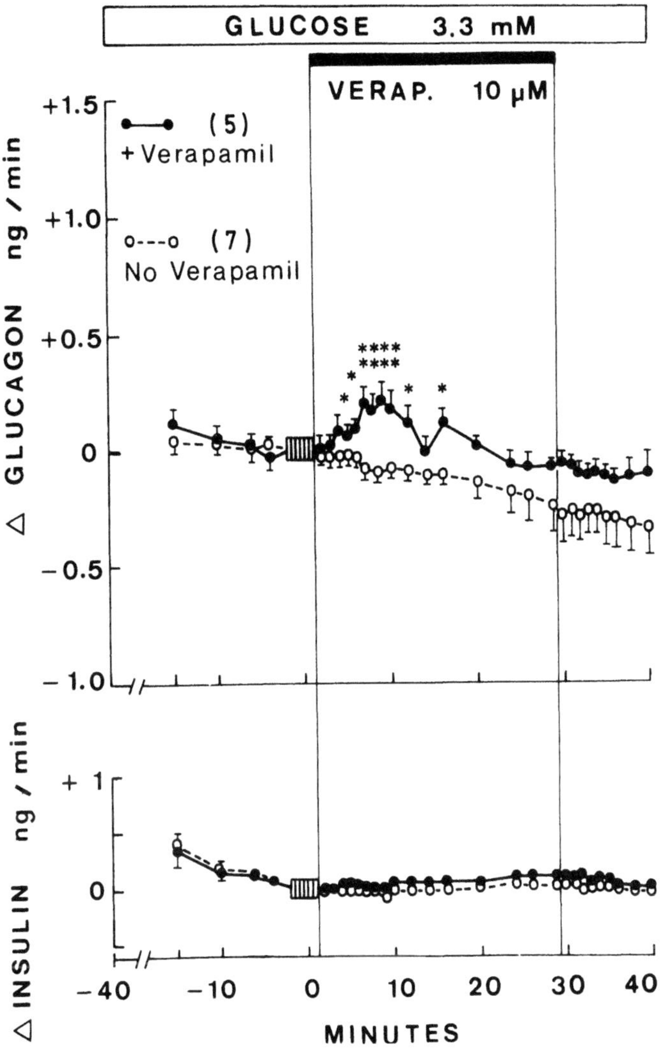

Fig. 4. The effects of verapamil (10 µM) upon glucagon and insulin release in the presence of a 3.3 mM concentration of glucose (closed circles). Presentation as in Fig. 1. Significant differences above the control secretory rates seen in the absence of verapamil (open circles) are indicated. (*) : P <0.05; (II) : P <0.02 or less.

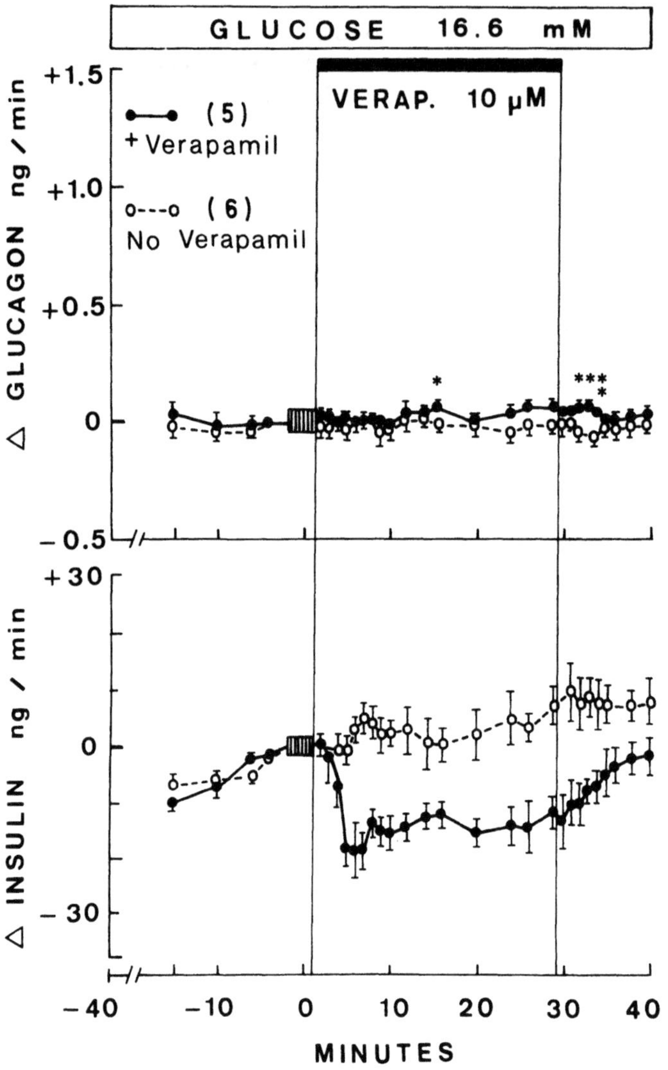

Fig. 5. The effects of verapamil (10 μM) upon glucagon and insulin release in the presence of a 16.6 mM concentration of glucose (closed circles). Presentation as in Fig. 1. significant differences above the control secretory rates seen in the absence of verapamil (open circles) are indicated as in Fig. 4.

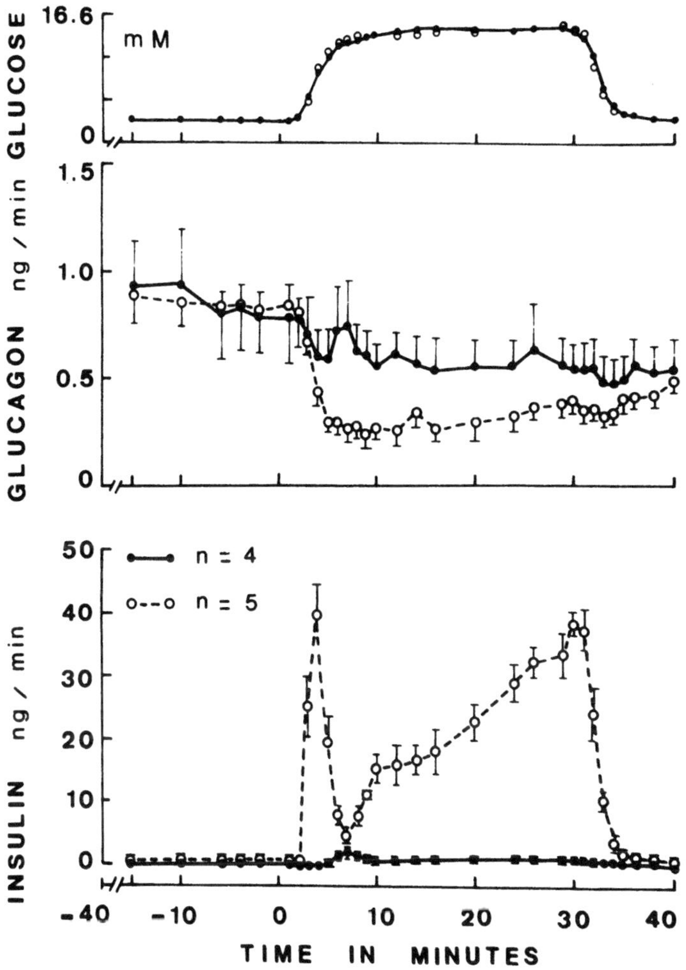

Fig. 6. The effect of a rise in the concentration of glucose from 3.3 to 16.6 mM applied late (40 min) during the period of calcium deprivation, on glucagon and insulin release (closed circles, calcium levels : 0.09 ± 0.01 mM, n = 4). The control experiments (open circles) were performed in the presence of normal calcium levels (1.94 ± 0.04 mM, n = 5). The mean secretory rates (± SEM) are shown together with the number of individual experiments in each group (n) (13, with permission).

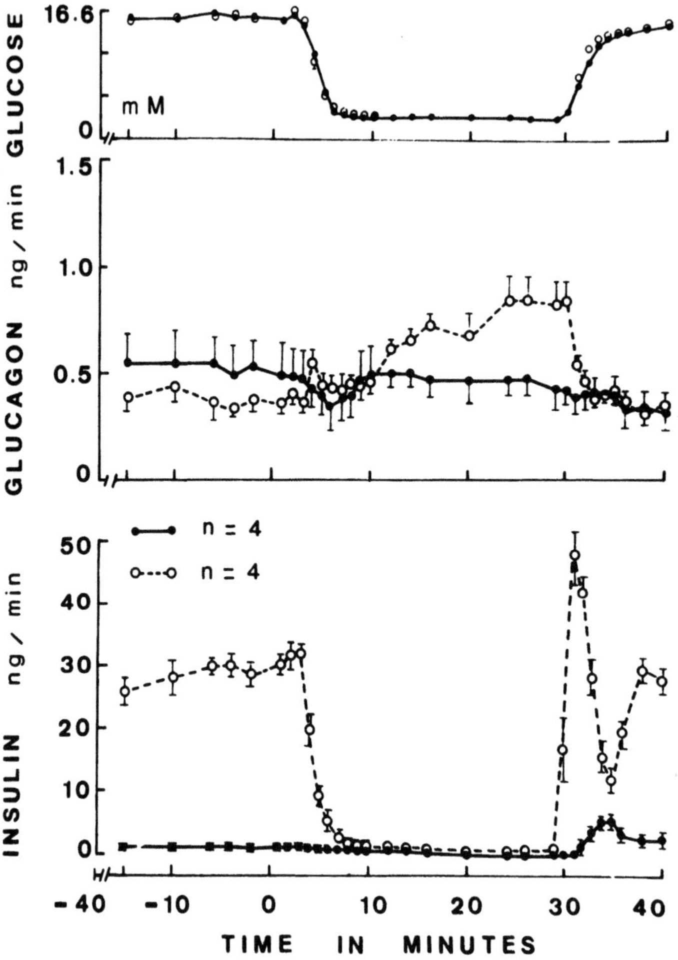

Fig. 7. The effect of a decrease in the concentration of glucose from 16.6 to 3.3 mM applied late (40 min) during the period of calcium deprivation, upon glucagon and insulin release (closed circles, calcium levels : 0.09 ± 0.01 mM, n = 4). The control experiments were performed in the presence of normal calcium levels (open circles, Ca^{2+} : 1.93 ± 0.03 mM, n = 4). Presentation as in Fig. 6 (13 with permission).

The arginine stimulus. In close confirmation of the results previously obtained by others (8), we have observed that a 10 mM arginine stimulus loses its ability to stimulate the release of glucagon, after a prolonged (40 min) as well as a short period of calcium deprivation (Fig. 8). Thus, as it is the case for glucose, a normal extracellular calcium concentration is a prerequisite for the usual positive glucagonotropic effect of arginine upon the A cell.

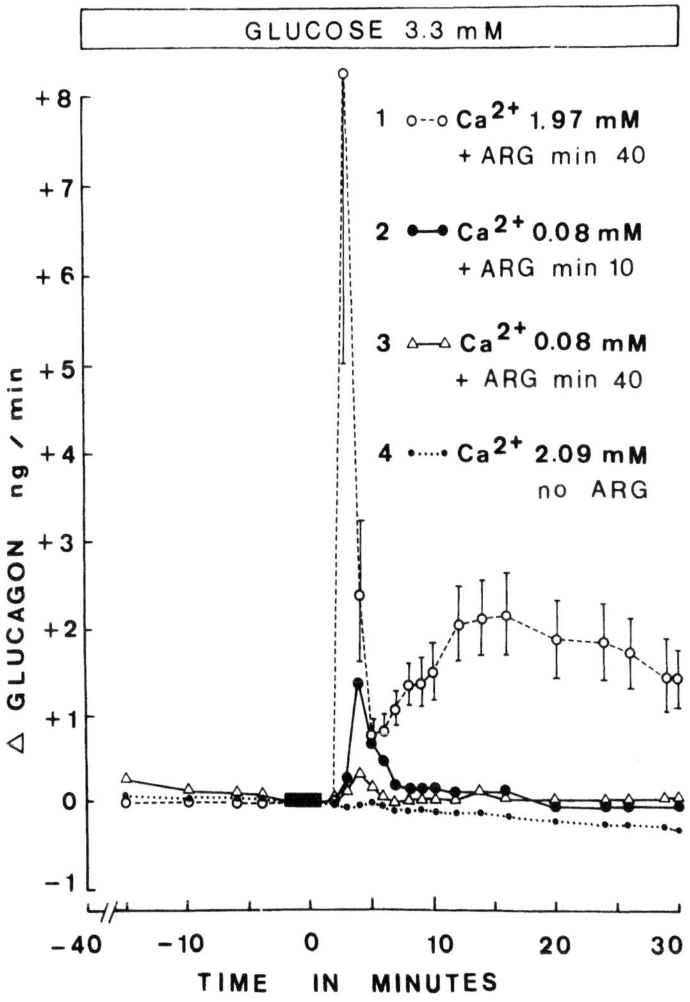

Fig. 8. Comparison on the effects of a 10 mM arginine infusion upon glucagon release, in the presence of a 3.3 mM concentration of glucose, either in the presence of normal calcium levels (curve no. 1, n = 6) or after a short (10 min, curve no. 2, n = 6) and prolonged period of calcium deprivation (40 min, curve no. 3, n = 5). Control experiments, without arginine infusion, are also indicated (curve no. 4, n = 5). Presentation as in Fig. 1. From Leclercq-Meyer et al., Hormone Res. 7, 348, 1976. (Reproduced from Hormone Research by permission of S. Karger AG, Basel).

The Calcium Dependency of Glucagon Release Evoked by Either Arginine or a Mixture of Fumarate, Glutamate and Pyruvate

The arginine stimulus. In the presence of a low 3.3 mM glucose concentration and a constant 10 mM arginine infusion, calcium deprivation provoked an abrupt reduction of glucagon release (Fig. 9). Upon restoration of normal calcium levels, there was a sudden increase in glucagon release, followed by a return to the secretory rates seen in the experiments in which the pancreas had not been submitted to calcium deprivation. These results confirm the calcium-dependency of arginine-induced glucagon release, first documented by Gerich et al (8). In a way, our experiments reveal that the situation created by the removal of calcium corresponds mainly, if not exclusively, to a suppression of the proper effect of arginine. Indeed, for about 20 min after removal of calcium, the secretory rate remained at a value (approximately 1 ng/min) close to that otherwise seen in the absence of arginine, but in the presence of glucose (3.3 mM), whether at normal (Fig. 3) or low (Fig. 1) calcium concentration. That calcium deprivation may affect the glucagonotropic capacity of arginine and glucose simultaneously even if through a different mechanism, is supported further by the next set of experiments.

In the presence of a high 16.6 mM glucose concentration, and of a constant 10 mM arginine infusion, the removal of calcium did not provoke inhibition, but rather, a transient enhancement of glucagon release (Fig. 10). It is well known that in the presence of normal calcium levels and high glucose, the glucagonotropic effect of arginine is notably reduced (7, 20; cf. also Figs. 9, 10, 15, and 16). It is therefore understandable that, under such an experimental condition, the predominant response of the A cell to calcium deprivation reflects the increase in output due to calcium deprivation and to high glucose concentration in the absence of arginine (cf. Fig. 2). Nevertheless, such a stimulation is of a smaller amplitude than that seen in Fig. 2, suggesting a combination of the opposite effects of calcium deprivation on arginine-induced glucagon secretion on one hand, and the release seen in the presence of glucose alone, on the other hand. Incidentally, the increase in glucagon release, due to calcium deprivation in the presence of arginine and high glucose concentration, confirms previous observations from this laboratory (12). Indeed, the 8.3 mM concentration of glucose which was used in combination with arginine, in the previous experiments, is as effective in inhibiting glucagon release as the higher 16.6 mM concentration (11).

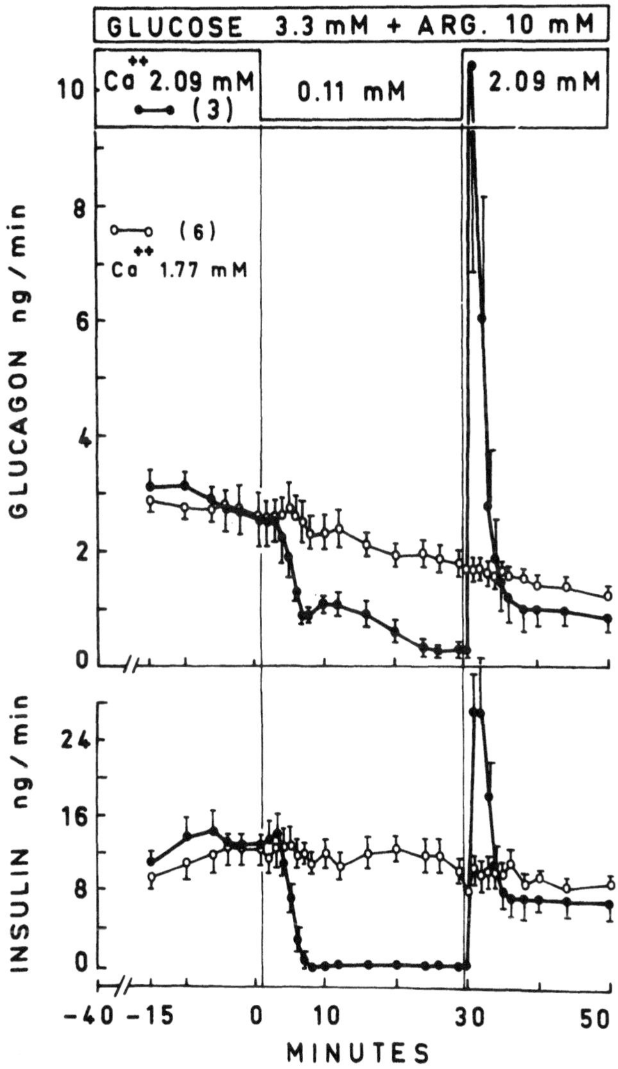

Fig. 9. The effects of calcium deprivation and restoration upon glucagon and insulin release, in the presence of 3.3 mM glucose and 10 mM arginine concentrations (closed circles). The control experiments were performed in the presence of normal calcium levels throughout (open circles: Ca^{2+} : 1.77 ± 0.04 mM, n = 6). Presentation as in Fig. 6.

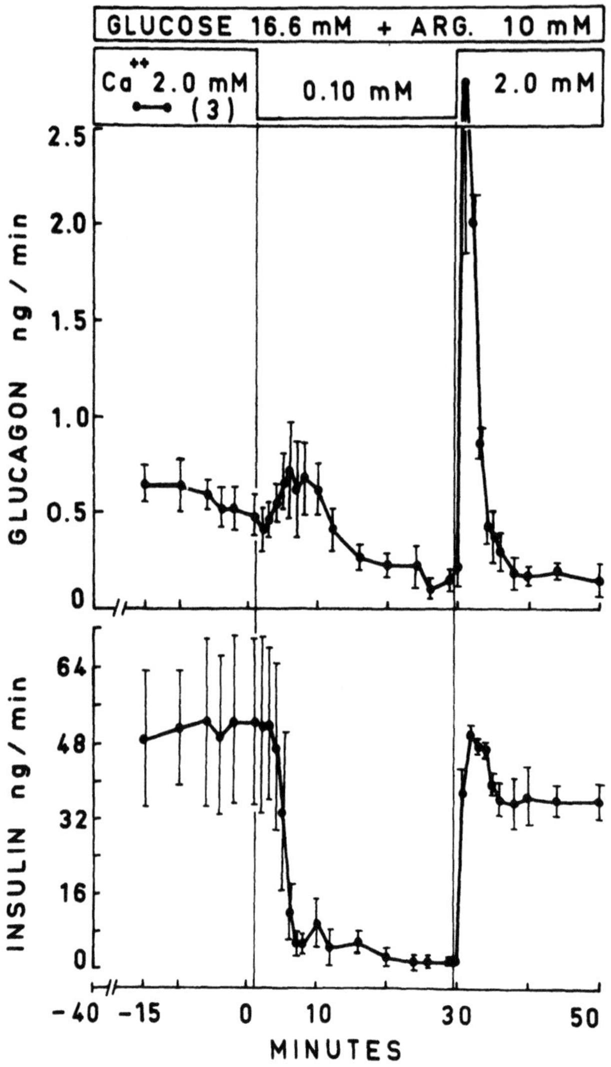

Fig. 10. The effects of calcium deprivation and restoration upon glucagon and insulin release, in the presence of 16.6 mM glucose and 10 mM arginine concentrations. Presentation as in Fig. 6.

The fumarate + glutamate + pyruvate (FGP) stimulus. We have recently observed that a FGP mixture has arginine-like effects in the perfused rat pancreas (unpublished results). Thus, like arginine, FGP evoked a biphasic response, which was inversely related to the environmental glucose concentration (compare the glucagon output during the equilibration period in Figs. 11 and 12, to those seen in Figs. 9 and 10; see also Fig. 17).

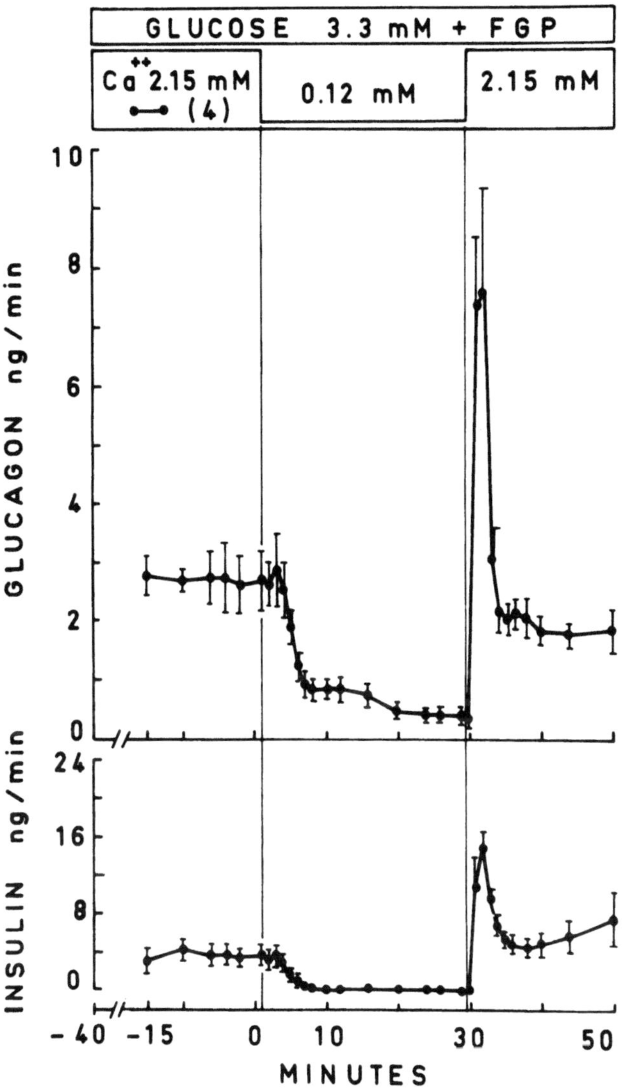

Fig. 11. The effects of calcium deprivation and restoration upon glucagon and insulin release, in the presence of 3.3 mM glucose and a mixture of "fumarate + glutamate + pyruvate" (FGP, 5 mM of each salt). Presentation as in Fig. 6.

In the presence of the FGP mixture, the response of the A cell to calcium deprivation closely duplicated that seen in the presence of arginine. Thus, in the presence of a low 3.3 mM concentration of glucose, an inhibition of glucagon release was observed, followed by a striking outburst of glucagon upon restoration of normal calcium levels (compare Fig. 11 to Fig. 9). In the presence of a high 16.6 mM concentration of glucose, calcium deprivation stimulated glucagon release

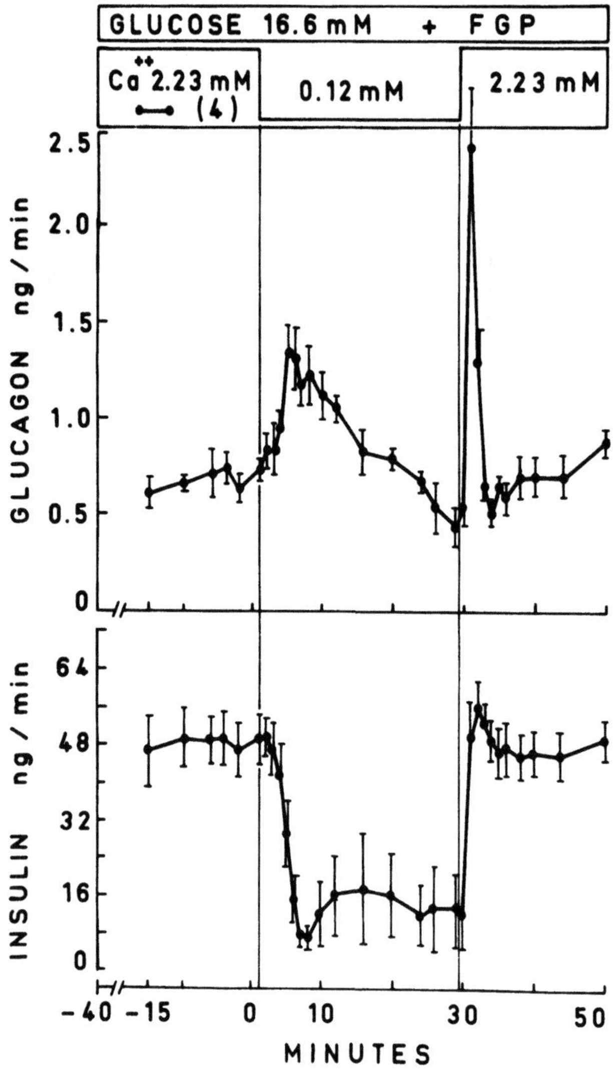

Fig. 12. The effects of calcium deprivation and restoration upon glucagon and insulin release, in the presence of 16.6 mM glucose and a mixture of "fumarate + glutamate + pyruvate" (FGP, 5 mM of each salt). Presentation as in Fig. 6.

even more than in the presence of arginine (Fig. 12, compare to Fig. 10). The close analogy between the results obtained with arginine and with the FGP mixture fully support the view that the overall effect of calcium deprivation upon the A cell is the result of two opposite actions: the inhibition of either arginine- or FGP-induced glucagon release and the enhancement of release usually seen in the presence of glucose alone, especially at high concentrations.

The Participation of Inward Transport of Calcium in Glucagon Release

The participation of inward transport of calcium in glucagon release was studied by means of perfusion experiments in which the medium contained verapamil at a 20 M concentration. Various stimuli, including glucose, arginine and the FGP mixture, were applied during the experiment, after a 40 min equilibration period. Some of these results have already been reported elsewhere (10).

The glucose stimulus. Raising the glucose concentration from 3.3 to 16.6 mM, clearly inhibited the rate of glucagon release, in the presence of verapamil as well as in its absence (Fig. 13). By contrast, the stimulatory effect of decreasing the concentration of glucose was blunted in the presence of verapamil (Fig. 14). These results suggest that the integrity of inward calcium transport is not necessary for the inhibitory action of glucose on glucagon release. On the other hand, the increase in glucagon release upon lowering the concentration of glucose seems to depend, at least in part, upon the integrity of such transport.

The arginine stimulus. In the presence of verapamil and of a low 3.3 mM glucose concentration, the response of the A cell to a 10 mM arginine stimulus, which normally is very pronounced, was strikingly modified (Fig. 15). Thus, a blunted first phase was followed by an inhibition of glucagon release below pre-stimulatory levels. As a matter of fact, such inhibition brought the rate of glucagon release to levels comparable to those found in the presence of high (16.6 mM) glucose concentrations and in the absence of arginine (cf. Fig. 3).

At high (16.6 mM) glucose concentration, the already reduced response of the A cell to arginine was further decreased by verapamil (Fig. 16). Thus, the late phase of glucagon release was virtually absent. The first phase, however, was still present.

The foregoing results suggest that the integrity of the inward transport of calcium represents an absolute requirement for arginine-induced glucagon release, in the presence of low as well as high glucose concentrations.

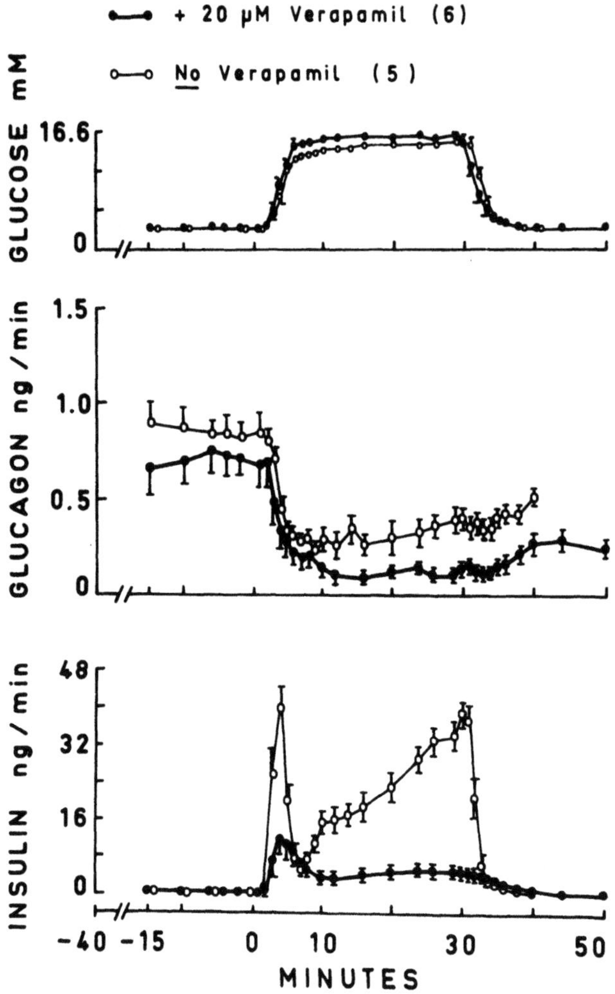

Fig. 13. The effect of a 20 μM verapamil infusion, maintained throughout the experiments, upon the release of glucagon and insulin in response to a rise in the concentration of glucose from 3.3 to 16.6 mM (closed circles). The control experiments were performed in the absence of verapamil (open circles). Presentation as in Fig. 6.

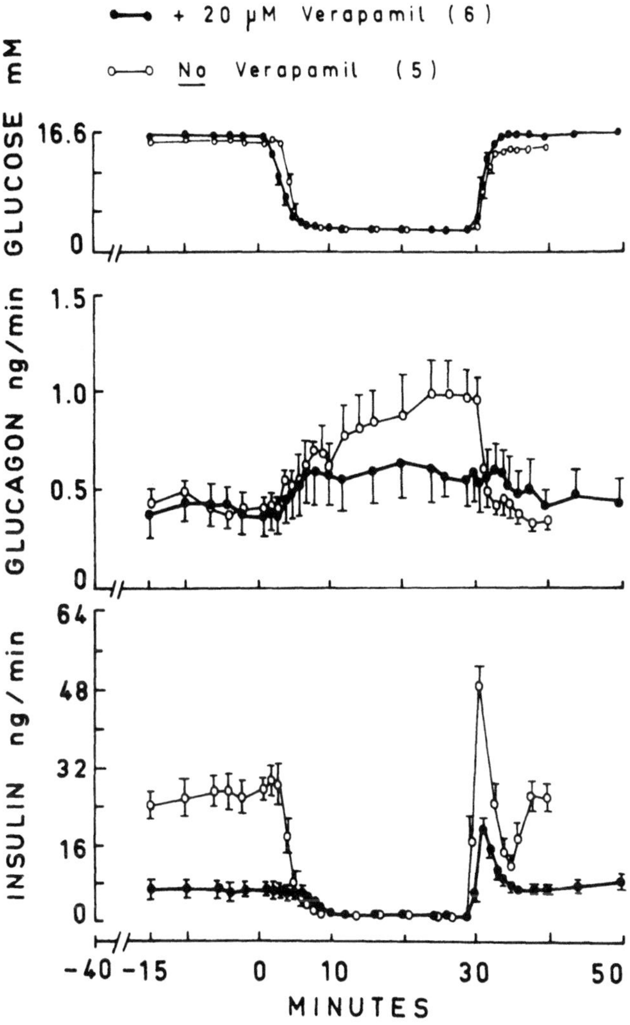

Fig. 14. The effect of a 20 μM verapamil infusion, maintained throughout the experiments, upon the release of glucagon and insulin in response to a decrease in the concentration of glucose from 16.6 to 3.3 mM (closed circles). The control experiments were performed in the absence of verapamil (open circles). Presentation as in Fig. 6.

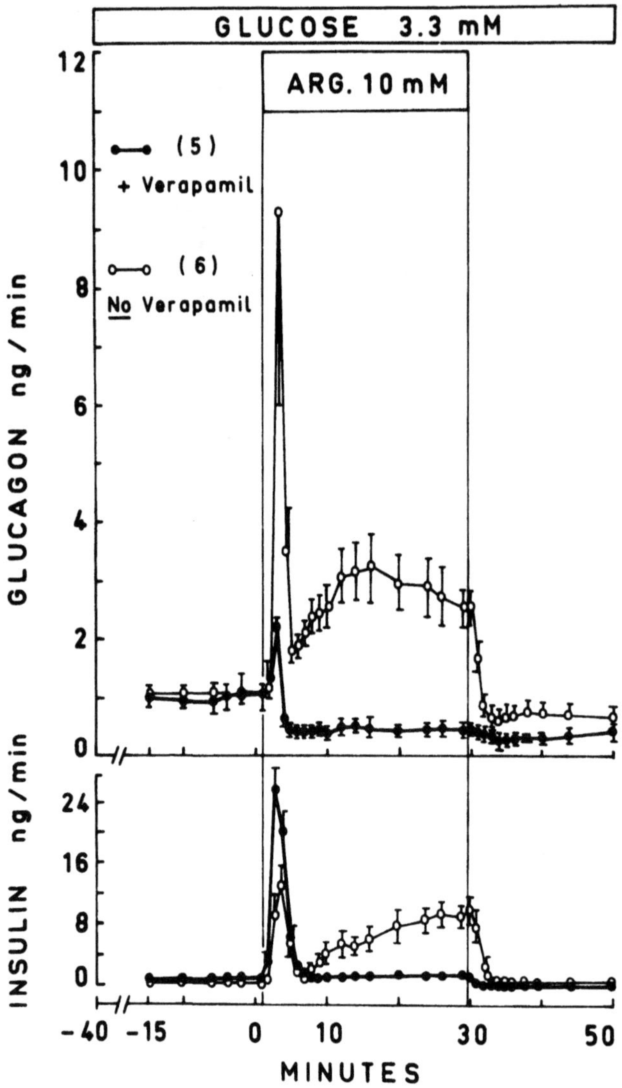

Fig. 15. The effect of a 20 μM verapamil infusion, maintained throughout the experiments, upon the release of glucagon and insulin, in response to 10 mM arginine, in the presence of a 3.3 mM concentration of glucose (closed circles). Presentation as in Fig. 6.

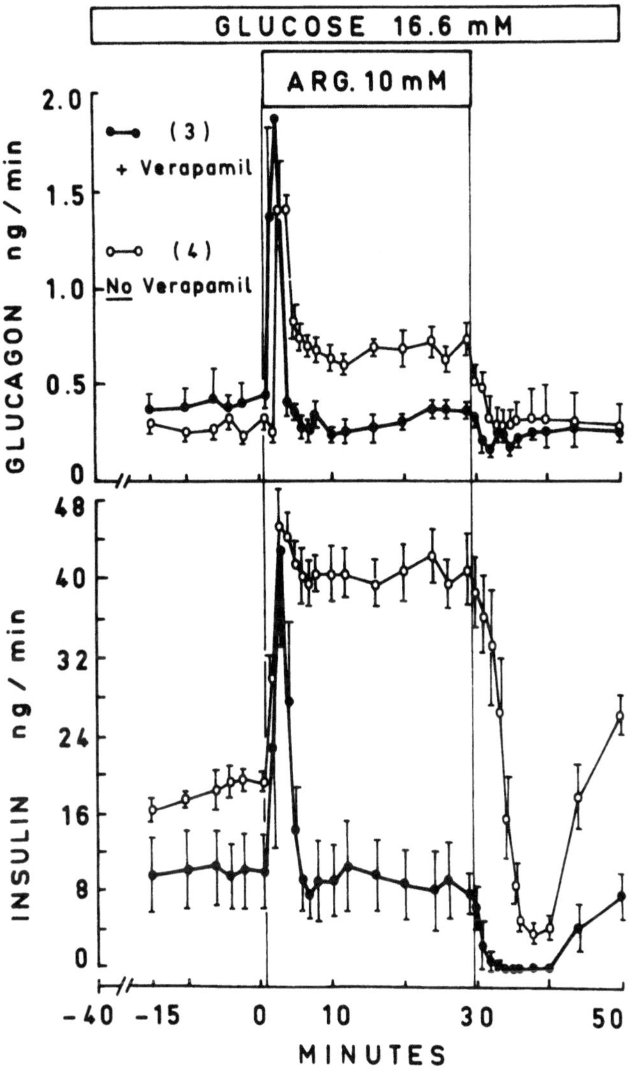

Fig. 16. The effect of a 20 μM verapamil infusion, maintained throughout the experiments, upon the release of glucagon and insulin, in response to 10 mM arginine, in the presence of a 16.6 mM concentration of glucose (closed circles). The control experiments were performed in the absence of verapamil (open circles). Presentation as in Fig. 6.

The fumarate + glutamate + pyruvate stimulus (FGP). So far, only two experiments have been performed in the presence of a 3.3 mM glucose concentration (Fig. 17).

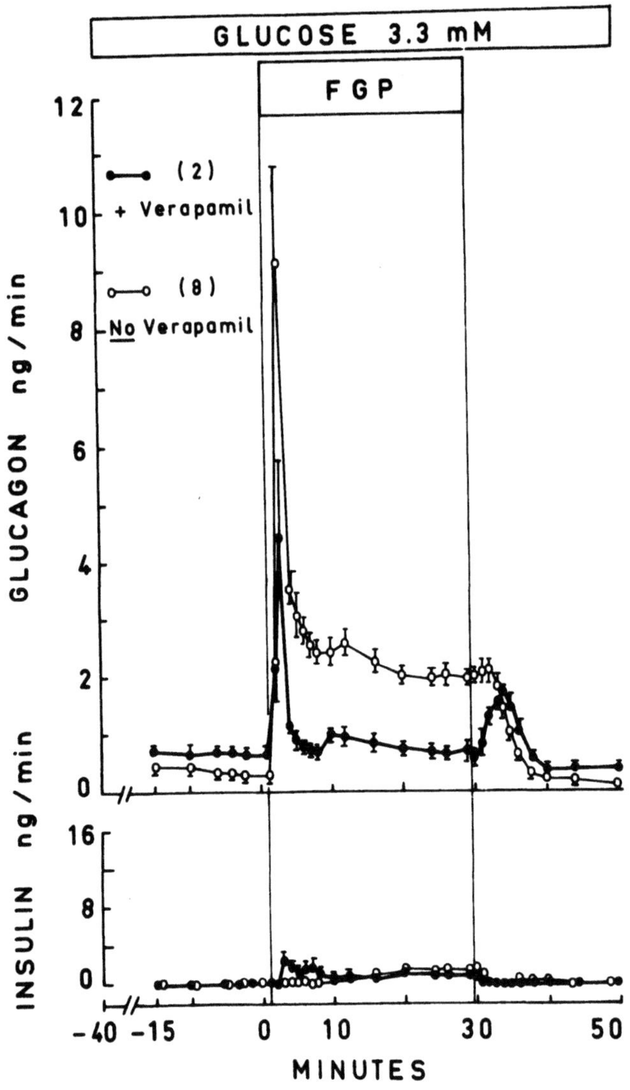

Fig. 17. The effect of a 20 μM verapamil infusion, maintained throughout the experiments, upon the release of glucagon and insulin, in response to a mixture of "fumarate + glutamate + pyruvate" (FGH, 5 mM of each salt), in the presence of a 3.3 mM concentration of glucose (closed circles). The control experiments were performed in the absence of verapamil (open circles). Presentation as in Fig. 6.

The results suggest that, as in the case of arginine, an inward transport of calcium is needed for the effectiveness of the FGP mixture on glucagon release. Indeed, both phases of the FGP-induced glucagon release were strongly reduced when the A cell was exposed to verapamil. No inhibition of release below prestimulatory levels was observed during the late phase, in contrast to what was seen with arginine (cf. Fig. 15).

A Working Hypothesis for the Role of Calcium in Glucagon Release

Any working hypothesis for the role of calcium in the secretory activity of the A cell may be only tentative at the present time. However, on the basis of data available in the literature and of our own findings, it seems increasingly evident that calcium may play the following three roles, at three distinct sites.

Intracellular calcium may act as a positive regulator of glucagon release. The first postulated role of calcium is the most easy to define, because it appears identical to that currently ascribed to this divalent cation in the process of insulin release. It is generally accepted that the concentration of calcium in some critical site, possibly the cytosol of the B cell, controls the rate of insulin release, whatever agent is used to either stimulate or inhibit secretion (17).

Calcium could play a comparable role in the A cell. The data obtained with arginine and with the FGP mixture seem to support such a postulate, for the following reasons:

(i) Arginine fails to stimulate glucagon release in the absence of calcium;

(ii) Verapamil, which by inhibiting calcium entry into the cell may lower the intracellular calcium concentration, markedly reduces the secretory response to either arginine or the FGP mixture. In the presence of verapamil, the rate of secretion may fall below prestimulatory levels;

(iii) During continuous administration of either arginine or the FGP mixture, removal of extracellular calcium, apparently suppresses the usual effect of these secretagogues.

(iv) The secretory response to arginine is increased in the presence of high (4 mM), as distinct from normal (2 mM) calcium concentrations (unpublished results).

Other secretagogues could operate through the postulated calcium-dependent pathway. These may include calcium itself, which has been shown to stimulate glucagon release (16; cf. the so-called "off-responses" in Figs. 1 and 2) cationic ionophores (1, 27); and Ba^{2+} which induces glucagon release (27) and is transported into the B cell through a verapamil-sensitive channel (24). To some extent, even the release of glucagon evoked by a lowering of the glucose concentration may be considered a classical calcium-dependent process (see below).

Other findings from the literature could fit this scheme. For instance, the recent observation that glucagon release from the perfused canine pancreas is positively related to calcium and inhibited by verapamil was documented using a perfusate supplemented with the FGP mixture (9). Similarly, a calcium-dependent modulation of the A cell, as reported by Fujimoto and Ensinck (6), might be expected in studies dealing with islet monolayer cultures, where the culture medium is supplemented with several glucagonotropic amino acids (22).

<u>Extracellular calcium may be a prerequisite for the recognition of other environmental factors.</u> The second postulated role of calcium is based on the observation that the A cell may require a sufficient concentration of extracellular calcium in order to recognize certain agents which normally control its secretory activity. This permissive role of extracellular calcium is most evident when concentration of glucose is high, as documented by the following observations:

(i) During constant exposure to high (16.6 mM) glucose concentration, the lowering of extracellular calcium results in a sustained and reversible enhancement of glucagon release (Fig. 2). The rate of secretion reached during the period of calcium deprivation is close to that found at a low glucose level, as if the high glucose concentration were no longer recognized by the B cell (Fig. 3).

(ii) During a period of calcium deprivation, a modification of the glucose concentration does not result in any marked changes in glucagon output. Thus, when the glucose concentration is raised, during calcium deprivation, the rate of secretion remains close to its initial level, as if the new high glucose level were not sensed by the A cell (Fig. 6). Similarly, when the glucose concentration is reduced, the rate of secretion also remains close to its initial level, probably because the A cell had already been derepressed during the initial equilibration period (Fig. 7).

(iii) In the presence of glucose (16.6 mM) with either arginine or the FGP mixture, the removal of extracellular calcium again results in an increase in glucagon output. Thus, despite the expected reduction in the glucagonotropic action of either arginine or the FGP mixture, the dominant phenomenon appears to be a relief from the normal repressing action exerted by the high glucose concentration.

These observations suggest that the presence of extracellular calcium is a prerequisite for the recognition of high glucose levels by the A cell. There are a number of ways by which the absence of calcium could impede such a recognition process. Thus, an extracellular calcium depletion may alter the structural, electric and functional coupling between adjacent cells in the endocrine pancreas (19) as it does in other tissues (15), and this, in turn, may result in an impaired identification of glucose. Alternately, the transport of glucose across the B cell membrane may be a calcium-dependent process.

Two additional points must be made. The first is that the extent to which the permissive role of extracellular calcium in the identification of environmental factors is restricted to the recognition of high glucose levels remains to be determined. The second is that for this recognition, it is the extracellular rather than intracellular amount of calcium which apparently represents the key permissive factor. Indeed, a rise in glucose concentration can inhibit glucagon release in the presence of verapamil (Fig. 13), while verapamil itself exerts little effect upon the rate of glucagon release at the high glucose concentration (Fig. 5).

Calcium may inhibit glucagon release. Taking into account the postulated roles for calcium outlined in previous sections, the question remains whether, under certain experimental conditions, calcium may also exert an inhibitory role upon glucagon release.

In trying to answer such a question, we have to consider the most puzzling set of results in the present study, namely that related to the influence of calcium upon the rate of glucagon release normally seen at low glucose concentration. Is such a release process, like that evoked by arginine, a classical "calcium-dependent" phenomenon?

In a way, it may be mediated through the accumulation of calcium in the A cell leading to the subsequent activation of the effector system controlling the exocytosis of secretory granules. Indeed, the increase in glucagon output evoked by a sudden lowering of the glucose concentration is reduced when the pancreas is simultaneously exposed to verapamil. To some extent, and taking into account the previous discussion, the failure of a sudden lowering in glucose concentration to affect glucagon release at a low extracellular calcium concentration could also be mentioned in support of the idea that glucagon release evoked by cataglycemia is a classical calcium-dependent process.

However, the situation is different when the pancreas is exposed during a prolonged time to low glucose concentrations. During such a prolonged exposure, a trend towards progressive reduction of glucagon release is often observed (Fig. 3, curve No. 1; 20, 25). This trend could indicate that the postulated cellular accumulation of calcium attributable to the low glucose level may somehow exert a negative feedback upon the secretory sequence. If such were the case, one would expect that a reduction in the supply of calcium to the A cell may enhance glucagon release at least transiently, by relieving the secretory machinery from the negative feed-back control. Indeed, upon either removal of extracellular calcium or addition of verapamil, a transient increase in glucagon output is invariably observed when the pancreas had been exposed for a prolonged time to a low concentration of glucose.

A last comment seems appropriate. When the pancreas is exposed for a prolonged period of time to a low glucose concentration, either the removal of extracellular calcium or the addition of verapamil fails to cause inhibition of release, even after the initial facilitation has worn off. Therefore, if the release of glucagon attributable to a low glucose level is a classical calcium-dependent process, A cell must be less dependent on the maintenance of a normal calcium influx after such a prolonged exposure to low glucose levels than under other experimental conditions.

In concluding this discussion, we wish to underline that we have purposely avoided, or at least restricted, speculations as to the intimate mode and site of action of calcium in the A cell. Calcium could affect a number of cellular events likely to be involved in the secretory sequence. These include membrane receptors, glucose transport and metabolism as well as cyclic AMP accumulation, the activity of the microtubular-microfilamentous system and the release of different hormones (insulin, somatostatin, HPP-like substance, catecholamines) capable of affecting the A cell when released by adjacent cells in the islets. It is reasonable to assume that the complex role of calcium upon glucagon release, as documented by the present report, will be better understood when more will be known about the effect of the cation upon these various processes.

Acknowledgements

This work was supported by grants 3.4527.75 and 3.4537.76 from the FRSM (Belgium). We thank Drs. G. Wald (Bayer; Brussels, Belgium), M. A. Root (Eli Lilly Co., Indianapolis, USA) and J. Schlichtkrull (Novo, Bagsvaerd, Denmark) for the generous gifts of Trasylol, glucagon and insulin used in these studies.

REFERENCES

1. Ashby, J. P. and Speake, R. N. (1975): Insulin and glucagon secretion from isolated islets of Langerhans. The effects of calcium ionophores. Biochem. J. 150, 89-96.

2. Devis, G., Somers, G., Van Obberghen, E. and Malaisse, W. J. (1975): Calcium antagonists and islet function. I. Inhibition of insulin release by verapamil. Diabetes 24, 547-551.

3. Edwards, J. C. (1973): A-cell metabolism and glucagon secretion. Postgr. Med. J. 49, 611-615.

4. Edwards, J. C. and Howell, S. L. (1973): Effects of vinblastine and colchicine on the secretion of glucagon from isolated guinea-pig islets of Langerhans. FEBS Letters 30, 89-92.

5. Fleckenstein, A. (1971): Specific inhibitors and promoters of calcium action in the excitation-contraction coupling of heart muscle and their role in the prevention or production of myocardial lesions. In: Calcium and the Heart. P. Harris and L. Opie, Eds., Academic Press, New York, pp. 135-188.

6. Fujimoto, W. Y. and Ensinck, J. W. (1976): Somatostatin inhibition of insulin and glucagon secretion in rat islet culture: reversal by ionophore A23187. Endocrinology 98, 259-262.

7. Gerich, J. E., Charles, M. A. and Grodsky, G. M. (1974): Characterization of the effects of arginine and glucose on glucagon and insulin release from the perfused rat pancreas. J. Clin. Invest. 54, 833-841.

8. Gerich, J. E., Frankel, B. J., Fanska, R., West, L., Forsham, P. H. and Grodsky, G. M. (1974): Calcium dependency of glucagon secretion from the in vitro perfused rat pancreas. Endocrinology 94, 1381-1385.

9. Hermansen, K. and Iversen, J. (1976): Calcium, glucose and glucagon release. Diabetologia 12, 398.

10. Leclercq-Meyer, V. and Marchand, J. (1976): Differential need for an inward calcium transport in the inhibition by glucose and stimulation by arginine of glucagon release. Studies with verapamil. Diabetologia 12, 405.

11. Leclercq-Meyer, V., Marchand, J., Leclercq, R. and Malaisse, W. J. (1976): Glucagon and insulin release by the in vitro perfused rat pancreas. Influence of the colloid composition of the perfusate. Diàbete Métab. 2, 57-65.

12. Leclercq-Meyer, V., Marchand, J. and Malaisse, W. J. (1973): The effect of calcium and magnesium on glucagon secretion. Endocrinology 93, 1360-1370.

13. Leclercq-Meyer, V., Marchand, J. and Malaisse, W. J. (1976): The role of calcium in glucagon release. Interactions between glucose and calcium. Diabetologia 12, 531-538.

14. Leclercq-Meyer, V., Rebolledo, O., Marchand, J. and Malaisse, W. J. (1975): Glucagon release: paradoxical stimulation during calcium deprivation. Science 189, 897-899.

15. Loewenstein, W. R. (1966): Permeability of membrane junctions. Ann. N. Y. Acad. Sci. 137, 441-472.

16. Lundquist, I., Fanska, R. and Grodsky, G. M. (1976): Direct calcium-stimulated release of glucagon from the isolated perfused rat pancreas and the effect of chemical sympathectomy. Endocrinology 98, 815-818.

17. Malaisse, W. J., Herchuelz, A., Levy, J., Somers, G., Devis, G., Ravazzola, M., Malaisse-Lagae, F. and Orci, L. (1975): Insulin release and the movements of calcium in pancreatic islets. In: Calcium Transport in Contraction and Secretion. E. Carafoli, F. Clementi, W. Drabikowski and A. Margreth, Eds. North Holland Publishing Company, Amsterdam, pp. 211-226.

18. Ohneda, A., Matsuda, K., Horigome, K., Ishii, S. and Yamagata, S. (1974): Effect of intrapancreatic administration of calcium upon glucagon secretion in dogs. Tohoku J. Exp. Med. 113, 301-311.

19. Orci, L., Blondel, B., Malaisse-Lagae, F., Ravazzola, M., Wollheim, C. and Renold, A. E. (1974): Cell motility and insulin release in monolayer cultures of endocrine pancreas. Diabetologia 10, 382.

20. Pagliara, A. S., Stillings, S. N., Hover, B., Martin, D. M. and Matschinsky, F. M. (1974): Glucose modulation of amino acid-induced glucagon and insulin release in the isolated perfused rat pancreas. J. Clin. Invest. 54, 819-832.

21. Ravazzola, M., Malaisse-Lagae, F., Amherdt, M., Perrelet, A., Malaisse, W. J. and Orci, L. (1976): Patterns of calcium localization in pancreatic endocrine cells. J. Cell Sci. 27, 107-117.

22. Rocha, D. M., Faloona, G. R. and Unger, R. H. (1972): Glucagon-stimulating activity of 20 amino acids in dogs. J. Clin. Invest. 51, 2346-2351.

23. Somers, G., Devis, G., Van Obberghen, E. and Malaisse, W. J. (1976): Calcium antagonists and islet function. II. Interaction of theophylline and verapamil. Endocrinology 99, 114-124.

24. Somers, G., Devis, G., Van Obberghen, E. and Malaisse, W. J. (1976): Calcium antagonists and islet function. VI. Effect of barium. Pflügers Arch. 365, 21-28.

25. Weir, G. C., Knowlton, S. D. and Martin, D. B. (1974): Glucagon secretion from the perfused rat pancreas. Studies with glucose and catecholamines. J. Clin. Invest. 54, 1403-1412.

26. Wollheim, C. B., Blondel, B., Renold, A. E. and Sharp, G. W. G. (1976): Stimulatory and inhibitory effects of cyclic AMP on pancreatic glucagon release from monolayer cultures and the controlling role of calcium. Diabetologia 12, 269-277.

27. Wollheim, C. B., Blondel, B., Renold, A. E. and Sharp, G. W. G. (1976): Calcium induced glucagon release in monolayer cultures of the endocrine pancreas. Studies with ionophore A23187. Diabetologia 12, 287-294.

INTERRELATIONSHIP OF CALCIUM AND SOMATOSTATIN IN THE SECRETION OF INSULIN AND GLUCAGON

G. M. Grodsky, I. Lundquist, R. Fanska and R. Pictet

Metabolic Research Unit and Department of Biochemistry and Biophysics,
University of California
San Francisco, California USA

Using the isolated perfused rat pancreas, previous experiments suggesting that glucagon secretion is calcium dependent were expanded. Spontaneous glucagon secretion was inhibited when residual calcium in the system was removed by EGTA. However, glucose at various concentrations poorly inhibited glucagon secretion unless calcium was present. Thus, both stimulation of glucagon release and its inhibition by glucose are calcium requiring. The overall effect of calcium, therefore, depends on the relative concentration of secretagogues vs. glucose. In addition, a mixed stimulation/inhibition at higher levels of calcium was noted.

The effect of somatostatin on insulin and glucagon secretion was evaluated in experiments in which calcium was perfused at various concentrations as single steps or staircases. Although increasing extracellular calcium increased insulin and glucagon secretion in the presence or absence of somatostatin, percent inhibition by somatostatin was not affected. Thus, somatostatin's action is relatively independent of extracellular calcium. In the perifused islet system, somatostatin inhibited ^{45}calcium efflux--suggesting that this inhibitor does influence intracellular calcium distribution. Comparison with the effects of dinitrophenol indicated that somatostatin does not inhibit by blocking cellular adenosine triphosphate production.

Since somatostatin inhibits various endocrine systems, it probably acts on a mechanism common to all. It has been reported that calcium, in part, reverses the effects of somatostatin on insulin release (2, 6), suggesting that somatostatin may act at the level of calcium metabolism. However, these studies did not establish the existence of a direct quantitative relationship between somatostatin and extracellular calcium. A similar mechanism for the inhibition of glucagon release by somatostatin requires the unequivocal demonstration that A cell activity is calcium dependent. In the isolated perfused rat pancreas we have shown that amino acid-stimulated glucagon secretion requires calcium and that calcium can stimulate glucagon secretion directly even in the absence of other secretagogues (7, 13).

However, apparently conflicting results have been reported by others who have observed that glucagon secretion is augmented during calcium deprivation (11, 12). Our current studies (14), summarized here, indicate that in addition to promoting glucagon release, calcium may act as a positive modulator of the inhibiting effect of glucose. Such a mixed action of calcium could explain the contradictory results in the literature, while supporting the conclusion that the A cell, like most endocrine cells, has an absolute calcium requirement.

The present studies indicate that changes of extracellular calcium ion have little influence on the effectiveness of somatostatin as an inhibitor of insulin and glucagon secretion. Direct measurements of ^{45}calcium efflux from perifused islets indicate that the polypeptide might act, at least partially, on intracellular calcium translocation.

MATERIALS AND METHODS

The methods for isolation and perfusion of the rat pancreas have been described in detail (9, 10). To minimize the contribution of glucagon secretion from the stomach, a modified preparation without the stomach and spleen and including only a part of the duodenum was used (7). Perfusion medium was a calcium-depleted Krebs-Ringer buffer containing 1% human serum albumin and 3% dextran, but no added calcium (total calcium by atomic absorption = 0.2 mEq/L). The flow rate was 4 ml/min. The 4 ml samples were collected in iced tubes containing 12 mg EDTA.

In all experiments, perfusion was begun at -30 min with calcium-depleted perfusate, either alone or containing glucose or arginine and glucose. When used, somatostatin*, at various concentrations, was introduced at -5 min and perfused continuously. Calcium, ranging from 1 to 15 mEq/L, was added at min 0 and perfused continuously or in a series of increasing steps.

---

*Kindly supplied by Dr. Roger Guillemin of the Salk Institute, San Diego, California USA.

For the ^{45}calcium efflux experiments, 100 rat islets, isolated by collagenase, were imbedded on thin solid agar films, which were then cultured for five days in medium M199 containing 20% fetal calf serum, antibiotics and mycostatin, as described (16). ^{45}Calcium (5 mCi/ml) was added 24 hr before use. Islets were removed manually from the agar and perifused by the Lacy technique (1). Control experiments with ^{3}H-sucrose showed that the washout of ^{45}calcium into the perfusate was complete by 20 min. Somatostatin was added during the second phase of glucose-stimulated insulin release or in the absence of glucose.

RESULTS

Figures 1 - 3 are part of a more detailed study (14). In the absence of calcium or glucose from the beginning of a perfusion experiment, glucagon in the effluent was elevated but then slowly decreased with time (Fig. 1). When glucose was present from the beginning of the perfusion, glucagon secretion was partially suppressed. Inhibition was incomplete, since the concentrations of glucose used (8.3 and 27.8 mM) would, in the presence of normal calcium levels, cause suppression to baseline within 5 min (13). At no time in the experiment did glucose, even at a 27.8 mM concentration, cause a paradoxical increase of glucagon secretion (12). A small increase of insulin release occurred with 27.8 mM glucose. The right panel of Figure 1 shows the results obtained when 0.2 mM EGTA was added to the perfusing fluid to minimize the effect of residual calcium in the medium and tissues. Glucagon secretion was already inhibited in the first sample (-15 min) and remained undetectable in the presence or absence of glucose. Insulin release at a 27.8 mM glucose concentration was also inhibited. All subsequent experiments were performed without EGTA because detectable hormone release was required.

As we previously reported (13, 14), calcium (0.5 mEq/L) alone caused a large transient release of glucagon followed by weaker responses at 1 mEq/L (Fig. 2). Relative insensitivity of the pancreas was noted at 3 mEq/L calcium

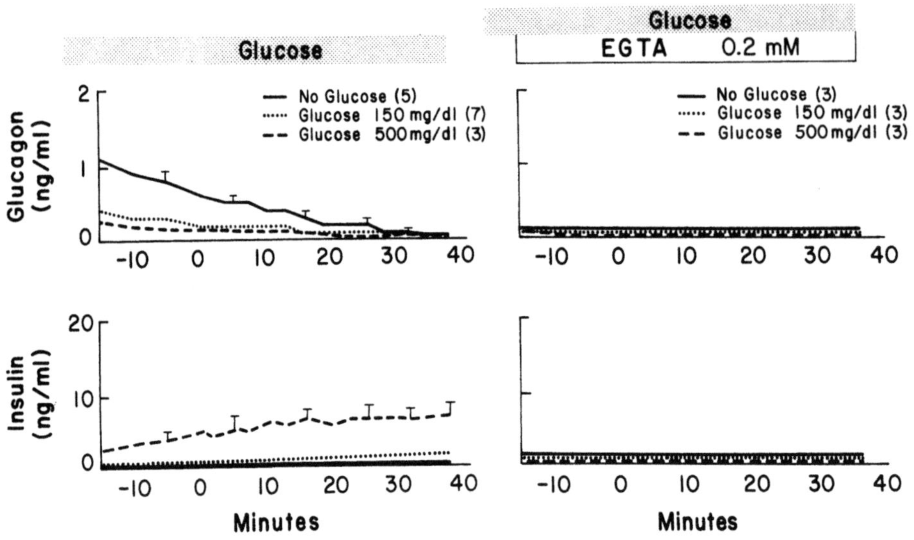

Fig. 1. Effect of glucose on insulin and glucagon secretion without added calcium. Experiments were begun at -30 min. Glucose (8.3 and 27.8 mM) caused significant suppression of glucagon secretion (p 0.05) at all points until min 5. Ave. ± SEM (14, with permission).

(to be discussed later), whereas further calcium steps again augmented release. As seen in the experiments containing glucose (8.3 mM) inhibition of glucagon secretion by glucose was rapidly enhanced by the addition of calcium, producing an effect directly opposite to the increased release caused by calcium alone (Fig. 2). Glucagon was reduced to undetectable levels within 2 min in all glucose experiments and remained suppressed at all subsequent calcium steps. Predictably, insulin secretion also occurred during the glucose/calcium-induced glucagon inhibition.

Because of the relative insensitivity of the A cell during the 3 mEq/L calcium step (Fig. 2), ancillary experiments were performed (Fig. 3). In the absence of glucose, 3 mEq/L calcium was added at the same time (min 14) as in Fig. 2, but without the previous calcium steps. A mixed action of calcium on glucagon release was demonstrated as indicated by a prompt inhibition followed by a large rise. We have not noted this initial inhibition with lower concentration of calcium

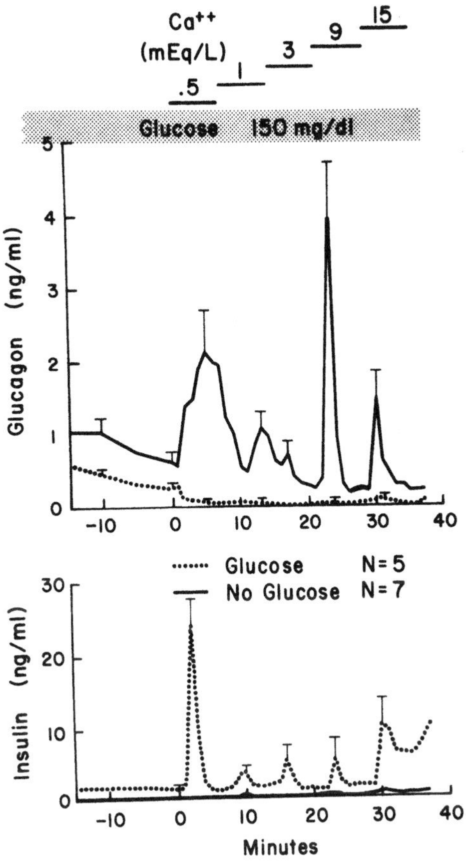

Fig. 2. Effect of increasing steps of calcium on glucagon and insulin secretion with (8.3 mM) and without glucose. At min 2, after addition of calcium, glucagon was undetectable. Ave. ± SEM (14, with permission).

(0.5 mEq/L), whether added initially at min 0 (Fig. 2) or at min 14 (not shown). The secondary rise in glucagon-secretion was more characteristic of that caused by calcium added as an initial stimulus, even though at a lower concentration (0.5 mEq/L), than of that elicited by 3 mEq/L added as a subsequent increment (Fig. 2). This pattern of hormone secretion is similar to that caused by glucosamine and suggests the occurrence of a mixed, overlapping stimulation and inhibition.

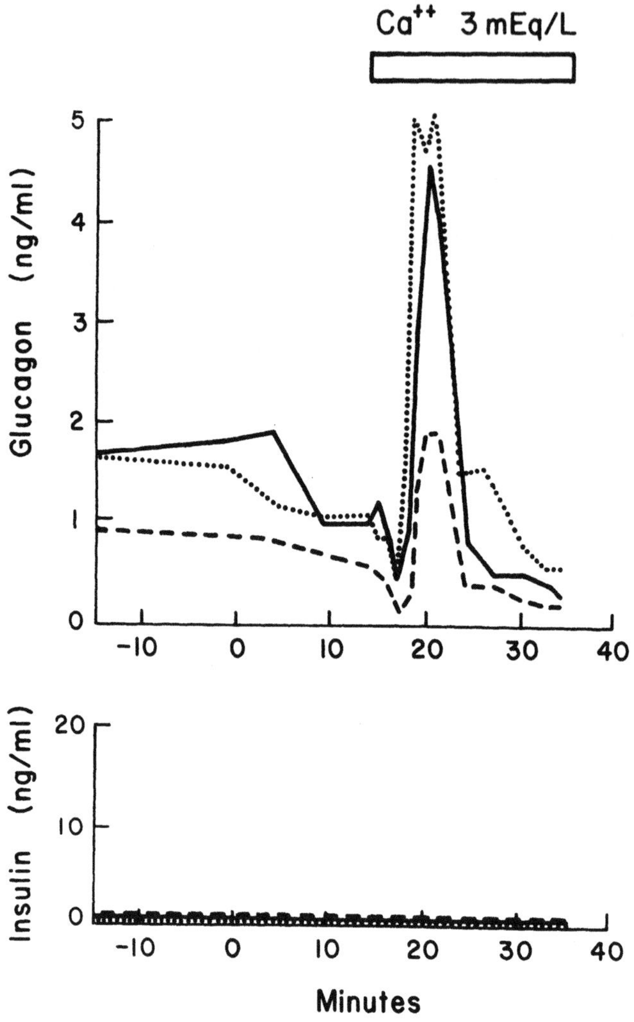

Fig. 3. Effect of calcium (3 mEq/L) on insulin and glucagon secretion. Experiments were begun at -30 min. Calcium was added as a single step at min 14, corresponding to the time of addition of this calcium concentration in Fig. 2. Each line is a separate experiment (14, with permission).

Figures 4 - 6 show the interrelationship between calcium and somatostatin on insulin and glucagon release during constant perfusion with different agents. To minimize variation, the four different experimental conditions for each study, i.e. calcium concentrations of 1 and 9 mEq/L with and without somatostatin (50 ng/ml), were performed on the same day and the hormone secretion patterns measured together on a subsequent day. Mean values for all experiments are given in the figures and the percentage of inhibition, calculated from paired data for each day, is shown in Table 1.

TABLE 1

PERCENT INHIBITION OF INSULIN AND GLUCAGON SECRETION BY SOMATOSTATIN*

Additional agents	Inhibition of release (%)** First phase	Second phase
	Insulin	
Glucose (16.6 mM) +		
Calcium (1 mEq/L)	59.3 ± 5.9	27.4 ± 9.7
Calcium (9 mEq/L)	69.5 ± 5.3	23.8 ± 10.8
Glucose (8.3 mM) +		
Calcium (1 mEq/L)	88.5 ± 5.9	87.8 ± 9.6
Calcium (9 mEq/L)	79.5 ± 8.5	81.3 ± 6.3
Arginine (19.2 mM) + Glucose (5.5 mM) +		
Calcium (1 mEq/L)	55.8 ± 9.6	51.8 ± 2.9
Calcium (9 mEq/L)	52.8 ± 4.3	39.7 ± 8.0
	Glucagon	
Arginine (19.2 mM) + Glucose (5.5 mM) +		
Calcium (1 mEq/L)	88.4 ± 2.5	90.7 ± 3.6
Calcium (9 mEq/L)	54.7 ± 7.1	75.7 ± 4.5

*Data (paired for a given day) taken from the experiments shown in Figs. 4-6.

**Mean ± S.E.M.

Figure 4 shows results of experiments using near-maximal concentrations of glucose (16.6 mM). In the absence of added calcium, but with some calcium remaining in the tissues (Fig. 1), somatostatin caused a prompt inhibition of glucose-stimulated insulin release. The addition of calcium caused a biphasic insulin release. Raising calcium from 1 to 9 mEq/L had little effect on the first phase of insulin release but consistently caused an increase in the second phase. In these high glucose experiments, somatostatin inhibited both phases of insulin release but had a greater inhibitory effect on the first phase. If the second phase of insulin release in the presence of somatostatin and of calcium (1 or 9 mEq/L) is compared with the "control" experiment in which calcium (1 mEq/L) alone was used, increasing the concentration of calcium appears to overcome somatostatin's inhibition (the rates of secretion by 30 min with 1 mEq/L calcium were indistinguishable from those with 9 mEq/L calcium and somatostatin). However, when somatostatin experiments are compared with their respective controls, the percent inhibition of both phases of insulin release by somatostatin does not appear to have been affected by extracellular calcium (Table 1).

Results of similar experiments using a half-maximal glucose concentration (8.3 mM) are summarized in Table 1. In contrast to the high glucose experiments, somatostatin now inhibited both phases equally and more effectively. Again, the percentage of inhibition by somatostatin of both phases of insulin release was not affected by extracellular calcium (Table 1).

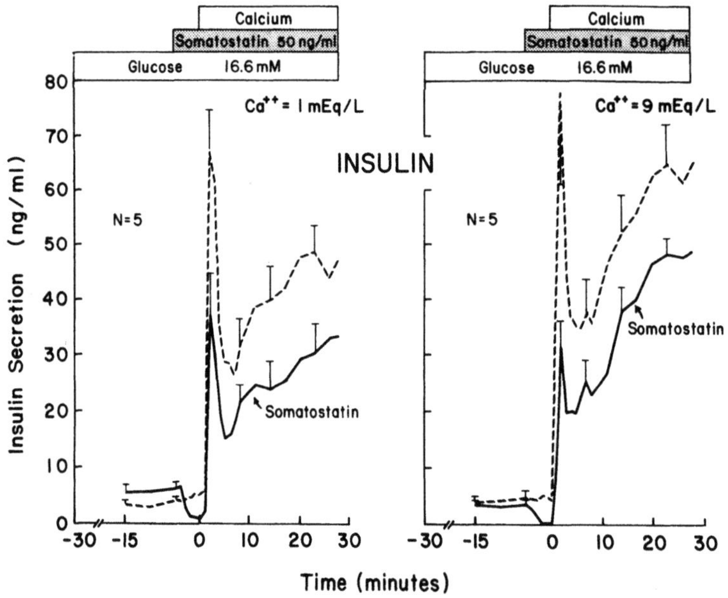

Fig. 4. Interrelationship of calcium and somatostatin on insulin secretion during stimulation with high glucose. Curves represent mean data. Paired results are summarized in Table 1.

With the moderate stimulus, arginine and glucose (Fig. 5), somatostatin inhibited both phases of insulin release equally, and gave results similar to those obtained using 8.3 mM glucose. However, somatostatin more effectively inhibited the action of combined arginine and glucose than that of glucose alone. Once again, the percent inhibition by somatostatin of both phases of insulin release was not affected by extracellular calcium (Table 1).

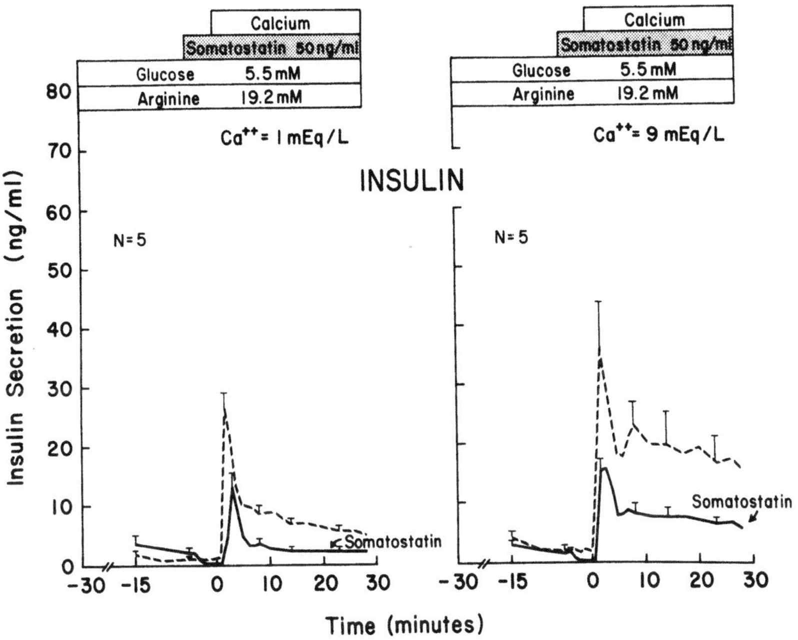

Fig. 5. Interrelationship of calcium and somatostatin during arginine/glucose-stimulated insulin secretion. Curves represent mean data. Paired results are summarized in Table 1.

Figure 6 shows the glucagon secretion during the arginine/glucose perfusion. As reported previously (7), somatostatin inhibits the secretion of glucagon more effectively than that of insulin in this system. In almost all experiments, calcium (9 mEq/L) caused multiple spikes (oscillations), possibly a reflection of the mixed stimulation and inhibition of glucagon release at high calcium concentrations. The percent inhibition by somatostatin at high calcium (9 mEq/L) was significantly less than at low calcium (1 mEq/L), indicating some reversal of somatostatin's inhibitory effect by calcium (Table 1). However, the difference was small and normalization was not achieved.

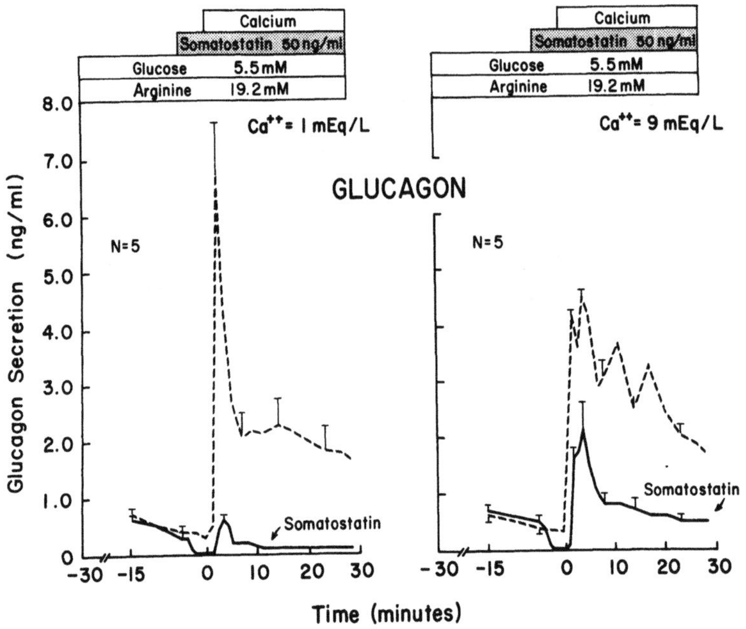

Fig. 6. Interrelationship of calcium and somatostatin during arginine/glucose-stimulated glucagon secretion. Curves are mean data obtained from the experiments shown in Fig. 5. Paired results are summarized in Table 1.

Experiments illustrated in Figs. 7 and 8 were carried out using the same arginine/glucose stimulus but several various additional calcium steps and somatostatin concentrations (5 and 200 µg/ml) straddling those used in the previous experiments. When compared with the results obtained without somatostatin, it was apparent that the percent inhibition of insulin or glucagon release by different concentrations of somatostatin was not affected by changing extracellular calcium. It is emphasized that at very high calcium (15 mM) glucagon secretion was inhibited.

<u>^{45}Calcium efflux.</u> As reported by others (5, 17), we found that somatostatin had little effect on insulin secretion from freshly isolated islets. When 5-day cultures were used, the response to somatostatin increased, although not consis-

tently and the concentrations required to inhibit were higher than those for the perfused pancreas.

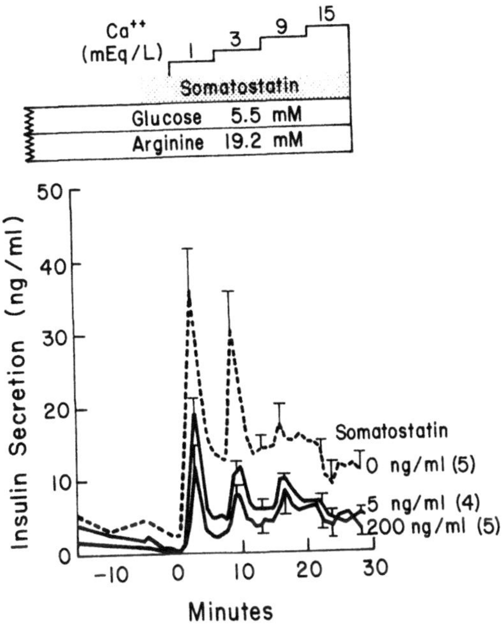

Fig. 7. Interrelationship of calcium, added in increasing steps, and of somatostatin on arginine/glucose-stimulated insulin release. Results with somatostatin were significantly ($p < 0.05$) below control values at all times after min 1.

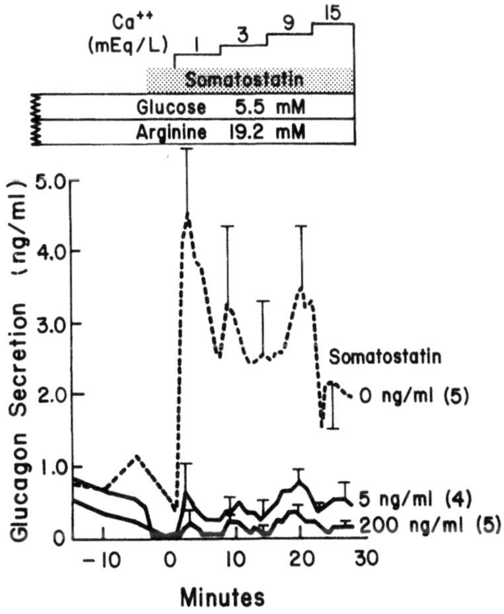

Fig. 8. Interrelationship of calcium, added in increasing steps, and of somatostatin on arginine/glucose-stimulated glucagon release. Results with somatostatin were significantly ($p < 0.01$) below control values at all times after min 1.

Figure 9 gives the ^{45}calcium efflux in those perifused islets where inhibition of the second phase of glucose-stimulated insulin release by somatostatin could be demonstrated. Inhibition of ^{45}calcium efflux occurred within 2 min after administration of somatostatin, corresponding to the decreased insulin release. Both ^{45}calcium efflux and insulin release returned to normal promptly after discontinuing the somatostatin.

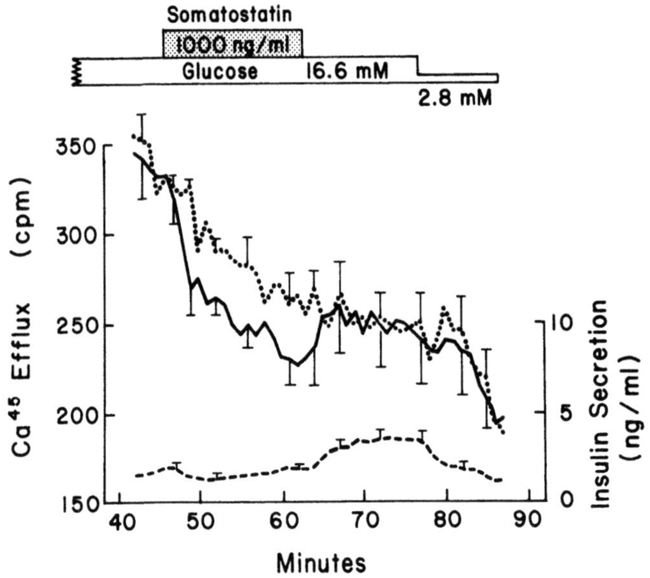

Fig. 9. Effect of somatostatin on ^{45}calcium efflux during glucose stimulation of cultured islets. Glucose was maintained at 16.6 mM from min 32. ^{45}Calcium efflux (control [N = 4] normalized to experimental data at min 47) = ········; ^{45}calcium efflux (+ somatostatin [N = 4]) = ______; insulin secretion (+ somatostatin [N = 4]) = -------.

Figure 10 shows that somatostatin still caused a prompt decrease in ^{45}calcium efflux in the absence of glucose or of changing insulin levels. In contrast, dinitrophenol caused a prompt increase in ^{45}calcium efflux.

DISCUSSION

The present studies provide further evidence that the A cell requires calcium. As with insulin (4, 8), the addition of calcium either in single steps or in mul-

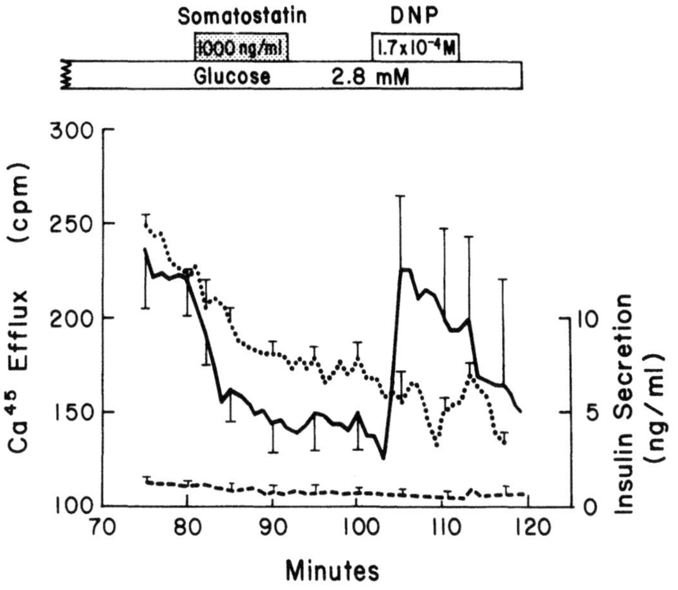

Fig. 10. Comparative effects of somatostatin and dinitrophenol on ^{45}calcium efflux from unstimulated cultured islets. In these experiments designed as controls for another study, the high glucose concentration (16.6 mM) was maintained from min 25 to min 65 and somatostatin was added at min 81. ^{45}Calcium efflux (control normalized to experimental data at min. 80) = ...; ^{45}calcium efflux (+ somatostatin and dinitrophenol = ———; insulin secretion (+ somatostatin and dinitrophenol) = -----. Each point represents 3 measurements.

tiple ascending steps, either alone or during arginine/glucose perfusion, always resulted in positive secretion of glucagon consistent with that previously observed (7-14). In addition, removal of residual calcium by EGTA inhibited spontaneous glucagon secretion.

However, the calcium ion has multiple effects on glucagon secretion: 1) its stimulation involves at least two processes that differ in their sensitivity to neural input (13); 2) in special conditions, elevated calcium may inhibit glucagon release (Figs. 3 and 8 and Ref. 18); and 3) calcium is required not only for A cell stimulation, but also for A cell inhibition by glucose. Whether this inhibition is the result of a direct effect of glucose/calcium on the A cell or of the concomitant increase in insulin release has been discussed elsewhere (14).

Initial studies showing that the inhibition of insulin release by somatostatin could be normalized by increased calcium (2) did not adequately recognize the effect of calcium alone. Using the same high glucose concentration (Fig. 4), we found that perfusing with calcium (at 9 mEq/L) "normalized" somatostatin's inhibition of the second phase, that is, it restored the secretion to values close to those obtained without somatostatin at lower calcium concentration. However, in all experiments, with high or low glucose or with arginine and glucose, the percent inhibition of insulin release was the same at high or low calcium, whether the ion was added as a single or repeated steps. Thus, the action of somatostatin on insulin secretion is relatively independent of extracellular calcium concentration, suggesting, although not proving, that its major action may not be related to uptake of extracellular calcium.

The inhibition of glucagon secretion by somatostatin also seems to be independent of extracellular calcium. Although increased calcium caused a significant decrease in percent inhibition in single-step experiments, the difference was small. Furthermore, the percent inhibition of glucagon secretion was constant at all calcium concentrations in the multiple-step experiments.

Our studies showing that somatostatin preferentially inhibits the first-phase of insulin release induced by glucose stimulation, agree with those of others (3). However, this is not a consistent characteristic of somatostatin's action, since both phases are equally inhibited at lower glucose concentrations or when arginine/glucose is the signal. This may explain why a preferential sensitivity is not consistently reported (5).

^{45}Calcium efflux. Somatostatin inhibited ^{45}calcium efflux from glucose-stimulated perifused cultured islets concurrently with its prompt inhibition of insulin release. If part of the calcium efflux represents calcium associated with insulin secretory granules (15) one would expect inhibition of 45calcium efflux by any inhibitor of insulin release. However, somatostatin also inhibited efflux in the absence of stimulatory levels of glucose when insulin release was marginal. Thus, somatostatin does influence calcium metabolism, although this effect may not be the direct or primary action of somatostatin on hormone release. Current concepts suggest that calcium is concentrated in a specific small cellular compartment that positively modulates insulin release. Since somatostatin is an inhibitor, the observed decreased efflux may indicate an action of somatostatin to increase calcium storage in islet organelles, thereby reducing its availability to the active compartment.

Since secretion mechanisms require chemical energy, the ubiquitous inhibitory action of somatostatin could be to block energy production. However, dinitrophenol caused increased calcium efflux from the cultured islets; thus, it is unlikely that somatostatin acts to inhibit hormone release by uncoupling adenosine triphosphate production.

Acknowledgements

Supported in part by Grant AM-01410 from the National Institute of Arthritis, Metabolism and Digestive Diseases, by the Kroc Foundation of Santa Ynez, California, the Juvenile Diabetes Foundation of New York City, and the American Diabetes Association. During the course of these investigations, Dr. Lundquist was supported by the Swedish Medical Research Council (04P-4289).

We are indebted to Florence Schmid, Michele Manning and Mary Ann Jones for technical help.

REFERENCES

1. Charles, M. A., Lawecki, J., Pictet, R. and Grodsky, G. M. (1975): Insulin secretion: interrelationships of glucose, cyclic adenosine 3',5'-monophosphate, and calcium. J. Biol. Chem. 250, 6134-6140.

2. Curry, D. L. and Bennett, L. L. (1974): Reversal of somatostatin inhibition of insulin secretion by calcium. Biochem. Biophys. Res. Commun. 60, 1015-1019.

3. Curry, D. L. and Bennett, L. L. (1976): Does somatostatin inhibition of insulin secretion involve two mechanisms of action? Proc. Natl. Acad. Sci. USA 73, 248-251.

4. Curry, D. L., Bennett, L. L. and Grodsky, G. M. (1968): Dynamics of insulin secretion by the perfused rat pancreas. Endocrinology 83, 572-584.

5. Efendic, S., Luft, R. and Grill, V. (1974): Effect of somatostatin on glucose induced insulin release in isolated perfused rat pancreas and isolated rat pancreatic islets. FEBS Lett. 42, 169-172.

6. Fujimoto, W. Y. and Ensinck, J. W. (1976): Somatostatin inhibition of insulin and glucagon secretion in rat islet culture: Reversal by ionophore

A23187. Endocrinology 98, 259-262.

7. Gerich, J. E., Frankel, B. J., Fanska, R., West, L., Forsham, P. H. and Grodsky, G. M. (1974): Calcium dependency of glucagon secretion from the in vitro perfused rat pancreas. Endocrinology 94, 1381-1385.

8. Grodsky, G. M. (1972): A threshold distribution hypothesis for packet storage of insulin. II. Effect of calcium. Diabetes 21 (Suppl. 2), 584 (abst.).

9. Grodsky, G. M., Batts, A. A., Bennett, L. L., Vcella, C., McWilliams, N. B. and Smith, D. F. (1963): Effects of carbohydrates on secretion of insulin from isolated rat pancreas. Am. J. Physiol. 205, 638-644.

10. Grodsky, G. M. and Fanska, R. (1975): The in vitro perfused pancreas. In: Methods in Enzymology, Vol. 39: Hormone Action, Part D: Isolated Cells, Tissues, and Organ Systems. J. G. Hardman and B. W. O'Malley, Eds. Academic Press, New York, pp. 364-372.

11. Leclercq-Meyer, V., Marchand, J. and Malaisse, W. J. (1973): The effect of calcium and magnesium on glucagon secretion. Endocrinology 93, 1360-1370.

12. Leclercq-Meyer, V., Rebolledo, O., Marchand, J. and Malaisse, W. J. (1975): Glucagon release: Paradoxical stimulation by glucose during calcium deprivation. Science 189, 897-899.

13. Lundquist, I., Fanska, R. and Grodsky, G. M. (1976): Direct calcium-stimulated release of glucagon from the isolated perfused rat pancreas and the effect of chemical sympathectomy. Endocrinology 98, 815-818.

14. Lundquist, I., Fanska, R. and Grodsky, G. M. (1976): Interaction of calcium and glucose on glucagon secretion. Endocrinology 99, 1304-1312.

15. Malaisse, W. J. (1973): Insulin secretion: Multifactorial regulation for a single process of release. Diabetologia 9, 167-173.

16. Pictet, R. and Forsham, P. H. (1975): Long-term islet culture. Diabetes 24, (Suppl. 2), 421 (abst.).

17. Turcot-Lemay, L., Lemay, A. and Lacy, P. E. (1975): Somatostatin inhibition of insulin release from freshly isolated and organ cultured rat islets of Langerhans in vitro. Biochem. Biophys. Res. Commun. 63, 1130-1138.

18. Wollheim, C. B., Blondel, B., Renold, A. E. and Sharp, G. W. G. (1976): Calcium induced glucagon release in monolayer culture of the endocrine pancreas. Studies with ionophore A23187. Diabetologia 12, 287-294.

DYNAMICS AND INTERRELATIONSHIPS OF IN VITRO SECRETION OF GLUCAGON AND INSULIN

S. Pek and T.-Y. Tai

Department of Internal Medicine, Division of Endocrinology and Metabolism
and The Metabolism Research Unit
The University of Michigan
Ann Arbor, Michigan USA

The dynamics as well as the magnitudes of the multiple phases of secretion of glucagon and insulin, studied in the perfused rat pancreas, failed to provide evidence for the existence of feedback relationships between the two pancreatic islet hormones. On the other hand, in response to common secretagogues, the release of glucagon, which consistently precedes that of insulin, may have a modifying influence on the function of the B cell.

Administered glucagon stimulates the release of insulin *in vivo* (3, 9) and *in vitro* (2, 10). An inhibitory effect of exogenous insulin on *in vivo* and *in vitro* release of glucagon has been documented (4, 5). The existence of a feedback regulatory system between endogenous glucagon and insulin remains to be established. We wish to review the results of some of our studies with the isolated, perfused rat pancreas in search for a relationship between dynamics of secretion of glucagon and that of insulin.

Pancreata were removed from overnight-fasted, male Sprague-Dawley rats which weighed 250-300 grams (8). The organ was perfused at 37°C via the aortic cannula with Krebs-Ringer bicarbonate buffer containing dextran (40 g/l) and bovine serum albumin (2 g/l). The perfusion flow rate was 2.5 ml/min. Portal venous effluent was collected in 12-60 second fractions. The effluent levels of glucagon and insulin were measured by radioimmunoassay.

The administration of L-arginine over a 15-minute period evoked a clearly biphasic release of both glucagon and insulin (Fig. 1). The phase-1 as well as phase-2 of glucagon release started 1-2 minutes earlier than the corresponding phases of insulin release. Thus, the time sequence and the concordance of the biphasic responses seen in these experiments would be consistent with the hypothesis that the secretion of glucagon promotes that of insulin.

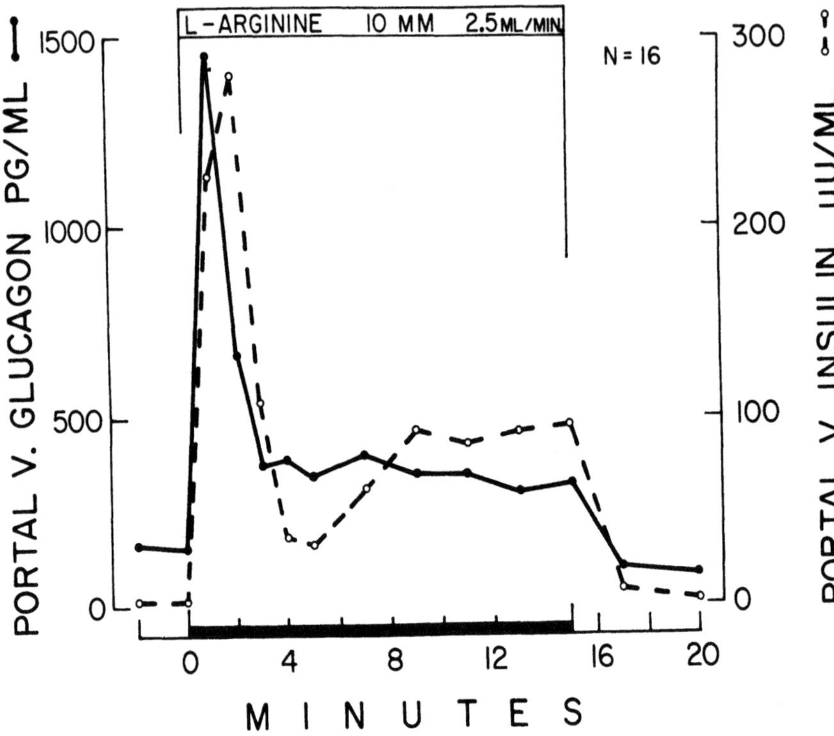

Fig. 1. Mean portal venous effluent levels of immunoreactive glucagon (solid line) and insulin (interrupted line) in response to 15 minute perfusions of 10 mM L-arginine in the presence of 5.6 mM D-glucose.

We have reported (8) that prostaglandin E-2 is a potent secretagogue of both islet hormones (Fig. 2). These experiments exemplify the fact that a major phase-2 secretion of glucagon in response to a secretagogue common to both islet hormones need not be accompanied by a phase-2 release of insulin.

In the isolated, perfused rat pancreas, L-leucine is a secretagogue common to both islet hormones, which is contrary to our earlier observations in man and dog that this amino acid stimulates insulin, but not glucagon release (1, 6). As shown in Fig. 3, leucine evoked a rapid phase-1 secretion of glucagon, followed by that of insulin. Yet, despite the fact that the perfusate concentration of glucose was low (a condition which usually augments glucagon release), in this set of experiments a phase-2 release occurred only for insulin.

These latter experiments with prostaglandin E-2 and leucine demonstrate that in response to secretagogues common to both hormones, phase-2 release of glucagon and insulin may be discordant, despite a concordant phase-1 release.

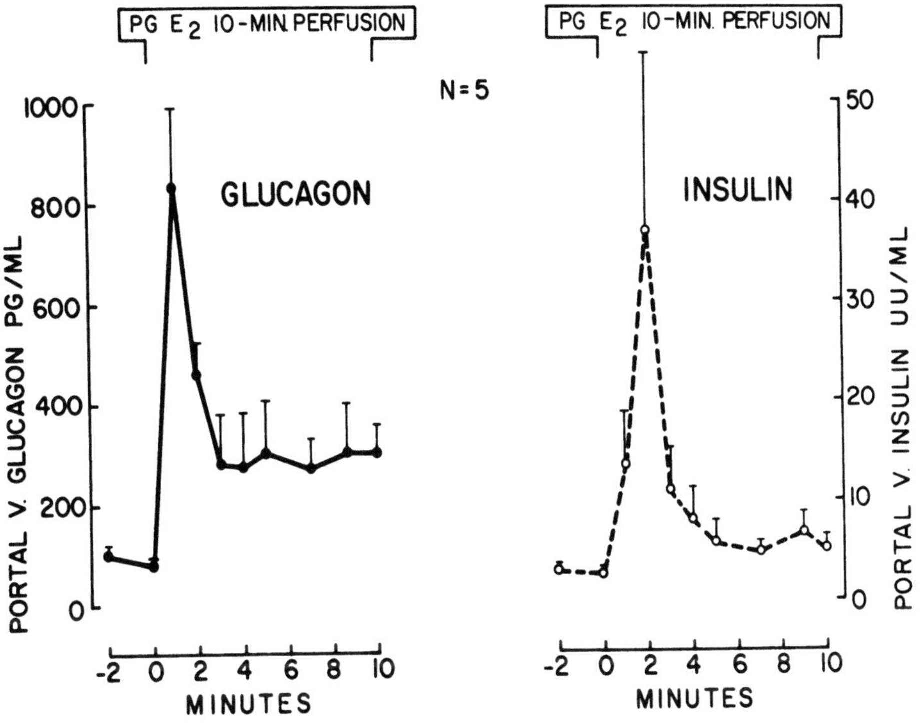

Fig. 2. Effluent levels of glucagon and insulin with 10 minute perfusions of 1.4 M prostaglandin E-2 in the presence of 5.6 mM D-glucose. Mean ± S. E. (8, with permission).

In response to compounds which stimulate the secretion of glucagon and insulin, the acute phase (phase-1) release occurs consistently for both hormones (7). We have analyzed the time course of this acute phase of hormone secretion in terms of a) the time of the first significant change from basal levels, and b) the time of "half-maximal" hormone concentrations calculated from a function derived from normalized response areas. Arginine, leucine, prostaglandin E-2, prostaglandin F-2-alpha, bovine growth hormone or isoproterenol were perfused over 60 seconds and the portal venous effluent was sampled at 12 second intervals over 120 seconds. As shown in Fig. 4 a-f, when the means of partial sums of response areas for each 12 second interval were plotted, a clear pattern of hor-

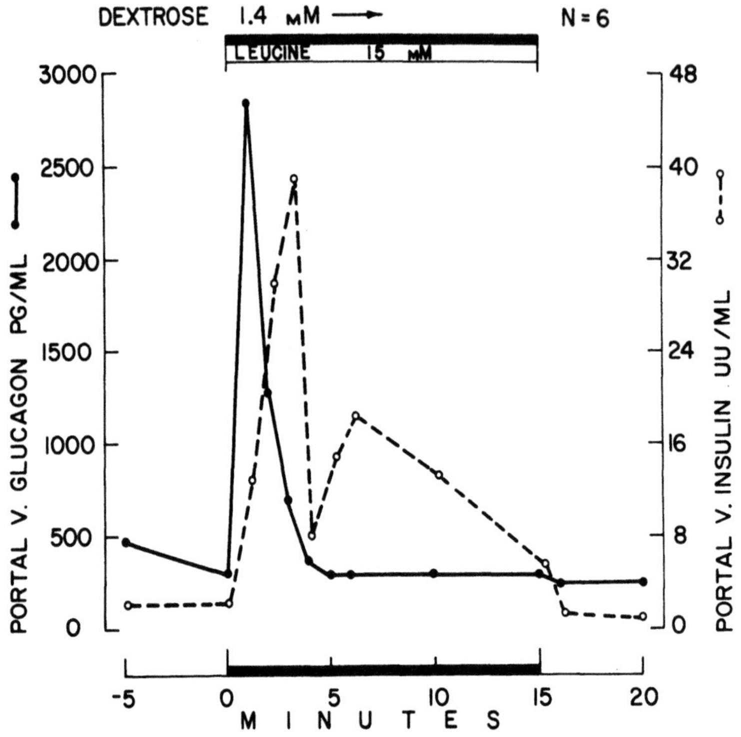

Fig. 3. Mean effluent levels of glucagon and insulin with 15 minute perfusions of 15 mM L-leucine in the presence of 1.4 mM D-glucose.

mone secretion emerged: during the entire period of observation, in response to all secretagogues tested, glucagon release preceded insulin release. These data would be consistent with the proposed role for endogenous glucagon as a modulator of the secretion of insulin. If the antecedent release of glucagon had a significant effect upon the release of insulin in these experiments, significant relationships ought to exist between the time course of secretion and the magnitude of secretion of the two hormones. As shown in Fig. 5, the mean "half-maximal" values for glucagon varied over a considerable range (29 to 49 seconds, mean time for all experiments 40 seconds). On the other hand, the mean "half-maximal" value for insulin occurred within a narrow time range of 6 seconds (64 to 70, mean time 68 seconds). Thus, no correlation could be demonstrated between the values for glucagon and insulin. We have also analyzed the values for the total 120-second response areas pertaining to 8 sets of experiments with 6 different secretagogues common to both hormones (Fig. 6). The magnitude of the secretory response in glucagon and insulin correlated for only two sets of experiments (prostaglandin F-2-alpha, and isoproterenol, 0.11 μM); the correlation between the average glucagon and insulin response areas for all experiments was not significant. These observations provide evidence that the antecedent release of glucagon is not the principal mediator of insulin release in response to stimuli common to both hormones.

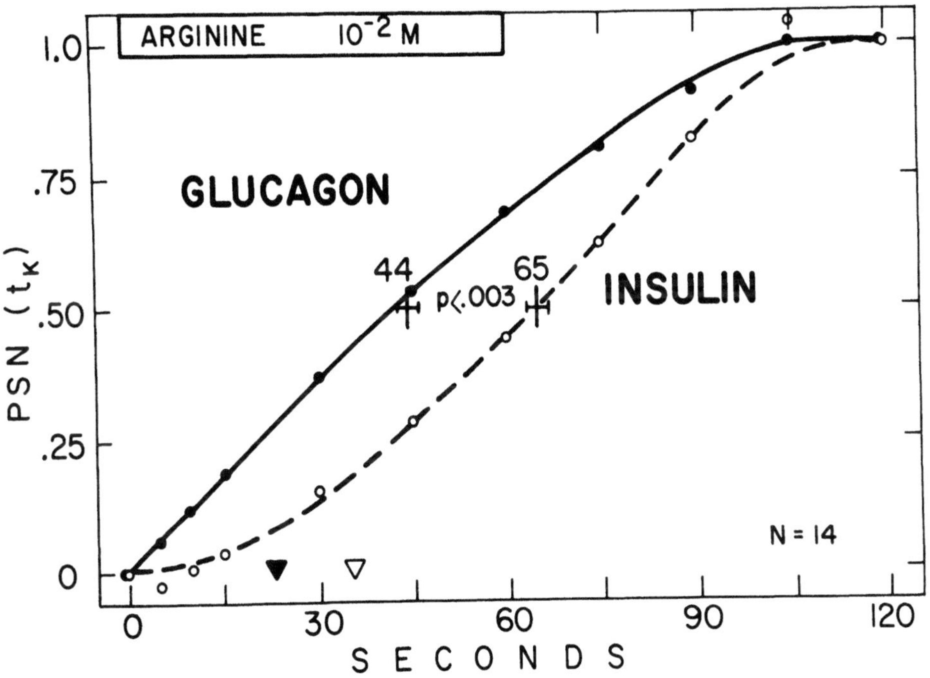

4-a

Fig. 4 a-f: Mean normalized partial sums (PSN) of the response areas for immunoreactive glucagon and insulin, as determined for each of the "k"th 12-second intervals (t_k) over a period of 120 seconds from the beginning of 60-second perfusions of various secretagogues. The mean ± S.E. of the time of "half-maximal" hormone concentrations are superimposed on each regression slope. The p-value refers to the significance of the difference between the "half-maximal" values for glucagon and insulin. The arrowheads signify the time of the first significant increase above basal levels. In all experiments the perfusate concentration of D-glucose was 5.6 mM. The type and concentration of the secretagogue are given in the upper left-hand corner of each figure (bGH = bovine growth hormone).

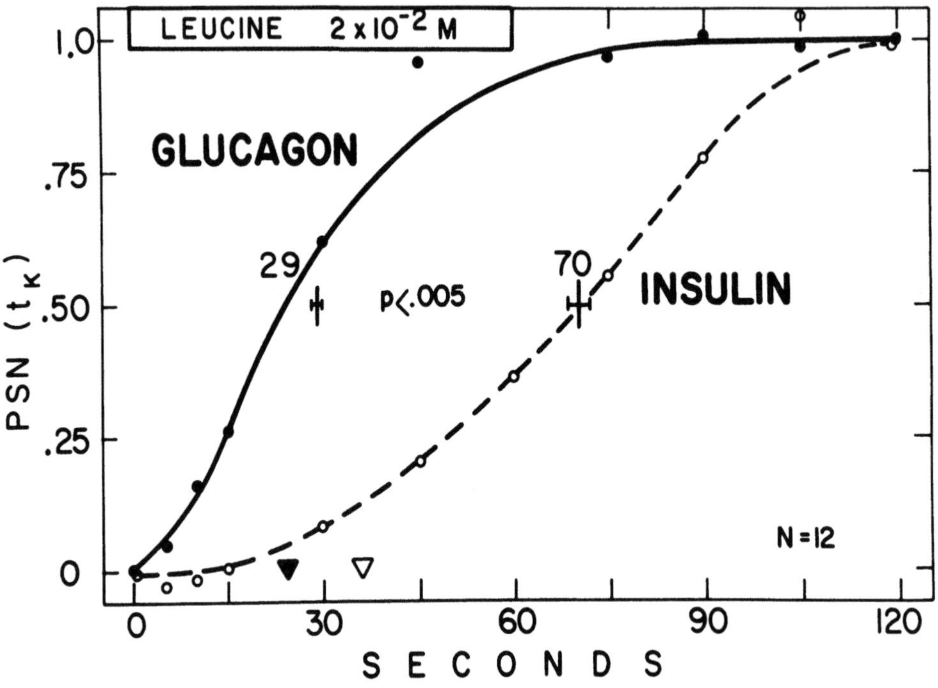
LEUCINE 2 x 10^-2 M
GLUCAGON
INSULIN
29
70
P<.005
N=12
PSN (t_K)
1.0
.75
.50
.25
0
0
30
60
90
120
SECONDS

4-b

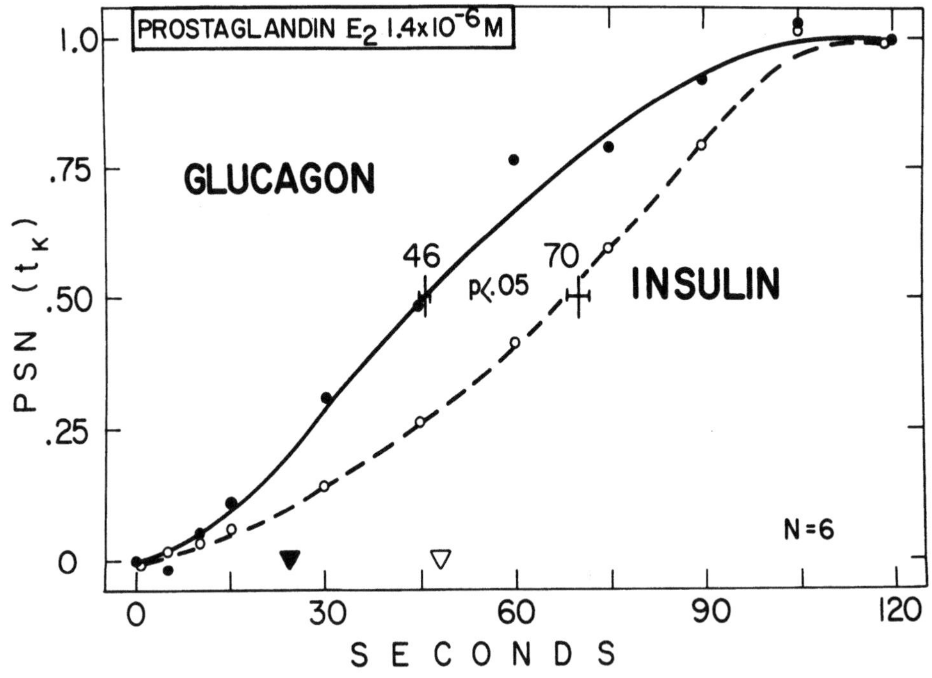
PROSTAGLANDIN E2 1.4x10^-6 M
GLUCAGON
INSULIN
46
70
P<.05
N=6
PSN (t_K)
1.0
.75
.50
.25
0
0
30
60
90
120
SECONDS

4-c

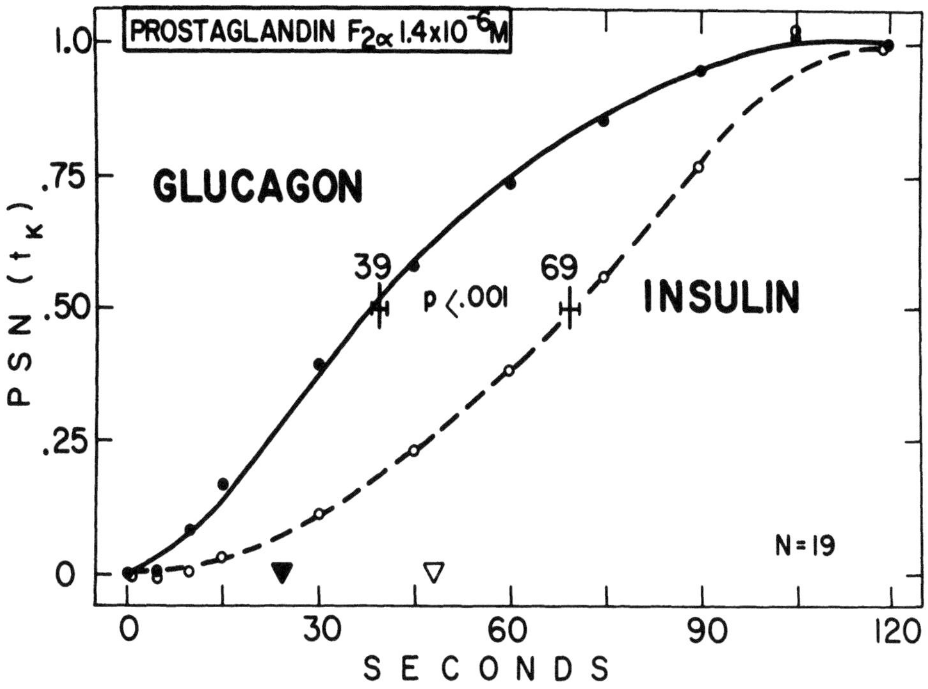
PROSTAGLANDIN F2α 1.4x10-6M
1.0
.75
.50
.25
0
PSN (tK)
GLUCAGON
INSULIN
39
69
p <.001
N=19
0
30
60
90
120
SECONDS

4-d

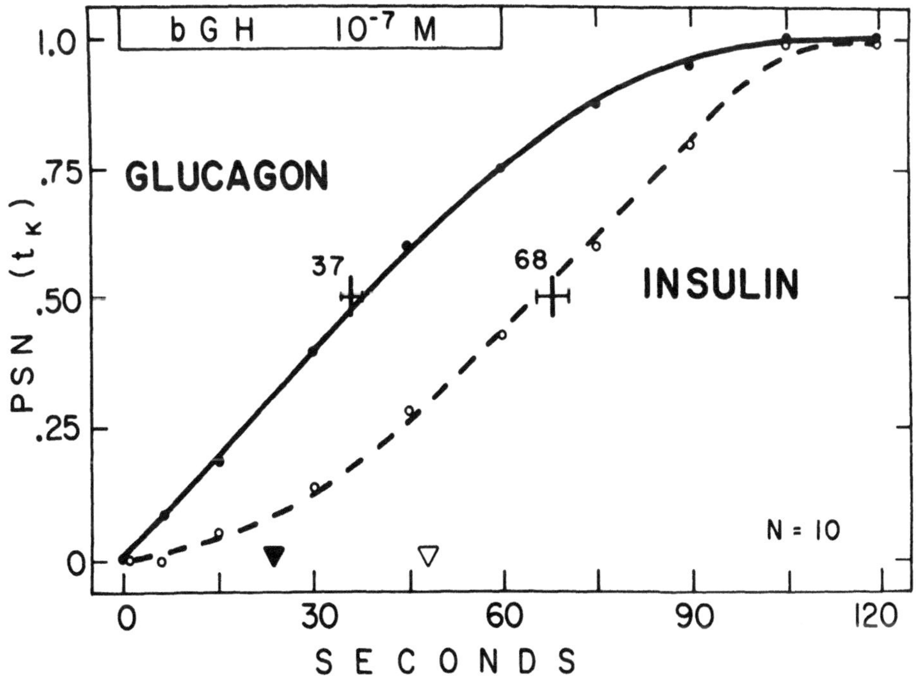
bGH 10-7 M
1.0
.75
.50
.25
0
PSN (tK)
GLUCAGON
INSULIN
37
68
N = 10
0
30
60
90
120
SECONDS

4-e

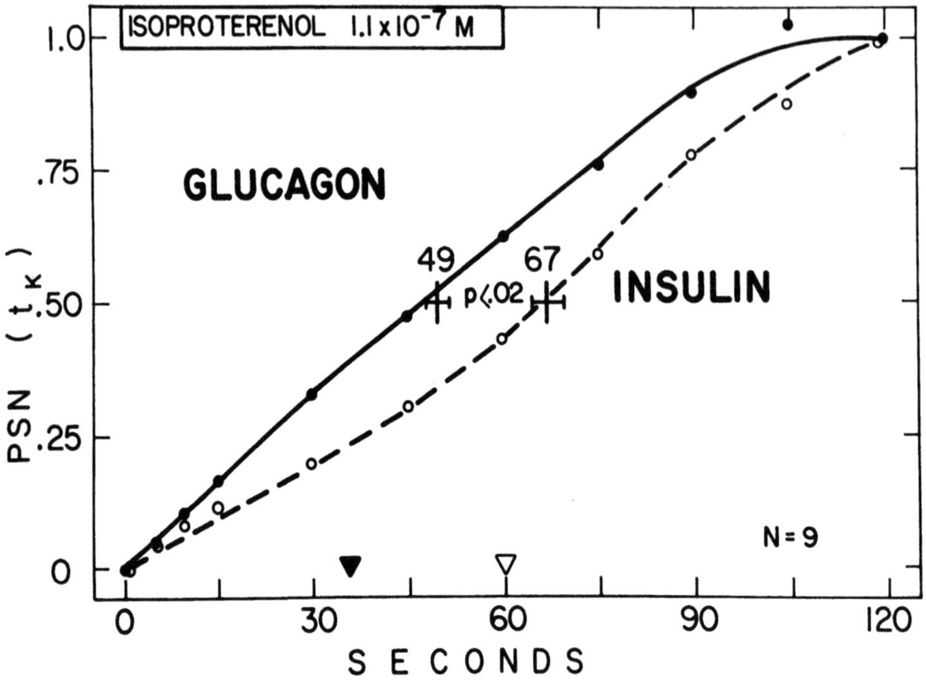

4-f

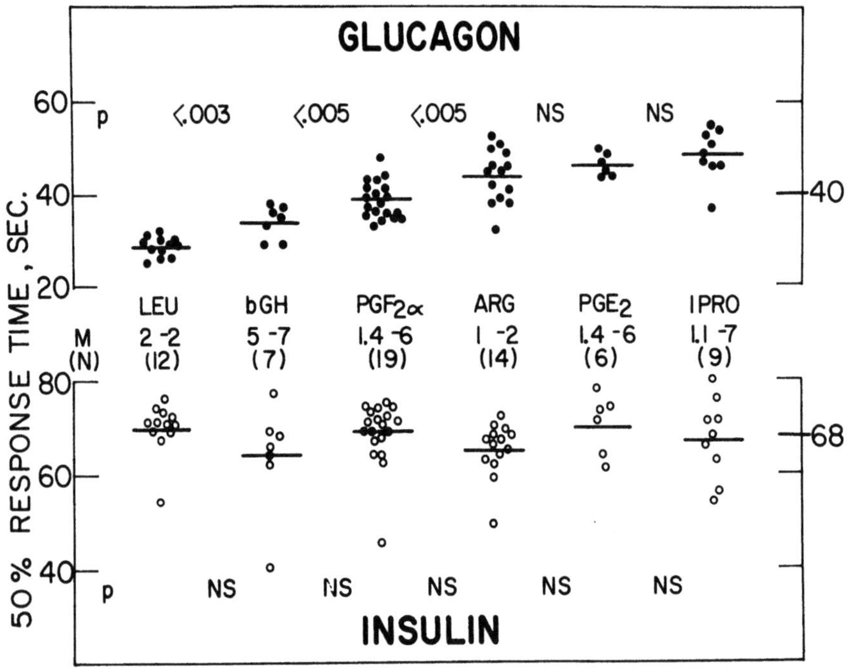

Fig. 5. Individual and mean time, in seconds, for "half-maximal" secretory responses (the time by which 50% of the total 120-second secretory response had occurred) in glucagon (upper panel) and insulin (lower panel) to 60-second perfusions of various secretagogues in the presence of 5.6 mM D-glucose: (from left to right) 20 mM L-leucine, 50 μM bovine growth hormone, 1.4 μM prostaglandin F-2-alpha, 10 mM L-arginine, 1.4 μM prostaglandin E-2 and 0.11 μM isoproterenol.

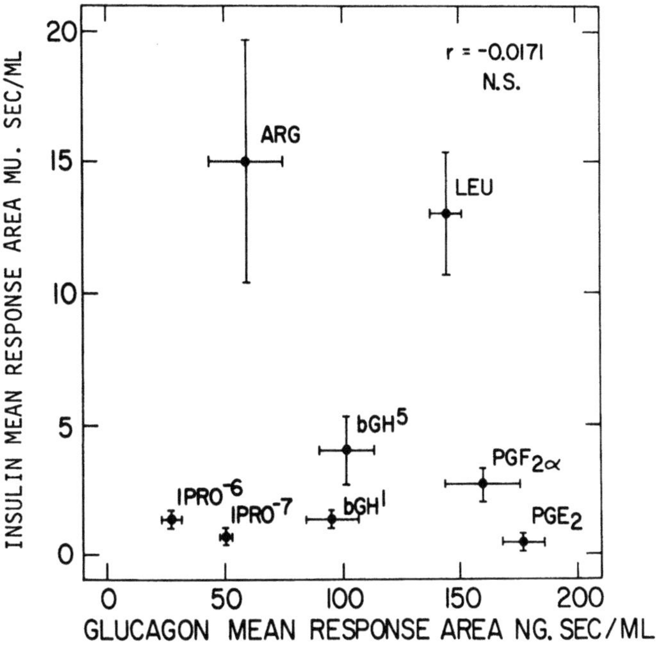

Fig. 6. Scatter plot of means ± S.E. of 120-second secretory response areas in glucagon (horizontal axis) versus those in insulin (vertical axis) to the six secretagogues depicted in Figs. 4 and 5. Note that isoproterenol (IPRO) and bovine growth hormone (bGH) have been administered at two different concentrations each: $IPRO^{-6}$ 1.1 x 10^{-6}M, $IPRO^{-7}$ 1.1 x 10^{-7}M, bGH^1 10^{-7}M and bGH^5 5 x 10^{-7}M.

Recently, we have observed a much less recognized phase of secretion of pancreatic islet hormones. Within 2-5 minutes after the termination of the perfusion of L-leucine, a short-lived secretion of insulin occurs only in the presence of glucose (Fig. 7). This secretory event is fairly specific for L-leucine; we have not observed the phenomenon with any of the other secretagogues discussed above. The post-leucine secretory response of insulin is not accompanied by a secretion of glucagon, and thus constitutes another example of discordant secretion of the two hormones.

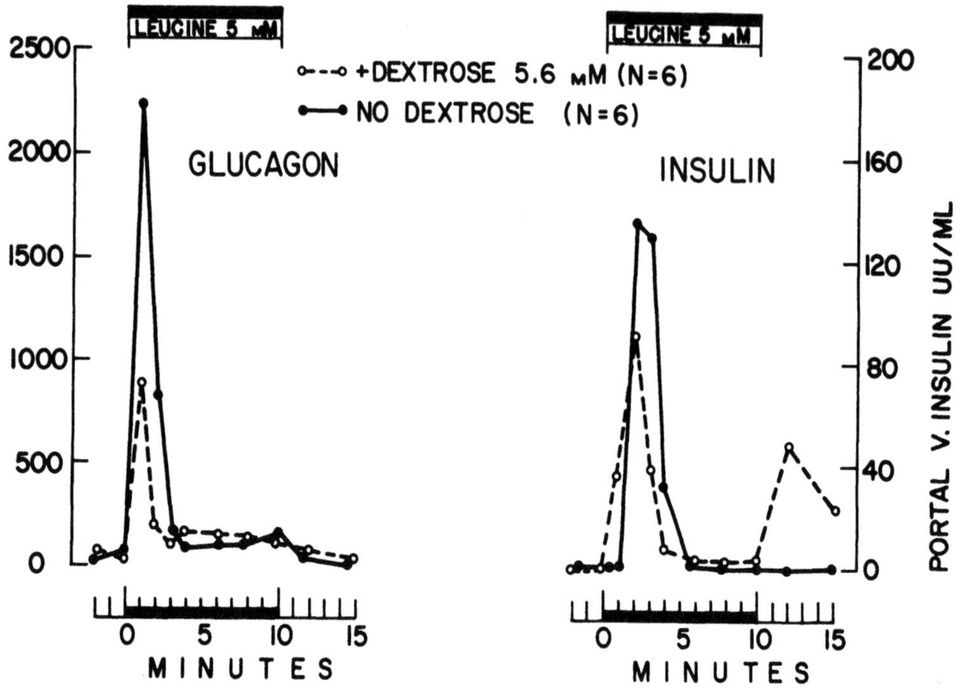

Fig. 7. Mean effluent levels of glucagon (panel on the left) and insulin before, during and following 10 minute perfusions of 5 mM L-leucine in the absence (solid lines) and presence (interrupted lines) of 5.6 mM D-glucose.

REFERENCES

1. Fajans, S. S., Quibrera, R., Pek, S., Floyd, J. C., Jr., Christensen, H. N. and Conn, J. W. (1971): Stimulation of insulin release in the dog by a non-metabolizable amino acid. Comparison with leucine and arginine. J. Clin. Endocrinol. Metab. 33, 35-41.

2. Grodsky, G. M., Bennett, L. L., Smith, D. F. and Schmid, F. G. (1967): Effect of pulse administration of glucose or glucagon on insulin secretion in vitro. Metabolism 16, 222-233.

3. Ketterer, H., Eisentraut, A. M. and Unger, R. H. (1967): Effect upon insulin secretion of physiologic doses of glucagon administered via the portal vein. Diabetes 18, 283-288.

4. Müller, W. A., Faloona, G. R. and Unger, R. H. (1971): The effect of experimental insulin deficiency on glucagon secretion. J. Clin. Invest. 50, 1992-1999.

5. Pagliara, A. S., Stillings, S. N., Haymond, M. W., Hover, B. A. and Matschinsky, F. M. (1975): Insulin and glucose as modulators of the amino acid-induced glucagon release in the isolated pancreas of alloxan and streptozotocin diabetic rats. J. Clin. Invest. 55, 244-255.

6. Pek, S., Fajans, S. S., Floyd, J. C,. Jr., Knopf, R. F. and Conn, J. W. (1969): Effects upon plasma glucagon of infused and ingested amino acids and of protein meals in man. Diabetes 18, Suppl. 1, 328.

7. Pek, S., Tai, T.-Y., Crowther, R. and Fajans, S. S. (1976): Glucagon release precedes insulin release in response to common secretagogues. Diabetes 25, 764-770.

8. Pek, S., Tai, T.-Y., Elster, A. and Fajans, S. S. (1975): Stimulation by prostaglandin E-2 of glucagon and insulin release from isolated rat pancreas. Prostanglandins 10, 493-502.

9. Samols, E., Marri, G. and Marks, V. (1965): Promotion of insulin secretion by glucagon. Lancet 2, 125-126.

10. Turner, D. S. and McIntyre, N. (1966): Stimulation by glucagon of insulin release from rabbit pancreas in vitro. Lancet 1, 351-352.

Acknowledgements

Supported in part by USPHS Grants AM-02244 and TI-AM-05001; by the Michigan Affiliate of the American Diabetes Association; the Department of Public Health of the State of Michigan; the Upjohn Company, Kalamazoo, Michigan and the Pfizer Company, New York, New York.

Dr. T.-Y. Tai was sponsored by the China Medical Board of New York, Inc., as a research scholar.

EFFECTS OF INSULIN ON THE GLUCAGON RELEASE, GLUCOSE UTILIZATION AND ATP CONTENT OF THE PANCREATIC A* CELLS OF THE GUINEA PIG

C.-G. Östenson, A. Andersson, S. E. Brolin, B. Petersson
and C. Hellerström

Department of Histology, University of Uppsala
Uppsala, Sweden

Previous studies have shown that insulin suppresses hyperglucagonemia in animals with experimental diabetes. This observation has raised the question whether the pancreatic A cell itself is insulin sensitive. The availability of A cell rich pancreatic islets opens the possibility of a more direct approach to this question. The effects of exogenous insulin on the glucagon release, glucose utilization and ATP content were studied in A cell rich islets from streptozotocin treated guinea pigs. Batches of islets were incubated in glucose concentrations of 3.3 mM, 16.7 mM or 16.7 mM plus 30 mU/ml insulin. The glucagon release was similar at both the low and high glucose concentrations but was strongly suppressed when insulin was added to the high-glucose medium. Glucose utilization increased with the glucose concentration in both normal and A cell rich islets. Addition of insulin caused a significant stimulation only in the latter islets. Similarly, the ATP concentration of the A cell rich islets increased when the glucose concentration was raised and displayed a further significant increase when insulin was added to the high-glucose medium. The present observations support the view that the A cell is sensitive to insulin and suggest that the suppressive effect of insulin on glucagon release is mediated via a stimulated glucose metabolism of the A cell.

INTRODUCTION

Unexpectedly high levels of circulating glucagon are found in both juvenile and adult onset human diabetics and in animals made diabetic with streptozotocin or alloxan (5, 6, 10, 31, 32). Administration of insulin has been shown to correct this high glucagon level in experimentally diabetic animals and, at least to some degree, also in human insulin-dependent diabetics (5, 27, 33). Reversal of the hyperglucagonemia in adult onset diabetics with normal or elevated insulin levels, requires, however, extremely high doses of insulin (32).

The above observations have led to the view that the A cell is sensitive to insulin although the possible mode of action is poorly understood. Previously it

*A_2, in the original manuscript

has been shown that the A cell can oxidize both glucose and fatty acids (7) and that both these compounds are able to suppress glucagon release (8, 9). Conversely, glucose depletion or inhibition of ATP formation in the A cell seems to enhance glucagon mobilization (8). The possibility that insulin causes suppression of glucagon release by facilitation of the glucose transport and metabolism within the A cell has been considered in this context. The availability of A cell rich islets (16, 26) makes it feasible to study this question in a more direct way than hitherto. In the present investigation we have examined the effects of insulin on the glucagon release, glucose utilization and ATP content of A cell rich islets isolated from the guinea pig.

MATERIALS AND METHODS

Chemicals. Streptozotocin was generously given by Dr. W. E. Dulin, The Upjohn Co., Kalamazoo, Michigan; Clinistix R were obtained from the Ames Co., Slough Bucks., U. K.; collagenase from Worthington Biochemical Corp., Freehold, N. J.; tissue culture medium (TCM) 199 and calf serum were supplied by Statens Bakteriologiska Laboratorium, Stockholm; beef insulin, 10 x cryst., was a gift of Novo Terapeutisk Laboratorium A/S, Bagsvaerd, Denmark; bovine plasma albumin (fraction V) was obtained from Armour Pharmaceutical Co., Eastbourne, U. K.; glucagon radioimmunoassay kits were obtained from Novo Research Institute, Bagsvaerd, Denmark; [5-^{3}H]-glucose (s. a., 2.4 Ci/mmol) was purchased from The Radiochemical Centre, Amersham, U.K.; fire-fly lantern extract and apyrase were purchased from Sigma Chemical Co., St. Louis, Missouri; ATP from Boehringer-Ingerheim, Germany. Other reagents were of the purest grade commercially available.

Preparation of A cell enriched islets: Ninety two male guinea pigs weighing about 200 g were used. Streptozotocin was injected i. p. in a single dose of 375 mg/kg b. w. to 76 of these animals. Immediately before the injection the drug was dissolved in citrate buffer, pH 4.5, at a concentration of 35 mg/ml. During the period following the injection the metabolic state of the animals was evaluated by measurements of their body weights, by the presence of urinary sugar and by assays of their blood glucose concentrations (15). Between 1-4 weeks after treatment animals were killed by decapitation and their pancreatic glands removed. A small piece of each gland was fixed in Bouin's solution, while the remaining part was treated with collagenase for islet isolation as described by Howell and Taylor (17). Islets were isolated in the same way from non-injected age- and weight-matched control animals.

Islets intended for measurements of glucose utilization were used immediately, whereas those to be used for determination of ATP or glucagon release were maintained in tissue culture (1-3). The latter islets were isolated aseptically and cultured for 24-48 hours in plastic Petri dishes containing 4 ml of TCM 199, 1 ml of calf serum, antibiotics (penicillin, 100 U/ml; streptomycin, 0.1 mg/ml) and 6.1 mM glucose.

Morphologic studies. The Bouin fixed pieces of pancreas from the streptozotocin injected animals were embedded in paraffin, cut in 4 or 7 μ thick sections and stained with aldehyde fuchsin trichrome (12).

Glucagon release. Groups of 15 A cell rich islets were incubated for 60 min in 250 μl of bicarbonate buffer (11), the gas phase consisting of 95% O_2:5% CO_2. Bovine plasma albumin (2 mg/ml) was also added. After incubation, samples of the medium were immediately frozen and stored at -20°C for later assay of glucagon, according to Heding (13), using bovine-porcine glucagon standards. Control studies showed that no correction was needed for the small amounts of glucagon, present in the insulin added to the medium. After incubation the dry weights of the islets were determined as described by Hellerström (14).

Glucose utilization. The utilization of glucose by the isolated islets was determined by the formation of [^{3}H]-water from [5-^{3}H]-glucose, essentially as described by Ashcroft et al (4). Groups of 15 islets each were incubated in small vessels (9), containing 20 μl of bicarbonate buffer (11). These vessels were placed inside scintillation vials containing 0.5 ml distilled water and sealed with a rubber membrane. [5-^{3}H]-glucose was added to give a final concentration of 3.3 or 16.7 mM with a specific activity of 30 mCi/mmol and 6 mCi/mmol, respectively. Insulin was dissolved directly in the incubation medium at a final concentration of 30 mU/ml. Islets were also incubated in 16.7 mM glucose without the addition of insulin. After 60 min of incubation at 37°C in a gas phase of 95% O_2:5% CO_2 the metabolism was stopped by the injection of 10 µl of 0.2 M HCl into the inner incubation vessel. The scintillation vials were then incubated at room temperature for another 20 h to allow the [^{3}H]-water formed by the islet cells to equilibrate with the water in the outer compartment. The radioactivity in the water was then measured by liquid scintillation spectrometry. After incubation the islets were recovered and their dry weight was determined (14).

Control samples without islets were used to correct for the initial [^{3}H]-water contained in the [5-^{3}H]-glucose. The recovery of [^{3}H]-water during the equilibration period was tested by measurements of the diffusion of known amounts of [^{3}H]-water. Under the conditions of the present experiments approximately 70% of the [^{3}H] activity was recovered.

ATP content. The effects of glucose and insulin on the ATP content of A cell rich islets were evaluated with the aid of the luciferine-luciferase technique described by Wettermark (34). Groups of 5-10 islets were incubated for 60 min in a bicarbonate buffered medium similar to that described above and with a gas phase of 95% O_2:5% CO_2. Immediately after incubation, the islets were frozen at -70°C, and freeze-dried. They were then carefully freed from adhering salt crystals and weighed on a quartz fiber balance (20) before being assayed for their ATP content.

Calculation and expression of results. The glucagon release rates were calculated as pg of glucagon released per µg of dry islet weight per hour. Glucose utilization rates were calculated as pmol of glucose utilized per µg of dry islet weight per hour, according to the formula of Ashcroft et al (4). The islet content of ATP was calculated as mmol per kg dry islet weight. The results in each experimental group were expressed as the mean ± S.E.M., each experiment representing the mean of 2-4 observations on islets obtained from the same animal. Statistical significances were evaluated with the aid of Student's t-test (30).

RESULTS

All streptozotocin injected guinea pigs showed a positive reaction for urinary sugar and a marked retardation of growth. Despite this, their mean blood glucose concentration at the time of death was only 138 ± 8 mg/100 ml (n = 15) as compared to 161 ± 10 mg/100 ml (n = 5) in the controls ($p < .05$). Light microscopic examination of pancreatic sections from the streptozotocin treated animals confirmed that their islets were composed mainly of A cells with only few remaining B cells. The latter observations confirmed that isolated islets from these animals were suitable for the present studies.

The rates of glucagon release from the isolated A cell rich islets are shown in Table 1. The highest release was found at a glucose concentration of 3.3 mM, but this was not significantly different from that at 16.7 mM glucose. By contrast, the addition of insulin to a medium containing the higher glucose concentration strongly inhibited the release of glucagon, which was now significantly lower than in either the low or the high-glucose media.

As seen in Table 2 the rate of glucose utilization, as reflected in the formation of [^{3}H]-water, increased markedly ($p < .001$) in both normal and A cell rich islets when the glucose concentration of the incubation medium was increased from 3.3 mM to 16.7 mM. At both these glucose concentrations there was, however, a significantly lower glucose utilization of the A cell rich islets ($p < .001$). Addition of insulin to the high-glucose medium affected the two groups of islets differently, in that the glucose utilization of the A cell rich islets increased by about 50 per cent ($p < .01$) whereas no significant effect was seen in the group of normal islets ($p > .05$). In these experimental conditions the glucose utilization of the A cell rich islets was in fact very similar to that of the normal islets (p >.05).

TABLE 1

GLUCAGON RELEASE FROM A CELL RICH ISLETS OF THE GUINEA PIG

Additions to the Medium	Number of Experiments	Glucagon Release (pg/μg · h)
Glucose (3.3 mM)	10	67.1 ± 6.6
Glucose (16.7 mM)	11	58.4 ± 6.9*
Glucose (16.7 mM) + Insulin (30 mU/ml)	10	36.2 ± 3.5**, ***

* p <.05 vs 3.3 mM glucose; ** p <.001 vs 3.3 mM glucose; ***p <0.01 vs 16.7 mM glucose

TABLE 2

GLUCOSE UTILIZATION EXPRESSED AS THE FORMATION OF TRITIATED WATER BY ISOLATED PANCREATIC ISLETS OF NORMAL AND STREPTOZOTOCIN INJECTED GUINEA PIGS

	Formation of [^{3}H] Water (pmole/μg . h)	
Additions to the Medium	Normal Islets	A Cell Rich Islets
Glucose (3.3 mM)	35.9 ± 3.4 (5)	23.7 ± 2.8 (13)
Glucose (16.7 mM)	94.4 ± 2.3 (5)	64.6 ± 5.8 (12)
Glucose (16.7 mM) + Insulin (30 mU/ml)	110.5 ± 8.8 (5)	97.2 ± 8.1 (11)

The ATP contents of the A cell rich islets are shown in Fig. 1. Incubation for 60 min at 16.7 mM glucose raised significantly the ATP content as compared with islets incubated in 3.3 mM glucose ($p < .001$). Addition of insulin to the incubation medium induced a further significant increase of the ATP content ($p < .001$). In fact, the stimulation by insulin was more pronounced than that observed when the glucose concentration was increased from 3.3 to 16.7 mM.

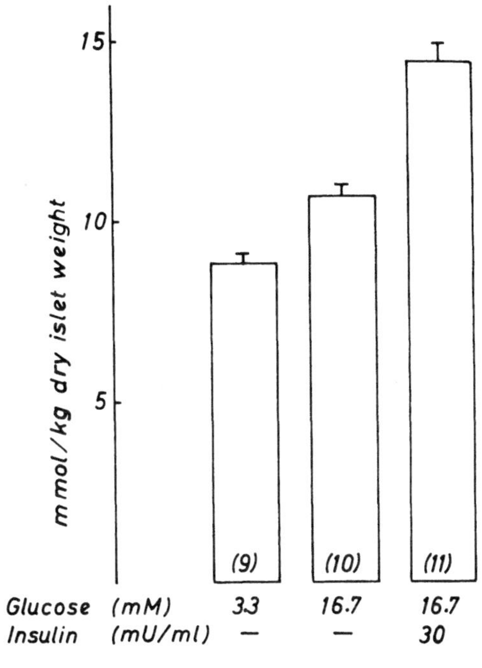

Fig. 1. The concentration of ATP in A cell rich islets which had been cultured for 24-48 hours in 6.1 mM glucose and subsequently incubated for 60 min in a bicarbonate buffer containing glucose and insulin as indicated in the figure. Results are given as mean values ± S.E.M. with the number of experiments within parentheses.

DISCUSSION

The present observations of marked growth retardation, a tendency to low blood glucose levels and heavy glycosuria in the streptozotocin treated guinea pigs well conform to previous findings in such animals (16, 26). The lack of hyperglycemia in the B cell deficient guinea pigs is particularly noteworthy and may signify a lowered threshold for glucose in the kidneys, which have been shown to be severely damaged by streptozotocin (26). This paradoxical blood sugar response means that the pancreatic A cell of these animals will remain in a relatively normoglycemic environment despite the insulin lack. Sustained hyperglycemia may alter considerably the glucagon response of the A cell (29) and the guinea pig may therefore be the animal of choice for induction of A cell rich islets. By contrast, the mouse and the rat develop a manifest diabetic state with very high blood sugar levels within a day or so after injection of streptozotocin (18, 28).

Previous histologic studies have shown that the proportion of A cells in the guinea pig islets will change from about 15 per cent in the normal animal to 60-80 per cent in the streptozotocin treated animal (16, 26). The islet glucagon concentration (per µg protein) is correspondingly doubled, whereas the concentration of insulin falls to about 10 per cent of that in the normal islets (16). These observations strongly support the view that our measurements of glucose utilization or ATP content of the A cell rich islets are representative for this cell type. On the other hand, the B cell deficient islets are considerably smaller than the normal ones making it necessary to express the values on a basis of islet weight.

The present finding of a marked inhibition of the in vitro glucagon release by insulin is in general agreement with previous reports on experimentally diabetic animals, in which insulin injections caused a prompt suppression of the circulating glucagon levels (23). However, several recent reports failed to confirm these results with in vitro techniques (24, 25). Relevant comparisons between these studies and the present findings are difficult due to differences in experimental design, the species used, the mode of islet preparation, etc. It should nevertheless be pointed out that the prompt decline in glucagon release in response to insulin reported by us was obtained with A cells which had been deprived of their surrounding B cells for several weeks before the experiment and then cultured for 1-2 days before the incubation with insulin. The sustained deficiency of insulin in the extracellular space may well have made the A cell more sensitive to acute exposure to this hormone.

More direct evidence for an insulin effect on the glucose metabolism of the A cell was obtained in the measurements of [^{3}H]-water formation by the islets. This technique was originally introduced in islet research by Ashcroft et al (4), who concluded that the rate of metabolic formation of [^{3}H]-water from [5-^{3}H]-glucose represents the combined rates of flow through the glycolytic and pentose cycle pathways. Glucose converted into glycogen is not included in this reaction, but it is conceivable that the capacity for glycogen formation in the A cell is lower than that in the B cell, in which less than 10 per cent of the metabolized glucose is converted into glycogen (4). We therefore conclude that formation of [^{3}H]-water from [5-^{3}H]-glucose represents a satisfactory measure of glucose utilization also by the islet A cell. It is noteworthy in this context that the figures for glucose utilization of the A cell rich islets are lower than those of the normal islets. In agreement with this, Lundqvist (21) reported a relatively low activity of hexokinase in the A cell of the duck, which might indicate that glucose phosphorylation is rate limiting for glycolytic flow in this cell type. On the other hand, the enzymatic capacity for ATP formation from triose phosphates is higher in the A cells of the guinea pig than in either the B cells or the exocrine pancreatic cells (22).

The present observation of a highly significant effect of insulin on both glucose utilization and ATP formation in the A cell strongly suggests that the suppressive effect of insulin on glucagon release is mediated via a stimulated glucose metabolism of the A cell. This notion also agrees with previous observations suggesting that suppression of ATP formation enhances glucagon release, and conversely that this process is inhibited by exposing the A cell to substrates such as glucose and fatty acids (7, 8). However, the precise molecular mechanisms involved in the action of insulin on the A cells and the reason why these cells become functionally stimulated by a decline in available energy-rich compounds remain to be clarified.

Acknowledgement

This work was supported by grants from the Swedish Medical Research Council, the Swedish Diabetes Association and the Medical Faculty of the University of Uppsala.

REFERENCES

1. Andersson, A. (1976): Tissue culture of isolated pancreatic islets. Acta Endocrinol. (Kbv.), 83, Suppl. 205, 283-294.

2. Andersson, A. and Hellerström, C. (1972): Metabolic characteristics of isolated pancreatic islets in tissue culture. Diabetes 21, Suppl. 2, 546-554.

3. Andersson, A., Petersson, B., Westman, J., Edwards, J., Lundqvist, G. and Hellerström, C. (1973): Tissue culture of A_2-cell rich pancreatic islets isolated from guinea pigs. J. Cell Biol. 57, 241-247.

4. Ashcroft, S. J. H., Weerasinghe, L. C. C., Bassett, J. M. and Randle, P. J. (1972): The pentose cycle and insulin release in mouse pancreatic islets. Biochem. J. 126, 525-532.

5. Braaten, J. T., Faloona, G. R. and Unger, R. H. (1974): The effect of insulin on the alpha cell response to hyperglycemia in long standing alloxan diabetes. J. Clin. Invest. 53, 1017-1021.

6. Buchanan, K. D. and Mawhinney, W. A. (1973): Glucagon release from isolated pancreas in streptozotocin-treated rats. Diabetes 22, 797-800.

7. Edwards, J. C., Hellerström, C., Petersson, B. and Taylor, K. W. (1972): Oxidation of glucose and fatty acids in normal and in A_2-cell rich pancreatic islets isolated from guinea pigs. Diabetologia 8, 93-98.

8. Edwards, J. C.and Taylor, K. W. (1970): Fatty acids and the release of glucagon from isolated guinea pig islets of Langerhans incubation in vivo. Biochim. Biophys. Acta 215, 310-315.

9. Gerich, J. E., Charles, M. A.and Grodsky, G. M. (1974): Characterization of the effects of arginine and glucose on glucagon and insulin release from the perfused rat pancreas. J. Clin. Invest. 54, 833-841.

10. Gerich, J. E., Lorenzi, M., Karam, J. H., Schneider, V., Forsham, P. H. (1975): Abnormal pancreatic glucagon secretion and postprandial hyperglycemia in diabetes mellitus. J. Am. Med. Assoc. 234, 159-165.

11. Gey, G. O. and Gey, M. K. (1936): Maintenance of human normal cells and tumor cells in continuous culture; preliminary report: cultivation of mesoblastic tumors and normal tissue and notes on methods of activation. Amer. J. Cancer 27, 45-76.

12. Gomori, G. (1950): Rapid one-step trichrome stain. Am. J. Clin. Path. 20, 661-664.

13. Heding, L. G. (1971): Radioimmunological determination of pancreatic and gut glucagon. Diabetologia 7, 10-19.

14. Hellerström, C. (1967): Effects of carbohydrates on the oxygen consumption of isolated pancreatic islets of mice. Endocrinology 81, 105-112.

15. Hjelm, M. and Verdier, C.-H. de (1963): A methodological study of the enzymatic determination of glucose in blood. Scand. J. Clin. Lab. Invest. 14, 415-429.

16. Howell, S. L., Edwards, J. C. and Whitfield, M. (1971): Preparation of B cell deficient guinea pig islets of Langerhans. Horm. Metab. Res. 3, 37-43.

17. Howell, S. L. and Taylor, K. W. (1968): Potassium ions and the secretion of insulin by islets of Langerhans incubated in vitro. Biochem. J. 108, 17-24.

18. Junod, A., Lambert, A. E., Stauffacher, W. and Renold, A. E. (1969): Diabetogenic action of streptozotocin: relationship of dose to metabolic response. J. Clin. Invest. 48, 2129-2139.

19. Keen, H., Field, J. B. and Pastan, I. (1963): A simple method for in vitro metabolic studies using small volumes of tissue and medium. Metabolism 12, 143-147.

20. Lowry, O. H., Roberts, N. R. and Chang, M.-L. W. (1956): Analysis of simple cells. J. Biol. Chem. 222, 97-107.

21. Lundqvist, G. (1972): Enzymatic studies of glucose phosphorylation in the glucagon producing cells of duck pancreas. Horm. Metab. Res. 4, 83-86.

22. Lundqvist, G. (1972): Enzymatic prerequisites for ATP formation from triose phosphates in the A_2 and B cells of the endocrine pancreas in the guinea pig. Horm. Metab. Res. 4, 237-241.

23. Müller, W. A., Faloona, G. R. and Unger, R. H. (1971): The effect of experimental insulin deficiency on glucagon secretion. J. Clin. Invest. 50, 1992-1999.

24. Oliver, J. R., Williams, V. L. and Wright, P. H. (1976): Studies on glucagon secretion using isolated islets of Langerhans of the rat. Diabetologia 12, 301-306.

25. Pagliara, A. S., Stillings, S. N., Haymond, M. W., Hover, B. A. and Matschinsky, F. M. (1975): Insulin and glucose as modulators of the amino acid induced-glucagon release in the isolated pancreas of alloxan and streptozotocin diabetic rats. J. Clin. Invest. 55, 244-255.

26. Petersson, B., Hellerström, C. and Gunnarsson, R. (1970): Structure and metabolism of the pancreatic islets of streptozotocin treated guinea pigs.

Horm. Metab. Res. 2, 313-317.

27. Raskin, P., Fujita, Y. and Unger, R. H. (1975): Effect of insulin-glucose infusions on plasma glucagon levels in fasting diabetics and non-diabetics. J. Clin. Invest. 56, 1132-1138.

28. Rerup, C. C. (1970): Drugs producing diabetes through damage of the insulin secreting cells. Pharmacol. Rev. 22, 485-518.

29. Segerström, K., Andersson, A., Lundqvist, G., Petersson, B. and Hellerström, C. (1976): Regulation of the glucagon release from mouse pancreatic islets maintained in tissue culture at widely different glucose concentrations. Diabète Metab. 2, 45-48.

30. Snedecor, G. W. (1956): In: Statistical Methods, 5th Ed., The Iowa State University Press, Ames, p. 35.

31. Unger, R. H., Aguilar-Parada, E., Müller, W. A. and Eisentraut, A. M. (1970): Studies of pancreatic alpha cell function in normal subjects. J. Clin. Invest. 49, 837-848.

32. Unger, R. H., Madison, L. L. and Müller, W. A. (1972): Abnormal alpha cell function in diabetics. Response to insulin. Diabetes 21, 301-307.

33. Weir, G. C., Knowlton, S. D., Atkins, R. F., McKennan, K. X. and Martin, D. B. (1976): Glucagon secretion of the perfused pancreas of streptozotocin-treated rats. Diabetes 25, 275-282.

34. Wettermark, G., Tegner, L., Brolin, S. E. and Borglund, E. (1970): Photokinetic measurements of the ATP and ADP levels in isolated islets of Langerhans. In: The Structure and Metabolism of the Pancreatic Islets, Vol. 16. S. Falkmer, B. Hellman and I.-B. Täljedal, Eds. Pergamon Press, Oxford, pp. 275-282.

EFFECT OF VASOACTIVE INTESTINAL PEPTIDE (VIP) AND GASTRIC INHIBITORY PEPTIDE (GIP) ON INSULIN AND GLUCAGON RELEASE BY PERIFUSED NEWBORN RAT PANCREAS

D. Bataille, C. Jarrousse, N. Vauclin, C. Gespach and G. Rosselin

Unité I.N.S.E.R.M. U.55, ERA C.N.R.S. n° 494
Hôpital Saint-Antoine, 75571 Paris Cedex 12, France

The effect of porcine vasoactive intestinal polypeptide (VIP) and of gastric inhibitory peptide (GIP) on insulin and glucagon secretion was studied using the perifused splenic portion of the 3 day-old rat pancreas. The B cells were markedly stimulated by glucagon (50 ng/ml) in the presence of glucose (16.7 mM), indicating that this preparation was suitable for testing the endocrine response to polypeptide hormone stimulation. VIP, at the concentration of 2 ng/ml, induced the release of insulin and glucagon in the presence of either 4 mM or 16.7 mM glucose. At high glucose concentration (16.7 mM), VIP was more effective at a concentration of 10 ng/ml than at a concentration of 2 ng/ml. However, at a concentration of 50 ng/ml, the effectiveness of VIP decreased sharply.

GIP, a well known insulin secretagogue, induced a sharp release of insulin and of glucagon at the two concentrations tested (2 and 10 ng/ml). On a molar basis and a high glucose concentration, GIP was more effective than VIP on the release of either hormone.

From these and other recently published data, we propose a hypothesis on the possible role of VIP in the physiology of the endocrine pancreas.

INTRODUCTION

It has been observed that a load of glucose is a more effective stimulus to insulin secretion when given orally than when injected intravenously (28). These data led to the suggestion that in response to oral glucose the gut releases an insulin secretagogue (11, 12, 28, 41), for which the name "incretin" was proposed (46). This concept is represented diagrammatically in Fig. 1. Several peptides, found in extracts of mammalian gut, have been considered incretin candidates. Among them are gastrin (16), cholecystokinin-pancreozymin (22), secretin (21), vasoactive intestinal peptide (VIP; 37), gastric inhibitory peptide (GIP; 6) and a material with glucagon-like immunoreactivity (GLI), not yet obtained in a bioactive pure form (1, 3, 4, 43). However, because of the high doses used to obtain an effect either *in vivo* or *in vitro*, and the difficulty of obtaining pure

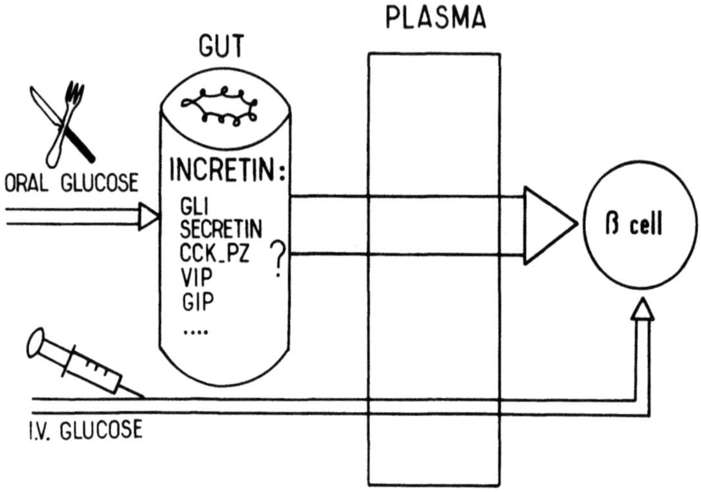

Fig. 1. Diagram representing the possible role of a gut factor ("incretin") in the greater insulinogenic effect of oral glucose as compared to intravenous glucose.

materials, there remains the possibility that the hormone release induced by most of these preparations may be due to contaminants. One exception appears to be GIP, whose effect on insulin secretion can be obtained at physiologic concentration (31) and whose plasma level rises after an oral glucose load (13). Thus, this peptide seems to best fulfill the requirements of being the effective insulin secretagogue or at least one of its components. On the other hand, few data are available on the possible effect of intestinal peptides on glucagon secretion.

The present paper deals with the effect of low doses of VIP and GIP on insulin and glucagon secretion by the perifused pancreas of the 3 day-old rat. This in vitro system has evolved from the incubated pancreas preparation, successfully used in our laboratory for the study of insulin and glucagon secretion and of insulin biosynthesis (14, 17-19). In view of the data presented in this paper and of other recent developments in this field, we propose a modification of the incretin hypothesis.

MATERIALS AND METHODS

Perifusion of newborn rat pancreas was performed as described elsewhere (45). Briefly, 3 day-old Wistar rats from our colony were killed by decapitation, the splenic part of the pancreas was removed (17) and immediately cut in 0.5 mm squares with a tissue chopper (McIlvain, Mickle Laboratories). The tissue from 3 pancreatic explants was placed in a small perifusion chamber, maintained at 37^oC, immediately washed with Krebs-Ringer bicarbonate buffer, pH 7.5, containing 5 mM Hepes* and 0.5% (w/v) bovine serum albumin (Fraction V, Pentex) and perifused with the same buffer by means of a multichannel peristaltic pump (Ismatec MP 13 GJ 10). The buffer had been previously saturated with a mixture of O_2 (95%) and CO_2 (5%) and was gassed during the whole experiment. The flow rate was 0.8 ml/min. After 45 minutes of equilibration, the same buffer containing the substance to be tested** was introduced using a 3 way-valve (Cheminert R 60 31V3). After a 10 minute period of stimulation, the valve was turned back to the first position, allowing the original buffer to flow through the perifusion chamber again. Five minute (4 ml) or one minute (0.8 ml) fractions were collected and kept at -20^oC until analyzed. Insulin and glucagon were measured by means of radioimmunoassays described previously (42), using talc to separate free from bound radioactivity (35). The cross-reactivities of VIP and GIP with the glucagon antibody (GN_3) were less than 0.1% and 0.3%, respectively.

---

*N_2-hydroxyethylpiperazine-N'_2-ethanesulfonic acid.

**VIP (lot 24/10/74) was a generous gift of V. Mutt, GIH Laboratory, Karolinska Institutet, Stockholm, Sweden; GIP was kindly provided by J. C. Brown, Department of Physiology, University of British Columbia, Vancouver, Canada; glucagon (lot B66) was obtained through the courtesy of J. Schlichtkrull, The Novo Research Institute, Copenhagen, Denmark.

RESULTS

The basal levels of insulin and glucagon release were stable during the 45-85 minutes of the experiment (Table 1).

TABLE 1

BASAL RELEASE OF INSULIN AND GLUCAGON OBSERVED AFTER 45 MINUTES OF EQUILIBRATION

Femtomoles of hormone released per minute and per explant at three different times. Ave. ± SEM of 3 experiments. The medium contained 16.7 mM glucose

	Minutes of Perifusion		
	45th	65th	85th
Insulin	259.5 ± 36	271.5 ± 74	233.3 ± 59
Glucagon	15.2 ± 3.8	14.5 ± 6.5	19 ± 6.8

The sensitivity of the perifusion system to peptide hormones was tested first using glucagon, a well known stimulant of insulin secretion. Fig. 2 shows the results of a typical experiment performed at high glucose concentration (16.7 mM) and in the presence of 18 amino acids at subphysiologic concentration (25 μM each).

Glucagon, at the concentration of 10 ng/ml, did not modify insulin output but when the concentration was raised to 50 ng/ml a clear-cut increase in insulin release was obtained and continued for several minutes after cessation of the stimulus. Thus, the system seemed suitable for the study of the effect of the glucagon-related peptides VIP and GIP on the secretory activity of the endocrine pancreas.

Figure 3 displays the results of an experiment using VIP at three concentrations, tested under the same conditions as those described in Fig. 2. As shown in the left panel VIP stimulated insulin release at concentrations as low as

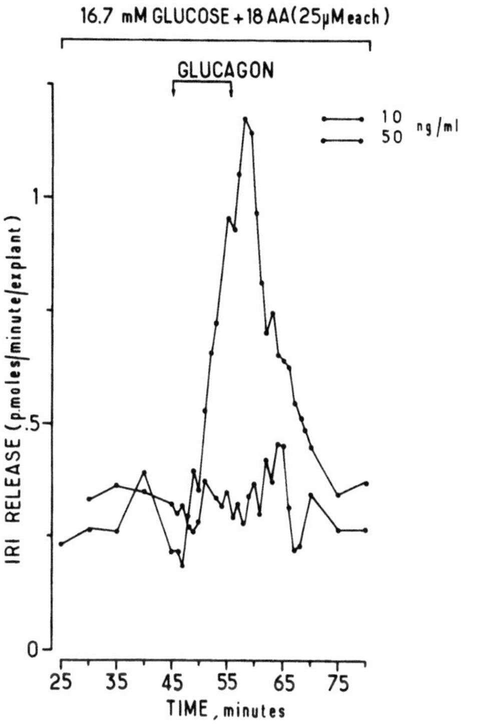

Fig. 2. Effect of two doses of glucagon (10 and 50 ng/ml) on insulin release by newborn rat pancreas perifused in the presence of glucose (16.7 mM) and of 18 amino acids (AA, 25 μM each). Glucagon was perifused between min 45 and 55.

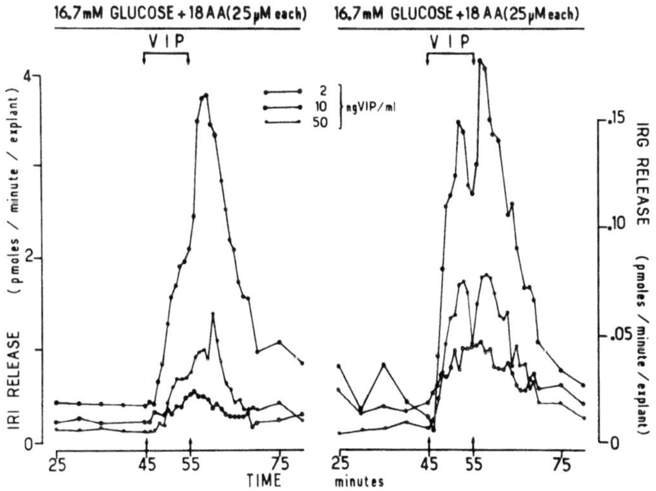

Fig. 3. Effect of three doses of VIP on the release of insulin (IRI left panel) and glucagon (IRG, right panel) by newborn rat pancreas perifused in the presence of glucose (16.7 mM) and of 18 amino acids (AA, 25 μM each). Conditions as in Fig. 2.

2 ng/ml. At a concentration of 10 ng/ml, the insulinogenic effect of VIP was markedly increased, but surprisingly, a still higher dose did not induce a greater release of insulin, but a smaller one. A possible explanation for this apparent paradox is given in DISCUSSION. The right panel of Fig. 3 shows the glucagon output measured in the same experiment: the same pattern was observed for the three doses of VIP. These data suggest that VIP acts on B and A cells at similar concentrations. At the most efficient dose, that is, 10 ng/ml for 10 minutes, VIP increased insulin and glucagon release, measured between 45 and 75 minutes, 4.5-fold and 3-fold respectively, relative to the basal level. Furthermore, as pointed out for the effect of glucagon (Fig. 2), the peaks of insulin and glucagon were obtained after the stimulus had been discontinued. This was also observed by others using a different system (15).

In order to check whether the presence of substrates could modulate the effect of VIP, we performed similar experiments in the absence of amino acids and at a normal (4 mM) glucose concentration. Fig. 4 shows the results of experiments in which the 18 AA were omitted, run simultaneously with the experiments described in Fig. 3.

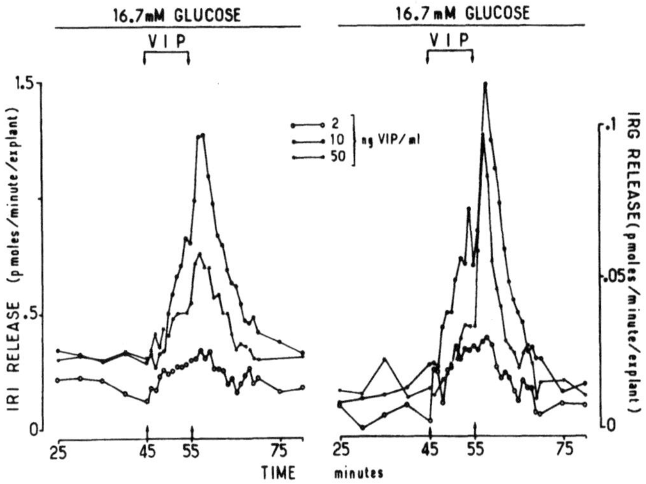

Fig. 4. Experiment run in parallel with that of Fig. 3. Conditions as for Fig. 3, except that the 18 amino acids (AA) were omitted.

Similar patterns of insulin and glucagon response were observed for each concentration of VIP in the absence of AA as were observed in their presence (Fig. 3), even though the insulin output was slightly lower when the AA were absent. It can be concluded that amino acids are not necessary for the effectiveness of VIP, even if the presence of low concentrations of AA seem to increase the responsiveness of the system, at least so far as insulin release is concerned. Fig. 5 depicts the results of a comparative experiment performed in the presence of the 18 AA (25 μmM each) and of a normal (4 mM, left panel) or a high glucose (16.7 mM, right panel) concentration. A high glucose concentration was not necessary for the effect of VIP on insulin secretion, neither was a low glucose concentration necessary for an effect on glucagon secretion.

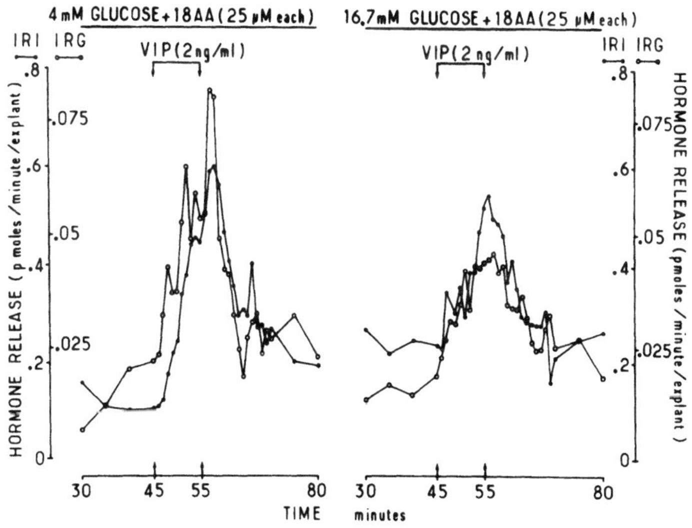

Fig. 5. Comparative effect of VIP (2 ng/ml) on insulin (IRI) and glucagon (IRG) release at normal (4 mM, left panel) and at high (16.7 mM, right panel) glucose concentration and in the presence of 18 amino acids (AA, 25 μM each). The two experiments were run simultaneously. Other conditions as described in the preceding figures.

Fig. 6 shows that GIP, at the concentration of 2 ng/ml and in the presence of high glucose concentration (16.7 mM) and of the 18 AA (25 μM each), was effective in stimulating both insulin and glucagon release. At the concentration of 10 ng/ml, the effect of GIP was still greater.

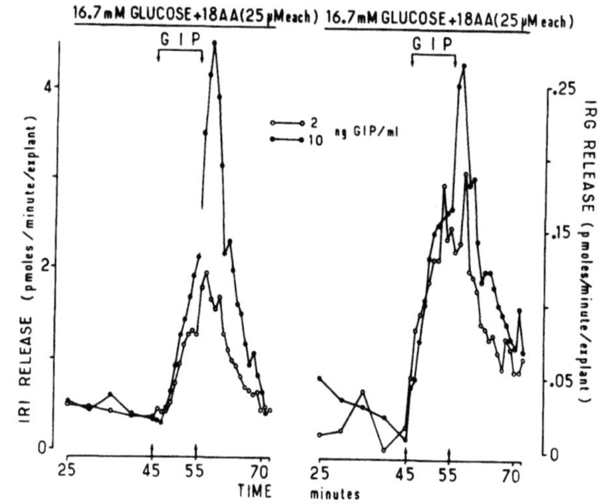

Fig. 6. Effect of two doses of GIP (2 and 10 ng/ml) on the release of insulin (IRI, left panel) and glucagon (IRG, right panel) by newborn rat pancreas, perifused in the presence of glucose (16.7 mM) and of 18 amino acids (AA, 25 µM each). All conditions were identical to those described in Fig. 3.

As observed when glucagon and VIP were used as stimulants, the hormonal release continued to rise after the end of the stimulation. Concerning this problem, several experiments show that the time required to obtain a maximal rise in hormonal release was not related to the length of stimulation but was an intrinsic characteristic of the system. This phenomenon suggests that the stimulus triggers a biochemical process within the target cells and that the process would then proceed according to its own dynamics.

DISCUSSION

This paper represents the first report concerning the in vitro insulinotropic effect of VIP, at concentrations as low as 2 ng/ml. A similar effect has been previously observed in the perfused cat pancreas and the perifused mouse pancreas, albeit using VIP in microgram amounts (39). On the other hand, no effect of VIP, in doses of 2 μg/kg, on blood insulin levels could be obtained in vivo (40). An effect of GIP on glucagon release was observed in adults at normal, but not at high glucose concentration (33). The fact that GIP is effective on glucagon release in our system, even at high glucose concentrations, may be related to the poor dependency of the A cell upon glucose in the neonate (18). There are no reports concerning the effect of VIP on glucagon secretion.

If we compare the effect of VIP and GIP on insulin and glucagon release, we can observe that at high glucose concentrations, both peptides stimulated the secretion of both hormones, when added in comparable amounts, but since the molecular weight of GIP is ≈5100, whereas that of VIP is ≈3300, GIP was more efficient than VIP, on a molar basis. Since no data are available on the concentration of VIP surrounding the islets in different physiological conditions, no conclusion can be drawn about the physiological meaning of this observation. Concerning the glucose-dependence of the insulinotropic effect, it has been observed that GIP needs a glucose concentration of at least 5-6 mM (31), whereas, in our system, VIP was effective at normal (4 mM) as well as high (16.7 mM) glucose concentrations.

One possible explanation for the paradoxical effect of a relatively high concentration (50 ng/ml) of VIP lies in its clear-cut secretin-like activity on the exocrine pancreas (23). Thus, VIP may induce a release of proteolytic enzymes that may destroy, at least partially, the peptide hormones used as stimulants, their receptors on the islet cell membrane or the secreted hormones. In this connection it should be noted that enzymatic digestion used to isolate the islets of Langerhans tends to suppress their responsiveness to polypeptides, such as secretin (44) and somatostatin (20, 29). In contrast, islets isolated by microdissection are sensitive to secretin and CCK-PZ. In addition, their responsiveness is lower when they are incubated with surrounding exocrine tissue than when they are free of exocrine cells (8). The effectiveness of polypeptide hormones in our system seems to be related, at least partially, to the low proteolytic activity of the newborn rat pancreas (18, 34).

The status of GIP as a gastrointestinal hormone and more precisely as a component of "incretin" seems to be well established (31). As far as VIP is concerned, the picture is not as clear, although several observations favor the concept that it has a role of this type. Thus, 1) the VIP content of the gut of a variety of species, including man, is very large (5); 2) VIP-containing cells have been observed in the intestinal mucosa by immunofluorescence (32); 3) the VIP content of the jejunum-ileum of the rat increases sharply at the time of weaning, that is, between the 14th and the 21st day of age (24), coincident with a similar rise in the gastrin content of the fundus (27); 4) VIP target tissues, such as liver (2, 9), fat cells (2, 3, 10) and the exocrine pancreas (7) contain VIP receptors related to the activation of adenylate cyclase (2, 7, 9, 10). By analogy with other well known peptide hormones, these data suggest that VIP may act through a classic hormonal mechanism, involving release into the blood, binding to specific sites on plasma membranes and activation of adenylate cyclase. On the other hand, no physiologic variations in radioimmunoassayable plasma VIP have been observed up to now. Furthermore, VIP, or a VIP-like material has been found in the central nervous system by radioimmunoassay (5, 38) and by radioreceptor-assay (5) and in nerve terminals (36) and nerves with cerebrovascular (25) and intestinal distribution (26) by radioimmunoassay and immunofluorescent techniques. These observations suggest that VIP may act as a "local hormone" or as a neurotransmitter substance, rather than as a hormone in the classic sense of the word.

On the basis of present knowledge about VIP, we propose the following conclusions regarding the relationship between VIP and the endocrine pancreas (see Fig. 7):

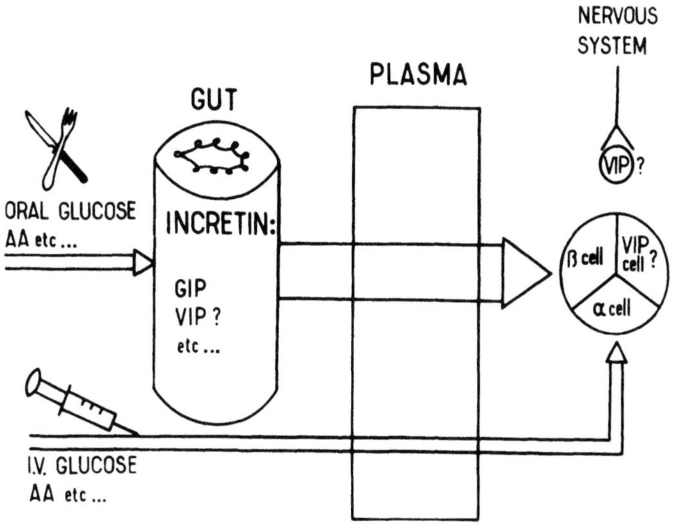

Fig. 7. Modified diagram showing the possible role of GIP and VIP in endocrine activity of the pancreas (See Fig. 1).

- VIP, like GIP, may be released from the gut after an oral load of nutrients and, through a classic hormonal mechanism, induce insulin and glucagon release.

- VIP, produced by specific cells in the pancreas (5, 26), may stimulate the secretory activity of surrounding A and B cells, without necessarily being released into the blood. A similar mechanism has been suggested for the effect of glucagon on insulin release on the basis of ultrastructural connections between A and B cells (30).

- The action of VIP on the endocrine pancreas may be aided by its dilator effect on the islet capillaries.

- VIP, or a closely related substance may be present mainly in the synaptosomes and may act as a mediator for the nervous control of A and B cell metabolism. Thus, VIP could be considered a neurohormone or a transmitter of nervous impulses. The latter view is gaining increasing experimental support but, in any case, VIP will have to be taken into account in any future investigation of the physiology of the endocrine pancreas.

Acknowledgements

We should like to express our thanks to Drs. V. Mutt, J. C. Brown and J. Schlichtkrull for providing the hormones and to C. Brunet and D. Lhenry for their help in the preparation of the manuscript. This work was supported by the Institut National de la Santé et de la Recherche Médicale (A.T.P. n° 16-75-39), by the Centre National de la Recherche Scientifique, and by a grant from the Fondation pour la Recherche Médicale Française.

REFERENCES

1. Bataille, D., Besson, J., Bastard, C., Laburthe, M. and Rosselin, G.: Specificity in hormone receptor interaction: Studies with insulin, glucagon and vasoactive intestinal peptide (VIP), In: Hormonal Receptors in Digestive Tract Physiology. North Holland Publ. Co. Amsterdam, p. 113-125.

2. Bataille, D., Freychet, P. and Rosselin, G. (1974): Interactions of glucagon, gut glucagon, vasoactive intestinal polypeptide and secretin with liver and fat cell plasma membranes: binding to specific sites and stimulation of adenylate cyclase. Endocrinology 95, 713-721.

3. Bataille, D., Rosselin, G. and Freychet, P. (1975): Interactions of glucagon, gut glucagon, vasoactive intestinal polypeptide and secretin with their membrane receptors. Israel J. Med. Sci. 11, 687-692.

4. Bataille, D., Rosselin, G., Freychet, P. and Mutt, V. (1975): Partial purification and characterization of active "enteroglucagon" from porcine jejunoileum. Diabetologia 11, 331 (abst.).

5. Besson, J., Laburthe, M., Bataille, D. and Rosselin, G. (in preparation).

6. Brown, J. C. (1971): A gastric inhibitory polypeptide. I. The amino acid composition and tryptic peptides. Can. J. Biochem. 49, 255-261.

7. Christophe, J. P., Conlon, T. P. and Gardner, J. D. (1976): Interaction of porcine vasoactive intestinal peptide with dispersed pancreatic acinar cells from the guinea pig. J. Biol. Chem. 251, 4629-4634.

8. Danielsson, Å. and Lernmark, Å. (1974): Effects of pancreozymin and secretin on insulin release and the role of exocrine pancreas. Diabetologia 10, 407-409.

9. Desbuquois, B. (1974): The interaction of vasoactive intestinal polypeptide and secretin with liver cell membranes. Eur. J. Biochem. 46, 439-450.

10. Desbuquois, B., Laudat, M. H. and Laudat, P. (1973): Vasoactive intestinal polypeptide and glucagon: stimulation of adenylate cyclase activity via distinct receptors in liver and fat cell membranes. Biochem. Biophys. Res. Commun. 53, 1187-1194.

11. Dupré, J. (1964): An intestinal hormone affecting glucose disposal in man. Lancet 2, 672-673.

12. Dupré, J. and Beck, J. C. (1966): Stimulation of release of insulin by an extract of intestinal mucosa. Diabetes 15, 555-559.

13. Dupré, J., Ross, S., Watson, D. and Brown, J. C. (1973): Stimulation of insulin secretion by gastric inhibitory polypeptide in man. J. Clin. Endocr. Metab. 37, 826-828.

14. Duran Garcia, S., Jarrousse, C. and Rosselin, G. (1976): Interaction of glucose, cyclic AMP, somatostatin and sulfonylureas on the ^{3}H-leucine incorporation into immunoreactive insulin. J. Clin. Invest. 57, 230-243.

15. Eloy, R., Garaud, J. C., Moody, A., Jaeck, D. and Grenier, J. F. (1975): Jejunal factor stimulating insulin release in the isolated perfused canine pancreas and jejunum. Horm. Metab. Res. 7, 461-467.

16. Gregory, R. A. and Tracy, H. J. (1961): The preparation and properties of gastrin. J. Physiol. 156, 523-543.

17. Jarrousse, C., Rançon, F., Rosselin, G. and Freychet, P. (1973): Sécrétion de l'insuline et du glucagon par le pancréas du rat nouveau-né: effet du glucose et de l'adénosine 3'-5' cyclique monophosphate. C. R. Acad. Sci. Paris (D) 276, 797-800.

18. Jarrousse, C. and Rosselin, G. (1975): Interaction of amino acids and cyclic AMP on the release of insulin and glucagon by newborn rat pancreas. Endocrinology 96, 168-177.

19. Jarrousse, C. and Rosselin, G. (1975): Regulation by glucose and cyclic nucleotides of the glucagon and insulin release induced by amino acids. Diabète et Métabolisme 1, 135-142.

20. Johnson, D. G., Ensinck, J. W., Koerker, D., Palmer, J. and Goodner, J. C. (1975): Inhibition of glucagon and insulin secretion by somatostatin in the rat pancreas perfused in situ. Endocrinology 96, 370-374.

21. Jorpes, J. E., Mutt, V., Magnusson, S. and Steele, B. B. (1962): Amino acid composition and N-terminal amino acid sequence of porcine secretin. Biochem. Biophys. Res. Commun. 9, 275-279.

22. Jorpes, J. E., Mutt, V. and Toezko, K. (1964): Further purification of cholecystokinin and pancreozymin. Acta Chem. Scand. 18, 2408-2410.

23. Konturek, S. J., Pucher, A. and Radecki, T. (1976): Comparison of vasoactive intestinal peptide and secretin in stimulation of pancreatic secretion. J. Physiol. 255, 497-509.

24. Laburthe, M., Bataille, D. and Rosselin, G. (in press): Vasoactive intestinal peptide (VIP): variation of the jejuno-ileal content in the developing rat as measured by radioreceptorassay. Acta Endocr.

25. Larsson, L. I., Edvinsson, L., Fahrenkrug, J., Hakanson, R., Owman, C., Schaffalitzky de Muckadell, O. and Sundler, F. (1976): Immunohistochemical localization of a vasoactive intestinal polypeptide (VIP) in cerebrovascular nerves. Brain Res. 113, 400-404.

26. Larsson, L. I., Fahrenkrug, J., Schaffalitzky de Muckadell, O., Sundler, F., Håkanson, R. and Rehfeld, J. F. (1976): Localization of vasoactive intestinal polypeptide (VIP) to central and peripheral neurons. Proc. Natl. Acad. Sci. USA 73, 3197-3200.

27. Lichtenberger, L. and Johnson, L. R. (1974): Gastrin in the ontogenic development of the small intestine. Amer. J. Physiol. 227, 390-395.

28. McIntyre, N., Holdsworth, C. D. and Turner, D. S. (1964): New interpretation of oral glucose tolerance. Lancet 2, 20-21.

29. Norfleet, W. T., Pagliara, A. S., Haymond, M. W. and Matschinsky, F. (1975): Comparison of alpha and beta cell secretory responses in islets isolated with collagenase and in isolated perfused pancreas of rats. Diabetes 24, 961-970.

30. Orci, L., Malaisse-Lagae, F., Ravazzola, M., Rouiller, C., Renold, A. E., Perrelet, A. and Unger, R. H. (1975): A morphological basis for intracellular communication between α and β cells in the endocrine pancreas. J. Clin. Invest. 56, 1066-1070.

31. Pederson, R. A. and Brown, J. C. (1976): The insulinotropic action of gastric inhibitory polypeptide in the perfused isolated rat pancreas. Endocrinology 99, 780-785.

32. Polak, J. M., Pearse, A. G. E., Garaud, J. C. and Bloom, S. R. (1974): Cellular localization of a vasoactive intestinal peptide in the mammalian and avian gastrointestinal tract. Gut 15, 720-724.

33. Rabinovitch, A. and Dupré, J. (1974): Effect of gastric inhibitory polypeptide in impure pancreozymin-cholecystokinin on plasma insulin and glucagon in the rat. Endocrinology 94, 1139-1144.

34. Robberecht, P., Dechodt-Lanckman, M., Camus, J., Bruylands, J. and Christophe, J. (1971): Rat pancreatic hydrolases from birth to weaning and dietary adaptation after weaning. Amer. J. Physiol. 221, 376-381.

35. Rosselin, G., Assan, R., Yalow, R. and Berson, S. A. (1966): Separation of antibody-bound and unbound peptide hormones labelled with iodine 131 by talcum powder and precipitated silica. Nature 216, 355-357.

36. Said, S. I., personal communication.

37. Said, S. I. and Mutt, V. (1972): Isolation from porcine intestinal wall of a vasoactive octacosapeptide related to secretin and to glucagon. Eur. J. Biochem. 28, 199-204.

38. Said, S. I. and Rosenberg, R. N. (1976): Vasoactive intestinal polypeptide: Abundant immunoreactivity in neural cell lines and normal nervous tissue. Science 192, 907-908.

39. Schebalin, M., Brooks, A. M., Said, S. I. and Makhlouf, G. M. (1974): The insulinotropic effect of vasoactive intestinal peptide (VIP): direct evidence from in vitro studies. Gastroenterology 66, 772 (abst.).

40. Turner, D. S., Etheridge, L., Marks, V., Brown, J. C. and Mutt, V. (1974): Effectiveness of the intestinal polypeptides IRP, GIP, VIP and motilin on

insulin release in the rat. Diabetologia 10, 459-463.

41. Unger, R. H. and Eisentraut, A. M. (1969): Enteroinsular axis. Arch. Intern. Med. 123, 261-266.

42. Unger, R. H., Eisentraut, A. M., McCall, M. S. and Madison, L. L. (1961): Glucagon antibodies and an immunoassay for glucagon. J. Clin. Invest. 40, 1280-1289.

43. Valverde, I., Rigopoulo, D., Marco, J., Faloona, G. R. and Unger, R. H. (1970): Characterization of glucagon-like immunoreactivity (GLI). Diabetes 19, 614-623.

44. Guidoux-Grassi, L. and Felber, J. P. (1968): Effect of secretin on insulin release by the rat pancreas. Diabetologia 4, 391-392.

45. Vauclin, N., Gespach, C., Bataille, D., Jarrousse, C. and Rosselin, G. (in press): Sécrétion d'insuline et de glucagon par le pancréas perifusé de rat nouveau-ne: Effet du peptide intestinal vasoactif ou VIP. C. R. Acad. Sci. Paris.

46. Zunz, E. and La Barre, J. (1928): Hyperinsulinémie consécutive à l'injection de solution de sécrétine non hypotensive. C. R. Soc. Biol. (Paris) 98, 1435-1438.

THE ROLE OF THE AUTONOMIC NERVOUS SYSTEM IN GLUCAGON SECRETION

A. Kaneto

Third Department of Internal Medicine, University of Tokyo
Tokyo, Japan

The conditions under which the sympathetic nervous system stimulates the secretion of glucagon have not yet been adequately defined. However, stressful situations, such as intensive physical exercise, birth, starvation, severe hypoglycemia, infection, extensive burns, rapid exsanguination, myocardial infarction and diabetes mellitus may be associated with heightened sympathetic activity. In these situations an increased supply of metabolic fuels in the form of endogenous glucose and free fatty acids would be advantageous and may be rapidly attained by direct neural stimulation of the liver and adipose tissue. This mobilization of nutrients is aided by an increased glucagon secretion in response to primary sympathetic nerve stimulation and to increased release of the catecholamines due to concomitant stimulation of the adrenal medulla.

The circumstances where the parasympathetic nervous system is stimulated are less clearly understood. However, an increase in plasma insulin has been reported in some humans during imaginary food ingestion under hypnosis. Visual and olfactory food stimuli have been known to result in a cephalic phase of insulin secretion, without demonstrable changes in plasma glucose, in obese young persons. The mechanisms involved in these types of hormonal response are not clear, but most probably are reflex in nature, requiring integrated activity of the parasympathetics through the vago-insulin and vago-glucagon systems.

Thus, under stress or under rhythmic physiologic phenomena, such as feeding, both divisions of the autonomic nervous system may play a role leading to changes in the endocrine function of the pancreas.

Since Langerhans (45) observed that autonomic nerve fibers were closely associated with the pancreatic islets, many histologists have confirmed his finding (58). Electron micrographs have shown autonomic nerve terminals in close proximity to both A and B cells (3, 21, 43, 51, 69, 76, 78). The A cells have been clearly demonstrated to be in close contact with cholinergic as well as adrenergic nerve endings (61).

More than one century after the piqûre experiments of Claude Bernard in dogs (4), Rosenberg and DiStefano (63) found that the microinjection of epinephrine into the fourth ventricle induced hyperglycemia in cats. They observed that the hyperglycemic response to intravenous epinephrine was inhibited by brain section distal to this region and by spinal cord transection at C_1-C_3, but not by midcollicular transection, and concluded that epinephrine hyperglycemia results in large part from activation of a hyperglycemic brain center. Their observations were extended by Ezdinli et al. (22). According to these authors, the interruption of cervical spinal cord pathways (C_2-C_7) resulted in a significant fall of the blood glucose level and inhibited the hyperglycemic response to epinephrine in dogs, without blocking glucagon hyperglycemia. These observations support the conclusion that the brain is involved in the regulation of the blood glucose level and that brain centers play a major role in mediating epinephrine hyperglycemia, thus leading to a revised explanation for the hyperglycemia of central nervous system origin. The release of epinephrine from the adrenal medulla during splanchnic nerve stimulation has long been known to cause hepatic glycogenolysis (13), nevertheless liver glycogen can be mobilized rapidly in response to stimulation of the hepatic sympathetic innervation, at frequencies within the physiological range, also in adrenalectomized animals (17-19). Ezdinli and Sokal (23) demonstrated that only glucagon can serve as a physiologic hepatic glycogenolytic agent and that the effect of moderate doses of epinephrine on the liver is an indirect one, possibly mediated through stimulation of glucagon secretion.

The central nervous system, which is critically dependent on glucose for its own metabolism, contains one or more centers sensitive to blood glucose concentration which initiate the reactions to hypoglycemia (64), thus playing a role in homeostasis. These reactions include stimulation of glucagon and epinephrine secretion via sympathetic pathways and a direct neural action on hepatic glycogenolysis (67, 68). Frohman and Bernardis (25) demonstrated that stimulation of the ventro-medial hypothalamus caused a rapid rise in plasma glucagon concentration in the rat and that this response persisted after removal of both adrenal glands. In the baboon a substantial rise in the plasma glucagon level occurs in response to stress or anxiety with such rapidity as to suggest a direct nervous pathway (5).

Role of the Sympathetic Nervous System in Glucagon Secretion

Morphologic changes have been observed in A and B cells following sympathetic nerve stimulation in the cat (66). Esterhuizen and Howell (20) demonstrated significant rises in the glucagon concentration in pancreatic effluent plasma in response to sympathetic nerve stimulation in certain cat preparations. The ultrastructure of the A cells of innervated islets was correlated with this increased secretory activity.

By means of measurement of blood flow and hormone concentration in the venous effluent from the pancreas, Marliss et al. (55) observed that electrical stimulation of the distal end of the discrete bundles of mixed nerve fibers along the superior pancreaticoduodenal artery in anesthetized dogs was followed by an increase in glucagon output. That the preparation behaved in physiologic fashion was confirmed by a fall in glucagon output and a rise in insulin output during hyperglycemia induced by intravenous glucose.

Bloom et al. (6) investigated the extent to which the splanchnic sympathetic innervation was implicated in the control of plasma glucagon concentration in the young calf. Stimulation of the peripheral ends of both splanchnic nerves caused an abrupt increase in plasma glucagon concentration in adrenalectomized calves. In spite of the accompanying hyperglycemia, insulin release was completely inhibited throughout the period of stimulation at all frequencies within the physiologic range tested, but rebounded immediately when the stimulation was discontinued. The authors concluded that tonic changes in sympathetic efferent activity are likely to modify plasma glucagon concentration in normal conscious calves. According to them, even though the glucagon release mechanism is extremely sensitive to stimulation via this pathway, there is as yet no evidence that the sympathetic system responds to fluctuations in plasma glucose concentration within the normal range.

Kaneto et al. (unpublished) observed that electrical stimulation of the cut peripheral end of the splanchnic nerve at the subdiaphragmatic level caused an increase in plasma glucagon concentration and plasma flow in the cranial pancreaticoduodenal vein, reflecting an augmentation of glucagon output from the pancreas in anesthetized dogs. During the stimulation a slow increase in insulin output followed the hyperglycemia, and it was not suppressed by atropine which had inhibited the basal output of both hormones. This suggests the insulin release cannot be ascribed to the stimulation of aberrant parasympathetic fibers in the splanchnic nerve.

The results of these experiments in vivo are in accord with the in vitro studies of Leclercq-Meyer et al (49) where a large increase in glucagon output by isolated pancreatic islets was observed when epinephrine was added to the preparation. Iversen (31) reported that isoproterenol, a β-adrenergic drug, stimulated glucagon as well as insulin release from the isolated, perfused canine pancreas. Loubatieres et al (53) documented that the β-adrenergic receptor involved in isoproterenol-provoked insulin secretion is of type β_2. Luyckx and Lefebvre (54) observed a rise in plasma glucagon in rats subjected to forced swimming. This exercise-induced glucagon rise was completely blocked by propranolol, a specific β-receptor blocking agent, but not by practolol, a specific blocker of the β_1 receptors. Kaneto et al (39) demonstrated that glucagon and insulin secretion stimulated by trimetoquinol, a selective receptor stimulant of the β_2 type, as well as by isoproterenol was totally blocked by propranolol, but not by practolol (Fig. 1).

It is well known that the α-adrenergic receptors are responsible for the inhibition of insulin release induced by epinephrine (60). Phenylephrine and methoxamine, highly specific α-receptor stimulants, suppressed basal output of glucagon and insulin into the cranial pancreaticoduodenal vein when infused intrapancreatically in anesthetized dogs, and that this inhibition was prevented by phentolamine, a specific α-receptor blocker (Fig. 2). These pharmacologic findings suggest that the glucagon secretion induced by stimulation of the sympathetic nervous system in some mammalian species is mediated by the preponderant action of the β_2-receptors over that of the α-receptors in the A-cell.

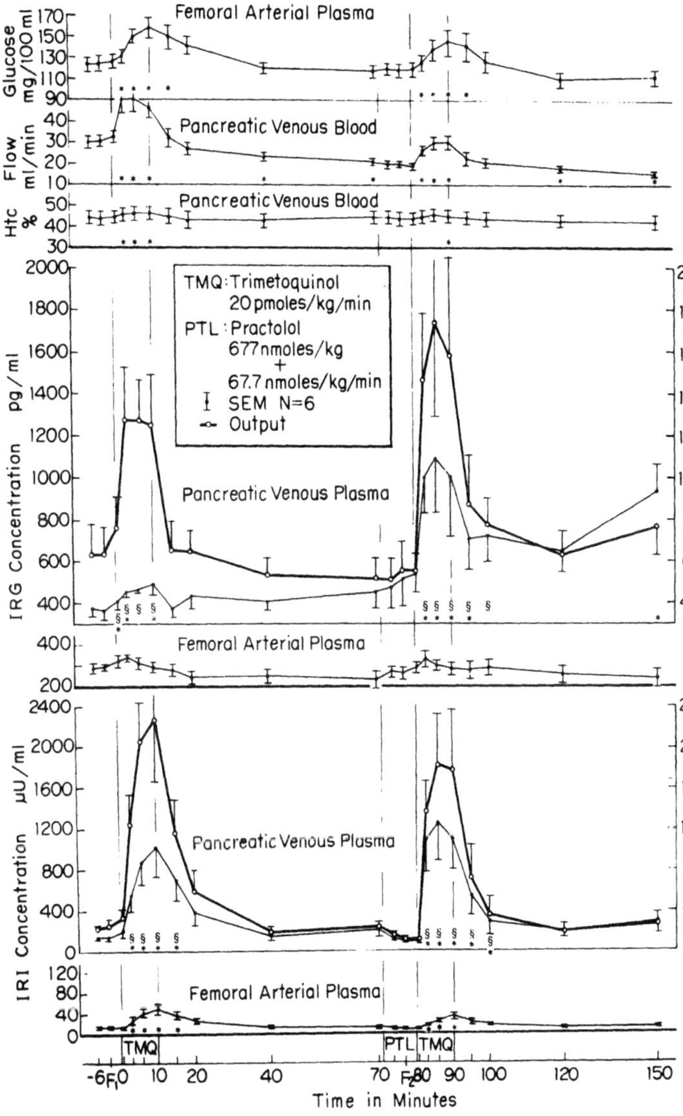

Fig. 1. Effect of practolol pretreatment on the responses of arterial plasma glucose, pancreatic venous blood flow, hematocrit and plasma IRG and IRI to the infusion of trimetoquinol (20 pmol/kg/min). * = P 0.05 vs pre-infusion values, designated F_1 or F_2. § = P <0.05 of IRG and IRI output, indicated by open circles.

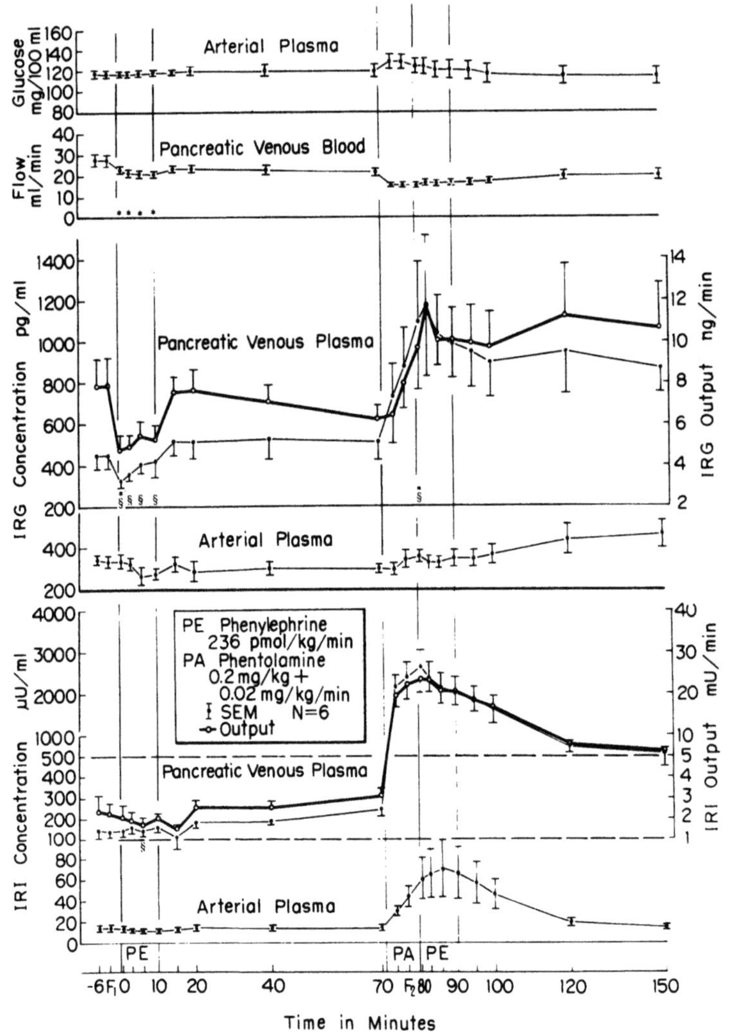

Fig. 2. Effect of phenylephrine infused intrapancreatically at a dose of 236 pmol/kg/min on the levels of arterial plasma glucose, pancreatic venous blood flow and plasma IRG and IRI. Symbols as in Fig. 1.

Role of the Parasympathetic Nervous System in Glucagon Secretion

Tyler and Kajinuma (70, 71) demonstrated that, in ducks, glucagon secretion was stimulated by the intravenous infusion of norepinephrine and inhibited by the infusion of acetylcholine. The opposite results were seen for insulin release. This reciprocal control of pancreatic A-cell function by the two neurochemical transmitters seems to suggest that, in birds, there exists a dual system of vago-insulin and sympatho-glucagon discharges.

Kaneto et al (37) noticed a delayed hyperglycemic response to intrapancreatic infusion of some acetylcholine derivatives which caused a prompt increase of insulin secretion. Such a seemingly paradoxical hyperglycemia also followed stimulation of the vagus nerve (26, 36).

Iversen (32) showed an increase of glucagon and insulin secretion during perfusion of the isolated canine pancreas with acetylcholine. Kaneto et al (39) demonstrated that intrapancreatic infusion of acetylcholine in anesthetized dogs caused an immediate increase in plasma glucagon and insulin and in plasma flow in the pancreatic vein which was totally prevented by atropine.

According to Bloom et al. (7), the administration of atropine alone to normal fasting calves produced a significant fall in plasma concentrations of both glucose and glucagon, without affecting that of insulin. In atropinized calves with cut splanchnic nerves, a small dose of insulin produced severe hypoglycemia accompanied by convulsions. In spite of the intensity of the hypoglycemic stimulus, the rise in plasma glucagon concentration was both delayed and diminished in these animals as compared to the intact calves or calves with only the splanchnic nerve sectioned. A significant increase in plasma glucagon concentration occurred in response to stimulation of the peripheral ends of the thoracic vagi in adrenalectomized calves with cut splanchnic nerves under barbiturate anesthesia. This was accompanied by a rise in mean plasma glucose concentration which correlated significantly with the glucagon response. These results seem to indicate that the cholinergic mechanism is more sensitive to hypoglycemia than the sympathetic system.

Kaneto et al (unpublished) reported that electrical stimulation of the dorsal vagus trunk at the diaphragmatic level caused a rapid increase in blood flow and in the concentration of plasma glucagon and insulin in the cranial pancreaticoduodenal vein of anesthetized dogs which was completely blocked by pretreatment with atropine.

Bloom et al. (8) showed that in fasting men a significant fall in plasma pancreatic glucagon was produced by the injection of atropine. Furthermore, the glucagon response to insulin hypoglycemia in patients with truncal vagotomy was significantly less than in preoperative controls or after selective vagotomy. These results suggest that also the vagus nerve is implicated in the regulation of glucagon release from the pancreas under resting conditions. Although the stimulus caused by vagal activation is widespread, making it hard to exclude secondary or indirect effects, these in vivo results as well as the in vitro data (32) strongly indicate the existence of a vago-glucagon system in some mammalian species.

When an alteration of the "internal milieu" associated with an increased release of insulin occurs, the glucose concentration would fall to a critical point if factors elevating the blood glucose level were not mobilized. The possible stimulation of insulin and glucagon secretion by the vagus could be an important mechanism for the maintenance of glucose homeostasis.

Role of Hypothalamic Polypeptides on Glucagon Secretion

Some polypeptides, such as somatostatin (11) and substance P (15) have been found in the hypothalamus, the gut and/or the pancreas. It has been shown that the systemic administration of two hypothalamic polypeptides, substance P and neurotensin (14), increases the plasma concentration of glucose and glucagon and reduces that of insulin in the rat (12), whereas the administration of somatostatin suppresses all of them (10, 16, 27, 33, 34, 44, 48, 56, 65, 77).

Kaneto et al (unpublished) observed that synthetic substance P and synthetic neurotensin increased plasma glucose, glucagon and insulin concentrations as well as plasma flow in the cranial pancreaticoduodenal vein, when infused intrapancreatically in anesthetized dogs (Figs. 3 and 4). While the response to substance P was quite rapid, the response to the same molar dose of neurotensin was somewhat delayed, and the latter was suppressed by propranolol. This suggests that the neurotensin-induced enhancement in glucagon and insulin output may be mediated through the β-adrenergic receptor system. These two polypeptides as well as other gut hormones, such as vasoactive intestinal polypeptide (VIP) (Kaneto, et al., unpublished) have the common property of increasing pancreatic venous plasma flow, raising the problem of whether the augmentation of hormonal output may be attributed to the circulatory increase per mass of islets rather than to a direct action on the A- and B-cells.

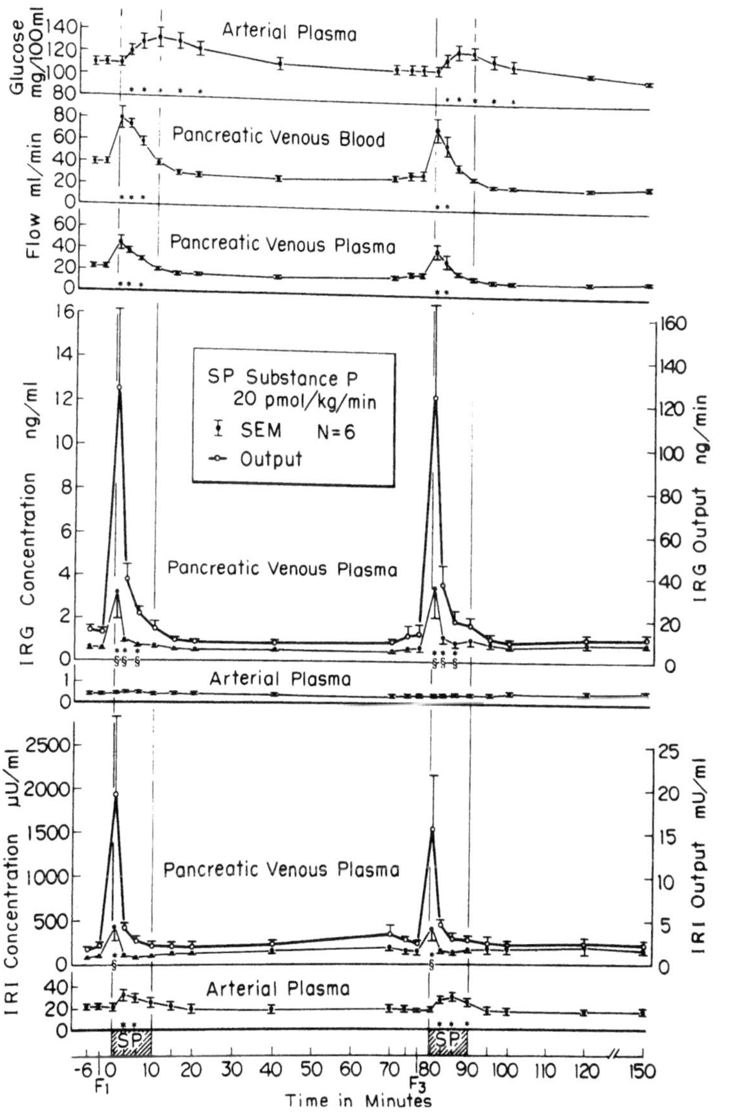

Fig. 3. (see following page for legend)

Fig. 3. (See previous page). Effect of substance P infused intrapancreatically at a dose of 20 pmol/kg/min on the levels of arterial plasma glucose, pancreatic venous blood and plasma flow and plasma IRG and IRI. Symbols as in Fig. 1.

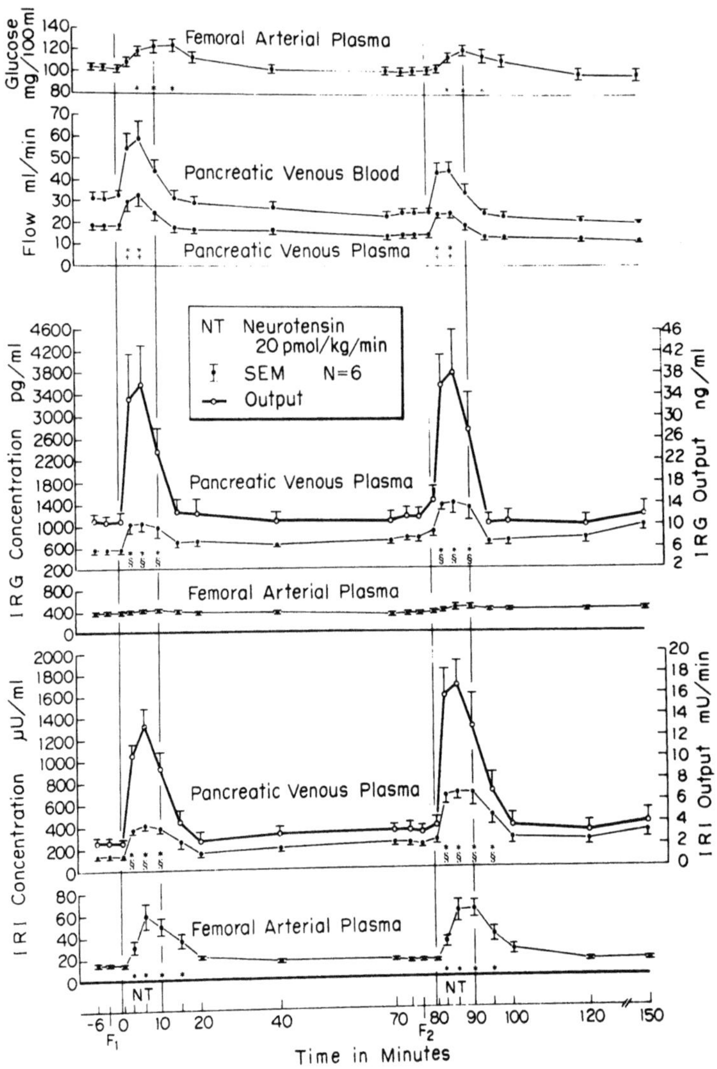

Fig. 4. Effect of neurotensin infused intrapancreatically at a dose of 20 pmol/kg/min on the levels of arterial plasma glucose, pancreatic venous blood and plasma flow, and plasma IRG and IRI. † = P <0.05 (pancreatic venous flow). Other symbols as in Fig. 1.

REFERENCES

1. Aguilar-Parada, E., Eisentraut, A. M. and Unger, R. H. (1969): Effects of starvation on plasma pancreatic glucagon in normal man. Diabetes 18, 717-723.

2. Allison, S. P., Hinton, P. and Chamberlain, M. J. (1968): Intravenous glucose-tolerance, insulin and free-fatty-acid levels in burned patients. Lancet 2, 1113-1116.

3. Bencosme, S. A. (1959): Studies on the terminal autonomic nervous system with special reference to the pancreatic islets. Lab. Invest. 8, 629-646.

4. Bernard, C. (1849): Chiens rendus diabétiques. Compt. Rend. Soc. Biol. 1 60-64.

5. Bloom, S. R., Daniel, P. M., Johnston, D. I., Ogawa, O. and Pratt, O. E. (1973): Release of glucagon induced by stress. Quart. J. exp. Physiol. 58, 99-108.

6. Bloom, S. R., Edwards, A. V. and Vaughan, N. J. A. (1973): The role of the sympathetic innervation in the control of plasma glucagon concentration in the calf. J. Physiol. 233, 457-466.

7. Bloom, S. R., Edwards, A. V. and Vaughan, N. J. A. (1974): The role of the autonomic innervation in the control of glucagon release during hypoglycemia in the calf. J. Physiol. 236, 611-623.

8. Bloom, S. R., Vaughan, N. J. A. and Russell, R. C. G. (1974): Vagal control of glucagon release in man. Lancet 2, 546-549.

9. Böttger, I., Schlein, E. M., Faloona, G. R., Knochel, J. P. and Unger, R. H. (1972): The effect of exercise on glucagon secretion. J. Clin. Endocrinol. Metab. 35, 117-125.

10. Brazeau, P. and Guillemin, R. (1974): Somatostatin: newcomer from the hypothalamus. N. Engl. J. Med. 290, 963-964.

11. Brazeau, P., Vale, W., Burgus, R., Ling, N., Butcher, M., Rivier, J. and Guillemin, R. (1973): Hypothalamic polypeptide that inhibits the secretion of immunoreactive pituitary growth hormone. Science 179, 77-79.

12. Brown, M. and Vale, W. (1976): Effects of neurotensin and substance P on plasma insulin, glucagon and glucose levels. Endocrinology 98, 819-822.

13. Cannon, W. B., McIver, M. A. and Bliss, S. W. (1924): Studies on the conditions of activity in endocrine glands. XIII. A sympathetic and adrenal mechanism for mobilising sugar in hypoglycemia. Am. J. Physiol. 69, 46-66.

14. Carraway, R. and Leeman, S. E. (1973): The isolation of a new hypotensive peptide, neurotensin, from bovine hypothalami. J. Biol. Chem. 248, 6854-6861.

15. Chang, M. M. and Leeman, S. E. (1970): Isolation of a sialogogic peptide from bovine hypothalamic tissue and its characterization as substance P. J. Biol. Chem. 245, 4784-4790.

16. DeVane, G. W., Siler, T. M. and Yen, S. S. C. (1974): Acute suppression of insulin and glucose levels by synthetic somatostatin in normal human subjects. J. Clin. Endocrinol. Metab. 38, 913-915.

17. Edwards, A. V. (1972): The sensitivity of the hepatic glycogenolytic mechanism to stimulation of the splanchnic nerves. J. Physiol. 220, 315-334.

18. Edwards, A. V. (1972): The hyperglycemic response to stimulation of the hepatic sympathetic innervation in adrenalectomized cats and dogs. J. Physiol. 220, 697-710.

19. Edwards, A. V. and Silver, M. (1970): The glycogenolytic response to stimulation of the splanchnic nerves in adrenalectomized calves. J. Physiol. 211, 109-124.

20. Esterhuizen, A. C. and Howell, S. L. (1970): Ultrastructure of the A cells of cat islets of Langerhans following sympathetic stimulation of glucagon secretion. J. Cell. Biol. 46, 593-598.

21. Esterhuizen, A. C., Spriggs, T. L. B. and Lever, J. D. (1968): Nature of islet cell innervation in the cat pancreas. Diabetes 17, 33-36.

22. Ezdinli, E. Z., Javid, R., Owens, G. and Sokal, J. E. (1968): Effect of high spinal cord section on epinephrine hyperglycemia. Am. J. Physiol. 214, 1019-1024.

23. Ezdinli, E. Z. and Sokal, J. E. (1966): Comparison of glucagon and epinephrine effects in the dog. Endocrinology 78, 47-54.

24. Felig, P., Wahren, J., Hendler, R. and Ahlborg, G. (1972): Plasma glucagon levels in exercising man. N. Engl. J. Med. 287, 184-185.

25. Frohman, L. A. and Bernardis, L. L. (1971): Effect of hypothalamic stimulation on plasma glucose, insulin and glucagon levels. Am. J. Physiol. 221, 1596-1603.

26. Frohman, L. A., Ezdinli, E. Z. and Javid, R. (1967): Effect of vagotomy and vagal stimulation on insulin secretion. Diabetes 16, 443-448.

27. Gerich, J. E., Lorenzi, M., Schneider, V., Karam, J. H., Rivier, J., Guillemin, R. and Forsham, P. H. (1974): Effects of somatostatin on plasma glucose and glucagon levels in human diabetes mellitus. N. Engl. J. Med. 291, 544-547.

28. Girard, J., Bal, D. and Assan, R. (1972): Glucagon secretion during the early postnatal period in the rat. Horm. Metab. Res. 4, 168-170.

29. Goldfine, I. D., Abraira, C., Gruenewald, D. and Goldstein, M. S. (1970): Plasma insulin levels during imaginary food ingestion under hypnosis. Proc. Soc. Exp. Biol. Med. 133, 274-276.

30. Hinton, P., Allison, S. P., Littlejohn, S. and Lloyd, J. (1971): Insulin and glucose to reduce catabolic response to injury in burned patients. Lancet 1, 767-769.

31. Iversen, J. (1973): Adrenergic receptors and the secretion of glucagon and insulin from the isolated, perfused canine pancreas. J. Clin. Invest. 52, 2102-2116.

32. Iversen, J. (1973): Effect of acetylcholine on the secretion of glucagon and insulin from the isolated, perfused canine pancreas. Diabetes 22, 381-387.

33. Iversen, J. (1974); Inhibition of pancreatic glucagon release by somatostatin in vitro. Scand. J. Clin. Lab. Invest. 33, 125-129.

34. Johnson, D. G., Ensinck, J. W., Koerker, d., Palmer, J. and Goodner, C. J. (1975): Inhibition of glucagon and insulin secretion by somatostatin in the rat pancreas perfused in situ. Endocrinology 96, 370-374.

35. Johnston, D. I. and Bloom, S. R. (1973): Plasma glucagon levels in the term human infant and effect of hypoxia. Arch. Dis. Childh. 48, 451-545.

36. Kaneto, A., Kajinuma, H., Hayashi, M. and Kosaka, K. (1973): Effect of splanchnic nerve stimulation on glucagon and insulin secretion. Endocrinologia Japon. 20, 609-617.

37. Kaneto, A., Kajinuma, H. and Kosaka, K. (1975): Effect of splanchnic nerve stimulation on glucagon and insulin output in the dog. Endocrinology 96, 143-150.

38. Kaneto, A., Kajinuma, H., Kosaka, K. and Nakao, K. (1968): Stimulation of insulin secretion by parasympathomimetic agents. Endocrinology 83, 651-658.

39. Kaneto, A. and Kosaka, K. (1974): Stimulation of glucagon and insulin secretion by acetylcholine infused intrapancreatically. Endocrinology 95, 676-681.

40. Kaneto, A., Kosaka, K. and Nakao, K. (1967): Effects of stimulation of the vagus nerve on insulin secretion. Endocrinology 80, 530-536.

41. Kaneto, A., Miki, E. and Kosaka, K. (1974): Effects of vagal stimulation on glucagon and insulin secretion. Endocrinology 95, 1005-1010.

42. Kaneto, A., Miki, E. and Kosaka, K. (1975): Effect of $beta_1$ and $beta_2$ adrenoreceptor stimulants infused intrapancreatically on glucagon and insulin secretion. Endocrinology 97, 1166-1173.

43. Kobayashi, S. and Fujita, T. (1969): Fine structure of mammalian and avian pancreatic islets with special reference to D cells and nervous elements. Z. Zellforsch. Mikrosk. Anat. 100, 340-363.

44. Koerker, D. J., Ruch, W., Chideckel, E., Palmer, J., Goodner, C. J., Ensinck, J. and Gale, C. C. (1974): Somatostatin: hypothalamic inhibitor of the endocrine pancreas. Science 184, 482-484.

45. Langerhans, P. (1869): Beitrage zur mikroskopischen Anatomie der Bauchspeicheldrüse. Thesis, Friedrich-Wilhelms-Universitat, Berlin, p. 1.

46. Laniado, S., Segal, P. and Esrig, B. (1973): The role of glucagon hypersecretion in the pathogenesis of hyperglycemia following acute myocardial infarction. Circulation 48, 797-800.

47. Lawrence, A. M. (1966): Radioimmunoassayable glucagon levels in man: effects of starvation, hypoglycemia and glucose administration. Proc. Natl. Acad. Sci. USA 55, 316-320.

48. Leblanc, H., Anderson, J. R., Sigel, M. B. and Yen, S. S. C. (1975): Inhibitory action of somatostatin on pancreatic α and β cell function. J. Clin. Endocrinol. Metab. 40, 568-572.

49. Leclercq-Meyer, V., Brisson, G. R. and Malaisse, W. J. (1971): Effect of adrenaline and glucose on release of glucagon and insulin in vitro. Nature (Lond.) 231, 248-249.

50. Lefèbvre, P. J., Luyckx, A. S. and Federspil, G. (1972): Muscular exercise and pancreatic function in rats. Israel J. Med. Sci. 8, 390-393.

51. Legg, P. G. (1967): The fine structure and innervation of the beta and delta cells in the islets of Langerhans of the cat. Z. Zellforsch. Mikrosk. Anat. 80, 307-321.

52. Lindsey, C. A., Faloona, G. R. and Unger, R. H. (1975): Plasma glucagon levels during rapid exsanguination with and without adrenergic blockade. Diabetes 24, 313-316.

53. Loubatières, A., Mariani, M. M., Sorel, G. and Savi, L. (1971): The action of β-adrenergic blocking and stimulating agents on insulin secretion. Characterization of the type of β-receptor. Diabetologia 7, 127-132.

54. Luyckx, A. S. and Lefèbvre, P. J. (1974): Mechanisms involved in the exercise-induced increase in glucagon secretion in rats. Diabetes 23, 81-93.

55. Marliss, E. B., Girardier, L., Seydoux, J., Wollheim, C. B., Kanazawa, Y., Orci, L., Renold, A. E. and Porte, D. Jr. (1973): Glucagon release induced by pancreatic nerve stimulation in the dog. J. Clin. Invest. 52, 1246-1259.

56. Mortimer, C. H., Carr, D., Lind, T., Bloom, S. R., Mallinson, C. N., Schally, A. V., Tunbridge, W. M. G., Yeomans, L., Coy, D. H., Kastin, A., Besser, G. M. and Hall, R. (1974): Effects of growth-hormone release-inhibiting hormone on circulating glucagon, insulin, and growth hormone in normal, diabetic, acromegalic, and hypopituitary patients. Lancet 1, 697-701.

57. Müller, W. A., Faloona, G. R., Aguilar-Parada, E. and Unger, R. H. (1970): Abnormal alpha-cell function in diabetes. N. Engl. J. Med. 283, 109-115.

58. Munger, B. L. (1972): The histology, cytochemistry and ultrastructure of pancreatic islet A-cells. In: Glucagon: Molecular Physiology, Clinical and Therapeutic Implications. P. J. Lefebvre and R. H. Unger, Eds. Pergamon Press, Oxford and New York, pp. 7-25.

59. Parra-Covarrubias, A., Rivera-Rodriguez, I. and Lamaraz-Ugalde, A. (1971): Cephalic phase of insulin secretion in obese adolescents. Diabetes 20, 800-802.

60. Porte, D. Jr. (1967): A receptor mechanism for the inhibition of insulin release by epinephrine in man. J. Clin. Invest. 46, 86-94.

61. Renold, A. E. (1971): The beta cell and its reponses. Diabetes 21, (Suppl. 2), 619-631.

62. Rocha, D. M., Santeusanio, F., Faloona, G. R. and Unger, R. H. (1973): Abnormal pancreatic alpha-cell function in bacterial infections. N. Engl. J. Med. 288, 700-703.

63. Rosenberg, F. J. and Distefano, V. (1962): A central nervous system component of epinephrine hyperglycemia. Am. J. Physiol. 203, 782-788.

64. Sakata, K., Hayano, S. and Sloviter, H. A. (1963): Effect on blood glucose concentration of changes in availability of glucose to the brain. Am. J. Physiol. 204, 1127-1132.

65. Sakurai, H., Dobbs, R. and Unger, R. H. (1974): Somatostatin-induced changes in insulin and glucagon secretion in normal and diabetic dogs. J. Clin. Invest. 54, 1395-1402.

66. Sergeyeva, M. A. (1940): Microscopic changes in the islands of Langerhans by sympathetic and parasympathetic stimulation in the cat. Anat. Rec. 77, 297-317.

67. Shimazu, T. and Amakawa, A. (1968): Regulation of glycogen metabolism in liver by the autonomic nervous system. II. Neural control of glycogenolytic enzymes. Biochim. Biophys. Acta. 165, 335-348.

68. Shimazu, T. and Amakawa, A. (1968): Regulation of glycogen metabolism in liver by the autonomic nervous system. III. Differential effects of sympathetic-nerve stimulation and of catecholamines on liver phosphorylase. Biochim. Biophys. Acta. 165, 349-356.

69. Stahl, M. (1963): Electronenmikroskopische Untersuchungen über die vegetative Innervation der Bauchspeicheldrüse. Z. Mikrosk. Anat. Forsch. 70, 62-102.

70. Tyler, J. M. and Kajinuma, H. (1972): Influence of beta-adrenergic and cholinergic agents in vivo on pancreatic glucagon and insulin secretion. Diabetes 21 (Suppl. 1), 332.

71. Tyler, J. M., Mialhe, P. and Kajinuma, H. (1972): Stimulation of pancreatic glucagon secretion by norepinephrine in vivo. Excerpta Medica No. 256, IV International Congress Endocrinology, p. 67.

72. Unger, R. H. (1971): Glucagon and the insulin : glucagon ratio in diabetes and other catabolic illness. Diabetes 20, 834-838.

73. Unger, R. H., Aguilar-Parada, E., Müller, W. A. and Eisentraut, A. M. (1970): Studies on pancreatic alpha cell function in normal and diabetic subjects. J. Clin. Invest. 49, 837-848.

74. Unger, R. H., Eisentraut, A. M., McCall, M. S. and Madison, L. L. (1962): Measurements of endogenous glucagon in plasma and the influence of blood glucose concentration upon its secretion. J. Clin. Invest. 41, 682-689.

75. Vendsalu, A. (1960): Studies on adrenaline and noradrenaline in human plasma. Acta Physiol. Scand. 49 (Suppl. 173), 1-123.

76. Watari, N. (1968): Fine structure of nervous elements in the pancreas of some vetebrates. Z. Zellforsch. Mikrosk. Anat. 85, 291-314.

77. Weir, G. C., Knowlton, S. D. and Martin, D. B. (1974): Somatostatin inhibition of epinephrine-induced glucagon secretion. Endocrinology 95, 1744-1746.

78. Winborn, W. B. (1963): Light and electron microscopy of the islets of Langerhans of the Saimiri monkey pancreas. Anat. Rec. 147, 65-93.

IV. MODE OF ACTION OF GLUCAGON

INTERACTIONS OF HEPATOCYTE MEMBRANE RECEPTORS WITH PANCREATIC AND GUT GLUCAGON

J. J. Holst

Department of Clinical Chemistry, Bispebjerg Hospital
Copenhagen, Denmark

The pancreas contains at least two molecular forms of glucagon. One of them behaves like true glucagon on gel filtration and radioreceptor analysis. The other has no or very low affinity for the hepatocyte receptors, contains the entire glucagon molecule elongated from the C-terminal threonine residue and may be the biosynthetic precursor of glucagon. Human and porcine gastrointestinal mucosa contain several major glucagon components, two of which have the immunologic characteristic of the pancreatic fractions. The other components, which do not cross-react with antisera specific for pancreatic glucagon, can be separated into four fractions by isoelectric focusing. Although one of these binds to glucagon receptors and has glycogenolytic activity, the physiologic role of the gut glucagons remains obscure.

INTRODUCTION

The biochemistry of the glucagon receptor and the associated adenylate cyclase of the rat liver cell membranes has been intensively studied, notably by Rodbell and coworkers (12). Porcine hepatocyte membranes also possess glucagon receptors with high affinity for glucagon (7.0×10^9 l/mol; 4). Indeed, porcine membranes isolated by means of an aqueous two-phase polymer system form the basis for a radio-receptor assay for glucagon, which is sensitive, precise, and does not cross-react with non-glucagon gastrointestinal peptides. The glucagon radio-receptor assay constitutes an important supplement to the methods used in the characterization of the components of the glucagon group of peptides; until now most members of this group have been included because of glucagon-like behaviour in various assay systems, and it is still uncertain which of the gastrointestinal peptides are really glucagon-like from a chemical point of view. Molecules may be glucagon-like in that they possess glucagon-like biologic activity and/or glucagon-like immunoreactivity. Since true glucagon (MW 3885) has no unique metabolic actions (unless one combines several systems and includes knowledge of molar potencies) no assay based on biologic activity can be specific. Other substances which are obviously not glucagon-like from a chemical point of view, may possess strong glucagon-like bioactivity (e.g. catecholamines). If a substance, however, is found to possess glucagon-like immunoreactivity, this means that the substance shares an antigenic determinant with true glucagon. The radioimmunoassay is therefore, by definition, highly specific, in that one and only one antigenic determinant is being detected. The antigenic determinant, however, may be part of widely different molecules, and obviously nothing can be deduced about the rest of the molecule from the binding reactions of only a part of the molecule. The characterizations of the glucagons must rest upon the immunologic reactions of these moieties, until they can be prepared in such amounts and to such purity to allow for chemical and biologic analyses. At the early stages of characterization, analyses with a radioreceptor assay which combines some of the chemical specificity of the radioimmunoassay with the biologic relevance of the bioassay represent a valuable shortcut.

The Pancreatic Type and the Gut Type Glucagons

Since the demonstration of extrapancreatic sources of a peptide which is indistinguishable from true glucagon (6, 13), the designation "pancreatic glucagon", which usually means a peptide measured in a so-called specific glucagon assay, has lost some of its meaning. The designation "pancreatic type glucagon" (PTG) shall therefore be proposed. It covers all peptides, including "true glucagon", which are recognized in assays which do not crossreact with purified gut glucagon-like immunoreactive materials or GLI (14). Since the antibodies which are used in such assays are probably directed against the C-terminal of the glucagon molecule (1), the pancreatic type glucagons presumably all share the same C-terminal. Peptides of gastrointestinal origin (including pancreatic) which are not recognized in the specific assays, can then be distinguished by the designation gut type glucagon (GTG). These GTS's may be measured in specific assays (2, 7), or more commonly, with the aid of antisera which crossreact strongly with gut extracts; such antisera which should preferably crossreact completely with the above mentioned GLI, will then measure the total glucagon-like immunoreactivity, from which the GTG can be calculated by subtracting the PTG content. The cross-reacting antisera are not necessarily directed against the N-terminal part of the glucagon molecule, but in some instances seem to require an almost intact molecule. An illustration of this phenomenon is given in Fig. 1, in which the reactions of purified tryptic digests of glucagon with the antisera commonly in use in this laboratory are shown.

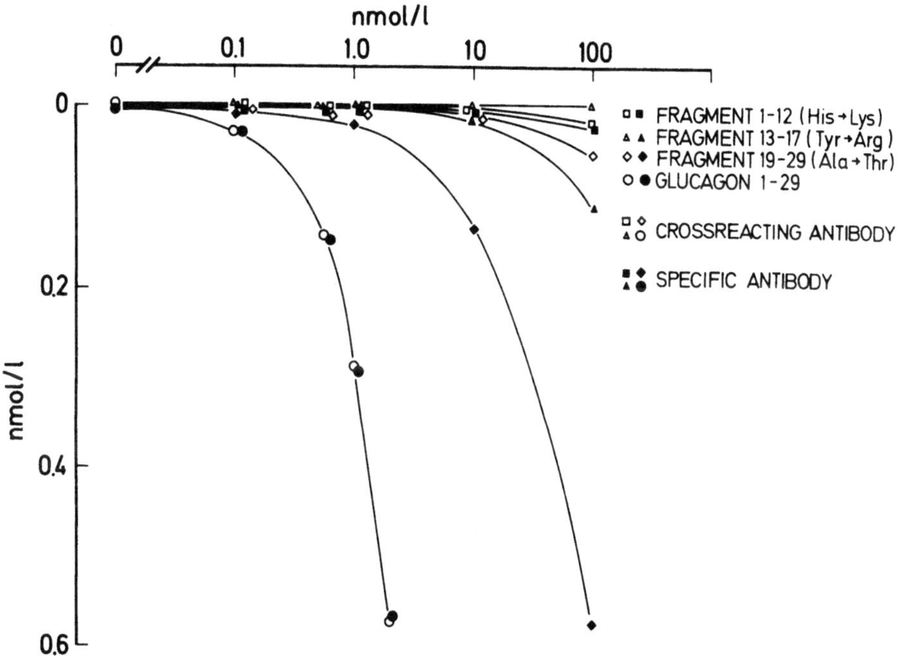

Fig. 1. Interaction of purified tryptic digests of glucagon in glucagon radioimmunoassays with crossreacting antibodies (measuring gut and pancreatic type glucagon, GTG + PTG, open symbols) or specific antibodies (measuring only PTG, closed symbols). The molar concentrations of glucagon and its fragments, added to the assay system, are indicated on the horizontal scale (logarithmic). The vertical scale represents the resulting inhibition of tracer binding in the assay, transformed into the concentration of glucagon required to cause equivalent inhibition. The tryptic fragments were gifts from Dr. A. Salokangas, National Institute for Medical Research, London, and L. G. Heding, Novo Research Institute, Copenhagen.

The Pancreatic Forms of Glucagon

In this laboratory the concentration of glucagon-like immunoreactivity in the human pancreas has been determined to range between 0.4 and 1.7 nmol-equivalents/g tissue (wet weight, 5 determinations). In two morphologically normal pancreata obtained at autopsy, the concentration was 0.9 and 1.1 nmol-equivalents/g. Fig. 2 shows the elution pattern of the PTG in an acid-ethanol extract of a human pancreas subjected to gel filtration on Sephadex G 50 in dilute acetic acid. The overwhelming majority of the PTG eluted at K_{av} 0.65, corresponding to the glucagon marker. However, small peaks can be seen at K_{av} 0.25 and 0.43. The peak at K_{av} 0.25 is a constant feature of the normal human pancreas, and was also found in three glucagon-producing tumors. Furthermore, a PTG-component of K_{av} 0.23 - 0.25 was also found in normal plasma (n = 3) and in plasma from patients with glucagon-producing tumors. We have not been able to identify PTG components of lower K_{av}. While the component of K_{av} 0.65 was identified by its full immunometric potency in the receptor assay (Fig. 2, lower panel), the other components did not interact with glucagon for binding to the receptors, at least, not in the concentrations reached in this and other similar experiments. When assayed with a crossreacting antiserum the elution patterns obtained in the above experiments was virtually identical to those obtained with the so-called specific antiserum. Gut type glucagon of pancreatic origin was therefore not identified.

The Gastrointestinal Forms of Glucagon

Descriptions of canine gastrointestinal glucagon forms may be found in previous publications (6, 8, 13, 16). The dog, however, differs markedly in this respect from man and pig (3, 5, 6) and here mainly porcine forms will be discussed.

The concentration of glucagon-like immunoreactivity in the porcine gastrointestinal mucosa increases from very low levels in the stomach (less than

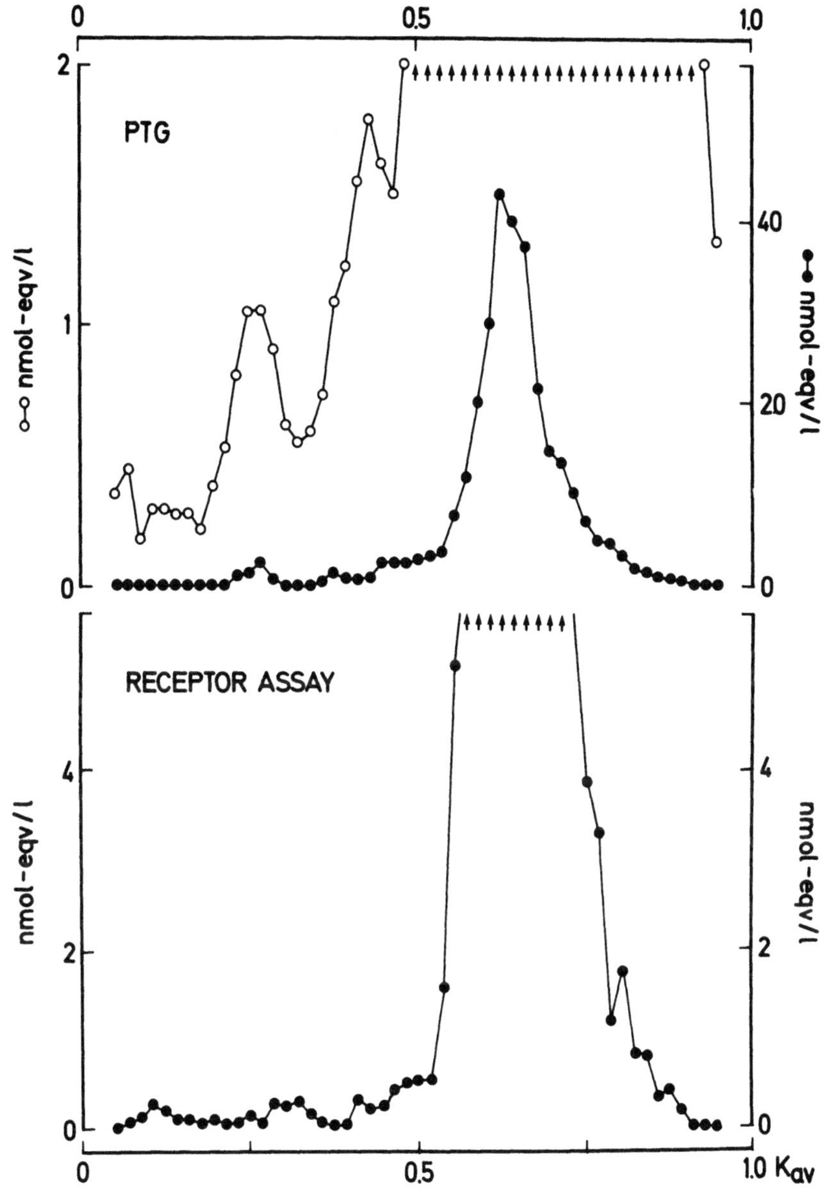

Fig. 2. Gel filtration of an acid ethanol extract of human pancreas. The extract of 20 g of the body of a human pancreas, removed at autopsy 10 hours after death, was applied to a 50 x 1000 mm Sephadex G-50 SF column, equilibrated and eluted with 0.5 mol/l acetic acid. The eluted fractions were analyzed for their concentration of PTG (upper panel, note the different scales on the two ordinates) and for receptor-active moieties, using the glucagon radioreceptor assay (4). Elution positions are indicated by the coefficient of distribution, K_{av}.

0.003 nmol/g tissue), with the exception of the so-called cardiac gland region (0.01 - 0.04 nmol/g), and in the duodenum (0.009 - 0.015 nmol/g) to concentrations of 0.3 - 0.5 nmol/g in the ileum and colon (5). All tissues contain PTG and GTG. The PTG dominates in the gastric and duodenal mucosa (20 - 80%) and accounts for 5 - 12% of the GLI in the lower intestinal mucosa.

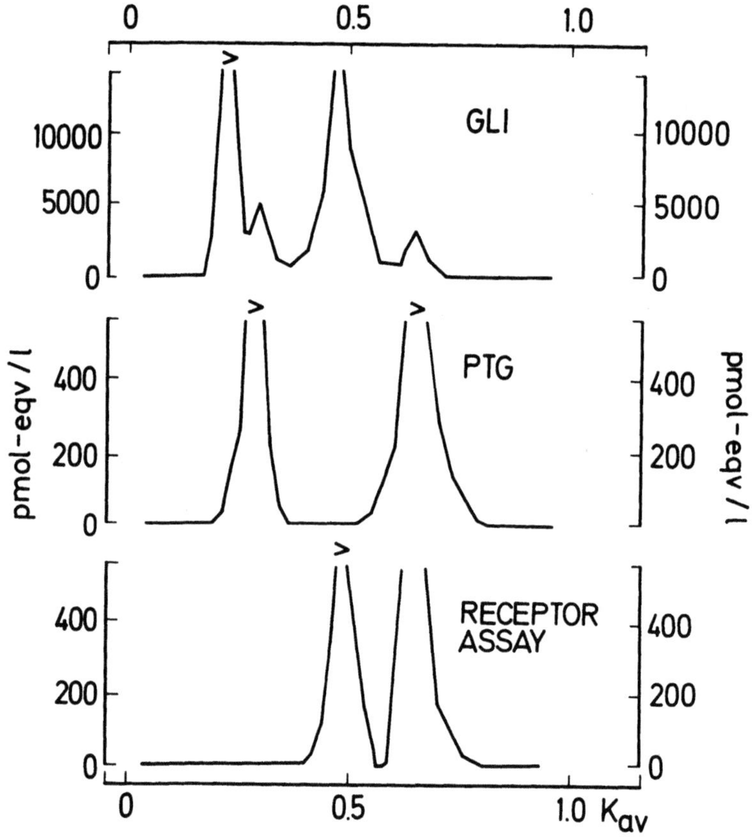

Fig. 3. Schematic distribution of GLI (upper panel), PTG (central panel), and receptor-active moieties, measured by the glucagon radioreceptor assay (lower panel) in extracts of porcine caudal intestinal mucosa, separated by gel filtration on Sephadex G-50 in 0.5 mol/l acetic acid. Elution positions are indicated by the coefficient of distribution, K_{av}.

By gel filtration on Sephadex G-50 SF in dilute acetic acid, the PTG may be resolved into two components (Fig. 3), one with a coefficient of distribution of 0.65, corresponding to the glucagon marker, and one eluting earlier with K_{av} = 0.30, suggesting a larger molecular size. The smaller form dominates in the proximal tissues, and the larger form dominates in the caudal tissues. When analyzed by radioreceptor assay, the smaller form reacts with a potency corresponding closely to its immunometric concentration, while the larger form does not react, even in concentrations exceeding 1 nmol-equivalents/l.

By gel filtration the GTG may be resolved into two major components (Fig. 3), one eluting with a K_{av} of 0.49 and one of 0.22 - 0.23, corresponding to the position of highly purified GLI-I (14). By re-filtration of large loads of gut extract, the larger peak may be resolved further into the major component of K_{av} 0.22 and a smaller component (which has no PTG-activity) of K_{av} = 0.38 (Fig. 4). Also by polyacrylamide-electrophoresis in 15% urea-containing gels at pH 8.6, the 0.22 component appears as a broad, apparently heterogenous peak (Fig. 5). Neither by refiltration nor by polyacrylamide-electrophoresis does the smaller peak reveal any heterogeneity. By isoelectric focusing in polyacrylamide gels in the nominal pH range of 3 to 10, the large GTG component may be resolved into three well defined peaks, with isoelectric pH's of 5.0, 6.5, and 6.9, respectively, none of which show any PTG activity (Fig. 6). Again, upon isoelectric focusing the small GTG component is homogenous, with almost all of the activity found at pH 8 (Fig. 7). This, however, may not be the proper pI of this component; the experiments were performed with cytochrome C as marker, and the run was stopped, when the marker was about to leave the gel tubes at the basic end. The GTG component followed closely after the cytochrome C, suggesting an even more basic pI. When analyzed by receptor-assay the smaller form reacted with a potency of $\frac{1}{50}$ to $\frac{1}{10}$ of its immunometric concentration, while the larger form never reacted in spite of concentrations exceeding 100 nmol-equivalents/l.

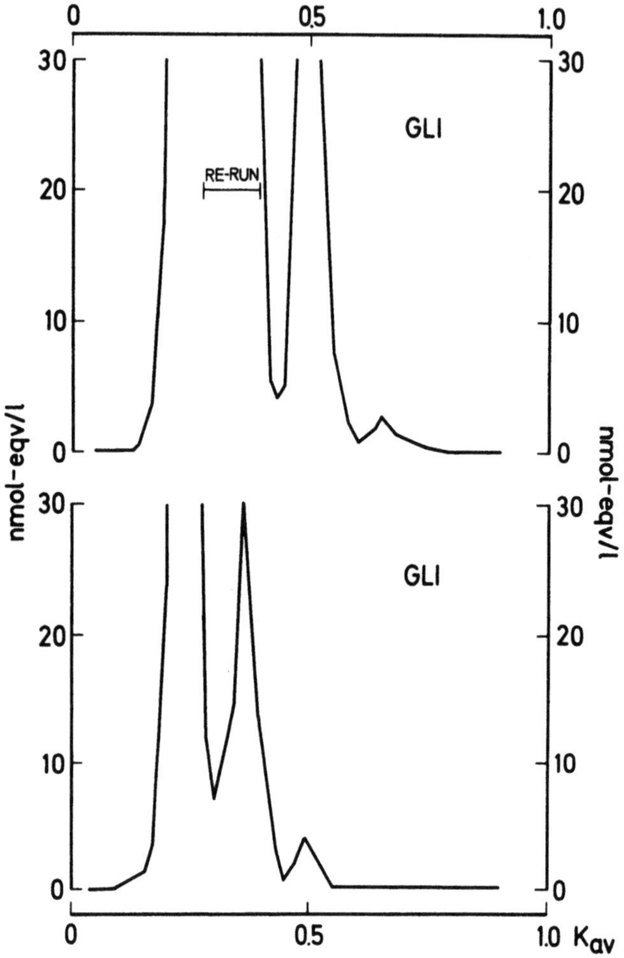

Fig. 4. Gel filtration of an extract of porcine intestinal mucosa on Sephadex G-50 (upper panel) and refiltration of a pool of the fractions eluted at the position of the large GTG (lower panel). Concentrations of GLI in the effluents are plotted against coefficient of distribution, K_{av}.

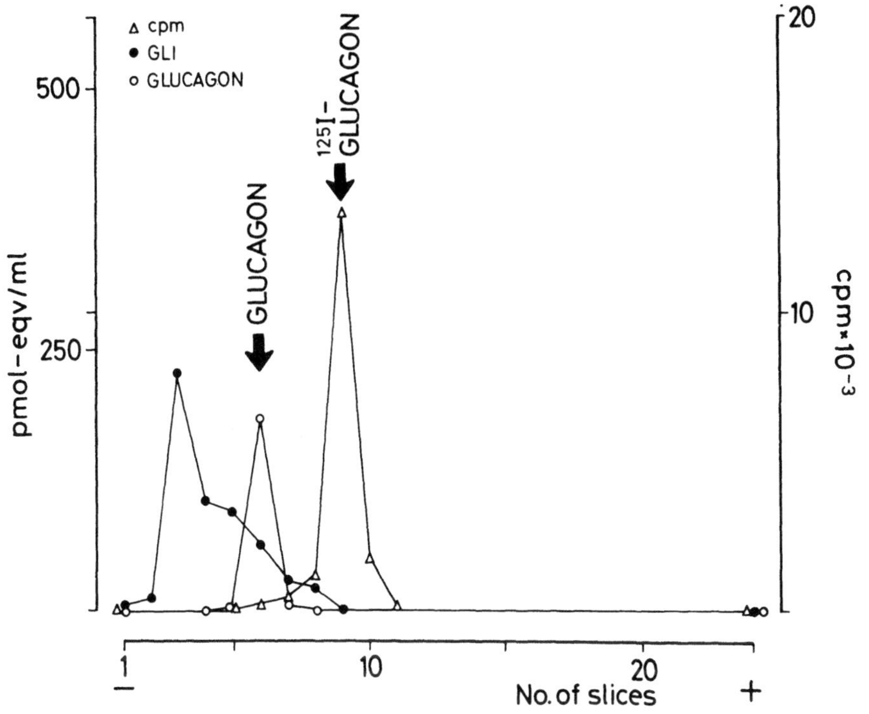

Fig. 5. Polyacrylamide-gel electrophoresis of large GTG from porcine intestinal mucosa, purified by gel filtration. Electrophoresis was performed in 15% urea-containing gels at pH 8.6, as previously described (4). The gels were sliced, eluted with assay buffer, and assayed for GLI concentration in the eluate. Also shown are the elution positions of true glucagon and ^{125}I-labelled glucagon.

In further experiments large loads of intestinal extracts were fractionated by gel filtration, and the two GTG peaks purified further by affinity chromatography on columns of Sepharose 4 B substituted with crossreacting glucagon antibodies (4). After affinity chromatography the non-glucagon protein content of the purified GTG components was too low to be detected by polyacrylamide-gel electrophoresis with subsequent staining with coomassie blue, in spite of loading of the gel with 1-2 nmol of GTG. The purified large GTG showed the same heterogeneity as before purification and still failed to react in the radioreceptor assay. The smaller component remained homogenous, and retained its ability to react in the

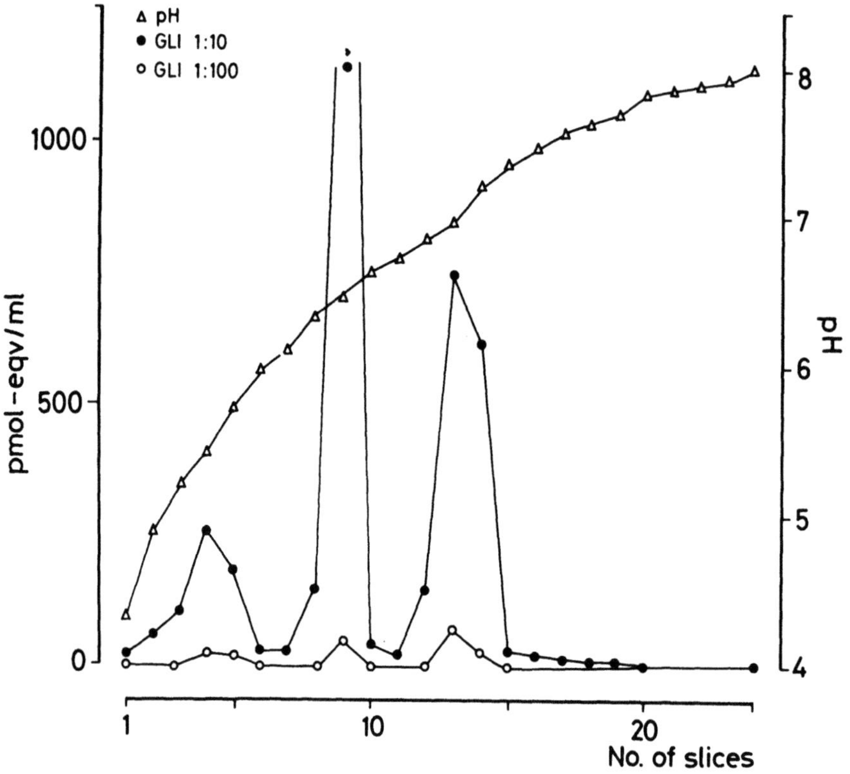

Fig. 6. Isoelectric focusing in polyacrylamide gels of large GTG from porcine intestinal mucosa purified by gel filtration. The focusing was performed according to Wrigley (17). The gels were sliced, eluted with assay buffer, and assayed for GLI concentration after further dilution as indicated in the figure. Control tubes were sliced and eluted with degassed distilled water for pH determination.

receptor assay. As the small GTG persistently competed with glucagon for binding to the glucagon receptor of the liver cell membranes, this was suggestive of a possible glucagon-like bioactivity of this component. The purified small GTG was therefore incubated with isolated hepatocytes from fed rats (10) in Krebs-Ringer bicarbonate buffer containing in addition 0.5% gelatine. After a 30 minute incubation the glucose concentration in the medium was measured and the results expressed as the per cent increase in glucose production (Table 1) over that observed without addition of hormones. It appears that the maximum effect of

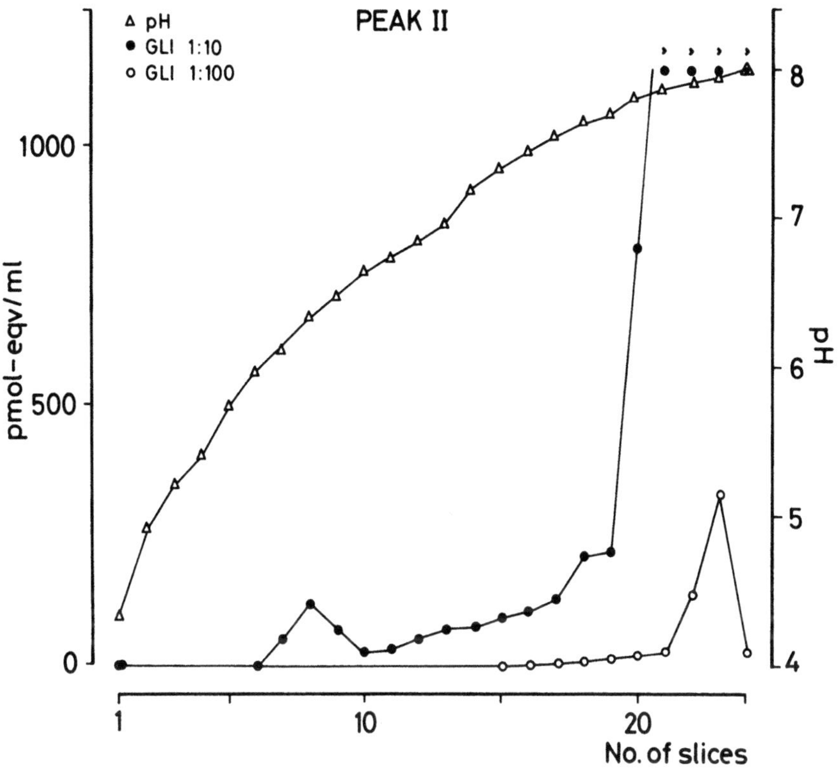

Fig. 7. Isoelectric focusing in polyacrylamide gels of small (K_{av} 0.49) GTG from porcine intestinal mucosa purified by gel filtration. For further explanations see legend to Fig. 6.

glucagon was reached at a concentration between 10^{-9} and 10^{-8} mol/l, while concentrations of small GTG of 10^{-8} mol/l elicited a glucose response which was equivalent to that noted with a glucagon concentration of 10^{-9} mol/l. The minimum effective concentration of glucagon in this system (not shown) was 10^{-12} mol/l, while minimum effective concentration of small GTG was close to 10^{-10} mol/l. Additions of GTG to maximum effective concentrations of glucagon did not increase glucose production further, and additions of GTG to the concentration of glucagon necessary to elicit 50% of the maximum glucose production, neither inhibited nor enhanced glucose production. This response is compatible with a role of small GTG as a partial agonist for the actions of glucagon on the receptor of the hepatocyte. A detailed description of these experiments will be given elsewhere (Holst et al, in preparation).

TABLE 1

EFFECT OF GLUCAGON AND OF PURIFIED, SMALL, GUT-TYPE GLUCAGON ON GLUCOSE PRODUCTION IN ISOLATED HEPATOCYTES

Added	Increase Over Basal, %	Added	Increase Over Basal, %
glucagon 10^{-8} M	255 ± 4 (4)	small GTG 10^{-8} M	202 ± 12 (7)
glucagon 10^{-9} M	221 ± 8 (4)	small GTG 10^{-9} M	135 ± 9 (7)
glucagon 10^{-10} M	198 (2)	small GTG 10^{-10} M	111 ± 5 (5)
glucagon $5 \cdot 10^{-11}$ M	188		
glucagon 10^{-9} M + small GTG 10^{-8} M	221	glucagon 5×10^{-11} +GTG 10^{-8}	191 (2)
glucagon 10^{-9} M + small GTG 10^{-9} M	240	glucagon 5×10^{-11} +GTG 10^{-9}	173 (2)
glucagon 10^{-10} M + small GTG 10^{-8} M	212	glucagon 5×10^{-11} +GTG 10^{-10}	173 (2)

Isolated hepatocytes from fed rats were incubated with glucagon and/or small GTG (K_{av} 0.49) in the concentrations indicated in the table. The glucose production was calculated from the glucose concentrations in the medium after 30 minutes of incubation and expressed as per cent of that observed without addition of hormones. Mean ± SEM. Number of experiments in parentheses.

Analysis of the amino acid composition of the purified small GTG showed the molecule to be composed of approximately 50 amino acids, its elution position on Sephadex G-50 being taken into account. The residues and their quantity represented the entire 29-amino-acid molecule of glucagon. In addition, one residue each of Ile, Val, Leu, Tyr, Phe, Axs, Ser; two residues each of Glx and Arg; two or three residues each of Pro and Ala, and three residues of Lys were found.

The number of glycine residues is uncertain because of a slight contamination of the purified preparation with glycine buffer during the affinity chromatography step.

Further experiments were performed with the highly purified GLI-I, a generous gift from F. Sundby and A. Moody, Novo Research Institute, Copenhagen, Denmark (14). GLI-I probably corresponds to our large GTG, the component with pI 6.9, since the behavior of the latter on immunological analysis, gel filtration and isoelectric focusing corresponds closely to that of GLI-I from Novo. GLI-I was analyzed by radioreceptor assay in concentrations up to 10^{-7} mol/l and always failed to interfere with glucagon for binding to the membranes. Furthermore, ^{125}I-labelled GLI-I (a gift from F. Sundby and A. Moody), a preparation which retained fully the immunological properties of unlabelled GLI-I, failed completely to bind to liver cell membranes under conditions where labelled glucagon was strongly bound. Finally, GLI-I did not stimulate glucose production in isolated hepatocytes, but may have been degraded to fragments with glycogenolytic activity (Holst et al, unpublished observations).

The human intestinal mucosa seems to contain much the same components of GLI as the porcine. The distribution of GTG and PTG in an extract of mucosa from a specimen of normal colon transversum obtained at surgery is shown in Fig. 8.

CONCLUSIONS

As discussed above, the glucagon family of peptides is composed of several members. The pancreas was found to contain at least two molecular forms, one which, according to its behaviour during gel filtration and radioreceptor analysis, must be true glucagon and another one of larger molecular size which has no, or very low, affinity for the glucagon receptor of the liver cells. These findings are in accordance with the results of Rigopoulou et al (14) on dog pancreas. The large form may be the biosynthetic precursor or prohormone of true glucagon (9). The "prohormone" fragment, analyzed and described by Tager and Steiner (15) con-

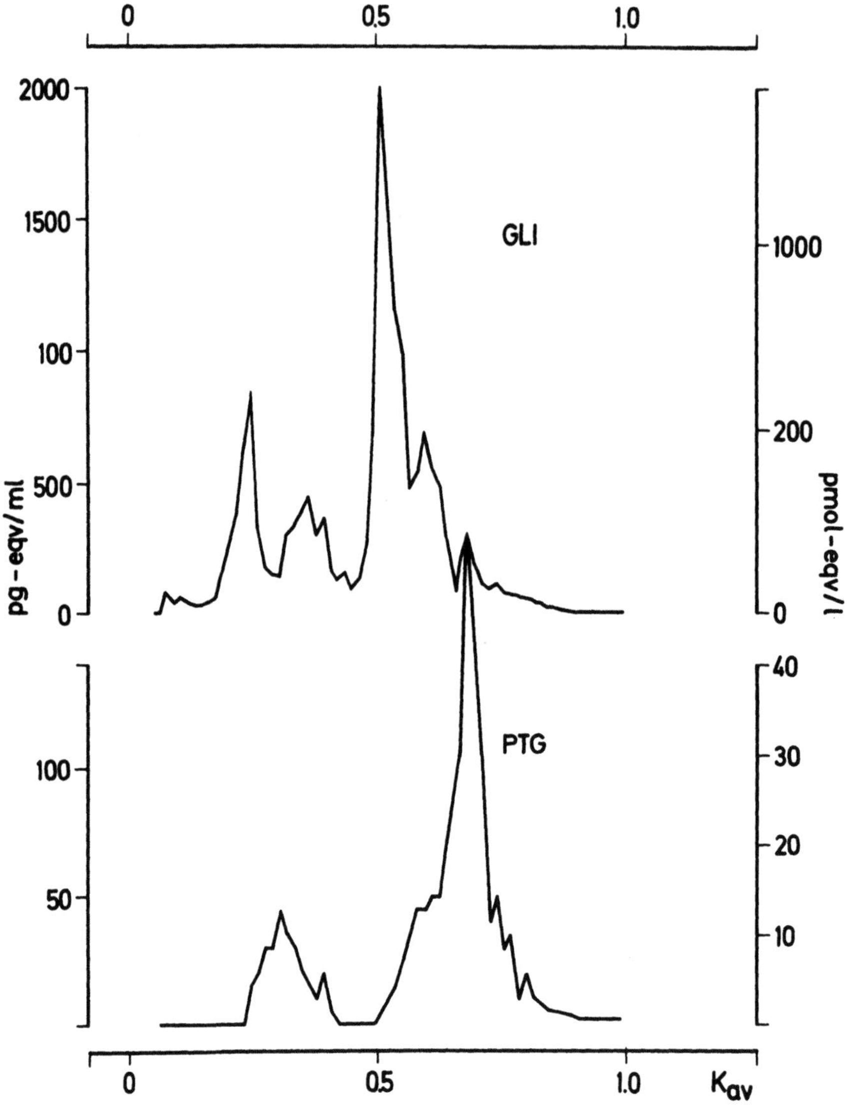

Fig. 8. Gel filtration of an extract of human intestinal mucosa (colon transversum, specimen obtained at surgery) on Sephadex G-50 in 0.5 mol/l acetic acid. Concentrations of GLI (upper panel) and PTG (lower panel) in the effluent are plotted against coefficient of distribution, K_{av}.

sisted of the entire glucagon molecule elongated from the C-terminal threonine residue. In view of the preference of the so-called specific antibodies for the C-terminus of glucagon, the "prohormone" of Tager and Steiner (15) and the large form, described here, could be identical. If so, the antibodies should be able to bind the C-terminal sequence even when hidden in a longer peptide chain.

We have not identified significant amounts of pancreatic type glucagon (PTG) components of even larger molecular weight, nor have we been able to identify such components in plasma, perhaps because our specific antibodies do not react with such forms.

The gastrointestinal mucosa of man and pig contains four major components: two gut type glucagon (GTG) forms and two PTG forms. Whether the PTG components which were found mainly in the low intestinal mucosa are identical to the pancreatic forms cannot be decided from the present investigations. The equipotent receptor activity of the intestinal K_{av} 0.65 component suggests that it may be identical to true glucagon. The large form (K_{av} 0.30) distinguished itself from the pancreatic counterpart in two respects: 1) it eluted later (0.30 vs 0.25) on gel filtration and it was the dominating form of the PTG's in the intestine. Like the pancreatic form it showed no affinity for the glucagon receptor.

The GTG's eluted by gel filtration in 2 or 3 positions. The GTG eluting at K_{av} 0.23, showed further heterogeneity, being separable into three components by isoelectric focusing, in agreement with the findings of Sundby et al (14). The isolation and purification of the major form (14) represents one of the important advances in the exploration of the GTG's. A physiologic role for this peptide, however, has not yet been defined. In its intact form it does not show any affinity for the glucagon receptor of the liver cell membrane, nor has it any receptor of its own. The smaller form, the basic peptide with K_{av} 0.49, does bind to glucagon receptor and has glycogenolytic activity, when tested on isolated hepatocytes from fed rats. Its relation to the other GLI forms remains undefined, and nothing is known about its possible physiologic role. In the fasting state the concentration of GTG in human plasma is very low (10-20 pmol-eqv./ml, n = 25), but the GTG concentration is significantly increased during the second to fifth hour after the ingestion of a mixed meal in normal subjects (unpublished results). Therefore, a hormonal role of the GTG's is still a possibility.

Acknowledgement

This study was supported by grants 512-5266, 512-3717, and 512-6602 from the Danish Medical Research Council.

REFERENCES

1. Assan, R. and Slusher, N. (1972): Structure/function and structure/immunoreactivity relationships of the glucagon molecule and related synthetic peptides. Diabetes 21, 843-855.

2. Bloom, S. R. (1972): An enteroglucagon tumor. Gut 13, 520-523.

3. Bloom, S. R., Bryant, M. G. and Polak, J. M. (1975): Distribution of gut hormones. Gut 16, 821.

4. Holst, J. J. (1975): A receptor-assay for glucagon: binding of enteroglucagon to liver plasma membranes. Diabetologia 11, 211-219.

5. Holst, J. J. (1977): Extraction, gel filtration pattern, and receptor binding of porcine gastrointestinal glucagon-like immunoreactivity. Diabetologia. In press.

6. Larsson, L.-I., Holst, J. J., Hakanson, R. and Sundler, F. (1975): Distribution and properties of glucagon immunoreactivity in the digestive tract of various mammals: an immunohistochemical and immunochemical study. Histochemistry 44, 281-290.

7. Moody, A. (1977): Gut glucagon-like immunoreactivity (GLI). In: Endocrinology of the Gut. N. S. Track, Ed. McMaster University Symposium, Hamilton, Canada, p. 84.

8. Morita, S., Doi, K., Yip, C. and Vranic, M. (1976): Measurement and partial characterization of immunoreactive glucagon in gastrointestinal tissues of dogs. Diabetes 25, 1018-1025.

9. Noe, B. D. (1976): Biosynthesis of glucagon. Metabolism 25 (Suppl. 1), 1339-1341.

10. Quistorff, B., Bondesen, S. and Grunnet, N. (1973): Preparation and biochemical characterization of parenchymal cells from rat liver. Biochim. Biophys. Acta, 320, 503-516.

11. Rigopoulou, D., Valverde, I., Marco, J., Faloona, G. and Unger, R. H. (1970): Large glucagon immunoreactivity in extracts of pancreas. J. Biol. Chem. 245,

496-501.

12. Rodbell, M. and Londos, C. (1976): Regulation of hepatic adenylate cyclase by glucagon, GTP, divalent cations, and adenosine. Metabolism 25 (Suppl. 1) 1347-1349.

13. Sasaki, H., Rubalcava, B., Baetens, D., Blázquez, E., Srikant, C. B., Orci, L. and Unger, R. H. (1975): Identification of glucagon in the gastrointestinal tract. J. Clin. Invest. 56, 135-145.

14. Sundby, F., Jacobsen, H. and Moody, A. (1976): Purification and characterization of a protein from porcine gut with glucagon-like immunoreactivity. Horm. Metab. Res. 8, 366-371.

15. Tager, H. S. and Steiner, D. F. (1973): Isolation of a glucagon-containing peptide: primary structure of a possible fragment of proglucagon. Proc. Nat. Acad. Sci. USA 70, 2321-2324.

16. Unger, R. H. and Orci, L. (1975): The essential role of glucagon in the pathogenesis of diabetes mellitus. Lancet 1, 14-16.

17. Wrigley, C. (1968): Gel electrofocusing. A technique for analyzing multiple protein samples by isoelectric focusing. Science Tools, 15, 17.

GLUCAGON, INSULIN AND THEIR HEPATIC RECEPTORS: AN ENDOCRINE PATTERN CHARACTERIZING HEPATOPROLIFERATIVE TRANSITIONS IN THE RAT

H. L. Leffert

Cell Biology Laboratory, Salk Institute for Biological Studies
San Diego, California USA

Hyperglucagonemia and hypoinsulinemia (absolute or relative) characterize at least four "different" hepatoproliferative states which include:

1. Regeneration after 70% hepatectomy;
2. Developmental transitions;
3. Lipotrope-deficiency induced proliferation; and
4. Combined hormonal infusion

During regeneration, glucagon receptor sites "disappear" from the liver, as revealed by labelled hormone binding studies and Scatchard analyses of the data. However, this is not detected during the early prereplicative phase of regeneration when numerous glucagon-dependent changes (particularly those involving altered hepatic VLDL metabolism) are known to occur. By contrast, isolated liver "membranes" from regenerating liver early in the prereplicative phase show an enhanced ability to bind labelled insulin. A common endocrine pattern, related to the physiology of pancreatic hormones and interactions with their receptors, may participate in controlling normal hepatocyte proliferation.

Our laboratory has sought to identify humoral factors regulating liver regeneration, in order to define mechanisms controlling the proliferation of normal animal cells (16). We have approached this problem, first, by defining different factors from biologic fluids which regulate cultured hepatocyte proliferation (8, 10, 12, 19) and, second, by determining the physiologic behaviour of these factors during various *in vivo* hepatoproliferative transitions (10, 14, 15, 17, 19). These studies have led to a working hypothesis postulating multifactorial involvement of hormones, nutrients, lipoproteins, and novel highly phosphorylated purine nucleotides (9). In this brief report, we will focus only upon glucagon and insulin as regulators of this process. Apparently, common endocrine patterns formed by these hormones and their receptor interactions facilitate or retard normal liver cell proliferation.

Hepatic Proliferation in Partially Hepatectomized Adult Rats

Partial hepatectomy stimulates hepatic DNA synthesis (within 12-16 hr) and cell division (within 24-30 hr) in a highly reproducible manner dependent upon the amount of liver removed. Regeneration is well-regulated: after a few days, initial cell numbers are restored and proliferative rates decline to the steady-state "quiescent" level (the fraction of cells labelled with [^{3}H]-thymidine, i.e. the labelling-index, is ca. 0.05%; for review see 3). In vitro studies with differentiated fetal rat hepatocytes implicated insulin and glucagon as regulators of DNA synthesis initiation (13, 21). Accordingly, pancreatic hormone physiology has been studied during the regenerative process.

Within minutes after partial hepatectomy, "dose-dependent" arterial hyperglucagonemia and hypoinsulinemia ensue (15). Non-specific surgical stimuli do not produce these alterations, which are sustained many hours after the initial "stimulus." Other laboratories have confirmed these findings (5, 20). Similar changes occur in portal blood, suggesting that a differential A and B cell secretion may be partly responsible for them. If glucagon and insulin are made "limiting," by prior abdominal evisceration of appropriately maintained rats, regeneration is markedly inhibited. Therefore, regeneration requires both peptides despite paradoxical changes in their blood levels (for discussion, see 17).

Preliminary hormone disappearance studies suggest that both ^{125}I-labelled glucagon and insulin are cleared less rapidly from the liver 0 to 3 hr after partial hepatectomy than from the liver of laparotomized control rats (17). Therefore, altered binding of these peptides to their hepatic receptors and/or altered turnover also could account for altered blood levels. Because we had reported earlier that the binding of glucagon to regenerating liver membrane fractions (hereafter, referred to as "membranes") is decreased, we decided to extend this type of study (15). Figures 1-3 show our preliminary results (details are given in 18).

Scatchard analysis of ^{125}I-labelled glucagon binding to hepatic membranes from anesthetized, laparotomized and 70% hepatectomized rats 24 hr post-operatively (Fig. 1) indicates that regenerating membranes contain ca. 50% fewer specific binding sites than are present in control preparations. No detectable differences in the apparent affinity of glucagon for its receptors is observed. Figure 2 shows results of similar studies with ^{125}I-labelled insulin which, for reasons discussed elsewhere (18) are more difficult to interpret. Computer analyses of these data indicate that an increase either in the number of sites (Fig. 2, left panel) and/or in the apparent affinity of these sites (Fig. 2, right panel) for insulin occurs in regenerating membranes. Like altered blood hormone levels, altered binding is somewhat dependent upon the extent of hepatectomy (Fig. 3, left panel). Interestingly, enhanced insulin binding to regenerating membranes is most pronounced during the pre-DNA synthetic phase (4-8 hr postoperatively) whereas diminished glucagon binding is not detected until after 24 hr (Fig. 3, right panel).

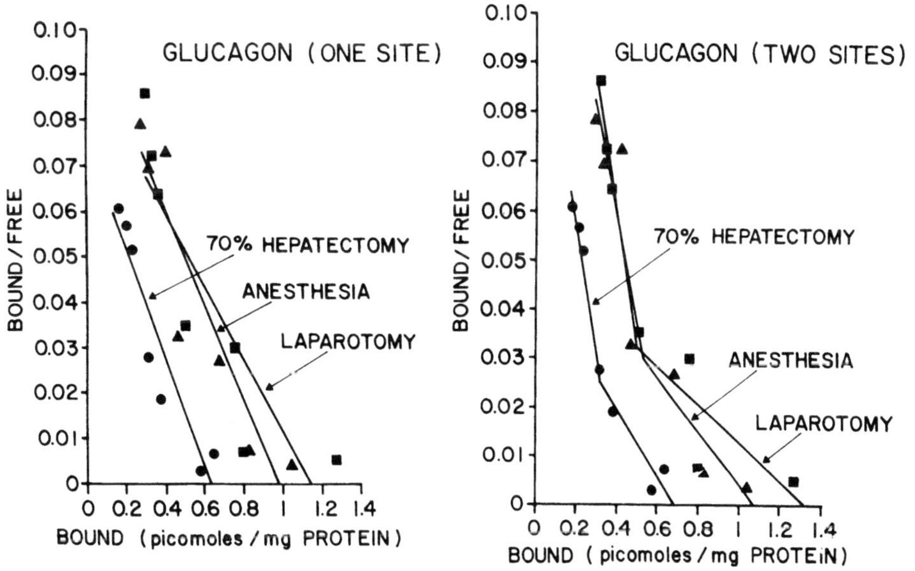

Fig. 1. Scatchard-type analyses of ^{125}I-labelled glucagon binding to hepatic plasma membrane fractions from adult rats 24 hr post-operatively. Full details are given in Leffert et al (18).

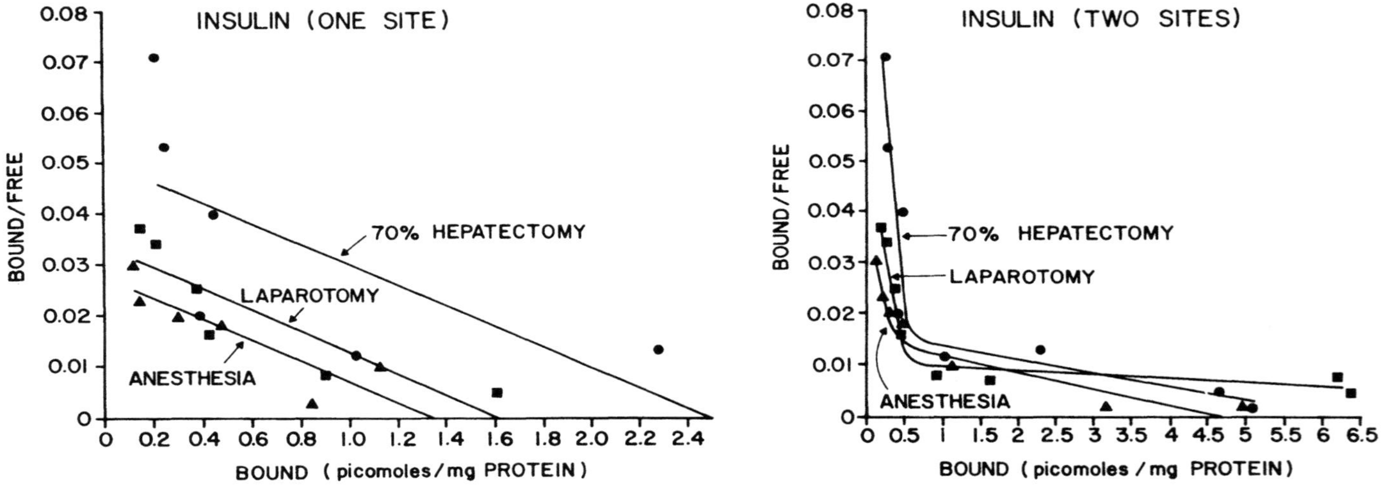

Fig. 2. Scatchard-type analyses of ^{125}I-labelled insulin binding to hepatic plasma membrane fractions. (see text, Fig. 1 legend and Leffert et al. [18] for details).

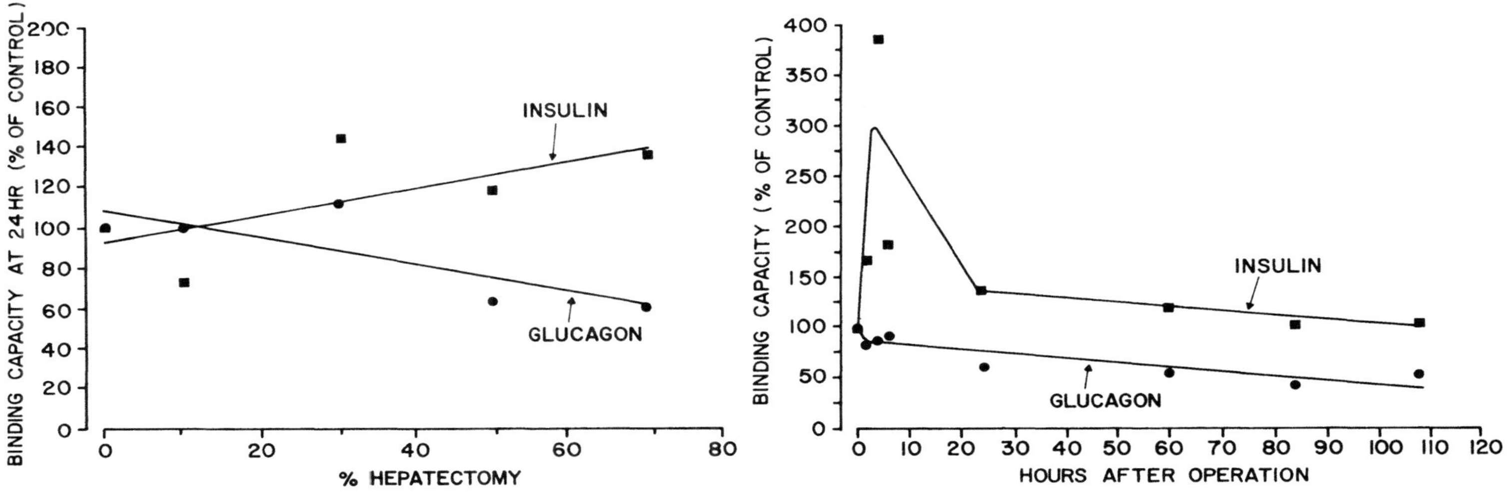

Fig. 3. Binding capacity of ^{125}I-labelled glucagon and insulin to liver membranes from partially hepatectomized rats. "One site" curves were used to obtain these data (see Leffert et al [18] for further details).

The combined results imply that the intrahepatic actions of glucagon and insulin are elevated early during the regenerative process; with time, these actions appear to decline. It would seem that dissimilar physicochemical mechanisms underlie these differential receptor-binding changes; in this regard, hepatic turnover mechanisms may prove to be quite important in these processes. Current models of hormone-receptor modulation, hormone resistant states, and negative cooperativity should be considered in attempts to design and interpret these types of experiments (18).

Hepatic Proliferation in Developing Rats

Hepatic proliferation rates decrease during development, reaching steady-state levels 50-60 days after birth (24). Concomitantly, plasma glucagon levels fall, insulin levels rise (the "tail-end" of this process is shown in Fig. 4, top and middle panels, respectively, where the zero-time represents 14-21 day-old weaned rats), and, ^{125}I-labelled glucagon binding to liver membranes continuously increases (4). Therefore, regenerating liver recapitulates ontogeny with respect to glucagon binding to its receptors (developmental studies require Scatchard analyses to further substantiate this point). These observations suggest that functional retrogression observed during regeneration, e. g. the resumption of fetal enzyme activity patterns, could be partly related to restoration of a "fetal-receptor-like" condition resulting, in turn, from sustained altered pancreatic blood hormone levels.

Hepatic Proliferation in Lipotrope-Deficient Rats

Feeding weaned rats choline-and methionine-free diet ("lipotrope-deficiency"; 23) arrests the maturational changes in plasma glucagon and insulin levels (Fig. 4, top and middle panels, respectively). This arrest is associated with a stimulation of hepatocyte DNA synthesis and with an increased labelling-index (Fig. 4, bottom panel). These changes are strikingly similar to those occurring shortly

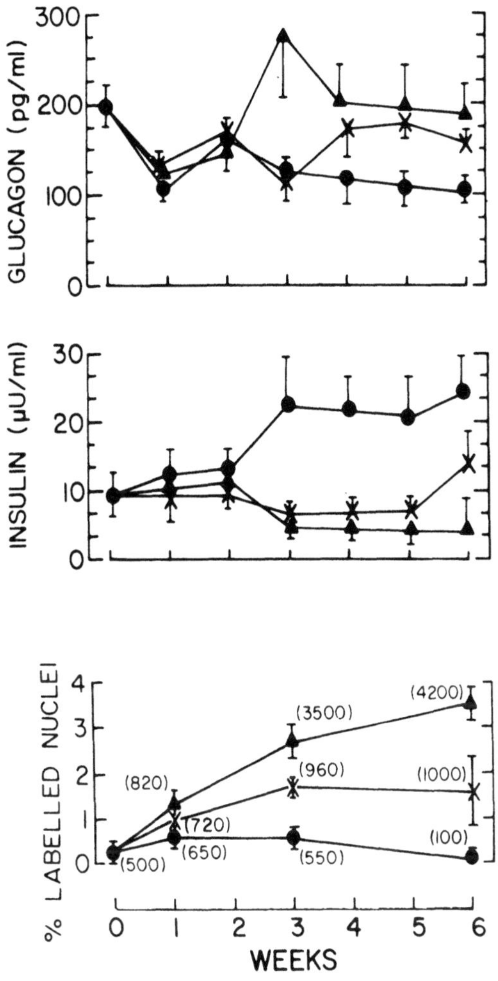

Fig. 4. Plasma levels of pancreatic hormones and liver DNA synthesis in lipotrope-deficient rats. Hormone measurements were made by routine procedures (15). Weaned rats were placed onto normal (circles), moderately (crosses) or severely-deficient (triangles) diets as described elsewhere (23). DNA synthesis was determined by measuring either [^{3}H]-thymidine uptake into isolated hepatic nuclei (numbers in parentheses) or by routine radioautographic procedures (% labelled nuclei = labelling index). Values for uptake = cpm/mg DNA/2 hr pulse (input doses were adjusted according to body weight).

after partial hepatectomy. Whether or not this is a specific response, restricted to liver, is presently unknown.

That labelling-indices in severely deficient rats reach only ca. 4%, compared to 20-25% levels obtained 24 hr post-hepatectomy, probably reflects the fact that additional hormonal and nutritional changes have not occurred (or, occured too slowly), presumably because their blood levels too are modulated by functional liver mass (16). This reasoning predicts that liver from lipotrope-deficient rats, "primed" by chronic hypoinsulinemic and hyperglucagonemic environments (Fig. 5, top and middle panels, respectively), should be more sensitive to superimposed proliferative stimuli. This seems true because moderately-deficient rats are significantly stimulated to synthesize DNA after 30% hepatectomy, in contrast to well-fed controls (Fig. 5, botton panel). Furthermore, severely-deficient rats (whose initial plasma insulin and glucagon levels are lower and higher, respectively, than those of the moderately-deficient group) respond to sub-maximal stimulation by synthesizing hepatic DNA nearly as well as do the well-fed 70% hepatectomized animals. This also is shown in Fig. 5 (bottom panel).

Recently, we have developed a proliferating adult hepatocyte primary culture system and have shown that, like in fetal liver cells, proliferation is hormonally controlled (18). Preliminary results suggest that cultured hepatocytes derived from lipotrope-deficient rats are less dependent upon certain hormones for proliferation (unpublished). Therefore, "priming" events may reflect the in vivo preactivation of cellular mechanisms by appropriate endocrine changes.

Hepatic Proliferation in Diabetic, "Fatty," Protein-Deficient, and TAGH-Infused Rats

Low insulin/glucagon (I/G) ratios and insulin-resistant states arising from still other environmental and genetic conditions are known to modify hepatic proliferation. For example, the liver of chronically diabetic rats manifests a slightly elevated rate of proliferation, that can be further stimulated by

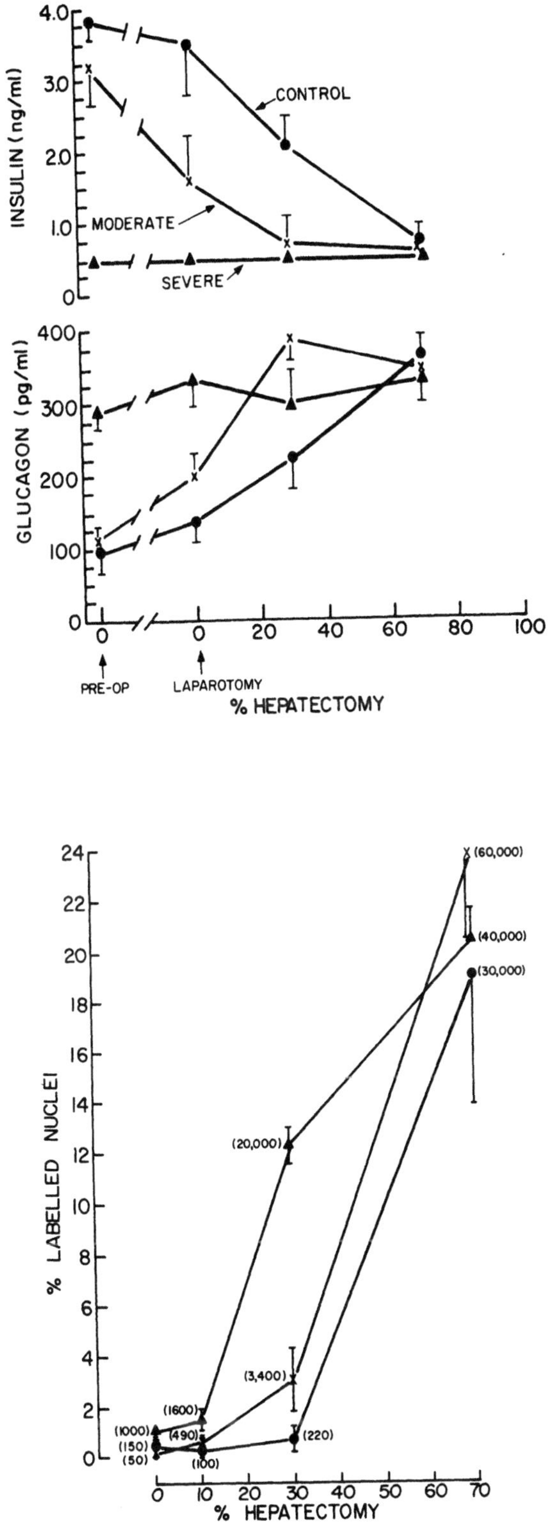

Fig. 5. Enchanced sensitivity of liver from lipotrope-deficient rats to synthesizing DNA after partial hepatectomy. Animals made deficient for 6 weeks were used and all measurements were as described in legend to Fig. 4. Preoperative blood samples were obtained 24 hr prior to experimental manipulations (see 15, for further details.).

peripheral insulin administration (28, 29). Recent studies with alloxanized mice indicate increased numbers of available insulin binding sites in their liver membranes. Conversely, insulin-resistant "Fatty" rats (30) show impaired DNA synthesis after 70% hepatectomy.

Protein-deficient rats, like diabetic animals, also have low plasma I/G ratios (1) and become "primed" for proliferative stimulation by protein or amino-acid repletion (11, 25). "Priming" mechanisms in these models may be different from those occurring during lipotrope-deficiency, inasmuch as protein-deficient rats respond poorly to partial hepatectomy. Plasma glucagon and insulin levels have not been measured, to our knowledge, in partially hepatectomized protein-deficient rats.

Peripheral infusion of pharmacologic doses of triiodo-L-thyronine (T_3), amino acids, glucagon and heparin (TAGH) also stimulates, somewhat non-specifically, hepatic proliferation (25). Certain results obtained with this experimental model have been discussed in detail elsewhere (16, 17, 19); notably, the observation that the omission of glucagon reduces DNA synthesis stimulation by 30-40%.

The Role of Glucagon in the Control of Hepatic Proliferation

Extrahepatic lipolytic activity of glucagon (favored by concomitant hypo-insulinemia) could be important in promoting hepatic DNA synthesis initiation (17, 26, 27). Mitochondrial energy metabolism, altered membrane structure and modulation of hormone receptor function are examples of growth-related processes which are modified by lipids (some of our studies have implicated PGE specifically, 17). In certain instances, glucagon reportedly stimulates parathyroid hormone secretion (2). Ablation-repletion studies suggest that parathyroid hormone also is required for DNA synthesis initiation after 70% hepatectomy (22). These few examples indicate that many of glucagon's growth promoting effects may be indirect.

The known direct effects of glucagon upon the liver include the stimulation of cyclic AMP formation of amino acid transport and, possibly, of specific histone phosphorylation and lysosomal activation. All these interrelated functions have been implicated in the regulation of growth (16) although much of the existing evidence is indirect and open to different interpretation.

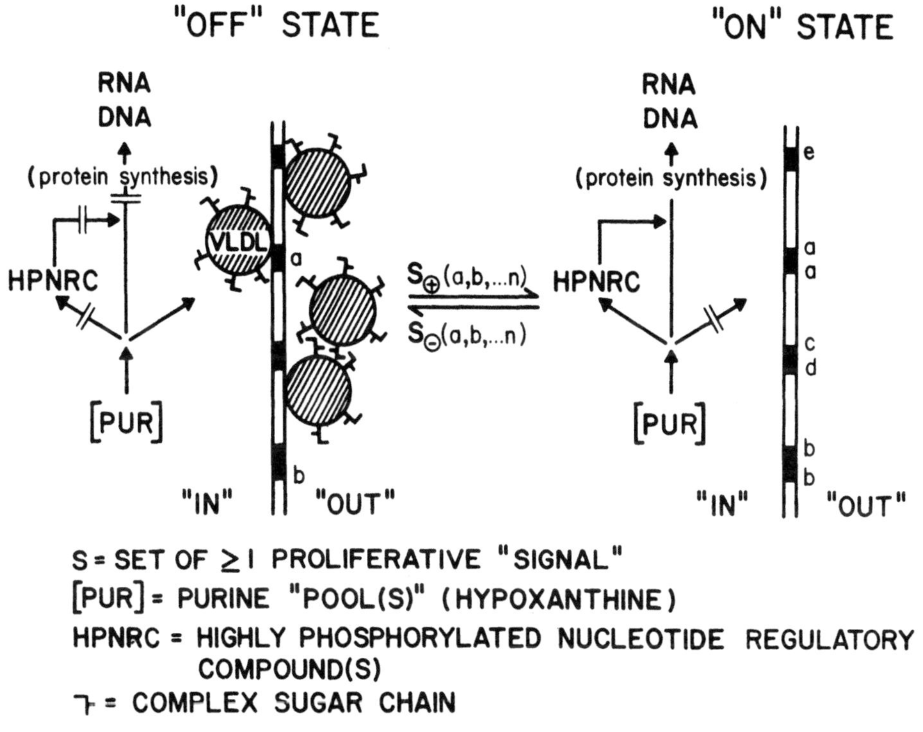

Fig. 6. A model for normal regulation of hepatoproliferative transitions. (See text and Koch and Leffert (9) for further details.

Figure 6 considers a still additional regulatory mode based upon glucagon's ability to suppress hepatic production of very low density lipoprotein (VLDL; 6). This has attracted our attention because VLDL specifically blocks the in vitro initiation of fetal hepatocyte DNA synthesis and reduces prereplicative protein synthesis rates (19). Blood VLDL levels specifically fall in a dose-dependent manner after partial hepatectomy (with kinetics expected from the rapid alterations of plasma glucagon and insulin) and, as mentioned above, impaired DNA synthesis is observed in 70% hepatectomized hyper-VLD-lipoproteinemic "Fatty" rats (17). In addition, VLDL blood levels increase during development but not during lipotrope-deficiency (18); and TAGH infusions rapidly clear circulating VLDL with glucagon and heparin both contributing to these effects (17). These and other findings (10) have led us to postulate that VLDL (and/or some metabolic event related to it) maintains a refractory "Off" state, i.e. the proliferative metabolic reactions required of DNA synthesis initiation occur at very low rates, if at all. In turn, if hormonal and nutritional conditions are altered so that VLDL formation is suppressed, then purine-mediated reactions regulating appropriate macromolecular syntheses can proceed (9).

The reasons and preliminary evidence for invoking purine action, especially that of a novel highly phosphorylated purine "regulatory" nucleotides (HPNRC, see Fig. 6) have been presented earlier (10, 17). As yet, no direct evidence links glucagon action to HPNRC formation but it is, perhaps, one of the more interesting directions in which to proceed in this complicated problem.

Acknowledgements

The collaboration of K. S. Koch, P. Newberne, A. Rogers, B. Rubalcava; the technical assistance of T. Moran; and the help of M. Brown are gratefully appreciated.

Supported by the National Cancer Institute (CA 14312 and CA 14195) and the National Science Foundation (BMS 74-05471).

REFERENCES

1. Anthony, L. E. and Faloona, G. (1974): Plasma insulin and glucagon levels in protein-malnourished rats. Metabolism 23, 303-306.

2. Avioli, L. V., Birge, S. J., Scott, S. and Shieber, W. (1969): Role of the thyroid gland during glucagon-induced hypocalcemia in the dog. Amer. J. Physiol. 216, 939-945.

3. Becker, F. F. (1973): Humoral aspects of liver regeneration. In: Humoral Control of Growth and Differentiation, Vol. 1. J. Lo Bue and A. S. Gordon, Eds. Academic Press, New York and London, pp. 249-256.

4. Blázquez, E., Rubalcava, B., Montesano, R., Orci, L. and Unger, R. (1976): Development of insulin and glucagon binding and the adenylate cyclase response in liver membranes of the prenatal, postnatal, and adult rat: evidence of glucagon resistance. Endocrinology 98, 1014-1023.

5. Bucher, N. L. R. and Swaffield, M. N. (1975): Regulation of hepatic regeneration in rats by synergistic action of insulin and glucagon. Proc. Nat. Acad. Sci. (USA) 72, 1157-1160.

6. Eaton, R. P. (1973): Hypolipemic action of glucagon in experimental endogenous lipemia in the rat. J. Lip. Res. 14, 312-318.

7. Kahn, C. R. (1976): Membrane receptors for hormones and neurotransmitters. J. Cell Biol. 70, 261-286.

8. Koch, K. S. and Leffert, H. L. (1974): Growth control of differentiated fetal rat hepatocytes in primary monolayer culture. VI. Studies with conditioned medium and its functional interactions with serum factors. J. Cell Biol. 62 780-791.

9. Koch, K. S. and Leffert, H. L. (1976): Control of hepatic proliferation: a working hypothesis involving hormones, lipoproteins, and novel nucleotides. Metabolism 25, 1419-1422.

10. Koch, K. S., Leffert, H. L. and Moran, T. (1976): Hepatic proliferation-control by purines, hormones and nutrients. In: Onco-Developmental Gene Expression. W. Fishman, Ed. Academic Press, New York, p. 21.

11. Leduc, E. H. (1949): Mitotic activity in liver of mouse during inanition followed by refeeding with different levels of protein. Amer. J. Anat. 84, 397-429.

12. Leffert, H. L. (1974); Growth control of differentiated fetal rat hepatocytes in primary monolayer culture. V. Occurrence in dialyzed fetal bovine serum of growth regulatory functions. J. Cell Biol. 62, 767-769.

13. Leffert, H. L. (1974): Growth control of differentiated fetal rat hepatocytes in primary monolayer culture. VII. Hormonal control of DNA synthesis and its possible significance to the problem of liver regeneration. J. Cell Biol. 62, 792-801.

14. Leffert, H. L. and Alexander, N. M. (1976): Thyroid hormone metabolism during liver regeneration in adult rats. Endocrinology 98, 1205-1211.

15. Leffert, H., Alexander, N. M., Faloona, G., Rubalcava, B. and Unger, R. (1975): Specific endocrine and hormonal changes associated with liver regeneration in adult rats. Proc. Nat. Acad. Sci. (USA) 72, 4033-4036.

16. Leffert, H. L. and Koch, K. S. (1977): Control of animal cell proliferation. In: Growth, Nutrition and Metabolism of Cells in Culture, Vol. III. G. H. Rothblatt and V. J. Cristofalo, Eds. Academic Press, New York, pp. 225-294.

17. Leffert, H. L., Koch, K. S. and Rubalcava, B. (1976): Present paradoxes in the environmental control of hepatic proliferation. Cancer Res. 36, 4250-4255.

18. Leffert, H. L., Koch, K. S., Rubalcava, B., Sell, S., Moran, T. and Boorstein, R.: Hepatocyte growth control: in vitro approach to problems of liver regeneration and function. J. Nat. Cancer Inst. Monograph Series. in press.

19. Leffert, H. L. and Weinstein, D. B. (1976): Growth control of differentiated fetal rat hepatocytes in primary monolayer culture. IX. Specific inhibition of DNA synthesis initiation by very low density lipoprotein and possible significance to the problem of liver regeneration. J. Cell Biol. 70, 20-32.

20. Morley, C. G. D., Kuku, S., Rubenstein, A. H. and Boyer, J. L. (1975): Serum hormone levels following partial hepatectomy in the rat. Biochem. Biophys. Res. Commun. 67, 653-661.

21. Paul, D. and Walter, S. (1975): Growth control in primary fetal rat liver cells in culture. J. Cell. Physiol. 85, 113-123.

22. Rixon, R. H. and Whitfield, J. F. (1976): The control of liver regeneration by parathyroid hormone and calcium. J. Cell Physiol. 87, 147-156.

23. Rogers, A. E. (1975): Various effects of a lipotrope deficient, high fat diet on chemical carcinogenesis in rats. Cancer Res. 35, 2469-2474.

24. Sell, S., Nichols, M., Becker, F. F. and Leffert, H. L. (1974): Hepatocyte proliferation and alpha 1-fetoprotein in pregnant, neonatal and partially hepatotectomized rats. Cancer Res. 34, 865-871.

25. Short, J., Armstrong, N. B., Kolitsky, M. A., Mitchell, R. A., Zemel, R. and Lieberman, I. (1974): Amino acids and the control of nuclear DNA replication in liver. In: Cold Spring Harbor Conferences on Cell Proliferation 1, 37-48..

26. Short, J., Brown, R. F., Husakova, J. R., Gilbertson, A., Zemel, R. and Lieberman, I. (1972): Induction of deoxyribonucleic acid synthesis in the liver of the intact animal. J. Biol. Chem. 247, 1757-1766.

27. Simek, J., Chmelar, V. L., Melka, J., Pazderka, J. and Charvat, A. (1967): Influence of protracted infusion of glucose and insulin on the composition and regeneration activity of liver after partial hepatectomy in rats. Nature 213, 910-911.

28. Starzl, T. E., Porter, K. A., Kashiwagi, N., Lee, I. Y., Russell, W. J. I. and Putnam, C. W. (1975): The effect of diabetes mellitus on portal blood hepatotrophic factors in dogs. Surg. Gynec. Obstetr. 140, 549-562.

29. Younger, L. R., King, J. and Steiner, D. F. (1966): Hepatic proliferative response to insulin in severe alloxan diabetes. Cancer Res. 26, 1408-1414.

30. Zucker, L. M. and Antoniades, H. N. (1972): Insulin and obesity in the Zucker genetically obese rat "fatty". Endocrinology 90, 1320-1330.

REEXAMINATION OF THE SECOND MESSENGER HYPOTHESIS OF GLUCAGON AND CATECHOLAMINE ACTION IN LIVER

J. H. Exton, A. D. Cherrington, N. J. Hutson
and F. D. Assimacopoulos-Jeannet

Laboratories for the Studies of Metabolic Disorders,
Howard Hughes Medical Institute,
Department of Physiology, Vanderbilt University Medical School
Nashville, Tennessee

Glucagon causes a rapid activation of cAMP-dependent protein kinase in rat liver parenchymal cells which correlates well with the accumulation of cAMP. Full activation of phosphorylase or inactivation of glycogen synthase is achieved with half-maximal activation of protein kinase. Epinephrine stimulates glycogen breakdown and gluconeogenesis in these cells mainly by mechanisms involving α-adrenergic receptors and not β-receptors. Activation of α-receptors results in rapid activation of phosphorylase and inactivation of glycogen synthase without accumulation of cAMP or activation of cAMP-dependent protein kinase. Activation of β-receptors causes a transient rise in cAMP and a short-lived activation of protein kinase with correspondingly little stimulation of glycogenolysis and gluconeogenesis.

Thus the actions of glucagon on glycogen metabolism and gluconeogenesis are fully in accord with the second messenger hypothesis of Sutherland, whereas the actions of epinephrine on these processes mainly involve non-cAMP mediated mechanisms.

INTRODUCTION

The second messenger hypothesis of hormone action which proposes that adenosine 3':5'-monophosphate (cAMP) acts as the intracellular mediator of the actions of a large number of hormones arose from the classic investigations of Sutherland and his associates into the mechanism of action of glucagon and epinephrine on hepatic glycogenolysis. According to the original scheme, interaction of certain hormones with specific receptors on the plasma membrane of target cells leads to activation of the membrane-bound enzyme adenylate cyclase which catalyzes the conversion of ATP to cAMP. The resulting intracellular accumulation of cAMP is then postulated to cause the physiological response through modifications of the activities of certain enzymes. Figure 1 illustrates present knowledge of the mechanisms by which a hormonally induced increase in cAMP leads to activation of glycogenolysis and inhibition of glycogen synthesis in muscle. A similar scheme is believed to operate in liver.

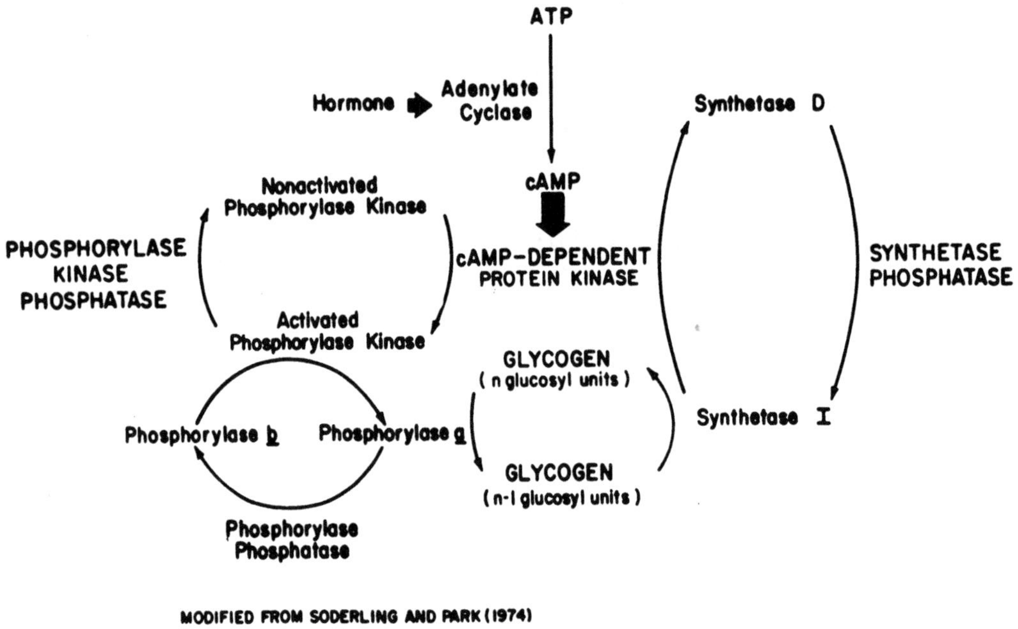

Fig. 1. Sequence of reactions by which epinephrine stimulates glycogen breakdown and inhibits glycogen breakdown in skeletal muscle (22).

The first step in cAMP action is generally thought to be activation of cAMP-dependent protein kinase. The holo form of this enzyme (R_2C_2) has a molecular weight of 165,000 and consists of a regulatory subunit dimer (R_2, M. W. 95,000) combined with two catalytic subunits (C, M. W. 35,000). The holoenzyme is inactive because the regulatory dimer exerts an inhibitory effect on the catalytic subunits. cAMP activates the holoenzyme by binding to the regulatory dimer causing it to dissociate from the catalytic subunits. The increase in "free" active catalytic subunits then leads to enhanced phosphorylation of a number of enzymes and other cellular proteins with ATP as the phosphoryl donor.

As shown in Figure 1, one of the substrates of the cAMP-dependent protein kinase in muscle and liver is phosphorylase kinase. Phosphorylation of this enzyme leads to an increase in its activity, i. e., enhanced conversion of phosphorylase from the b form to the more active a form and hence an acceleration of glycogen breakdown. cAMP-dependent protein kinase also phosphorylates the active I form of glycogen synthase converting it to the inactive D form resulting in a reduction in glycogen synthesis. The end result of these changes is enhanced lactate production in muscle and increased glucose output in liver.

Recently, however, observations in hepatocytes and the perfused liver have raised questions concerning the role of cAMP in the activation of hepatic glycogenolysis and gluconeogenesis by glucagon and catecholamines (3, 4, 6, 10, 14, 17, 19, 20, 25). Okajima and Ui (17) have reported that infusion of low doses of epinephrine and glucagon in vivo causes glycogenolysis without an increase in cAMP. Birnbaum and Fain (3) have made similar observations in hepatocytes. It has also been found that the β-adrenergic antagonist propranolol reduces or abolishes the accumulation of cAMP due to epinephrine in hepatocytes and the perfused liver without affecting the stimulation of glycogenolysis or gluconeogenesis (3, 10, 14, 19, 20, 25). Furthermore, the β-adrenergic agonist isoproterenol has been observed to have little or no effect on glycogenolysis or gluconeogenesis although it is as effective as epinephrine in raising cAMP (6, 14, 19, 25). On the other hand, the α-adrenergic antagonists phenoxybenzamine and phentolamine markedly inhibit the effects of catecholamines on hepatic glycogen metabolism and gluconeogenesis without reducing cAMP levels (6, 10, 14, 20), and the α-agonist phenylephrine activates glycogenolysis and gluconeogenesis at concentrations which do not elevate cAMP (3, 6, 10, 20).

These findings prompted us to reexamine the role of cAMP in the activation of hepatic glycogenolysis and gluconeogenesis by glucagon and catecholamines. In addition we have studied the role of cAMP-dependent protein kinase in these effects. Homogeneous preparations of rat liver parenchymal cells have been employed to avoid possible complications arising from changes in cAMP in non-parenchymal cells as may occur in preparations such as the perfused liver.

METHODS

Isolated liver parenchymal cells were prepared from fed or 24 hr-fasted rats by a modification (10) of the method of Berry and Friend (2). They were suspended in Krebs-Henseleit bicarbonate buffer containing 1.5% gelatin and incubated with shaking at 37° in Erlenmeyer flasks gassed with O_2-CO_2 (95:5). A preliminary incubation for 10 min was routinely employed before addition of hormones or other agents. Glucose and [^{14}C] glucose were measured as described elsewhere (10). cAMP was determined by the protein binding assay (8), cAMP-dependent protein kinase by the method of Cherrington et al. (4) and phosphorylase kinase by the method of Vandenheede et al. (26). Phosphorylase a was measured by the method of Stalmans and Hers (23) and glycogen synthase according to Thomas et al. (24) after preparation of cell homogenates according to Hutson et al. (10). Cell viability was routinely checked at the beginning and at the end of the experiments by measuring trypan blue exclusion. Experiments in which viability was less than 90% were discarded. The wet weight of packed cells was measured as described earlier (10). Glucagon and α- and β-agonists and antagonists were obtained from sources noted previously (10).

RESULTS

Effects of glucagon on glycogen metabolism and lactate gluconeogenesis in hepatocytes. As shown in Figure 2, a physiologic level of glucagon (10^{-10}M) significantly increased glucose production and [^{14}C] glucose synthesis from [^{14}C] lactate measured over 30 min in hepatocytes from fed rats. Almost 3-fold increases in both parameters were observed with maximally effective concentrations of the hormone (10^{-9}M) and the two dose response curves were indistinguishable. Measurements of glycogen and calculations from the specific radioactivity of

[^{14}C] lactate showed that in the presence or absence of the hormone, glucose release was largely attributable to glycogen depletion. Thus, glucose output was a valid index of net glycogenolysis.

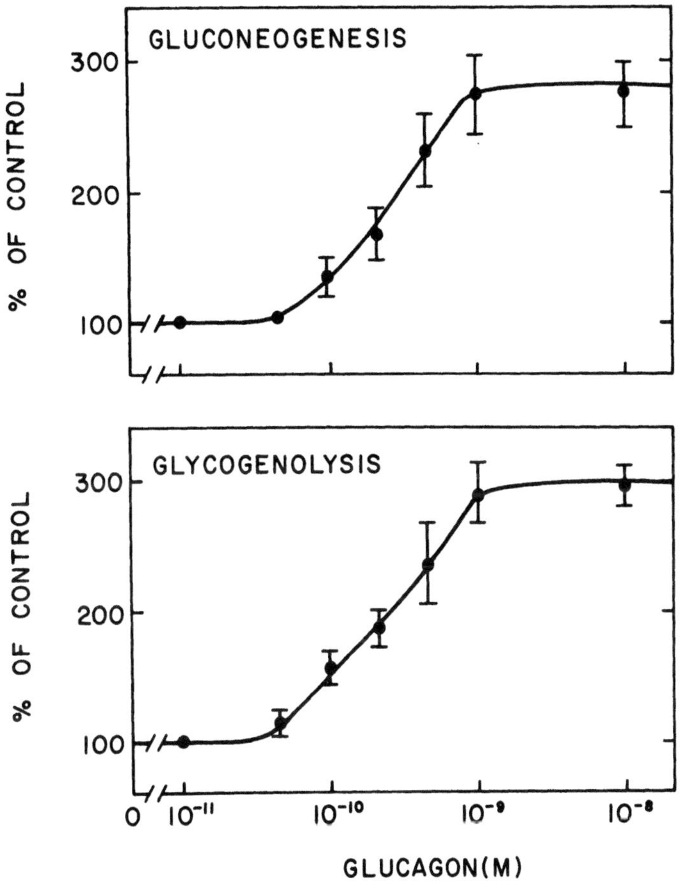

Fig. 2. Dose response curves for glucagon on glucose release and [^{14}C] glucose synthesis from [^{14}C] lactate in hepatocytes from fed rats. For experimental details, see (10).

For methodologic reasons, it was difficult to discern how rapidly the activation of glucose release and gluconeogenesis occurred, but significant increases in glucose and [^{14}C] glucose were observed at 5 min. Early changes in glycogenolysis are more readily detected by measurement of phosphorylase a levels.

Figure 3 and Figure 4 (lower panel) show that 10^{-10}M - 10^{-9}M glucagon caused abrupt increases in phosphorylase a followed by slow declines. Activation of the enzyme was discernible at 1 min (Fig. 3). In hepatocytes incubated with glucose or obtained from fasted rats, physiologic levels of glucagon also rapidly promoted the conversion of glycogen synthase from the active I form to the inactive D form (10).

Effects of glucagon on cAMP and cAMP-dependent protein kinase in hepatocytes. If cAMP and cAMP-dependent protein kinase mediate the actions of glucagon on glycogen metabolism and gluconeogenesis, they should show changes within 1 min and with concentrations of the hormone as low as 10^{-10}M. Figure 3 shows that this low level of glucagon produced a significant increase in cAMP and in the protein kinase activity ratio (-cAMP/+cAMP). The effects were maximal at 1 min and were diminished at 10 min. The maximal increases in these parameters preceded that in phosphorylase a.

Figure 4 shows that similar results were obtained when higher, but physiologic concentrations of the hormone were tested. There was a striking similarity between the time courses of the changes in cAMP and protein kinase consistent with the nucleotide being the primary modulator of the enzyme. The changes in phosphorylase with 10^{-10}M and 5 x 10^{-10}M glucagon followed similar time courses to the cAMP and protein kinase changes, but were proportionally greater. Phosphorylase was maximally activated by 5 x 10^{-10}M and 10^{-9}M glucagon at 2 and 5 min, even though these concentrations did not maximally affect cAMP or protein kinase. The same was true for 10^{-9}M hormone at 15 min.

In other experiments (not shown), glucagon activated phosphorylase kinase in the hepatocytes in agreement with Vandenheede et al. (26). An effect was seen within 1 min and with hormone concentrations within the physiologic range.

Analysis of Figure 4 and many other experiments not shown indicates that maximal activation of phosphorylase, i. e., a 3-fold rise in phosphorylase a may be achieved with half-maximal activation of protein kinase and a cAMP level of

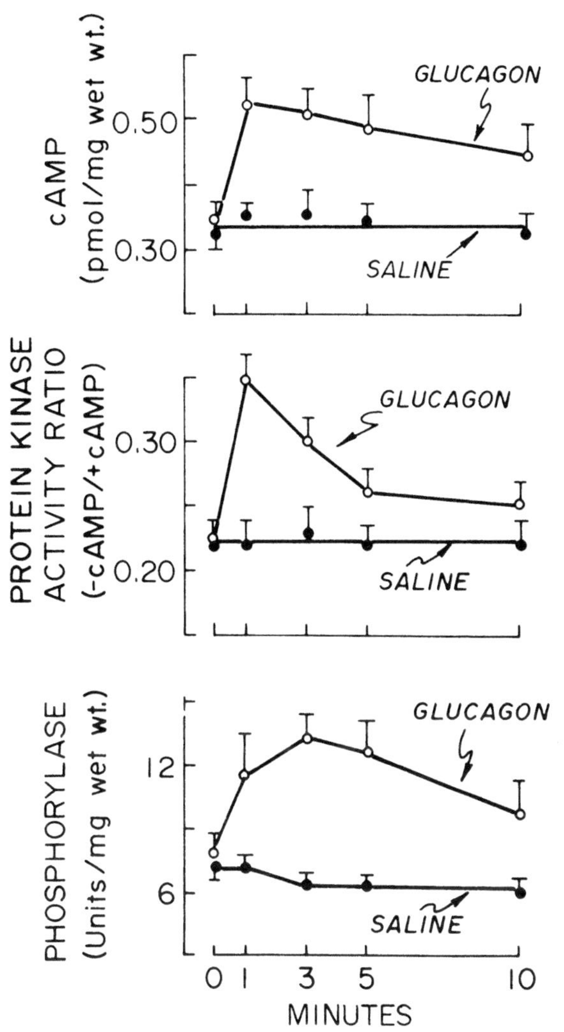

Fig. 3. Time courses of changes in cAMP, protein kinase activity ratio and phosphorylase a in hepatocytes incubated with 10^{-10}M glucagon. For assay procedures see (10).

less than 1 pmol per mg of cells. More detailed dose response curves for the effects of glucagon on the level of cAMP and the protein kinase activity ratio measured 1 min after exposure of hepatocytes to the hormone are shown in Figure 5. It is seen that half-maximal activation of protein kinase occurred at 5 x 10^{-10}M hormone, but this concentration was much lower than that which induced half-maximal accumulation of cAMP (2 x 10^{-9}M). Thus glucagon is capable of promoting cAMP increases far greater than those required for full biologic responses, as noted previously (7).

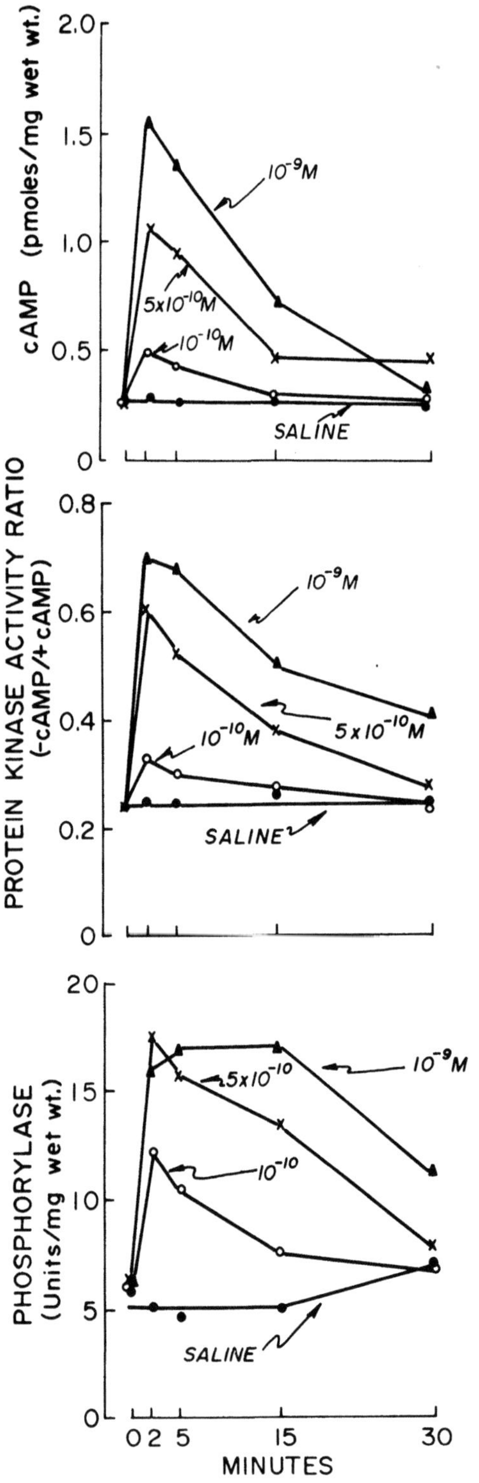

Fig. 4. Time courses of changes in cAMP, protein kinase activity ratio, and phosphorylase a in hepatocytes incubated with different concentrations of glucagon.

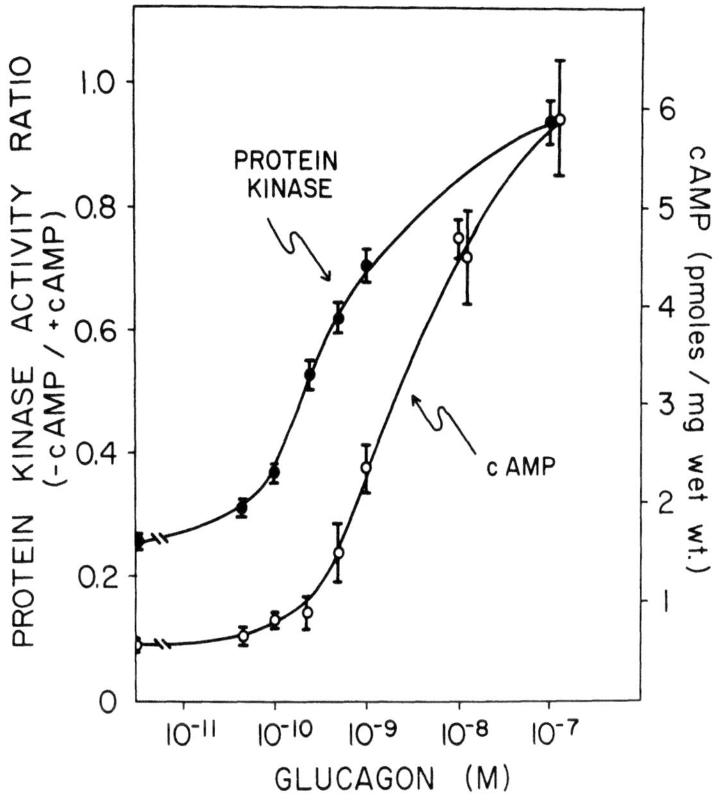

Fig. 5. Dose response curves for glucagon on the cAMP level and protein kinase activity ratio in hepatocytes from fed rats. Samples were taken 1 min after exposure to the hormone.

Effects of epinephrine on glycogen metabolism and lactate gluconeogenesis in hepatocytes. As illustrated in Figure 6, 10^{-6}M and 10^{-5}M epinephrine cause rapid increases in glycogenolysis in hepatocytes from fed rats. These concentrations also activated [^{14}C] glucose synthesis from [^{14}C] lactate after a delay of 5 min. The effect of epinephrine on glycogenolysis was attributable to rapid phosphorylase activation since glycogen synthase was not inactivated under these conditions (Fig. 7). As noted previously for glucagon, epinephrine caused rapid inactivation of glycogen synthase in hepatocytes previously incubated with glucose or obtained from fasted rats (10). Significant effects of epinephrine on glucose release, gluconeogenesis, phosphorylase activation and glycogen synthase (under appropriate conditions) were observed at 10^{-7}M concentration and maximal changes occurred at 10^{-5}M (10; Fig. 8).

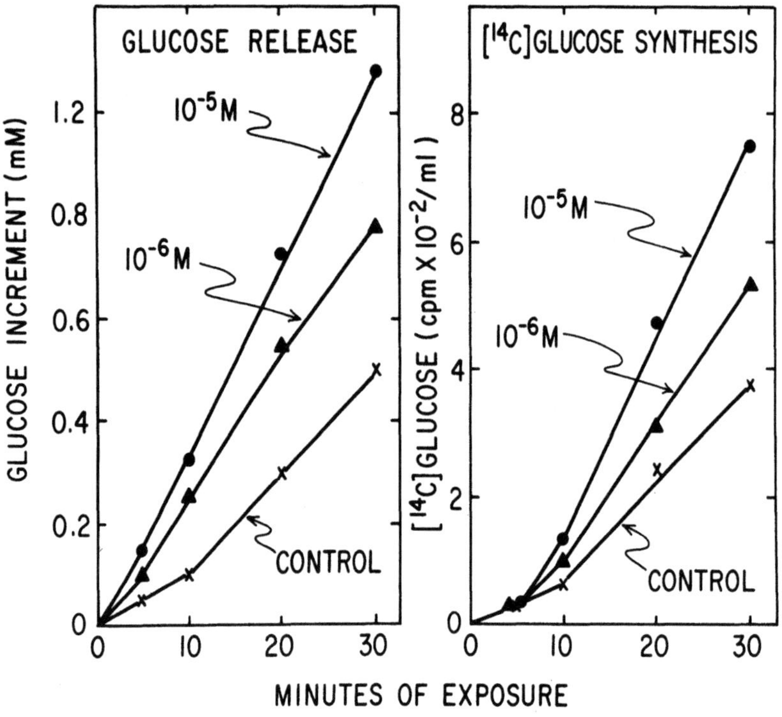

Fig. 6. Time courses of glucose release and [^{14}C] glucose synthesis in hepatocytes from fed rats incubated with 5 mM [^{14}C] lactate.

<u>Effects of epinephrine on cAMP and cAMP-dependent protein kinase in hepatocytes.</u> In accord with observations in the perfused liver (7), epinephrine (10^{-7}M - 10^{-5}M) caused rapid transient increases in cAMP in hepatocytes (Fig. 9). Peak elevations were seen at 1-3 min and thereafter the nucleotide declined such that there were no significant increases at 10 min. The transient nature of the cAMP response probably reflects developing refractoriness of the β-adrenergic system since readdition of epinephrine at 10 min provokes no further cAMP in-

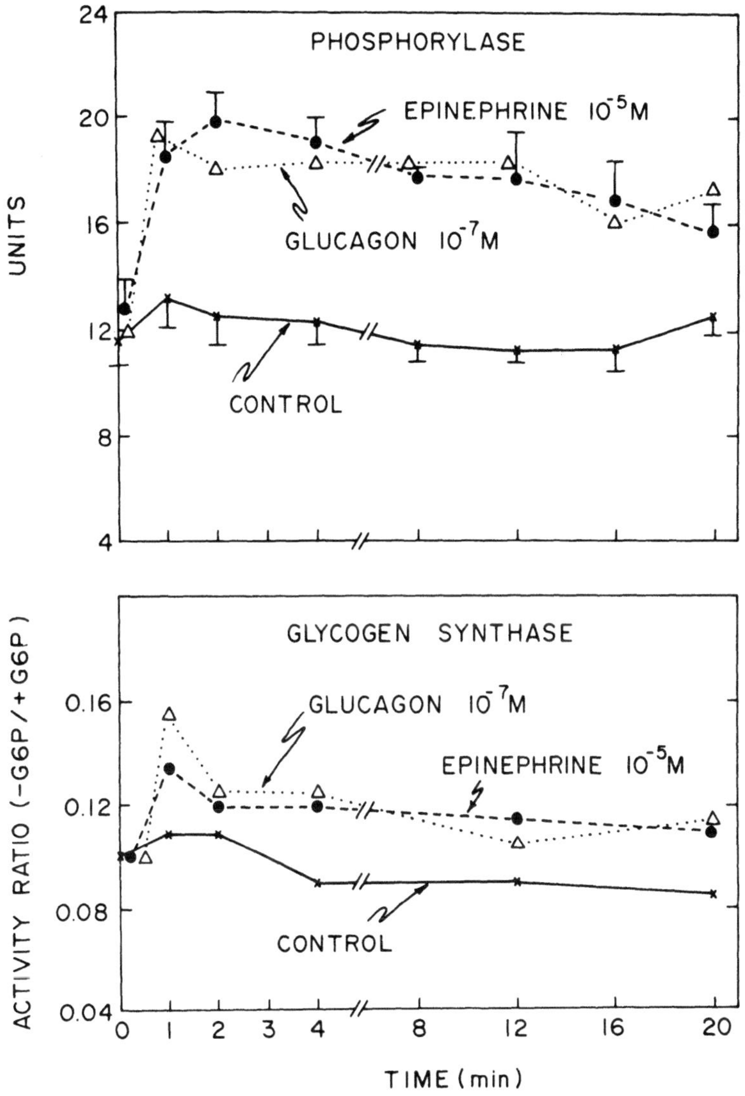

Fig. 7. Time courses of changes in phosphorylase a and glycogen synthase activity ratio (-G6P/+G6P) in hepatocytes from fed rats incubated with maximally effective concentrations of glucagon and epinephrine. For experimental details see (10).

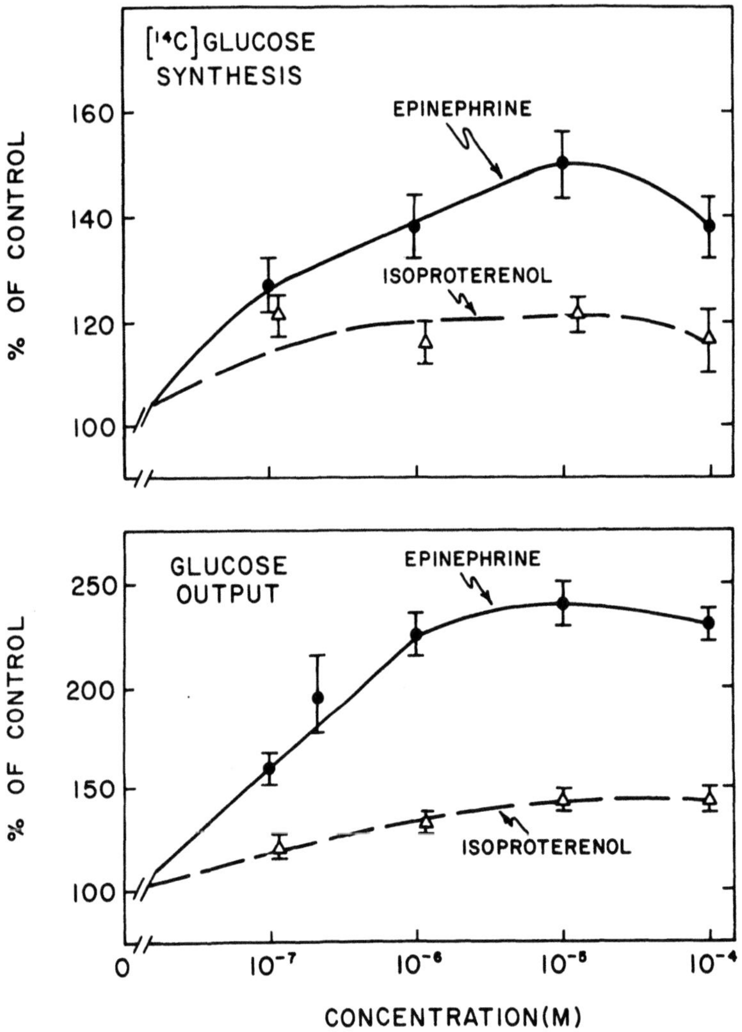

Fig. 8. Dose response curves for epinephrine and isoproterenol on glucose release and [^{14}C] glucose synthesis from [^{14}C] lactate. Experimental conditions were as in Fig. 2.

crease, whereas addition of glucagon does (data not shown). As expected from the cAMP response, epinephrine produces a transient activation of protein kinase which is maximal at 1 min (data not shown).

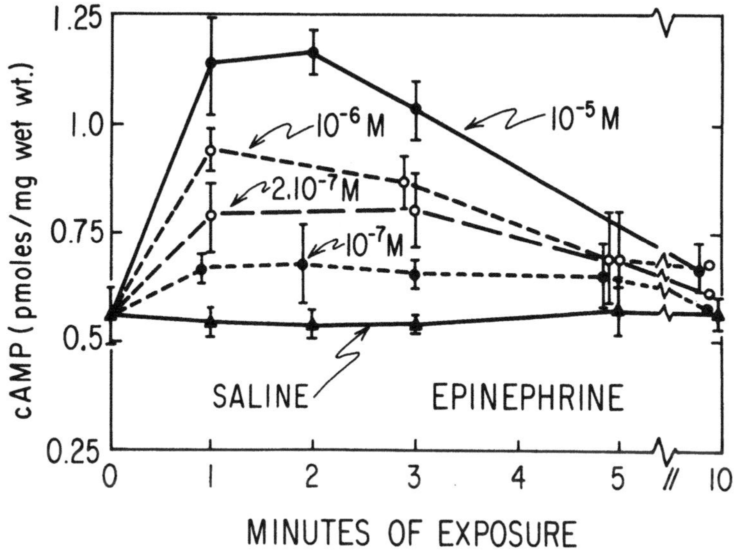

Fig. 9. cAMP levels in hepatocytes incubated with different concentrations of epinephrine.

Effects of adrenergic antagonists on the actions of epinephrine in hepatocytes. In accord with the general finding that the cAMP response to epinephrine is mediated by β-adrenergic receptors, the β-adrenergic antagonist propranolol abolished the cAMP increase observed with epinephrine (Fig. 10 and Fig. 11, upper panel). As also expected, the α-adrenergic blockers phentolamine and phenoxybenzamine did not modify the nucleotide response (Fig. 10 and Fig. 11, upper panel). However, the effects of the adrenergic antagonists on glycogenolysis and gluconeogenesis were unexpected and contrary to the hypothesis that cAMP mediated the actions of epinephrine on these processes. As illustrated in Fig. 11 (middle and lower panels), the β-blocker had minimal or no effects on the activation of glucose release and [^{14}C] glucose synthesis from [^{14}C] lactate by epinephrine,

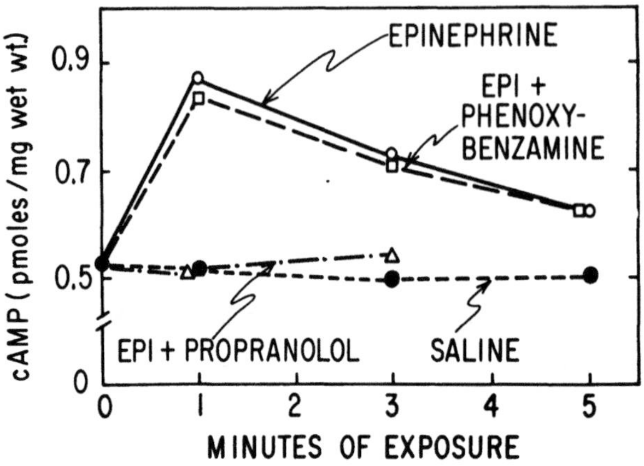

Fig. 10. Effects of the β-adrenergic blocker propranolol and the α-adrenergic blocker phenoxybenzamine on the cAMP increase in hepatocytes incubated with 10^{-5}M epinephrine. Blocker concentrations were 10^{-5}M.

whereas the α-blockers produced significant inhibitions. Thus the data indicated a major role for α-adrenergic receptors in catecholamine activation of glycogenolysis and gluconeogenesis, and a minimal role for the β-receptor-cAMP system. Similar conclusions were reached when the effects of α- and β-adrenergic antagonists on epinephrine activation of phosphorylase and inactivation of glycogen synthase were examined (10). The conclusion that an alternative mechanism not involving cAMP exists for catecholamine activation of glucose release is also indicated by comparing Figures 6, 7 and 9. These figures illustrate that the activation of phosphorylase and glycogenolysis by 10^{-5}M epinephrine persists far beyond the point at which cAMP levels return to baseline.

<u>Effects of the α-adrenergic agonist phenylephrine on hepatocytes.</u> To further establish the existence of an α-adrenergic-mediated mechanism for phosphorylase activation in hepatocytes not involving cAMP, the effects of the specific α-agonist phenylephrine were examined. As illustrated in Figure 12, this agent rapidly increased phosphorylase <u>a</u> and activated glycogenolysis in cells from fed

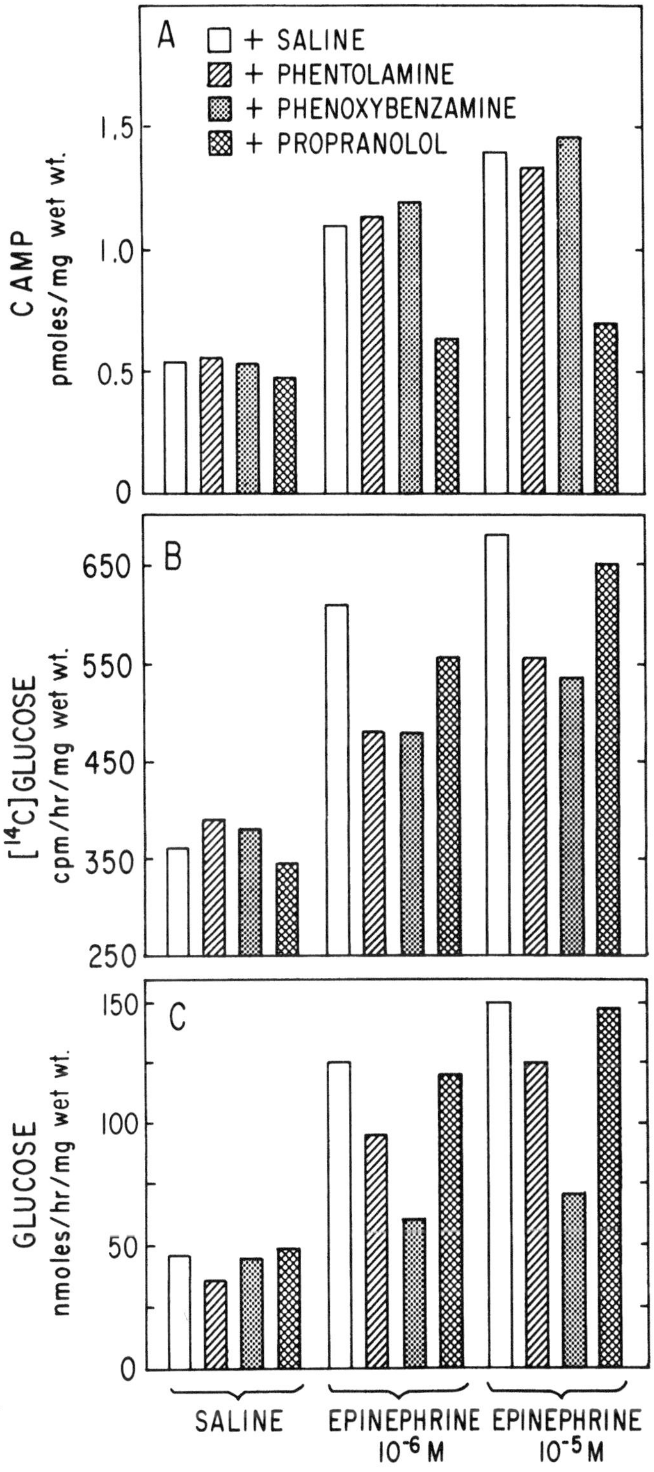

Fig. 11. Effects of 10^{-5}M adrenergic blockers on cAMP levels, glucose released and [^{14}C] glucose synthesis from [^{14}C] lactate in hepatocytes from fed rats incubated with epinephrine. cAMP was measured at 1 min and glucose and [^{14}C] glucose were measured at 30 min.

rats. It also inactivated glycogen synthase in cells from fasted rats and stimulated gluconeogenesis (10). All these effects were specifically due to activation of α-adrenergic receptors as shown by studies with adrenergic blockers (10).

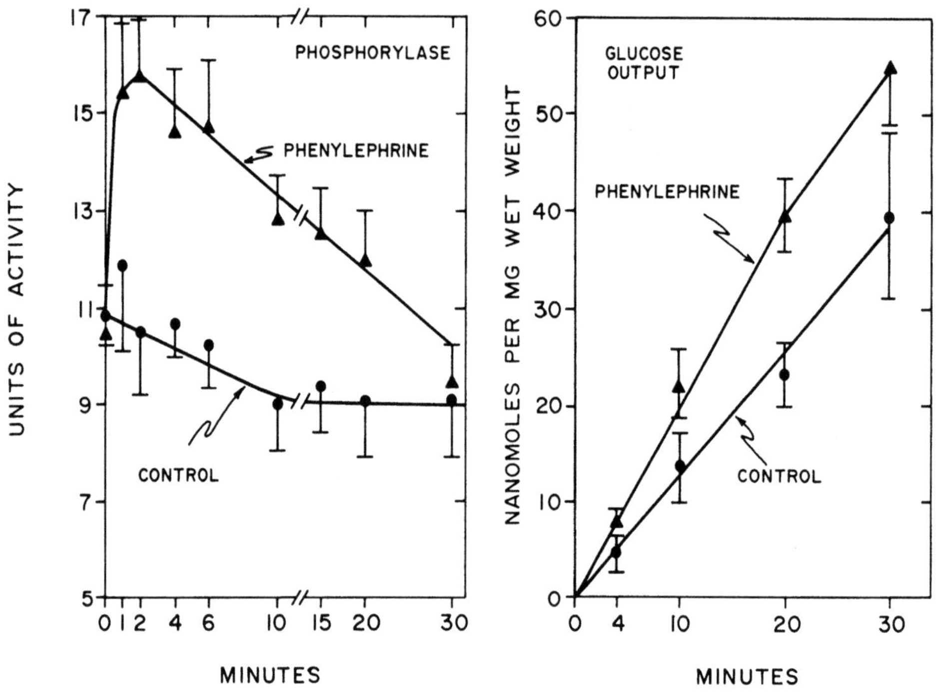

Fig. 12. Time courses of phenylephrine activation of phosphorylase and glucose release in hepatocytes from fed rats. Phenylephrine concentration was 10^{-5}M.

Concentrations of phenylephrine (10^{-6}M and 10^{-5}M) which were maximal or near-maximal for glycogenolysis and gluconeogenesis (10) induced little or no change in cAMP levels (Fig. 13). In contrast, concentrations of glucagon (10^{-10}M and 2 x 10^{-10}M) which induced changes in glucose release and [^{14}C] glucose synthesis from [^{14}C] lactate equivalent to 10^{-6}M and 10^{-5}M phenylephrine, respectively (4), clearly promoted the accumulation of cAMP (Fig. 13).

The effect of a maximally effective concentration of phenylephrine (10^{-5}M) on protein kinase measured at 1 min is shown in Figure 14. The agonist produced a

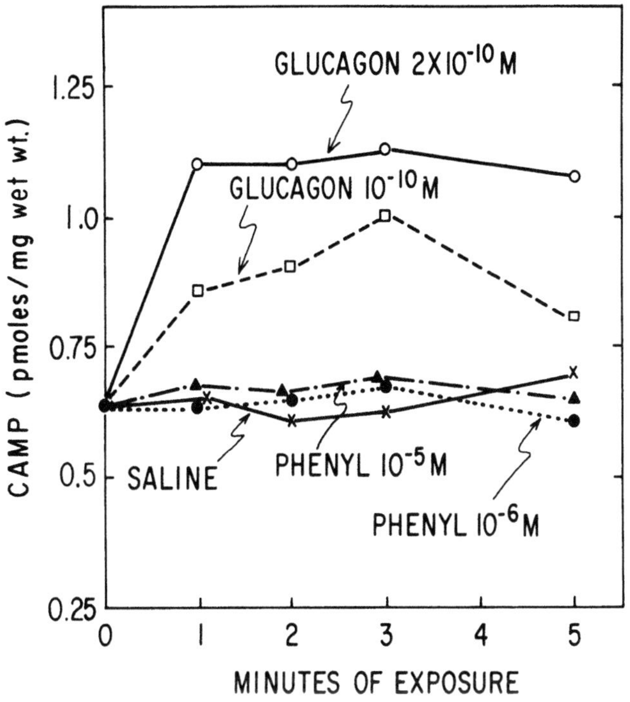

Fig. 13. Comparison of the effects of glycogenolytically equivalent concentrations of phenylephrine and glucagon on hepatocyte cAMP levels.

small increase in the activity ratio of the enzyme which was significant by paired t analysis. In comparison, glucagon induced much greater activation of the enzyme (Fig. 14) in line with its larger effect on cAMP accumulation. For example, the concentration of glucagon (2 x 10^{-10}M) which was glycogenolytically equivalent to 10^{-5}M phenylephrine induced a 6-fold greater increase in the kinase activity ratio. When the hepatocytes were incubated with propranolol or phenoxybenzamine and the effects of 10^{-5}M phenylephrine on protein kinase tested, the small activation of the enzyme was abolished by the β-blocker and unaffected by the α-blocker (4). Thus these results confirm the conclusion that α-adrenergic activation of phosphorylase is not associated with a rise in cAMP or activation of cAMP-dependent protein kinase.

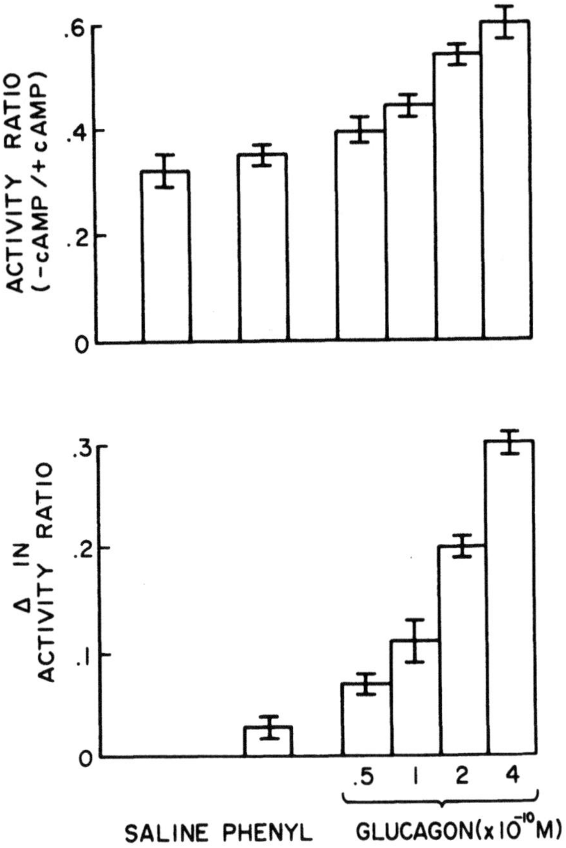

Fig. 14. Comparison of the effects of 10^{-5}M phenylephrine and different concentrations of glucagon on the protein kinase activity ratio of hepatocytes measured at 80 sec.

Since the effects of phenylephrine on phosphorylase and glycogenolysis are depressed when the calcium content of hepatocytes is decreased by incubation in Ca-free medium containing EGTA, whereas those of glucagon on these parameters are unaffected (1, 5), it has been postulated that calcium is involved in the α-adrenergic activation of phosphorylase (1, 5). In support of this hypothesis, it has been found that phosphorylase is activated by an increase in medium Ca^{++} and by the divalent cation ionophore A23187 in a calcium-dependent manner (1, 5, 18).

Furthermore, phenylephrine has been found to stimulate ^{45}Ca influx and exchange in hepatocytes to a greater degree than a glycogenolytically equivalent concentration of glucagon (1). A likely site of Ca action in the phosphorylase activation pathway is phosphorylase kinase, since this enzyme is inactivated by EGTA and reactivated by micromolar concentrations of Ca (13, 21).

Effects of the β-adrenergic agonist isoproterenol on hepatocytes. It has been reported by many workers that the β-adrenergic agonist isoproterenol is a much less effective glycogenolytic and gluconeogenic agent than epinephrine in rat liver preparations (6, 7, 9, 11, 14, 19, 20, 25). It has also been noted that isoproterenol is as effective as epinephrine in promoting the accumulation of cAMP (6, 7, 19, 20, 25). Obviously, the relatively weak biological activity of isoproterenol can now be attributed to the lack of α-adrenergic activity. However, the minimal changes in glucose release and gluconeogenesis observed with the agonist have led to the view that an inhibitory metabolite or other factor is produced during its action which inhibits the effects of the increased cAMP (16).

Figure 8 compares the relative potencies of isoproterenol and epinephrine on glycogenolysis and gluconeogenesis and Figure 15 shows the time courses of cAMP changes in hepatocytes incubated with several concentrations of isoproterenol. Comparison of Figure 15 with Figure 9 confirms the view that the β-agonist is as effective as epinephrine in increasing cAMP, but, like epinephrine, its effect is transient, being no longer significant at 10 min. Figure 16 illustrates the changes in cAMP, protein kinase activity ratio and phosphorylase a in hepatocytes incubated with a maximally effective concentration of isoproterenol (10^{-5}M). cAMP accumulation was similar to that seen in Figure 15 and the time course of activation of protein kinase generally followed the changes in cAMP. There was a 2-fold increase in phosphorylase a at 1 min followed by a slow decline. As expected, the changes in this enzyme were less in magnitude and duration than those seen with 10^{-5}M epinephrine (cf. Fig. 7).

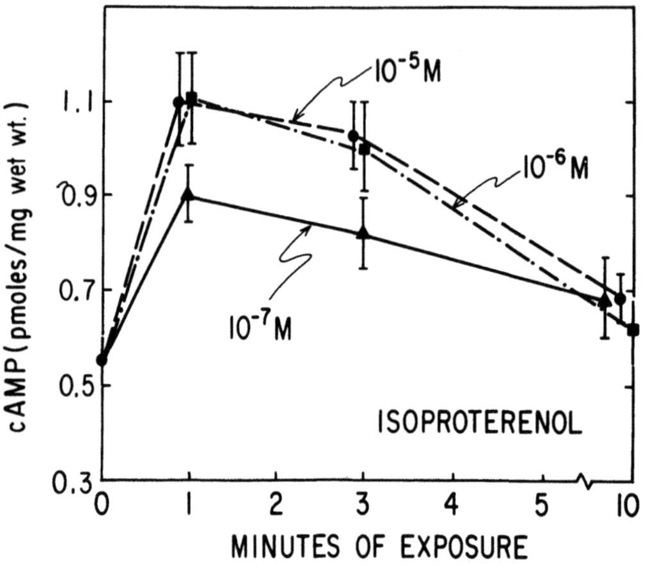

Fig. 15. cAMP levels in hepatocytes incubated with different concentrations of isoproterenol.

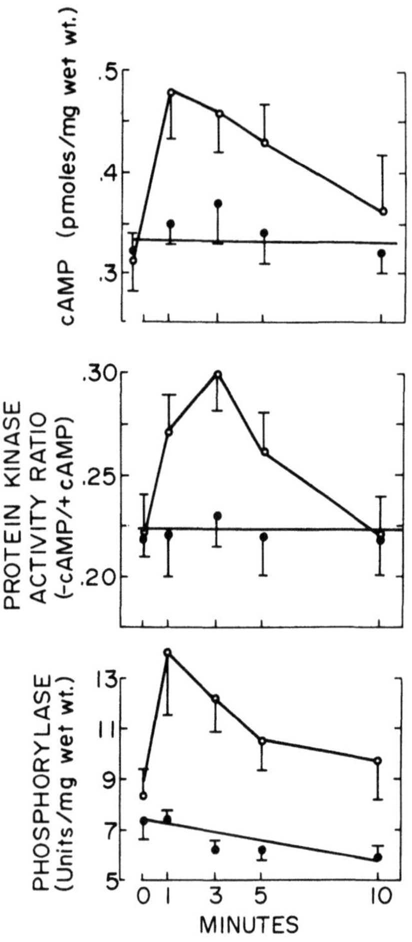

Fig. 16. Time courses of changes in cAMP, protein kinase activity ratio and phosphorylase _a_ in hepatocytes incubated with 10^{-5}M isoproterenol.

Comparison of Figures 3 and 16 indicates that the changes in cAMP, protein kinase and phosphorylase induced by 10^{-5}M isoproterenol are similar in degree and duration to those induced by 10^{-10}M glucagon. Thus there is no evidence of any impairment of the linkages between cAMP, protein kinase and phosphorylase with isoproterenol. Furthermore, Figure 2 indicates that 10^{-10}M glucagon caused only 30-60% increases in glycogenolysis and gluconeogenesis over a 30 min period, and these are comparable to those induced by 10^{-5}M isoproterenol (Fig. 8).

DISCUSSION

The results of the present study with glucagon are in complete agreement with the scheme shown in Figure 1. They show that cAMP accumulation and protein kinase activation precede the changes in phosphorylase and glycogen synthase, and that concentrations of glucagon which significantly activate phosphorylase and inactivate glycogen synthase also increase cAMP and activate protein kinase. In addition the present findings are consistent with the hypothesis that cAMP and cAMP-dependent protein kinase also mediate the effects of glucagon on gluconeogenesis.

No definitive explanations can be offered for the dissociation between cAMP and glycogenolysis in livers infused with low doses of glucagon reported by Okajima and Ui (17) and the similar discrepancy between cAMP and phosphorylase in hepatocytes incubated with glucagon noted by Birnbaum and Fain (3). However, it should be noted that low concentrations of the hormone induce small changes in cAMP (Fig. 5), but large changes in phosphorylase and glycogenolysis (Fig. 3) so that suboptimal assay procedures may fail to detect the changes in cAMP induced by physiological concentrations of glucagon. In addition, the significant delay in tissue fixation for cAMP which occurred in the study of Birnbaum and Fain (3) may have led to a diminution of the nucleotide increase because of phosphodiesterase activity.

In contrast to the situation with glucagon, the present findings represent a radical departure from the classical view of how epinephrine stimulates glycogenolysis in liver. They indicate that the effect is mediated mainly by α-adrenergic receptors and that the β-adrenergic receptor-cAMP system plays little role. They also point to a significant α-adrenergic component in the activation of gluconeogenesis by epinephrine in agreement with findings in hepatocytes from fasted rats (25). It could be argued that this situation is due to a derangement of the adrenergic receptors resulting from the procedures involved in preparation of the hepatocytes. However, data from the perfused liver (6, 11, 19, 20) also indicate a major role for α-adrenergic receptors in epinephrine action. Furthermore, the cAMP changes with epinephrine in the hepatocytes (Fig. 9) indicate that the β-adrenergic receptors are as responsive as in the perfused liver (7), and experiments with exogenous cAMP (data not shown) indicate that glycogenolysis and gluconeogenesis are as sensitive to the nucleotide as in the perfused liver. The sensitivity of the hepatocytes employed in the present study to epinephrine also closely resembles that reported by other workers.

It is possible that the predominance of α-adrenergic receptors as mediators of catecholamine-induced glycogenolysis in the perfused rat liver and rat hepatocytes represents a species difference. The classical studies of Sutherland, Rall and coworkers were carried out in cat, dog and rabbit liver in which β-receptors appear to be more important than α-receptors in catecholamine-induced glycogenolysis. The relative roles of α- and β-adrenergic receptors in the actions of catecholamines in liver of most other species have not been explored systematically (9, 12). The situation in guinea pig liver resembles that in rat liver: both α- and β-receptors are present and activation of either leads to glucose release.

The conclusion that the β-receptor-cAMP system plays a minor role in the stimulation of glucose output in rat hepatocytes is based largely on experiments with adrenergic antagonists. First, the α-antagonist phenoxybenzamine inhibits epinephrine-induced glycogenolysis, gluconeogenesis, phosphorylase activation and glycogen synthase inactivation without altering cAMP levels (10; Fig. 10 and 11). Second, the β-antagonist propranolol, almost completely abolishes the rise in cAMP with epinephrine, but does not alter glycogenolysis, gluconeogenesis, phosphorylase activation or glycogen synthase inactivation (10; Fig. 10 and 11). The specificities of phenoxybenzamine and propranolol for their respective receptors has been established in our experimental system, and it has been shown that neither agent interferes with the actions of glucagon or exogenous cAMP (10). The α-blockers phentolamine and dihydroergotamine produce similar effects to phenoxybenzamine and the β-blockers sotalol and dichloroisoproterenol resemble propranolol in their actions (10).

Additional evidence that mechanisms other than a rise in cAMP are involved in the glycogenolytic action of epinephrine comes from the present and previous studies. For example, the stimulation of glucose release or rise in phosphorylase a induced by the catecholamine persists beyond the point at which cAMP levels return to the baseline (Fig. 6 and 7, cf. Fig. 9). Comparison of the effects of low concentrations of glucagon and epinephrine on cAMP levels and glycogenolysis also suggests that the metabolic effects of the catecholamine are greater than can be attributed to the small rise in cAMP. For example, 10^{-7}M epinephrine produced an activation of glycogenolysis which was similar to that caused by 2 x 10^{-10}M glucagon, yet the increase in cAMP induced by the catecholamine was much less than that induced by glucagon (Fig. 8, cf. Fig. 2 and Fig. 9, cf. Fig. 13). Similar quantitative discrepancies between the effects of epinephrine on cAMP levels and glycogenolysis or phosphorylase levels have been reported in the perfused liver (6, 20) and a lack of correlation between changes in cAMP levels and gluconeogenesis has been noted in hepatocytes from fasted rats (14, 25).

The most extreme dissociations between catecholamine effects on carbohydrate metabolism and cAMP are seen with the α-adrenergic agonist phenylephrine and the β-adrenergic agonist isoproterenol. Phenylephrine produces significant stimulation of glycogenolysis, phosphorylase and gluconeogenesis, and inactivation of glycogen synthase, but causes little or no increase in cAMP in the perfused liver or hepatocytes (4, 6, 20; Fig. 12 and 13). On the other hand, isoproterenol increases cAMP, but produces minimal activation of glycogenolysis and gluconeogenesis (Fig. 8). The present studies showing these dissociations are of particular significance since they have been carried out in preparations of parenchymal cells and not in heterogeneous systems such as the perfused liver or liver slices. As stated earlier, the possibility exists that, in the latter systems, cAMP changes in nonparenchymal cells can lead to spurious results.

The lack of effects of α-blockade on cAMP levels in the presence of epinephrine and the inability of phenylephrine to elevate cAMP in hepatocytes indicate that activation of the α-receptor leads to increased glycogenolysis and gluconeogenesis by mechanisms not involving a rise in total cellular cAMP. As described in the RESULTS section, evidence has been obtained in support of the hypothesis that α-agonists activate phosphorylase by increasing the cytosolic concentration of Ca^{++} ions which leads to activation of phosphorylase kinase.

Another aspect of the present study which needs comment is the suggestion that cAMP generated by activation of the β-adrenergic receptor is less effective than that accumulated in response to glucagon. An example is the low glycogenolytic and gluconeogenic activity of isoproterenol. However, careful comparison of the metabolic effects of a maximally effective concentration of the β-agonist with those produced by an equivalent concentration of glucagon (10^{-10}M) fails to reveal a discrepancy between cAMP accumulation and glycogenolysis or gluconeogenesis. The same thing is true when the effects of adrenergic antagonists on the actions of epinephrine are carefully examined (4).

The existence of two apparently independent mechanisms for activation of hepatic glycogenolysis and gluconeogenesis by catecholamines mediated by separate receptors probably has survival value. Sympathetic nervous stimulation of liver glucose output is importantly involved in the classical "fight or flight" response of the organism to danger and it is understandable that a dual system could be of advantage. The fact that both α- and β-adrenergic systems mediate the same end response is unusual since these systems generally cause opposing effects in most tissues (15). The presence of both α- and β-receptors in the liver of certain species has also caused much confusion concerning their classification by pharmacologists (9, 12). More important than these considerations is the fact that the existence of α-adrenergic control of relatively well defined biochemical sequences such as phosphorylase activation and glycogen synthase inactivation in liver provides a tool for the elucidation of the chemical and enzymatic nature of α-adrenergic mechanisms.

Acknowledgements

Supported by Grants 5 P01 AM 07462, 1 R01 AM 18660 and 1 P17-AM 17026 from the National Institutes of Health, U. S. Public Health Service.

J. H. Exton is a Senior Investigator, Howard Hughes Medical Institute.

A. D. Cherrington is recipient of a Medical Research Council of Canada Fellowship and a Solomon A. Berson Award from the American Diabetes Association.

F. D. Assimacopoulos-Jeannet is an NIH International Postdoctoral Fellow. Present address: Laboratoires de Recherches Médicales, University of Geneva, Geneva, Switzerland.

REFERENCES

1. Assimacopoulos-Jeannet, F. D., Blackmore, P. F. and Exton, J. H. (1977): Studies on the α-adrenergic activation of hepatic glucose output. III. Studies on the role of calcium in the α-adrenergic activation of phosphorylase. J. Biol. Chem. Submitted for publication.

2. Berry, M. N. and Friend, D. S. (1969): High-yield preparation of isolated rat liver parenchymal cells. J. Cell Biol. 43, 506-520.

3. Birnbaum, M. J. and Fain, J. N. (1977): Activation of protein kinase and glycogen phosphorylase in isolated rat liver cells by glucagon and catecholamines. J. Biol. Chem. (in press).

4. Cherrington, A. D., Assimacopoulos, F. D., Harper, S. C., Corbin, J. D., Park, C. R. and Exton, J. H. (1976): Studies on the α-adrenergic activation of hepatic glucose output. II. Investigation of the roles of adenosine 3':5'-monophosphate and adenosine 3':5'-monophosphate-dependent protein kinase in the actions of phenylephrine in isolated hepatocytes. J. Biol.

Chem. 251, 5209-5218.

5. DeWulf, H. and Keppens, S. (1975): Is calcium the second messenger in liver for cyclic AMP-independent glycogenolytic hormones? Arch. Intern. Physiol. Biochim. 84, 159-160.

6. Exton, J. H. and Harper, S. C. (1975): Role of cyclic AMP in the actions of catecholamines on hepatic carbohydrate metabolism. Adv. Cyclic Nucleotide Res. 5, 519-532.

7. Exton, J. H., Robison, G. S., Sutherland, E. W. and Park, C. R. (1971): Studies on the role of adenosine 3',5'-monophosphate in the hepatic actions of glucagon and catecholamines. J. Biol. Chem. 246, 6166-6177.

8. Gilman, A. G. (1970): A protein binding assay for adenosine 3',5'-cyclic monophosphate. Proc. Natl. Acad. Sci. USA 67, 305-312.

9. Hornbrook, K. R. (1970): Adrenergic receptors for metabolic responses in the liver. Fed. Proc. 29, 1381-1385.

10. Hutson, N. J., Brumley, F. T., Assimacopoulos, F. D., Harper, S. C. and Exton, J. H. (1976): Studies on the α-adrenergic activation of hepatic glucose output. I. Studies on the α-adrenergic activation of phosphorylase and gluconeogenesis and inactivation of glycogen synthase in isolated rat liver parenchymal cells. J. Biol. Chem. 251, 5200-5208.

11. Jakob, A. and Diem, S. (1976): Metabolic responses of perfused rat livers to alpha- and beta-adrenergic agonists, glucagon and cyclic AMP. Biochim. Biophys. Acta 404, 57-66.

12. Jenkinson, D. H. (1973): Classification and properties of peripheral adrenergic receptors. Brit. Med. Bull. 29, 142-147.

13. Khoo, J. C. and Steinberg, D. (1975): Stimulation of rat liver phosphorylase kinase by micromolar concentrations of Ca^{2+}. FEBS Letters 57, 68-72.

14. Kneer, N. M., Bosch, A. L., Clark, M. G. and Lardy, H. A. (1974): Glucose inhibition of epinephrine stimulation of hepatic gluconeogenesis by blockade of the α-receptor function. Proc. Natl. Acad. Sci. 71, 4523-4527.

15. Moran, N. C. (1975): Adrenergic receptors. In: Handbook of Physiology. Section 7: Endocrinology, Vol. 6. Adrenal Gland. H. Blaschko, G. Sayer and A. D. Smith, Eds. American Physiological Society, Washington, D. C., pp. 447-472.

16. Newton, N. E. and Hornbrook, K. R. (1972): Effects of adrenergic agents on carbohydrate metabolism of rat liver: activities of adenyl cyclase and glycogen phosphorylase. J. Pharm. Exp. Ther. 181, 479-488.

17. Okajima, F. and Ui, M. (1976): Lack of correlation between hormonal effects on cyclic AMP and glycogenolysis in rat liver. Arch. Biochem. Biophys. 175, 549-557.

18. Pointer, R. H., Butcher, F. R. and Fain, J. N. (1976): Studies on the role of cyclic guanosine 3':5'-monophosphate and extracellular Ca^{++} in the regulation of glycogenolysis in rat liver cells. J. Biol. Chem. 251, 2987-2992.

19. Saitoh, Y. and Ui, M. (1976): Stimulation of glycogenolysis and gluconeogenesis by epinephrine independent of its beta-adrenergic function in perfused rat liver. Biochem. Pharmacol. 25, 841-845.

20. Sherline, P., Lynch, A. and Glinsmann, W. H. (1972): Cyclic AMP and adrenergic receptor control of rat liver glycogen metabolism. Endocrinology 91, 680-690.

21. Shimazu, T. and Amakawa, A. (1975): Regulation of glycogen metabolism in liver by the autonomic nervous system. VI. Possible mechanism of phosphorylase activation by the splanchnic nerve. Biochim. Biophys. Acta 385, 242-256.

22. Soderling, T. R. and Park, C. R. (1974): Recent advances in glycogen metabolism. In: Advances in Cyclic Nucleotide Research, Vol. 4. P. Greengard and G. A. Robison, Eds. Raven Press, New York, pp. 283-333.

23. Stalmans, W. and Hers, H. G. (1975): The stimulation of liver phosphorylase b by AMP, fluoride and sulfate. Eur. J. Biochem. 54, 341-350.

24. Thomas, J. A., Schlender, K. K. and Larner, J. (1968): A rapid filter paper assay for UDP glucose-glycogen glucosyltransferase, including an improved biosynthesis of UDP-^{14}C-glucose. Anal. Biochem. 25, 486-499.

25. Tolbert, M. E. M., Butcher, F. R. and Fain, J. N. (1973): Lack of correlation between catecholamine effects on cyclic adenosine 3':5'-monophosphate and gluconeogenesis in isolated rat liver cells. J. Biol. Chem. 248, 5686-5692.

26. Vandenheede, J. R., Keppens, S. and DeWulf, H. (1976): The activation of liver phosphorylase b kinase by glucagon. FEBS Letters 61, 213-217.

GUANYL NUCLEOTIDE REGULATION OF THE LIVER GLUCAGON-SENSITIVE ADENYLYL CYCLASE SYSTEM

L. Birnbaumer and T. L. Swartz

Department of Cell Biology, Baylor College of Medicine
Houston, Texas

Properties of the GMP-P(NH)P-activated glucagon sensitive adenylyl cyclase from rat liver plasma membranes were studied with respect to reversal of nucleotide imidodiphosphate analog activation. It was found that incubation of the analog-activated state with excess GTP (the natural effector) in the presence of substrate (ATP), divalent cation (Mg^{++}), a chelator (EDTA), the reaction product (cAMP), and an ATP-regeneration system, resulted in reversal of activity to that seen with GTP alone. The reversed state was demonstrated to be both sensitive to re-stimulation by GMP-P(NH)P and to stimulation by glucagon, and was therefore not the result of GTP-induced inactivation of the enzyme. Examination of the time course of GTP-induced reversal of GMP-P(NH)P activation, as well as of GMP-P(NH)P-induced re-stimulation of reversed activity revealed that guanyl nucleotide effects occur with progressively increasing lag times, suggesting the existence or participation of one or more co-factors. These findings are inconsistent with the mechanism of GMP-P(NH)P activation involving the irreversible formation of an enzyme-P(NH)P derivative and suggest a more complex mechanism than the involvement of a simple two step isomerization reaction.

Glucagon exerts its action by stimulating adenylyl cyclase activity on the membrane surface of its target tissues. Description of its mode of action, therefore, resides in the elucidation of the molecular mechanisms involved in hormonal activation of the membrane-bound enzyme. This enzyme is extremely complex and to date we do not know how hormones affect its activity. However, there are clues. The most important of these clues stems from the finding in 1970 that guanyl nucleotides play a key regulatory role in hormonal stimulation of adenylyl cyclases. On the following pages we shall review some of the basic findings relating to guanyl nucleotide regulation of adenylyl cyclases, some of the models that have been proposed to explain their mode of action, and our most recent experiments with GMP-P(NH)P, the imidodiphosphate analog of GTP, which has been assumed to activate adenylyl cyclase in an irreversible and persistent manner, not only in liver membranes but also in all membranes of eukaryotic cells studied to date.

As membrane-bound enzyme systems, adenylyl cyclases are no exception in that they are complex both in a regulatory and a structural sense. The substrate for these enzymes is ATP·Mg^{++}. Kinetic studies throughout the last 10 years suggest that the enzyme may contain an allosteric site for Mg^{++} whose affinity for the divalent cation is increased upon hormonal stimulation. Activity can also be supported by Mn^{++}. On the other hand, Ca^{++} appears to be inhibitory to catalytic activity and, in some systems such as the one found in heart muscle and described by Drummond and Duncan (6), it can be shown to interact competitively with Mg^{++} in excess of ATP, i.e. at the putative allosteric site. In liver membranes, however, it has been proposed (14, 15, 21, 28, 29) that the regulation of the activity of this enzyme, rather than depending on the interaction of Mg^{++} with an allosteric site, depends on the affinity changes of the catalytic site for inhibitory protonated free substrate, thus explaining a strong requirement for Mg^{++} in excess of ATP for enzymatic activity and a decreased Mg^{++} dependency for activity seen upon hormonal stimulation. Adenylyl cyclases are dependent not only on divalent cations for their activity, but also on phospholipids which surround it in the membranes. These phospholipids exert a specific regulatory role in hormonal stimulation. Key findings to this effect reside in the fact that if membranes are treated with phospholipases or detergents, hormonal stimulation is selectively lost and can be restored totally, as shown by Levey in heart (11, 12), or partially, as shown by Pohl et al. (19) in liver membranes, upon readdition of specific phospholipids.

The complexity of the enzyme system can be further emphasized by some unpublished experiments from our laboratory indicating that reagents such as parachloromercuribenzoate (PCMB) which react with sulfhydryl groups, can totally inactivate adenylyl cyclase activity. This inactivated system can be reactivated by treatment with mercaptoethanol or Cleland's reagent, thus suggesting the existence of key -SH groups in adenylyl cyclases. However, addition of mercaptoethanol or Cleland's reagent per se has also been found to strongly inhibit hormonal stimulation, especially in LH-sensitive adenylyl cyclase derived from rabbit corpora lutea, clearly suggesting the existence of -SS- bridges essential for proper hormonal stimulation. Taken together, these experiments indicate that proper hormonal regulation and activity of adenylyl cyclases are likely to depend on the natural state of both -SS- bridges and -SH groups. In spite of all these difficulties, adenylyl cyclases are undoubtedly excellent model systems to study hormone actions in cell free systems. In fact, they are the only cell free systems where, to date, clear-cut hormonal effects can be demonstrated unequivocally.

From physiologic studies one expects hormonal action in cell free systems, especially peptide and biogenic amine action, to be a reversible phenomenon. In isolated membranes it has been shown clearly that this reversibility is preserved. Figure 1 presents an experiment carried out with the glucagon-sensitive adenylyl cyclase of liver plasma membranes in which this reversibility was tested. It can be seen that the addition of 4×10^{-9}M glucagon to an ongoing adenylyl cyclase reaction results in an immediate increase of cAMP accumulation. This stimulation of enzyme activity by glucagon can be reversed by post addition of DH-glucagon (des-his^{1}-glucagon), a glucagon analog that competitively interacts with glucagon receptor, but cannot cause the conformational change that is necessary to stimulate the activity of the catalytic unit of the system (14, 15, 23-25). That both stimulation of adenylyl cyclase and its reversal are truly the expression of reversible interactions was demonstrated by the fact that readdition of excess native glucagon resulted in final re-stimulation of the system. The reversibility thus demonstrated is a key feature of adenylyl cyclase systems and should be considered in any purification and reconstitution experiments. It is on this aspect of regulation that we have centered our attention lately.

In 1970 Dr. Martin Rodbell and one of us (LB) carried out the experiment shown in Figure 2 in which we tested the effects of glucagon and GTP on liver membrane adenylyl cyclase that was being assayed using the synthetic substrate AMP-P(NH)P, the imidodiphosphate analog of ATP. It was found that, when either

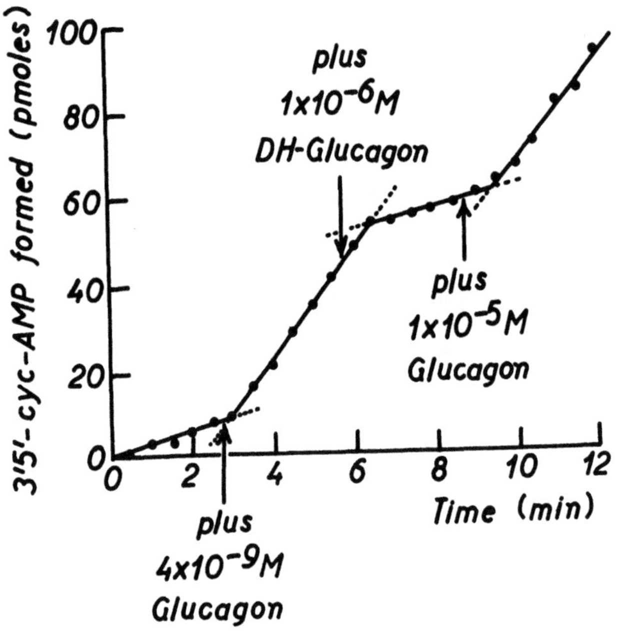

Fig. 1. Reversibility of stimulatory action of glucagon and of inhibitory action of dehistidine-glucagon on rat liver membrane adenylyl cyclase. The reactions were carried out in the presence of 3.2 mM ATP, 5.0 mM $MgCl_2$, 1.0 mM EDTA, 1.0 mM cAMP, ATP-regenerating system, 25 mM Tris-HCl, pH 7.5, and the indicated additions. For further details see Birnbaumer et al (3).

glucagon or GTP was added at the beginning of the reaction, both of these agents were capable of stimulating adenylyl activity. Combination of GTP and glucagon (10^{-5}M and 10^{-6}M respectively) resulted in greater activity than that obtained with either agent alone. However, when glucagon or GTP was added 5 min after the initiation of the reaction, a surprising result was obtained. Neither one of the agents by itself was capable of stimulating or enhancing the accumulation of cAMP beyond that of control. The combination of the two, however, restored the activity to that seen when they were added at time zero of the reaction. This experiment presented us with several findings. First, it shows unequivocal evidence for an important regulatory role of guanyl nucleotides and suggested an obligatory role for GTP in the action of glucagon. The reverse statement was also suggested by the data, i.e. that under the conditions of the assay, glucagon was essential for the GTP-mediated stimulation of adenylyl cyclase. Second, it can be seen that while GTP is capable of stimulating adenylyl cyclase in "fresh" membranes, i.e. when added at time zero, it is incapable of doing so, at least within the period tested, when added to "aged" membranes, i.e. 5 min after initiation of incubation. Third, glucagon, which is also incapable of stimulating activity in "aged" membranes, can do so if GTP is also added. In an initial interpretation of these results, we postulated (24, 25) that endogenous guanyl nucleotides, present in fresh membranes, were lost during the first 5 min of incubation in the absence of an ATP regenerating system, thus explaining the necessity of having to add GTP to obtain a glucagon effect in "aged" membranes. We did not speculate, however, on what the reason might be for which GTP addition at the late time did not cause any immediate changes in enzymatic activity. As will be shown below, this may be due to the fact that "aging" results in a progressive decrease of the rate at which nucleotides affect the enzyme system and that this decrease in rate of action can be partially if not totally reversed by hormones.

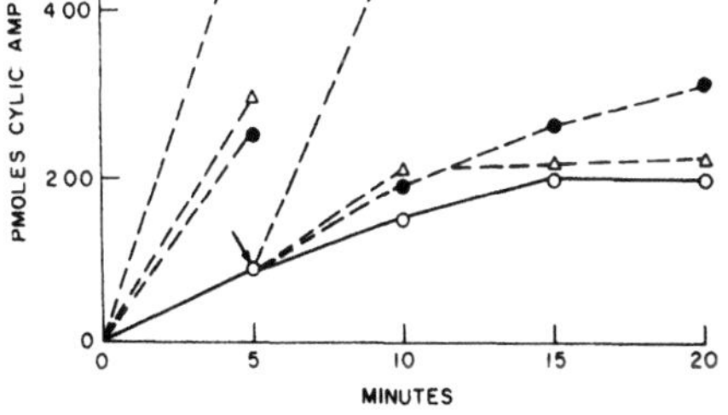

Fig. 2. Effect of glucagon and GTP on adenylyl cyclase activity in liver membranes determined in the absence of regenerating system and with AMP-P(NH)P as substrate. Note requirement of both GTP and glucagon for sustained hormonal stimulation if added after 5 min of incubation for basal activity (24, with permission).

Further studies carried out since then, not only in the laboratory of Rodbell and collaborators (7, 16, 22, 26, 27), but also in the laboratories of, among others, Lefkowitz (8-10), Aurbach (1, 33, 34), Levitski (13, 32), and Helmreich (18, 20) have confirmed that guanyl nucleotides indeed play a central role in the regulation of adenylyl cyclase activity. Studies with GTP were complicated, however, by the fact that this nucleotide is rapidly degraded by membrane nucleotidases. To circumvent this difficulty, the imidodiphosphate analog of GTP, GMP-P(NH)P, was pioneered by Rodbell and collaborators (17) and shown to be a potent stimulator of eukaryotic adenylyl cyclases. It has been and is being used widely to probe regulatory features of adenylyl cyclases and their stimulation by hormones. Using this analog, it was soon found that in addition to being more effective in stimulating basal adenylyl cyclase activity than GTP, it seems to stimulate in an irreversible manner that persists even after extensive washing of membranes (5, 29, 31, 34). Indeed, stimulation by GMP-P(NH)P has been found to persist even after solubilization of membranes (10, 18, 31).

The difference in activities obtained when liver plasma membranes are exposed to either GTP or GMP-P(NH)P is exemplified in Figure 3. It can be seen that basal activity under assay conditions that include ATP regenerating system is not very stable, and tends to fall off as the incubation time progresses. Addition of GTP, however, leads to a very rapid stimulation of basal activity, about two-fold in the experiment presented in Figure 3, which then remains constant for as long as 30 min, the longest time tested in these experiments. The stimulation obtained by GTP is rapid and, at the concentration used (1×10^{-5}M) does extrapolate back to the point of addition of the nucleotide. Addition of GMP-P(NH)P, on the other hand, stimulates adenylyl cyclase activity more than GTP, as seen by the steady state rate of cAMP accumulation, but appears slowly and exhibits a lag in the order of 5-10 min, varying somewhat from membrane preparation to membrane preparation. Similar studies to these, showing similar results, have been published by others before (2, 4, 17, 29). Thus, one very important characteristic of the ac-

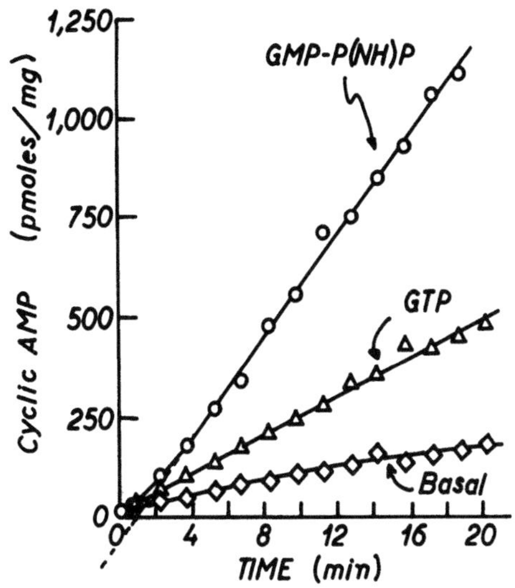

Fig. 3. Effect of GMP-P(NH)P (10 μM) and of GTP (10 μM) on adenylyl cyclase activity in liver membranes. The reaction was carried out in the presence of 0.1 mM [α-^{32}P] ATP (900 cpm/pmole), 5.0 mM $MgCl_2$, 1.0 mM EDTA, 1.0 mM [^{3}H] cAMP (10,000 cpm), an ATP regenerating system, 25 mM BTP buffer, pH 7.5 at 30°. The reactions were started by addition of 10 μl (18.5 μg) of partially purified liver membranes to 40 μl of reagent, and terminated at various times by addition of 100 μl of stopping solution (40 mM ATP, 10 mM cAMP and 1% sodium dodecyl sulfate) followed by immediate boiling for 2.5 min. The [^{32}P] cAMP formed and the [^{3}H] cAMP added as a recovery marker were isolated by a slight modification (4) of the method of Salomon et al (30).

tion of GMP-P(NH)P is that of exhibiting a lag for the activation of the adenylyl cyclase. Rodbell (22) found by testing the action of GMP-P(NH)P on adenylyl cyclase in fat cell membranes, that the initial rate of cAMP accumulation after nucleotide addition is less than that obtained under basal conditions where no nucleotide had been added, and that the rate of cAMP accumulation increased drastically (well above that of control) only after a short lag of 1-2 min. This distinct initial activity indicates that slow effects of the nucleotide are not due to slow binding, but rather to slow isomerization of the enzyme system from a conformation having low activity (equal or lower than that of control) to one that is more active. Using rat liver plasma membrane, Rodbell and collaborators demonstrated that addition of glucagon will also accelerate the rate at which

GMP-P(NH)P can activate the adenylyl cyclase of these membranes. This is shown in Figure 4, where among other features, it can be seen that the addition of glucagon (2 μM) together with GMP-P(NH)P resulted in abolishment of the lag characteristic of GMP-P(NH)P action in the absence of hormones.

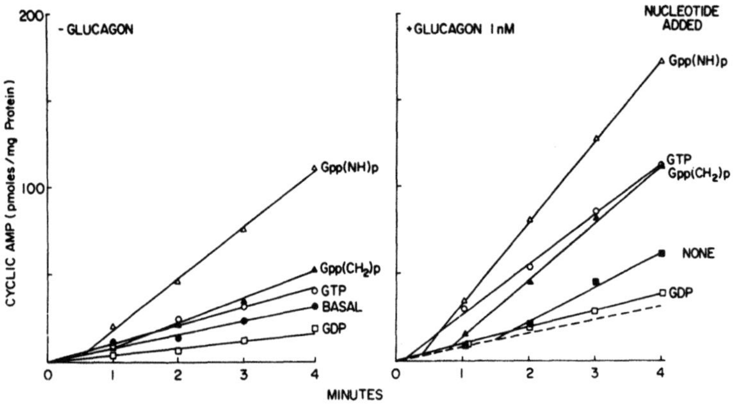

Fig. 4. Guanyl nucleotide effects on liver adenylyl cyclase activity in presence and absence of glucagon. All guanyl nucleotides were 0.1 mM. The assay medium contained 25 mM Tris-HCl, pH 7.5, 5 mM $MgCl_2$, 0.1 mM [^{32}P] AMP-P(NH)P (29, with permission).

Based on this kind of results and on detailed kinetic analysis of the substrate and divalent cation dependency of the liver-glucagon sensitive system as a function of pH, Rodbell and collaborators proposed a model called "Tri-State Model" for the action of nucleotides on adenylyl cyclase by hormones. This model is summarized in Table 1. In summary, it states that adenylyl cyclase system normally exists in a basal state, E, which upon nucleotide addition changes its conformation to an E'_N state that has an altered catalytic capacity and either an altered affinity for the free protonated (inhibitory) substrate or, if one wants to adapt this model to one where the enzyme has an allosteric site for magnesium, an altered affinity for magnesium at the allosteric site. These two alterations

TABLE 1

REGULATION OF ADENYLYL CYCLASE AS PROPOSED BY RENDELL et al. (21)

1. In the presence of GMP-P(NH)P:

$$E + N \underset{\uparrow\ \text{rapid}}{\rightleftharpoons} E'_N \underset{\uparrow\ \text{"slow" (lag)}}{\rightleftharpoons} E''_N$$

a. Very slowly reversible

a. Accelerated by hormone

b. Sometimes increased by hormone towards E'' so that:

$$(E''_N/E'_N)_{\text{without hormone}} < (E''_N/E'_N)_{\text{with hormone}}$$

2. In the presence of GTP:

$$E + N \underset{\uparrow\ \text{rapid}}{\rightleftharpoons} E'_N \underset{\uparrow\ \text{rapid}}{\rightleftharpoons} E''_N \quad \text{(no lag or lag very short)}$$

rapidly reversible

increased by hormone towards E''_N

are such that their combination results in either no apparent change of cAMP accumulation or even in a decreased rate of accumulation such as seen in the fat cell system. The formation of the E'_N state is then followed by a slow isomerization of this state into an E''_N state which now has not only the altered (increased) catalytic capacity but also a lowered affinity for the inhibitory protonated substrate (or increased affinity for the allosteric binding of divalent cation), resulting in a notable increase in the rate of cAMP production, i. e. in a stimulated state of the enzyme system. The "Tri-State Model" accounts for the persistence of adenylyl cyclase activity which is obtained with GMP-P(NH)P and seen after washing of membranes, by assuming that the initial binding of nucleotide to the enzyme giving the E'_N state is very tight. Although no specific experiments were done in this respect, this theory does assume that the initial interaction is reversible, occurring with a very slow rate constant for the dissociation of nucleotide from the E'_N state.

While Rodbell and collaborators account for the strange kinetics obtained with GMP-P(NH)P by proposing the existence of three discrete states of the enzyme (E, E'_N, and E''_N) and suggest that the effect of occupied hormone receptor on this enzyme system limits itself to modifying the rate constants with which the enzyme E'_N isomerizes to the E''_N (14, 15, 21, 28, 29), Cuatrecasas et al. (5), proposed a different mechanism by which the adenylyl cyclase system becomes persistently activated by GMP-P(NH)P. By testing activity in fat cell membranes and liver membranes they essentially corroborated and confirmed the kinetic properties described by Rodbell's group. In addition, in the fat cell membrane system, they observed that if enough time is allowed after addition of GMP-P(NH)P, the activity obtained approaches very closely that seen in the presence of the combination of GMP-P(NH)P and hormone (isoproterenol in this case). This can be seen in Figure 5. Also shown is the fact that the addition of isoproterenol together with GMP-P(NH)P increases the rate by which GMP-P(NH)P activates that adenylyl cyclase system. Based mainly on the fact that very little, if any, further activity was obtained by inclusion of a hormone with GMP-P(NH)P into the assay if steady state of activation is tested for, they postulated the catalytic unit of adenylyl cyclase systems to exist in three states: E (basal, essentially inactive), Enzyme·GTP or Enzyme GMP-P(NH)P (a Michaelis complex, with low catalytic capacity), and Enzyme-PP (derived from Enzyme·GTP complex) or Enzyme-P(NH)P (derived from the GMP-P(NH)P-driven Michaelis complex). Both derivative forms of the enzyme are assumed to be highly active, possibly of equal potency. The model continues to suggest that while Enzyme-PP form is either unstable or enzymatically hydrolyzed with formation of the basal state, E, the Enzyme-P(NH)P is incapable of reverting to the basal E state. As a consequence, Enzyme-P(NH)P accumulates and "total" activation results in due time when the system is exposed to GMP-P(NH)P. On the other hand, when GTP is added, the enzyme system is in constant turnover and the activity is representative of steady state levels of Enzyme-PP. As in the "Tri-State Model", also here hormones are assumed not to interact directly with adenylyl cyclase but rather to increase the rate of Enzyme-P(NH)P or Enzyme-PP formation from their respective Michaelis complexes. This model is presented in schematic form on Table 2.

Interest in the mechanism by which GMP-P(NH)P affects adenylyl cyclase systems stems not only from the facts that it stimulates activity, that it mimics many or perhaps all of the effects of GTP, the natural regulator, and that by studying its mode of action one might learn the mechanism of nucleotide regulation and gain insight into the mechanism of action of hormones, but also from the fact that the activity obtained after GMP-P(NH)P treatment of membranes is remarkably stable towards heat inactivation and towards inactivation upon solubilization from membranes with detergents. Moreover, adenylyl cyclase systems solubilized with nonionic detergents often retain their capacity to respond to GMP-P(NH)P stimulation while losing their capacity to respond to hormonal stimulation. Thus, a general strategy one might want to use in purifying adenylyl cyclase systems and their components is to (a) treat membranes with GMP-P(NH)P, thereby activating the adenylyl cyclase system, (b) solubilize the enzyme using nonionic detergents such as Lubrol, (c) purify the stabilized active enzyme using classical solution procedures developed for soluble enzymes, and finally, (d) reconstitute membrane particles and test whether or not a hormonally sensitive adenylyl cyclase system can be obtained.

This strategy, however, will work only if a way can be found of reversing the active state obtained with GMP-P(NH)P back to a basal state susceptible to regulation. Since this is a very important problem for those of us interested in the purification of components of adenylyl cyclase systems and in the elucidation of the mechanisms by which hormones activate these adenylyl cyclases, we decided to investigate the mode of action of GMP-P(NH)P on the glucagon-sensitive adenylyl cyclase of rat liver plasma membranes and to search for ways of reversing

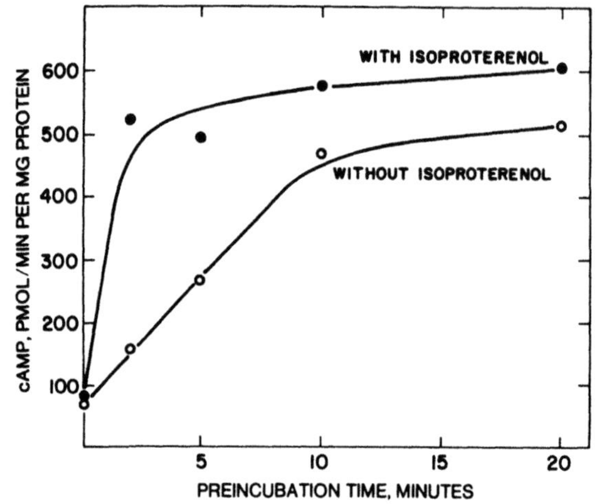

Fig. 5. Appearance of GMP-P(NH)P activation of fat cell membrane adenylyl cyclase as a function of time. Membranes were incubated for various times at 30°C with GMP-P(NH)P (0.1 mM) in the presence or absence of isoproterenol (20 μM). Membranes were then washed twice and assayed for adenylyl cyclase activity (5, with permission).

TABLE 2

REGULATION OF ADENYLYL CYCLASE AS PROPOSED BY CUATRECASAS et al. (5)

1. In the presence of GTP:

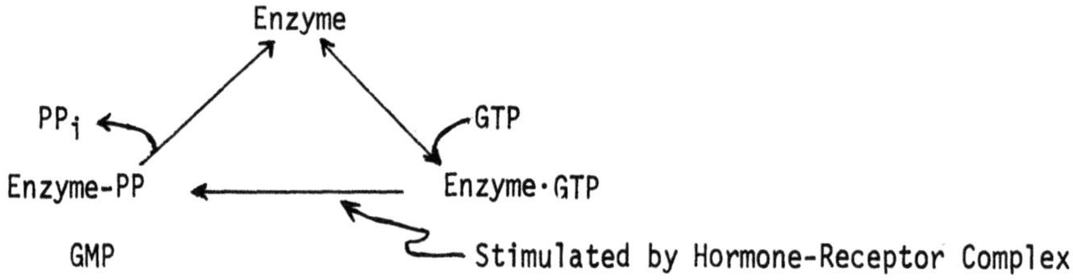

2. In the presence of GMP-P(NH)P:

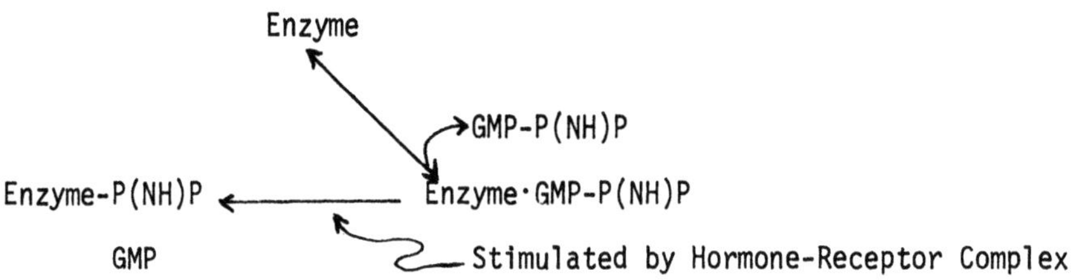

GMP-P(NH)P activated adenylyl cyclase back to its basal state. We initiated our experiments by designing a slightly different assay to test for "persistent" activation by GMP-P(NH)P. Rather than stopping the effect of the GTP analog by dilution and a single or double washing cycle, as has been done by others (5, 31, 34), we decided to make use of the finding made by others (2, 8, 18, 29, 33) and confirmed in our laboratory that GTP, when added simultaneously with GMP-P(NH)P, inhibits in a competitive manner the action of the analog. We reasoned that, since GMP-P(NH)P activates the enzyme more than GTP, we should be able to test for the degree of activation obtained with GMP-P(NH)P under various conditions and times of incubation by stopping the effect of the analog by the simple addition of excess GTP to the assay medium, followed immediately by assay of adenylyl cyclase activity. In this way, we would be able to avoid the somewhat cumbersome procedures involved in diluting, centrifuging, and resuspending membranes, for these treatments are always associated with variable losses of membrane protein that often make it difficult to interpret the results correctly. Figure 6 shows the results of an experiment such as the one just described.

It can be seen that post-addition of GTP at varying times after initiation of GMP-P(NH)P action on membranes, does have the expected result, i.e. it causes a slow progressive increase of activity, as seen when cAMP accumulation is tested between 10 and 16 min of incubation. This result is consistent with an apparent loss of reversibility of GMP-P(NH)P stimulation and appearance by 10 min of a state of activity that is "persistent", for it is not significantly inhibited by a further 5 min incubation in the presence of excess GTP. Note the similarity between these results and those obtained by Cuatrecasas et al. (5) with the washing procedure (shown in Fig. 5). The experiment shown in Figure 6 was carried out in the absence of glucagon and presented us with a puzzle, for the rate at which GMP-P(NH)P induced the appearance of an "irreversibly stimulated" activity was found to be a linear function of time. This result is surprising since one would have expected the appearance of the active form to depend on abundancy of

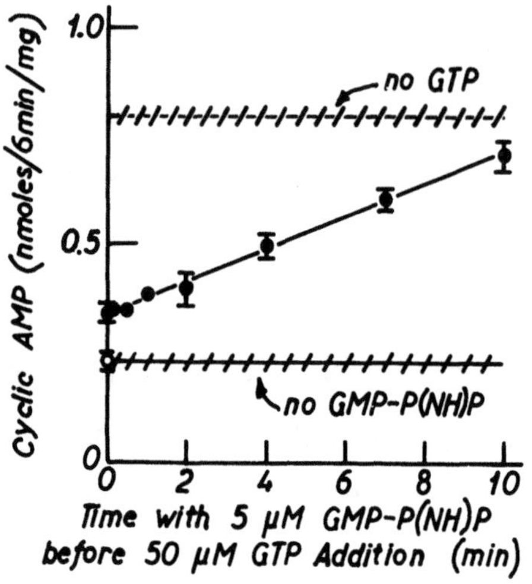

Fig. 6. Time course of inhibition by GTP of GMP-P(NH)P activation of liver membrane adenylyl cyclase. Liver membranes were incubated in complete incubation medium as described for Figure 3, including 5 μM GMP-P(NH)P but using unlabeled substrate. At various times after the initiation of the reaction, 50 μM GTP was added and the reaction was further incubated to a total of 10 min. After this time, [α-^{32}P] ATP was added and the assay was terminated 6 min later. Each point represents the mean ± SD of triplicate determinations.

inactive precursor and, hence, follow a rate function of the general type $1-e^{-kt}$, rather than being linear.

We decided to explore this phenomenon further by studying the time courses of adenylyl cyclase activity under the various incubation conditions shown in the experiment of Figure 6. The results were surprising and indicated to us that contrary to general assumption, the state of activity induced by GMP-P(NH)P is not of a persistent nature. Figure 7 shows the effect of post-addition of GTP at 40 sec after initiating the stimulatory action of GMP--P(NH)P. We found that after a short, but definite, lag period, GTP addition resulted in complete reversal of the GMP-P(NH)P-stimulated activity to that seen with GTP alone. In this experiment, the lag lasted about 1-2 min.

Figure 8 shows a similar experiment, but one in which GTP was added 2 min after the initiation of the stimulatory action of GMP-P(NH)P. Again, the effect

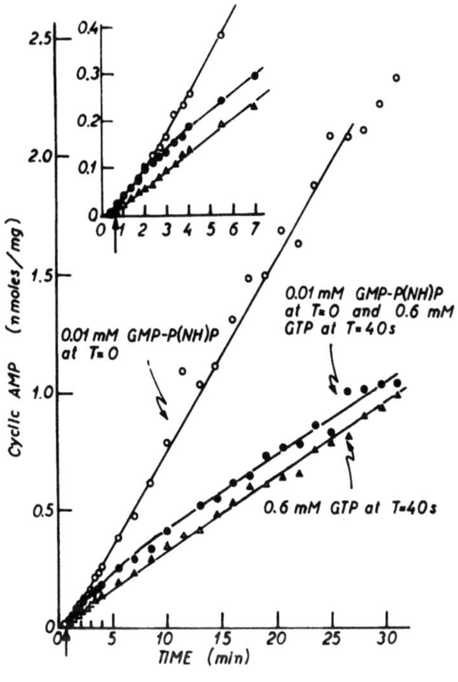

Fig. 7. Time course of rat hepatic adenylyl cyclase activity stimulated by GMP-P(NH)P (10 μM) at t = 0, GTP (0.6 mM) at t = 0, and reversion of GMP-P(NH)P stimulated activity by GTP addition to reactions 40 sec after initiation. Expanded scale inset indicates that reversal by GTP added at 40 sec is not immediate.

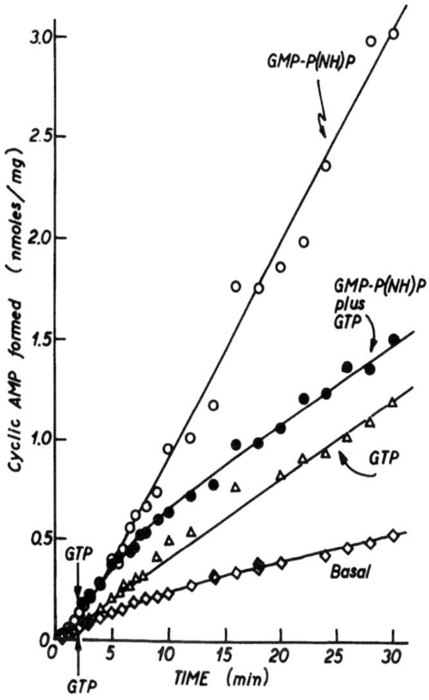

Fig. 8. Time course similar to Figure 7; however, GTP was added to GMP-P(NH)P activated adenylyl cyclase at 2 min rather than 40 sec. Also time course of basal activity is shown.

of GMP-P(NH)P was found to be reversed by GTP. One difference was observed, however: the lag required for GTP to reverse GMP-P(NH)P stimulation became longer. It was now in the order of 2-4 min. Post-addition of GTP after 6 min of GMP-P(NH)P stimulation (Figure 9) also resulted in complete reversal of GMP-P(NH)P stimulated activity, but now the lag was in the order of 6 min. Thus, the kinetic properties of the liver plasma membrane system appear to be such that GMP-P(NH)P stimulates adenylyl cyclase after a short lag period leading to an active state that can be reversed with GTP to an activity that corresponds to that obtained with GTP alone. The reversal requires, however, a finite time to become effective and this increases as the time between initiation of GMP-P(NH)P action and addition of GTP is increased. In other experiments not shown here, post-addition of GTP 30 min after initiation stimulation with GMP-P(NH)P still reverted the activity back to that seen with GTP alone, the lag period being about 10 min.

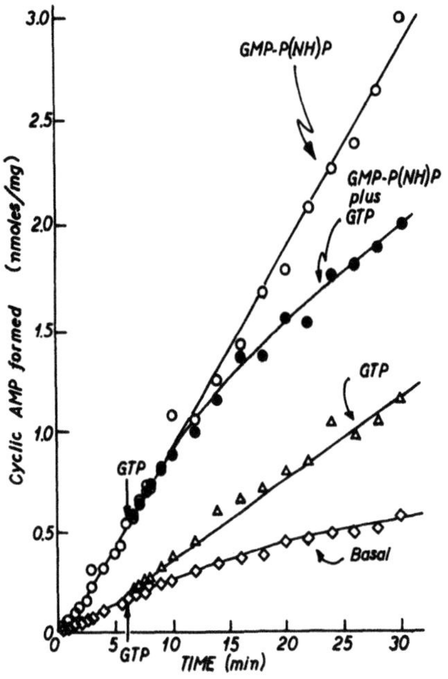

Fig. 9. Time course similar to Figure 8 with GTP added to reaction at 6 min.

We tested whether indeed we were dealing with a truly reversible system in which GTP and GMP-P(NH)P simply appear to compete for the same site and determined the effect of addition of GMP-P(NH)P to an enzyme system that had first been exposed to the analog and then "stopped" with GTP. The result of two such experiments, one in which post-addition of GTP had been made at 2 min, is shown in Figure 10. It can be seen that in each instance, a further addition of a ten-fold excess of GMP-P(NH)P over GTP resulted, after a given lag period, in restimulation of the enzymatic activity. Interestingly, this lag period appears to be longer the later the GMP-P(NH)P is added in a manner that resembles to some extent the prolongation of the lag period seen with the lateness of GTP addition for the initial reversal of GMP-P(NH)P action.

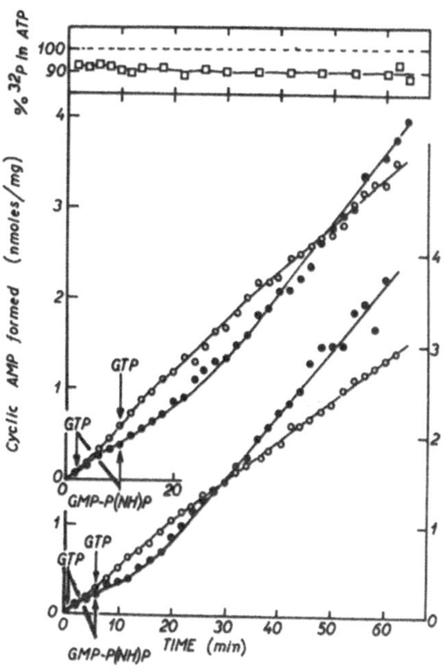

Fig. 10. Effect of post-addition of GTP and GMP-P(NH)P (alone or in combination) to GMP-P(NH)P-stimulated adenylyl cyclase activity. Each curve represents one incubation from which 50 1 aliquots were removed at the indicated times and subjected to analysis for cAMP formed using a modification (4) of the method of Salomon et al (30). All curves represent incubations for adenylyl cyclase activity to which 10 μM GMP-P(NH)P was added at time = 0. Open symbols represent incubations to which a single addition of GTP was made to give a final concentration of 100 μM after either 6 min (lower panel) or 10 min (upper panel) of incubation. Closed symbols represent incubations to which were added first GTP to give a final concentration of 100 μM (after 40 sec in the lower panel and after 2 min in the upper panel) and then GMP-P(NH)P to give a final concentration of 1000 μM at time = 6 min (lower panel) or 10 min (upper panel).

The results may mean that under normal adenylyl cyclase assay conditions, i.e. in the presence of ATP (0.1 mM in our experiments) and an ATP regenerating system (creatine phosphate and creatine kinase), the action of GMP-P(NH)P on adenylyl cyclase is essentially reversible. It was now of interest to determine whether during these various manipulations of the adenylyl cyclase system, i.e. stimulation with GMP-P(NH)P and then reversal of that stimulation with GTP, the glucagon-sensitive adenylyl cyclase from these liver plasma membranes had preserved its hormonal responsiveness. It was found (Figure 11) that addition of glucagon alone or together with GMP-P(NH)P to membranes that had been initially exposed to GMP-P(NH)P alone or exposed to GMP-P(NH)P and then to GTP, still preserved their sensitivity to the hormone. Furthermore, the rate of re-stimulation of the enzyme whose stimulation by GMP-P(NH)P had been reversed by post-addition of GTP seemed to be increased by the presence of glucagon (compare first and second panel of Figure 11). Thus, two effects of glucagon can be seen after all these manipulations: (a) stimulation of total activity, and (b) an increase in the rate at which stimulation by the nucleotide GMP-P(NH)P occurred.

It is our interpretation of the experiments presented above that in liver plasma membranes and perhaps also other enzyme systems, some of which we are now in the process of testing, the stimulatory action of guanyl nucleotides as well as the stimulatory action of hormones are reversible phenomena. The apparent irreversibility observed by several other laboratories after extensive washing of GMP-P(NH)P-treated membranes is probably a reflection of the combined effect of slow dissociation of the analog from the stimulatory site and a decrease in the rate or effectiveness at which the obviously complex mechanism by which GMP-P(NH)P and GTP modify the activity of adenylyl cyclase is operative. The simplest interpretation for this decreased effectiveness resulting in slow down of the rate at which GTP reverses GMP-P(NH)P stimulation if added late or of the rate at which GMP-P(NH)P stimulates at late times, is that the effect of guanyl nucleotides requires the presence of a factor to be operative. This putative factor, which we

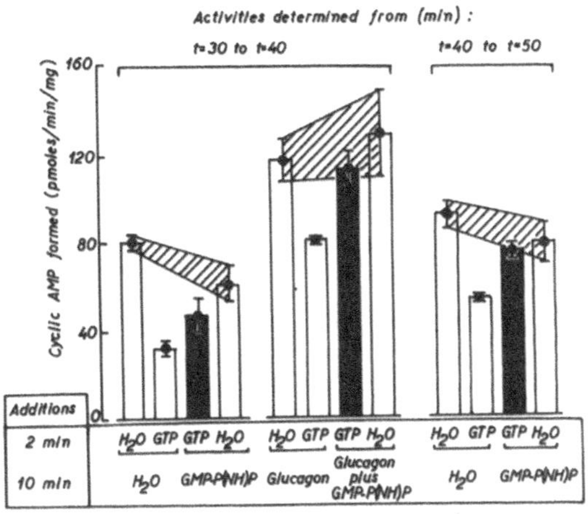

Fig. 11. Glucagon and GMP-P(NH)P sensitivity of adenylyl cyclase after previous stimulation by GMP-P(NH)P and reversal of that stimulation by GTP. The enzyme was stimulated with GMP-P(NH)P (10 μM) at t = 0. Post-additions were as indicated to give final concentrations of 100 μM GTP, 1.0 mM GMP-P(NH)P, and 3.0 μM glucagon. Adenylyl cyclase activity between 30 and 40 min, and 40 and 50 min was followed by addition at 30 or 40 min of labeled ATP (20 Ci/mmole) and stopping the reactions 10 min later. Bars represent mean ± SD of triplicate determinations. Note that the solid bars represent GTP-reversed activity which was restimulated either by GMP-P(NH)P alone or by the combination of GMP-P(NH)P and glucagon.

may call coupling factor, may be either a cofactor or an enzyme, and at the present time we have no information to speculate further. But the results are consistent with the existence of a factor, that upon incubation of membranes at 30° under adenylyl cyclase assay conditions, either leaches out of the membranes or becomes slowly inactivated, thereby slowing down the effects of nucleotides on the enzyme system. Since treatment of membranes with nucleotides under conditions that lead to decreased rates at which nucleotides affect adenylyl cyclase did not appear to influence the susceptibility of these membranes to respond to hormone receptor, it may be speculated that the putative factor with which guanyl nucleotides have to interact to exert their action on adenylyl cyclase systems, is separate from both the hormone receptor and the catalytic unit of the system and, therefore, susceptible to isolation. It should be mentioned in this context that

Pfeuffer and Helmreich (18) have recently reported that it is possible to isolate from pigeon erythrocyte membranes solubilized with Lubrol PX a guanyl nucleotide binding protein that is capable of restoring guanyl nucleotide sensitivity to a guanyl nucleotide insensitive, soluble adenylyl cyclase preparation. We are currently in the process of testing whether fractions obtained from salt extracts of membranes contain a factor that may restore to incubated liver plasma membranes the rapidity with which GMP-P(NH)P is capable of activating their adenylyl cyclase.

SUMMARY

In conclusion, we believe that the data presented in this paper are inconsistent with the model for guanyl nucleotide action and hormonal regulation of adenylyl cyclase suggested by Cuatrecasas et al. (5), for they clearly demonstrate that GMP-P(NH)P affects the enzyme system in a reversible manner. Furthermore, these data suggest that the enzyme system is more complex than had been assumed by Rodbell and collaborators (14, 15, 21, 28, 29) when proposing a "Tri-State Model", where the slow activation by GMP-P(NH)P is the simple result of an isomerization reaction, for the rate at which this isomerization occurs should not be varying with incubation time. It is more likely that hormone sensitive adenylyl cyclases are complex molecular systems composed of at least three types of component: the hormone receptor, the guanyl-nucleotide-affected coupler, and the catalytic unit. It is to be hoped that with the realization that GMP-P(NH)P-stabilized enzymes can be reversed to their basal state in a relatively easy manner by displacing the analog from the system with GTP, future purification studies, coupled to identification of components involved, will provide an answer to the as yet unresolved problem of how peptide hormones and catecholamines in general, and glucagon in particular, alter the activity of membrane bound adenylyl cyclases.

Acknowledgement

This work was supported in part by grants from the National Institutes of Health (AM-19318 and HD-07495) and the Clayton Foundation.

REFERENCES

1. Bilezikian, J. P. and Aurbach, G. D. (1974): The effects of nucleotides on the expression of β-adrenergic adenylate cyclase activity in membranes from turkey erythrocytes. J. Biol. Chem. 249, 157-161.

2. Birnbaumer, L., Nakahara, T. and Yang, P.Ch. (1974): Studies on receptor-mediated activation of adenylyl cyclases. II. Nucleotide and nucleoside regulation of the activities of the renal medullary adenylyl cyclase and their stimulation by neurohypophyseal hormones. J. Biol. Chem. 249, 7857-7866.

3. Birnbaumer, L., Pohl, S. L. and Rodbell, M. (1972): The glucagon-sensitive adenylate cyclase system in plasma membranes of rat liver. VII. Hormonal stimulation: reversibility and dependence on concentration of free hormone. J. Biol. Chem. 247, 2038-2043.

4. Bockaert, J., Hunzicker-Dunn, M. and Birnbaumer, L. (1976): Hormone-stimulated desensitization of hormone-dependent adenylyl cyclase: Dual action of luteinizing hormone on pig graafian follicle membranes. J. Biol. Chem. 251, 2653-2663.

5. Cuatrecasas, P., Jacobs, S. and Bennet, V. (1975): Activation of adenylate cyclase by phosphoramidate and phosphonate analogs of GTP: Possible role of covalent enzyme-substrate intermediates in the mechanism of hormonal activation. Proc. Nat. Acad. Sci. USA, 72, 1739-1743.

6. Drummond, G. I. and Duncan, L. (1970): Adenyl cyclase in cardiac tissue. J. Biol. Chem. 245, 976-983.

7. Harwood, J. P., Low, H. and Rodbell, M. (1973): Stimulatory and inhibitory effects of guanyl nucleotides on fat cell adenylate cyclase. J. Biol. Chem. 248, 6239-6245.

8. Lefkowitz, R. J. (1974): Stimulation of catecholamine-sensitive adenylate cyclase by 5'guanylyl-imidodiphosphate. J. Biol. Chem., 249, 6119-6124.

9. Lefkowitz, R. J. (1975): Guanosine triphosphate binding sites in solubilized myocardium. J. Biol. Chem. 250, 1006-1011.

10. Lefkowitz, R. J. and Caron, M. G. (1975): Characteristics of 5'-guanyl imidodiphosphate-activated adenylate cyclase. J. Biol. Chem. 250, 4418-4422.

11. Levey, G. S. (1971): Restoration of norepinephrine responsiveness of solubilized myocardial adenylate cyclase by phosphatidylinositol. J. Biol. Chem. 246, 7405-7410.

12. Levey, G. S. (1971): Restoration of glucagon responsiveness of solubilized myocardial adenyl cyclase by phosphatidylserine. Biochem. Biophys. Res. Commun. 43, 109-111.

13. Levitski, A., Sevilla, N. and Steer, M. L. (1976): The regulatory control of β-receptor dependent adenylate cyclase. J. Supramol. Str. 4, 405-418.

14. Lin, M. C., Salomon, Y., Rendell, M. and Rodbell, M. (1975): The hepatic adenylate cyclase system. II. Substrate binding and utilization and the effects of magnesium ion and pH. J. Biol. Chem. 250, 4246-4252.

15. Lin, M. C., Wright, D. W., Hruby, V. J. and Rodbell, M. (1975): Structure-function relationships in glucagon: Properties of highly purified des-his^{1}-, monoiodo-, and des-asn^{28}, thr^{29} (homoserine lacton^{27})-glucagon. Biochemistry 14, 1559-1563.

16. Londos, C. and Rodbell, M. (1975): Multiple inhibitory and activating effects of nucleotides and magnesium on adrenal adenylate cyclase. J. Biol. Chem. 250, 3459-3465.

17. Londos, C., Salomon, Y., Lin, M. C., Harwood, J. P., Schramm, M., Wolff, J. and Rodbell, M. (1974): 5'-guanylylimidodiphosphate, a potent activator of adenylate cyclase systems in eukaryotic cells. Proc. Nat. Acad. Sci. USA, 71, 3087-3090.

18. Pfeuffer, T. and Helmreich, E. J. M. (1975): Activation of pigeon erythrocyte membrane adenylate cyclase by guanylnucleotide analogues and separation of a nucleotide binding protein. J. Biol. Chem. 250, 867-876.

19. Pohl, S. L., Krans, H. M. J., Kozyreff, V., Birnbaumer, L. and Rodbell, M. (1971): The glucagon-sensitive adenyl cyclase system in plasma membranes of rat liver. VI. Evidence for a role of membrane lipids. J. Biol. Chem. 246, 4447-4454.

20. Puchwein, G., Pfeuffer, T. and Helmreich, E. J. M. (1974): Uncoupling of catecholamine activation of pigeon erythrocyte membrane adenylate cyclase. J. Biol. Chem. 249, 3232-3240.

21. Rendell, M., Salomon, Y., Lin, M. C., Rodbell, M. and Berman, M. (1975): The hepatic adenylate cyclase system. III. A mathematical model for the steady state kinetics of catalysis and nucleotide regulation. J. Biol. Chem. 250, 4235-4260.

22. Rodbell, M. (1975): On the mechanism of activation of fat cell adenylate cyclase by guanine nucleotides. J. Biol. Chem. 250, 5826-5834.

23. Rodbell, M., Birnbaumer, L., Pohl, S. L. and Krans, H. M. J. (1971): Regulation of glucagon action and its receptor. In: Structure-Activity Relationships of Protein and Polypeptide Hormones. M. Margoulis and F. C. Greenwood, Eds. Excerpta Medica Int. Congr. Ser. No. 241, Vol. 1, pp 199-211.

24. Rodbell, M., Birnbaumer, L., Pohl, S. L. and Krans, H. M. J. (1971): The glucagon-sensitive adenyl cyclase system in plasma membranes of rat liver. V. An obligatory role of guanylnucleotides in glucagon action. J. Biol. Chem. 246, 1877-1882.

25. Rodbell, M., Birnbaumer, L., Pohl, S. L. and Sundby, F. (1971): The reaction of glucagon with its receptor: Evidence for discrete regions of activity and binding in the glucagon molecule. Proc. Nat. Acad. Sci. USA 68, 909-913.

26. Rodbell, M., Lin, M. C. and Salomon, Y. (1974): Evidence for interdependent action of glucagon and nucleotides on the hepatic adenylate cyclase system. J. Biol. Chem. 249, 59-65.

27. Rodbell, M., Lin, M. C., Salomon, Y., Londos, C., Harwood, J. P., Martin, B. R., Rendell, M. and Berman, M. (1974): The role of adenine and guanine nucleotides in the activity and response of adenylate cyclase systems to hormones: Evidence for multi-site transition states. Acta Endocr. 77, 11-37.

28. Rodbell, M., Lin, M. C., Salomon, Y., Londos, C., Harwood, J. P., Martin, B. R., Rendell, M. and Berman, M. (1975): Role of adenine and guanine nucleotides in the activity and response of adenylate cyclase systems to hormones: Evidence for multiple transition states. In: Advances in Cyclic Nucleotide Research. G. I. Drummond, P. Greengard and G. A. Robinson, Eds. Raven Press, New York, Vol. 5, pp. 3-30.

29. Salomon, Y., Lin, M. C., Londos, C., Rendell, M. and Rodbell, M. (1975): The hepatic adenylate cyclase system. I. Evidence for transition states and structural requirements for guanine nucleotide activation. J. Biol. Chem. 250, 4239-4260.

30. Salomon, Y., Londos, C. and Rodbell, M. (1974): A highly sensitive adenylate cyclase assay. Anal. Biochem. 58, 541-548.

31. Schramm, M. and Rodbell, M. (1975): A persistent active state of the adenylate cyclase system produced by the combined actions of isoproterenol and guanylyl imidodiphosphate in frog erythrocyte membranes. J. Biol. Chem. 250, 2232-2237.

32. Sevilla, N., Steer, M. L. and Levitzki, A. (1976): Synergistic activation of adenylate cyclase by guanylyl imidophosphate and epinephrine. Biochemistry 15, 3493-3499.

33. Spiegel, A. M. and Aurbach, G. D. (1974): Binding of 5'guanylyl-imidodiphosphate to turkey erythrocyte membranes and effects on beta-adrenergic-activated adenylate cyclase. J. Biol. Chem. 249, 7630-7636.

34. Spiegel, A. M., Brown, E. M., Fedak, S. A., Woodward, C. J. and Aurbach, G. D. (1976): Holocatalytic state of adenylate cyclase in turkey erythrocyte membranes: Formation with guanylylimidodiphosphate plus isoproterenol without effect on affinity of β-receptor. J. Cyclic Nucleotide Res. 2, 47-56.

A PEPTIDE INHIBITOR OF GLUCAGON-RESPONSIVE ADENYLATE CYCLASE IN LIVER

D. C. Lehotay, G. S. Levey, J. M. Canterbury, L. A. Bricker and E. Ruiz

Division of Endocrinology and Metabolism, Department of Medicine,
University of Miami School of Medicine
Miami, Florida USA

The effects of a peptide inhibitor of adenylate cyclase produced by the isolated, perfused rat liver under hypocalcemic conditions were studied. This inhibitory peptide non-competitively abolished the activation of adenylate cyclase in particulate preparations of rat liver by glucagon, epinephrine, and parathyroid hormone, as well as cyclic AMP production in glucagon-stimulated rat liver slices. Higher concentrations of inhibitor also decreased basal adenylate cyclase activity and its fluoride-responsiveness. The data suggest that this substance is normally present in liver, but is released only under hypocalcemic conditions. The peptide does not crossreact in the radioimmunoassay of glucagon, insulin, or parathyroid hormone. Its inhibitory effects are not duplicated by somatostatin, angiotensin I, renin substrate, or the desoctadecapeptide of insulin.

INTRODUCTION

It has been observed that the hypocalcemic perfusion of the rat liver *in situ* causes the release of a small molecular weight substance which inhibits basal adenylate cyclase activity in rat renal cortex. We have partially purified this material and found it to be a heat and acid-stable peptide which also inhibits hormone-stimulated rat renal cortical adenylate cyclase in a dose-related, non-competitive manner (5). This report describes the effect of this inhibitor on glucagon-, and other hormone-stimulated adenylate cyclases in liver homogenates and slices, and provides some additional information about its probable site of action on the enzyme.

MATERIALS AND METHODS

Liver Perfusions. Male Sprague-Dawley rats (200-250g) were anesthetized with sodium pentobarbital (5 mg/100 g) and surgically prepared for liver perfusion *in situ*, as described previously (5).

Gel Filtration. Two ml samples of perfusate were centrifuged and the supernatants applied to calibrated Bio-Gel P-10 columns (1.5 x 50 cm) equilibrated with bovine serum albumin in the eluting buffer, 0.15 M ammonium acetate, pH 5.5. Selected fractions were pooled, lyophilized and reconstituted in 400 μl of 0.01 N acetic acid for testing in the adenylate cyclase assay.

Adenylate cyclase assay. Rat liver or renal cortex was homogenized in cold, 0.25 M sucrose and centrifuged at 12,000 g for ten minutes at $4^{\circ}C$. The particles were resuspended in cold sucrose. Adenylate cyclase activity was assessed by the method of Krishna, Weiss and Brodie (4) as described in detail elsewhere (5). Inhibitor was added at a concentration of 10 or 15 μl per 100 μl of enzyme containing 400-600 μg of protein. An ATP regenerating system was not required.

Tissue slices technique. Slices were cut freehand. Twenty to fifty milligrams were suspended in 0.5 ml Krebs-Ringer Tris-albumin buffer, pH 7.4, containing theophylline (10 mM), preincubated at $37^{\circ}C$ for 5 minutes, the inhibitor and/or hormone were added and the samples were incubated for 3 minutes. The reaction was stopped by the addition of 0.1 ml 30% trichloroacetic acid. Cyclic AMP was measured by the Gilman assay (1), and expressed as pmol of cyclic AMP per mg of tissue protein.

RESULTS

Table 1 shows the effect of the inhibitor on glucagon-stimulated adenylate cyclase in liver. The inhibitor, at a concentration of 15 μl per 100 μl of particulate enzyme preparation, almost completely abolished the activation of adenylate cyclase by the hormone, but did not impair fluoride-activation. However, higher concentrations of inhibitor decreased both basal and fluoride-stimulated activity.

A concentration response curve for glucagon-activation of the adenylate cyclase in the presence and absence of inhibitor, as shown in Fig. 1, indicates that the inhibition was non-competitive with respect to the hormone. When the

TABLE 1

EFFECT OF THE INHIBITOR ON GLUCAGON-STIMULATED RAT LIVER ADENYLATE CYCLASE

	pmoles of cyclic 3',5'-AMP accumulated/mg protein
Control	42 ± 4*
Control + Inhibitor**	39 ± 3
Glucagon (1 x 10^{-5} M)	466 ± 23
Glucagon (1 x 10^{-5} M) + Inhibitor	118 ± 12

* Each value is the mean ± SEM of 9 samples.

** When present, 15 µl of inhibitor, prepared as described in Material and Methods, was added to each 100 µl of particulate fraction containing 400-600 µg protein; the final inhibitor concentration was 3 µl per flask or 45 µl/ml incubation media.

concentration of inhibitor was lowered to 5 µl, a 54% decrease in glucagon-activation, a 50% decrease in PTH-activation, and a 60% decrease in epinephrine activation of liver adenylate cyclase were observed (Table 2). The inhibitor also abolished the accumulation of cyclic AMP in liver slices incubated with glucagon: control, 4.3 ± 0.1; control + inhibitor, 5.5 ± 0.5; glucagon, 8.0 ± 0.4; glucagon + inhibitor 5.8 ± 0.2 pmoles/3 min/mg protein (mean ± SEM, n = 4).

The purified inhibitor was tested in a number of radioimmunoassay systems, and no cross-reactivity was observed with either insulin, glucagon, or with parathyroid hormone. Furthermore, angiotensin I, renin substrate, the desoctatetradeca-peptide from the C-terminal end of the B chain of insulin and the sulfhydryl and cyclic forms of somatostatin were without effect.

DISCUSSION

Most studies of adenylate cyclase have focused on activation of the enzyme,

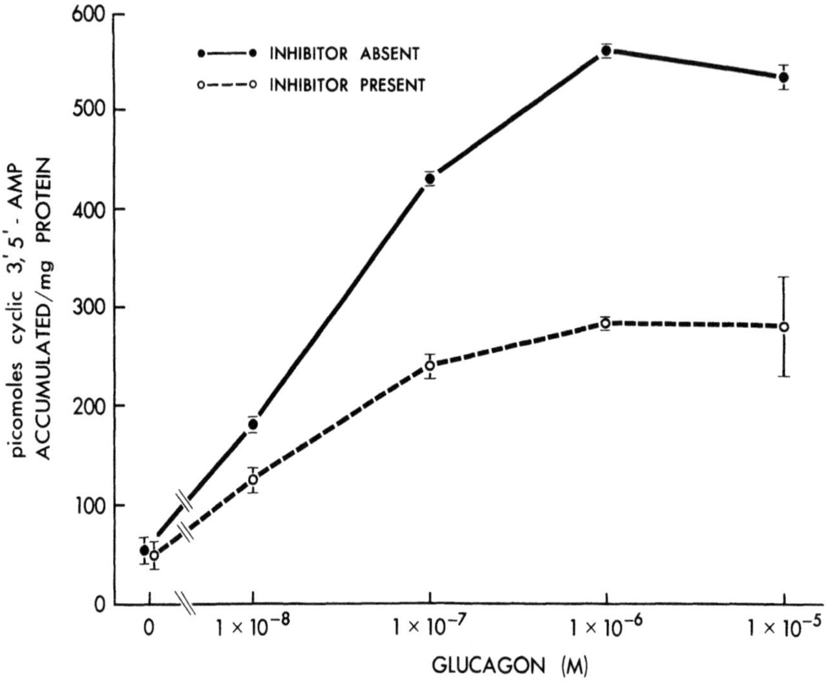

Fig. 1. Glucagon concentration response curve in the presence and absence of inhibitor. The source of adenylate cyclase was rat liver. Each value represents the mean ± SEM of 9 samples. The concentration of inhibitor was noted in the legend to Table 1.

TABLE 2

EFFECT OF LOWER CONCENTRATIONS OF INHIBITOR ON GLUCAGON, PTH, AND EPINEPHRINE ACTIVATION OF LIVER ADENYLATE CYCLASE

	Picomoles cyclic 3',5'-AMP accumulated/mg protein*		
	(-) Inhibitor	(+) Inhibitor	% Decrease
Control	90 ± 3	80 ± 10	--
Glucagon (1 x 10^{-5} M)	550 ± 90	290 ± 100	54%
PTH (2 x 10^{-6} M)	170 ± 30	120 ± 3	50%
Epinephrine (1 x 10^{-4} M)	140 ± 3	100 ± 5	60%

* Each value represents the mean ± SEM of 3 samples. The final inhibitor concentration was 1 μl per incubation.

particularly by hormone activators. Comparatively little attention has been directed to inhibitors of adenylate cyclase and/or its product, cyclic AMP. Ho and Sutherland have extensively described the formation of an inhibitor in rat fat cells stimulated by epinephrine, ACTH, or glucagon, which can be found circulating in rat plasma (2,3). The factor has a molecular weight of about 500 to 1000 and, in contrast to the peptide inhibitor in the present report, is alkali-stable, acid-labile, and to some extent soluble in lipid solvents.

The data obtained from the present studies show that the peptide inhibitor of adenylate cyclase released by rat livers perfused with hypocalcemic medium is a potent non-competitive inhibitor of the glucagon-, epinephrine-, and parathyroid hormone-responsive rat liver adenylate cyclase. When the inhibitor concentration was decreased to an amount which resulted in approximately 50 percent inhibition for one hormone, it inhibited the stimulation by the other hormones to a comparable extent. These data suggest that the inhibitor acts on the catalytic site of adenylate cyclase.

The precise chemical structure, the mode of synthesis and the physiologic or pathophysiologic role of the inhibitory peptide remains to be elucidated. However, it seems evident that when the peptide inhibitor is purified, it will serve as a useful investigational probe to thoroughly define which actions of glucagon are mediated by cyclic AMP and which are not, thereby increasing our understanding of glucagon action at the cellular level.

Acknowledgements

This investigation was supported in part by Grants 1 R01 HL 23715-05 and AM 26768 from the National Institutes of Health and by the Damon Runyon Cancer Fund, DRG 1120.

G. S. Levey is an Investigator at the Howard Hughes Medical Institute.

Purified bovine PTH was obtained from Inolex Laboratories. Glucagon was a gift of the Eli Lilly Laboratories. Epinephrine was obtained from Sigma Chemical Co., Bio-Gel P-10 from Bio-Rad and alpha-labeled ^{32}P-ATP from International Chemical and Nuclear Corporation.

REFERENCES

1. Gilman, A. G. (1970): A protein binding assay for adenosine 3',5'-cyclic monophosphate. Proc. Nat. Acad. Sci. USA 67, 305-312.

2. Ho, R. J. and Sutherland, E. W. (1971): Formation and release of a hormone antagonist by rat adipocytes. J. Biol. Chem. 246, 6822-6827.

3. Ho, R. J. and Sutherland, E. W. (1974): Inhibition of adenylate cyclase by a factor isolated from human plasma. Fed. Proc. 33, 480 (abst.)

4. Krishna, G., Weiss, B. and Brodie, B. B. (1968): A simple sensitive method for the assay of adenyl cyclase. J. Pharmacol. Exp. Ther. 163, 379-385.

5. Levey, G. S., Lehotay, D. C., Canterbury, J. M., Bricker, L. A. and Meltz, G. J. (1975): Isolation of a unique peptide inhibitor of hormone-responsive adenylate cyclase. J. Biol. Chem. 250, 5730-5733.

MODULATION OF NUCLEAR PROTEIN KINASE ACTIVITY BY GLUCAGON

W. K. Palmer and D. A. Walsh

Department of Biological Chemistry, School of Medicine
University of California
Davis, California USA

Protein kinase (PK) activity associated with the nuclei isolated from perfused rat liver was increased 3-4 times above control when the perfusate contained 7 x 10^{-9} M glucagon. Maximal incorporation of phosphate from ATP into histone f_{2b} was observed 5 minutes following the addition of hormone to the perfusate. The increased activity seen as a result of glucagon could be inhibited to control levels with purified regulatory subunit from cyclic AMP-dependent protein kinase and by the heat-stable protein kinase inhibitor. The control PK activity associated with nuclei isolated from unstimulated liver was not affected appreciably by these two inhibitory proteins. The substrate preference for control nuclei was phosvitin while the stimulated nuclei preferred histone as phosphate acceptor. The phosvitin activity in control and stimulated nuclei was not changed with hormone. There was a non-specific increase in the nuclear-associated protein kinase with the addition of purified catalytic subunit of cyclic AMP-dependent protein kinase to liver homogenates. When livers were stimulated with low doses of glucagon (3.4 x 10^{-10}M) a spike of cyclic AMP formation occurred at 5 minutes returning to control (0.5 nmoles/gram) by 15 minutes. The nuclear-associated PK activity remained elevated 2-3 fold even after the cyclic AMP content had returned to control levels. Evidence indicates that with glucagon stimulation, the catalytic subunit of cytoplasmic cyclic AMP-dependent protein kinase becomes associated with the nucleus *in vivo*. This phenomenon can be mimicked with the addition of cyclic AMP and free catalytic subunit *in vitro*.

INTRODUCTION

Investigators have reported (17, 25, 26) a correlation between increased enzyme induction and protein synthesis and elevations in the intracellular levels of adenosine 3',5'-cyclic monophosphate (cyclic AMP). There have also been reports that demonstrate increased phosphorylation of specific nuclear histone (16) and non-histone (14) proteins in response to hormones whose mechanism of action is thought to be mediated through the induction of cyclic AMP. Cyclic AMP, on the other hand, seems to influence intracellular events in mammalian tissues through the activation of a cyclic AMP-dependent protein kinase. This enzyme has been postulated to have two distinct subunits: a regulatory subunit (R) which binds cyclic AMP, and a catalytic subunit (C) which becomes active when dissociated from the holoenzyme (RC) complex by cyclic AMP (2). We have demonstrated that N^6-butyryl-cAMP binds primarily to the R subunit of a cAMP-dependent protein kinase in perfused liver (4).

Hepatic cyclic AMP-dependent protein kinase seems to be found primarily in the cytoplasmic fraction (6), and although there have been protein kinases isolated from purified hepatic nuclei, the influence of cyclic AMP on these enzymes is equivocal and if a stimulation of activity does occur it is only marginal (21, 22).

If, as has been hypothesized, cyclic AMP effects are mediated through a cyclic AMP-dependent protein kinase and this enzyme is located in the cytoplasm, by what mechanism does cyclic AMP mediate nuclear events? In an earlier communication (18) we reported that fifteen minutes following the administration of glucagon to perfused rat liver preparations, there was a significant increase in nuclear-associated protein kinase activity compared to activities measured in nuclei isolated from perfused livers not receiving hormone. Concommitant with the increased nuclear-associated activity there was a significant decrease in total cytoplasmic protein kinase activity following glucagon stimulation. The elevated enzyme activity associated with the hormone stimulated nuclei was inhibited to control activity values with the heat-stable cyclic AMP-dependent protein kinase inhibitor and purified regulatory subunit from rabbit skeletal muscle. All preliminary findings indicated that following the hormonally induced elevation in cyclic AMP there was an association of the cytoplasmic cyclic AMP-dependent protein kinase catalytic subunit with the nuclear fraction. Whether this phenomenon occurred in the tissue or as an artifact of the technique used to prepare the tissue was not determined. In this paper we further characterize the phenomenon of increased nuclear-associated protein kinase activity occurring in response to elevated cyclic AMP levels. We also attempt to determine indirectly if this response occurs in vivo or if there is a non-specific association of protein kinase with subcellular organelles associated with elevated concentrations of cyclic AMP.

EXPERIMENTAL PROCEDURE

Materials. Glucagon was obtained from Sigma Chemical Company; bromsulphalein was purchased from Fischer Scientific Products; purified cyclic AMP-dependent protein kinase and preparations of rabbit and bovine muscle catalytic and regulatory subunits, as well as partially purified cyclic AMP binding proteins were gifts of Drs. J. Beavo, P. Bechtel, and E. G. Krebs; protein kinase inhibitor was prepared by the technique of Walsh et al. (24); histone fractions were prepared by Method 1 of Johns (10). All other chemicals utilized in this investigation were obtained commercially.

Animal Preparation and Perfusion. Male Sprague-Dawley rats (120-250 g) were fasted the evening preceding surgery. Animals were anesthetized with sodium pentobarbital (6 mg/100 g body weight) and the livers perfused by the technique of Hems et al (9) with the addition of 10-15 mg of bromsulphalein (BSP) to 110 ml of perfusate. The removal of this compound from the circulation is done selectively by the liver and its removal from the perfusion medium was employed as an indicator of tissue viability.

Two livers were perfused simultaneously. Following a 30-min equilibration period, glucagon (7×10^{-9}M, unless otherwise indicated) was added to one of the perfusion systems (GS), while no addition was made to a parallel control (NS) system. Blood samples were removed throughout the entire perfusion period for determination of BSP content. At various times following hormone addition, NS and GS livers were removed and rinsed in cold 0.9% NaCl. Simultaneously, a small piece of hepatic tissue (approx. 50 mg) was frozen in liquid nitrogen for subsequent analysis for cyclic AMP concentration.

Preparation of Tissue. Fresh tissue was homogenized immediately and the nuclei were isolated by the technique of Chauveau et al. (5). Unless otherwise stated, nuclei were washed once in a solution containing 100 mM 2-(N-morpholino)-ethane sulphonic acid (MES) buffer, 15 mM magnesium chloride, and 6 mM EGTA (pH 6.9). Washed nuclei were resuspended in a volume of washing buffer equal to one ml per gram of liver and homogenized in a tight fitting Dounce (glass:glass) homogenizer.

Assays. Protein kinase was assayed by a modification of the procedure of Reimann et al. (19), as described elsewhere (18). Cyclic AMP was measured by a modification of the technique of Brostrom and Kon (1) employing hydroxylapatite to bind the cyclic AMP binding protein. Tissue was extracted in 0.1 N HCl.

Using this extraction procedure our standard recoveries were approximately 90% and the assay was linearly related to the amount of tissue utilized. The amount of cyclic AMP binding protein in cytoplasmic and nuclear fractions was measured by the technique of Gilman (8). BSP was measured spectrophotometrically by the method of Seligson et al. (20). Livers deemed viable by their removal of BSP from the perfusate were utilized for further experimentation.

RESULTS

Time Course. The relative cyclic AMP-independent protein kinase activity associated with nuclei isolated from glucagon stimulated livers and expressed as a function of time following hormone addition to the recirculating perfusate is illustrated in Fig. 1. Five minutes following the addition of hormone, the enzyme activity measured using histone f_{2b} as substrate was approximately five times greater in stimulated nuclei than in simultaneously perfused controls. The nuclear-associated activity then declined and returned to control values thirty minutes following hormone addition. A lag time of approximately 45 seconds occurred between the time the glucagon was added to the perfusate and when it reached the liver. If glucagon was added again at thirty minutes, a rapid increase in the nuclear-associated protein kinase activity to a level approximately two times greater than control activities was noted five minutes following the second hormone addition.

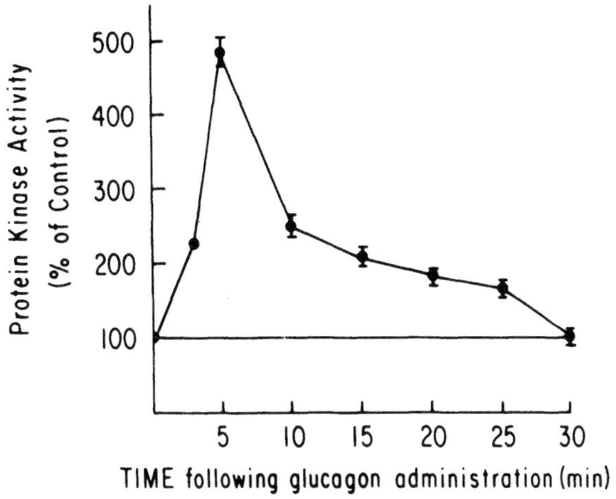

Fig. 1. Nuclear-associated cyclic AMP-independent protein kinase activity measured at various times following glucagon administration to the isolated perfused rat liver preparation.

Inhibition of Activity. Figs. 2 and 3 show a typical titration curve obtained when increasing amounts of heat-stable protein kinase inhibitor or purified rabbit skeletal muscle regulatory subunit were added to the assay. At all time points (Fig. 1) the addition of these two proteins inhibited nuclear-associated cyclic AMP-dependent protein kinase activity of GS livers to control values. In some instances these two compounds also inhibited a small fraction of the activity associated with NS nuclei.

To determine if something was present in the control nuclei that would inhibit the binding of the R or inhibitor proteins to the enzyme, equal amounts of nuclear extracts were combined and assayed with increasing amounts of R and heat-stable inhibitor. The activity of the combined extracts was midway between that of the control and the stimulated nuclear enzyme activities and could be titrated to control levels by either regulatory subunit or heat-stable protein kinase inhibitor (data not shown).

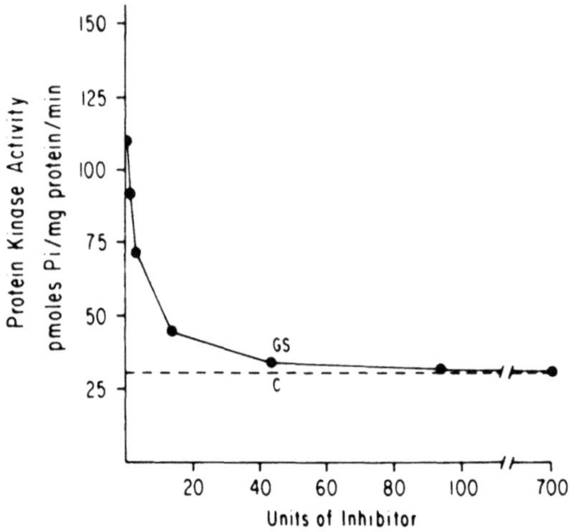

Fig. 2. Nuclear-associated cyclic AMP-independent protein kinase activity measured in the presence of increasing amounts of heat stable inhibitor of protein kinase. Closed circles (—•—) denote enzyme activity in nuclei isolated from glucagon stimulated livers. Broken line (-----) denotes activity of nuclear-associated enzyme from control livers measured in the presence of inhibitor.

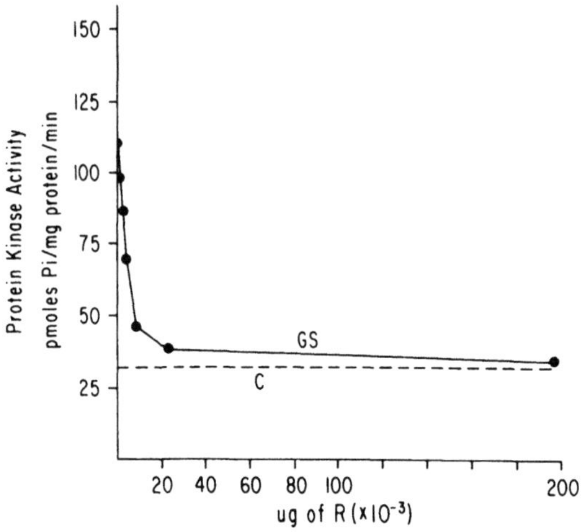

Fig. 3. Nuclear-associated cyclic AMP-independent protein kinase activity measured in the presence of increasing amounts of protein kinase regulatory subunit. Circles (—•—) denote enzyme activity in nuclei isolated from glucagon stimulated livers. Broken line (-----) denotes activity of nuclear-associated enzyme from control livers measured in the presence of the R subunit.

Substrate Specificity. The specificity of the enzyme for particular substrates can give some indication of the species of protein kinase being investigated. Chen and Walsh (6) have shown that protein kinase activity associated with liver nuclei phosphorylates phosvitin or casein in preference to histones, while the catalytic subunit from cytoplasmic cyclic AMP-dependent protein kinase uses histone as substrate in preference to casein or phosvitin.

Measurement of cytoplasmic and nuclear protein kinase activities in control livers showed that phosvitin is the preferred substrate for the nuclear enzyme, while the cytoplasmic enzyme has greater activity toward most of the histone fractions. However, following glucagon stimulation, there is a shift in the nuclear kinase specificity from that shown in control nuclei to a pattern similar to that seen in cytoplasmic enzyme.

In Vitro Stimulation of Nuclear-Associated Protein Kinase. The next series of experiments were designed to determine if the phenomenon we were dealing with actually occurred in the intact liver or was only an artifact of the nuclear isolation procedure. First we had to ascertain if these results could be duplicated *in vitro* in the presence of elevated levels of cyclic AMP. To do this we homogenized one liver in the sucrose homogenization medium containing 10^{-6}M cyclic AMP. The control liver was homogenized without the cyclic nucleotide. Protein kinase activity was assayed in nuclear and cytoplasmic fractions. The enzyme activity associated with the nuclei of the cyclic AMP-stimulated liver was approximately 5 x greater than that of control nuclei.

The influence of adding rabbit skeletal muscle catalytic subunit to the homogenate of GS and NS livers can be seen in Fig. 4. When catalytic subunit was added to the homogenate of either GS or NS perfused rat livers there was an equivalent increase in the nuclear-associated protein kinase activity above that measured when C was not added. The presence or absence of hormone stimulation did not seem to make any difference in the amount of activity associated with the nuclei. No detectable difference in cytoplasmic specific activity was evident when NS and GS homogenates plus C were compared.

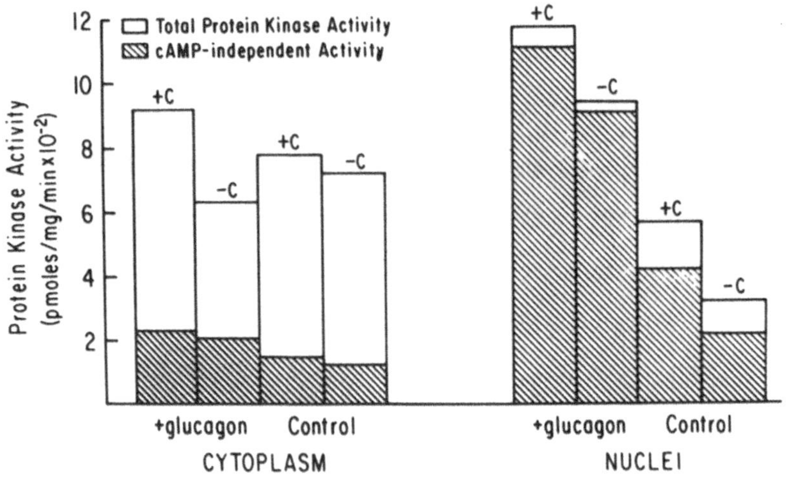

Fig. 4. Cytoplasmic and nuclear-associated protein kinase activity measured in glucagon-stimulated and control livers. Total protein kinase activity is the activity measured in the presence of 2 x 10^{-5}M cyclic AMP, while cyclic AMP-independent activity is that measured when cyclic AMP was not present. Columns denoted by +C represent the addition of approximately 10^6 units of purified skeletal muscle catalytic subunit to one-half of the liver homogenates. The -C columns represent experiments in which no catalytic subunit was added to the other half of the homogenate.

When the time-course of hepatic cyclic AMP content and that of nuclear-associated protein kinase at low levels (3.4 x 10^{-10}M) of glucagon stimulation were plotted together, we found the results reported in Fig. 5. While nuclear-associated protein kinase remained elevated for up to 30 minutes following hormone administration the level of cyclic AMP was reduced to control levels within the same time period.

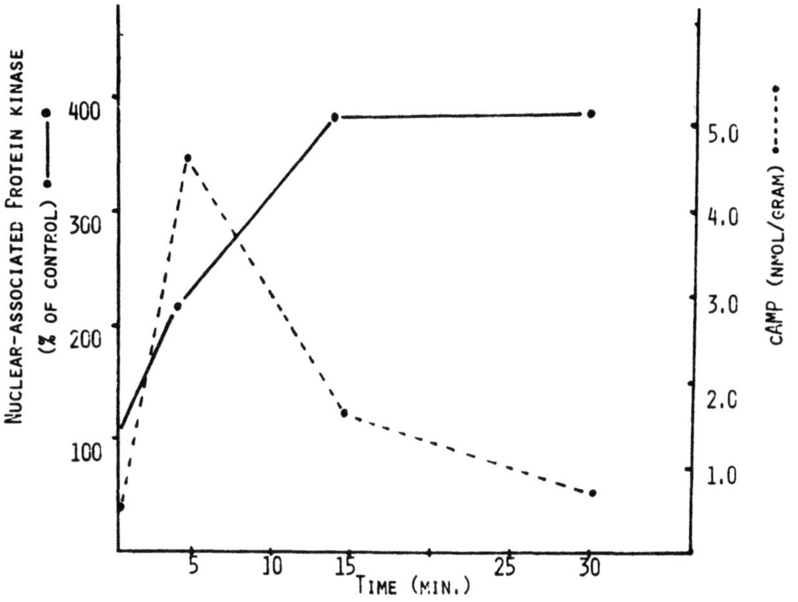

Fig. 5. Nuclear-associated protein kinase activity (—•—) and cyclic AMP concentrations (---o---) in glucagon-stimulated livers at various times following hormone administration. Glucagon was administered in a bolus directly above the liver. Concentration is that computed for final perfusate volume.

DISCUSSION

Numerous studies (3, 11, 15, 18) have implicated protein kinase translocation as a possible mechanism for relating the cytoplasmic enzyme to events which occur in the nucleus or other organelles in concert with increased cyclic AMP titers. Our data support the contention that nuclei isolated from glucagon-stimulated livers have a significantly higher protein kinase activity toward histone than nuclei isolated from non-stimulated livers. The characterization of the increased activity was that utilized by Walsh, et al (23) who outlined criteria for cyclic AMP-dependent protein kinase. While the protein kinase associated with nuclei isolated from control livers was not influenced by the heat-stable inhibitor of protein kinase or by the regulatory subunit, the hormone-induced activity was titrated to control values with either of these proteins. The specificity of binding of these proteins for the catalytic subunit of the cylic AMP-dependent protein kinase has been outlined. The substrate-specificity of the nuclear-associated kinase from control livers had the characteristics of a casein or phosvitin kinase, while the activity of the nuclear enzyme from glucagon-stimulated tissue had the characteristic substrate specificity of the cyclic AMP-dependent protein kinase. When we stimulated perfused livers with dibutyryl cyclic AMP (3) and determined the sedimentation characteristics of the nuclear protein kinase, the enzyme species that increased with stimulation sedimented at 3.9S. This size is similar to that of purified catalytic protein kinase (19).

The critical question regarding the physiologic value of these data has been asked by Corbin et al. (12). They found that the (C) subunit sediments non-specifically with particulate fractions of hormone-stimulated heart. The results of our studies indicate the same phenomenon occurs in liver. Cyclic AMP addition to hepatic homogenates mimics the in situ effects. The addition of purified skeletal muscle protein kinase to glucagon-stimulated and non-stimulated liver homogenates also supports the lack of specificity of protein kinase catalytic subunit association with nuclei. However, the fact that this enzyme does associate with the particulate fraction does not, in itself, negate a physiologic action providing the association occurs in vivo prior to homogenization and not as a result of cellular disruption. The results illustrated in Fig. 5 suggest that the association of the enzyme with the hepatic nuclei occurs within the tissue. Thirty minutes following hormonal stimulation, nuclear protein kinase is elevated more than 3-fold, while the cyclic AMP content is at control levels. These data infer that the hormone causes a dissociation of the cyclic AMP-dependent protein kinase. The catalytic subunit, in turn, associates with the nucleus in situ. The fact that there is no alteration in the amount of cyclic AMP binding protein associated with stimulated nuclei (3) is supportive evidence for the in vivo dissociation of the cyclic AMP-dependent protein kinase (7, 13).

Fig. 6 illustrates a working model summarizing a system by which glucagon may indirectly modify nuclear protein kinases. The hormone-induced stimulation of

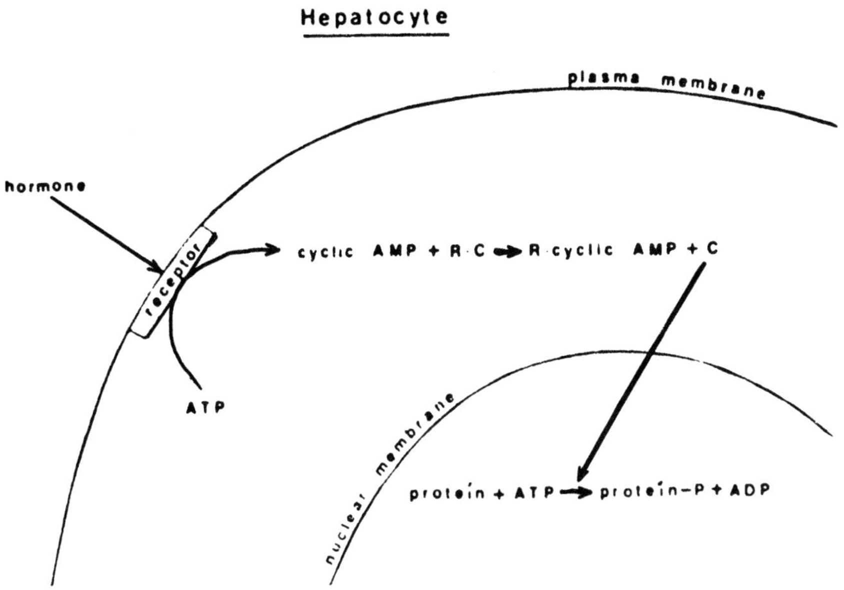

Fig. 6. Schematic representation of the mechanism by which glucagon may influence nuclear protein kinase.

cyclic AMP causes the dissociation of the nucleotide-dependent protein kinase. The catalytic subunit of the dissociated enzyme associates with or translocates into the nucleus, where, in the presence of ATP-Mg, phosphorylation of nuclear histone or nonhistone proteins may occur.

Acknowledgements

This study was supported in part by Grant AM 13613 from the National Institutes of Health.

Present address for Dr. Palmer: Biokinetics Research Laboratory, Pearson Hall, Temple University, Philadelphia, Pennsylvania 19122.

Dr. D. A. Walsh is an Established Investigator of the American Heart Association.

REFERENCES

1. Brostrom, C. O. and Kon, C. (1974): An improved protein binding assay for cyclic AMP. Anal. Biochem. 58, 459-468.

2. Brostrom, M. A., Reimann, E. M., Walsh, D. A. and Krebs, E. G. (1970): A cyclic 3',5'-AMP-stimulated protein kinase from cardiac muscle. Advan. Enzyme Regul. 8, 191-203.

3. Castagna, M., Palmer, W. K. and Walsh, D. A. (1975): Nuclear protein kinase activity in perfused rat liver stimulated with dibutyryl-adenosine cyclic 3',5'-monophosphate. Euro. J. Biochem. 55, 193-199.

4. Castagna, M., Palmer, W. K. and Walsh, D. A. (1977): Metabolism of [^{3}H] N^6, O^2-dibutyryl cyclic adenosine 3',5-'monophosphate and macromolecular interactions of the products in perfused rat liver. Arch. Biochem. Biophys. (submitted)

5. Chauveau, J., Moule, Y. and Rouiller, C. (1956): Isolation of pure and unaltered liver nuclei: morphology and biochemical composition. Exp. Cell Res. 11, 317-321.

6. Chen, L. J. and Walsh, D. A. (1971): Multiple forms of hepatic adenosine 3',5'-monophosphate-dependent protein kinase. Biochemistry 10, 3614-3621.

7. Cherrington, A. D., Assimacopoulos, F. D., Harper, S. C., Corbin, J. D., Park, C. R. and Exton, J. H. (1976): Studies on the α-adrenergic activation of hepatic glucose output. II. Investigation of the roles of adenosine 3',5'-monophosphate and adenosine 3',5'-monophosphate-dependent protein kinase in the actions of phenylephrine in isolated hepatocytes. J. Biol. Chem. 251, 5209-5218.

8. Gilman, A. G. (1970): A protein binding assay for adenosine 3',5'-cyclic monophosphate. Proc. Natl. Acad. Sci. USA 67, 305-312.

9. Hems, R., Ross, B. D., Berry, M. N. and Krebs, H. A. (1966): Gluconeogenesis in perfused rat liver. Biochem. J. 101, 284-292.

10. Johns, E. W. (1964): Studies on histones. 7. Preparative methods for histone fractions from calf thymus. Biochem. J. 92, 55-59.

11. Jungman, R. A., Hiestand, P. C. and Schweppe, J. S. (1974): Mechanism of action of gonadotropin. IV. Cyclic adenosine monophosphate-dependent translocation of ovarian cytoplasmic cyclic adenosine monophosphate-binding protein and protein kinase to nuclear acceptor sites. Endocrinology 94, 168-183.

12. Keely, S. L., Corbin, J. D. and Park, C. R. (1975): On the question of translocation of heart cAMP-dependent protein kinase. Proc. Natl. Acad. Sci. USA 72, 1501-1504.

13. Keeley, S. L., Cordin, J. D. and Park, C. R. (1975): Regulation of adenosine 3',5'-monophosphate-dependent protein kinase. Regulation of the heart enzyme by epinephrine, glucagon, insulin and 1-methyl-3-isobutylxanthine. J. Biol. Chem. 250, 4832-4840.

14. Kleinsmith, L. J. (1975): Phosphorylation of non-histone proteins in the regulation of chromosome structure and function. J. Cell Physiol. 85, 459-476.

15. Korenman, S. G., Bhalla, R. C., Sanborn, B. M. and Stevens, R. H. (1974): Protein kinase translocation as an early event in the hormonal control of uterine contraction. Science 183, 430-432.

16. Langan, T. A. (1969): Phosphorylation of liver histone following the administration of glucagon and insulin. Proc. Natl. Acad. Sci. USA 64, 1276-1283.

17. Miller, J. P., Beck, A. H., Simon, L. N. and Meyer, R. B. Jr. (1975): Induction of hepatic tyrosine aminotransferase in vivo by derivatives of cyclic adenosine 3',5'-monophosphate. J. Biol. Chem. 250, 426-431.

18. Palmer, W. K., Castagna, M. and Walsh, D. A. (1974): Nuclear protein kinase activity in glucagon-stimulated perfused rat livers. Biochem. J. 143, 469-471.

19. Reimann, E. M., Walsh, D. A. and Krebs, E. G. (1971): Purification and properties of rabbit skeletal muscle adenosine 3',5'-monophosphate-dependent protein kinase. J. Biol. Chem. 246, 1986-1995.

20. Seligson, D., Marino, J. and Dodson, E. (1957): Determination of sulfobromophthalein in serum. Clin. Chem. 3, 638-645.

21. Siebert, G., Ord, M. G. and Stocken, L. A. (1971): Histone phosphokinase activity in nuclear and cytoplasmic cell fractions from normal and regenerating rat liver. Biochem. J. 122, 721-725.

22. Takeda, M., Yamamura, H. and Ohga, Y. (1971): Phosphoprotein kinases associated with rat liver chromatin. Biochem. Biophys. Res. Comm. 42, 103-110.

23. Walsh, D. A. and Ashby, C. D. (1973): Protein kinases: aspects of their regulation and diversity. Recent Progr. Hormone Res. 29, 329-359.

24. Walsh, D. A., Ashby, C. D., Gonzalez, C., Calkins, D., Fischer, E. H. and Krebs. E. G. (1971): Purification and characterization of a protein inhibitor of adenosine 3',5'-monophosphate-dependent protein kinase. J. Biol. Chem. 246, 1977-1985.

25. Wicks, W. D., Barnett, C. A. and McKibbin, J. B. (1974): Interaction between hormones and cyclic AMP in regulating specific hepatic enzyme synthesis. Fed. Proc. 33, 1105-1111.

26. Wicks, W. D., Kee, K. L. and Kenney, F. T. (1969): Induction of hepatic enzyme synthesis in vivo by adenosine 3',5'-monophosphate. J. Biol. Chem. 244, 6008-6013.

COMBINED INFLUENCE OF INSULIN AND GLUCAGON UPON HEPATOCYTES IN RATS WITH LIVER LOSS OR VIRAL INJURY

N. L. R. Bucher

The John Collins Warren Laboratories of Harvard University
at the Massachusetts General Hospital
Boston, Massachusetts USA

Insulin and glucagon act in synergy to promote hepatic regeneration in eviscerated rats and to enhance survival in mice with viral hepatitis. Relatively low doses of insulin and high doses of glucagon are required, in keeping with the low portal venous blood levels of insulin and high levels of glucagon in normal rats following partial hepatectomy.

The means by which these effects are implemented are unknown. Possible roles of cyclic purine and cyclic pyrimidine nucleotides, and newly discovered highly phosphorylated nucleotides are currently under investigation.

The enrichment of portal venous blood with factors that promote survival and proliferation of hepatocytes was suggested by Rous and Larrimore in 1920 (20), subsequently confirmed, and studied with increasing intensity by numerous investigators as transplantation of livers in animal and human subjects developed into a reality. The extensive work of Starzl and his associates in this area not only emphasized the requirement for portal venous blood, but singled out insulin as its most important hepatotrophic component; although proposing that other portal venous constituents may also contribute to this function, they have recently expressed doubts concerning the role of glucagon (23-26). The influence of insulin and glucagon were not assessed individually, however, either by these or other investigators, under conditions in which other portal venous constituents were eliminated (7, 28, 24-26). For example, glucagon of gastrointestinal origin (21) was largely overlooked. Consequently, we adapted for rats the procedure of portal splanchnic evisceration first used to explore portal blood factors in hepatic regeneration by Price _et al._ in dogs (14). Following resection of the gastrointestinal tract, pancreas and spleen, rats were maintained in good condition for up to three days by continuous intravenous infusion of a mixture of electrolytes, supplemented with nutrients and hormones as indicated. The livers were supplied with blood solely by the hepatic artery, the portal vein having been sectioned and ligated. Cervical esophagostomy to prevent the debilitating effects and high mortality caused by aspiration pneumonitis, and large doses of subcutaneous five percent glucose solution at the time of surgery to prevent acute hypoglycemia reduced the operative mortality to negligible levels (3).

Partial hepatectomy in such animals induced only sluggish regenerative activity, as evidenced by the rate of incorporation of ^{3}H-thymidine into DNA in the liver remnant. DNA synthesis, thus evaluated, was found negligible at 24 hours, an interval when normally regenerating livers exhibit peak activity; by 48 hours the rate had risen to only 10-20 percent of the normal maximum (3). The persistence of trace amounts of enteric or pancreatic hormones in such animals could not be entirely excluded, but as these hormones tend to be rapidly eliminated from the body their presence in appreciable amounts seemed unlikely; radioimmunoassay of plasma insulin showed minimal or undetectable concentrations.

Addition to the infusion mixture of either insulin or glucagon in increasing amounts had little effect upon the rate of DNA synthesis determined at 24 hours post hepatectomy. In all experiments insulin was supplemented with glucose, and glucagon with an amino acid mixture. Control animals received the nutrients without hormones. When the two hormones were combined in various ratios, the results were striking (Table 1). Although there was considerable variability among rats due to the stressful nature of the procedure, in responsive animals the DNA synthetic rate was restored completely to normal (4).

It has been established for a number of years that hepatic regeneration is regulated by humoral agents. For example, we found that when blood is cross circulated between pairs of rats connected only by intravenous polyethylene cannulas, partial hepatectomy in one partner initiates a rise in DNA synthesis in the liver of the other partner (15). We have been completely unable to reproduce this effect, however, by continuous administration of insulin and glucagon to nonhepatectomized rats, whether normal or eviscerated, at a variety of dosage levels

TABLE 1

EFFECT OF CONTINUOUS INFUSION OF INSULIN AND GLUCAGON ON INCORPORATION OF ^{3}H-THYMIDINE INTO HEPATIC DNA AT 24 HOURS AFTER PARTIAL HEPATECTOMY IN EVISCERATED RATS

Rats were infused with the dosages of insulin and glucagon indicated, supplemented with glucose and amino acid, starting immediately after evisceration and partial hepatectomy, as previously described (4). ^{3}H-thymidine (50 μCi, specific activity 6 Ci/mmol) was injected intravenously one hour before the rats were killed

Group No.	No. of rats	Insulin U/kg/24 hr	Glucagon mg/kg/24 hr	DNA dpm/10 μg ± SEM
1	6	0	0	161 ± 37
2	4	2	0	468 ± 176
3	7	4	0	200 ± 52
4	5	0	1	376 ± 155
5	4	0	2	357 ± 158
6	4	2	1	1948 ± 243
7	6	2	2	3381 ± 325
8	10	4	2	2030 ± 338

and even when infused directly into the portal vein. These and other findings suggest that the two pancreatic hormones although highly active promoters cannot serve by themselves as primary initiators of hepatic regeneration, and that additional blood borne agents, some of which may not be of enteric or pancreatic origin, are involved in physiological control of hepatocyte proliferation.

The hormone doses given to eviscerated rats were empirical. To learn what concentrations of these hormones are actually supplied to the liver under physiologic conditions we determined levels of insulin and glucagon in portal venous blood of normal rats at intervals after partial hepatectomy (5). Portal rather than systemic blood was examined because of the avid uptake of hormones, especially insulin, by the liver (25). Almost immediately after partial hepatectomy portal venous insulin content fell whereas glucagon rose abruptly (Fig. 1). The low insulin levels are in keeping with the modest doses of insulin that we routinely infused, only a small fraction of the maintenance dose employed in treatment of rats made diabetic with alloxan (29). It appears that though insulin is essential for hepatic regeneration in vivo, low levels suffice. On the other hand, relatively high levels of glucagon seem to be required. The doses that we considered optimal for eviscerated rats were in the pharmacologic rather than physiologic range, as judged by the excessively high portal venous blood levels found by radioimmunoassay in normal rats infused with similar amounts. Perhaps physiologic doses would be effective under less drastic experimental conditions.

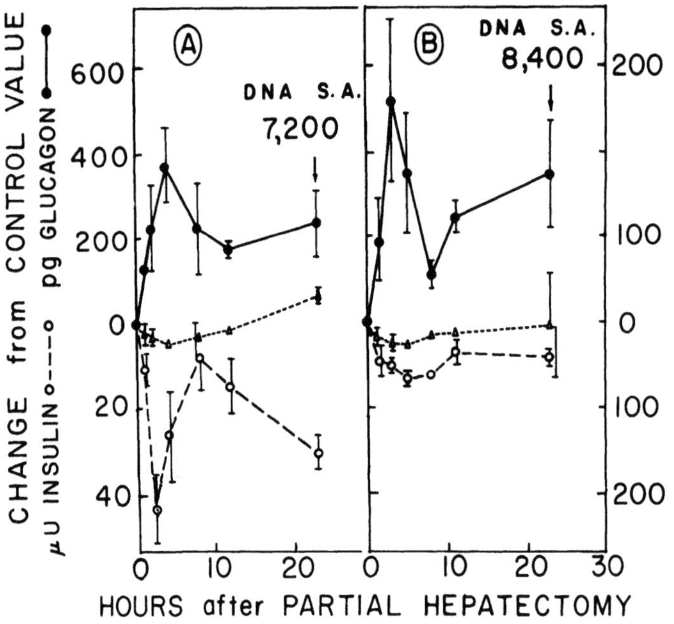

Fig. 1. Changes in concentrations of immunoreactive glucagon, immunoreactive insulin, and glucose in portal venous blood at intervals after partial hepatectomy. DNA specific activities were determined at 24 hours, following one hour incorporation of ^{3}H-thymidine. Number of rats: six per point in Group A and three rats, sampled sequentially from indwelling portal venous cannulas without anesthesia, for curves in Group B. Prehepatectomy (zero hours) concentrations of insulin were approximately 50 µU/ml and glucagon 125 pg/ml.

Influence of Insulin and Glucagon on Rat Hepatocytes

To find whether insulin and glucagon would influence the outcome of a liver deficit induced by a different means and in a different animal species we instigated a collaborative study with Drs. Farivar, Wands, and Isselbacher. Dr. Wands had found that young mice given overwhelming doses of A-59 murine hepatitis virus developed fulminant hepatitis, with massive liver necrosis; 100 percent of the animals died within four days. Continuous intraperitoneal infusion of pancreatic hormones into such animals was begun immediately after infection with the virus, in mixtures of electrolytes and nutrients similar to those employed in the eviscerated rat studies. The results strongly supported the previous findings; the two hormones in combination more than doubled the mean survival time and resulted in complete recovery in 40 percent of the mice, whereas insulin by itself had much less effect (6 percent recovery), and glucagon none (8).

When insulin and glucagon dosages were varied, it appeared that although reduction of insulin by 90 percent only slightly impaired its effectiveness, high doses of glucagon, i. e., amounts that approached the limits of tolerance, maximally promoted mouse survival. This parallels the observation in rats that low levels of blood insulin and high levels of glucagon are associated with active regeneration.

The promotion of survival in mice by the dual hormone treatment probably stemmed in part from enhancement of liver regeneration, since DNA synthesis was shown by autoradiography to be highly active in the hepatocytes of the treated animals. This lacks definitive proof, which would hinge upon comparison with untreated controls, because no controls survived long enough for DNA synthesis to get underway. In addition, the mouse model raised a new point: the combined insulin-glucagon treatment afforded a significant measure of protection against viral hepatocellular injury as evidenced by the severity of cellular necrosis, which after 24 hours was only half as extensive as in the hormone deprived control animals. The nature of this hormone induced resistance to viral damage is unknown, but it appears not to be directly concerned with DNA replication which begins much later, i. e., not before about 48 hours after infection with the virus.

The principal point to emerge from this work so far is that in two animal models, with acute liver loss or injury, respectively, insulin and glucagon appear to act in synergy, profoundly influencing certain aspects of hepatocyte function, including DNA replication. This contrasts with other aspects of hepatocyte function, including metabolism of lipid and carbohydrate moieties where these two hormones often exert opposing effects.

Although liver is known to be a prime target of pancreatic hormone action, our results, based upon whole animal studies, do not prove that the observed functional changes resulted from direct action of insulin and glucagon upon the hepatocytes. Recently, however, Friedman and his associates showed that isolated adult rat hepatocytes in monolayer culture were modestly stimulated to synthesize DNA by either insulin or epidermal growth factor; when combined, these two agents acted in synergy to increase the DNA labeling index by several fold. Addition of glucagon caused considerable further enhancement (19). These observations favor a direct interaction of hormones with hepatocytes, although the remote possibility remains that epithelioid and fibroblastic types of cells also present in these cultures could have mediated the hormonal influence.

The way in which the synergism of insulin and glucagon is implemented is unknown; among a number of possible mechanisms are those involving cyclic nucleotides. Cyclic AMP and possibly cyclic GMP levels are influenced by the pancreatic hormones and, indeed, the effects of glucagon are largely mediated by cyclic AMP (6, 16). In addition cyclic UMP and cyclic CMP have recently been isolated from rat liver, cyclic CMP reportedly undergoing a 50-200 fold increase in regenerating liver (1, 2). We have consequently initiated a study of cyclic nucleotides, in normal and regenerating rat liver, employing a novel procedure involving in vivo labeling with $^{32}P_i$ which permits quantitative determination of all four compounds from a single tissue sample. So far no differences from previously reported findings regarding cyclic AMP and cyclic GMP have emerged (9, 11-13, 22, 27):

cyclic AMP levels rose during the prereplicative phase of liver growth and cyclic GMP did not change significantly. Cyclic UMP and cyclic CMP, however, were not detectable by this technique, indicating that even if present in 1/100th of the amount previously reported, their turnover rates must be extremely low, as no $^{32}P_i$ was incorporated during a two-hour labeling period. This suggests that the role of cyclic pyrimidine nucleotides is of little importance at least in the early prereplicative phase of liver regeneration. These findings are preliminary and no studies have yet been carried out in hormone-treated animals.

In addition to the cyclic purine and pyrimidine derivatives we have sought other possible growth-associated nucleotides, resulting in the discovery of two previously undetected highly phosphorylated compounds which both contain modified adenosine as the nucleotide (17, 18). One of these two nucleotides, which differs from the other only by containing one less phosphate group, appears preferentially in livers that have been regenerating for more than six hours; this suggests that like insulin and glucagon (4) it may not be concerned with the very earliest prereplicative steps in growth initiation. As yet there is no direct evidence for a hormone-highly phosphorylated nucleotide interaction. Similar observations have subsequently emerged from the studies of Leffert and his associates in fetal liver cultures (10). We are currently investigating the identity and function of these compounds.

Many additional means through which insulin and glucagon may exert their mutually complementary hepatotrophic influences also remain to be explored, as well as their modes of interaction with various other putative growth regulatory agents.

Acknowledgements

This is publication No. 1516 of the Cancer Commission of Harvard University. This work was supported by USPHS grants CA 02146-23 and 1 R01 AM 19423-01.

REFERENCES

1. Bloch, A. (1975): Isolation of cytidine 3',5'-monophosphate from mammalian tissues and body fluids and its effects on leukemia L-1200 cell growth in culture. In: Advances in Cyclic Nucleotide Research, Vol. 5, G. I. Drummond, P. Greengard and G. A. Robison, Eds. Raven Press, New York, pp. 331-338.

2. Bloch, A. (1975): Uridine 3',5'-monophosphate (cyclic UMP). I. Isolation from liver extracts. Biochem. Biophys. Res. Comm. 64, 210-218.

3. Bucher, N. L. R. and Swaffield, M. N. (1973): Regeneration of liver in rats in the absence of portal splanchnic organs and a portal blood supply. Can. Res. 33, 3189-3194.

4. Bucher, N. L. R. and Swaffield, M. N. (1975): Regulation of hepatic regeneration in rats by synergistic action of insulin and glucagon. Proc. Nat. Acad. Sci. USA 72, 1157-1160.

5. Bucher, N. L. R. and Weir, G. C. (1976): Insulin, glucagon, liver regeneration and DNA synthesis. Metab. 25 (Suppl. 1), 1423-1425.

6. Cuatrecasas, P. (1975): Hormone receptors--their function in cell membranes and some problems related to methodology. In: Advances in Cyclic Nucleotide Research, Vol. 5, G. I. Drummond, P. Greengard and G. A. Robison, Eds. Raven Press, New York, pp. 79-104.

7. Duguay, L. R., Lee, S. and Orloff, M. J. (1976): Regulation of liver regeneration by the pancreas. Gastroenterology 70, 980 (abst.).

8. Farivar, M., Wands, J. R., Isselbacher, K. J. and Bucher, N. L. R. (1976): Effect of insulin and glucagon on fulminant murine hepatitis. New Engl. J. Med. 295, 1517-1519.

9. Fausto, N. and Butcher, F. R. (1976): Cyclic nucleotide levels in regenerating liver. Biochim. Biophys. Acta 428, 702-706.

10. Koch, K. S., Leffert, H. L. and Moran, T. (1976): Hepatic proliferation control by purines, hormones and nutrients. In: Onco-developmental Gene Expression, W. H. Fishman and S. Sell, Eds. Academic Press, New York, pp. 21-33.

11. MacManus, J. P., Braceland, B. M., Youdale, T. and Whitfield, J. F. (1973): Adrenergic antagonists and a possible link between the increase in cyclic adenosine 3',5'-monophosphate and DNA synthesis during liver regeneration. J. Cell. Physiol. 82, 157-164.

12. MacManus, J. P., Franks, D. J., Youdale, T. and Braceland, B. M. (1972): Increases in rat liver cyclic AMP concentrations prior to initiation of DNA synthesis following partial hepatectomy or hormone infusion. Biochem. Biophys. Res. Comm. 49, 1201-1207.

13. MacManus, J. P., Whitfield, J. F., Boynton, A. L. and Rixon, R. H. (1975): Role of cyclic nucleotides and calcium in the positive control of cell proliferation. In: Advances in Cyclic Nucleotide Research, Vol. 5, G. I. Drummond, P. Greengard and G. A. Robison, Eds. Raven Press, New York, pp. 719-734.

14. Max, M. H., Price, J. B., Jr., Takeshige, K. and Voorhees, A. B., Jr. (1972): The role of factors of portal origin in modifying hepatic regeneration. J. Surg. Res. 12, 120-123.

15. Moolten, F. L. and Bucher, N. L. R. (1967): Regeneration of rat liver: transfer of humoral agent by cross circulation. Science 158, 272-274.

16. Park, C. R. and Exton, J. H. (1972): Glucagon and the metabolism of glucose. In: Glucagon, P. J. Lefèbvre and R. H. Unger, Eds. Pergamon Press, New York, pp. 77-108.

17. Rapaport, E. and Bucher, N. L. R. (1976): Two new adenine nucleotides in normal and regenerating rat liver. In: Onco-developmental Gene Expression, W. H. Fishman and S. Sell, Eds. Academic Press, New York, pp. 13-20.

18. Rapaport, E. and Bucher, N. L. R. (1975): Two new nucleotides in normal and regenerating rat liver. J. Cell Biol. 67, 352a.

19. Richman, R. A., Claus, T. H., Pilkis, S. J. and Friedman, D. L. (1976): Hormonal stimulation of DNA synthesis in primary cultures of rat hepatocytes. Proc. Nat. Acad. Sci. USA 73, 3589-3593.

20. Rous, P. and Larimore, L. D. (1920): Relation of the portal blood to liver maintenance: A demonstration of liver atrophy conditional on compensation. J. Exper. Med. 31, 609-632.

21. Sasaki, H., Rubalcava, B., Baetens, D., Blázquez, E., Srikant, C. B., Orci, L. and Unger, R. H. (1975): Identification of glucagon in the gastrointestinal tract. J. Clin. Invest. 56, 135-145.

22. Short, J., Tsukada, K., Rudert, W. A. and Lieberman, I. (1975): Cyclic adenosine 3',5'-monophosphate and the induction of deoxyribonucleic acid synthesis in liver. J. Cell Biol. 250, 3602-3606.

23. Starzl, T. E., Marchioro, T. L., Rowlands, D. T., Jr., Kirkpatrick, C. H., Wilson, W. E. C., Rifkind, D. and Waddell, W. R. (1964): Immunosuppression after experimental and clinical homotransplantation of the liver. Ann. Surg. 160, 411-439.

24. Starzl, T. E., Porter, K. A., Kashiwagi, N. and Putnam, C. W. (1975): Portal hepatotrophic factors, diabetes mellitus and acute liver atrophy, hypertrophy and regeneration. Surg. Gynecol. Obstet. 141, 843-858.

25. Starzl, T. E., Porter, K. A. and Putnam, C. W. (1975): Intraportal insulin protects from the liver injury of portacaval shunt in dogs. Lancet 2, 1241-1242.

26. Starzl, T. E., Porter, K. A., Watanabe, K. and Putnam, C. W. (1976): Effects of insulin, glucagon, and insulin/glucagon infusions on liver morphology and cell division after complete portacaval shunt in dogs. Lancet 1, 821-825.

27. Thrower, S. and Ord, M. G. (1974): Hormonal control of liver regeneration. Biochem. J. 144, 361-369.

28. Whittemore, A. D., Kasuya, M., Voorhees, A. B., Jr. and Price, J. B., Jr. (1975): Hepatic regeneration in the absence of portal viscera. Surgery 77, 419-426.

29. Younger, L. R., King, J. and Steiner, D. F. (1966): Hepatic proliferative response to insulin in severe alloxan diabetes. Cancer Res. 26, 1408-1414.

V. METABOLIC EFFECTS AND PHYSIOLOGIC ROLE OF GLUCAGON

INSULIN AND GLUCAGON (PANCREATIC AND EXTRAPANCREATIC) INTERACTIONS AND THE REGULATION OF GLUCOSE TURNOVER IN PHYSIOLOGY AND IN THE DIABETIC STATE

M. Vranic, C. Yip, K. Doi, L. Lickley, S. Morita and G. Ross

Department of Physiology and Banting and Best Department of Medical Research
University of Toronto
Toronto, Canada

Dynamics of the in vivo effect of glucagon on glucose production. Glucagon has the unique property of acutely increasing glucose turnover in the normal dog without changing blood glucose levels. This occurs because when glucagon is infused, or released, a synchronous release of insulin increases peripheral glucose utilization. The effect of glucagon on the liver is unopposed by plasma insulin increments (up to 80 µU/ml) and the increase in glucose production is of the same magnitude as the increase in glucose utilization. The release pattern of the two hormones is such that the plasma glucagon/insulin ratio changes continuously. Such a change of the hormonal ratio is necessary because the effect of glucagon on the liver is transient, which is best demonstrated by maintaining insulin concentrations at a constant level in depancreatized dogs. The transient effect of glucagon suggests that an increase in glucagon release is important for acute, rather than chronic, adjustments of glucose turnover. On the other hand, glucagon has a role in the control of basal glucose production, as evidenced by experiments in which selective glucagon deficiency was induced by concurrent infusions of insulin and somatostatin. In studying the characteristics of glucagon effect, it was necessary to apply a non-steady state tracer method. Validation studies are discussed and it is demonstrated that by using these methods non-steady turnover rates can be measured within a 10% error.

Extrapancreatic glucagon and glucose turnover. In assessing the role of glucagon in the regulation of glucose production both pancreatic and extrapancreatic glucagon must be considered since the endogenous release of extrapancreatic glucagon in depancreatized dogs has a marked effect on glucose production. Furthermore, immunologic and biochemical similarities between extrapancreatic and pancreatic glucagon suggest that suppression or stimulation of extrapancreatic glucagon in depancreatized dogs can be used as a model for assessing the role of immunoreactive glucagon in diabetes, particularly because depancreatized dogs can continue to secrete extrapancreatic glucagon practically indefinitely. The highest concentration of extrapancreatic glucagon was found in oxyntic mucosa of the stomach both in adult and newborn dogs. A component which is immunologically pure and apparently identical to pancreatic glucagon was isolated from extracts of this mucosa following gel filtration in acid and alkaline columns and ion exchange chromatography. This purified fraction yielded one sharp peak on electrophoresis which reacted equally with antisera against pancreatic or gut glucagon. The control of secretion of pancreatic and extrapancreatic glucagon differs because arginine stimulates the release of the latter only when plasma IRI is subnormal; exercise stimulates the release of pancreatic, but not of extrapancreatic, glucagon.

The role of glucagon in the diabetic state. The following evidence suggests that glucagon plays an important role in the acute development of glucose overproduction and hyperglycemia when insulin levels in plasma are low, but still measurable. When both insulin and glucagon were suppressed by somatostatin in normal dogs, hyperglycemia and overproduction of glucose did not occur. In contrast, selective insulin deficiency, in depancreatized dogs which had normal plasma concentrations of extrapancreatic IRG, or in normal dogs concurrently infused with somatostatin and glucagon, induced promptly overproduction of glucose with hyperglycemia. In depancreatized dogs, somatostatin arrested, and arginine accelerated, development of hyperglycemia due to overproduction of glucose. These results with the administration of somatostatin can be attributed to the increased sensitivity of liver to insulin, when glucagon has been suppressed to below basal levels. Under such conditions (increased liver sensitivity), the portal-peripheral insulin gradient may be less important than under normal conditions and, therefore, it is speculated that suppression of glucagon below its normal level might have potential value as an adjuvant in the management of diabetic patients.

In order to find out whether glucagon also plays a role in the control of glucose homeostasis in animals totally deprived of insulin, the effect of somatostatin was also assessed in depancreatized dogs kept without insulin and food for 4 days. These dogs had elevated plasma IRG, increased gluconeogenesis (2 - 2 1/2 times) and marked hyperglycemia. Somatostatin suppressed IRG and gluconeogenesis decreased by 40%, indicating that both hyperglucagonemia and insulinopenia played a role in the control of gluconeogenesis. Surprisingly, however, plasma glucose levels did not change because somatostatin, for unexplained reasons, also suppressed the rate of glucose disappearance.

INTRODUCTION

The effect of insulin on the regulation of hepatic glycogenolysis and gluconeogenesis is opposed by a number of factors, one of which is glucagon, originating from pancreatic or extrapancreatic A-cells. In the periphery, the effect of insulin appears unopposed by glucagon and the counter regulatory effect of other hormones are less well characterized than at the level of the liver. The effects of insulin and glucagon are such that a simultaneous increment of both hormones can increase the flux of glucose without affecting its plasma concentration. For example, an arginine infusion (equivalent to a protein meal) stimulates the secretion of glucagon with an initial increase of glucose production by the liver (Ra), virtually unopposed by the simultaneous release of insulin. However, the latter stimulates peripheral glucose uptake (Rd). This coordinated interaction between the two hormones brings about a matched increase in production and utilization of glucose (9, 10). In this paper some of the previously published work describing the dynamics of the insulin and glucagon effects will be reviewed.

The tracer method used to measure glucose turnover in and out of steady state will be discussed. It will be shown that glucagon of extrapancreatic origin has immunologic and biologic characteristics similar to those of the pancreatic hormone and the role of extrapancreatic glucagon in regulating glucose production in depancreatized dogs will be considered in some detail. In view of the biologic activity of both pancreatic and extrapancreatic glucagon, the question whether suppression of glucagon below its normal level might play a role in alleviating the abnormalities of glucose metabolism in the diabetic state will be considered.

Methods of Calculating Glucose Production, Utilization and Clearance

Since the work described in this paper deals with measurements of glucose turnover, the validity and rationale of these methods will be described before outlining the glucagon experiments (for more detailed reviews of the tracer methods, see also 21, 34, 41).

When the flux of a metabolite is controlled only at its site of production, the concentration changes in blood reflect its turnover, and therefore measurements of blood concentration of such a substance can yield information about turnover dynamics. Under such conditions the rate of disappearance (uptake) of the metabolite is linearly proportional to its concentration; e. g., despite a changing concentration its clearance from the blood remains constant. Such is the case for most of the hormones and for some metabolites. For example, in the resting state, the rate of lipolysis (which is hormonally and metabolically regulated) will determine the plasma concentration and the rate of uptake of plasma FFA, because this rate is solely determined by mass action (4). This, however, does not apply to glucose, because the production rate by the liver and its disappearance rate are controlled by separate mechanisms. Thus when both production and utilization of glucose increase or decrease concurrently, this may result in a marked change in turnover, with only a small or no effect on plasma glucose concentration. This occurs, for example, during prolonged fasting (16), exercise (46), chronic administration of steroids (33), or growth hormone (3). This is also true for the glucagon-insulin system, where the two hormones can interact in the control of glucose production, but only insulin has a marked effect on glucose utilization. For example, arginine infusions do not induce changes in the concentration of glucose either in normal or in depancreatized insulin-infused dogs, but in the former glucose turnover increases markedly due to the simultaneous increase in the secretion of insulin and glucagon, while in the latter, the turnover does not change because the supply of insulin and glucagon is unaltered (10, 11).

In the experiments described in this paper, glucose turnover was measured by injecting a priming dose of glucose labeled with ^{14}C or ^{3}H followed by a constant infusion. Thus, a plateau of specific activity was attained at an early time point. In the steady state, when the concentration and turnover of glucose are constant, the rate of glucose production (rate of appearance, Ra) equals Ra*/SA, where Ra* equals the rate of infusion of labeled glucose (dpm/min) and SA equals the specific activity of plasma glucose, or the ratio of the concentrations of labeled to unlabeled glucose in the plasma (DPM/ g). In non-steady state conditions, when the turnover changes with or without a change in glucose concentration, other equations have to be used. Among them are those based on the principles introduced by Steele (39) and further elaborated by De Bodo _et al_ (17). This principle assumes a one-compartment (extracellular volume) glucose system into which glucose is delivered mainly by the liver, and from which glucose disappears to be taken up by the various tissues of the body. The assumption and the derivation of the equations describing such a simple system have been described elsewhere (34, 35, 39). The rate of production is calculated as $Ra = \frac{Ra^*}{SA} - N \frac{\frac{dSA}{dt}}{SA}$, where N indicates the body glucose mass (glucose pool), calculated as V · C, where V is the volume of distribution of glucose, and C its concentration. The more the system approaches steady state ($\frac{dSA}{dt}$ approaches 0) the smaller is the influence exerted by the second part of the equation on the value of Ra. When the rate of glucose delivery changes rapidly, the newly formed glucose will not intermix promptly with the whole extracellular compartment: therefore Steele added a correction factor (p = 0.5) to the equation which reduces the volume of distribution by one-half (V x p). When these assumptions were made it was not known what errors would be encountered under physiologic conditions. It was therefore necessary to assess the validity of such an approach. For this purpose we used the following procedure: a substance (which is not produced in the body, or whose endogenous production is suppressed) is infused at a known rate and its appearance in the blood is calculated by the tracer method; the two rates, the known and the calculated ones, are then compared. One example of the validation procedure is shown in Fig. 1. The polysaccharide inulin was used in 10 experiments as a test substance in conscious dogs (34). Radioactive inulin was infused at a constant rate (primed tracer infusion) while unlabeled inulin was infused concurrently, but at rates which were intermittently increased and decreased. The upper panel shows that the concentration of labeled inulin in plasma declined, indicating a progressively increasing clearance rate of the polysaccharide; the middle panel illustrates the plasma concentrations of unlabeled inulin, changing as a consequence of the changing infusion rate of unlabeled inulin; the lower panel indicates the measured (heavy line) and the calculated (circles) delivery of inulin. From the concentration of labeled and unlabeled inulin, its specific activity and change with time were calculated. The rate of inulin appearance was calculated (open circles) by using the equation shown previously. The best result was obtained, by correcting V by a factor of 0.65 rather than 0.5 as found also by Cowan and Hetenyi (15). The figure shows that the rates based on these calculations approximated the known rate of inulin delivery very closely. It was concluded that this tracer method yields an error of approximately 10%. If greater

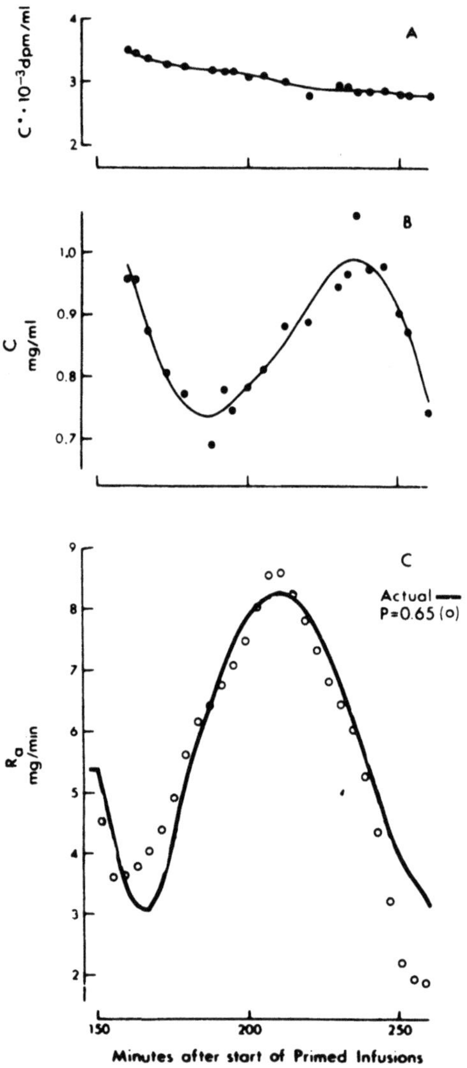

Fig. 1. Validation of non-steady state techniques for measuring turnover in a normal dog. ^{14}C-inulin was infused at a constant rate. Unlabeled inulin was infused at varying rates shown in the lower panel. Upper panel (A) illustrates the concentration of ^{14}C-inulin in plasma, and the middle panel (B) the concentrations of unlabeled inulin. Lower panel (C) shows that the rate of appearance of inulin (o) calculated by a non-steady state method (pool fraction, p = 0.65) yielded results which were very similar to the known rate of inulin infusion. Adapted from Radziuk et al. (34).

accuracy is desired, more complex equations based on multicompartment analysis can be used (34, 35). In order to study more closely the validity of these equations for the glucose system the following approach was used. Labeled and unlabeled glucose were infused into normal dogs (17 experiments) (35). Labeled glucose was infused at a constant rate and unlabeled glucose at varying rates, mimicking changes in glucose production rate under physiologic conditions, such as during and after an infusion of glucagon (11). The rate of unlabeled glucose infusion was sufficient to suppress endogenous glucose production so that the known rate of infusion of glucose could be compared to the tracer-calculated rate of glucose inflow into the system, using the non-steady state equation. Although the glucose system differs from the inulin system because its clearance changes markedly with changes in glucose concentration (stimulation of insulin secretion) the conclusions were the same as those for the inulin system, namely the method yielded accurate results with an error of 10%.

The rate of glucose production equals the rates of glycogenolysis and gluconeogenesis. When it is desirable to study the rate of gluconeogenesis by the outlined methods, one has to devise some experimental condition in which glycogenolysis is minimal. One such condition is prolonged fasting. If ^{14}C is used as a tracer, corrections are made for the recycling of labeled glucose fragments (mainly via lactate and alanine). Such corrections are not needed when 3-^{3}H or 6-^{3}H glucose is used, because this label does not recycle (1, 22, 24, 25). With 2-^{3}H glucose Ra is somewhat greater because in addition to the newly produced glucose a futile cycle (occurring in the liver) is measured as well. The rate of glucose utilization (rate of disappearance, Rd) which in the postabsorptive state equals the rate of peripheral glucose utilization, is calculated by applying the mass conservation argument, namely $\frac{dN}{dt} = Ra - Rd$, e. g. the accumulation rate of glucose in plasma equals the difference between its appearance and disappearance rates.

The rate of glucose utilization is governed by the concentration of glucose in the plasma (mass effect) and by factors which can accelerate or inhibit glucose uptake at a given concentration of plasma glucose. In order to separate the effects of glucose from other factors it is necessary to normalize Rd for a standard glucose concentration in plasma. This is achieved by dividing Rd by the prevailing plasma glucose concentration (g) at that time. This ratio, Rd/g is called the metabolic clearance (M) of glucose. Under most physiologic conditions, the factor which regulates M is insulin. When the concentration of insulin in plasma is kept constant in depancreatized dogs, despite changes in glucose concentration induced by glucagon, M does not change. This is illustrated in Fig. 2. Glucose concentration was maintained at normal levels in depancreatized dogs by a constant insulin infusion. A glucagon infusion increased glucose production by the liver and, consequently, glucose concentration in plasma rose; this increase in glucose concentration increased Rd, but M did not change. (M can also be described by the equation M = VK, where K is the fractional disappearance rate of glucose). When insulin was infused at rates higher than basal, the rate of change of K or M was determined by the concentration of insulin at various times. A simple mathematical equation was found to describe this relationship (44). Thus, whenever it is desirable to differentiate the effects of glucose concentration on Rd from the effects of other factors (mainly insulin), the clearance rates are calculated in addition to the glucose disappearance rates.

Surgical Procedures

In order to study the role of glucagon-insulin interactions in regulating glucose homeostasis, experiments were performed either in normal or in depancreatized dogs. It was desirable to be able to maintain normoglycemia and normal glucose turnover in depancreatized dogs during the control period. This could be achieved by one of the following approaches. In the first, the dogs underwent a 2-stage operation. During the first stage, two-thirds of the pancreas was

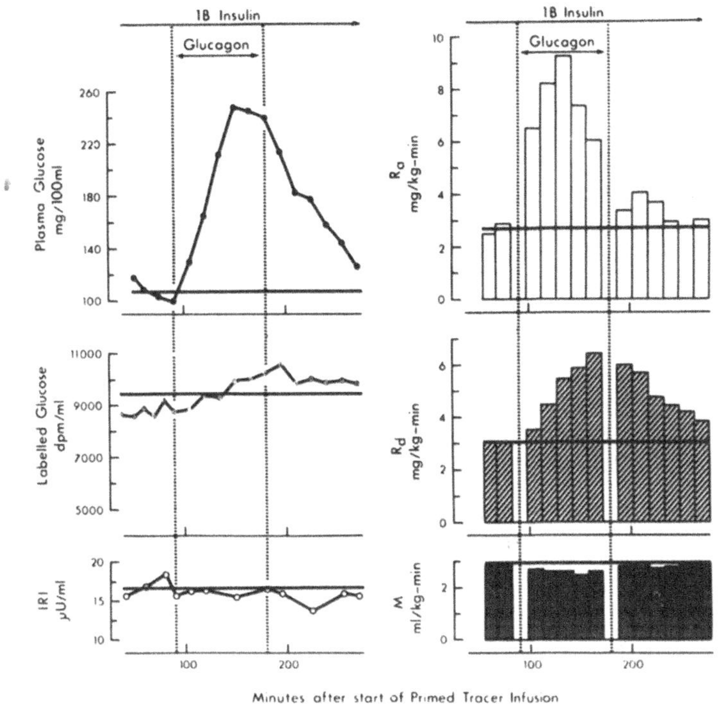

Fig. 2. Effects of glucagon infusion (t = 90 to t = 180 min) into a depancreatized dog maintained on a basal insulin infusion (250 µU/kg/min) on the plasma concentration of labeled and unlabeled glucose and insulin and on the rate of production (Ra), disappearance (R_d) and metabolic clearance (M) of glucose. Glucagon induced an increase in R_a and therefore glucose concentration rose. The increase in R_d was due only to a rise in glucose concentration and therefore M (R_d divided by glucose concentration) remained unchanged. Therefore, M can be used to differentiate between the effects of glucose and other factors which control R_d in vivo. Adapted from Radziuk _et al_. (34).

removed and the remaining third (uncinate process) was relocated with its blood supply under the skin of the dog. At the same time a cannula was implanted into the portal vein and kept patent. The uncinate process secreted sufficient insulin to maintain normoglycemia in the fasting state and the dogs were essentially aglycosuric when fed. A week later, the uncinate process was removed under local anesthesia and a basal intraportal insulin infusion (200-250 µU/kg/min) was started immediately to replace the endogenous hormone (46). In the second approach, the dogs were totally depancreatized, a portal cannula was introduced and the dogs were maintained on long acting pork insulin for a week. Forty-eight hours prior to the experiment, long acting insulin was replaced with crystalline insulin and, on the day of the experiment, the hyperglycemic dogs were given sufficient insulin intraportally to normalize the plasma glucose levels, which were thereafter maintained with a constant insulin infusion which could be varied as desired (43).

Dynamics of the Effect of Glucagon on Glucose Production In Vivo

In this section, we will describe three characteristics of the role of glucagon in the maintenance of glucose homeostasis: 1) the ability of glucagon to increase markedly glucose turnover at a time when glucose levels are unchanged; 2) the predominant effect of glucagon on the liver when both insulin and glucagon rise concurrently, and 3) the waning effect of glucagon increments on glucose production by the liver.

1) When glucagon was infused into 6 normal dogs for 90 minutes (29 ng/kg/min during the last hour of infusion), the concentration of glucose rose minimally and only transiently. Measurement of glucose concentration would indicate that glucagon infusions capable of generating plasma immunoreactive glucagon (IRG) concentrations up to 1200 pg/ml (9) have little effect on glucose metabolism. However, the tracer method revealed that glucose turnover increased by 2 1/2 times. This dissociation between glucose concentration and its turnover was due to glucagon-stimulated insulin release which caused a simultaneous 2 1/2-fold increase of the glucose disappearance rate and of its metabolic clearance (11). In order to find out how much insulin is necessary to bring about such a coordinated response, glucagon infusion was given to 11 conscious depancreatized dogs which, at the same time, received intraportal infusions of insulin at basal rate (B = 240 μU kg/min) or multiples of this rate. Such a basal rate had been shown

previously to maintain normal and steady turnover and concentrations of glucose in depancreatized dogs. Fig. 3 illustrates that when an insufficient amount of insulin was infused together with glucagon marked hyperglycemia resulted. However, the glycemic response to glucagon was similar to that seen in normal dogs when insulin was infused at rates near 12 x B.

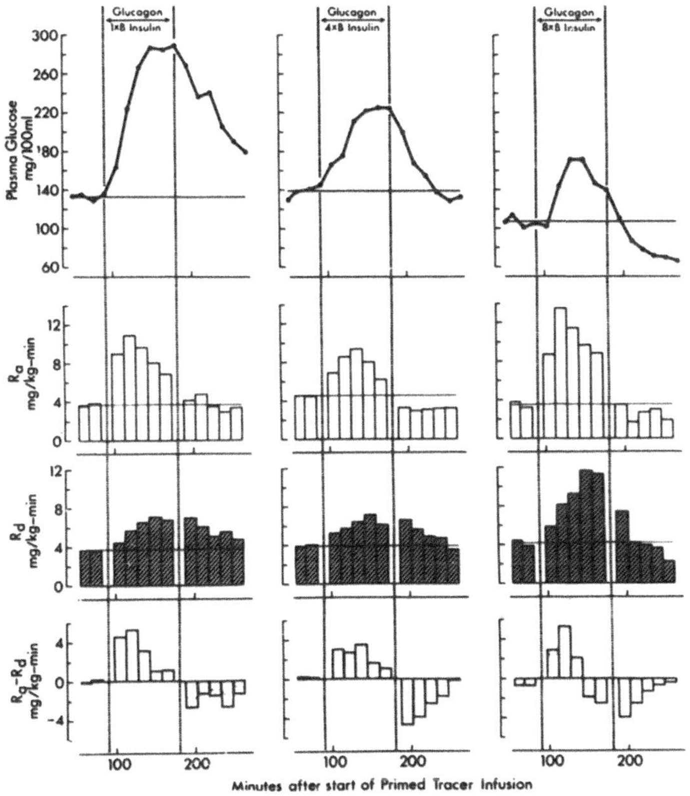

Fig. 3a

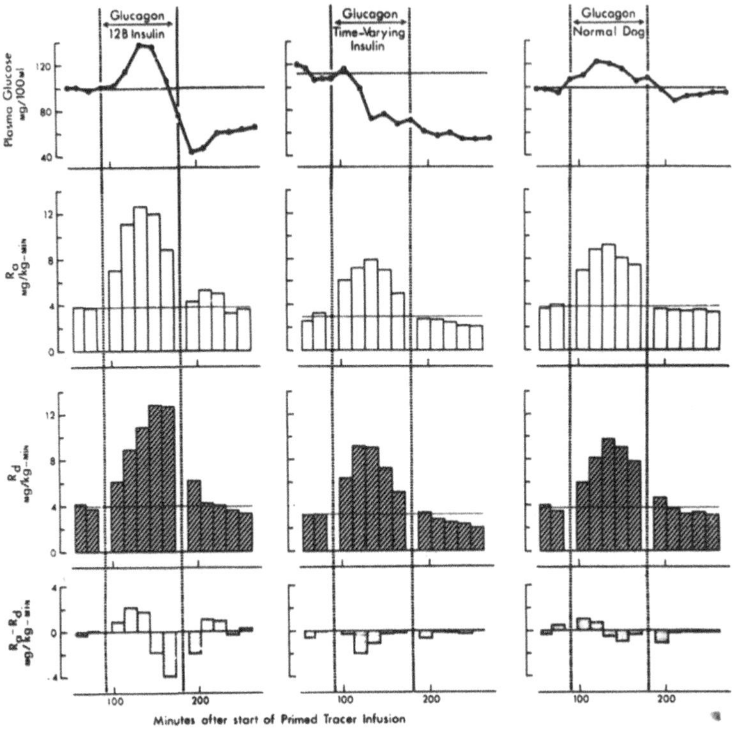

Fig. 3b

Figs. 3a and 3b. Effect of glucagon infusion (13-29 ng/kg/min) on the average concentration and rate of production and utilization of glucose in 6 normal and in 10 depancreatized dogs. All depancreatized dogs were maintained on a basal intraportal insulin infusion (R = 240 μU/kg/min) during the first and second control periods. In three depancreatized dogs, basal infusion was continued throughout (1 x B); in others, during glucagon infusion, insulin delivery was increased to a constant rate of 4 x B (2 dogs), 8 x B (1 dog) and 12 x B (2 dogs). In 2 dogs the rate of infusion varied (16 x B for 20 min, 8 x B for 30 min and 4 x B for 40 min). The lowest panel shows the difference between glucose production and utilization for each time point. Adapted from Cherrington and Vranic (11).

In subsequent experiments arginine was infused into normal dogs and glucose turnover increased markedly while glucose concentration did not change (10). That this was due solely to the arginine-stimulated release of insulin and glucagon is suggested by the fact that, in depancreatized insulin-infused dogs, arginine did not stimulate either insulin or glucagon release and glucose turnover did not change (Fig. 4). However, a normal increase in glucose turnover was obtained in these dogs when insulin and glucagon were added to the infusion of arginine in amounts capable of producing plasma concentrations of the hormones similar to those seen in normal dogs during arginine infusions (9). These data also indicated a physiologic role of multiphasic insulin release in the control of glucose metabolism, while rejecting the hypothesis that a given insulin:glucagon ratio controls glucose production by the liver, because in order to maintain a steady level of glucose production it was necessary to change this ratio continuously.

2) Fig. 3 and 5 also demonstrate that even when the rate of insulin infusion was as high as 2,888 μU/kg/min, the glucagon-induced increment of the rate of glucose production was not affected by insulin, suggesting that under these conditions, the effect of glucagon on the liver was predominant. Insulin affected glucose concentrations only by its peripheral effect which was proportional to the dose of insulin used. Due to such a dynamic relationship between insulin and glucagon, glucose turnover can be increased markedly with only a minimal effect on glucose concentration. This unique property of glucagon could be an important physiologic mechanism to minimize hyperglycemia, whenever glucose fluxes are increased. This is in contrast to other acute regulators of glucose production, such as catecholamines, which stimulate glucose fluxes, but at markedly hyperglycemic levels.

3) Figs. 2 and 3 reveal that the effect of glucagon waned with time, irrespective of whether or not insulin concentration was elevated during the infusion of glucagon. This characteristic of the glucagon effect has been discussed

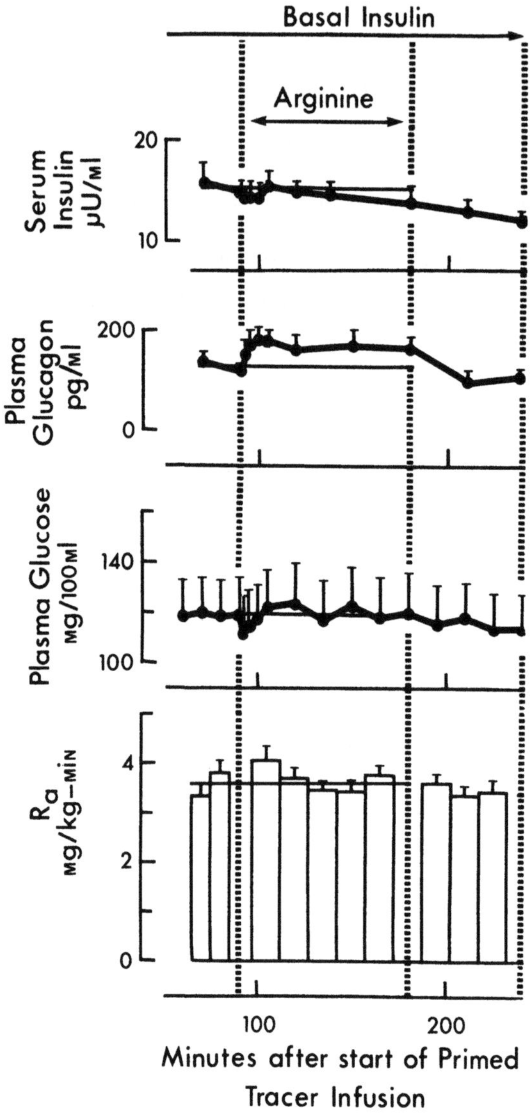

Fig. 4. Effect of arginine infusion (12.5 mg/kg/min) in 6 depancreatized dogs maintained normoglycemic with a basal infusion of insulin. Arginine did not appreciably affect either one of the parameters studied. Adapted from Cherrington et al. (9).

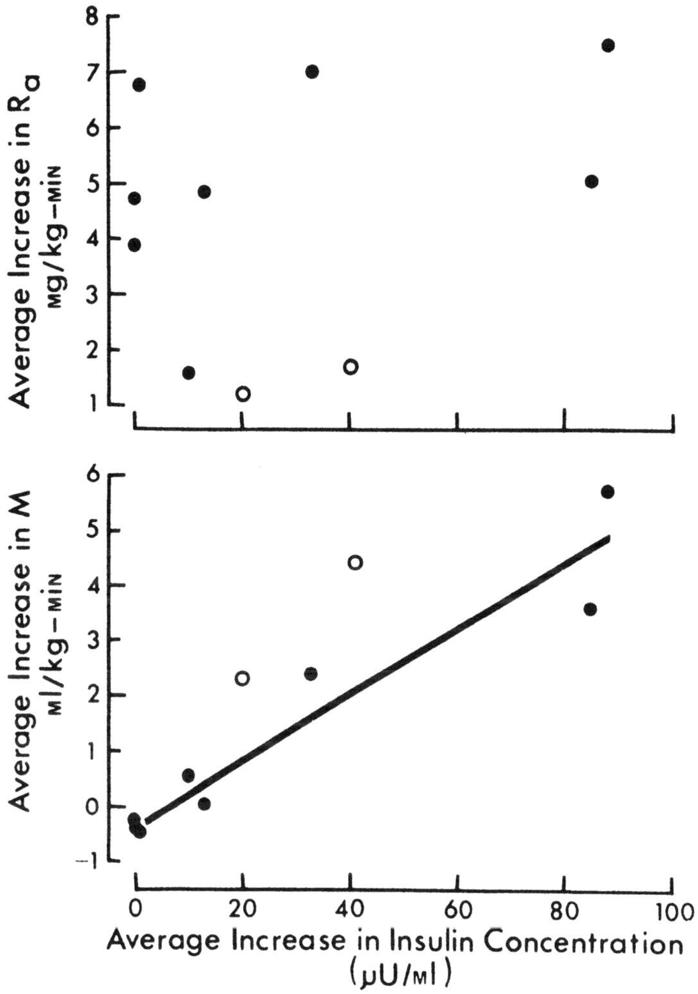

Fig. 5. Correlations calculated from data in Fig. 3, demonstrating that glucagon-induced increments of glucose production were not affected by the amount of insulin given to the depancreatized dogs; however, a highly significant positive correlation was found between plasma insulin concentrations and the metabolic clearance rate of glucose. ● constant insulin infusion; o varying pattern of insulin infusion. Adapted from Cherrington and Vranic (11).

in detail in previous papers (11, 12).

An evanescent effect of glucagon after protein feeding has also been observed recently in diabetic patients (18, 47). These observations suggest that the physiologic role of glucagon is to increase the glucose production rate by the liver acutely, and not to sustain increased glucose production for a prolonged period. A sustained increase of gluconeogenesis might be the function of other gluconeogenic hormones. The transient effect of glucagon on the liver would protect this organ against a sudden depletion of its glycogen stores and would provide an efficient mechanism for transient changes in glucose production, whenever metabolically desirable.

Extrapancreatic Glucagon and Glucose Turnover

When we designed a protocol for the study of the role of insulin and glucagon in the regulation of glucose turnover in dogs, we planned to use depancreatized animals as a model of glucagon deficiency. However, much to our surprise we found that depancreatized insulin-treated dogs had normal basal concentrations of IRG measured with an antibody considered to be specific for pancreatic glucagon (9). Furthermore, when insulin infusion was discontinued for a week, plasma IRG rose progressively, reaching a mean value of 986 ± 18 pg/ml (46). In order to explain this finding, it was necessary to exclude the possibility that the measured plasma IRG did not originate from residual pancreatic tissue, that it did not reflect a peculiarity of the antibody used; and that this plasma IRG was immunologically similar to the pancreatic hormone. These possibilities were excluded because 1) no immunoreactive insulin (IRI) could be detected in plasma using an assay sensitive to less than 1 μU/ml; 2) no pancreatic tissue was found at autopsy; 3) a progressive rise of IRG could be detected using 4 different antibodies specific for pancreatic glucagon; 4) the immunologic dilution curves of sera of depancreatized dogs were parallel to those of the pancreatic standard. Finally, appreciable amounts of IRG (measured by the antibody 30-K) could be detected in mucosa of the upper stomach of normal and depancreatized dogs (32, 44). Indeed, increased concentrations of plasma IRG, following pancreatectomy, were found also by other investigators (30, 31). In the dog the main source of immunoreactive glucagon is thought to be the oxyntic mucosa of the stomach which, among all gastrointestinal tissues, yields extracts with the highest glycogenolytic activity (39), the highest content of IRG in adult and newborn dogs (20, 32, 44) and which contains A-cells (5) and produces one major IRG component which has a molecular weight similar to pancreatic glucagon (5, 27, 32). Finally, IRG from stomach extracts has been chromatographed and measured with 2 antisera: 20-K and K-4023 (an antiserum that crossreacts strongly with glucagon-like immunoreactivity (GLI). IRG was eluted as one major peak corresponding to pancreatic glucagon from acid biogel P-30 columns (Fig. 6), and from the same gel in 0.05M NH_4CO_3 (Fig. 7). The IRG was resolved by polyacrylamide disc gel electrophoresis in urea at pH 8.7 into three immunologic components (Fig. 8), including a fast minor component, identified as desamido-glucagon and a slow component which reacted more strongly with K-4023 and presumably contained fragments of glucagon. The IRG peak obtained from gel filtration was therefore re-chromatographed on a DEAE Sephadex A-25 column (Fig. 9). The major IRG peak eluted as pancreatic glucagon. Electrophoresis of this peak showed only one immunoreactive component reacting equally well with 30-K and K-4023. These observations suggest that the IRG present in the stomach of the dog is similar if not identical to pancreatic glucagon. The question whether the biologic activities of gastric glucagon are identical to those of pancreatic glucagon can only be answered by using purified stomach glucagon. Studies on the in vitro biologic activity of partially purified extracts of stomach mucosa could be misleading, since such extracts might contain other biologically active materials which either potentiate or inhibit the effect of gastric glucagon. Isolated stomach glucagon has not yet been tested in vivo and therefore it is not yet known whether pancreatic and stomach glucagon are identical in their biologic effects. It was possible, nevertheless, to assess

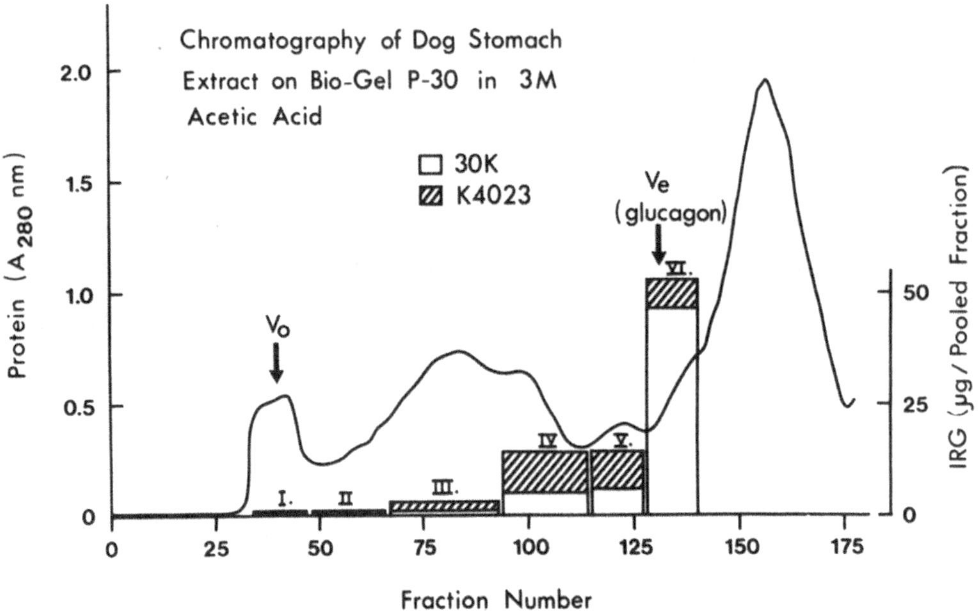

Fig. 6. Gel filtration of acid ethanol extract of mucosal scrapes from the gastric fundi of five normal dogs. The ether-alcohol precipitates obtained from the acid ethanol extract were dissolved in 15 ml of 3-M acetic acid. The material applied to a column (5 cm. x 100 cm.) of Bio-Gel P-30 (100-200 mesh) was equilibrated with 3-M acetic acid. The column was eluted with 3-M acetic acid at a flow rate of about 18 ml/hr. Fractions of 8 ml were collected, pooled as indicated and assayed for IRG with antiserum 30-K and K-4023. The elution volume (Ve) of glucagon was determined by using ^{125}I-glucagon. Vo = void volume. Reproduced from Morita *et al.* (32).

whether the *in vivo* release of extrapancreatic glucagon affects glucose production by the liver in a way similar to pancreatic glucagon. Fig. 10 illustrates one of 6 experiments designed to study this problem: a depancreatized dog was maintained normoglycemic by an intraportal infusion of insulin, then insulin was discontinued for one hour. Arginine induced a marked release of extrapancreatic IRG in the dog partially deprived of insulin, resulting in hyperglycemia. This could be accounted for entirely by a marked increase in the rate of glucose production. Thus, extrapancreatic glucagon appears to have a pronounced effect on glucose production *in vivo*. Arginine induced the release of IRG also in the

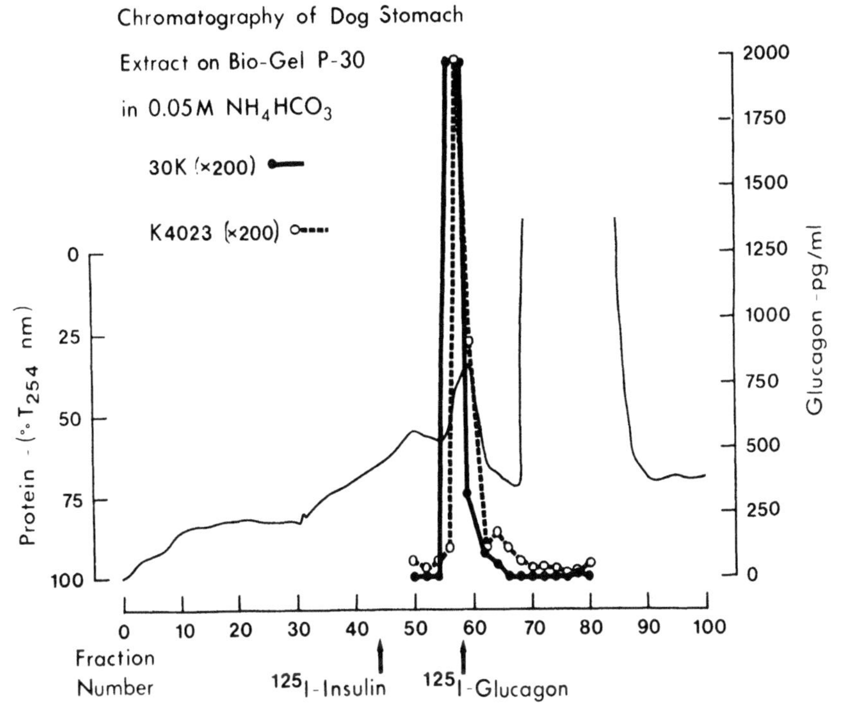

Fig. 7. Gel filtration of pooled fraction VI obtained as shown in Fig. 4. Fraction VI was reduced in volume by ultrafiltration at 5°C and was then lyophilized. The dry material was dissolved in 4 ml of 0.05-M NH_4HCO_3 and chromatographed on a column (2.5 cm. x 100 cm.) of Bio-Gel P-30 (100-200 mesh) equilibrated to 0.05-M NH_4HCO_3. The column was eluted with the same buffer at a rate of 15 ml/hr. Fractions of 7.5 ml were collected and assayed for IRG with antiserum 30-K (●——●) or K-4023 (o-----o). Ve of glucagon was determined using ^{125}I-glucagon. Reproduced from Morita et al. (32).

isolated stomach of the dog (28). The regulation of secretion of extrapancreatic glucagon, however, differs from that of pancreatic glucagon, since when insulin infusion was maintained, arginine did not affect either the release of extrapancreatic glucagon or glucose production (Fig. 4). Thus, the arginine test can be used to explore the presence of extrapancreatic glucagon only under conditions of partial insulin deprivation; it should not be used as a criterion to decide whether depancreatized man can secrete extrapancreatic glucagon when insulin levels are unknown (6). The other difference we have noted between glucagon and

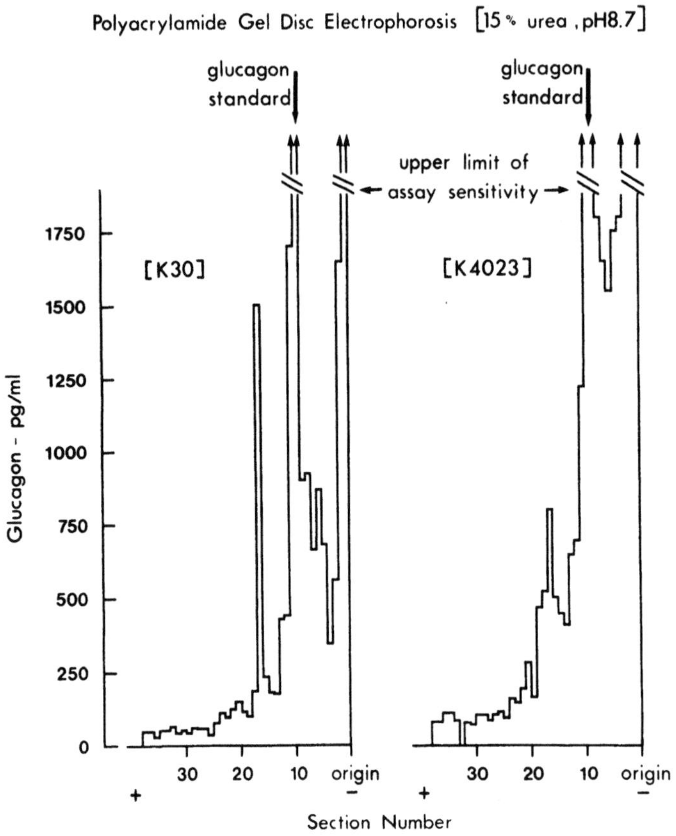

Fig. 8. Polyacrylamide disc gel electrophoresis of an aliquot of the partially purified immunoreactive material obtained as shown in Fig. 5. Electrophoresis was carried out in 15 percent gel containing 4-M urea at pH 8.7. Standard porcine crystalline glucgon was electrophoresed in parallel in a separate gel tube and stained. The unstained gel containing the sample was cut into 1.5 mm sections. Each section was eluted with 0.1 ml of the immunoassay diluent. The eluate from each section was assayed for IRG with antiserum 30-K and 4-4023 as indicated. Reproduced from Morita et al (32).

extrapancreatic glucagon was that strenuous exercise which stimulates release of IRG in normal dogs, does not stimulate the release of extrapancreatic glucagon in the depancreatized dog (43). Extrapancreatic glucagon was present in depancreatized dogs up to 5 years after removal of the pancreas, suggesting that the gastrointestinal tract can secrete physiologic amounts of IRG indefinitely (42).

In conclusion, the similarities between the immunologic and biologic effects of pancreatic and gastrointestinal glucagon suggest that depancreatized dogs can be used as a model to investigate the role of immunoreactive glucagon in the control of glucose turnover in the diabetic state.

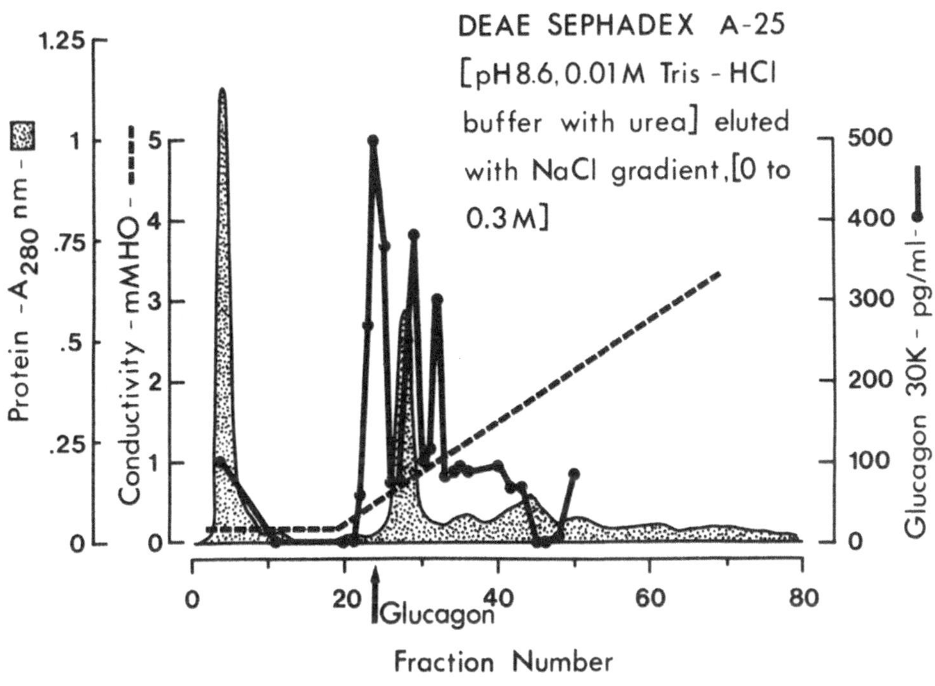

Fig. 9. Ion-exchange chromatography of gastric IRG obtained after gel filtration on Bio-Gel P.30 in 3M acetic acid, followed by gel filtration on Bio-Gel P-30 in 0.05 M NH_4HCO_3. DEAE-Sephadex A-25 was equilibrated with 0.01 M Tris-HCl, pH 8.6 containing 7M urea and was packed in a 1.5 x 30 cm column. The column was eluted at (13 ml/hr) with NaCl in a gradient of increasing concentration up to 0.3M. Fractions of 4.4 ml each were collected and protein content was measured at 280 nm. Each fraction was assayed for IRG using 30-K antiserum. The arrow indicates the pancreatic glucagon marker. From Vranic et al. (44).

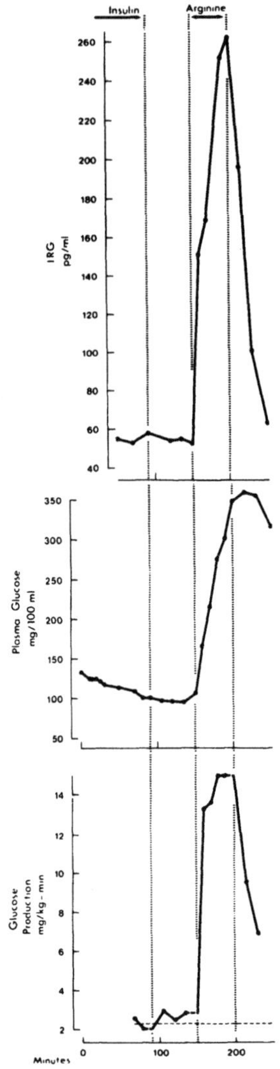

Fig. 10. Effect of arginine infusion (12.5 mg/kg/min) in a depancreatized dog maintained on basal insulin infusion during the control period, but deprived of insulin thereafter. Arginine given during the period of insulin-deprivation induced a marked increase in plasma IRG, and in plasma glucose. Hyperglycemia could be accounted for solely by increased hepatic glucose production.

The Role of Pancreatic or Extrapancreatic Glucagon in the Regulation of Glucose Turnover

In order to investigate the role of glucagon in the development of the diabetic state, we compared the effects of acute insulin deficiency induced by somatostatin (0.8 μg/kg/min) infusions in normal dogs when both insulin and glucagon were suppressed, or when normal plasma IRG was maintained by infusions of exogenous glucagon, or of insulin deficiency induced by total pancreatectomy, when extrapancreatic plasma IRG was within normal range. In addition we infused somatostatin and arginine into depancreatized dogs deprived of insulin for 1-2 hours, to examine the effect of the suppression or stimulation of glucagon on the concentration and turnover of glucose. We have shown previously that sudden cessation of endogenous insulin supply (clamping of the blood supply to a pancreatic remnant transplanted under the skin) or of exogenous insulin supply (cessation of an intraportal infusion of insulin in depancreatized dogs) resulted, within an hour, in hyperglycemia, due primarily to overproduction, but also to decreased metabolic clearance of glucose (46). In a recent study, cessation of basal insulin infusion in depancreatized dogs resulted in a 70% decrease of plasma IRI, within one hour, while extrapancreatic IRG did not change (133 ± 44, and 159 ± 38 pg/ml). Glucose production increased by 60%, metabolic clearance decreased by 37%, and plasma glucose increased by 50% (43). In contrast, somatostatin (0.8 μg/kg/min) which, in 5 normal dogs, decreased plasma IRI by 67 ± 4%, within one hour, did not induce hyperglycemia. Instead plasma glucose decreased temporarily (8 ± 1%), because Ra fell transiently (26 ± 3%), while glucose clearance did not decrease significantly. This difference between normal and depancreatized dogs could be attributed essentially to a 37 ± 4% decrease in plasma IRG. Indeed when glucagon (4.2 μg/kg/min) was infused with somatostatin, the results were similar to those obtained in depancreatized dogs deprived of insulin: plasma glucose increased by 52 ± 7%, Ra by 69 ± 12%, while M fell by 19 ± 7% (45). Fig. 11 compares glucose concentration and glucose production in depancreatized and in normal dogs infused with somatostatin. Similar effects of somatostatin on glucose turnover in normal dogs were observed also by Cherrington et al. (8) and by Altszuler et al. (2). These results emphasize the importance of glucagon in the early development of diabetes and indicate also (as discussed in the previous section) that pancreatic and extrapancreatic glucagon have a similar role in the control of glucose turnover. It cannot be excluded, however, that some metabolic effects of somatostatin could be exerted also through a mechanism

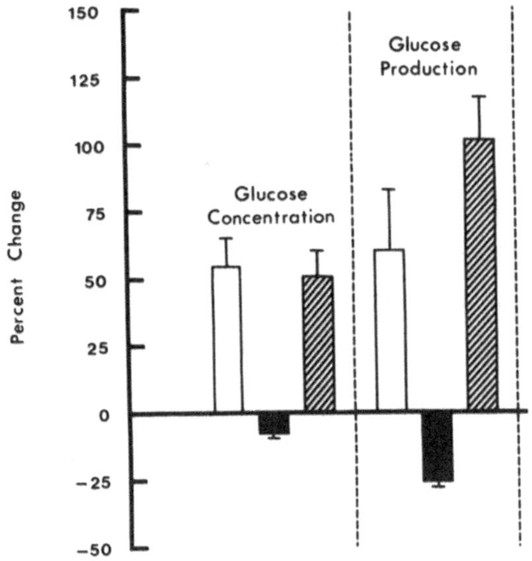

Fig. 11. The effects of selective insulin deficiency induced either by pancreatectomy (□), or by the concurrent infusion of somatostatin and glucagon (▨) are compared to those of the combined deficiency of insulin and glucagon induced by the infusion of somatostatin alone (■). Maximal percent changes (MEAN ± SE) in glucose concentration and rate of production from the control periods.

□ Pancreatectomy (Insulin $\downarrow$)

Normal Dogs
■ Somatostatin (Insulin and Glucagon $\downarrow$)
▨ Somatostatin + Glucagon (Insulin $\downarrow$)

which does not involve glucagon suppression. However, when glucagon was infused concurrently with somatostatin to normal dogs, the effects of pancreatectomy were restored, indicating that under the conditions studied nonpancreatic effects of somatostatin did not play a major role in the control of glucose production. The fact that the decrease of Ra in normal dogs infused with somatostatin was transient, suggests that initially the effect of glucagon deficiency prevailed and glucose production decreased. By the end of somatostatin infusion, however, Ra was normal, reflecting a balance of both insulin and glucagon deficiency. In order to find out the effects of selective glucagon deficiency, an infusion of somatostatin was given concurrently with an intraportal infusion of insulin to 8 normal dogs. Normal plasma IRI levels were maintained while plasma IRG fell. Ra was suppressed by 27 ± 6%, but this suppression was not transient; instead it was sustained until the end of the infusion period. These data suggest that basal glucagon plays a role in controlling about 1/3 of the basal glucose production (29).

The role of IRG in the rise of glucose production following insulin deprivation was explored further in 4 depancreatized dogs (Table 1). A basal insulin infusion maintained normal plasma concentrations of IRI, IRG and glucose and a normal glucose turnover for 90 minutes. When the infusion of insulin was discontinued for 140 minutes, plasma IRI progressively declined to a low, but still measurable level of 4.7 ± 0.6 μU/ml. During the first hour of insulin deprivation plasma IRG was unchanged, but glucose concentration and production rose. During the second hour of insulin deprivation, somatostatin was infused: IRG fell below the limits of our assay and the diabetogenic effect of insulin deprivation was promptly arrested. Cessation of somatostatin infusion induced a rebound of IRG to normal levels and glucose production and concentration rapidly increased. Thus, the presence of IRG, but not hyperglucagonemia is necessary for the induction of hyperglycemia. Arginine (12.5 mg/kg/min) brought about a three-fold increase in plasma IRG, and further increments in the production and concentration of glucose. Such a marked acceleration of the diabetic state was also noted when arginine was given in the second hour of insulin deficiency. Thus, the effects of suppression and stimulation of IRG secretion suggest that immunoreactive glucagon plays a major role in the development of hyperglycemia when plasma IRI is still detectable, and that glucose production reflects both glycogenolysis and gluconeogenesis. These data are consistent with the hypothesis that in the presence of endogenous or exogenous insulin the suppression of glucagon below its

TABLE 1

EFFECT OF SOMATOSTATIN (0.6 μg/kg/min) AND ARGININE (12.5 mg/kg/min)ON IMMUNOREACTIVE INSULIN (IRI), GLUCAGON (IRG) AND GLUCOSE AND ON THE RATE OF GLUCOSE PRODUCTION IN FOUR DEPANCREATIZED DOGS*

		No Insulin			
		(1)	(2) Somatostatin	(3)	(4) Arginine
IRI (μU/ml)	14.4 ± 1.9	6.8 ± 0.4	6.1 ± 0.4	5.2 ± 0.2	4.7 ± 0.6
IRG (pg/ml)	51 ± 8	60 ± 14	26 ± 5**	40 ± 10	127 ± 32
Glucose concentration (mg/100 ml)	116 ± 10	167 ± 32	170 ± 30	220 ± 44	340 ± 37
Glucose production (mg/kg/min)	5.2 ± 0.6	7.9 ± 2.0	6.3 ± 0.5	9.3 ± 1.7	15.1 ± 2

*Depancreatized dogs were initially maintained normoglycemic with a constant basal intraportal infusion of insulin (200-250 μU/kg/min) for 90 min. Insulin supply was withdrawn for the remaining 200 min of the experiment: (1) 60 min of no treatment, (2) 50 min of somatostatin infusion, (3) 50 min of no treatment, (4) 40 min of arginine infusions. Mean ± S.E. of the last value of each period.

**IRG values were within the sensitivity limits of the assay.

Reproduced from Vranic *et al* (45).

basal values may be helpful in the control of glycemia.

The next aim was to find out whether plasma IRG plays some role in the control of gluconeogenesis in depancreatized dogs chronically deprived of insulin. Five totally depancreatized dogs (fasted for 1 or 4 days) were deprived of insulin for 4 days until IRI could no longer be detected in their plasma. In such dogs the contribution of glycogenolysis to hepatic glucose production is expected to be minimal. The diabetic state in the 5 dogs was characterized by elevated extrapancreatic IRG (255 ± 38 pg/ml), elevated GLI measured by antibody K-4023 (841 ± 76 pg/ml), hyperglycemia (388 ± 21 mg/100 ml) and overproduction of glucose (11.9 ± 1.0 mg/kg/min). Somatostatin infusions (0.5 μg/kg/min) for 6 hr decreased plasma IRG to 56 ± 7 and GLI to 270 ± 21 pg/ml. Both glucose production and utilization decreased by 36% (from 11.9 mg/kg/min to 7.7 ± 0.8 mg/kg/min) but were still above normal values. Since both Ra and Rd decreased proportionally, hyperglycemia remained essentially unchanged (335 ± 11 mg/100 ml): another example of the dissociation between concentration and turnover data. If one considers that the glucose production (measured by 2-^{3}H-glucose) of fasted, depancreatized, insulin-controlled dogs was 5.2 ± 0.6 mg/kg/min (Table 1) and that of insulin deficient hyperglucagonemic dogs was 11.9 mg/kg/min, it appears that an increment of 6.7 mg/kg/min had been induced by the abnormalities of the 2 hormones. Glucagon suppression decreased the overproduction by 4.2 mg/kg/min (more than half of the increment). This strongly supports the concept that glucagon plays a role in the control of gluconeogenesis (13), at least in the diabetic state. The cause of the decrease in rate of glucose disappearance, at a time when plasma glucose was virtually unchanged, is unknown. Further work is needed to find out whether this is due to a somatostatin-mediated effect which became apparent only in the state of chronic and total insulin deprivation.

Since plasma IRG is elevated at birth in the rat, dog, and humans (7, 19, 20, 37) we have infused somatostatin (0.8 or 1.6 μg/kg/min) to newborn pups to find out the relative roles of glucagon and insulin in the maintenance of basal glucose production. However, somatostatin did not turn out to be an adequate tool to answer this question, because it did not affect either the plasma concentration of the 2 pancreatic hormones, or the concentration and turnover of glucose.

Is There a Rationale for the Search of a Specific Glucagon Suppressant as an Adjuvant in the Treatment of Diabetes?

The data described in the previous section suggest that elevated IRG plays a role in the control of gluconeogenesis, but is not important for the maintenance of hyperglycemia in dogs which are chronically insulinopenic. On the other hand, in the presence of low plasma IRI, hyperglycemia and overproduction of glucose were alleviated when plasma IRG was suppressed below its normal level. Such data support the contention that partial glucagon suppression could be useful in the management of diabetes. It is conceivable that diabetics with a reduced insulin reserve could control their metabolism without insulin if glucagon were partially suppressed and that in other patients insulin requirements could be decreased. A decrease of insulin requirements in insulin dependent patients per se, however, may not be so important since an adequate administration of insulin may suppress secretion of glucagon anyway. In order to rationalize the search for glucagon suppressants, we have suggested a hypothesis, schematized in Fig. 12 (45). In normal subjects insulin is delivered into the portal vein and a gradient is established between the portal and peripheral blood so that the liver is exposed to higher insulin concentrations than the peripheral tissues; this gradient is particularly marked during food ingestion, when insulin secretion is enhanced. The physiologic significance of this gradient may be related to the fact that in the liver the effect of insulin is counteracted by a number of glycogenolytic and gluconeogenic factors, while the peripheral action of insulin is less subject to counterregulation. One of the hormones antagonistic to insulin in the liver is glucagon, which may be most important for the control of glucose production during fasting and for the prevention of insulin induced hypoglycemia following a protein

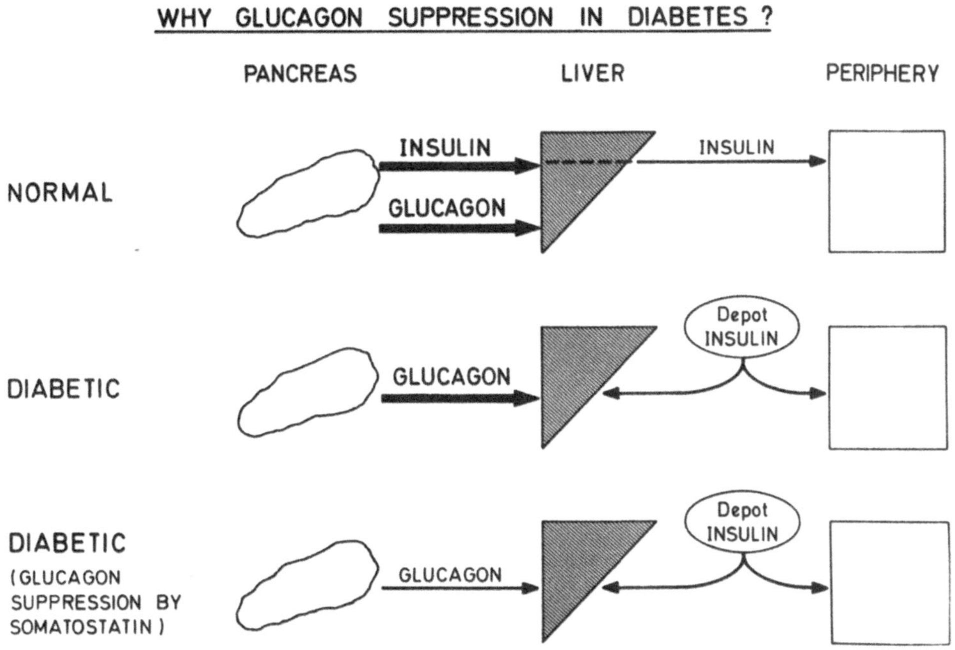

Fig. 12. The scheme illustrates the porto-peripheral gradient of insulin concentrations in normal subjects, and the absence of the gradient in insulin injected diabetics. It is hypothesized that in diabetics the suppression of glucagon below its basal level could help normalize the relationship between insulin and glucagon at the liver site.

meal. In insulin dependent diabetic patients, prolonged periods of fasting must be avoided. In addition, these patients cannot respond to protein ingestion by releasing insulin. Therefore, it is conceivable that in such patients an adequate control of glucose metabolism could be achieved with subnormal concentrations of plasma glucagon.

In the diabetic patient, insulin is delivered through a peripheral route and the insulin concentration gradient between the liver and periphery is absent. Therefore, either the periphery is exposed to larger than normal insulin concentrations, or the liver is exposed to IRI concentrations which are low in relation to the portal IRG levels (Fig. 12, middle panel). Suppression of IRG below its normal level would make the liver more sensitive to insulin while not changing the sensitivity of the peripheral tissues. Thus, even when a concentration gradient between portal and peripheral blood is absent, a partial suppression of glucagon could generate a sensitivity gradient for insulin. This concept is hypothetical and does not take into account other metabolic abnormalities of diabetes. It provides, however, some rationale for the search of glucagon suppressants which, unlike somatostatin, might be more specific for glucagon. One aspect of glucagon suppression, however, has not yet been explored. It was found that during exercise in the depancreatized dogs, long-acting insulin is mobilized from its subcutaneous depot at an accelerated rate. Thus, in contrast to controls, where exercise depresses insulin production, the diabetic animals are exposed to elevated plasma IRI which suppresses glucose production by the liver and hypoglycemia results (26). Similar observations were also made in human diabetics (48). The question to be explored is whether IRG suppression below normal levels would make the liver so sensitive to IRI fluctuations to create problems with insulin-induced hypoglycemia following exercise.

Acknowledgements

We are indebted to Mrs. N. Kovecevic, P. Eng, and Mrs. J. Wilkins, for their excellent assistance and we acknowledge the following generous gifts: Somatostatin, from Dr. R. Deghengi (Ayerst Research Laboratories, Montreal); Festal tablets (Hoechst Pharmaceutical Company); Dextrostix and Reflectance Metre (Ames Co. Ltd.) and PZI and crystalline pork insulin (Connaught Laboratories).

This work was supported by the Medical Research Council of Canada (grant MT-2196 and MT-3034), Canadian Diabetes Association, Banting Foundation, and the C. H. Best Foundation.

Dr. K. Doi and Dr. S. Morita were recipients of the Connaught and C. Stevens postdoctoral fellowships of the University of Toronto, respectively. Mr. G. Ross was an M.Sc. student in the Department of Physiology.

REFERENCES

1. Altszuler, N., Barkai, A., Bjerknes, C., Gottlieb, B. and Steele, R. (1975): Glucose turnover values in the dog obtained with various species of labelled glucose: Am. J. Physiol. 229, 1662-1667.

2. Altszuler, N., Gottlieb, B. and Hampshire, J. (1976): Interaction of somatostatin, glucagon and insulin on hepatic glucose output in the normal dog. Diabetes 25, 116-121.

3. Altszuler, N., Rathgeb, I., Winkler, B., De Bodo, R. C. and Steele, R. (1968): The effects of growth hormone on carbohydrate and lipid metabolism in the dog. Ann. N. Y. Acad. Sci. 148, 441-468.

4. Armstrong, D. T., Steele, R., Altszuler, N., Dunn, A., Bishop, J. S. and De Bodo, R. C. (1961): Regulation of free fatty acid turnover. Am. J. Physiol. 201, 9-15.

5. Baetens, D., Ruffer, C., Srikant, B., Dobbs, R. E., Unger, R. H. and Orci, L. (1976): Identification of glucagon producing A-cells in dog gastric mucosa. J. Cell Biol. 69, 455-464.

6. Barnes, A. J. and Bloom, S. (1976): Pancreatectomised man: a model for diabetes without glucagon. Lancet 1, 219-220.

7. Blázquez, E., Sugase, T., Blázquez, M. and Foà, P. P. (1974): Neonatal changes in the concentration of rat liver cyclic AMP and of serum glucose, free fatty acids, insulin, pancreatic and total glucagon in man and in the rat. J. Lab. Clin. Med. 83, 957-967.

8. Cherrington, A. D., Chiasson, J. L., Liljenquist, J. E., Jennings, A. S., Neller, K. and Lacy, W. W. (1976): The role of glucagon in maintaining basal glucose production. J. Clin. Invest. 58, 1407-1418.

9. Cherrington, A. D., Kawamori, R., Pek, S. and Vranic, M. (1974): Arginine infusion in dogs. Model for the roles of insulin and glucagon in regulating glucose turnover and free fatty acid levels. Diabetes 23, 805-815.

10. Cherrington, A. and Vranic, M. (1973): Effect of arginine on glucose turnover and plasma free fatty acids in normal dogs. Diabetes 22, 537-543.

11. Cherrington, A. and Vranic, M. (1974): Effect of interaction between insulin and glucagon on glucose turnover and FFA concentration in normal and depancreatized dogs. Metabolism 23, 729-744.

12. Cherrington, A., Vranic, M., Fono, P. and Kovacevic, N. (1972): Effect of glucagon on glucose turnover and free fatty acids in depancreatized dogs maintained on matched insulin infusions. Canad. J. Physiol. Pharmacol. 50, 946-954.

13. Chiasson, J. L., Cook, J., Liljenquist, J. E. and Lacy, W. W. (1974): Glucagon stimulation of gluconeogenesis from alanine in the intact dog. Am. J. Physiol. 227, 19-23.

14. Chiasson, J. L., Liljenquist, E., Sinclair-Smith, B. C. and Lacy, W. W. (1975): Gluconeogenesis from alanine in normal postabsorptive man. Intrahepatic stimulatory effect of glucagon. Diabetes 24, 574-584.

15. Cowan, J. S. and Hetenyi, C., Jr. (1971): Glucoregulatory responses in normal and diabetic dogs recorded by a new tracer method. Metabolism 20, 360-372.

16. Cowan, J. S., Vranic, M. and Wrenshall, G. A. (1969): Effects of preceding diet and fasting on glucose turnover in normal dogs. Metabolism 18, 319-330.

17. De Bodo, R. C., Steele, R., Altszuler, N., Dunn, A. and Bishop, J. (1963): On the hormonal regulation of carbohydrate metabolism. Studies with C^{14} glucose. Rec. Progr. Horm. Res. 19, 445-488.

18. Felig, F., Wahren, J. and Hendler, R. (1976): Influence of physiological hyperglucagonemia on basal and insulin inhibited splanchnic glucose output in normal man. J. Clin. Invest. 58, 76-765.

19. Girard, J. R., Cuendet, G. S., Marliss, E. B., Kerman, A., Rientart, M. and Assan, R. (1973): Fuels, hormones and liver metabolism at term and during the early postnatal period in the rat. J. Clin. Invest. 52, 3190-3200.

20. Hetenyi, G., Jr., Kovacevic, N., Hall, S. E. H. and Vranic, M. (1976): Plasma glucagon in pups, decreased by fasting, unaffected by somatostatin or hypoglycemia. Am. J. Physiol. 231, 1377-1382.

21. Hetenyi, G., Jr. and Norwich, K. H. (1974): Validity of rates of production and utilization of metabolites as determined by tracer methods in intact animals. Fed. Proc. 33, 1841-1848.

22. Issekutz, B. Jr., Allen, M. and Borkow, I. (1972): Estimation of glucose turnover in the dog with glucose-2-T and glucose-U-^{14}. Am. J. Physiol. 222, 710-712.

23. Jennings, A. S., Cherrington, A. D., Chiasson, J. L., Liljenquist, J. E. and Lacy, W. W. (1975): The fine regulation of basal hepatic glucose production. Clin. Res. 23, 323A.

24. Katz, J. and Dunn, A. (1967): Glucose-2-T as a tracer for glucose metabolism. Biochemistry 6, 1-5.

25. Katz, J., Dunn, A., Chenoweth, M. and Golden, S. (1974): Determination of synthesis, recycling and body mass of glucose in rats and rabbits in vivo with ^{3}H and ^{14}C-labelled glucose. Biochem. J. 142, 171-183.

26. Kawamori, R. and Vranic, M. (1977): Mechanism of exercise-induced hypoglycemia in depancreatized dogs maintained on long-acting insulin. J. Clin. Invest. 59, 331-337.

27. Larsson, L. I., Holst, J., Hakanson, R. and Sundler, F. (1975): Distribution and properties of glucagon immunoreactivity in the digestive tract of various mammals: an immuno-histochemical and immunochemical study. Histochemistry 44, 281-290.

28. Lefèbvre, P. J., Luyckx, A. S., Brassinne, A. H. and Nizet, A. H. (1976): Glucagon and gastrin release by the isolated perfused dog stomach in response to arginine. Metabolism 25 (Suppl. 1), 1477-1479.

29. Lickley, L. and Vranic, M. (1976): Effects of selective glucagon deficiency. Clin. Res. 24, 660A.

30. Mashiter, K., Harding, P. E., Chou, M., Mashiter, G. D., Stout, J., Diamond, D. and Field, J. B. (1975): Persistent pancreatic glucagon but not insulin response to arginine in pancreatectomized dogs. Endocrinology 96, 678-693.

31. Matsuyama, T. and Foà, P. P. (1974): Plasma glucose, insulin, pancreatic and enteroglucagon levels in normal and depancreatized dogs. Proc. Soc. Exp. Biol. Med. 147, 97-102.

32. Morita, S., Doi, K., Yip, C. and Vranic, M. (1976): Measurement and partial characterization of immunoreactive glucagon in gastrointestinal tissues of dogs. Diabetes 25, 1018-1025.

33. Ninomiya, R., Forbath, N. F. and Hetenyi, G. Jr. (1965): Effects of adrenal steroids on glucose kinetics in normal and diabetic dogs. Diabetes 14, 729-739.

34. Radziuk, J., Norwich, K. H. and Vranic, M. (1974): Measurement and validation of nonsteady turnover rates with application to the inulin and glucose systems. Fed. Proc. 33, 1855-1864.

35. Radziuk, F., Vranic, M. and Norwich, K. H. (1974): Experimental validation of tracer determined nonsteady glucose turnover rates and a functional relationship between glucose clearance and insulin levels. Fed. Proc. 33, 276.

36. Sasaki, H., Rubalcava, B., Baetens, D., Blázquez, E., Srikant, C. B., Orci, L. and Unger, R. H. (1975): Identification of glucagon in the gastrointestinal tract. J. Clin. Invest. 56, 135-175.

37. Oh, W., Fisher, D. A. (1974): Spontaneous and amino acid stimulated glucagon secretion in the immediate postnatal period. J. Clin. Invest. 53, 1159-1166.

38. Steele, R. (1959): Influence of glucose loading and of injected insulin on hepatic glucose output. Ann. N. Y. Acad. Sci. 82, 420-430.

39. Sutherland, E. W. and de Duve, C. (1948): Origin and distribution of hyperglycemic glycogenolytic factor of the pancreas. J. Biol. Chem. 175, 663-674.

40. Unger, R. and Orci, L. (1976): Physiology and pathophysiology of glucagon. Physiol. Rev. 56, 778-826.

41. Vranic, M. (1974): An overview, tracer methodology and glucose turnover. Fed. Proc. 33, 1837-1840.

42. Vranic, M., Engerman, R., Doi, K., Morita, S. and Yip, C. (1976): Extrapancreatic glucagon in the dog. Metabolism 25 (Suppl. 1), 1469-1473.

43. Vranic, M., Kawamori, R., Pek, S., Kovacevic, N. and Wrenshall, G. A. (1976): The essentiality of insulin and the role of glucagon in regulating glucose utilization and production during strenuous exercise in dogs. J. Clin. Invest. 57, 245-255.

44. Vranic, M., Pek, S. and Kawamori, R. (1974): Increased "glucagon immunoreactivity" in plasma of totally depancreatized dogs. Diabetes 23, 905-912.

45. Vranic, M., Ross, G., Doi, K. and Lickley, L. (1976): The role of glucagon-insulin interactions in control of glucose turnover and its significance in diabetes. Metabolism 25 (Suppl. 1), 1375-1380.

46. Vranic, M. and Wrenshall, G. A. (1968): Matched rates of insulin infusion and secretion and concurrent tracer-determined rates of glucose appearance and disappearance in fasting dogs. Can. J. Physiol. Pharmacol. 46, 383-390.

47. Wahren, J., Felig, P. and Hagenfeldt, L. (1976): Effect of protein ingestion on splanchnic and leg metabolism in normal man and in patients with diabetes mellitus. J. Clin. Invest. 57, 987-999.

48. Zinman, B., Murray, F. T., Vranic, M., Albisser, A. M., Leibel, B. S., McClean, P. A. and Marliss, E. B. (1977): Glucoregulation during moderate exercise in insulin treated diabetes. J. Clin. Endocrinol. Metab. In Press.

REGULATION OF HEPATIC GLUCONEOGENESIS BY GLUCAGON IN THE RAT

M. S. Ayuso-Parrilla and R. Parrilla

Department of Metabolism, Instituto G. Marañon, C.S.I.C.
Velázquez 144, Madrid-6, Spain

Glucagon (2×10^{-9} M) increased the rates of gluconeogenesis, ureogenesis and ketogenesis by the perfused isolated rat liver. Glucagon to insulin ratios of 2.6 reversed all the metabolic effects induced by the administration of glucagon alone.

The changes in the total hepatic content of intermediary metabolites were compatible with an interaction of glucagon with the gluconeogenic sequence at two different sites: that is, between pyruvate and phosphoenolpyruvate and at the level of the reactions catalyzed by phosphofructokinase and fructose diphosphatase. Calculation of the intracellular metabolite distribution demonstrates that glucagon produces a significant rise in the intramitochondrial concentration of oxaloacetate, suggesting an activation of pyruvate carboxylase, the first nonequilibrium step in the gluconeogenic sequence. Since the cytosolic concentration of oxaloacetate is below the Km of phosphoenolpyruvate carboxykinase for this metabolite, the finding of a decreased cytosolic level of oxaloacetate in conjunction with an accelerated gluconeogenic flux implies that phosphoenolpyruvate carboxykinase is activated also by glucagon.

Biosynthetic processes from L-alanine, as judged by the total hepatic oxygen uptake, display saturation type kinetics with a "Vmax" of about 2.0 mM. Thus, in the isolated perfused rat liver, the effect of glucagon in raising the intracellular concentration of L-alanine from 8.5 to 10.3 mM and the stimulation of gluconeogenesis were fully independent of one another.

Glucagon is known to stimulate gluconeogenesis _in vivo_ and _in vitro_ from a variety of substrates (4, 6, 10, 12, 20). The significance of this action of glucagon became even more apparent when it was found that in situations like fasting or diabetes, characterized by an accelerated glucose output, the plasma glucagon levels are elevated, contributing to an increase in net glucose output from the liver in concert with a decrease in insulin secretion. Apparently, the molar ratio of these hormones, rather than their absolute concentrations, is the crucial factor in the regulation of hepatic gluconeogenesis in the rat (13). Thus, a molar ratio of glucagon to insulin above 2.6 brings about a significant acceleration in the gluconeogenic flux. The mechanism for this hormonal effect remains unclear. Recent work seems to localize the site of action of these hormones in the gluconeogenic sequence on a metabolic step between pyruvate and phosphoenolpyruvate (5, 27), that is, on a step involved in the movement of metabolites across the mitochondrial membrane. In the rat pyruvate carboxylation takes place exclusively within the mitochondria (21, 24), while the formation of phosphoenolpyruvate is a virtually exclusive cytosolic event (21). On these grounds, it was considered highly desirable to explore whether the intracellular distribution of intermediary metabolites was affected by hormonal treatment since some of the gluconeogenic enzymes respond readily to variations in the concentration of their substrates within the physiologic range.

It has also been suggested that the stimulation of gluconeogenesis from amino acids by glucagon may be due to increased amino acid transport into the hepatocyte (11). Although results reported herein seem to indicate that this effect is independent of the stimulation of gluconeogenesis, at least in the perfused isolated liver, the possibility exists that this mechanism may play an important role in the regulation of gluconeogenesis *in vivo*.

EXPERIMENTAL PROCEDURE

Male Wistar albino rats, weighing 190 to 200 g were used. The animals were purchased when they weighed approximately 120 g and were kept under standard conditions of food intake, light and temperature. They were fasted for 24-36 hours prior to experiments. The livers were perfused with hemoglobin-free Krebs-Henseleit bicarbonate buffer (9), containing fraction V bovine serum albumin (4%), in an apparatus equipped with a disc oxygenator (8). The outflow was in contact with oxygen (Clark type) and pH electrodes and was recirculated at a rate of 28 ml/min (7). In all the experiments the livers were equilibrated for 30 minutes with basal medium followed, when indicated, by the addition of substrates (10 mM) and hormones. The perfusion was then continued for an additional 60 minutes. At this time the livers were frozen in situ using aluminum clamps precooled in liquid nitrogen (29). Representative portions of the livers were freeze dried and extracted with 35 volumes of perchloric acid (8%, w/v). Acid extracts were brought to pH 6.0 with potassium carbonate and immediately used for metabolite assays. Samples of perfusate were taken every ten minutes, deproteinized in cold perchloric acid (6%, w/v) and brought to pH 6.0 with potassium carbonate, for the determination of metabolites. Metabolic rates were calculated over 60-90 minute intervals. Metabolites were measured fluorimetrically according to previously described methods (2,28).

Most of the reagents were obtained from Sigma Chemical Co. (St. Louis, Missouri, USA). Enzymes were purchased from Boehringer (Mannheim, Germany).

RESULTS

Glucagon increased ureogenesis, ketogenesis and glucose production from L-lactate or L-alanine by the perfused isolated rat liver (Tables 1 and 2). The rates of gluconeogenesis from these two substrates differed significantly. Although about 80% more glucose was produced from L-lactate than from L-alanine, the degree of stimulation of gluconeogenesis from either substrate by glucagon was about 35%. The proportion of substrate utilized to glucose produced was also remarkably constant under all the situations studied: between 2.2 and 2.4. Thus, most of the glucose produced was accounted for by the substrate utilized, indicating that glycogenolysis was negligible.

Insulin reversed all glucagon effects (Tables 1 and 2). It is interesting to notice that the glucagon to insulin ratio used in these experiments was similar to the one observed in the portal vein in normal fed animals.

Effect of glucagon and of glucagon plus insulin on the hepatic content of intermediary metabolites. The changes in the concentration of gluconeogenic substrates was studied 60 minutes after the addition of the hormones. Figs. 1 and 2 show crossover plots of these substrates in livers perfused with L-lactate and L-alanine, respectively. From the study of these plots it becomes apparent that glucagon may interact with the gluconeogenic pathway at two different sites; that is, between pyruvate and phosphoenolpyruvate and also at the level of the reac-

TABLE 1

EFFECTS OF GLUCAGON AND OF GLUCAGON PLUS INSULIN ON LACTATE METABOLISM BY THE PERFUSED ISOLATED RAT LIVER

Metabolite Change	Control livers	Glucagon treated livers	Glucagon + Insulin treated livers
		μmoles/100 g body wt/h	
Glucose production	123 ± 10	165 ± 8	120 ± 10
Lactate used	287 ± 16	375 ± 24	264 ± 20
Lactate used/glucose formed	2.3 ± 0.2	2.2 ± 0.05	2.3 ± 0.2
Lactate accounted for as glucose	246 ± 20	330 ± 16	240 ± 20
Pyruvate formed	-7.9 ± 0.8	-6.5 ± 0.7	-9.4 ± 1
Ketone bodies formed	7 ± 0.5	10 ± 0.9	6 ± 0.6
Lactate unaccounted for	41.9	41.5	27.4
Urea formed	9 ± 0.7	16 ± 1.3	13 ± 0.9

Livers from fasted rats were pre-perfused for 30 minutes as described in Methods. At this time sodium L-lactate (10 mM initial concentration) and hormones were added and the livers perfused for 60 minutes. Metabolic rates were calculated over the last 30 minutes of perfusion. The results are means of at least eight experiments ± S. E.

tions catalyzed by phosphofructokinase and fructose diphosphatase. The patterns of change were similar with both substrates, and were not modified significantly by the combined use of glucagon and insulin. Hence, there was a good correlation between the observed changes in metabolite patterns induced by glucagon and the increase in the gluconeogenic flux.

Concentration of tricarboxylic acid cycle (TCA) intermediates. Glucagon treatment increased the concentration of most TCA intermediates, except oxaloacetate, which decreased regardless of which substrate was used (Figs. 3 and 4; Tables 3 and 4). When L-alanine was the carbon source, glucagon decreased α-ketoglutarate and increased aspartate. Probably, these changes were secondary to the rise in the intracellular concentration of L-alanine brought about by the hormone (Table 4).

TABLE 2

EFFECTS OF GLUCAGON AND OF GLUCAGON PLUS INSULIN ON ALANINE METABOLISM BY THE PERFUSED ISOLATED RAT LIVER

Metabolite Change	Control livers	Glucagon treated livers	Glucagon + Insulin treated livers
		μmoles/100 g body wt/h	
Glucose production	67 ± 4.4	95 ± 3.4	75 ± 4.4
Alanine used	168 + 19	212 ± 7	188 ± 16
Alanine used/glucose formed	2.4 ± 0.2	2.2 ± 0.3	2.4 ± 0.2
Alanine accounted for as glucose	133 ± 8.8	191 ± 6.8	149 ± 8.8
Lactate production	18.8 ± 1.5	15.6 ± 1.2	19.8 ± 1.4
Pyruvate production	2.7 ± 0.1	2.6 ± 0.2	2.0 ± 0.1
Alanine unaccounted for	12.5	2.7	16.6
Urea production	74 ± 5.4	98 ± 8.1	59 ± 9

The conditions of the experiment were the same as described in Table 1, with the difference that L-alanine (10 mM initial concentration) was added instead of sodium L-lactate. The results are means of at least eight experiments ± S.E.

Glucagon did not affect significantly the concentration of acetyl-CoA and of total soluble CoA, but produced a statistically significant decrease in free CoA. This fall in the concentration of CoA probably indicates a hormone-induced increase in the formation of CoA derivatives. The presence of insulin reversed most of the described changes (Figs. 3 and 4).

Adenine nucleotides content and oxygen utilization. Glucagon caused a significant decrease in the [ATP]/[ADP] ratio (Tables 5 and 6). It is interest-

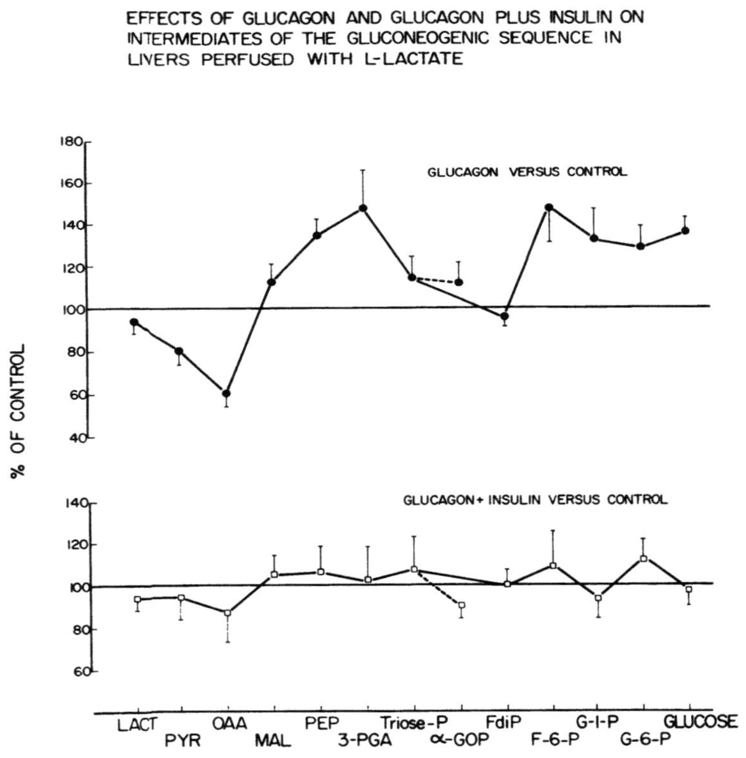

Fig. 1. Each point is the mean value of eight experiments and the vertical bars the standard error of the mean. (---) Connects glycerol 3-phosphate which is not in the direct gluconeogenic sequence (14).

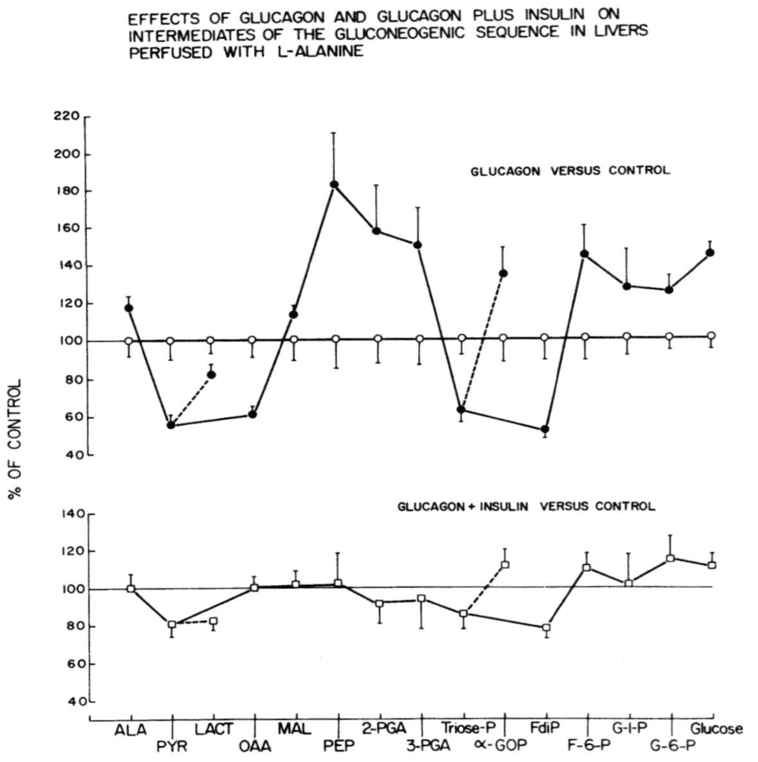

Fig. 2. Each point is the mean value of eight experiments and the vertical bars the standard error of the mean. (---) Connects lactate and glycerol 3-phosphate which is not in the direct sequence (15).

ing to notice that in the case of livers perfused with L-lactate this effect was caused by a rise in the ADP concentration, while it was due to a significant decrease in ATP when L-alanine was the substrate. This effect once again disappeared when insulin was present.

TABLE 3

EFFECTS OF GLUCAGON AND OF GLUCAGON PLUS INSULIN ON THE STEADY STATE CONCENTRATION OF TRICARBOXYLIC ACID CYCLE INTERMEDIARY METABOLITES AND CoA DERIVATIVES IN LIVER PERFUSED WITH 10 mM L-LACTATE AS SUBSTRATE

	Control livers	Glucagon treated livers	Glucagon + Insulin treated livers
		nmoles/g dry weight	
Malate	1186 ± 161	1334 ± 101*	1249 ± 112
Oxaloacetate	18.7 ± 2.4	11.2 ± 1.2***	16.3 ± 3
Citrate	2118 ± 266	2518 ± 376**	2202 ± 244
Isocitrate	91 ± 6	132 ± 8***	83 ± 8
α-Ketoglutarate	1763 ± 125	2009 ± 242**	1766 ± 168
Glutamate	9481 ± 372	10297 ± 470	8981 ± 1107
Aspartate	1902 ± 155	1515 ± 84**	1766 ± 179
Alanine	3730 ± 350	3120 ± 70	2610 ± 230**
Ammonia	2470 ± 265	2500 ± 300	2550 ± 325
CoA	412 ± 29	315 ± 14*	332 ± 19
Acetyl-CoA	347 ± 30	336 ± 17	335 ± 60
Total acid-soluble CoA	1400 ± 207	1269 ± 95	1342 ± 232

The conditions of the experiment were the same as described in Table 1. The results are means of at least eight experiments ± S. E. *p <0.1; **P <0.05; ***P <0.01, by t test.

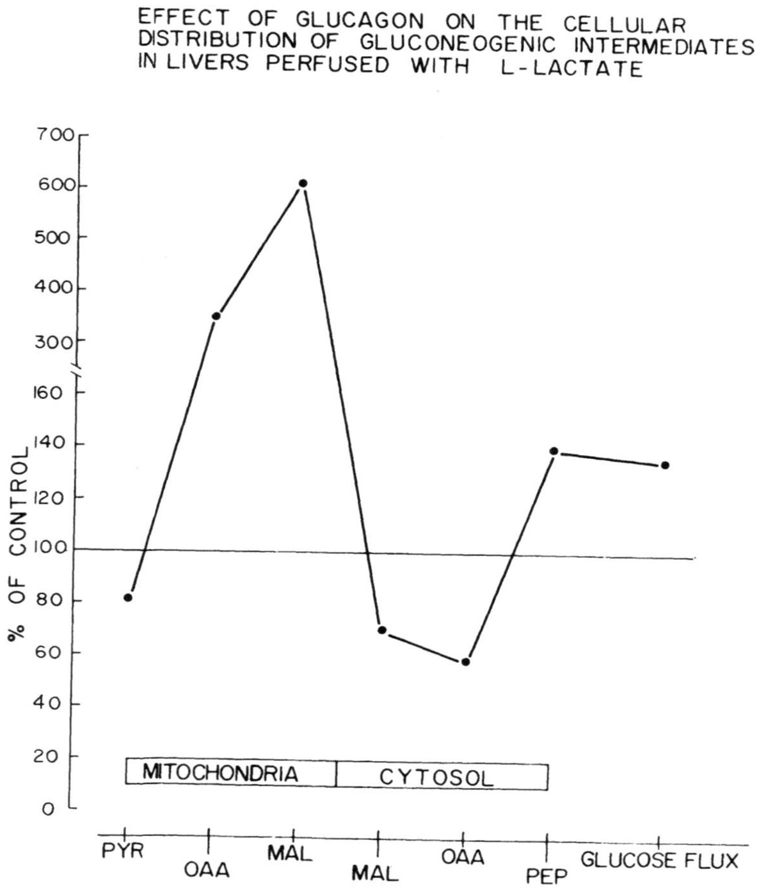

Fig. 3. The calculations for the intracellular distribution of metabolites were described elsewhere (14).

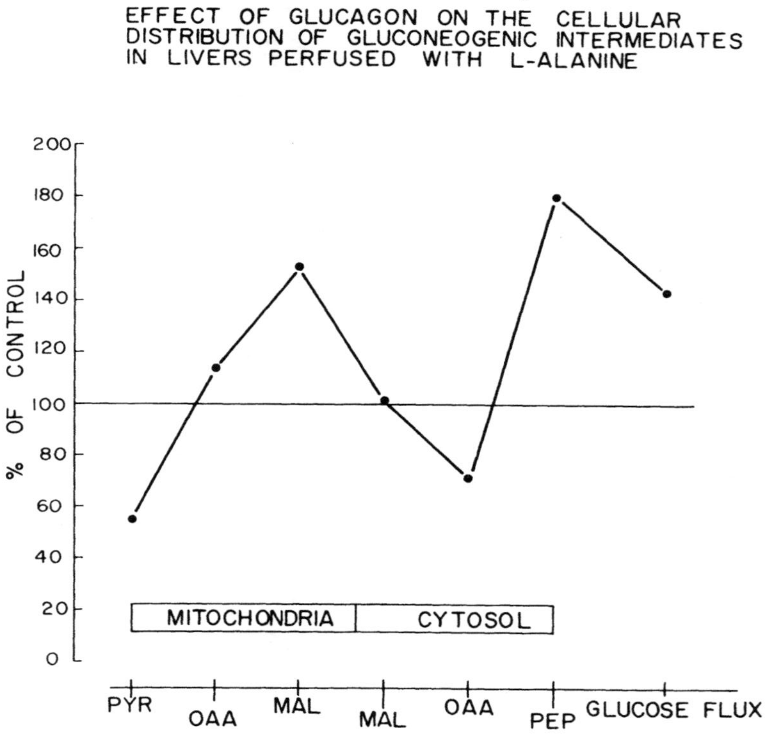

Fig. 4. The calculations for the intracellular distribution of metabolites were described elsewhere (14).

TABLE 4

EFFECTS OF GLUCAGON AND OF GLUCAGON PLUS INSULIN ON THE STEADY STATE CONCENTRATION OF TRICARBOXYLIC ACID CYCLE INTERMEDIARY METABOLITES AND CoA DERIVATIVES IN LIVERS PERFUSED WITH 10 mM L-ALANINE AS SUBSTRATE

	Control livers	Glucagon treated livers	Glucagon + Insulin treated livers
		nmoles/g dry weight	
Malate	616 ± 65	697 ± 33	625 ± 65
Oxaloacetate	18.1 ± 1.7	13.2 ± 0.8***	18.3 ± 1.1
Citrate	1227 ± 181	1363 ± 117	1281 ± 156
Isocitrate	36 ± 4	39 ± 4	41 ± 3.5
α-Ketoglutarate	529 ± 30	460 ± 35**	614 ± 58**
Glutamate	10172 ± 960	12090 ± 895**	11071 ± 1171
Aspartate	13075 ± 1290	14682 ± 2223*	10475 ± 549
CoA	418 ± 43	330 ± 14**	333 ± 22
Acetyl-CoA	348 ± 22	336 ± 17	326 ± 35
Total acid-soluble CoA	1350 ± 109	1340 ± 179	1386 ± 123

The conditions of the experiment were the same as described in Table 2. The results are means of at least eight experiments ± S. E. *P <0.1; **P <0.05; ***P <0.01, by t test.

The adenylate kinase mass action ratio was remarkably constant in all the situations studied and its calculated value was very close to the equilibrium constant determined in vitro (3).

Glucagon stimulated oxygen utilization and the presence of insulin reversed this effect also. The increment in oxygen uptake accounts for the extra energy consumed for the increased glucose and urea production, assuming a P:O ratio of 3 (*Tables 1, 2, 5 and 6*).

TABLE 5

EFFECTS OF GLUCAGON AND OF GLUCAGON PLUS INSULIN ON THE CONTENT OF ADENINE NUCLEOTIDES AND OXYGEN CONSUMPTION OF LIVERS PERFUSED WITH 10 mM L-LACTATE AS SUBSTRATE

	Control livers	Glucagon treated livers	Glucagon + Insulin treated livers
		nmoles/g dry weight	
ATP	7740 ± 628	7422 ± 390	8287 ± 562
ADP	2333 ± 153	2899 ± 218*	2622 ± 110
AMP	1595 ± 72	1790 ± 82	1804 ± 156
Total adenine nucleotides	11668 ± 698	12123 ± 494	12172 ± 174
[ATP]/[ADP]	3.36 ± 0.3	2.6 ± 0.2*	3.2 ± 0.3
$[ATP][AMP]/[ADP]^2$	2.10 ± 0.15	1.87 ± 0.24	2.01 ± 0.2
Δ Oxygen uptake (μatoms/100 g body wt/h)	276	339	300

The conditions of the experiment were the same as described in Table 1. The results are means of at least eight experiments ± S. E. *$P < 0.05$, by t test.

Cellular distribution of metabolites. The aforementioned effects of glucagon were accompanied by important changes in the intracellular distribution of intermediary metabolites. Tables 7 and 8 show how most of the metabolites, whose distribution was calculated, were redistributed after glucagon treatment, so that their mitochondrial:cytosolic concentration gradients were increased. The only exceptions were citrate and isocitrate in livers perfused with L-alanine. As expected, insulin also reversed all these changes almost completely. The changes in the intracellular distribution of oxaloacetate during glucagon stimulation of gluconeogenesis deserve particular attention. During gluconeogenesis from lactate the carbon actually seems to leave the mitochondria as aspartate (1, 16), however,

TABLE 6

EFFECTS OF GLUCAGON AND OF GLUCAGON PLUS INSULIN ON THE CONTENT OF ADENINE NUCLEOTIDES AND OXYGEN CONSUMPTION OF LIVERS PERFUSED WITH 10 mM L-ALANINE AS SUBSTRATE

	Control livers	Glucagon treated livers	Glucagon + Insulin treated livers
		nmoles/g dry weight	
ATP	8418 ± 65	6231 ± 340*	7224 ± 528
ADP	2746 ± 99	2612 ± 235	3407 ± 528
AMP	1599 ± 185	1687 ± 214	1634 ± 146
Total adenine nucleotides	12763 ± 632	10530 ± 583	12265 ± 562
[ATP]/[ADP]	2.93 ± 0.4	2.51 ± 0.2*	1.35 ± 0.2
$[ATP][AMP]/[ADP]^2$	1.7 ± 0.2	1.7 ± 0.3	1.3 ± 0.2
Δ Oxygen uptake (μatoms/100 g body wt/h)	230	316	235

The conditions of the experiment were the same as described in Table 2. The results are means of at least eight experiments ± S. E. *P <0.01 by t test.

for comparative purposes, this representation may be adequate. The most striking finding here is that in both situations, regardless of the source of carbon being utilized, glucagon produced a rise in the intramitochondrial concentration of oxaloacetate. This finding, in conjunction with a decrease in pyruvate levels, suggests that glucagon administration induced an activation of pyruvate carboxylase, the first non-equilibrium step in the gluconeogenic pathway. A decrease in the cytosolic concentration of oxaloacetate in conjunction with a rise in phosphoenolpyruvate levels is also compatible with the glucagon activation of this enzymatic step.

TABLE 7

EFFECTS OF GLUCAGON AND OF GLUCAGON PLUS INSULIN ON THE MITOCHONDRIAL:CYTOSOLIC CONCENTRATION GRADIENTS OF INTERMEDIARY METABOLITES OF LIVERS PERFUSED WITH 10 mM L-LACTATE

	Control livers	Glucagon treated livers	Glucagon + Insulin treated livers
Malate	0.78	6.96	4.32
Oxaloacetate	0.01	0.061	0.036
Citrate	2.16	4.71	5.36
Isocitrate	1.23	6.17	2.39
α-Ketoglutarate	0.013	0.054	0.022
Glutamate	0.22	1.79	0.53
Aspartate	0.17	2.07	0.88

Metabolite gradients have been calculated using their mM concentrations.

TABLE 8

EFFECTS OF GLUCAGON AND OF GLUCAGON PLUS INSULIN ON THE MITOCHONDRIAL:CYTOSOLIC CONCENTRATION GRADIENTS OF INTERMEDIARY METABOLITES OF LIVERS PERFUSED WITH L-ALANINE

	Control livers	Glucagon treated livers	Glucagon + Insulin treated livers
Malate	1.91	2.78	2.00
Oxaloacetate	0.039	0.065	0.037
Citrate	29.93	17.14	18.42
Isocitrate	14.79	6.94	9.55
α-Ketoglutarate	0.032	0.19	0.032
Glutamate	0.40	1.98	0.50
Aspartate	0.53	1.0	0.57

Metabolite gradients have been calculated using their mM concentrations.

Glucagon acceleration of amino acids transport and its relationship to the stimulating effect on the gluconeogenic flux. As can be seen in Table 9, glucagon treatment produced an enhancement of the alanine uptake by the liver cells, as previously reported (11). This effect was not present when insulin was also supplied to the liver. Thus, it is tempting to correlate the activation of gluconeogenesis with the increased supply of fuel to the gluconeogenic machinery. However, as can be seen in Table 10, this may not be the case. The stimulation of gluconeogenesis by L-alanine was accompanied by an elevation of the steady state levels of pyruvate. Since pyruvate carboxylase responds readily to variations in the concentration of this ketoacid within the physiologic range, these results suggest that the stimulation of gluconeogenesis by L-alanine was the result of a rise in the steady state levels of pyruvate. In contrast with these results, the addition of glucagon was accompanied by a significant decrease in the concentration of pyruvate (Table 8), suggesting that, at least under our experimental conditions, factors other than the stimulation of amino acid transport must be sought in order to explain the stimulation of gluconeogenesis. This point of view seems to be supported also by the evidence shown in Fig. 5 which indicates that in rat liver, biosynthetic processes are virtually saturated at a substrate concentration of L-alanine above 2 mM. This implies that the observed effect of glucagon in raising the intracellular concentration of alanine from 8.5 to 10 mM cannot possibly affect the rate of glucose synthesis.

DISCUSSION

Glucagon interaction with the gluconeogenic pathway. The experiments reported here, in agreement with previous publications (4, 14, 15, 27), seems to indicate the existence of two main sites of glucagon interaction with the gluconeogenic pathway; between pyruvate and phosphoenolpyruvate and at the level of the reactions catalyzed by phosphofructokinase and fructose diphosphatase. The second site would explain the observation that glucagon stimulates gluconeo-

TABLE 9

EFFECT OF GLUCAGON ON THE HEPATIC STEADY STATE CONCENTRATION OF SOME KETOACIDS DURING GLUCONEOGENESIS FROM L-ALANINE

Additions to the perfusate	None	10 mM L-alanine	10 mM L-alanine + glucagon
		nmoles/g dry wt	
Pyruvate	106 ± 12	788 ± 85	445 ± 48
α-Ketoglutarate	1150 ± 160	529 ± 30	460 ± 35
Oxaloacetate	13 ± 3	18 ± 1.7	13 ± 0.8
Glucose production μmoles/100 g body wt/h	12 ± 0.9	67 ± 4	95 ± 3

The experimental conditions were as described in Table 2. Oxaloacetate was calculated from the aspartate aminotransferase equilibrium reaction. Values are means of at least eight experiments ± S. E.

TABLE 10

EFFECT OF GLUCAGON ON THE INTRACELLULAR CONCENTRATION OF L-ALANINE IN ISOLATED LIVERS PERFUSED WITH L-ALANINE AS SUBSTRATE

Additions to the perfusate	L-Alanine Concentration Intracellular	Perfusion Fluid	Ratio IN/OUT
	mM		
L-Alanine (10 mM)	8.5	3.5	2.4
L-Alanine (10 mM) + Glucagon	10.3	3.2	3.2

The experiment was performed as described in Table 2. For the calculation of the hepatic intracellular concentration of L-alanine, the extracellular water content was taken to be 0.2 ml/g wet weight (11) and the ratio dry weight/wet weight was taken to be 4.94 ± 0.06. Mean of 87 determinations ± S. E.

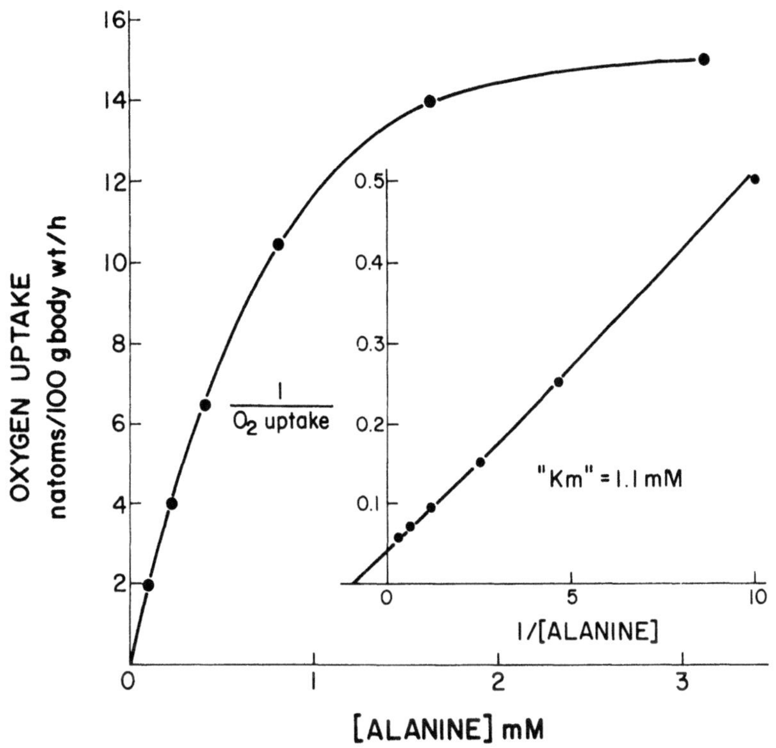

Fig. 5. The experiment was carried out with a non-recirculating perfusion system. L-alanine was added to the perfusion medium as to reach the indicated concentrations. Oxygen concentration in the liver outflow was determined with a Clark type oxygen electrode and continuously recorded. The insert shows a double reciprocal plot of the data.

genesis from fructose (25). However, on the basis of the known kinetic properties of both enzymes, it is difficult to ascertain how glucagon may act at this site. The observed effect is compatible with a phosphofructokinase inhibition and/or a fructose diphosphatase activation. Phosphofructokinase is inhibited by ATP and citrate (19), but glucagon did not affect the concentration of these metabolites sufficiently to affect this enzymatic step (Tables 5 and 8). On the other hand, fructose diphosphatase is known to be inhibited by AMP (23) and glucagon increased the levels of this nucleotide (Tables 5 and 6). Thus, at present, the glucagon effect cannot be explained on the basis of the known kinetics of the isolated purified enzymes. The possibility cannot be excluded that glucagon may have changed the kinetic behaviour of these enzymes or, perhaps, that it may have produced changes in enzyme effectors at sites which are not reflected in their total hepatic content.

The other apparent site of interaction of glucagon with the gluconeogenic pathway, between pyruvate and phosphoenolpyruvate, involves two main non-equilibrium enzymatic steps: pyruvate carboxylation and phophoenolpyruvate synthesis. In the rat, the former is an intramitochondrial reaction (21, 24) while the latter is a virtually exclusive cytosolic step (21). Since the mitochondrial membrane is almost impermeable to oxaloacetate, the carbon is transported to the cytosol as malate or aspartate (Figs. 6 and 7), depending on the degree of reduction of the gluconeogenic precursor (1, 13). A glucagon effect that decreases total oxaloacetate content seems to be in conflict with the increase in flux at either of these two non-equilibrium steps. A decrease in oxaloacetate in conjunction with a rise in phosphoenolpyruvate is compatible with an activation of phosphoenolpyruvate carboxykinase. Actually, unless the enzyme is activated it would be difficult to explain an increase in flux through this step, since the cytosolic oxaloacetate concentration is below the Km of phosphoenolpyruvate carboxykinase for this metabolite and the observed glucagon effect of decreasing its concentration (Tables 3 and 4) should result in a decreased rather than in the observed increased flux. On the other hand, an increase in flux through phosphoenolpyruvate carboxykinase cannot be clearly understood if the prior non-equilibrium step, pyruvate carboxylation, was not simultaneously activated; however, the observed metabolite patterns are not compatible with such an activation (Figs. 1 and 2). The calculation of the intracellular metabolite distribution was found to be extremely helpful in understanding this apparent contradiction. When the concentration of intermediary metabolites was determined at each of the compartments where the two enzymes are located (Figs. 3 and 4), it was found that the glucagon effect, regardless of the carbon source utilized, was accompanied by an

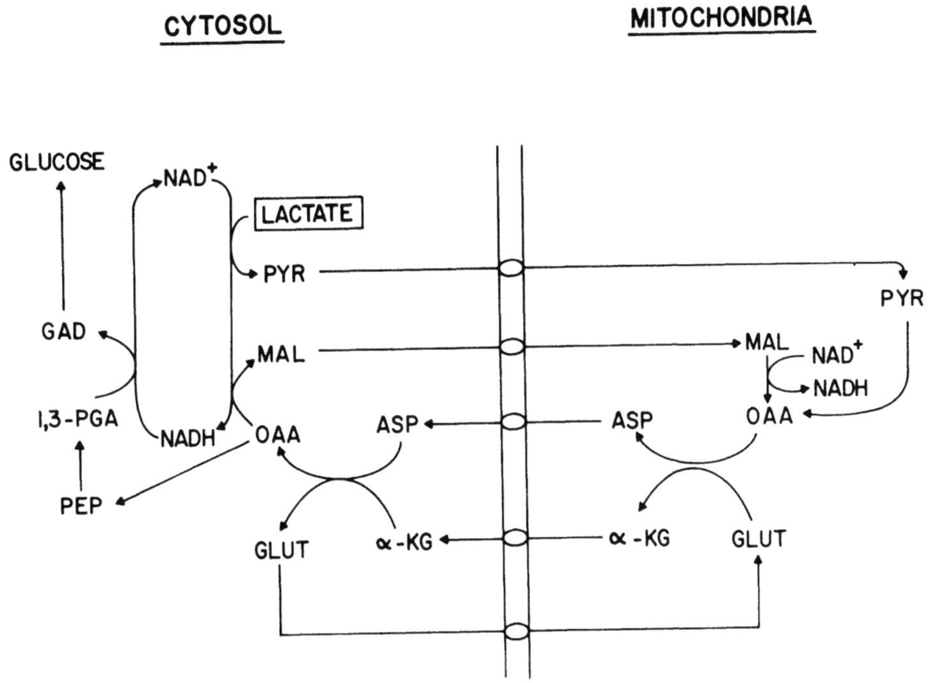

Fig. 6. Schematic representation of flow of intermediates across the hepatic mitochondrial membrane during gluconeogenesis from L-lactate.

increased mitochondrial concentration of oxaloacetate. This increase is compatible with an activation of pyruvate carboxylase, the first non-equilibrium step of the gluconeogenic path. Apparently, glucagon seems to have multiple sites of action on this pathway. An activation of pyruvate carboxylase, not clearly demonstrated before, is very important for a better understanding of the glucagon mechanism of activation of the gluconeogenic flux. The question of how pyruvate carboxylase was activated by glucagon does not seem to be easy to explain. Acetyl-CoA is a positive modulator of this enzyme (22, 27), but glucagon did not alter its concentration (Tables 3 and 4). Glucagon probably acts by somehow increasing acetyl CoA at the enzyme site, removing some inhibitor or perhaps by raising the concentration of some other positive modulator not yet identified.

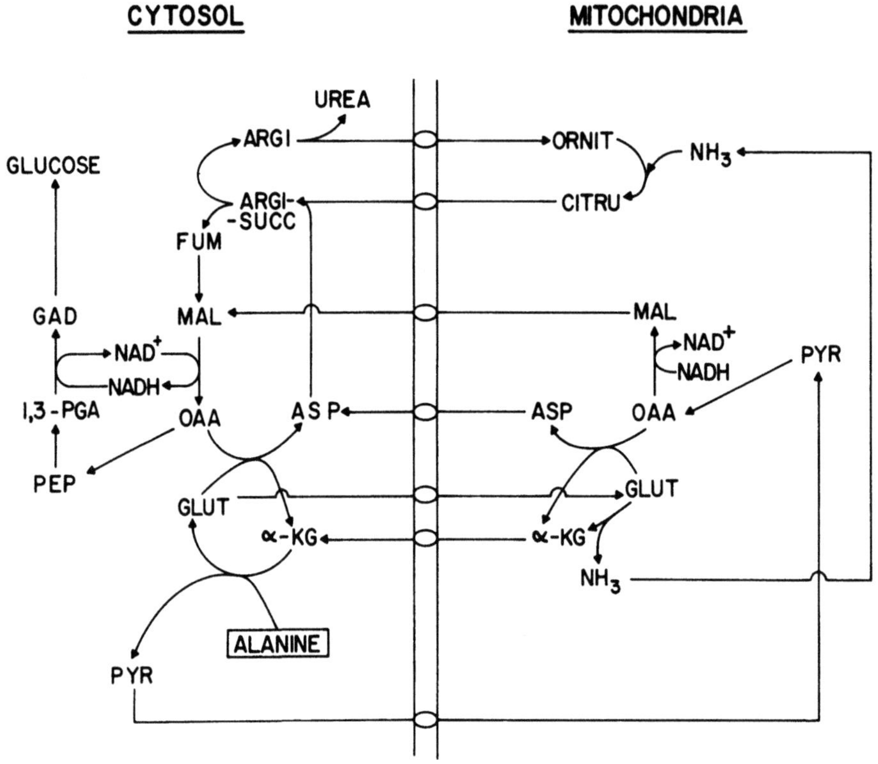

Fig. 7. Schematic representation of flow of intermediates across the hepatic mitochondrial membrane during gluconeogenesis from L-alanine.

The effect of glucagon was accompanied by drastic changes in the cellular distribution of intermediary metabolites leading to a rise in their mitochondrial: cytosolic gradients (Tables 7 and 8). As seen in the schemes of Figs. 6 and 7, in the case of gluconeogenesis from L-lactate the mitochondrial oxaloacetate leaves the mitochondria as aspartate while in the case of L-alanine most probably part of the carbon is transferred as malate and part as aspartate. At present the real driving force for the translocation of these metabolites is not clearly understood, however, a rise in their gradients may account for the observed increased rate of carbon transfer across the mitochondrial membrane. Since glucagon shifts the nicotinamide nucleotide couple to a more reduced state, as judged by the rise in the ratios lactate to pyruvate, α-glycerophosphate to dihydroxyacetone phosphate, citrate to α-ketoglutarate and glutamate to α-ketoglutarate, among others (Figs. 1 and 2; Tables 3 and 4), the cellular redistribution of intermediary metabolites may be the consequence of the increased cellular state of reduction, in an attempt to maintain the redox gradient across both cellular compartments.

The "Km" for alanine utilization by rat liver (Fig. 5) was found to be very close to the normal plasma L-alanine concentration (17) which explains why variations in plasma L-alanine levels within the physiologic range could affect the hepatic gluconeogenic activity. Nevertheless under our experimental conditions, glucagon stimulation of gluconeogenesis coexisted with oversaturated intracellular concentration of L-alanine indicating that both effects of glucagon: increased amino acid transport and stimulation of the gluconeogenic pathway, are fully independent of one another. However, one cannot exclude the possibility that under certain physiologic situations characterized by decreased plasma L-alanine levels, like starvation (17), glucagon may play a role also through an increase in L-alanine transport.

Glucagon-insulin interaction. Antagonistic actions of glucagon and insulin on several metabolic parameters, have been described by several authors. However, the importance of the relative concentration of both hormones for hepatic metabolism has been recognized only very recently (13-15, 18). It is remarkable that glucagon to insulin ratios such as those usually found in normal fed animals completely reversed all the metabolic effects of glucagon. This implies that no conclusions about the physiologic action of these hormones can be drawn unless their combined effect is considered.

The striking effect of insulin in reversing virtually all the metabolic changes induced by glucagon seems to indicate that both hormones share a common step which is the driving force for all these changes.

Acknowledgements

The authors acknowledge the technical assistance of Mr. T. Fontela and Mr. F. Martín.

This work has been supported in part by grants from Comisión Asesora para el Desarrollo de la Investigación, Lilly Indiana de España and Essex (España) S. A.

REFERENCES

1. Anderson, J. H., Nicklas, W. J., Blank, B., Refino, C. and Williamson, J. R. (1971): Transfer of carbon and hydrogen across the mitochondrial membrane in the control of gluconeogenesis. In: Regulation of Gluconeogenesis. H. D. Söling and B. Willms, Eds. Academic Press, New York, pp. 293-313.

2. Bergmeyer, H. U.. (1975): Methods of Enzymatic Analysis. Academic Press, New York.

3. Eggleston, L. V. and Hems, R. (1952): Separation of adenosine phosphates by paper chromatography and the equilibrium constant of the myokinase system. Biochem. J. 52, 156-160.

4. Exton, J. H., Jefferson, L. S., Butcher, K. W. and Park, C. R. (1966): Gluconeogenesis in the perfused liver. The effects of fasting, alloxan diabetes, glucagon, epinephrine, adenosine 3',5'-monophosphate and insulin. Amer. J. Physiol. 40, 709-715.

5. Exton, J. H. and Park, C. R. (1969): Control of gluconeogenesis in liver. III. Effects of L-lactate, pyruvate, fructose, glucagon, epinephrine, and adenosine 3',5-monophosphate on gluconeogenic intermediates in the perfused rat liver. J. Biol. Chem. 244, 1424-1433.

6. Garcia, A., Williamson, J. R. and Cahill, G. F., Jr. (1966): Studies on the perfused rat liver. II. Effect of glucagon on gluconeogenesis. Diabetes 15, 188-193.

7. Goodman, M. N., Parrilla, R. and Toews, C. J. (1973): Influence of fluorocarbon emulsions on hepatic metabolism in perfused rat liver. Amer. J. Physiol. 225, 1384-1388.

8. Kerstens, P. J. (1969): La perfusion du foie isolée. Bruxelles, Editions Arscia, pp. 26-38.

9. Krebs, H. A. and Henseleit, K. (1932): Untersuchungen uber die Harnstoffbildung im Tierkoerper. Z. Physiol. Chem. 210, 33-66.

10. Mallette, L. E., Exton, J. H. and Park, C. R. (1969): Control of gluconeogenesis from amino acids in the perfused rat liver. J. Biol. Chem. 244, 5713-5723.

11. Mallette, L. E., Exton, J. H. and Park, C. R. (1969): Effects of glucagon on amino acids transport and utilization in the perfused rat liver. J. Biol. Chem. 244, 5724-5728.

12. Miller, L. L. (1960): Glucagon; a protein catabolic hormone in the isolated perfused rat liver. Nature 185, 248.

13. Parrilla, R., Goodman, M. N. and Toews, C. J. (1974): Effects of glucagon: insulin ratios on hepatic metabolism. Diabetes 23, 725-731.

14. Parrilla, R., Jimenez, M. I. and Ayuso-Parrilla, M. S. (1975): Glucagon and insulin control of gluconeogenesis in the perfused isolated rat liver. Effects on cellular metabolite distribution. Eur. J. Biochem. 56, 375-383.

15. Parrilla, R., Jimenez, M. I. and Ayuso-Parrilla, M. S. (1976): Cellular redistribution of metabolites during glucagon and insulin control of gluconeogenesis in the isolated perfused rat liver. Arch. Biochem. Biophys. 174, 1-12.

16. Parrilla, R., Ohkawa, K., Lindros, K. O., Zimmerman, U. P., Kobayashi, K. and Williamson, J. R. (1974): Functional compartmentation of acetaldehyde oxidation in rat liver. J. Biol. Chem. 249, 4926-4933.

17. Parrilla, R. and Toews, C. J. (1976): The effect of starvation in the rat on metabolite concentrations in blood, liver, skeletal muscle and kidney. Amer. J. Physiol. In press.

18. Parrilla, R., Toews, C. J. and Goodman, M. N. (1972): The influence of the glucagon:insulin ratio on hepatic metabolism. Diabetes 21, suppl. 1, 52.

19. Passonneau, J. V. and Lowry, O. H. (1964): The role of phosphofructokinase in metabolic regulation. Adv. Enzyme Regul. 2, 265-274.

20. Ross, B. D., Hems, R. and Krebs, H. A. (1967): The rate of gluconeogenesis from various precursors in the perfused rat liver. Biochem. J. 102, 942-951.

21. Scrutton, M. C. and Utter, M. F. (1968): The regulation of glycolysis and gluconeogenesis in animal tissues. Ann. Rev. Biochem. 37, 249-302.

22. Struck, E., Ashmore, J. and Wieland, O. (1965): Stimulierung der Gluconeogenese durch langkettige Fettsäuren und Glukagon. Biochem. Z. 343, 107-110.

23. Taketa, K. and Pogell, B. H. (1965): Allosteric inhibition of rat liver fructose 1,6 diphosphatase by adenosine 5'-monophosphate. J. Biol. Chem. 240, 651-662.

24. Utter, M. F.and Scrutton, M. C. (1969): Pyruvate carboxylase. In: Current Topics in Cellular Regulation. B. L. Hörecker and E. R. Stadtman, Eds., Academic Press, New York, pp. 253-296.

25. Veneziale, C. M. (1971): Gluconeogenesis from fructose in isolated rat liver stimulation by glucagon. Biochem. 10, 3443-3447.

26. Williamson, J. R. (1969): Role of lipolysis in the gluconeogenic action of glucagon. In: Metabolic Regulation and Enzyme Action. A. Sols and S. Grisolia, Eds. Academic Press, New York, pp. 107-113.

27. Williamson, J. R., Browning, E. T., Thurman, R. G. and Scholz, R. (1969): Inhibition of glucagon effects in perfused rat liver by (+) decanoylcarnitine. J. Biol. Chem. 244, 5055-5064.

28. Williamson, J. R. and Corkey, B. E. (1968): Assays of intermediates of the citric acid cycle and related compounds by fluorometric enzyme methods. In: Methods in Enzymology, Vol. 13. J. M. Lowenstein, Ed. Academic Press, New York, pp. 434.

29. Wollemberger, A., Ristau, O. and Schoffa, G. (1970): Eine einfache Technik der extrem schnellen Abkülung grösserer Gewebestücke. Pflüger's Arch. ges. Physiol. 270, 399-412.

HORMONAL REGULATION OF HEPATIC PROTEIN METABOLISM IN THE RAT IN VIVO. CONTROL BY GLUCAGON

R. Parrilla, A. Martin-Requero, J. Perez-Diaz, and M. S. Ayuso-Parrilla

Department of Metabolism, Instituto G. Marañón, C.S.I.C.
Velazquez 144, Madrid-6, Spain

The ureogenic effect of glucagon in the rat *in vivo* was accompanied by a decrease in the rate of hepatic protein synthesis and by a significant and progressive increase of proteolysis, expressed as a rise in hepatic valine levels. The effect on protein synthesis was maximal five minutes after hormone administration and decreased progressively thereafter to reach control values after 20 minutes. On the contrary, ureogenesis and proteolysis increased progressively throughout the 20 minutes of the experiment suggesting a close relationship between the two processes.

The effect of glucagon in decreasing hepatic protein synthesis is localized at the level of the elongation and/or the termination step of polypeptide synthesis. Postmitochondrial supernatants from livers of control and glucagon-treated rats have a similar activity in incorporating radioactive precursors and therefore, we conclude that the hormone does not act directly on the components of the protein synthesis machinery. A close correlation has been observed between the decrease in the rate of protein synthesis, the decrease in the hepatic content of adenine nucleotides and the shift of the nicotinamide nucleotide system to a more reduced state. It is suggested that the effect of the hormone is due to the decreased availability of high energy phosphorylated compounds. Since the effect of glucagon on lipid mobilization was similar to the observed effects on protein synthesis, it seems logical to conclude that the primary effect of glucagon is to increase the supply of fatty acids to the liver. The oxidation of these fatty acids would bring about a shift in the [NADH]/[NAD$^+$] ratio to a more reduced state resulting in a decrease in the [ATP]/[ADP] ratio through the cytosolic equilibrium reactions to which the two both types of nucleotides are linked.

The idea that glucagon could operate as a catabolic hormone was initially suggested by Tybergheim (39) who observed an increase in plasma urea levels following the administration of glucagon, *in vivo*. This observation was corroborated by Miller (22) who, using an isolated rat liver preparation, was able to demonstrate a direct stimulation of urea production by glucagon. Based on these observations, it was postulated that glucagon could have an important regulatory role on protein catabolism *in vivo*. This point of view was supported further by observations on the ureogenic (29, 34, 41) as well as proteolytic effects of the hormone (5, 20, 45).

The mechanism by which glucagon accelerates protein catabolism remains obscure. Some authors (5) have suggested a primary effect of glucagon on proteolysis, thus increasing amino acid availability. The observations that glucagon is a potent inducer of autophagic vacuoles in rat liver (2) and that it increases the sensitivity of the lysosomes to osmotic and mechanical shock (9) support this point of view.

The possibility that glucagon would increase amino acid availability through a decrease in the rate of protein synthesis has not been thoroughly explored. It is well known that glucagon stimulates the synthesis of some hepatic enzymes such as tryrosine transaminase, phosphoenolpyruvate carboxykinase and serine dehydratase (13, 16, 42, 46), however, there is little information regarding the effect of glucagon on the synthesis of hepatic proteins in general. Pryor and Berthet (30) observed that glucagon inhibits the incorporation of amino acids into hepatic proteins in rat liver slices but, to our knowledge, no demonstration of an effect in vivo has ever been reported. Recently, the validity of experiments designed to determine the rate of protein synthesis in vivo from labeled precursors has been seriously questioned because the intracellular amino acid distribution does not seem to be homogeneous (1, 12, 24). Thus, it is difficult to determine the specific activity of amino acids serving as precursors for protein synthesis. This situation is further complicated when the effect of a hormone like glucagon is studied because its proteolytic effect may change the intracellular amino acid pool (5, 45).

The experiments reported herein are an attempt to elucidate whether the ureogenic effect of glucagon in vivo is mediated by an increase amino acid supply due to inhibition of protein synthesis, increased proteolysis or a combination of both. For the determination of rates of protein synthesis, in vivo, an approach originally described by Haschemeyer (15) and Mathews et al. (21) and modified by Scornik (33) has been used; it is based on the in vivo determination of the average transit time of a radioactive amino acid along the messenger RNA. Our results seem to demonstrate that the administration of glucagon in vivo had an inhibitory effect on hepatic protein synthesis acting mainly, if not exclusively, on the peptide chain elongation and/or termination steps. This inhibition does not seem to be the result of a direct effect of the hormone on the protein synthetic machinery, but rather to be a consequence of a change in the steady state concentration of the adenine nucleotides subsequent to an increased mobilization and oxidation of fatty acids.

EXPERIMENTAL PROCEDURES

Animals. Male Wistar albino rats, weighing 200 ± 20 g, were used. The animals were fed ad libitum until the experiments were started.

Determination of the incorporation of amino acids into hepatic proteins. The animals were anesthetized with Nembutal® (40 mg/Kg) ten minutes before surgery. In some of them, 200 μg of glucagon dissolved in 0.2 ml of saline were injected intraperitoneally and the portal vein was exposed through a midline abdominal incision for blood sampling, at the times indicated. Control rats received a similar volume of saline solution. In other rats, glucagon was infused into the inferior vena cava at a rate of 26 μg/min for the first five minutes and 10 μg/min thereafter. The controls received a similar volume of saline. Valine was used as a precursor for protein synthesis since it has been shown that this amino acid is not significantly metabolized by the liver (23). Twenty μCi of [H^3] valine dissolved in 0.2 ml of saline were slowly injected (30 seconds) into the portal vein. After the isotope injection, liver biopsies were taken in such a way that only two liver biopsies were taken from each rat (at 0.5 and 1.5, or at 1 and 2 minutes, respectively). In order to prevent hemorrhage, the syringe used for the isotope administration was not removed from the portal vein during the experiment. For the same reason the hepatic lobules were ligated immediately before the biopsies were taken. The liver samples were immediately frozen, using aluminum clamps precooled in a dry ice-methanol mixture, and stored in liquid nitrogen until they were processed.

To determine the incorporation of radioactivity into total protein, the liver biopsies were homogenized with three volumes of 0.3 M sucrose. To 0.02 ml of the homogenate, 3 ml of 10% perchloric acid were added. After centrifugation at 8000 rpm for 10 minutes in a refrigerated centrifuge, the pellet was washed twice with 10% perchloric acid, resuspended in 3 ml of 10% perchloric acid and heated for 30 minutes at 90°C. The precipitate was then washed with a mixture of alcohol, ether and chloroform (2:2:1), with acetone and with ether and stored at room temperature. The dried protein residue was dissolved in 1 ml of hyamine hydroxide and transferred to a vial containing 10 ml of toluene-based scintillation mixture. The radioactivity was determined in a Packard scintillation spectrometer.

The incorporation of radioactive amino acids into nascent chains was determined as follows: 1 ml of homogenate was diluted with 1 ml of 0.1 M Tris-HCl containing 2 mM magnesium acetate, pH 7.6, at 0°C. After centrifugation at 5000 rpm for 15 minutes at 4°C, sodium deoxycholate (2% final concentration) was added to the supernatant fluid and this was placed on top of a tube containing 6 ml of a solution consisting of 1 M sucrose and 1 mM magnesium acetate and was centrifuged at 150,000 x g for 3 h in a Beckman ultracentrifuge, at 0°C. The pellet containing the ribosomes was resuspended in 1 ml of distilled water at 0°C and 4 ml of this suspension were treated as described above for the determination of radioactivity in total proteins.

Determination of the liver polyribosomal profiles. The polyribosomal profiles were obtained from homogenates of biopsies taken as described above and quickly chilled by immersion in cold 50 mM triethanolamine buffer, pH 7.3, containing 5 mM $MgCl_2$, 25 mM KCl and 7 mM mercaptoethanol. Glucagon, when administered, was injected directly into the portal vein immediately after a small liver biopsy was taken to serve as control. The polyribosomes were purified through a discontinuous sucrose gradient and analyzed as described elsewhere (5).

Measurement of the protein synthetic activity in vitro. For these experiments, small liver biopsies were taken from anesthetized rats which had been injected five minutes earlier with either 0.2 ml of saline or 200 μg of glucagon dissolved in 0.2 ml of saline. The biopsies were immediately immersed in ice cold buffer containing: 20 mM Tris-HCl, pH 7.4; 0.3 M sucrose; 20 mM KCl; 3.5 mM $MgCl_2$; 3.5 mM $CaCl_2$ and 0.7 mM EDTA and homogenized with one volume of the same buffer. The homogenates were centrifuged for 15 minutes at 15,000 rpm in a Sorvall refrigerated centrifuge and the postmitochondrial supernatant fluids were filtered through a Sephadex G-25 column (1.5 x 15 cm) equilibrated with the buffer described above, and immediately used for the in vitro experiments since they are inactivated rapidly (31).

The incubation mixture (final volume, 0.5 ml) contained: 2 mM ATP; 0.8 mM GTP; 3 mM magnesium acetate; 150 mM KCl; 5 mM Tris-HCl, pH 7.8; 0.1 mM of all the natural amino acids except valine; 0.02 mM [U-^{14}C] valine (100 μCi/ mole); 2 mM phosphoenolpyruvate; 20 μg/ml of pyruvate kinase and 2-2.5 mg of protein from the postmitochondrial supernatant. After incubation at 37°C, 100 μl aliquots of supernatant were taken at the times indicated. Two ml of 10% trichloroacetic acid and 1 mg of bovine albumin were added to each aliquot. After heating at 90°C for 15 minutes, each sample was filtered through a Millipore membrane and washed with five 2 ml portions of 5% trichloroacetic acid. The filters were dried and placed in vials containing 15 ml of toluene-based scintillation mixture.

Analytical techniques. For the determination of intermediary metabolites, the frozen biopsies were freeze dried and then extracted with 35 volumes of 8% perchloric acid, the pH was adjusted to 6.0 with potassium carbonate and analyzed immediately, fluorometrically or spectrophotometrically, according to the expected concentrations of each metabolite, as described elsewhere (6, 25, 44). For the determination of plasma amino acid content, blood samples were taken with heparinized syringes from the aortic bifurcation. 0.2 ml aliquots from eight plasma samples, were pooled and deproteinized with sulphosalicylic acid. The acid extract was used directly for ion exchange chromatography in a Jeol automatic amino acid analyzer. For the determination of the hepatic amino acid content, 1 ml of the homogenates from six livers were pooled and deproteinized with 6 ml of 12% perchloric acid. After neutralization with 3 M K_2CO_3, the extracts were freeze dried and then stored at room temperature. For the measurement of valine concentrations the dried residues were resuspended in HCl, the pH adjusted to 2.5 and the extracts were chromatographed in a Jeol automatic amino acid analyzer. For the determination of the specific activity of the valine, 100 μl aliquots of the final solutions were placed on Millipore filters, and, after drying, the radioactivity of the filters was determined in a Packard scintillation spectrometer. Urea was measured using the diacetyl monoxime colorimetric method. Proteins were determined by the method of Lowry and plasma free fatty acids by the method of Dole (10).

RESULTS

The variations in plasma urea levels after glucagon administration are shown in Fig. 1. There was a progressive increase in urea levels during the 20 minutes of the experiment. This increase in plasma urea levels was paralleled by an increased availability of amino acids, as indicated by the progressive increase in hepatic valine concentration after glucagon administration (Fig. 2). As valine is an amino acid which is not significantly metabolized by the liver (23), variations in its intracellular concentration can be taken as an index of the net hepatic proteolytic activity (4, 25), provided that a decrease in the extracellular content of the amino acid does not account for its intracellular increase. This seems to be the case as indicated by the magnitude of the variations in total valine plasma concentration (Table 1).

TABLE 1

EFFECT OF GLUCAGON ON THE VALINE CONTENT OF PLASMA AND LIVER TISSUE

Time after glucagon administration (min)	Valine content in: Plasma	Tissue
	μmol/rat	
0	1.0	2.5
20	0.6	8.3
Δ	-0.4	5.8

For these calculations it was assumed that a 200 g rat has 12 ml of circulating blood and 7 g of liver.

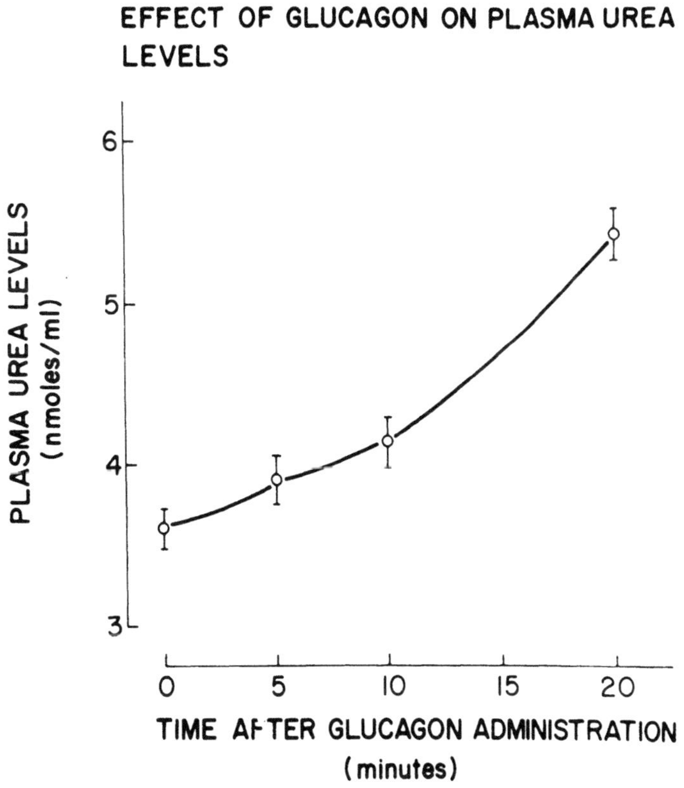

Fig. 1. Glucagon, 200 μg, was injected intraperitoneally as a single dose. The experimental details are described elsewhere (3). Each point represents the mean value of 16 experiments ± S. E.

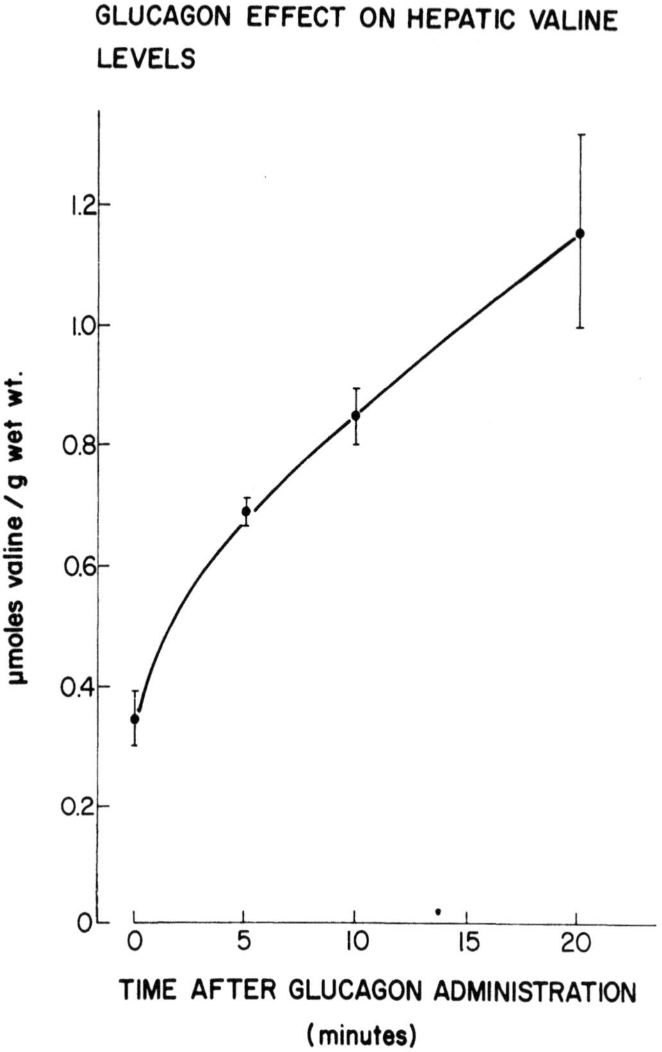

Fig. 2. Glucagon, 200 μg, was injected intraperitoneally as a single dose. The concentration of valine was determined in pools of six liver homogenates. The experimental details are described elsewhere (3). Each point represents the mean value of 4 pools ± S. E.

The effect of glucagon on protein synthesis was studied by determining the polypeptide chains completion time in vivo. This procedure assumes that after its injection, the radioactive precursor amino acid will meet the ribosomes in the middle of the translation of a message of average size. When a full cycle is completed the whole peptide on the ribosomes is completely labeled, while only one-half of the chains that have been terminated and released are labeled. In this way the ratio between the radioactivity incorporated into nascent peptides in the polyribosomes (n) and the radioactivity incorporated into total peptides (t) will be reduced by 50%. The time required to reach this 50% value is taken as the average polypeptide chains completion time (3, 33). The validity of this approach is largely based on the assumption that the specific activity of the precursor does not change during the two minutes of the experiment. That this was actually the case under our experimental conditions has been demonstrated by determining the specific activity of acid soluble valine at each experimental time. Table 2 shows the results of these determinations carried out in five different pools, each made up of aliquots from the perchloric acid supernatants of six livers. There were no statistically significant variations in the specific activity of valine during the time of the experiment.

TABLE 2

TIME COURSE OF THE SPECIFIC ACTIVITY OF VALINE

Time after valine administration (min)	Valine specific activity cpm/nmol
0.5	40 ± 7
1.0	33 + 3
1.5	35 ± 7
2.0	37 ± 16

The specific activity of valine was determined chromatographically using a Jeol automatic amino acid analyzer. The results are the means of five determinations carried out in pools made with aliquots from the acid soluble supernatants of six different livers, ± S.E.

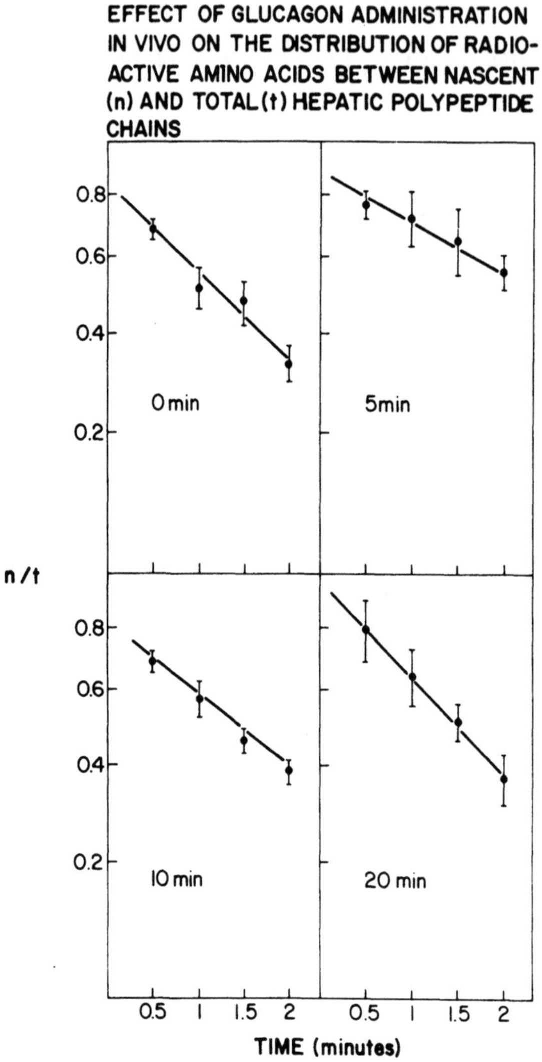

Fig. 3. Glucagon, 200 μg, was injected intraperitoneally as a single dose at the indicated times before the injection of the radioactive amino acid. The experimental procedure was described elsewhere (3). Each point represents the average value of 10-14 experiments ± S. E.

Figure 3 shows the rate of decay of the hepatic n/t value at different times after glucagon administration. As can be seen, the slope of the decay at 5 minutes was lower than in control livers, while 20 minutes after glucagon administration, the slope was similar to that of the control. These changes in the slope, 5 minutes after glucagon administration, indicate a slower ribosomal transit time along the messenger RNA. Figure 4 shows the values of the polypeptide chains completion time calculated using the results from the experiments reported in Fig. 3. The completion time in livers from control rats was 90 seconds. Five minutes after glucagon administration this value rose to 146 seconds and 20 minutes after the hormonal treatment it had returned to the control value.

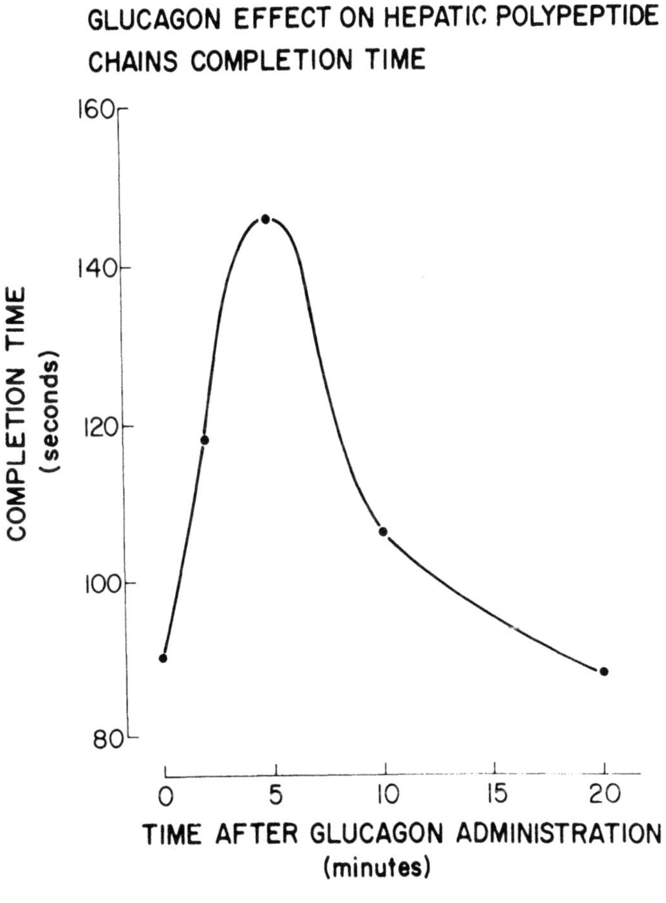

Fig. 4. The polypeptide chains completion time was calculated from the data of Fig. 3. The details of the experimental procedure are described elsewhere (3).

The transient effect of glucagon may be related to its rapid destruction by the organism. This point of view was confirmed by experiments in which glucagon was administered to the animal by continuous intravascular infusion. Under these conditions (Fig. 5), the effect of glucagon on the decay of the hepatic n/t value was sustained. Similar results were obtained whether glucagon was administered by intravascular or intraperitoneal infusion. The completion time calculated from these data (Fig. 6) was found to be 130 seconds, that is 40% higher than in livers from control animals.

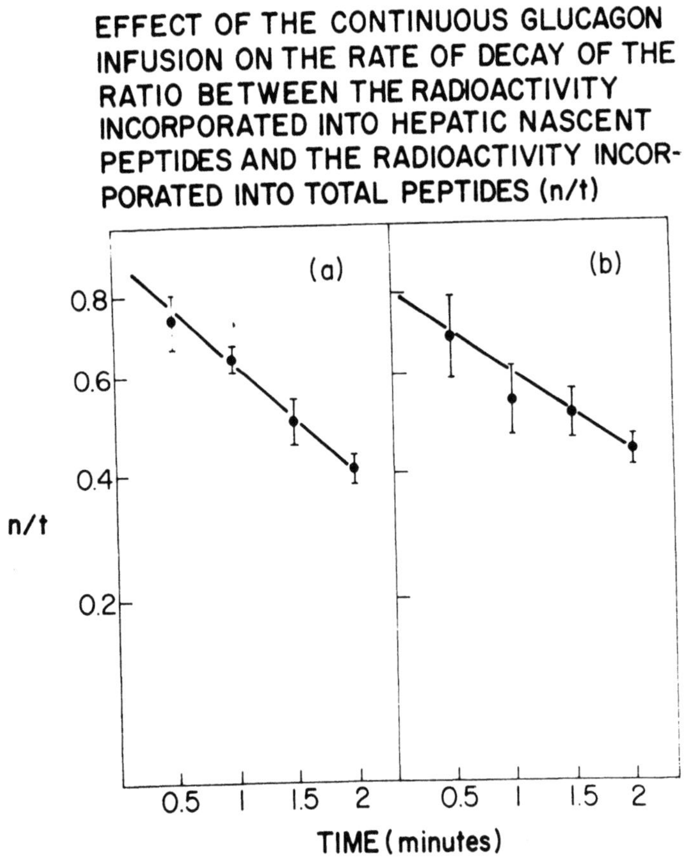

Fig. 5. Glucagon was infused continuously into the inferior vena cava of anesthetized animals. Further details of the experimental procedure are described elsewhere (3). a) 20 minutes of saline infusion; b) 20 minutes of glucagon infusion. Each point represents the average value of at least 8 experiments ± S. E.

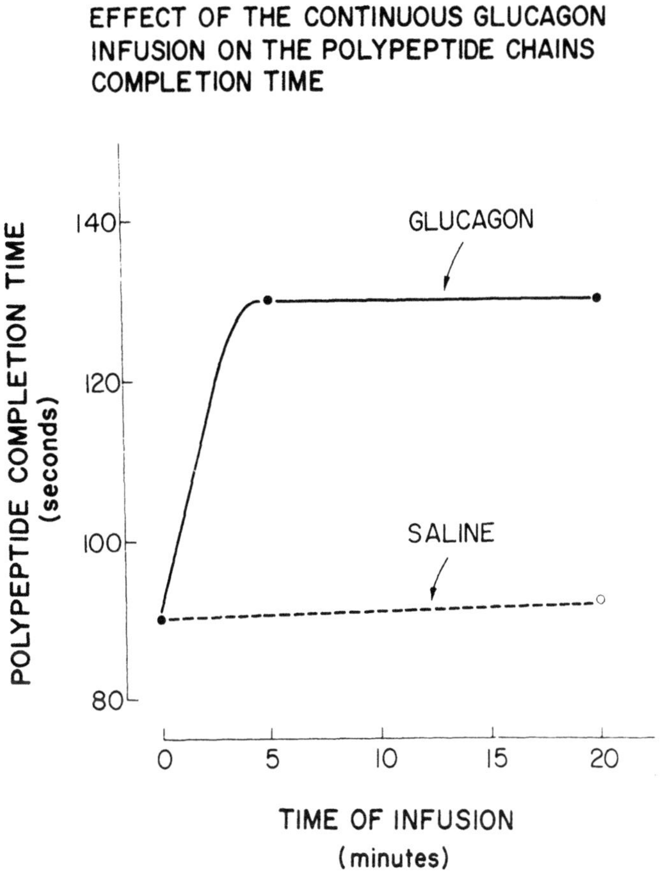

Fig. 6. The polypeptide chain completion times were calculated from the data in Fig. 5, as described elsewhere (3).

The ribosomal transit time reflects the rate of the peptide chain elongation step exclusively, since it is defined as the time needed by a ribosome to translate a messenger RNA of average size. It is then independent of the number of ribosomes engaged in the process, and thus independent of the initiation step of protein synthesis. In order to elucidate whether the observed inhibition of the elongation step was also accompanied by any significant effect on the initiation step, the hepatic polyribosomal profiles from control and glucagon-treated rats were studied. As shown in Fig. 7, five minutes after glucagon treatment the

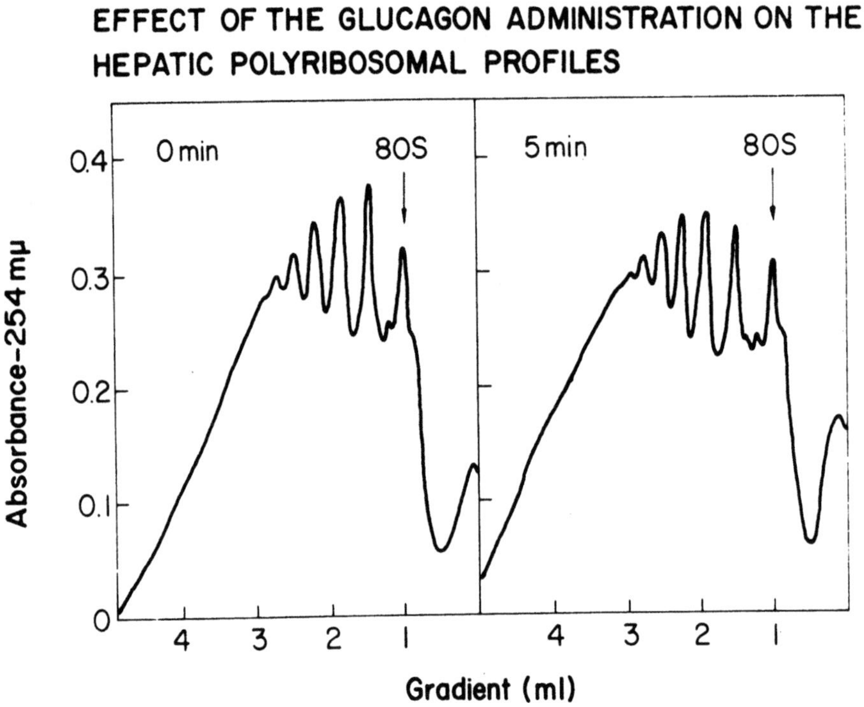

Fig. 7. Control and glucagon-treated liver biopsies were taken from the same animal, as described in Experimental Procedures. The experiment has been repeated 6 times.

hepatic polyribosomes displayed a higher state of aggregation. This finding is further emphasized in Table 3, in which it can be seen that the ratio of the area under the polyribosomes peaks to the area under monomers plus dimers peaks, that is the polyribosomal fraction or the ribosomal state of aggregation, was increased by 20% five minutes after glucagon administration. Since a block in the initiation step of polypeptide synthesis equal or greater than the inhibition of the elongation step would result in an unaltered or a broken polyribosomal profile, respectively (8, 18), these results also support the aforementioned point of view, namely, that glucagon acts preferentially, if not exclusively on the elongation and/or the termination step of polypeptide synthesis.

TABLE 3

EFFECT OF GLUCAGON ON THE HEPATIC POLYRIBOSOMAL FRACTION

Time after glucagon administration (min)	Polyribosomes / Monomers + Dimers	Percentage of control
0	2.2 ± 0.09	---
5	2.6 ± 0.10	119

The hepatic polyribosomal fraction was calculated by dividing the area under the polyribosome peaks by the area under the monomer plus dimer peaks. Each value represents the mean of five experiments ± S.E. Further details are described elsewhere (3).

AMINO ACIDS INCORPORATION INTO PROTEIN BY HEPATIC POSTMITOCHONDRIAL SUPERNATANTS FROM CONTROL AND GLUCAGON TREATED RATS

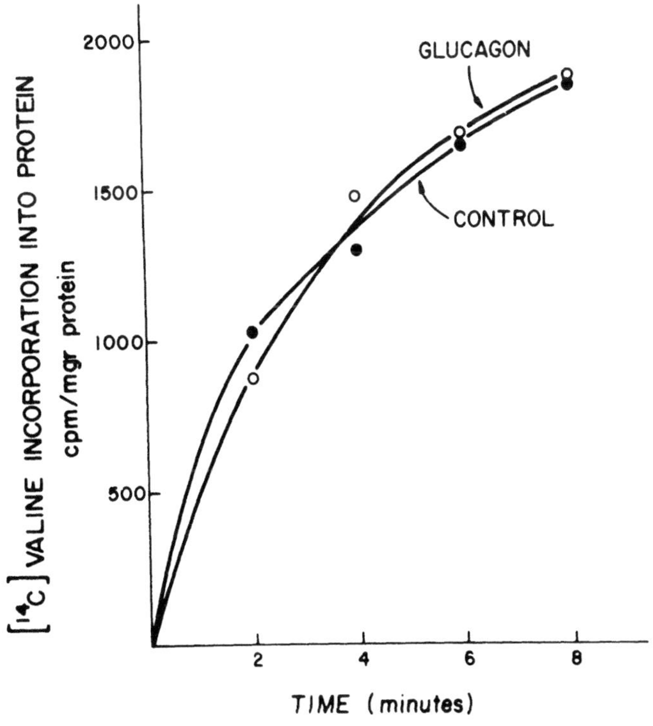

Fig. 8. Glucagon, 200 μg, was injected intraperitoneally in a single dose. Postmitochondrial supernatants were obtained from livers of control rats and of rats 5 min after treatment with glucagon. For further details see Experimental Procedure.

This effect of glucagon on hepatic protein synthesis does not seem to be the result of a direct action of the hormone on the protein synthetic machinery. This is suggested by the fact that protein synthesis in vitro was the same whether we used postmitochondrial supernatants of livers biopsied 5 minutes after glucagon-treatment or after the injection of saline (Fig. 8). Postmitochondrial supernatants, although they are rather crude preparations, offer the advantage of having greater activity than fractionated systems (7, 31).

The effect of glucagon on the hepatic polypeptide chain completion time was paralleled by changes in the hepatic content of adenine nucleotides (Table 4).

TABLE 4

EFFECT OF GLUCAGON ON THE HEPATIC CONTENT OF ADENINE NUCLEOTIDES

	Time after glucagon administration (min)		
	0	5	20
		nmoles/g dry wt	
[ATP]	14755 ± 291	13161 ± 317*	13674 ± 207***
[ADP]	4007 ± 130	4281 ± 104	4310 ± 131
[AMP]	1553 ± 75	1757 ± 105	1662 ± 223
Total adenine nucleotides	20316 ± 345	19239 ± 408	19574 ± 223
[ATP]/[ADP]	3.7 ± 0.1	3.0 ± 0.1**	3.3 ± 0.1***
Energy charge	0.82 ± 0.004	0.79 ± 0.006**	0.81 ± 0.009***

Each figure represents the mean value of 11-13 experiments ± S.E.

By t test: *P <.02; **P <.01; ***P = not significant.

The ATP content was significantly decreased (P <.05) five minutes after glucagon administration. At 20 minutes, although the ATP concentration remained below the control levels, the difference was not statistically significant. The [ATP]/[ADP] ratio and the energy charge displayed similar patterns of change; that is, five minutes after glucagon injection, they were decreased to a degree similar to that observed in untreated rat livers at 20 minutes. When glucagon was infused continuously its initial effect on the adenine nucleotide content (Table 5) was maintained throughout the 20 minutes of the experiment, as did its effect on the polypeptide chains completion time (Fig. 6).

TABLE 5

EFFECT OF THE CONTINUOUS INFUSION OF GLUCAGON ON THE HEPATIC CONTENT OF ADENINE NUCLEOTIDES

Time of infusion (min)	20	5	20
Additions to the saline	none	glucagon	glucagon
		nmoles/g dry wt	
[ATP]	15080 ± 488	12257 ± 752**	12706 ± 219**
[ADP]	4131 ± 135	4507 ± 142	4589 ± 205
[AMP]	1617 ± 152	1525 ± 103	1773 ± 206
Total adenine nucleotides	20829 ± 280	18290 ± 946	19070 ± 431
[ATP]/[ADP]	3.7 ± 0.2	2.6 ± 0.1***	2.7 ± 0.1***
Energetic charge	0.82 ± 0.01	0.79 ± 0.006*	0.79 0.006**

The figures are mean values of at least six livers ± S.E.

By t test: *P <.05; **P <.01; ***P <.001

TABLE 6

EFFECT OF GLUCAGON ON THE HEPATIC STATE OF REDUCTION

	Time after glucagon administration (minutes)		
	0	5	20
		nmoles/g dry wt	
[Lactate]	4334 ± 603	4776 ± 510	5286 ± 359
[Pyruvate]	204 ± 21	154 ± 19	207 ± 12
[Lactate]/[Pyruvate]	21 ± 3	28 ± 2*	22 ± 1
[β-hydroxybutyrate]	1416 ± 54	1894 ± 114**	1738 ± 128
[Acetoacetate]	7419 ± 339	7423 ± 524	8098 ± 431
[β-hydroxybutyrate]/[Acetoacetate]	0.19 ± 0.01	0.27 ± 0.03*	0.22 ± 0.02
Total ketone bodies	8837 ± 353	9317 ± 464	9836 ± 439

Each figure represents the mean value of at least 6 experiments ± S.E.
By t test: *P <.05; ** P <.001.

The changes in the hepatic content of the adenine nucleotides were accompanied by significant changes in the hepatic state of reduction, as expressed by the metabolite couples linked to cytosolic (lactate dehydrogenase) or mitochondrial (β-hydroxybutyrate dehydrogenase) nicotinamide nucleotide-dependent dehydrogenases. Due to the exclusive cytoplasmic location of lactate dehydrogenase the metabolite ratio of lactate to pyruvate reflects the cytosolic state of reduction. Seemingly, due to the virtually exclusive mitochondrial location of β-hydroxybutyrate dehydrogenase in the rat liver, the ratio of β-hydroxybutyrate to acetoacetate indicates the state of reduction of the intramitochondrial compartment. As shown in Table 6, the administration of glucagon resulted in a more reduced state in both cellular compartments at five minutes, thereafter returning

TABLE 7

EFFECT OF THE CONTINUOUS INFUSION OF GLUCAGON ON THE HEPATIC REDOX STATE

Time of infusion (min) Additions to the saline	20 none	5 Glucagon	20 Glucagon
		nmoles/g dry wt	
[Lactate]	6361 ± 1419	11537 ± 1029*	5577 ± 367
[Pyruvate]	287 ± 47	230 ± 41	159 ± 28*
[Lactate]/[Pyruvate]	25 ± 3	49 ± 5	40 ± 5**
[β-hydroxybutyrate]	2064 ± 132	2467 ± 312	2354 ± 124
[Acetoacetate]	5204 ± 159	5627 ± 376	5521 ± 427
[β-hydroxybutyrate]/[Acetoacetate]	0.39 ± 0.03	0.44 ± 0.04	0.46 ± 0.02
Total ketone bodies	7268 ± 44	8094 ± 591	8379 ± 932

The figures are mean values of at least six livers ± S.E.

By t test: *P <.05; **P <.01

to approximately control values. In contrast, when glucagon was infused continuously, these changes in the hepatic redox state were sustained throughout the experiment (Table 7). The action of glucagon in raising the [NADH]/[NAD^+] ratio is probably the consequence of the increased fatty acid oxidation secondary to the lipolytic effect of the hormone (14, 35, 40). This point of view is supported by the data shown in Fig. 9 indicating that the intraperitoneal administration of glucagon produced a transient rise in the plasma free fatty acid level with peak values coinciding with its maximal effect on the state of reduction and the decrease in the [ATP]/[ADP] ratio (Table 4 and 6, respectively).

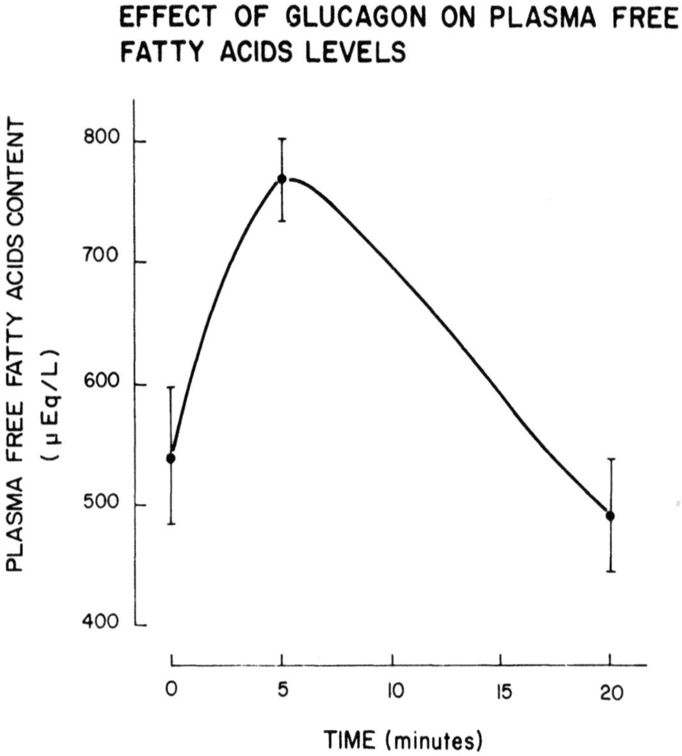

Fig. 9. Glucagon, 200 μg, was injected intraperitoneally in a single dose and the animals were bled at the indicated times. Each point represents the average value of at least 8 experiments ± S. E.

DISCUSSION

The observed effects of glucagon on the rates of protein synthesis and proteolysis (Figs. 2 and 4) allow one to conclude that its ureogenic effect is the result, at least in part, of the increased availability of amino acids secondary to accelerated proteolysis and not to decreased utilization for protein synthesis. This clear dissociation between the glucagon effects on protein synthesis and on proteolysis seems to indicate that they are mediated by different mechanisms.

Glucagon was found to inhibit the rate of hepatic protein synthesis by about 40%. This calculation has been made assuming a concentration of ribosomes of

3 nmoles/g (21) which, using a completion time of polypeptide chains of 90 and 146 seconds, yields a rate of protein synthesis of 120 and 75 nmoles of protein/g of liver/hour for livers of control and five minute glucagon-treated rats, respectively. The fact that the effect of glucagon lasts less than 20 minutes may be the consequence of the rapid degradation of the hormone. This is indicated by the observation that the continuous intravascular infusion of glucagon produced maximal effects throughout the experimental period (Fig. 6).

Glucagon stimulates insulin secretion in vivo as well as in vitro (32). For this reason, even though it is not yet known whether insulin prevents the described effects of glucagon on protein synthesis, as has been shown with some other metabolic parameters (26-28), the possibility cannot be excluded that the transient effect of glucagon was the result of a rise in plasma insulin levels, shifting the glucagon to insulin ratio to control values.

The effect of glucagon seems to be exerted exclusively on the elongation step of protein synthesis. This conclusion is based on the fact that the ribosomal state of aggregation was increased after glucagon administration (Fig. 7 and Table 3).

A glucagon inhibition of general hepatic protein synthesis seems to conflict with the known effects of the hormone in inducing several hepatic enzymes (13, 16, 42, 46). However the two processes are different from a mechanistic point of view: enzyme induction is a very slow process compared with the acute effects herein reported and it can take place also in the absence of substantial variations in the rate of total protein synthesis.

The mechanism by which glucagon inhibits protein synthesis does not seem to be a direct effect on the protein synthetic machinery, since postmitochondrial supernatants from livers of control and of glucagon-treated rats showed the same rate of radioactive valine incorporation into protein. Since glucagon is known to increase the hepatic concentration of cyclic AMP (11) and since a rise in this nucleotide is known to enhance the phosphorylation of ribosomal proteins (19), the fact that the hepatic polyribosomes from animals treated with glucagon did not show any functional difference in vitro, compared to those from control rats, indicates that the phosphorylation of ribosomal proteins mediated by glucagon does not play a regulatory role in protein synthesis in vivo. It is possible, however, that this phosphorylation of ribosomal proteins plays a role in the regulation of the synthesis of some specific proteins which are known to be induced by glucagon (13, 42, 46). On the other hand, the fact that the glucagon effect did not last 20 minutes after its administration (Fig. 4) together with the observation that under similar conditions the elevation of cyclic AMP levels persisted (38), also supports the view that the inhibition of hepatic protein synthesis by the hormone is not related to a cyclic AMP-mediated phosphorylation of ribosomal proteins.

Since the effect of glucagon does not seem to be mediated by a direct effect on the protein synthesis machinery, it seems reasonable to assume that its effect could be the result of some cellular metabolic perturbation brought about by the hormone. We investigated the possibility that glucagon could affect the rate of hepatic protein synthesis through changes in the content of adenine nucleotides because of the observation that glucagon decreases the concentration of ATP in the perfused isolated rat liver (28). On the other hand, previous work (5), summarized in Fig. 10, indicated that a decrease in the hepatic content of ATP within the physiologic range resulted in a preferential, if not exclusive, inhibition of peptide chain elongation and/or termination of protein synthesis. Actually, it is not ATP but GTP which is directly involved in these steps of protein synthesis. However, since GTP and ATP are in equilibrium through a potent nucleotide transphosphorylase, variations in the cellular ATP content might affect peptide chain elongation by altering the availability of GTP. On these grounds, it seems plausible to consider that the effect of glucagon in decreasing protein synthesis may be mediated through a decrease in the [GTP]/[GDP] ratio secondary to the decrease in the [ATP]/[ADP] ratio (Tables 4 and 5). The fact that the variations in the [ATP]/[ADP] ratio are parallel to the effect of the hormone on hepatic protein synthesis seem to support this point of view.

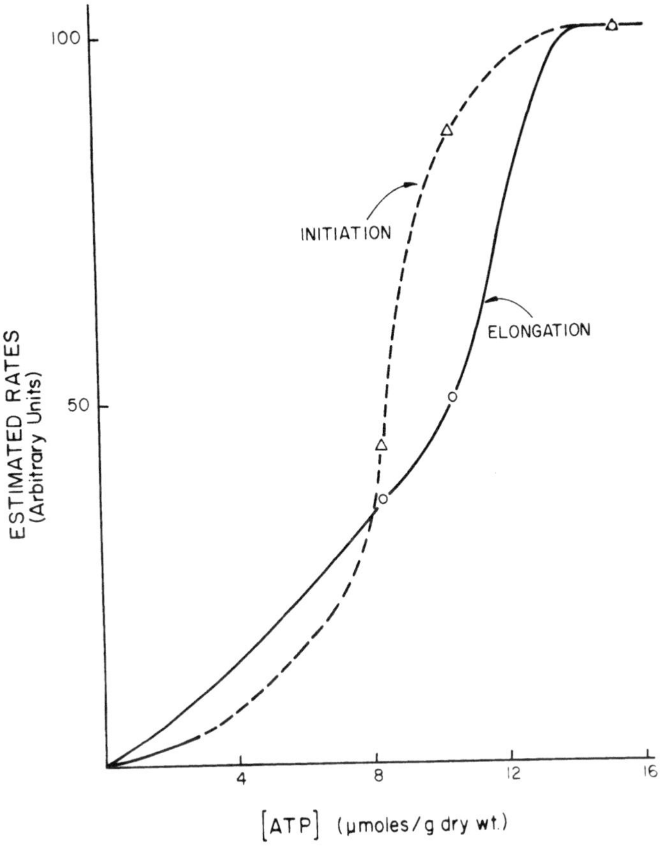

Fig. 10. Hypothetical relationship between the hepatic adenine nucleotide content and rates of peptide chain initiation and elongation.

If the glucagon effect on hepatic protein synthesis is related to its capacity to decrease the availability of high energy phosphorylated compounds, it seems pertinent to question by which mechanism(s) the hormone acts to decrease the concentration of these compounds. It is important to notice the close relationship, observed in all the situations studied, between the variations in adenine nucleotide content and changes in the state of reduction of the nicotinamide nucleotide system. Both types of nucleotide systems are in equilibrium through cytosolic reactions in such a way that an increase in the [NADH]/[NAD^+] ratio results in a decrease of the [ATP]/[ADP] ratio and vice versa (17). The decrease in ATP content after glucagon treatment could also be explained by the enhanced utilization of high energy phosphate compounds for gluconeogenesis. However, considering the observed similarity between the patterns of fatty acid mobilization and the changes in the state of reduction and adenine nucleotide content, we are more inclined to believe that the most important effect of the hormone is to mobilize endogenous lipids. All other effects would be secondary to the virtually unlimited capacity of the liver to oxidize fatty acids. Once more, the mechanism of glucagon action appears to be intimately related to its lipolytic effect (27, 28, 36, 37, 43).

Acknowledgements

The authors acknowledge the technical assistance of Mr. T. Fontela and Mr. F. Martin.

The work was supported in part by grants from the Comision Asesora para el Desarrollo de la Investigacion; Lilly Indiana de España and Essex (España) S. A.

A. M. R. and J. P. D. are recipients of research fellowships from the Spanish Ministry of Education and Science.

REFERENCES

1. Airhardt, J., Vidrich, A. and Khairallah, E. A. (1974): Compartmentation of free amino acids for protein synthesis in rat liver. Biochem. J. 140, 539-548.

2. Ashford, J. P. and Porter, K. R. (1962): Cytoplasmic components in hepatic cell lysosomes. J. Cell Biol. 12, 198-202.

3. Ayuso-Parrilla, M. S., Martin-Requero, A., Perez-Diaz, J. and Parrilla, R. (1976): Role of glucagon on the control of hepatic protein synthesis and degradation in the rat in vivo. J. Biol. Chem., in press.

4. Ayuso-Parrilla, M. S. and Parrilla, R. (1973): Glucagon effect on rat liver protein synthesis in vivo. Biochem. Biophys. Res. Commun. 52, 582-587.

5. Ayuso-Parrilla, M. S. and Parrilla, R. (1975): Control of hepatic protein synthesis. Differential effects of ATP levels on the initiation and elonga-

tion steps. Eur. J. Biochem. 55, 593-599.

6. Bergmeyer, H. U. (1965): Methods of Enzymatic Analysis. Academic Press, New York.

7. Clemens, M. J. and Pain, V. M. (1974): Hormonal requirements for acute stimulation of rat liver polysome formation by amino acid feeding. Biochim. Biophys. Acta 361, 345-358.

8. Colombo, B., Felicetti, L. and Baglioni, C. (1966): Inhibition of protein synthesis in reticulocytes by antibiotics. .. Effect on polysomes. Biochim. Biophys. Acta 119, 109-119.

9. Deter, R. L. and DeDuve, C. (1967): Influence of glucagon, an inducer of cellular autophagy, on some physical properties of rat liver lysosomes. J. Cell. Biol. 33, 437-449.

10. Dole, V. P. and Meinertz, H. (1960): Microdetermination of long chain fatty acids in plasma and tissues. J. Biol. chem. 235, 2595-2599.

11. Exton, C. R. and Park, J. H. (1972): Glucagon and the metabolism of glucose. In: Glucagon. Molecular Physiology, Clinical and Therapeutic Implications. P. J. Lefêbvre and R. H. Unger, Eds. Pergamon Press, New York, pp. 77-150.

12. Fern, E. B. and Garlick, P. J. (1974): The specific radioactivity of the precursor pool for estimates of the rate of protein synthesis. Biochem. J. 134, 1127-1130.

13. Greengaard, O. (1969): The hormonal regulation of enzymes in prenatal and postnatal rat liver. Biochem. J. 115, 19-24.

14. Hagen, J. H. (1961): Effects of glucagon on the metabolism of adipose tissue. J. Biol. Chem. 236, 1023-1027.

15. Haschemeyer, A. E. V. (1969): Rates of polypeptide chain assembly in liver in vivo: relation to the mechanism of temperature acclimation in Opsanus tau. Proc. Nat. Acad. Sci. USA 62, 128-135.

16. Jost, J. P., Hsie, A. W., Hughes, S. D. and Ryan, L. (1970): Role of cyclic adenosine 3'-5'-monophosphate in the induction of hepatic enzymes. Kinetics of the induction of rat liver serine dehydratase by cyclic adenosine 3'-5'-monophosphate. J. Biol. Chem. 245, 351-357.

17. Krebs, H. A. (1971): Gluconeogenesis and redox state. In: Regulation of Gluconeogenesis. H. D. Söling and B. Willms, Eds. Academic Press, New York, pp. 114-117.

18. Lin, S. Y., Mosteller, R. D. and Hardesty, B. (1966): The mechanism of sodium fluoride and cycloheximide inhibition of hemoglobin biosynthesis in the cell free reticulocyte system. J. Mol. Biol. 21, 51-69.

19. Loeb, J. E. and Blatt, C. (1970): Phosphorylation of some rat liver ribosomal proteins and its activation by cyclic AMP. FEBS Letters 10, 105-108.

20. Mallette, L. E., Exton, J. H. and Park, C. A. (1969): Effects of glucagon on amino acid transport and utilization in the perfused rat liver. J. Biol.

Chem. 244, 5724-5728.

21. Mathews, R. W., Oronsky, A. and Haschemeyer, A. E. V. (1973): Effect of thyroid hormone on polypeptide chain assembly kinetics in liver protein synthesis in vivo. J. Biol. Chem. 248, 1329-1333.

22. Miller, L. L. (1960): Glucagon: a protein catabolic hormone in the isolated perfused rat liver. Nature 185, 248.

23. Mortimore, G. E. and Mondon, C. E. (1970): Inhibition by insulin of valine turnover in liver. Evidence for a general control of proteolysis. J. Biol. Chem. 245, 2375-2383.

24. Mortimore, G. E., Woodside, K. H. and Henry, J. E. (1972): Compartmentation of free valine and its relation to protein turnover in perfused rat liver. J. Biol. Chem. 247, 2776-2784.

25. Parrilla, R. and Goodman, M. N. (1974): Nitrogen metabolism in the isolated perfused rat liver. Nitrogen balance, redox state and rates of proteolysis. Biochem. J. 138, 341-348.

26. Parrilla, R., Goodman, M. N. and Toews, C. J. (1974): Effect of glucagon: insulin ratios on hepatic metabolism. Diabetes 23, 725-731.

27. Parrilla, R., Jimenez, M. I. and Ayuso-Parrilla, M. S. (1975): Glucagon and insulin control of gluconeogenesis in the perfused isolated rat liver. Effects on cellular metabolite distribution. Eur. J. Biochem. 56, 375-383.

28. Parrilla, R., Jimenez, M. I. and Ayuso-Parrilla, M. S. (1976): Cellular redistribution of metabolites during glucagon and insulin control of gluconeogenesis in the isolated perfused rat liver. Arch. Biochem. Biophys. 174, 1-12.

29. Penhos, J. C., Wu, C. H., Daunas, J., Reitman, M. and Levine, R. (1966): Effect of glucagon on lipids and on urea formation by the perfused rat liver. Diabetes 15, 740-748.

30. Pryor, J. and Berthet, J. (1960): The action of adenosine 3',5'-monophosphate on the incorporation of leucine into liver proteins. Biochim. Biophys. Acta 43, 556-557.

31. Richardson, A., McGown, E., Henderson, L. M. and Swan, P. B. (1971): In vitro amino acid incorporation by the postmitochondrial supernatant from rat liver. Biochim. Biophys. Acta 254, 468-477.

32. Samols, E., Tyler, J. M. and Marks, V. (1972): Glucagon:insulin interrelationships. In: Glucagon. Molecular Physiology, Clinical and Therapeutic Implications. P. J. Lefèbvre and R. H. Unger, Eds. Pergamon Press, New York, pp. 151-173.

33. Scornik, O. (1974): In vivo rate of translation by ribosomes of normal and regenerating liver. J. Biol. Chem. 249, 3876-3883.

34. Sokal, J. E. (1966): Effect of glucagon on gluconeogenesis by the isolated perfused rat liver. Endocrinology 78, 538-548.

35. Steinberg, D., Shafrir, E. and Vaughan, M. (1959): Direct effect of glucagon

on release of unesterified fatty acid from adipose tissue. Clin. Res. 7, 250.

36. Struck, E., Ashmore, J. and Wieland, O. (1965): Stimulierung der Gluconeogenese durch langkettige Fettsaüren und Glukagon. Biochem. Z. 343, 107-110.

37. Struck, E., Ashmore, J. and Wieland, O. (1966): Effects of glucagon and long chain fatty acids on glucose production by isolated rat liver. Adv. Enzyme Regul. 4, 219-224.

38. Taunton, O. D., Steifel, F. B., Greene, H. L. and Herman, R. H. (1974): Rapid reciprocal changes in rat hepatic glycolytic enzyme and fructose diphosphatase activities following insulin and glucagon injection. J. Biol. Chem. 249, 7228-7239.

39. Tybergheim, J. (1953): Action du glucagon sur le métabolisme des proteines. Arch. internat. Physiol. 61, 104-107.

40. Weinges, K. F. (1961): Der Einfluss von Glukagon und Insulin auf den Stoffwechsel der nichtveresterten Fettsaüren am isolierten Fettgewebe der Ratte in vitro. Klin. Wschr. 39, 293-298.

41. Weinstein, J., Klausner, H. A. and Heimberg, M. (1973): The effect of concentration of glucagon on output of triglyceride, ketone bodies, glucose and urea by the liver. Biochim. Biophys. Acta 296, 300-309.

42. Wicks, W. D. (1969): Induction of hepatic enzymes by adenosine 3',5'-monophosphate in organ culture. J. Biol. Chem. 244, 3941-3950.

43. Williamson, J. R. (1970): Role of lipolysis in the gluconeogenic action of glucagon. In: Metabolic Regulation and Enzyme Action. A. Sols and S. Grisolia, Eds. Academic Press, pp. 107-113.

44. Williamson, J. R. and Corkey, B. E. (1968): Assays of intermediates of the citric acid cycle and related compounds by fluorometric enzyme methods. In: Methods in Enzymology, Vol. 13. J. M. Lowenstein, Ed. Academic Press, New York, pp. 434-513.

45. Woodside, K. H., Ward, W. F. and Mortimore, G. E. (1974): Effects of glucagon on general protein degradation and synthesis in perfused rat liver. J. Biol. Chem. 249, 5458-5463.

46. Yeung, D. and Oliver, I. T. (1968): Induction of phosphopyruvate carboxylase in neonatal rat liver by adenosine 3',5'-cyclic monophosphate. Biochemistry 7, 3231-3239.

THE ROLE OF GLUCAGON IN AMINO ACID HOMEOTASIS

G. F. Cahill, Jr. and T. T. Aoki

Elliott P. Joslin Research Laboratory and the Department of Medicine,
Harvard Medical School and the Peter Bent Brigham Hospital
Boston, Massachusetts USA

The uptake of amino acids by the liver, the plasma level of glucagon and the insulin to glucagon ratio have an important role in the amino acid homeostasis of man. Evidence that glucagon may be more important as an anabolic rather than a catabolic hormone is discussed.

In glucose homeostasis, insulin plays the leading role, but exercise, glucagon, catecholamines, hepatic innervation, adrenal steroids, and possibly growth hormone also participate. Compounding these interrelations are the variable contributions of different organs such as liver, gut and kidney, which may contribute glucose at one time (e. g., in fasting man) and remove glucose at another. In contrast, brain and muscle always remove this fuel. It is not surprising, therefore, that in such pathologic states as labile or "brittle" diabetes, non-pancreatic tumor-induced hypoglycemia or ketotic hypoglycemia in children, the abnormality in glucose homeostasis has not been clearly identified. Is it under-production or overutilization?

Amino acid homeostasis is much more complicated than glucose homeostasis because instead of a single substance there are at least two dozen different amino acids, because every organ in the body participates in their regulation and because this regulation involves amino acid transport in which red cells (3, 5), and probably other circulating elements may play active roles. In this regard, it is of interest that one amino acid, tryptophan, is variably bound to plasma proteins in competition with circulating free fatty acids (15). Thus, the three branched-chain amino acids are oxidized mainly by muscle (13, 17) and adipose tissue (23); alanine appears to be removed almost uniquely by the liver, whereas glutamine is removed by gut and kidneys, and to a lesser extent by the liver. The kidney is important in the uptake of glycine, lysine and proline, and appears to be the only organ contributing serine to the circulation (20). The splanchnic bed produces glutamate and the peripheral tissues remove it (18), returning glutamine to the blood as a second contribution of $-NH_2$ from the periphery, in addition to the alanine produced by muscle by transamination of pyruvate. Alanine, in turn, is removed by the splanchnic bed, and, in the fasted state, sent back to muscle as glucose. The precise role of glucagon in amino acid metabolism is difficult to define. Most data suggest that glucagon expedites the catabolism of the amino acids derived from a protein meal, as originally hypothesized by Unger and collaborators (24), while preventing a fall in the concentration of circulating glucose, as the serum insulin levels increase. Thus, the ingestion of protein leads to an increase in the blood levels of insulin, glucagon and of certain amino acids, particularly the branched-chain amino acids. The levels of the non-essential amino acids are little affected, probably due to the glucagon-enhanced trapping of these amino acids by the liver. In fact, it is surprising how effective this system is, particularly since the amount of these amino acids provided by the ingested meal may be several times greater than that present in the extracellular fluids.

On the other hand, the concentration of branched-chain amino acids increases dramatically, perhaps even trebling after a large protein meal (4). This selective increase is probably due in part to the ability of glucagon to initiate hepatic proteolysis (16) and to the inability of the liver to oxidize the branched-chain amino acids. Of these, leucine is a strong insulin, but not glucagon (22) secretogogue and, furthermore, may directly stimulate protein synthesis in muscle. Thus, leucine acts not only as an essential structural growth factor, but promotes growth also through its capacity to release insulin and its unique direct (probably allosteric) effect in promoting protein synthesis and inhibiting proteolysis (12). In addition to their role in muscle, the branched-chains are excellent precursors for the synthesis of fatty acids in adipose tissue, which, like muscle, export the residual amino group as alanine.

Since the flux of non-essential amino acids to the liver is increased without much change in their peripheral concentration, it is the operational "Km" of amino acid trapping by the liver which appears to be altered. Certainly, portal levels are somewhat increased above peripheral levels (8-10), but the striking stability of the glycogenic amino acids in the peripheral circulation attest to the increased efficiency of their removal. It must be emphasized that only the liver and, to a lesser extent, the kidney, can catabolize the glucogenic amino acids since those not entering the metabolic pathway as pyruvate (e. g., alanine) must be converted to phosphoenolpyruvate before they can be converted to pyruvate and hence, to acetyl CoA for terminal oxidation to CO_2. Thus, muscle, brain and adipose tissue, due to their very low or effectively absent phosphoenolpyruvate carboxykinase activity are incapable of significant catabolism of the amino acids entering the tricarboxylic acid cycle.

Furthermore, not only is alanine not taken up but it is produced. Glycine, on the other hand, appears to be neither removed nor produced in the periphery and serine is produced in the kidney, probably from glycine and alanine.

One may ask why should the body maintain such a finely regulated degree of amino acid homeostasis in the extracellular fluid (except for the branched-chain amino acids) particularly since, with the exception of the red cell, the glucogenic amino acids are maintained at a much higher concentration in the intracellular than in the extracellular fluid. The answer probably lies in the equilibrium between intracellular amino acids and their keto derivatives such as oxaloacetate, pyruvate and α-ketoglutarate, which, in many instances, share the same metabolic fate in other important pathways. Thus, the concentration of glucogenic amino acids must be finely regulated both in cells and in extracellular fluids, lest any gross perturbation in a given amino acid be reflected in a similar change in its ketoanalogue, with resulting disturbance in carbohydrate metabolism and citric acid cycle. Again, the branched-chain amino acids do not appear to abide by this rule and neither do certain other amino acids, such as tyrosine and tryptophan, which induce their own degradation systems whenever their intake increases and their concentration rises. The ketoanalogues of the branched-chain amino acids do not participate in other metabolic pathways except for sharing co-factors for fat oxidation, such as NAD^+ and CoASH. Thus, the catabolism of these amino acids is simply a function of their concentration. If insulin levels are high, they are used by adipose tissue for lipogenesis, or by muscle and, probably, by kidney as fuel. In fact, the absolute level of the branched-chain amino acids is linearly related to the overall rate of amino acid catabolism and, pari passu, of nitrogen excretion. Thus, in trauma (7), diabetic ketoacidosis (2), early starvation (1), and protein feeding (4, 5) the level of branched-chain amino acids and the nitrogen excretion are high, while, in prolonged starvation or after the intake of protein-sparing carbohydrate, they are low.

This parallelism between the level of branched-chain amino acids and nitrogen excretion does not hold following glucagon infusion (6) or in patients with glucagonoma (14). Here all amino acids, glucogenic as well as branched-chain are strikingly reduced, in spite of a normal amino acid flux. In fact, in a subject with glucagonoma (14), following a protein meal ingested under experimental conditions, the amino acid tolerance increased and the branched-chain and other essential amino acids did not reach normal levels, let alone increase, as they do in normal subjects (4). Augmented hepatic trapping could explain this hypoaminoacidemia. In addition, the apparently intact insulin secreting mechanism together with normal tissue responsiveness to this hormone probably accounts in part for the low levels of branched-chain amino acid associated with this tumor: ↑ glucagon → ↑ glucose → ↑ insulin → ↓ amino acids. Only the skin shows obvious signs of suffering from the apparent amino acid deficiency.

In summary, by altering the overall "Km" for amino acid uptake by the liver, glucagon, the ratio of insulin to glucagon or some function of this ratio help maintain amino acid homeostasis during protein ingestion. In man, when the carbohydrate intake is high and the protein intake is low or nil, the levels of alanine, pyruvate and lactate are high, yet the total daily nitrogen excretion is as low as 2 grams or about 0.2 mole. This means that the splanchnic alanine and glutamine extraction for gluconeogenesis is 10% or less, in spite of alanine concentrations approaching 1 mM. In this state, insulin levels are high and glucagon levels are low. Conversely, after a 4 to 5-day fast, the concentration of alanine is about 0.2 mM and the nitrogen excretion reaches 12 grams or 0.8 mole, which, in terms of alanine equivalents, would necessitate the extraction of all the alanine going through liver and of 3 to 4 times as much other amino acids. In fact, measurements have shown that 75 to 80% of the alanine is extracted across the splanchnic bed (11). In this state, insulin is low and glucagon is normal or slightly elevated (19). Under similar fasting conditions, the release of alanine from the muscles of the forearm is accelerated (21) when compared to the pre-fasting and more prolonged fasting states, when nitrogen excretion is greatly decreased. Thus, in the fed state, the primary role of glucagon is to enhance amino acid trapping and metabolism by the liver (and possibly also by the kidney) and to maintain glucose production via gluconeogenesis and glycogenolysis, should the meal fail to provide an amount of carbohydrate sufficient to satisfy requirements of the central nervous system. If the meal does contain some carbohydrate, glucagon secretion is suppressed, a most logical homeostatic phenomenon.

In the fasting state, glucagon has a less well-defined role, although it may act as a "catabolic" hormone. Studies with somatostatin have shown transient hypoglycemia as glucagon levels fall. In addition, glucagon, increasing the rate of alanine trapping across the splanchnic bed, probably has an essential "permissive" role in the maintenance of glucose levels in early fasting. However, as fasting progresses, glucagon levels return to normal post-absorptive values, as does the extraction of alanine. The evidence that glucagon is a catabolic hormone in the total organism is meager. In fact, the infusion of glucagon in physiologic amounts results in a decrease of amino acid levels accompanied by a slight but significant decrease in urea nitrogen excretion (14). This unexplained phenomenon suggests that glucagon may not be a catabolic but rather an anabolic hormone. The early embryologic development of the A cells and the demonstration of glucagon in the circulation of the embryo are in agreement with this hypothesis. This view is also supported by the observation that the A cells derive from the gut endoderm suggesting a direct nutritionally-related function for the hormone.

In summary, glucagon is intricately related to amino acid homeostasis and, although it expedites amino acid trapping, ureogenesis and gluconeogenesis by the liver, it appears that it may be more important as an anabolic rather than a catabolic hormone as far as total body economy is concerned.

Acknowledgements

This work was supported in part by USPHS Grants AM 05077 and AM 15191. T. T. Aoki, M. D. is Investigator, Howard Hughes Medical Institute.

REFERENCES

1. Aguilar-Parada, E., Eisentraut, A. M. and Unger, R. H. (1969): Effects of starvation on plasma pancreatic glucagon in normal man. Diabetes 18, 717-723.

2. Aoki, T. T., Assal, J. P., Manzano, F. M., Kozak, G. P. and Cahill, G. F., Jr. (1975): Plasma and cerebrospinal fluid amino acid levels in diabetic ketoacidosis before and after corrective therapy. Diabetes 24, 463-467.

3. Aoki, T. T., Brennan, M. F., Müller, W. A., Moore, F. D. and Cahill, G. F., Jr. (1972): Effect of insulin on muscle glutamate uptake. J. Clin. Invest. 51, 2889-2894.

4. Aoki, T. T., Brennan, M. F., Müller, W. A., Soeldner, J. S., Alpert, J. S., Saltz, S. B., Kaufmann, R. L., Tan, M. H. and Cahill, G. F., Jr. (1976): Amino acid levels across normal forearm muscle and splanchnic bed after a protein meal. Am. J. Clin. Nutr. 29, 340-350.

5. Aoki, T. T., Müller, W. A., Brennan, M. F. and Cahill, G. F., Jr. (1973): Blood cell and plasma amino acid levels across forearm muscle during a protein meal. Diabetes 22, 768-775.

6. Aoki, T. T., Müller, W. A., Brennan, M. F. and Cahill, G. F., Jr. (1974): Effect of glucagon on amino acid and nitrogen metabolism in fasting man. Metabolism 23, 805-814.

7. Aoki, T. T., Vignati, L. and Cahill, G. F., Jr. (unpublished data).

8. Elwyn, D. H.: Distribution of amino acids between plasma and red blood cells in the dog. Fed. Proc. 25, 854-861.

9. Elwyn, D. H., Launder, W. J., Parikh, H. C. and Wise, E. M. (1972): Roles of plasma and erythrocytes in interorgan transport of amino acids in dogs. Am. J. Physiol. 222, 1333-1342.

10. Elwyn, D. H., Parikh, H. C. and Shoemaker, W. C. (1968): Amino acid movements between gut, liver, and periphery in unanesthetized dogs. Am. J. Physiol. 215, 1260-1275.

11. Felig, P., Owen, O. E., Wahren, J. and Cahill, G. F., Jr. (1969): Amino acid metabolism during prolonged starvation. J. Clin. Invest. 48, 584-592.

12. Fulks, R. M., Li, J. B. and Goldberg, A. L. (1975): Effects of insulin, glucose, and amino acids on protein turnover in rat diaphragm. J. Biol. Chem. 250, 290-298.

13. Goldberg, A. L. and Odessey, R. (1972): Oxidation of amino acids by diaphragms from fasted rats. Am. J. Physiol. 223, 1384-1391.

14. Horton, E. S. (personal communication).

15. Lipsett, D., Madras, B. K., Wurtman, R. J. and Munro, H. N. (1973): Serum tryptophan level after carbohydrate ingestion: selective decline in non-albumin-bound tryptophan coincident with reduction in serum free fatty acids. Life Sci. 12, 57-64.

16. Mallette, L. E., Exton, J. H. and Park, C. R. (1969): Effects of glucagon on amino acid transport and utilization in the perfused rat liver. J. Biol. Chem. 224, 5724-5728.

17. Manchester, K. L. (1965): Oxidation of amino acids by isolated rat diaphragm and the influence of insulin. Biochem. Biophys. Acta 100, 295-298.

18. Marliss, E. B., Aoki, T. T., Pozefsky, T., Most, A. S. and Cahill, G. F., Jr. (1971): Muscle and splanchnic glutamine and glutamate metabolism in post-absorptive and starved man. J. Clin. Invest. 50, 814-817.

19. Marliss, E. B., Aoki, T. T., Unger, R. H., Soeldner, J. S. and Cahill, G. F., Jr. (1970): Glucagon levels and metabolic effects in fasting man. J. Clin. Invest. 49, 2256-2270.

20. Pitts, R. F., Damain, A. C. and MacLeod, M. B. (1970): Synthesis of serine by rat kidney in vivo and in vitro. Am. J. Physiol. 219, 584-589.

21. Pozefsky, T., Tancredi, R. G., Moxley, R. T., Dupre, J., Tobin, J. D. (1976): Effects of brief starvation on muscle amino acid metabolism in non-obese man. J. Clin. Invest. 57, 444-449.

22. Rocha, D. M., Faloona, G. R. and Unger, R. H. (1972): Glucagon-stimulating activity of 20 amino acids in dogs. J. Clin. Invest. 51, 2346-2351.

23. Rosenthal, J., Angel, A. and Farkas, J. (1974): Metabolic fate of leucine: a significant sterol precursor in adipose tissue and muscle. Am. J. Physiol. 226, 411-418.

24. Unger, R. H., Ohneda, A., Aguilar-Parada, E. and Eisentraut, A. M. (1969): The role of aminogenic glucagon secretion in blood glucose homeostasis. J. Clin. Invest. 48, 810-822.

THE ROLES OF GLUCAGON AND CYCLIC ADENOSINE 3',5'-MONOPHOSPHATE IN THE REGULATION OF LIPID BIOSYNTHESIS

J. J. Volpe and J. C. Marasa

Departments of Pediatrics and of Neurology and Neurosurgery (Neurology)
Washington University School of Medicine
St. Louis, Missouri USA

The role of glucagon and cyclic AMP in the long-term regulation of fatty acid synthesis was studied. A marked suppression of the induction of hepatic fatty acid synthetase by glucagon was observed in developing animals weaned to a fat-free diet. This effect of glucagon was similar to that observed when the hormone is administered to animals refed a fat-free diet after a fast. Although it is probable that cyclic AMP mediates these long-term effects of glucagon, tissue levels of the mononucleotide did not correlate with the activities of fatty acid synthetase. The mechanism(s) by which glucagon-induced elevations of hepatic cyclic AMP might lead to the suppression of the synthesis of fatty acid synthetase is discussed but remains unknown.

There is no uniform or obligatory relationship between intracellular cyclic AMP levels and rates of fatty acid synthesis or activities of the lipogenic enzymes. Thus, long-term studies with C-6 glial cells defined conditions which were characterized by increased rates of fatty acid synthesis but unchanged cyclic AMP levels and by unchanged rates of fatty acid synthesis but increased cyclic AMP levels. The nature of the specific inducer of changes in intracellular cyclic AMP may be critical in determining whether an effect on lipogenesis will result.

The information currently available on the role of glucagon and cyclic AMP in the short-term regulation of fatty acid synthesis is reviewed. Although some work suggests that these effectors lead to an inhibition of fatty acid synthesis that is mediated at the level of acetyl-CoA carboxylase, there are directly conflicting data. Moreover, the need for the use of very high levels of exogenous cyclic AMP to produce the short-term effects raises serious questions about the physiologic significance of the observed responses.

The information currently available on the role of glucagon and cyclic AMP in the short and long-term regulation of cholesterol synthesis is reviewed. A definite decrease in the rate of cholesterol synthesis and the activity of HMG-CoA reductase results in liver in hours after the injection of glucagon. Analogous effects can be produced by injection of dibutyryl cyclic AMP. The mechanism(s) of these relatively long-term effects is unknown. In short-term experiments utilizing several _in vitro_ systems cyclic AMP has been shown to lead to a decrease in cholesterol synthesis and HMG-CoA reductase in minutes. However, very high concentrations of the mononucleotide are required, and the effect may be caused by a competitive inhibition with enzyme substrate. Thus, the physiologic significance of the effect has been questioned. Moreover, studies with perfused rat liver demonstrate that the marked increase in hepatic cyclic AMP levels induced in minutes by glucagon is not accompanied by any change in the rate of cholesterol synthesis or the activity of HMG-CoA reductase.

Abbreviations: Cyclic AMP = Cyclic adenosine 3',5'-monophosphate

INTRODUCTION

The role of glucagon in the regulation of fatty acid and cholesterol synthesis in the living animal is unresolved. Most investigators currently conclude that glucagon causes a decrease in lipid synthesis and that this decrease is mediated by an increase in cellular levels of cyclic AMP. These conclusions are derived from experiments that generally are not free of alternative interpretations. In the following discussion we will review the current information on the roles of glucagon and cyclic AMP in the regulation of lipid biosynthesis. Because the mechanisms of short-term and long-term regulation of fatty acid synthesis are almost certainly different, we will discuss these regulations separately. Regulation of cholesterol synthesis is not so readily divisible but also will be discussed separately.

Fatty Acid Synthesis

The *de novo* biosynthesis of fatty acids in mammalian tissues is the result of the sequential action of two enzymes systems; acetyl-CoA carboxylase and fatty acid synthetase (see 54 for review). Acetyl-CoA carboxylase catalyzes the first committed step in cytosolic fatty acid biosynthesis, the biotin-dependent carboxylation of acetyl-CoA to form malonyl-CoA:

$$ATP + HCO_3^- + CH_3COSCoA \rightarrow {}^-OOCCH_2COSCoA + ADP + P_i$$

Malonyl-CoA is then utilized for the synthesis of fatty acids by fatty acid synthetase:

$$CH_3COSCoA + 7HOOCCH_2COSCoA + 14NADPH + 14H^+ \rightarrow CH_3CH_2(CH_2CH_2)_6CH_2COOH$$
$$+ 7CO_2 + 14\ NADP^+\ 8\ CoASH + 6H_2O.$$

Acetyl-CoA carboxylase may be involved particularly in the short-term regulation of fatty acid synthesis, and fatty acid synthetase (as well as acetyl-CoA carboxylase) in long-term regulation.

Short-term regulation. A role for glucagon and/or cyclic AMP in the short-term regulation of fatty acid synthesis has been suggested by several recent observations. In short-term experiments glucagon was associated with a marked depression of hepatic fatty acid synthesis, measured by incorporation of radioactive acetate or glucose in vitro by liver slices, in vivo by perfused liver preparations, or in isolated hepatocytes (17, 22, 26, 30, 37, 42). Particularly interesting was the demonstration by Klain and Weiser (30) of an approximately 60% reduction in fatty acid synthesis from glucose 15 min after infusion of glucagon and a 75% reduction after 30 min and simultaneous reductions in acetyl-CoA carboxylase activity of 55% and 70%. No change occurred in activities of fatty acid synthetase, glucose-6-phosphate dehydrogenase, NADP-malic dehydrogenase, isocitric dehydrogenase or citrate cleavage enzyme. Moreover, later work from the same laboratory (29) demonstrated that 3 hours after parenteral administration to rats of mannoheptulose, a compound known to cause a rapid fall in plasma insulin and a rise in glucagon, fatty acid synthesis from glucose and the activity of acetyl-CoA carboxylase (but not fatty synthetase) was reduced approximately three-fold. These experiments strongly suggested that short-term regulation of fatty acid synthesis by glucagon was mediated by alterations of acetyl-CoA carboxylase.

In contrast to these observations are those of Raskin et al (41) with perfused rat liver which show no change in the rate of fatty acid synthesis from acetate or octanoate after infusion of glucagon for 30 min. Differences in the size of the animal, dose of glucagon, etc. between these experiments and those noted above were not striking. The use of labeled octanoate as precursor for fatty acids afforded the opportunity to correct for endogeneous dilution of the cystolic acetyl-CoA pool involved in lipogenesis; this was not possible when acetate or glucose was utilized, as in the studies noted above. Thus, clearly the contention that glucagon exerts a short-term inhibition of fatty acid synthesis is not proven conclusively.

A variety of in vitro and in vivo studies have suggested that cyclic AMP exerts short-term control over fatty acid synthesis (1, 2, 6, 7, 22, 46). A possible role for acetyl-CoA carboxylase in this regulation was suggested by the observations of Allred and Roehrig (3) that lipogenesis, measured by the incorporation of tritium from ^3H_2O by liver slices, was reduced by 75% after a 30 min preincubation with 1 mM dibutyryl cyclic AMP and that this depression was accompanied by an approximately 60% reduction in the activity of acetyl-CoA carboxylase. Inhibition of acetyl-CoA carboxylase by approximately 50% could be attained by preincubation of tissue homogenates with 1 mM dibutyryl cyclic AMP. Other nucleotides or related compounds were not tested for such an effect on acetyl-CoA carboxylase.

In contrast to these observations are those of Rous (43) and Raskin et al. (41). Rous' data suggest that short-term inhibition of hepatic fatty acid synthesis from glucose, acetate or alanine by dibutyryl cyclic AMP is related to dilution of the cytosolic acetyl-CoA pool that results from the glycogenolytic effect of the cyclic nucleotide. Thus, when hepatic glycogen was depleted by one injection of dibutyryl cyclic AMP a second injection, 30 min later, caused no inhibition in hepatic fatty acid synthesis from acetate. No effect on hepatic acetyl-CoA carboxylase activity was noted 60 min after the intraperitoneal injection of dibutyryl cyclic AMP (50 mg/100 g body wt). Utilizing the perfused rat liver, Raskin et al (41) noted no change in fatty acid synthesis from octanoate after infusion of dibutyryl cyclic AMP (0.5 mM) or glucagon in a concentration (10^{-6}M) that caused a 50-fold increase in the level of hepatic cyclic AMP. Thus, clearly the contention that cyclic AMP exerts a short-term inhibition on fatty acid synthesis is not proven conclusively.

Worthy of particular emphasis is the fact that in the in vitro studies a clearly inhibitory effect of cyclic AMP on fatty acid synthesis requires concentrations of 10^{-4} to 10^{-3}M. These concentrations are several orders of magnitude higher than those usually observed in liver (18). Moreover, careful comparative studies of compounds with structures related to cyclic AMP have not been performed, and the possibility of physiologically irrelevant, competitive effects with substrates (e.g. ATP) has not been ruled out (see Regulation of Cholesterol

Synthesis below). If any short-term effect of cyclic AMP (or glucagon) on acetyl-CoA carboxylase is proven, the best possibilities for the mechanism of the effect are as follows. Acetyl-CoA carboxylase exists in protomeric (inactive) and polymeric (active) forms. The equilibrium between these forms is shifted to the active polymeric form by citrate and to the inactive protomeric form by long-chain acyl-CoA derivatives. A glucagon and cyclic AMP-sensitive lipase has been reported in liver (13, 38) and thus elevations of long-chain acyl-CoA derivatives could be the consequence of these two regulatory effectors. A second possibility would be that cyclic AMP activates a protein kinase that phosphorylates and inactivates the enzyme. The data of Kim and coworkers (8-10) suggest that acetyl-CoA carboxylase can be regulated by a phosphorylation-dephosphorylation system; however, the system was not sensitive to cyclic AMP *in vitro*. A third possibility would be that cyclic AMP interacts with a regulatory protein via a mechanism similar to that described for this mononucleotide in bacterial systems (16, 27, 56).

Long-term regulation. The most conclusive demonstration that glucagon is associated with long-term inhibition of fatty acid synthesis and of the enzymes thereof, utilized tube feeding techniques and the *in vivo* ^3H_2O method for measuring fatty acid synthesis (49). Thus, when starved animals were refed a high carbohydrate, fat-free diet for 48 hours, the incorporation of tritium from ^3H_2O into hepatic fatty acids was reduced by 75-80%. Similarly, under these conditions of equal caloric content the activities of acetyl-CoA carboxylase and fatty acid synthetase were reduced by 70-80%. The decrease in synthetase activity observed in glucagon-treated animals allowed to refeed *ad libitum* (33) is less impressive than in those tube fed. That the glucagon effect occurred at the level of enzyme synthesis was shown by isotopic-immunochemical studies with hepatic fatty acid synthetase (49). Thus, the synthesis of hepatic synthetase was reduced by approximately 60% in the animals injected with glucagon. Recently liver cells in culture were utilized to show that glucagon causes an inhibition of synthesis of another lipogenic enzyme, malic enzyme, and that the effect is at a post-transcriptional level (23). The induction by refeeding of the hepatic activity of still another lipogenic enzyme, glucose-6-phosphate dehydrogenase, is inhibited by glucagon (44), and this effect also occurs at the level of enzyme synthesis (19).

The possibility that the long-term effects of glucagon on the hepatic activities of the lipogenic enzymes are mediated by cyclic AMP is suggested by the production of similar effects after the injection of dibutyryl cyclic AMP into the intact animal (33, 44) and the addition of cyclic AMP to cultured liver cells (24). Whether these long-term effects of glucagon and cylic AMP are direct ones or occur indirectly as a consequence of one or more of the many metabolic effects associated with this hormone and cyclic nucleotide remains to be determined. This question is discussed in more detail below (see Discussion).

Cholesterol Synthesis

Long-term regulation. The effects of glucagon and several other hormones on cholesterol synthesis and the activity of HMG-reductase, the major rate-limiting enzyme in the biosynthetic pathway, have been studied in a number of laboratories over the past several years. Particularly important recent data, obtained by Porter and his coworkers (14, 31, 32, 39) indicate that the levels of hepatic HMG-CoA reductase and of cholesterol synthesis can be regulated by the relative amounts of stimulatory hormones, e. g. insulin and thyroid hormone, and of inhibitory hormones, e. g. glucagon and glucocorticoids. Thus, hepatic HMG-CoA reductase activity increased 2-7-fold after subcutaneous administration of insulin into normal or diabetic animals. The maximum effect occurred in 2-3 hours and could be elicited in animals that did not have access to food. Animals made diabetic exhibited a marked decrease in hepatic HMG-CoA reductase activity after 7 days (values approximately 5% of those of normal animals; 14). Moreover, the marked diurnal variation in enzymatic activity was abolished in the diabetic animal and was restored by insulin administration. These data suggest that insulin is necessary for the diurnal rise in HMG-CoA reductase activity. The fact that plasma insulin exhibits a diurnal variation (28), associated with feeding and similar to that of reductase activity, supports the notion that insulin is intimately involved in the diurnal change of reductase activity. Glucagon and perhaps other hormones also may be involved. Thus, the restorative effect of insulin on hepatic HMG-CoA reductase in diabetic animals is markedly suppressed by simultaneous administration of glucagon (14). The diurnal rise in reductase activity in liver of normal animals also can be suppressed markedly by administration of glucagon or cyclic AMP. Since glucagon exhibits a diurnal variation opposite to that of insulin (25), the data lead to the conclusion that the relative concentrations of insulin and glucagon play important roles in the regulation of the diurnal variation of hepatic HMG-CoA reductase activity.

Short-term regulation. The aforementioned data suggest a role for cyclic AMP in the relatively long-term regulation of HMG-CoA reductase and cholesterol biosynthesis. Recent experiments suggest the possibility for a role of the cyclic mononucleotide in short-term regulation as well. Utilizing rat liver slices, Bricker and Levey (6) demonstrated that incubation with cyclic AMP, 5×10^{-3}M, led to a more than 80% decrease in conversion of acetate to cholesterol. Only a minimal effect was noted at a concentration of 5×10^{-4}M. Similar effects have been observed with comparable concentrations of dibutyryl cyclic AMP with liver slices (2, 15) and isolated hepatocytes (7). Utilizing cell-free preparations (10000 x *g* supernatant solution), Bloxham and Akhtar (5) demonstrated a 10-fold reduction in cholesterol biosynthesis with 5 mM cyclic AMP in the incubation medium. The inhibition was noted when acetate, mevalonate, squalene or lanosterol was used as precursor. No effect was seen with cholest-7-en-β-ol as precursor. Thus, inhibition of the conversion of lanosterol to this sterol was suggested as the site of the cyclic AMP effect. Somewhat different data were obtained by Gibson and coworkers (4), who also utilized cell-free preparations (5000 x *g* supernatant solutions) of rat liver. Thus, these workers observed a 5-6 fold reduction in cholesterol biosynthesis from acetate when 2.5 mM cyclic AMP was added to the incubation. However, no effect on conversion of mevalonate to cholesterol was noted, and this suggested that HMG-CoA reductase was the enzyme affected. Indeed preincubation of liver slices, isolated cells or the 5000 x g supernatant solution for 60 min in the presence of 5 x 10^{-3} cyclic AMP resulted in decreased activities of HMG-CoA reductase, although the maximal effect on the enzyme was only an approximately 2-fold reduction. Moreover, utilizing isolated hepatocytes prepared by improved techniques, Edwards (15) demonstrated that the increase in HMG-CoA reductase activity noted in these cells upon addition of 10% rat serum was inhibited by at least 90% when cells were incubated for 3 hours in 3 x 10^{-4}M dibutyryl cyclic AMP. Gibson and coworkers suggested the possibility that reductase activity is regulated by a cyclic AMP-dependent phosphorylation-dephosphorylation mechanism because of the following findings: (1) washed microsomes preincubated with ATP, Mg^{++} and a cytosolic fraction exhibited a 10-fold reduction in HMG-CoA reductase activity; (2) addition of 10^{-6} cyclic AMP to the ATP and Mg^{++} accentuated the inhibition; and (3) a second preincubation with a separate cytosolic fraction resulted in some reversal of the inactivation. The physiologic significance of these observations with cyclic AMP was questioned by recent studies (11, 41). Although Raskin (41) confirmed the effects of high concentrations of cyclic AMP on cholesterol synthesis by liver slices, in perfusion experiments the addition of glucagon sufficient to cause a 50-fold increase in hepatic cyclic AMP levels resulted, in 30 min, in no significant change in cholesterol synthesis from acetate or octanoate or in HMG-CoA reductase activity. Moreover, Rudney and coworkers (11) examined the inhibitory effect of ATP and Mg^{++} in more detail and found that inhibition of HMG-CoA reductase by these factors varied widely from zero to complete. They could not demonstrate a requirement for a cytosolic fraction. Most important, utilizing specific antibody to precipitate the enzyme after incubation with ATP labeled uniformly with ^{14}C in the adenine portion and ^{32}P in the terminal phosphate, these workers were unable to demonstrate labeling of the protein. These data led to the conclusion that the mechanism of inhibition in the presence of ATP does not involve phosphorylation or adenylization of the enzyme and to the suggestion that competition between enzyme substrate and ATP might be responsible for the inhibition. Thus, at present, a physiologic role for cyclic AMP in the short-term regulation of cholesterol synthesis is unproven.

OBJECTIVES

The objectives of the work described below were to determine: (a) whether the marked increase in hepatic fatty acid synthetase activity that occurs at the time of weaning could be suppressed by the administration of glucagon, in the same way that activity is suppressed in animals refed after a fast; (b) whether the hepatic synthetase activity of animals recently weaned or refed after a fast is related to the tissue concentration of cyclic AMP at the time of sacrifice; and (c) whether there is a correlation between the intracellular levels of cyclic AMP and the rates of fatty acid synthesis and activities of fatty acid synthetase and acetyl-CoA carboxylase, under conditions of short and long-term regulation, in a cultured mammalian cell line, i. e. C-6 glial cells, in which lipogenesis is sensitive to the amount of exogenous lipid in the medium.

METHODS

Rats of the Sprague-Dawley strain were utilized. The fat-free diet was formulated according to Wooley and Sebrell (55). Preparation of liver extracts for enzyme assays was performed as described previously (48). Preparation of liver extracts for measurement of cyclic AMP was carried out according to Gilman (20). Preparation and administration of glucagon was performed as described previously (49). Fatty acid synthetase was assayed spectrophotometrically as described previously (48). One unit of synthetase activity is defined as the amount required to catalyze the oxidation of 1 nmole of NADPH per min at 37°C. Acetyl-CoA carboxylase was assayed by measuring the recovery of acid-stable radioactivity after incubation with $H^{14}CO_3^-$ (36). One unit of carboxylase activity is defined as the amount required to catalyze the fixation of 1 nmole of $H^{14}CO_3^-$ per min at 37°C. Cyclic AMP was measured by a modified cyclic AMP binding assay (35), based on the procedure of Gilman (20), with minor modifications. Protein was measured by the method of Lowry et al (34). The methods of C-6 glial cell culture and the preparation of extracts for enzymatic assays were described previously (50-52). Preparation of cellular extracts for the measurement of cylic AMP was carried out according to Clark et al. (12). Synthesis of fatty acids from [2-^{14}] acetate was determined as described previously (50). Statistical significance was determined by Student's t test (45).

RESULTS

Effect of glucagon on fatty acid synthetase in liver of animals weaned to a fat-free diet. To determine whether the marked increase in hepatic fatty acid synthetase activity that occurs at the time of weaning (47) could be suppressed by the administration of glucagon, late suckling animals were studied (Table 1). The pups were removed from their mother and provided a fat-free diet for 72 hours, during which time glucagon was administered in the total daily doses shown. In the three days after weaning hepatic synthetase activity rose dramatically, i.e. approximately 19-fold. This marked rise in activity was suppressed by the concurrent administration of glucagon. With the lower dose hepatic activity was approximately 75% of that in the animal that did not receive glucagon, and with the higher dose hepatic activity was only approximately 35% of that in the control. This inhibitory effect of glucagon on hepatic fatty acid synthetase in weaned animals is very similar in qualitative and quantitative characteristics to the effect of the hormone on hepatic synthetase in animals allowed to refeed after a fast (49).

TABLE 1

EFFECT OF GLUCAGON ON FATTY ACID SYNTHETASE IN LIVER OF ANIMALS WEANED TO A FAT-FREE DIET

Condition	Glucagon Dose (mg/100 g/day)	Fatty Acid Synthetase (units/mg protein)
Suckling	-	1.4 ± 0.2
Weaned	-	26.0 ± 1.7
Weaned	0.1	19.4 ± 2.0
Weaned	0.6	9.4 ± 0.8

Six of 8 animals of a litter were removed from their mother at 17 days of age. The 2 animals not weaned, i.e. "suckling", were sacrificed and hepatic synthetase activity determined. The 6 weaned animals, divided into 3 pairs, were fed a fat-free diet for 72 hours. During this period one pair was injected subcutaneously twice daily with glucagon in a dose of 0.05 mg/100 g body weight, one pair with 0.3 mg/100 g body weight, and one pair with 0.9% NaCl. At 20 days of age hepatic synthetase activity was determined. Values for enzymatic activities are means ± S.E.M. obtained from determinations of two levels of enzyme from each animal. Similar results were obtained in a separate experiment with another litter.

Relation of hepatic fatty acid synthetase and cyclic AMP in weaned or starved-refed animals. We next asked whether there was a discernible relation between the hepatic synthetase activity and cyclic AMP level in weaned (Table 2) or starved-refed animals (Table 3). Late suckling animals were weaned, fed a fat-free diet and injected with glucagon, as described above. A portion of each liver was assayed for either fatty acid synthetase activity or cyclic AMP level (Table 2). The marked rise in hepatic synthetase activity was again apparent in the weaned animals. In contrast, there was little difference in the hepatic concentrations of cyclic AMP among the various groups. Indeed, in the weaned animals only the difference between those given the higher dose of glucagon and the control animals was significant ($p < 0.05$), and the increase in the glucagon-injected animals was only 25%. In the starved-refed animals, although hepatic concentra-

TABLE 2

RELATION OF HEPATIC FATTY ACID SYNTHETASE AND CYCLIC AMP IN ANIMALS WEANED TO A FAT-FREE DIET

Condition	Glucagon Dose (mg/100 g/day)	Fatty Acid Synthetase (units/mg protein)	Cyclic AMP (pmol/mg protein)
Suckling	-	1.3 ± 0.2	1.67 ± 0.22
Weaned	-	20.3 ± 1.7	1.80 ± 0.16
Weaned	0.1	17.0 ± 1.3	2.09 ± 0.18
Weaned	0.6	7.7 ± 0.9	2.24 ± 0.20

This experiment was performed in an identical manner to that described in the legend of Table 1. However, one-half of each liver was utilized for assay of synthetase activity and one-half for determination of the concentration of cyclic AMP. Values are means ± S.E.M. obtained from determinations of two levels of extract from each animal. Similar results were obtained in a separate experiment with another litter.

tions of cyclic AMP did not differ, synthetase activities in the glucagon-injected animals were decreased by 35% and 52% in those given the low and high dose, respectively (Table 3). It should be emphasized that these data re: cyclic AMP were obtained 4 hours after the last injection of hormone and indicate only that there was no elevation of the concentration of cyclic AMP observable at the time of sacrifice of the animals. Transient increases of hepatic cyclic AMP levels would be expected after each injection. Indeed, one litter of 17-day-old rats was weaned to a fat-free diet and at 20 days of age injected subcutaneously with glucagon in a dose of 0.3 mg/100 g body weight. Pairs of animals were sacrificed at various times thereafter, and the hepatic levels of cyclic AMP at time zero, 15, 30, 60 and 120 min thereafter were 1.92, 4.05, 2.84, 2.06 and 1.84 pmol/mg protein, respectively.

TABLE 3

RELATION OF HEPATIC FATTY ACID SYNTHETASE AND CYCLIC AMP IN ANIMALS REFED A FAT-FREE DIET

Condition	Glucagon Dose (mg/100 g/day)	Fatty Acid Synthetase (units/mg protein)	Cyclic AMP (pmol/mg protein)
Fasted	-	3.1 ± 0.4	1.39 ± 0.15
Refed	-	82.8 ± 7.2	1.78 ± 0.24
Refed	0.1	53.2 ± 6.4	1.84 ± 0.33
Refed	0.6	39.5 ± 3.1	1.99 ± 0.24

Eight 100 g rats were fasted for 48 hours. At that time 2 animals, i.e. "fasted", were sacrificed and one-half portions of the liver utilized for assay of synthetase activity and for determination of cyclic AMP as described in the legend of Table 2. The remaining 6 animals were fed the fat-free diet and injected with glucagon as described in the legend of Table 1. After 48 hours hepatic synthetase activity and cyclic AMP concentration were determined. Values are means ± S.E.M. as described in the legend of Table 2. Similar results were obtained in a separate experiment.

Relation of fatty acid synthesis, fatty acid synthetase, acetyl-CoA carboxylase and cyclic AMP in C-6 glial cells upon removal of serum from the culture medium. To determine whether the increases in fatty acid synthesis and in the activity of fatty acid synthetase and acetyl-CoA carboxylase observed in C-6 glial cells upon removal of serum from the culture medium (50) were related to changes in intracellular cyclic AMP, these several parameters were examined in parallel (Table 4). Thus, after cells were grown initially in 10% serum, serum was removed from the culture medium and growth allowed to proceed for 24 hours. An approximately 17-fold increase in the rate of fatty acid synthesis from acetate was accompanied by a more than 2-fold increase in the specific activities of fatty acid synthetase and acetyl-CoA carboxylase. Despite these impressive changes in lipogenesis, no alteration in intracellular concentrations of cyclic AMP was observed. Thus, these data define an apparent dissociation between fatty acid

TABLE 4

RELATION OF FATTY ACID SYNTHESIS, FATTY ACID SYNTHETASE, ACETYL-CoA CARBOXYLASE AND CYCLIC AMP IN C-6 GLIAL CELLS UPON REMOVAL OF SERUM FROM THE CULTURE MEDIUM

Culture Medium	Fatty Acid Synthesis (cpm/mg protein x 10^{-3})	Fatty Acid Synthetase (units/mg protein)	Acetyl-CoA Carboxylase (units/mg protein)	Cyclic AMP (pmol/mg protein)
10% Serum	1.1 ± 0.2	2.0 ± 0.1	1.0 ± 0.2	28.7 ± 2.4
No Serum	19.0 ± 1.6	4.4 ± 0.3	2.5 ± 0.3	32.1 ± 2.8

Cells were transferred to a series of 25-cm^2 flasks and grown in 10% calf serum for 72 hours. At that time the flasks were divided into 3 groups, and for each group the medium was changed so that one-half of the flasks contained 10% calf serum and one-half no serum. After 24 hours, one of the following determinations was made on the cells of each group: 1) fatty acid synthesis from [2-^{14}C] acetate, 2) activities of fatty acid synthetase and acetyl-CoA carboxylase, and 3) concentration of cyclic AMP. Values are means ± S.E.M. obtained from determinations of two levels of extracts from each of two flasks.

synthesis and intracellular cyclic AMP.

Relation of fatty acid synthesis, fatty acid synthetase, acetyl-CoA carboxylase and cyclic AMP in C-6-glial cells treated with norepinephrine. We next evaluated the relation of lipogenesis and cyclic AMP in C-6 glial cells under conditions expected to be associated with an increase in cyclic AMP concentrations, e.g. after exposure to norepinephrine (Table 5). Catecholamines cause an increase in adenyl cyclase activity (40) and cyclic AMP levels (21) in cultured C-6 glial cells. To eliminate complicating factors that might be associated with the use of serum, we studied the effect of norepinephrine in the chemically defined medium without serum. The catecholamine caused no significant change in any of the parameters of lipogenesis. In contrast, cells treated with norepinephrine exhibited 4-fold higher concentrations of cyclic AMP than untreated cells. Thus, a clear dissociation between lipogenesis and intracellular cyclic AMP is again apparent.

DISCUSSION

The objectives of this work, as outlined in the Introduction, have been accomplished as follows: first, the marked increase in hepatic activity of fatty acid synthetase that occurs at the time of weaning is markedly suppressed by concurrent administration of glucagon. This effect of glucagon shares the qualitative and quantitative characteristics of the effect of the hormone on hepatic lipogenesis in starved-refed animals. Second, there is no clear relation between the hepatic synthetase activity and cyclic AMP level at the time of sacrifice of animals weaned or starved-refed. Third, in C-6 glial cells there is a readily defined dissociation between intracellular levels of cyclic AMP and parameters of lipogenesis, i.e. fatty acid synthesis from acetate and the activities of fatty acid synthetase and acetyl-CoA carboxylase.

The data presented in this and previous reports indicate that the administration of glucagon, particularly in high doses, is associated with a distinct, long-

TABLE 5

RELATION OF FATTY ACID SYNTHESIS, FATTY ACID SYNTHETASE, ACETYL-CoA CARBOXYLASE AND CYCLIC AMP IN C-6 GLIAL CELLS TREATED WITH NOREPINEPHRINE

Additions	Fatty Acid Synthesis (cpm/mg protein x 10^{-3})	Fatty Acid Synthetase (units/mg protein)	Acetyl-CoA Carboxylase (units/mg protein)	Cyclic AMP (pmol/mg protein)
None	21.4 ± 2.7	5.0 ± 0.8	2.8 ± 0.6	12.6 ± 1.7
Norepinephrine, 10^{-5}M	19.1 ± 3.2	4.6 ± 0.5	2.6 ± 0.5	50.9 ± 4.2

Cells were transferred to a series of 25-cm^2 flasks and grown in 10% calf serum for 72 hours. At that time the flasks were divided into 3 groups as described in the legend of Table 4, except that for each group the medium was changed so that one-half of the flasks contained no serum plus norepinephrine, 10^{-5}M, and one-half no serum with no additions of hormone. After 24 hours, rates of fatty acid synthesis, activities of fatty acid synthetase and acetyl-CoA carboxylase, and concentrations of cyclic AMP were determined, as described in the legend of Table 4.

term inhibition of fatty acid synthesis in mammalian liver. The effect is mediated by a decrease in synthesis of fatty acid synthetase and probably other lipogenic enzymes. The mechanism(s) by which glucagon produces a decrease in synthesis of hepatic fatty acid synthetase is unknown. It is likely that intracellular cyclic AMP is involved but the manner of this involvement is unclear. The recent studies of malic enzyme in isolated liver cells by Goodridge and Adelman (23) suggest that glucagon leads to a decrease in malic enzyme synthesis by acting at the level of translation or cytoplasmic messenger RNA processing. Cyclic AMP mimicked the effect of glucagon. These observations certainly do not prove that glucagon and/or cyclic AMP produced the effect directly. Indeed it is quite possible that glucagon and the cyclic mononucleotide produce the response indirectly as a consequence of one or more of their well-recognized metabolic effects, particularly those relating to carbohydrate metabolism. For example, glucagon, as well as cyclic AMP, leads to a drastic reduction in the conversion of glucose to CO_2 by liver slices (37). Such an effect would be expected to lead to a decrease in the hepatic concentration of intermediates of the glycolytic pathway at the triose phosphate step or beyond. We have presented data previously that suggest a major role for such intermediates in the regulation of synthesis of hepatic fatty acid synthetase (53, 54).

Although the aforementioned data suggest that cyclic AMP may mediate the effects of certain specific hormones on lipogenesis, it is clear that there is no uniform, obligatory relationship between intracellular cyclic AMP and lipid synthesis. Utilizing a mammalian cell line (C-6 glial cells) that contains an actively regulated pathway for *de novo* fatty acid synthesis, we demonstrated marked changes in the flux and enzymatic activities of this pathway that were unaccompanied by a comparable change in intracellular cyclic AMP concentrations. Moreover, norepinephrine induced a marked change in intracellular cyclic AMP concentrations that was unaccompanied by any change in the flux and enzymatic activities of the *de novo* pathway. Clearly, however, these experiments in no way could detect important changes in intracellular compartmentalization, transport or binding of cyclic AMP. We believe that insight into such changes will be needed to define the nature of the relationship of cyclic AMP and lipogenesis.

Acknowledgements

Supported by Grant 1-R01-HD-07464-04 from the National Institutes of Health.

J. J. Volpe is recipient of a Research Career Development Award 1-K4-HD70608 from the National Institutes of Health.

REFERENCES

1. Akhtar, M. and Bloxham, D. P. (1970): The co-ordinated inhibition of protein and lipid biosynthesis by adenosine 3':5'-cyclic monophosphate. Biochem. J. 120, 11P.

2. Allred, J. B. and Roehrig, K. L. (1972): Inhibition of hepatic lipogenesis by cyclic-3',5'-nucleotide monophosphates. Biochem. Biophys. Res. Commun. 46, 1135-1139.

3. Allred, J. B. and Roehrig, K. L. (1973): Inhibition of rat liver acetyl coenzyme A carboxylase by N^6,O^2-dibutyryl cyclic adenosine 3':5'-monophosphate in vitro. J. Biol. Chem. 248, 4131-4133.

4. Beg, Z. H., Allman, D. W. and Gibson, D. C. (1973): Modulation of 3-hydroxy-3-methylglutaryl coenzyme A reductase activity with cAMP and with protein fractions of rat liver cytosol. Biochem. Biophys. Res. Commun. 54, 1362-1369.

5. Bloxham, D. P. and Akhtar, M. (1971): Studies on the control of cholesterol biosynthesis: the adenosine 3':5'-cyclic monophosphate-dependent accumulation of a steroid carboxylic acid. Biochem. J. 123, 275-278.

6. Bricker, L. A. and Levey, G. S. (1972): Evidence for regulation of cholesterol and fatty acid synthesis in liver by cyclic adenosine 3',5'-monophosphate. J. Biol. Chem. 247, 4914-4915.

7. Capuzzi, D. M., Rothman, V. and Margolis, S. (1974): The regulation of lipogenesis by cyclic nucleotides in intact hepatocytes prepared by a simplified technique. J. Biol. Chem. 249, 1286-1294.

8. Carlson, C. A.and Kim, K.-H. (1973): Regulation of hepatic acetyl coenzyme A carboxylase by phosphorylation and dephosphorylation. J. Biol. Chem. 248, 378-380.

9. Carlson, C. A. and Kim, K.-H. (1974): Regulation of hepatic acetyl coenzyme A carboxylase by phosphorylation and dephosphorylation. Arch. Biochem. Biophys. 164, 478-489.

10. Carlson, C. A.and Kim, K.-H. (1974): Differential effects of metabolites on the active and inactive forms of hepatic acetyl CoA carboxylase. Arch. Biochem. Biophys. 164, 490-501.

11. Chow, J. C., Higgins, M. J. P. and Rudney, H. (1975): The inhibitory effect of ATP on HMG-CoA reductase. Biochem. Biophys. Res. Commun. 63, 1077-1084.

12. Clark, R. B., Gross, R., Su, Y.-F. and Perkins, J. P. (1974): Regulation of adenosine 3':5'-monophosphate content in human astrocytoma cells by adenosine and the adenine nucleotides. J. Biol. Chem. 249, 5296-5303.

13. Claycomb, W. C. and Kilsheimer, G. S. (1969:: Effect of glucagon, adenosine-3',5'-monophosphate and theophylline on free fatty acid release by rat liver slices and on tissue levels of coenzyme A esters. Endocrinology 84, 1179-1183.

14. Dugan, R. E., Ness, G. C., Lakshmanan, M. R., Nepokroeff, C. M. and Porter, J. W. (1974): Regulation of hepatic β-hydroxy-β-methylglutaryl coenzyme A reductase by the interplay of hormones. Arch. Biochem. Biophys. 161, 499-504.

15. Edwards, P. A. (1975): The influence of catecholamines and cyclic AMP on 3-hydroxy-3-methylglutaryl coenzyme A reductase activity and lipid biosynthesis in isolated rat hepatocytes. Arch. Biochem. Biophys. 170, 188-203.

16. Emmer, M., De Crombrugghe, B., Pastan, I. and Perlman, R. (1970): Cyclic AMP receptor protein of E. coli: Its role in the synthesis of inducible enzymes. Proc. Natl. Acad. Sci. USA 66, 480-487.

17. Exton, J. H., Corgin, J. G. and Harper, S. C. (1972): Control of gluconeogenesis in liver. V. Effects of fasting, diabetes, and glucagon on lactate and endogenous metabolism in the perfused rat liver. J. Biol. Chem. 247, 4996-5003.

18. Exton, J. H., Robison, G. A., Sutherland, E. W. and Park, C. R. (1971): Studies on the role of adenosine 3',5'-monophosphate in the hepatic actions of glucagon and catecholamines. J. Biol. Chem. 246, 6166-6177.

19. Garcia, D. R. and Holten, D. (1975): Inhibition of rat liver glucose-6-phosphate dehydrogenase synthesis by glucagon. J. Biol. Chem. 250, 3960-3965.

20. Gilman, A. G. (1970): A protein binding assay for adenosine 3':5'-cyclic monophosphate. Proc. Natl. Acad. Sci. USA 67, 305-312.

21. Gilman, A. G. and Nirenberg, M. (1971): Effect of catecholamines on the adenosine 3':5'-cyclic monophosphate concentrations of clonal satellite cells of neurons. Proc. Natl. Acad. Sci. 68, 2165-2168.

22. Goodridge, A. G. (1973): Regulation of fatty acid synthesis in isolated hepatocytes prepared from the livers of neonatal chicks. J. Biol. Chem. 248, 1924-1931.

23. Goodridge, A. G. and Adelman, T. G. (1976): Regulation of malic enzyme synthesis by insulin, triiodothyronine, and glucagon in liver cells in culture. J. Biol. Chem. 251, 3027-3032.

24. Goodridge, A. G., Garay, A. and Silpananta, P. (1974): Regulation of lipogenesis and the total activities of lipogenic enzymes in a primary culture of hepatocytes from prenatal and early postnatal chicks. J. Biol. Chem. 249, 1469-1475.

25. Harris, R. A., Rivera, E. R., Villemez, C. L. and Quackenbush, F. W. (1967): Mechanism of suckling-rat hypercholesteremia. II. Cholesterol biosynthesis and cholic acid turnover studies. Lipids 2, 137.

26. Haugaard, E. S. and Stadie, W. C. (1953): The effect of hyperglycemic-glycogenolytic factor and epinephrine on fatty acid synthesis. J. Biol. Chem. 200, 753-757.

27. Jost, J.-P. and Richenberg, H. V. (1971): Cyclic AMP. Ann. Rev. Biochem. 40, 741-774.

28. Kaul, L. and Berdanier, C. D. (1972): Diurnal rhythms of serum insulin and hepatic NADP-linked dehydrogenases in meal-fed rats. Fed. Proc. 31, 669.

29. Klain, G. J. and Meikle, A. W. (1974): Mannoheptulose and fatty acid synthesis in the rat. J. Nutrition 104, 473-477.

30. Klain, G. J. and Weiser, P. C. (1973): Changes in hepatic fatty acid synthesis following glucagon injections in vivo. Biochem. Biophys. Res. Commun. 55, 76-83.

31. Lakshmanan, M. R., Dugan, R. E., Nepokroeff, C. M., Ness, G. C. and Porter, J. W. (1975): Regulation of rat liver β-hydroxy-β-methylglutaryl coenzyme A reductase activity and cholesterol levels of serum and liver in various dietary and hormonal states. Arch. Biochem. Biophys. 168, 89-95.

32. Lakshmanan, M. R., Nepokroeff, C. M., Ness, G. G., Dugan, R. E. and Porter, J. W. (1973): Stimulation by insulin of rat liver β-hydroxy-β-methylglutaryl coenzyme A reductase and cholesterol-synthesizing activities. Biochem. Biophys. Res. Commun. 50, 704-710.

33. Lakshmanan, M. R., Nepokroeff, C. M. and Porter, J. W. (1972): Control of the synthesis of fatty-acid synthetase in rat liver by insulin, glucagon, and adenosine 3',5' cyclic monophosphate. Proc. Natl. Acad. Sci. USA 69, 3516-3519.

34. Lowry, O. H., Rosebrough, M. J., Farr, A. L. and Randall, R. J. (1951): Protein measurement with the Folin phenol reagent. J. Biol. Chem. 193, 265-275.

35. Lust, W. D., Dye, E., Deaton, A. V. and Passonneau, J. V. (1976): A modified cyclic AMP binding assay. Anal. Biochem. 72, 8-15.

36. Martin, D. B. and Vagelos, P. R. (1962): The mechanism of tricarboxylic acid cycle regulation of fatty acid synthesis. J. Biol. Chem. 237, 1787-1792.

37. Meikle, A. W., Klain, G. J. and Hannon, J. P. (1973): Inhibition of glucose oxidation and fatty acid synthesis in liver slices from fed, fasted and fasted-refed rats by glucagon, epinephrine, and cyclic adenosine-3',5'-monophosphate. Proc. Soc. Exp. Biol. Med. 143, 379-381.

38. Menahan, I. A. and Wieland, O. (1969): The role of endogenous lipid in gluconeogenesis and ketogenesis of perfused rat liver. Eur. J. Biochem. 9, 182-188.

39. Ness, G. C., Dugan, R. E., Lakshmanan, M. R., Nepokroeff, C. M., and Porter, J. W. (1973): Stimulation of hepatic β-hydroxy-β-methylglutaryl coenzyme A reductase activity in hypophysectomized rats by L-tri-iodothyronine. Proc. Natl. Acad. Sci. USA 70, 3839-3842.

40. Opler, L. A. and Makman, M. H. (1972): Mediation by cyclic AMP of hormone-stimulated glycogenolysis in cultured rat astrocytoma cells. Biochem. Biophys. Res. Commun. 46, 1140-1145.

41. Raskin, P., McGarry, J. D. and Foster, D. W. (1974): Independence of cholesterol and fatty acid biosynthesis from cyclic adenosine monophosphate concentration in the perfused rat liver. J. Biol. Chem. 249, 6029-6032.

42. Regen, D. M. and Terrell, E. B. (1968): Effects of glucagon and fasting on acetate metabolism in perfused rat liver. Biochim. Biophys. Acta 170, 95-111.

43. Rous, S. (1970): Effect of dibutyryl cAMP on the enzymes of fatty acid synthesis and of glycogen metabolism. FEBS Letters 12, 45.

44. Rudack, D., Davie, B. and Holten, D. (1971): Regulation of rat liver glucose-6-phosphate dehydrogenase levels by adenosine 3'-5'-monophosphate. J. Biol. Chem. 246, 7823-7824.

45. Snedecor, G. W. (1965): Statistical Methods Applied to Experiments in Agriculture and Biology. Iowa State College Press, Ames.

46. Tepperman, H. M. and Tepperman, J. (1972): Mechanism of inhibition of lipogenesis by 3',5' cyclic AMP. In: Insulin Action. I. B. Fritz, Ed. Academic Press, New York, p. 543.

47. Volpe, J. J. and Kishimoto, Y. (1972): Fatty acid synthetase of brain: Development, influence of nutritional and hormonal factors and comparison with liver enzyme. J. Neurochem. 19, 737-753.

48. Volpe, J. J., Lyles, T. O., Roncari, D. A. K. and Vagelos, P. R. (1973): Fatty acid synthetase of developing brain and liver. Content, synthesis, and degradation during development. J. Biol. Chem. 248, 2502-2513.

49. Volpe, J. J. and Marasa, J. C. (1975): Hormonal regulation of fatty acid synthetase, acetyl CoA carboxylase and fatty acid synthesis in mammalian adipose tissue and liver. Biochim. Biophys. Acta 380, 454-472.

50. Volpe, J. J. and Marasa, J. C. (1975): Regulation of palmitic acid synthesis in cultured glial cells: effects of lipid on fatty acid synthetase, acetyl-CoA carboxylase, fatty acid and sterol synthesis. J. Neurochem. 25, 333-340.

51. Volpe, J. J. and Marasa, J. C. (1976): Regulation of palmitic acid synthesis in cultured glial cells: effects of glucocorticoid on fatty acid synthetase, acetyl-CoA carboxylase, fatty acid and sterol synthesis. J. Neurochem. 27, 841-846.

52. Volpe, J. J. and Marasa, J. C. (1976): Long term regulation by theophylline of fatty acid synthetase, acetyl-CoA carboxylase and lipid synthesis in cultured glial cells. Biochim. Biophys. Acta 431, 195-205.

53. Volpe, J. J. and Vagelos, P. R. (1974): Regulation of mammalian fatty acid synthetase. The roles of carbohydrate and insulin. Proc. Natl. Acad. Sci. USA 71, 889-893.

54. Volpe, J. J. and Vagelos, P. R. (1976): Mechanisms and regulation of biosynthesis of saturated fatty acids. Physiol. Rev. 56, 339-417.

55. Wooley, J. G. and Sebrell, W. H. (1945): Niacin (nicotinic acid), an essential growth factor for rabbits fed a purified diet. J. Nutr. 29, 191-199.

56. Zubay, G., Schwartz, D. and Beckwith, J. (1970): Mechanism of activation of catabolite-sensitive genes: A positive control system. Proc. Natl. Acad. Sci. USA 66, 104-110.

LOWERING OF PLASMA LIPIDS, THE MAJOR EFFECT OF REPEATED GLUCAGON ADMINISTRATION IN RATS

F. Gey, H. Georgi and E. Buhler

Research Departments, F. Hoffmann-La Roche and Co. Ltd.
4002 Basel, Switzerland

The most prominent long lasting effect of repeated injection of commercial pancreatic glucagon (0.1-30 mg/kg s.c. twice daily) in fed and starving rats is a depression of the major plasma lipids. This lipid-lowering effect of glucagon becomes maximal after 2-3 days and persists for at least 4 weeks. After withdrawal of glucagon, the lipid levels, slowly return to normal. Plasma lipid depression occurs at glucagon doses (0.1 mg/kg) which have no comparable long-lasting effect on the level of other plasma components, such as glucose, FFA, α-amino nitrogen and urea, as well as on urinary excretion of urea. Dose-frequency curves suggest that lipid depression could also be obtained by the frequent release of small pulses of endogenous glucagon. On a molar basis, glucagon is more potent than other lipid-lowering hormones, like thyroxine and oestradiol. The injection of glucagon leads to a marked increase of the glucagon/insulin ratio in plasma, in spite of a considerable insulin release. Experiments in streptozotocin-treated rats with fully developed diabetes demonstrate that the lipid depression by glucagon does not depend on insulin. In consequence, a role of glucagon in the physiologic regulation of plasma lipids in the rat is suggested.

Recent studies in rats (6, 8, 20, 23, 38, 39, 41) have shown a reduction of plasma cholesterol or triglycerides by single or repeated injection of moderate to excessive doses of glucagon, i.e. 0.3 to 10 mg/kg. In men, including patients with hyperlipemia, a depression of plasma lipids has also been observed after daily injections of 0.07 - 5 mg glucagon (1, 2, 10, 13, 35, 42, 43). However, several questions have remained open. Among them:

- What are the relative changes of the major plasma lipids?
- What is the relative magnitude of the glucagon-induced changes in carbohydrate, lipid, and nitrogen metabolism?
- What are the effects of timing and dosage?
- Are the hypolipemic effects of glucagon dependent upon a decrease of body weight?
- What is the mechanism of the hypolipemic effect of glucagon?
- How does glucagon compare with other hypolipemic hormones?

- What is the role of the glucagon/insulin ratio?
- What is the physiologic significance of the hypolipemic action of glucagon?

The present paper attempts to deal with some of these questions.

METHODS

Randomized groups of pathogen-free albino rats of Wistar origin received s. c. injections of increasing doses of pancreatic glucagon at different intervals prior to decapitation (if not otherwise stated, at 8 a.m. and 4 p.m., with the last injection 3 hours prior to decapitation). Commercial glucagon (Novo and, in some cases, Lilly) gave the same results as highly purified pancreatic glucagon (Lilly, almost insulin-free). Control rats were injected with an equal volume of saline. In the *in vivo* studies each point in the following figures and tables represents mean ± S. E. of 3-6 experiments, each with 5-9 animals. Most data were obtained in female rats of about 150 g body weight, but the essentially similar results were obtained in males and in younger and older females.

Animals with fully developed streptozotocin diabetes were obtained as follows: streptozotocin (80 mg/kg) was injected i.v. under ether anesthesia. The animals received insulin therapy (five days, 1.2 U/kg i.m. twice a day; 0.6 U/kg twice a day for 2 additional days and, finally, a single dose of 0.6 U/kg for 1 day). Animals with plasma glucose levels of 500-700 mg/100 ml, accompanied by marked polydipsia and glycosuria for 1 weeks following insulin withdrawal, were chosen for the experiment.

Conventional photometric or fluorimetric techniques were used for the determination of cholesterol (4), lipid phosphorus (27), glucose (37), α-amino nitrogen (34), and urea (31) in heparinized plasma and of triglycerides in hydrocarbon extracts (40), free fatty acids in heptane extracts (29). Radioimmunoassays were used for plasma insulin (IRI; Insulin RIA kit IM 78, Amersham, GB) and plasma glucagon (18, using Unger's antibody 30 K). These measurements were kindly carried out by Dr. W. A. Müller, Geneva.

In preparation for liver perfusion, rats were starved for 2 days and subsequently fed ad libitum from 7 to 11 a. m. for 7 days. Prior to liver perfusion in situ (17), the liver was flushed with 10 ml Krebs-Ringer buffer, pH 7.4 and with 25 ml of medium. The perfusion was carried out at a constant flow rate and constant pressure and was divided into 3 periods: one hour for equilibration, a second hour for control observations, i. e. for the measurement of net secretion of cholesterol, triglycerides, glucose and urea into the perfusion medium, and one hour with the addition of glucagon (half of the dose in the first 5 minutes and the second half during the remaining minutes). The livers were perfused with pooled serum obtained from groups of 50 rats fed as described above.

RESULTS

Following the first and the fifth injection of a moderate dose of glucagon (0.5 mg/kg) the following immediate changes were observed in the plasma of fed and starving animals (Fig. 1); the FFA rose almost immediately, but decreased shortly thereafter; there was a slight rise in glucose (the secondary decrease in glucose and FFA may be explained at least in part, by the reactive insulin secretion); the α-amino nitrogen showed the expected marked decrease lasting about 2 hours; plasma and urinary urea did not change significantly.

In contrast to these well-known short term effects of glucagon, after one or 5 injections, the basal levels of all major plasma lipids was markedly decreased. In addition to this long term effect there was a further acute lipid depression 15 to 180 minutes after the last injection (Fig. 1).

Measurements made at the 3 hour intervals after each injection (Fig. 2) revealed that glucagon had a cumulative effect on the major plasma lipids, which became maximal after two days and lasted for about 3-8 hours after the last administration. The present studies extend previous observations (6, 8, 20, 22, 23, 38, 39, 41) and show that all major plasma lipids, i.e. cholesterol, triglycerides and phospholipids respond similarly to glucagon. In addition, they show that

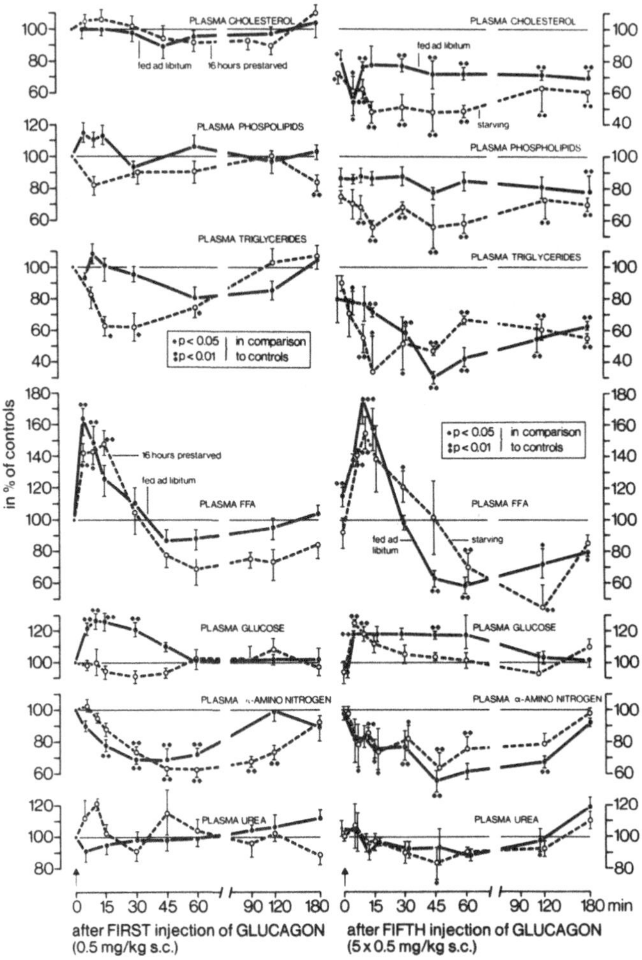

Fig. 1. Short-term changes of various plasma parameters after the first and fifth injection of glucagon (0.5 mg/kg, s.c.) in fed and starved rats.

glucagon is almost as potent in fed rats as in starving animals, indicating that the glucagon-induced lipid depression cannot be ascribed to an alteration of the caloric balance.

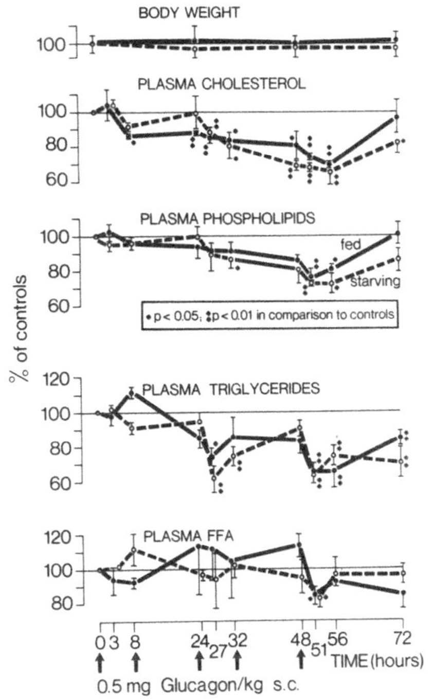

Fig. 2. Long lasting cumulative depression of the major plasma lipids in fed and starved rats following 1 to 5 injections of glucagon (0.5 mg/kg, s.c.).

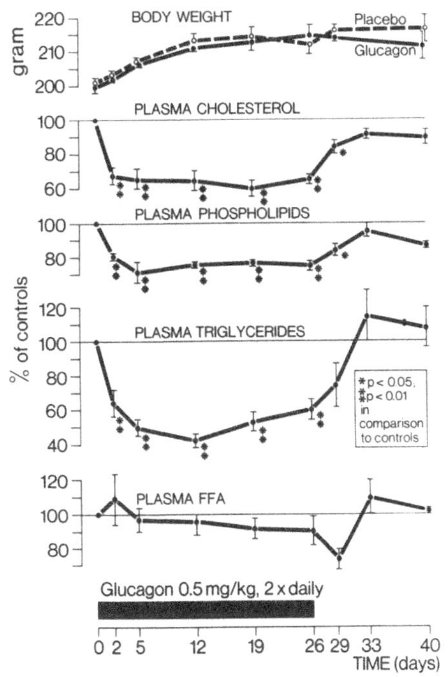

Fig. 3. Persistent depression of the major plasma lipids in fed rats by glucagon administration (0.5 mg/kg, s.c. twice a day for 26 days).

When the administration of glucagon was continued for 1 month (Fig. 3) the lipid lowering effect persisted as long as glucagon was administered, regardless of changes in body weight, as suggested by other authors (6, 8, 41). The lipid response-curves to 2 daily glucagon injections for 2 days was dose dependent (Fig. 4). Excessive amounts, as 10-30 mg/kg, depressed the major plasma lipids to about 20-40% of the control levels. The minimal effective dose was approximately 0.1 mg glucagon/kg, corresponding to 29 nmoles/kg, or an amount that produced no changes in other biochemical parameters (Fig. 5) lasting longer than 3 hours, except a slight increase of plasma urea. Thus, the most prominent persistent effect of repeated low doses of glucagon (0.1 mg/kg) in the normal rat was hypolipemia.

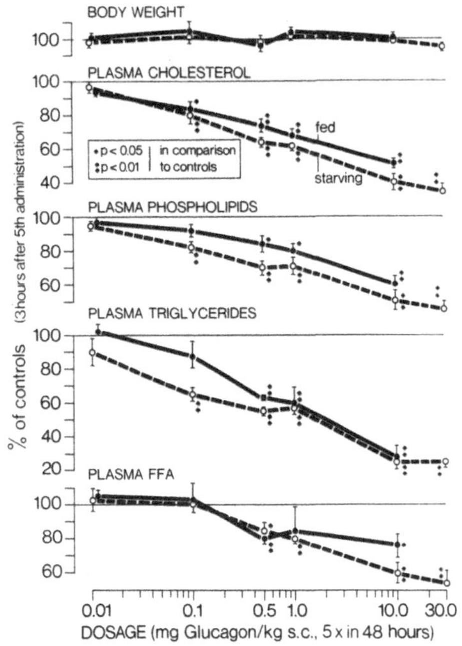

Fig. 4. Dose-response curves of plasma lipids in fed and starved rats after 5 s.c. injections of glucagon. The lipids were measured 3 hours after the last injection.

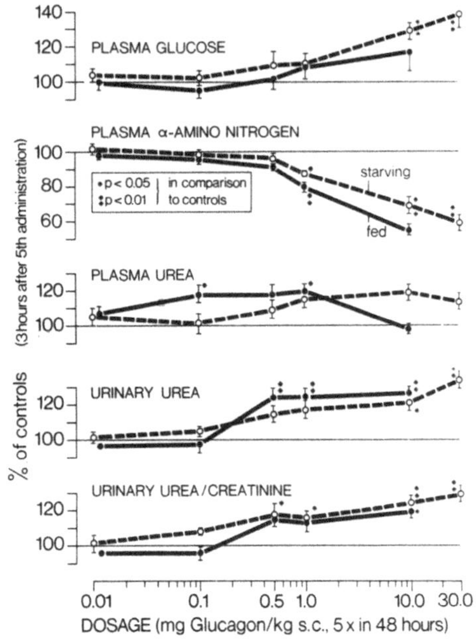

Fig. 5. Dose-response curves of non-lipid plasma parameters and urinary urea in fed and starved rats after 5 s.c. injections of glucagon. The plasma values were measured 3 hours after the last injection, the urine was collected for the whole experimental period (51 hours).

It may be argued that even this dose is in the pharmacologic, rather than in the physiologic range and, for this reason, dose-frequency studies were made. In these studies, the total 2 day dosage was divided into 3, 5, 9 or 17 portions (Fig. 6). Since subdividing the dose into 9 or 17 portions resulted in an increased efficacy, it may be assumed that the total daily dosage of glucagon can be

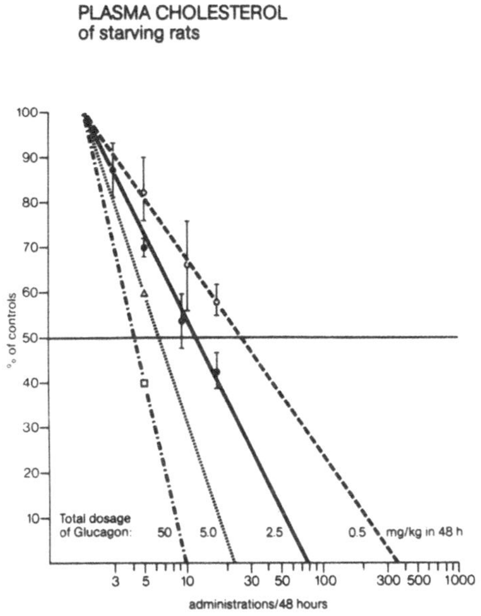

Fig. 6. Dose-frequency curves for plasma cholesterol of starved rats injected 3 to 17 times with 4 different doses of glucagon in 48 hours. The measurements were carried out 3 hours after the last injection.

reduced by increasing the frequency of administration. For instance, a 50% depression of plasma cholesterol could be achieved by a total dose of 2.5 mg/kg, divided into 12 subdoses or by a total dose of 0.5 mg/kg, divided into 26 subdoses. The number of subdoses for each total dose of glucagon may be seen in Fig. 7 where all points capable of reducing plasma lipid by 50% were obtained from Fig. 6. This plot suggests that a 50% depression of plasma cholesterol may be expected from a total amount of glucagon as little as 1000 ng/kg subdivided into 250 doses of 4 ng/kg given every 11 1/2 minutes for 2 days. If this kind of extrapolation were correct, one would have to conclude that exogenous glucagon can depress plasma lipids in minute amounts. This conclusion is acceptable considering the fact that a total daily dosage of 5000 ng/kg in the rat corresponds to about 8% of the total pancreatic glucagon content (roughly 1500 ng/250 g rat or 6000 ng/kg; 19, 22, 45). Almost identical dose-frequency curves were obtained for the depression of plasma triglycerides by glucagon.

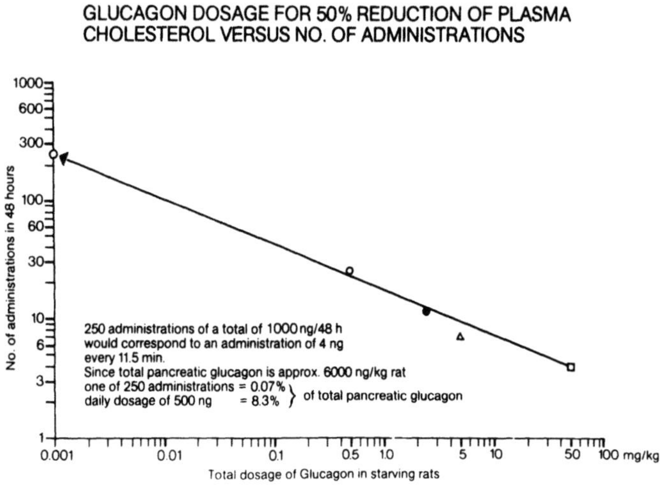

Fig. 7. Doses of glucagon required to cause a 50% reduction of plasma cholesterol obtained from Fig. 6, plotted against the number of injections. Extrapolation indicates that a dose as little as 1 µg/kg subdivided into 250 subdoses should result in 50% depression of plasma cholesterol as well.

In conclusion, glucagon may exert a strong and lasting hypolipemic effect in physiologic amount, provided it is released by the pancreas in numerous small pulses. Indeed, on a molar basis, glucagon is the most potent of several lipid-lowering hormones in starving rats (Fig. 8). Thus, glucagon achieves about the same maximal effect as oral oestradiol, but is about 10 times more potent if doses required to obtain a 50% depression are compared. D- and L-thyroxine (s. c. as well as per os) are about 100 times less potent than glucagon in starving rats and lack any activity in fed rats. With regard to plasma triglyceride depression, glucagon obviously exceeds any other hormone.

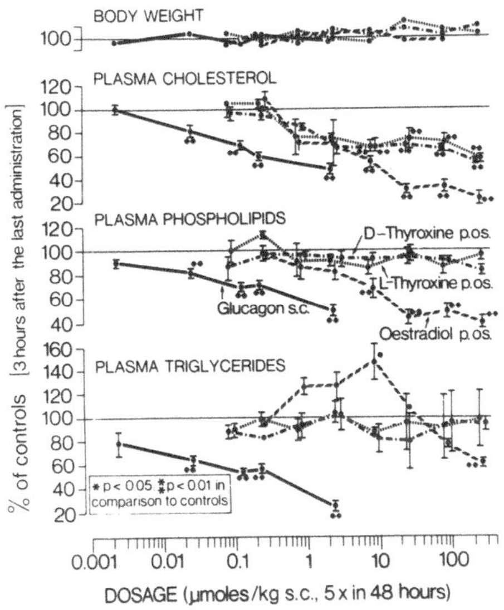

Fig. 8. Comparison of the hypolypemic effects of glucagon, oestradiol, D- and L-thyroxine on a molar basis. The hormones were given 5 times in 48 hours, the measurements were carried out 3 hours after the last administration. Starved rats.

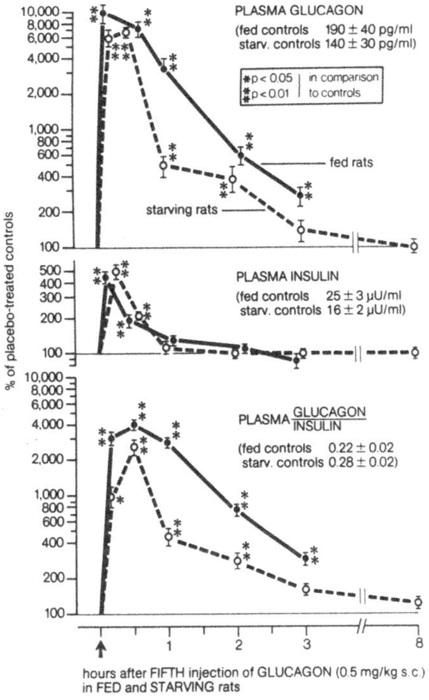

Fig. 9. Plasma glucagon, insulin and glucagon/insulin ratio after the fifth s. c. injection of 0.5 mg/kg of glucagon in fed and starved rats.

Since glucagon is a well-known insulin secretagogue, the glucagon to insulin ratio may be an important factor. Under the present experimental conditions, an injection of 0.5 mg glucagon/kg caused a 60- to 100-fold increase in plasma glucagon, and about a 5-fold increase in plasma insulin, or about a 28- to 40-fold increase in the glucagon/insulin ratio (Fig. 9). This rise could be of particular importance at the cellular receptor sites, e. g. in the liver. Indeed, an alteration of the glucagon/insulin ratio may have a role in the regulation of plasma lipids in genetically hyperlipemic obese rats (11), in man (10) and in the action of lipid-lowering drugs, such as nicotinic acid (21, 30, 44), clofibrate (9, 14, 15) and halofinate (12).

In spite of the glucagon-induced release of insulin the hypolipemic effect of glucagon does not appear to depend upon the presence of insulin, as indicated by experiments in streptozotocin-diabetic rats (Fig. 10). Under our experimental conditions, the pancreas of these animals lost more than 95.5% of its insulin and the injection of glucagon failed to increase plasma insulin, as it did in normal non-diabetic animals. Nevertheless, in the diabetic rats, in spite of their hypertriglyceridemia, repeated injections of a standard dose of glucagon decreased plasma lipids to a similar degree as in the non-diabetic controls.

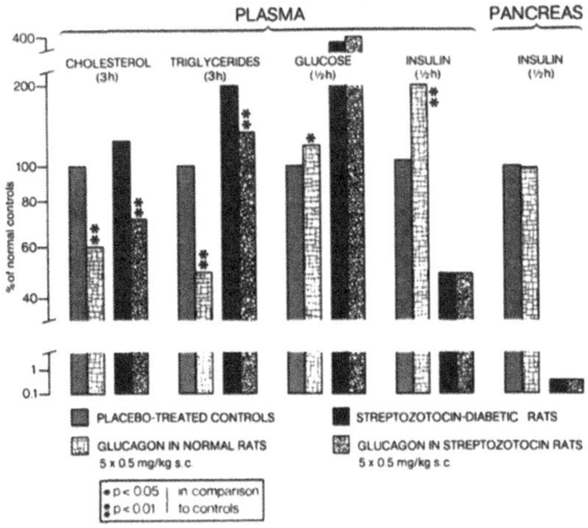

Fig. 10. Effect of 5 glucagon injections (0.5 mg/kg s.c., in 48 hours) in normal rats and in rats rendered diabetic with streptozotocin. The lipids were measured 3 hours, glucose and insulin one-half hour after the last injection.

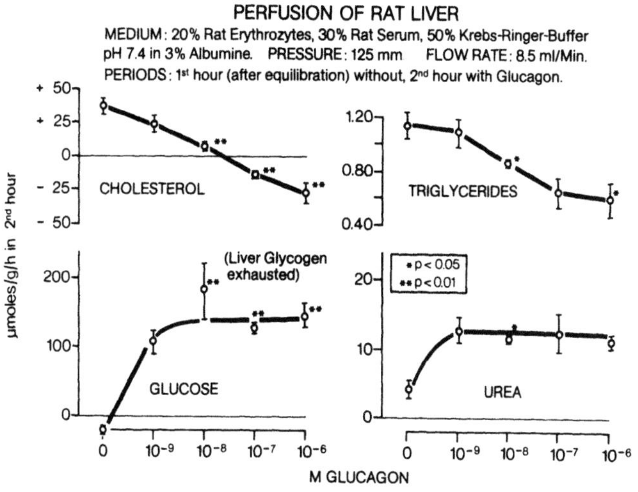

Fig. 11. Isolated perfused liver of fed rats: Effect of the concentration of glucagon in the perfusion medium on the hepatic secretion of cholesterol, triglycerides, glucose and urea. Ave. ± SEM of 3-4 experiments. Medium: 20% rat erythrocytes; 20% rat serum; 50% Krebs-Ringer buffer with 3% albumin, pH 7.4; perfusion pressure, 125 mm Hg; flow rate, 8.5 ml/min; after equilibration, the perfusion was carried out for 1 hour without and for 1 hour with glucagon.

With regard to the mechanism of the plasma lipid-lowering action of glucagon in rats, a major transfer of lipids into blood cells is unlikely, since under the present experimental conditions the total cholesterol and triglycerides content of these tissues is not altered significantly by glucagon (22), in agreement with the observation of others (6, 38, 41). Therefore, and because of experiments in vitro (3, 5, 24, 25, 28, 32, 33), in the perfused liver (16, 23, 26, 36, 46) and in vivo (6, 8), it has been suggested that glucagon suppresses lipid and/or lipoprotein synthesis in the liver. Perfusion of the isolated liver of fed rats with a serum-rich medium at constant pressure and flow (Fig. 11) demonstrated that glucagon at 10^{-9} M concentration decreased the secretion of cholesterol and triglycerides and increased hepatic glucose output and urea formation. Increasing the concentration enhanced the effects of glucagon, and when the concentration reached 10^{-7} M, cholesterol was taken up from the perfusion medium. This reduced secretion of triglycerides agrees with a previous report (46) although in our study, the livers were about 10 to 100 times more sensitive to glucagon. The finding of a dose-dependent blockade of cholesterol secretion and cholesterol re-uptake is the first demonstration of this effect in perfused rat livers. This has been suggested as an alternative explanation of experiments with H^3-cholesterol-labelled lipoproteins in the intact rat (6).

DISCUSSION

For a long time it was assumed that the mutual antagonism between the short-term effects of pancreatic glucagon and insulin is essential for the homeostasis of blood glucose and FFA. However, recent evidence, supplied by studies in diabetic patients, casts doubt on the primary importance of glucagon in glucose homeostasis. The study of other metabolic parameters in many research centers has demonstrated that the continuous infusion of minute amounts of glucagon (0.1 mg/daily) alters the concentration of some amino acids and of α-amino nitrogen in plasma and the excretion of urea in man (7). The present results obtained in rats together with some data obtained in man (10), demonstrate that the repeated pulse administration of glucagon alters the major plasma lipids more markedly than the metabolism of α-amino nitrogen and urea, suggesting that the overall metabolic effect of glucagon may depend not only on its total amount, but also on the manner in which it becomes available to the receptors (continuous steady state level versus pulsing peaks). In addition, the ratio of glucagon to its "antagonists", e. g. insulin, remains to be studied in more detail.

In conclusion, a direct role of glucagon in the physiologic regulation of the major lipids of rat plasma may be proposed on the basis of the following facts:

- a depression of plasma lipids is the most prominent long-lasting, dose-dependent effect of repeated injection of pancreatic glucagon (Figs. 1, 4 5);
- the injection of glucagon results in a rise of the glucagon/insulin ratio

in plasma (Fig. 9), and the depression of plasma lipids does not depend on insulin (Fig. 10);

- upon repeated injections, the lipid-lowering effect of pancreatic glucagon is superior to that of other "typical" lipid-lowering hormones, such as oestradiol and thyroxine (Fig. 8);
- dose-frequency curves (Fig. 6, 7) suggest that a relatively small fraction of the total glucagon content of the pancreas could exert a hypolipemic effect, if it were in multiple small pulse fractions;
- if the hypolipemic action of glucagon were a physiologic phenomenon, it would lend support to the idea that glucagon is a "catabolic" hormone and has a role in the regulation of metabolism in the post-absorptive state.

Several papers in this Symposium have demonstrated that plasma, gut and other tissues contain considerable amounts of proteins with structures and immunoreactive properties similar to those of pancreatic glucagon. The present data on the hypolipemic effect of pancreatic glucagon suggests that their hypolipemic potentials should be investigated.

REFERENCES

1. Amatuzio, D. S., Grande, F. and Wada, S. (1962): Effect of glucagon on the serum lipids in essential hyperlipemia and in hypercholesterolemia. Metabolism 11, 1240-1249.

2. Aubry, F., Marcel, Y. L. and Davignon, J. (1974): Effects of glucagon on plasma lipids in different types of primary hyperlipoproteinemia. Metabolism 23, 225-238.

3. Berthet, J. (1958): Action du glucagon et de l'adrénaline sur le métabolisme des lipides dans le tissu hépatique. In: Proceedings, IV International Congress of Biochemistry, Vienna. Pergamon Press, London, p. 107.

4. Block, W. D., Jarrett, K. J., Jr. and Levine, J. B. (1966): Use of a single color reagent to improve the automated determination of serum total cholesterol. In: Automation in Analytical Chemistry: Technicon Symposium 1965. L. T. Skeggs Jr., Ed. Mediad Incorporated, New York, pp. 345-347.

5. Bloxham, D. P. and Akhtar, M. (1971): Studies on the control of cholesterol biosynthesis: the adenosine 3',5'-cyclic monophosphate-dependent accumulation of a steroid carboxylic acid. Biochem. J. 123, 275-278.

6. Byers, S. O., Friedman, M. and Elek, S. R. (1975): Further studies concerning glucagon-induced hypocholesterolemia. Proc. Soc. Exp. Biol. Med. 149, 151-157.

7. Cahill, G. F. Jr. and Aoki, T. T. (1977): The role of glucagon in amino acid homeostasis. This volume, p. 487-494.

8. Eaton, R. P. (1973): Hypolipemic action of glucagon in experimental endogenous lipemia in the rat. J. Lipid Res. 14, 312-318.

9. Eaton, R. P. (1973): Effect of clofibrate on arginine-induced insulin and glucagon secretion. Metabolism 22, 763-767.

10. Eaton, R. P. (1977): Glucagon and lipoprotein regulation in man. This volume, p. 533-550.

11. Eaton, R. P., Conway, M. and Schade, D. S. (1976): Endogenous glucagon regulation in genetically hyperlipemic obese rats. Am. J. Physiol. 230, 1336-1340.

12. Eaton, R. P., Oase, R. and Schade, D. S. (1976): Altered insulin and glucagon secretion in treated genetic hyperlipemia: a mechanism of therapy? Metabolism 25, 245-249.

13. Eaton, R. P. and Schade, D. S. (1973): Glucagon resistance as a hormonal basis for endogenous hyperlipemia. Lancet 1, 973-974.

14. Eaton, R. P. and Schade, D. S. (1973): Clofibrate suppression of insulin/glucagon secretory ratio in man. Clin. Res. 21, 273.

15. Eaton, R. P. and Schade, D. S. (1974): Effect of clofibrate on arginine-stimulated glucagon and insulin secretion in man. Metabolism 23, 445-454.

16. Exton, J. H., Corbin, J. G. and Harper, S. C. (1972): Control of gluconeogenesis in liver. V. Effects of fasting, diabetes, and glucagon on lactate and endogenous metabolism in the perfused rat liver. J. Biol. Chem. 247, 4996-5003.

17. Exton, J. H. and Park, C. R. (1967): Control of gluconeogenesis in liver. I. General features of gluconeogenesis in the perfused livers of rats. J. Biol. Chem. 242, 2622-2636.

18. Faloona, G. R. and Unger, H. (1974): Glucagon. In: Methods of Hormone Radioimmunoassay. B. M. Jaffe and H. R. Behrmann, Eds. Academic Press, New York, pp. 317-330.

19. Foà, P. P. (1968): Glucagon. Rev. Physiol. Biochem. Exper. Pharmacol. 60 141-219.

20. Friedman, M., Byers, S. O., Rosenman, R. H. and Elek, S. (1971): Effect of glucagon on blood-cholesterol levels in rats. Lancet 2, 464-466.

21. Gerich, J. E., Langlois, M., Schneider, V., Karam, J. H. and Noacco, C. (1974): Effects of alterations of plasma free fatty acid levels on pancreatic glucagon secretion in man. J. Clin. Invest. 53, 1284-1289.

22. Gey, K. F. and Georgi, H. Unpublished data.

23. Gorman, C. K., Salter, J. M. and Penhos, J. C. (1967): Effects of glucagon on lipids and glucose in normal and eviscerated rats and on isolated perfused rat livers. Metabolism 16, 1140-1157.

24. Haugaard, E. S. and Haugaard, N. (1954): The effect of hyperglycemic-glycogenolytic factor on fat metabolism of liver. J. Biol. Chem. 206, 641-645.

25. Haugaard, E. S. and Stadie, W. C. (1953): The effect of hyperglycemic-glycogenolytic factor and epinephrine on fatty acid synthesis. J. Biol. Chem. 200, 753-757.

26. Heimberg, M., Weinstein, I. and Kohout, M. (1969): The effects of glucagon, dibutyryl cyclic adenosine 3',5'-monophosphate, and concentration of free fatty acid on hepatic lipid metabolism. J. Biol. Chem. 244, 5131-5139.

27. Kraml, M. (1966): A semi-automated determination of phospholipids. Clin. Chim. Acta. 13, 442-448.

28. Lakshmanan, M. R., Nepokroeff, C. M. and Porter, J. W. (1972): Control of the synthesis of fatty acid synthetase in rat liver by insulin, glucagon, and adenosine 3',5'-cyclic monophosphate. Proc. Nat. Acad. Sci. USA 69, 3516-3519.

29. Lorch, E. and Gey, K. F. (1966): Photometric "titration" of free fatty acids with the technicon autoanalyzer. Anal. Biochem. 16, 244-252.

30. Marks, V., Frizel, D., Twycross, R. G. and Buchanan, K. D. (1971): Effect of β-pyridylcarbinol on glucose tolerance, plasma glucagon, insulin, and growth hormone in man. In: Metabolic Effects of Nicotinic Acid and its Derivatives. K. F. Gey and L. A. Carlson, Eds. Hans Huber Verlag, Bern, pp. 961-976.

31. Marsh, W. H., Fingerhut, B. and Kirsch, E. (1957): Determination of urea nitrogen with the diacetyl method and an automatic dialyzing apparatus. Amer. J. Clin. Path. 28, 681-688.

32. Meikle, A. W., Klain, G. J. and Hannon, J. P. (1973): Inhibition of glucose oxidation and fatty acid synthesis in liver slices from fed, fasted and fasted-refed rats by glucagon, epinephrine and cyclic adenosine-3',5-monophosphate. Proc. Soc. Exp. Biol. Med. 143, 379-381.

33. Nepokroeff, C. M., Lakshmanan, M. R., Ness, G. C., Dugan, R. E. and Porter, J. W. (1974): Regulation of the diurnal rhythm of rat liver β-hydroxy-β-methyl-glutaryl coenzyme A reductase activity by insulin, glucagon, cyclic AMP and hydrocortisone. Arch. Biochem. Biophys. 160, 387-396.

34. Palmer, D. W. and Peters, T. Jr. (1969): Automated determination of free amino groups in serum and plasma using 2, 4, 6-trinitrobenzene sulfonate. Clin. Chem. 15, 891-901.

35. Paloyan, E., Dumbrys, N., Gallagher, T. F. Jr., Rodgers, R. E.and Harper, P. V. (1962): The effect of glucagon on hyperlipemic states. Fed. Proc. 21, 200.

36. Penhos, J. C., Wu, C. H., Daunas, J., Reitman, M. and Levine, R. (1966): Effect of glucagon on the metabolism of lipids and on urea formation by the perfused rat liver. Diabetes 15, 740-748.

37. Roche Diagnostica Procedure: System CentrifiChem®, Prod. No. 1405. (Glucose HK/G6P-DH).

38. Rothfeld, B., Margolis, S., Varady, A. Jr. and Karmen, A. (1974): Effects of glucagon on cholesterol and triglyceride deposition in tissues. Biochem. Med. 10, 122-125.

39. Rothfeld, B., Pare, W. P., Margolis, S., Karmen, A., Varady, A. Jr. and Isom, K. E. (1974): Effect of glucagon and stress on cholesterol metabolism and deposition in tissues. Biochem. Med. 11, 189-193.

40. Royer, M. E. and Ko, H. (1969): A simplified semiautomated assay for plasma triglycerides. Anal. Biochem. 29, 405-416.

41. Salter, J. M. (1960): Metabolic effects of glucagon in the Wistar rat. Amer. J. Clin. Nutr. 8, 535-539.

42. Schade, D. S. and Eaton, R. P. (1975): Modulation of fatty acid metabolism by glucagon in man. I. Effects in normal subjects. Diabetes 24, 502-509.

43. Schade, D. S. and Eaton, R. P. (1975): Modulation of fatty acid metabolism by glucagon in man. II. Effects in insulin-deficient diabetics. Diabetes 24, 510-515.

44. Stauffacher, W., Assan, R., Gey, K. F. and Renold, A. E. (1971): Effects of β-pyridylcarbinol on levels of immunoreactive insulin and glucagon in pancreas and blood of intact rats. In: Metabolic Effects of Nicotinic Acid and its Derivatives. K. F. Gey and L. A. Carlson, Eds. Hans Huber Verlag, Bern, pp. 987-994.

45. Unger, R. H. (1972): Glucagon and glucagon immunoreactivity in plasma and pancreatic tissues. In: Glucagon. Molecular Physiology, Clinical and Therapeutic Implications. P. J. Lefèbvre and R. H. Unger, Eds. Pergamon Press, New York, pp. 205-211.

46. Weinstein, I., Klausner, H. A. and Heimberg, M. (1973): The effect of concentration of glucagon on output of triglyceride, ketone bodies, glucose, and urea by the liver. Biochim. Biophys. Acta 296, 300-309.

GLUCAGON AND LIPOPROTEIN REGULATION IN MAN

R. P. Eaton

University of New Mexico School of Medicine
Albuquerque, New Mexico

We have proposed that reduced glucagon activity may participate in the induction and/or maintenance of acquired as well as genetically determined endogenous hyperlipemia. To critically consider this hypothesis, the physiologic response to glucagon must be evaluated in a variety of clinical conditions in which the glucagon-lipid axis may be altered. Moreover, in this process, the critical role of free fatty acid availability, simultaneous insulin secretion, and concurrent production of the counter-regulatory hormones cortisol, growth hormone and catecholamines must be considered. Therefore, in the present review, the lipoprotein response to glucagon is examined in normal subjects, insulin-deficient diabetic subjects, insulin-resistant obese subjects and glucagon-resistant hyperlipemic subjects. In addition, the alterations in glucagon secretion induced by "estrogen hyperlipemia" and clofibrate "hypolipemia" are considered in relation to the hypothesis that glucagon deficiency may lead to hyperlipemia, while glucagon excess may reduce plasma lipids. A basic function of glucagon in augmenting free fatty acid availability to the liver while simultaneously "shifting" hepatic fatty acid metabolism into oxidative pathways of ketogenesis and away from synthetic pathways of lipoprotein production, is suggested as the critical axis of altered metabolism in the many pathologic conditions in which disturbed lipid physiology is recognized.

Since the original clinical observations of Amatuzio _et al_ in 1962 (1), a plasma hypolipemic response to glucagon has been recognized. However, the physiologic significance of this effect in normal man and the implications for lipoprotein regulation during pathologically altered glucagon secretion are not clear. It has been proposed that glucagon excess should result in a state of reduced plasma lipid levels, while the opposite state of glucagon deficiency should result in hyperlipemia (10). Such a concept might apply only if other recognized factors, critical to lipoprotein regulation, remain unaltered. Thus, lipoprotein production depends upon (a) the availability of plasma free fatty acids (FFA) as precursors for the triglyceride component of the lipoprotein, (b) a promoting action of insulin upon hepatic protein and lipid synthesis and (c) a modulating action of the "stress" hormones, growth hormone, catecholamines, and cortisol, on lipolytic FFA generation and lipoprotein synthesis in the adipocyte and the hepatocyte (9). The interpretation of any clinical experimental situation in which glucagon may be exerting an effect upon lipoprotein regulation may only be examined accurately when these three factors are considered.

FFA Availability

In the absence of FFA supply to the liver, hepatic triglyceride (TG) synthesis is reduced and lipoprotein production limited. Similarly, in the presence of elevated concentrations of FFA, hepatic triglyceride synthesis is increased and hepatic production of lipoprotein augmented (18). Since glucagon has a direct lipolytic action in the adipocyte, causing plasma FFA concentration to rise, a substrate-product response might be expected to result in elevated hepatic production of lipoprotein. However, experimental evidence suggests that, in addition, glucagon causes a shift of FFA from synthetic pathways of triglyceride formation, to oxidative pathways of ketone generation (19). This "glucagon shift", depicted in Fig. 1, results in ketonemia with a reciprocal hypolipemia in response to the hormone-induced increase of available FFA. Such series of events may account for the metabolic picture of fasting, chronic exercise and starvation, in which hyperglucagonemia, ketonemia, hypolipemia, and elevated FFA are characteristic. On the other hand, if a brisk insulin secretion in response to glucagon evolves, then the lipolytic response may be blunted or reversed. When this situation happens, as it may in the obese patient (21), the availability of FFA may decrease and it may superimpose itself upon the hepatic glucagon shift, resulting in an exaggerated lipid lowering action (22). In this situation, the net plasma triglyceride reduction is a combination of two events: the reduction of FFA available for hepatic lipoprotein synthesis and a direct hepatic action of glucagon shifting FFA away from lipid synthesis and into ketone generation. Thus, when comparing the response to glucagon in two human populations, the availability of FFA to the liver must always be controlled.

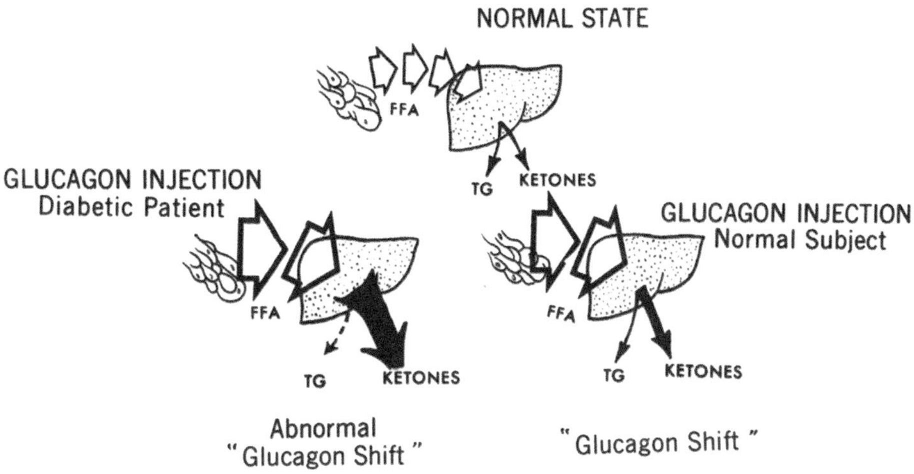

Fig. 1. Diagram showing that glucagon increases the availability of free fatty acid (FFA) from the adipocyte and induces a "shift" of hepatic FFA metabolism from the synthesis of triglyceride (TG) to the production of ketones.

Insulin Modulation

The insulin secretory response to an increase in plasma glucagon levels can be expected to exert a direct counter-regulatory effect on the adipocyte and the hepatocyte. Insulin is a potent antilipolytic hormone and can inhibit the glucagon stimulation of FFA release, creating a deficiency of FFA available to the liver for lipoprotein synthesis. In this activity insulin can be considered a hypolipemic hormone. However, insulin also acts directly on the hepatocyte to enhance lipoprotein production and release, in opposition to the actions of glucagon, as shown in Fig. 2. In this respect, insulin must be considered a hyperlipemic hormone (17). In addition, insulin enhances the degradation of newly synthesized lipoproteins (2) and thus it will accelerate lipid clearance and reduce plasma lipoprotein concentration. Which of these actions will predominate in any metabolic situation requires careful appraisal. Clearly, changes in FFA availability, in hepatic glucagon shift, or in the rate of peripheral FFA disposal, must be examined in any situation in which glucagon and insulin secretion are altered, before reaching any conclusion concerning the relative actions of these two hormones. The relative importance of insulin secretion during the experimental challenge of glucagon in three human populations is illustrated in Fig. 3. In the obese group, the intravenous injection of glucagon is followed by marked hyperinsulinemia, in contrast with the secretory response observed in normal weight patients, and in the absence of insulin response in ketosis-prone diabetic subjects (21). Thus, only in the obese patients was enough insulin secreted to overpower the action of glucagon on the adipocyte, resulting in a marked reduction in plasma FFA available for lipoprotein production. In the normal and diabetic patients, the injected glucagon prevailed and a lipolytic response ensued, increasing the amount of FFA available to the hepatocyte.

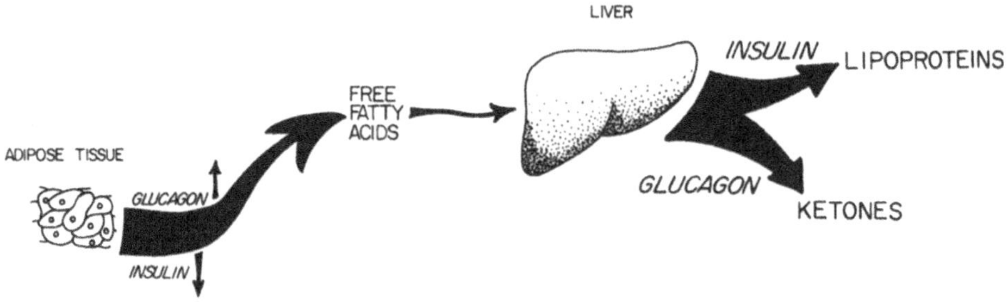

Fig. 2. Insulin reduces FFA availability and directs them toward the production of lipoprotein whereas glucagon increases FFA availability and directs them toward the production of ketones.

Modulation by Stress Hormones (growth hormone, catecholamines, and cortisol)

As illustrated in Fig. 4, glucagon and insulin do not act alone in the regulation of lipoprotein physiology, but must be considered within a larger hormone milieu. Catecholamines are commonly secreted in association with glucagon under conditions of stress, and are potent lipolytic agents increasing FFA

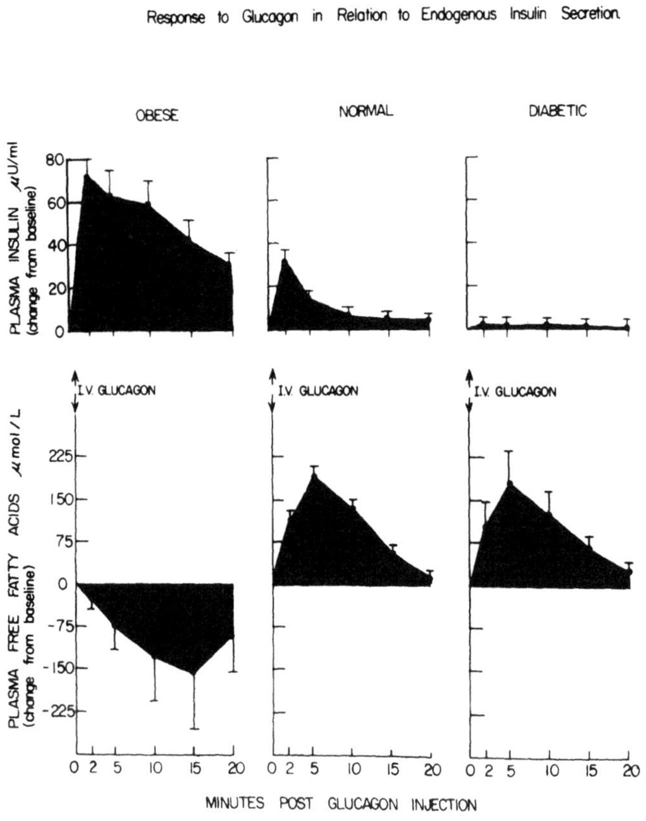

Fig. 3. Insulinogenic response to exogenous glucagon in obese, normal and diabetic subjects. It is apparent that the FFA response to glucagon (lower panel) is inversely related to the simultaneous insulinogenic response.

availability for hepatic lipoprotein production. Moreover, catecholamine infusion not only results in marked elevation in FFA (24), which is exaggerated in insulin deficiency, but also results in an intrahepatic shift of FFA metabolism from conversion to triglycerides (6, 13) to ketone generation (9). Thus, in examining the glucagon effect on lipoproteins in any patient population, attention to simultaneous catecholamine secretion must be given.

A second stress hormone, growth hormone, has also been implicated in the regulation of adipocyte FFA mobilization and subsequent ketogenesis (11), but its

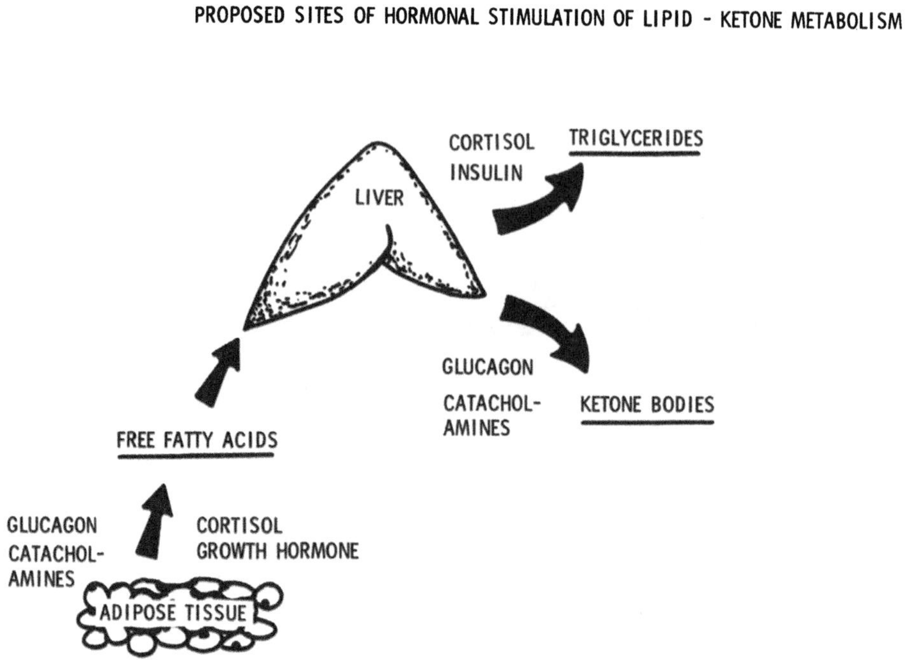

Fig. 4. Multiple interrelated regulatory effects of insulin, glucagon, catecholamines, cortisol and growth hormone on FFA availability and on their conversion to ketones or triglycerides in the liver.

role in lipoprotein physiology remains to be defined. A reciprocal elevation in plasma triglyceride concentration with a reduction in plasma cholesterol levels has been reported with growth hormone administration in ateleotic growth-hormone deficient dwarfs (12). Furthermore, untreated growth hormone deficient dwarfs are characterized by marked hypercholesterolemia, while acromegalic patients with excess growth hormone may have reduced plasma cholesterol levels (15). While the mechanism of these effects of growth hormone upon lipoprotein metabolism remains to be defined, it may be important that growth hormone is both an insulin secretagogue and a potentiator of glucagon secretion (16).

The last stress hormone, cortisol, is also capable of mobilizing adipocyte FFA, thus providing substrate for lipid production in the liver. Moreover, cortisone appears to share with insulin the capacity to augment hepatic lipoprotein production (3, 9). In vivo, this effect may be effectively neutralized by the steroid potentiation of glucagon secretion (14). Nevertheless, any interpretation of glucagon effects upon lipoprotein regulation must take into consideration the simultaneous secretion and effects of these counter-regulatory stress hormones.

Recognizing the limits imposed by the above considerations, we examined the dose response to exogenous glucagon in normal male subjects (19). As shown in Fig. 5, doses of 0.5 μg/Kg or greater resulted in similar FFA responses accompanied by a linear shift into betahydroxybutyrate generation. However, the plasma triglyceride response was complex: the rise in plasma lipid concentration was parallel to the rise in plasma FFA with hormone dosages of 0.1 and 0.5 μg/Kg; while greater doses induced a glucagon shift as indicated by a lipid lowering action in concert with a still greater rise in ketones. Using the 1 μg/Kg dose, we examined the parallel response in diabetic subjects whose insulin secretion failed to respond to the injection of glucagon (20). As shown in Fig. 6, we obtained a lipolytic response comparable to that seen in the normal subjects. However, in the absence of the restraining action of insulin, the hepatic glucagon shift was exaggerated as indicated by a four-fold augmentation of plasma ketone levels and a two-fold greater reciprocal depression of plasma triglyceride. This action of glucagon is seen schematically in Fig. 1. Extension of these observations to obese patients with "resistance" to insulin (as suggested by three-fold elevations in plasma immunoreactive insulin concentration) is shown in Fig. 7. Using a dose of glucagon of 2 μg/Kg in order to overcome the adipocyte resistance, we obtained a rise in FFA greater than that seen in subjects with normal weight (22). However, the degree of ketonemia was comparable in both groups, suggesting that obese patients generate less ketone bodies relative to the availability of FFA (Fig. 7). Moreover, the reciprocal reduction in plasma triglyceride concentration was greater in the obese than in the non-obese subjects, suggesting that

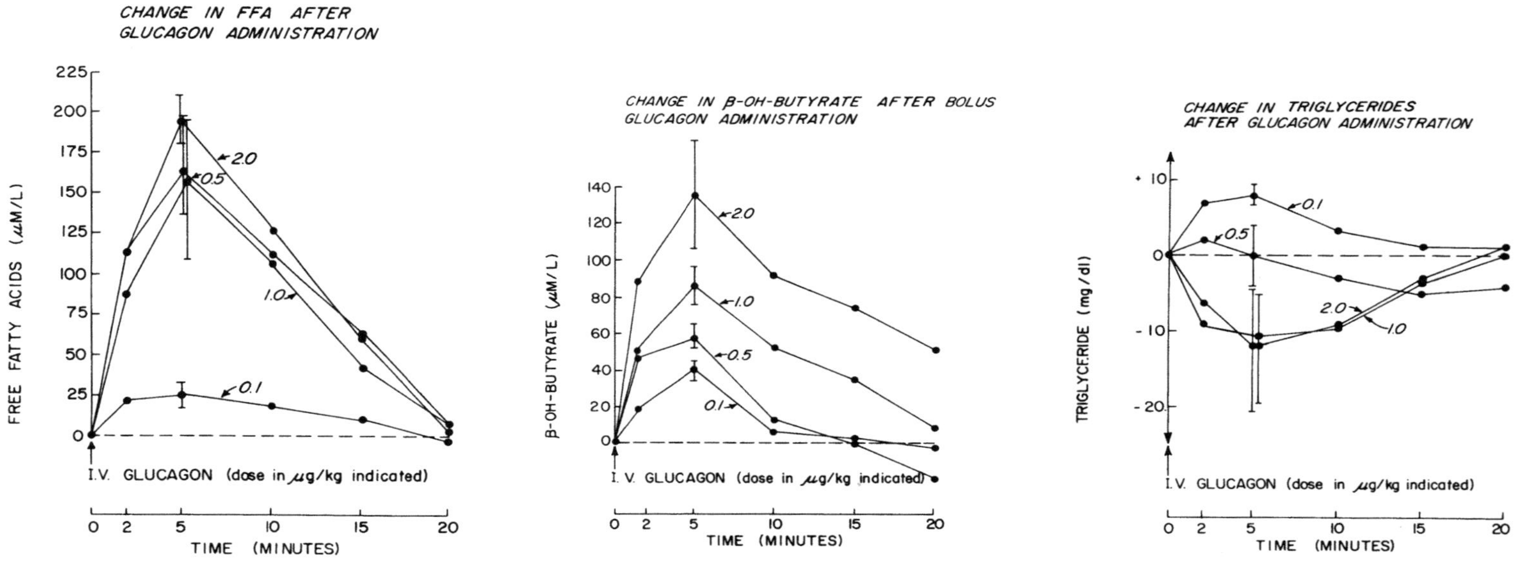

Fig. 5. Effect of increasing doses of intravenous glucagon on plasma free fatty acid and their hepatic conversion to ketone bodies or to triglycerides, in normal male subjects, interpreted according to the scheme shown in Fig. 1.

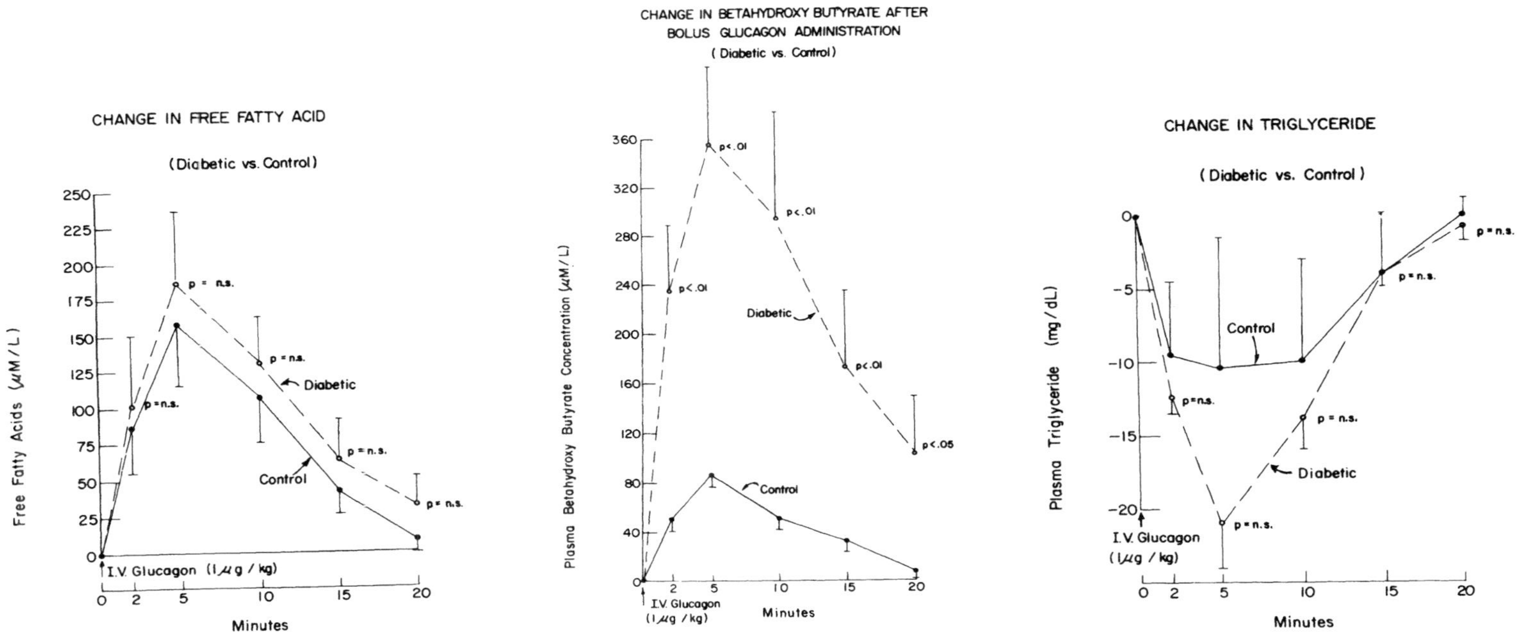

Fig. 6. Effect of intravenous glucagon in normal subjects on free fatty acid availability and their hepatic conversion to ketone bodies or to plasma triglycerides, in insulin-deficient diabetic patients and interpreted according to the scheme shown in Figs. 1 and 2.

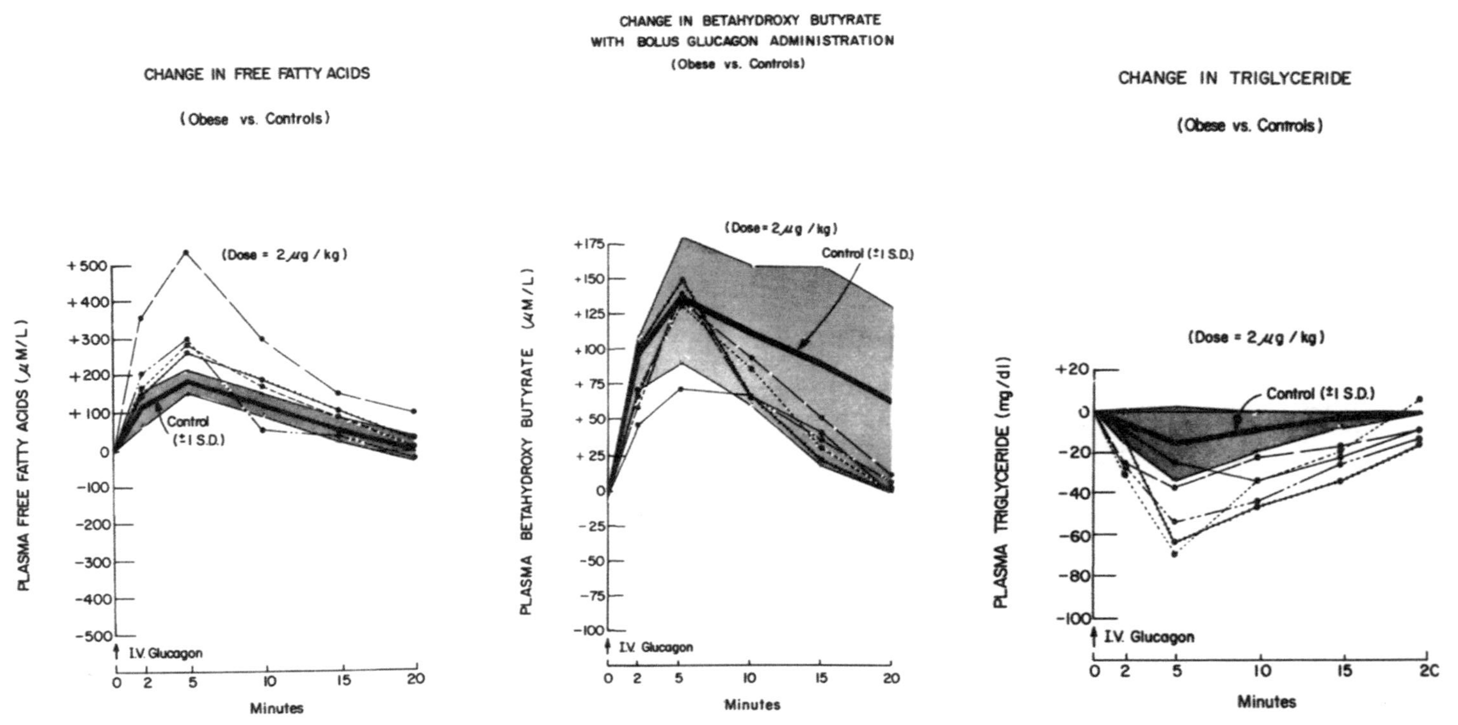

Fig. 7. Effect of intravenous glucagon on free fatty acid availability and on their hepatic conversion to ketone bodies and to triglycerides in insulin-resistant obese subjects and in normal controls, interpreted according to the scheme shown in Fig. 1.

insulin resistant obese subjects share the exaggerated glucagon shift characteristic of insulin deficient diabetics.

To examine the effects of glucagon upon lipoprotein concentration when physiologic instead of pharmacologic dosages were employed, we investigated six normal and six insulin-deficient diabetic subjects. Glucagon was infused to increase the plasma concentration from basal levels of 33 ± 10 pg/ml in normals and 64 ± 10 pg/ml in diabetics to 250 pg/ml. A 30 min control infusion of saline was given to all subjects. In normal subjects, plasma FFA were unaltered by saline infusion, but were elevated by the infusion of glucagon from basal levels of 704 ± 83 μmol/L to 1,111 ± 185 μmol/L ($p < .05$), in association with a rise in insulin levels from 13 μU/ml to 17 ± 2 μU/ml ($p < .05$). As plotted in the upper portion of Fig. 8, plasma triglyceride concentration was reduced by 7-15 mg% by either saline or glucagon infusion. The increased availability of FFA during glucagon infusion, associated with a decline in plasma triglyceride concentration, is consistent with a direct lowering effect of glucagon on plasma lipids either at the level of the hepatocyte or at peripheral sites. The data obtained in normal man are insufficient to establish that this action of glucagon is physiologic. However, the data obtained in insulin-deficient diabetic patients provide further support for a lipid-lowering action of glucagon at physiologically attainable levels of the hormone. In these patients, plasma FFA levels did not change during either saline or glucagon infusion, remaining at basal values of 900 ± 120 μmol/L. However, as shown in the lower portion of Fig. 8, a reduction in plasma triglyceride concentration of 10-20 mg% ($p < .05$) was obtained during the glucagon, but not during the saline infusion. Our results are consistent with the recent observation that the infusion of physiologic amounts of glucagon did not alter triglyceride concentration in non-diabetic subjects (23), and suggest that the stimulation of insulin secretion during glucagon infusion in normal subjects inhibits the lipid lowering action of glucagon at the hepatocyte. The absence of endogenous insulin secretion in our diabetic subjects may thus

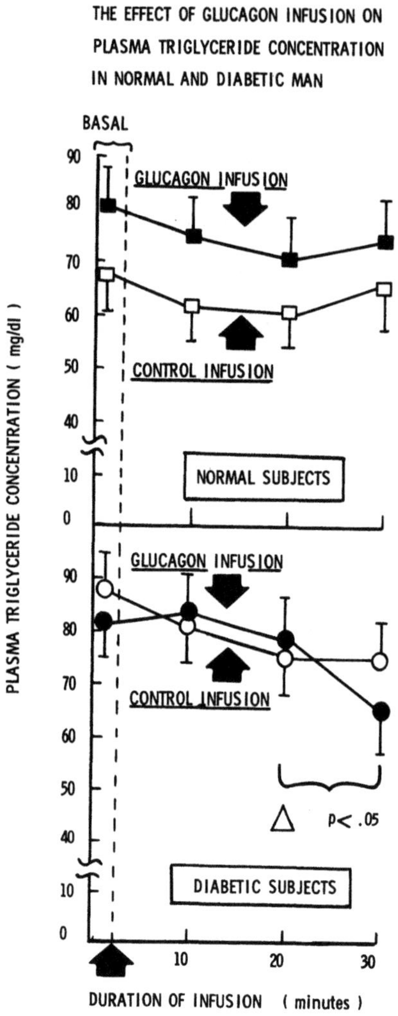

Fig. 8

Effect of "physiologic" elevations of plasma glucagon in normal subjects (upper panel) and in insulin-deficient diabetic subjects (lower panel). Control = saline infusion.

have allowed the hypolipemic activity of glucagon to manifest itself. Indeed, in the relatively insulinopenic duck, glucagon has been reported to have a marked triglyceride lowering effect which is dependent upon the presence of the liver (5). In this bird, immediately upon cessation of glucagon administration, a marked rebound hypertriglyceridemia appears which closely parallels the sustained FFA elevation resulting from the earlier glucagon injection. These investigations focus upon a modulating hypolipemic role for glucagon which can only be carefully assessed by consideration of simultaneous FFA availability and insulin secretion. Potential simultaneous influences exerted by cortisol, growth hormone, or catecholamines have not been evaluated in these preliminary studies.

In patients with "primary" hyperlipoproteinemia, unrelated to diabetes, to obesity, or to any recognizable alteration in counter-regulatory hormone secretion, we have identified an occasional patient with markedly elevated levels of immunoreactive plasma glucagon. In such patients we have suggested (7) that resistance to the lipid-lowering actions of glucagon may participate in the generation and/or maintenance of the endogenous hyperlipemia. Since our original observations, we have examined a large number of patients with Type IV hyperlipoproteinemia, but have found elevated immunoreactive glucagon only in three additional patients. To test the possibility that the proposed resistance to glucagon could be overcome, these patients were treated with increasing amounts of glucagon and the lipid lowering response was evaluated. The results in a patient with Type III hyperlipoproteinemia are shown in Fig. 9. At a glucagon dosage of 2.0 μg/Kg, the patient demonstrated a brisk lipolytic response with an appropriate rise in plasma FFA. Unlike the response observed in normal, obese and diabetic subjects, this rise in FFA concentration was associated with no apparent conversion to plasma ketones, but with a marked conversion to plasma triglycerides. Apparently, the glucagon shift did not occur in this patient. However, with the overwhelming glucagon dosage of 5.0 μg/Kg, causing an even greater rise in the availability of FFA, an appropriate glucagon shift was observed, as indicated by the marked rise in the concentration of plasma ketones associated with the expected reduction in plasma triglycerides (See Fig. 9). Thus, this patient with endogenous hyperlipoproteinemia of Type III, appears to have glucagon resistance, as indicated by the elevated plasma immunoreactive glucagon levels and the fact that a hypolipemic response could be obtained also with high doses of the hormone.

Having examined the lipoprotein response to glucagon in normal subjects, insulin deficient diabetic subjects, insulin-resistant obese subjects, and gluca-

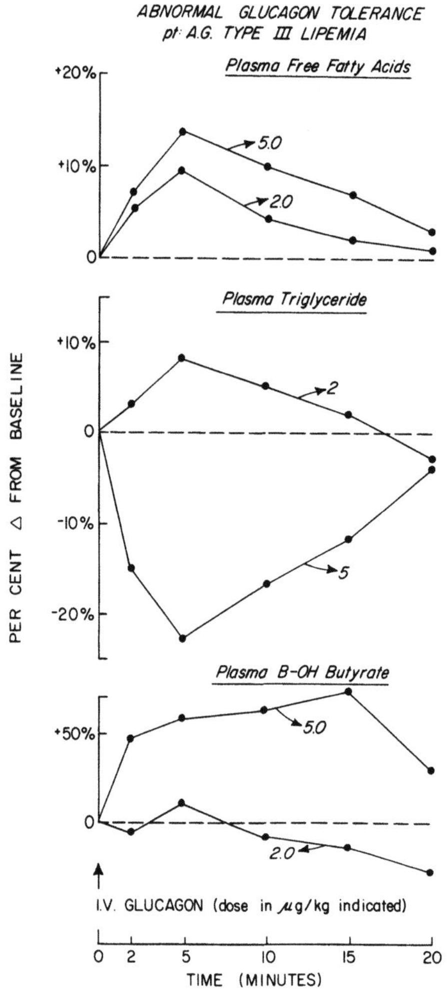

Fig. 9. Ability of high intravenous doses of glucagon (5.0 μg/kg) to overcome the "resistance" to lower doses (2.0 μg/kg) in a patient with Type III hyperlipoproteinemia and elevated basal immunoreactive glucagon levels. With the higher dose, a decrease of plasma triglycerides and an increase of plasma ketones was observed, indicating a normal "glucagon shift".

gon-resistant hyperlipemic subjects, we next investigated glucagon secretion in subjects with drug-induced lipemia. Using oral contraceptives as our model, we observed a 70% suppression of arginine-stimulated glucagon secretion in women given a standard combination steroid containing 0.08 mg mestranol and 1.0 mg norethindrone (4). This reduction in aminogenic glucagon secretion observed also following the oral administration of ethinyl estradiol (0.1 mg/day), but not of norethindrone alone (10.0 mg/day), was associated with a significant hyperlipemia, in agreement with the concept that the latter may have been due, in part, to estrogen-induced insufficiency of a hypolipemic hormone (4).

To pursue the hypothesis that reduced glucagon activity may be the common cause of drug (estrogen), carbohydrate- (dietary) and genetically- (Type IV-III) induced hyperlipemia, we examined the effect of a typical hypolipemic drug upon glucagon secretion. For this purpose, normal male volunteers, age 20-40 years, were given standard doses of clofibrate. Glucagon secretion was either unaltered, or minimally increased in individual patients, however, there was a significant 46% decrease in aminogenic insulin secretion (8), creating a glucagon excess, relative to insulin. Thus, clofibrate may well act, at least in part, by correcting a fundamental deficiency of glucagon relative to insulin.

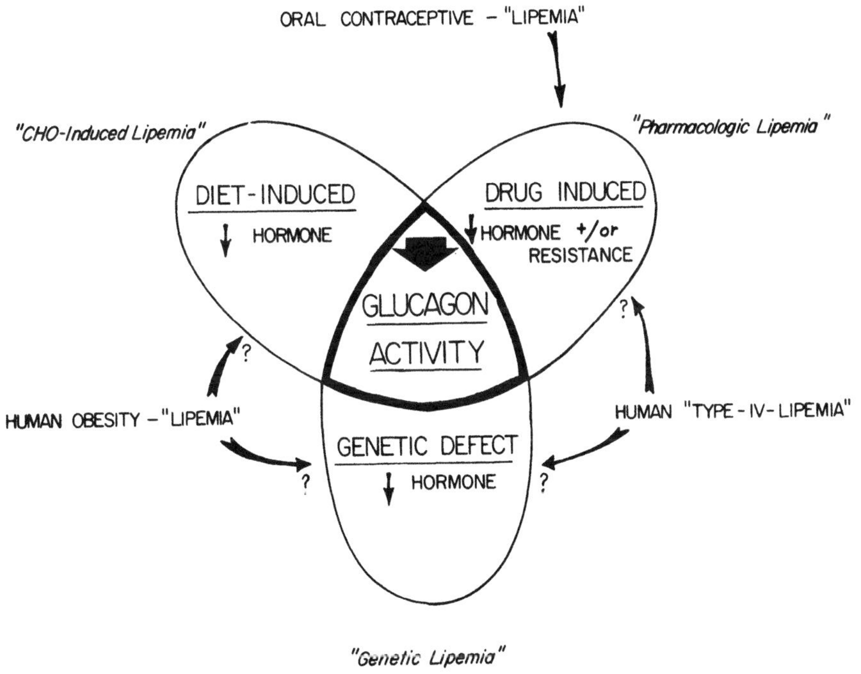

Fig. 10. Hypothesis proposing that a reduction in net glucagon activity may be the final common pathway mediating or participating in the development and/or maintenance of endogenous hyperlipemia in man. It is suggested that this hormonal mechanism may be initiated by diet, drug or genetic influences.

Fig. 10 summarizes the hypothesis based on these data, that a relative deficiency of glucagon has a pathogenetic role in the development of acquired and genetically determined hyperlipemia. While the picture is incomplete and the supporting data indirect, the concept of a hormonal basis for some forms of human lipemia represents a reasonable basis for further investigation and therapeutic trials.

Acknowledgement

The author wishes to express his appreciation to Ms. Jean Nichols for her excellent secretarial assistance in the preparation of this manuscript.

REFERENCES

1. Amatuzio, D. S., Grande, F. and Wada, S. (1962): Effect of glucagon on the serum lipids in essential hyperlipemia and in hypercholesterolemia. Metabolism 11, 1240-1249.

2. Bagdade, J. D., Porte, D., Bierman, E. (1967): Diabetic lipemia. N. Eng. J. Med. 276, 427-433.

3. Bagdade, J. D., Yee, E., Alberts, J. and Pykalisto, O. J. (1976): Glucocorticoids and triglyceride transport: Effects on triglyceride secretion rates, lipoprotein lipase, and plasma lipoproteins in the rat. Metabolism 25, 533-542.

4. Beck, P., Eaton, R. P., Arnett, D. M. and Alsever, R. N. (1975): Effect of contraceptive steroids on arginine-stimulated glucagon and insulin secretion in women: I-Lipid physiology. Metabolism 24, 1055-1065.

5. De Oya, M., Prigge, W. F. and Grande, F. (1971): Suppression by hepatectomy of glucagon-induced hypertriglyceridemia in geese. Proc. Soc. Exp. Biol. Med. 136, 107-110.

6. Eaton, R. P., Berman, M. and Steinberg, D. (1969): Kinetic studies of plasma free fatty acid and triglyceride metabolism in man. J. Clin. Invest. 48, 1560-1579.

7. Eaton, R. P. and Schade, D. S. (1973): Hypothesis: "Glucagon Resistance": An hormonal basis for endogenous hyperlipemia. Lancet 1, 973-974.

8. Eaton, R. P. and Schade, D. S. (1974): The effect of clofibrate on arginine stimulated glucagon and insulin secretion in man. Metabolism 23, 445-454.

9. Eaton, R. P. and Schade, D. S. (1977): The modulation and implications of the counter-regulatory hormones, glucagon, catecholamines, cortisol, and growth hormone in diabetes, obesity, and hyperlipemia. In: Adv. Modern Nutr. 2, 341-372.

10. Eaton, R. P., Schade, D. S. and Conway, M. (1974): Hypothesis: Decreased glucagon activity: A mechanism for both genetic and acquired hyperlipemia. Lancet 2, 1545-1548.

11. Felig, P., Marliss, E. B. and Cahill, G. F., Jr. (1971): Metabolic response to human growth hormone during prolonged starvation. J. Clin. Invest. 50, 411-421.

12. Friedman, M., Byers, S. O., Rosenman, R. H., Li, C. G. and Neuman, R. (1974): Effect of subacute administration of human growth hormone on various serum lipid and hormone levels of hypercholesterolemic and normocholesterolemic subjects. Metabolism 23, 905-912.

13. Heimberg, M., Fizette, N. B. and Klausner, H. (1964): The action of adrenal hormones on hepatic transport of triglycerides and fatty acids. J. Amer. Oil Chem. Soc. 41, 774-779.

14. Marco, J., Calle, C., Roman, D., Diaz-Fierros, M., Villanueva, M. L. and Valverde, I. (1973): Hyperglucagonism induced by glucocorticoid treatment in man. N. Engl. J. Med. 288, 128-132.

15. Merimee, T. J., Hollander, W., Fineberg, S. E. (1972): Studies of hyperlipidemia in the HGH-deficient state. Metabolism 21, 1053-1061.

16. Pek, S., Fajans, S. S., Floyd, J. C. and Knopf, R. F. (1973): Clinical conditions associated with elevated plasma levels of glucagon. Proceedings, International Diabetes Federation, VIII Congress, pp. 207-213.

17. Reaven, E. and Reaven, G. M. (1974): Mechanisms for development of diabetic hypertriglyceridemia in streptozotocin-treated rats. Effect of diet and duration of insulin deficiency. J. Clin. Invest. 54, 1167-1178.

18. Sailer, S., Sandhofer, F. and Braunsteiner, H. (1966): Umsatzraten für freie Fettsauren und Triglyceride im Plasma bei essentieller Hyperlipämie. Klin. Wschr. 44, 1032-1036.

19. Schade, D. S. and Eaton, R. P. (1975): Modulation of fatty acid metabolism by glucagon in man. Diabetes 24, 502-509.

20. Schade, D. S. and Eaton, R. P. (1975): Modulation of fatty acid metabolism by glucagon in man. II. Effects in insulin deficient diabetes. Diabetes 24, 510-515.

21. Schade, D. S. and Eaton, R. P. (1975): The contribution of endogenous insulin secretion to the ketogenic response to glucagon in man. Diabetologia 11, 555-559.

22. Schade, D. S. and Eaton, R. P. (1977): Modulation of the catabolic activity of glucagon by endogenous insulin secretion in obese man. Acta Diabetol. Latina (in press).

23. Sherwin, R. S., Fisher, M., Hendler, R. and Felig, P. (1976): Hyperglucagonemia and blood glucose regulation in normal, obese, and diabetic subjects. N. Engl. J. Med. 294, 455-561.

24. Willms, B., Böttcher, M., Wolters, V., Sakamoto, N. and Söling, H. D. (1969): Relationship between fat and ketone metabolism in obese and non-obese diabetics and non-diabetics during norepinephrine infusion. Diabetologia 5, 88-96.

METABOLIC FUELS IN THE EMBRYO AND THE NEWBORN

E. Blázquez

Departamento de Bioquimica, Facultad de Medicina Universidad Complutense
and Instituto G. Marañon, Velazquez, 144
Madrid-6, Spain

Although insulin and glucagon can be detected on the 11th day of gestation in the rat fetal pancreas, the secretion and biologic action of these hormones do not reach maturity until after the 30th day of postnatal life. During the perinatal life, the B and A cells are relatively insensitive to glucose, glucose tolerance is abnormal and the insulinogenic response to glucose is decreased. In spite of this, the hypersecretion of insulin by the fetal pancreas during the last days of pregnancy and the increased glucagon release immediately after birth, favor the storage and subsequent mobilization of nutrient. Although glucagon inhibits anabolic processes, such as the synthesis of hepatic DNA and the deposition of hepatic glycogen and although in the fetus these processes proceed at an extremely rapid rate, the concentration of plasma glucagon on the 21st day of fetal life is as high as it is in the adult rat. This apparent paradox can be explained by the delayed development of the hepatocyte glucagon receptors and therefore of the adenylate cyclase response to this hormone. On the 15th day of fetal life, the binding of I^{125}- glucagon by liver membranes was only 1% of the adult level, increased to 23% by the 21st day, but like the adenylate cyclase response to glucagon, did not reach maturity until after the 30th day of postnatal life. In contrast, insulin binding was 11% of the adult level on the 15th day of gestation, and 45% on the 21st day, reaching full maturity by the 30th postnatal day. These findings suggest that sequential development of the hepatic receptors for insulin and glucagon and the hypersecretion of these hormones during the perinatal age play an important role in fetal metabolism and in the disposal of nutrients in the newborn.

Although insulin and glucagon have been detected in the pancreas of the rat fetus as early as the 11th day of gestation (26) and in fetal plasma on day 19 (8), their fetal secretory and biosynthetic processes and their effects on the developing target organs may differ from those occurring in the adult. In fully mature animals, insulin and glucagon are believed to regulate the disposition of nutrients in harmony with their availability and with the energy requirements. These can vary extensively. For example, in pregnancy the maternal metabolism can assume a catabolic mode in order to provide enough nutrients for the anabolic needs of the fetus. This difference between maternal and fetal metabolism is made possible, in part, by the fact that insulin and glucagon cannot cross the placental barrier and thus create two independent endocrine environments (8, 18, 19). On the other hand, maternal glucose reaches the fetus freely and constitutes its major source of energy. Accordingly, the enzymes required for hexose metabolism appear early in development (10, 17), lipogenesis is active (29) and from the 17th day of gestation glycogen stores increase rapidly. This metabolic situation reverses immediately after delivery, when the newborn are deprived of transplacental nutrition and are exposed to the danger of neonatal hypoglycemia, especially before they start nursing. In contrast to the fetus, the neonate utilizes little glucose (2, 31) and subsists on a milk diet in which carbohydrates represent only 8% of the caloric intake, while fats, representing 69% of the caloric intake (12), become the principal source of energy. This dietary change is reflected in a decrease of the respiratory quotient from 1 to 0.7, in an increase of circulating glycerol, FFA and ketone bodies (11, 21, 25) and in a decrease in the activity of the enzymes of glycolysis and lipogenesis (29, 30) that occur shortly after birth.

A new shift in metabolic activities, characterized by a sudden change in growth rate, by an increase in the stores of tissue glycogen and in lipogenesis occurs approximately on the 20th postnatal day (29, 30). This shift to a pattern resembling that of the adult, may also be conditioned by the diet since from about the 20th day, the rats begin to eat the standard laboratory fare rich in carbohydrates and poor in fats.

Factors other than the diet, of course, may control the sequential changes between a metabolism based primarily on glucose and one based primarily on the utilization of fat. Among them are the levels of insulin and glucagon, a bihormonal unit, which regulate the flux of major nutrients to the tissues in response to metabolic needs and whose rate of secretion varies significantly, especially during the perinatal period (6, 8, 9, 14, 16-18, 20, 22, 23, 27).

As illustrated in Fig. 1, fetal serum insulin concentrations increase markedly during the last days of pregnancy, decline immediately after birth and remain low during the nursing period. In contrast, a sharp rise in the concentration of serum glucagon occurs as early as 15 minutes after birth and continues for the next 20 days.

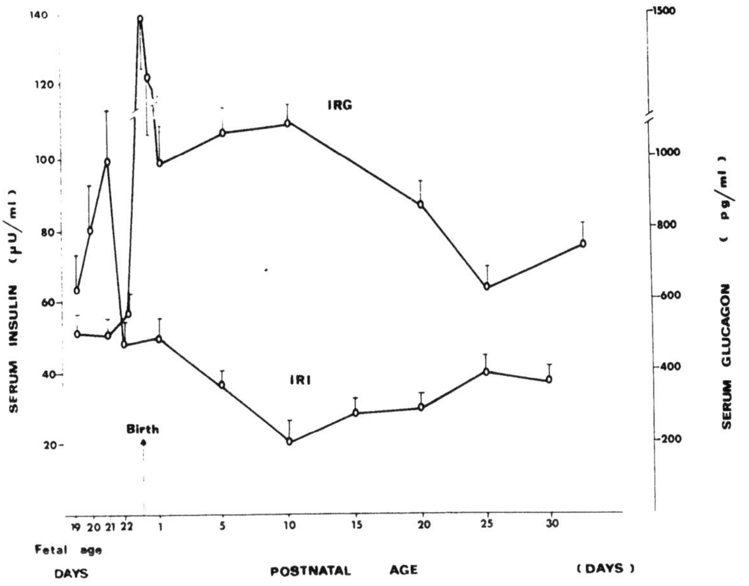

Fig. 1. Serum immunoreactive glucagon (IRG) and immunoreactive insulin levels (IRI) in fetal and postnatal rats.

These hormonal changes appear to control the prenatal accumulation of hepatic glycogen and its rapid breakdown and, thus, help prevent hypoglycemia in the immediate postnatal period of fasting. In addition, they help maintain a relatively high rate of gluconeogenesis throughout the period of nursing, when a large percentage of the necessary glucose must derive from amino acids (3, 4).

Despite the important role played by insulin and glucagon during the development of the rat, the mechanisms involved in their secretion and their ability to control the disposal of a glucose load do not reach full maturity until after the third week of life. As shown in Fig. 2, in 5 and 10 day old rats, glucose tolerance is impaired and the insulinogenic response is significantly lower than that of adult animals. Similarly, the activity of the A cells of the fetal and neonatal pancreas may differ from that of the adult, as indicated by the observations that glucagon secretion by the fetal pancreas in vitro is stimulated by high as well as by low concentrations of glucose (9), that the plasma glucagon level of newborn infants does not correlate with plasma glucose (8, 23) and that the release of glucagon by the pancreas of the newborn rat is relatively insensitive to varying concentrations of glucose (13).

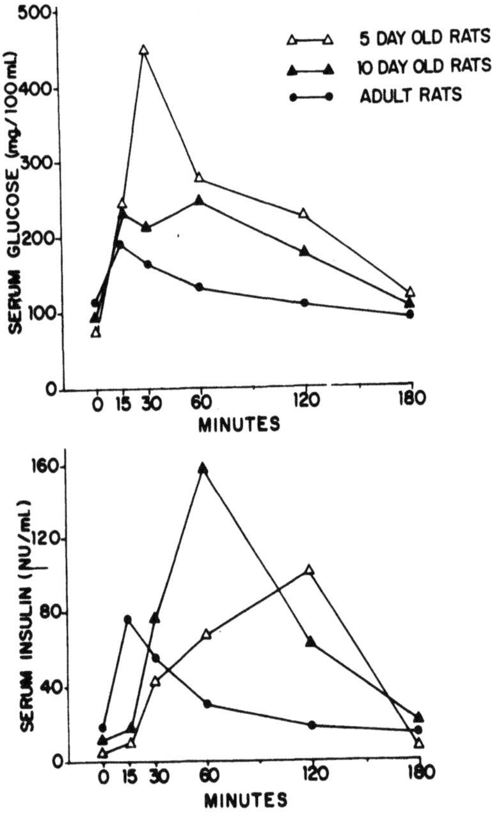

Fig. 2. Serum glucose and insulin after the oral administration of glucose (2g/Kg. body weight) to postnatal and adult rats.

Thus, the maturation of the mechanisms for insulin and glucagon release appears to be delayed until after birth. Nevertheless, the pancreatic islets of fetal and nursing rats are able to synthesize proinsulin and insulin to the same or greater extent than the islets of adult animals (1, 5; Fig. 3), while the fractionation of fetal rat plasma immunoreactive glucagon on Biogel P-10 columns gives a pattern similar to that of adult animals (unpublished results).

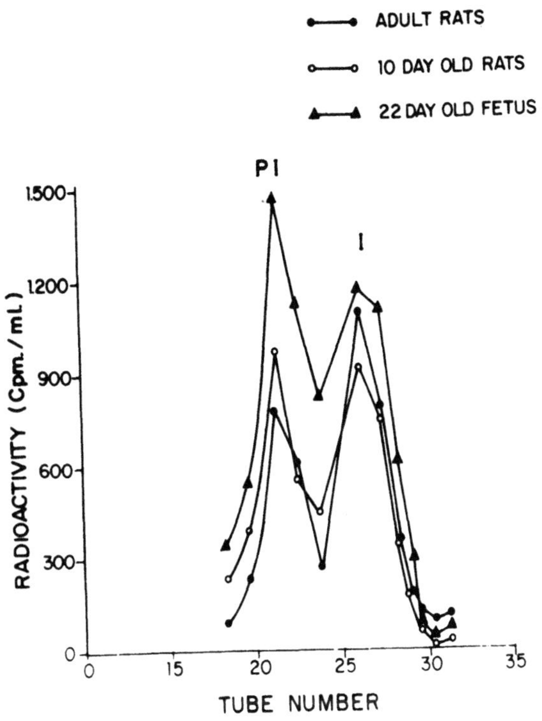

Fig. 3. Chromatographic patterns of partially purified islet proteins. Pancreatic islets were incubated during 4 hr. with L-[4,5-^{3}H] leucine and 3 mg/ml of glucose. PI: Proinsulin. I: Insulin

The synthesis of hepatic DNA and the deposition of liver glycogen proceed at a rapid rate during fetal development. Glucagon inhibits both of them and, yet, on the 21st day of the rat's fetal life, plasma glucagon is as high as it is in the adult rat. In order to determine if the paradoxically high fetal liver glycogen in the presence of adult levels of plasma glucagon reflects a diminished response of the fetal liver to glucagon, the adenylate cyclase response to this hormone was measured throughout the perinatal period (7). Rat liver membranes were harvested at various times before and after birth and were partially purified using the method of Neville (24). Adenylate cyclase was determined by the method of Salomon et al. (27) before and after incubation of the membranes with various concentrations of glucagon. Fig. 4 illustrates the response of the adenylate cyclase activity of liver membranes of fetal, postnatal and adult rats to increasing doses of glucagon. The results indicate that the response of the enzyme to glucagon during fetal life and during the first month of postnatal life is markedly lower than that characteristic of the adult liver. In order to test the possibility that this unresponsiveness may be due to decreased hormone binding by the receptors on the hepatocyte membrane, I^{125}-labeled glucagon and I^{125}-labeled insulin at 10^{-9}M concentration were incubated with liver membranes for 20 minutes and free and bound hormones were separated by microfiltration. As shown in Fig. 5, up to the time of birth the binding of glucagon was minimal, increased substantially during the first 10 days of life, but did not reach adult levels until after the 30th postnatal day. In contrast, insulin binding was substantial on the 15th prenatal, changed very little until the 5th postnatal day and increased thereafter reaching levels very close to adult values by the 20th day.

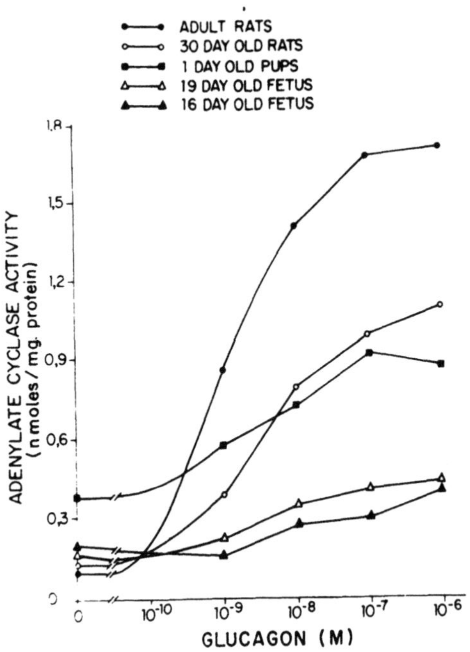

Fig. 4. Dose response curves of glucagon on adenylate cyclase activity in liver membranes of prenatal, postnatal and adult rats.

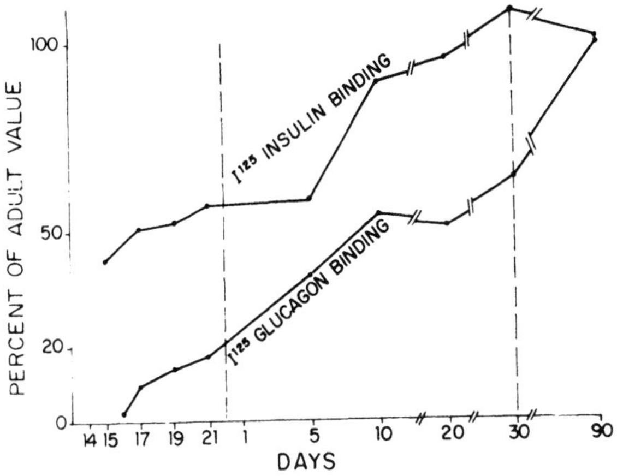

Fig. 5. Comparison of glucagon I^{125} and insulin I^{125} binding to liver membranes of prenatal, postnatal and adult rats.

To exclude the possibility that these results might have been due to the inclusion of membranes from hematopoietic elements, which become less abundant as gestation progresses, rather than to changes in the binding characteristics of the hepatocyte membrane, experiments were conducted with membranes derived from highly purified suspensions of hepatocyte from fetal, nursing and adult rats. As shown in Fig. 6, the results revealed a pattern of hormone binding similar to that observed with membrane preparations obtained from mixed cell suspensions, with the exception of a greater insulin binding by the hepatocytes of nursing rats, at the time when the circulating insulin levels were very low. In comparison, the binding of insulin and glucagon by the other types of cells found in the liver (hematopoietic elements, Kupfer cells and circulating erythrocytes) was negligible, (Fig. 7).

The results of these studies suggest the following conclusions: a) the reduced fetal adenylate cyclase response to glucagon appears to be the result of a paucity of glucagon receptors or of a decrease in their affinity; b) the resulting "glucagon resistance", at a time when liver binding sites for insulin are available, could be a factor in determining the high rate of hepatic DNA synthesis

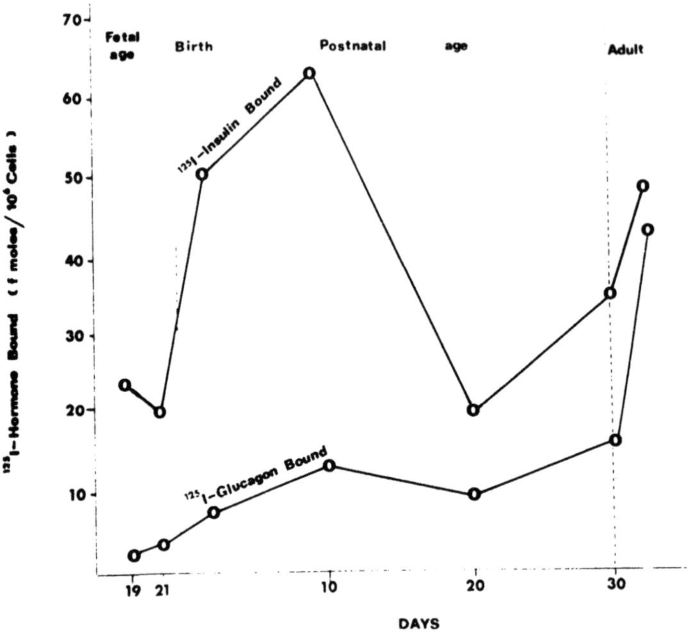

Fig. 6. Insulin I^{125} and glucagon I^{125} binding to isolated purified hepatocytes of prenatal, postnatal and adult rats.

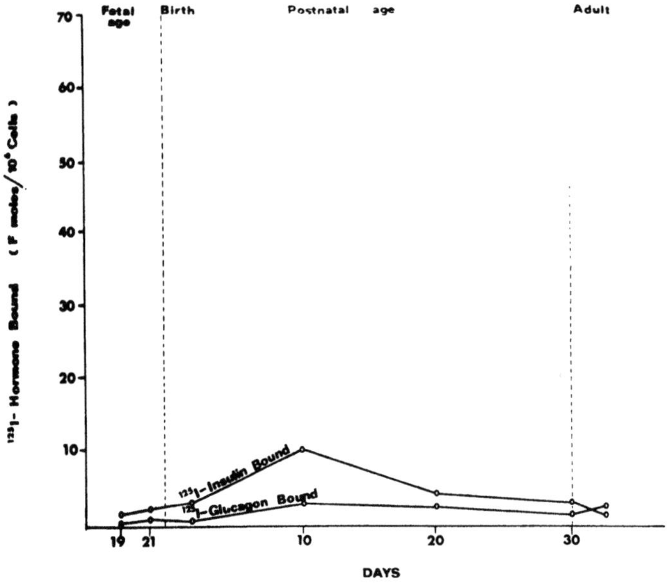

Fig. 7. Insulin I^{125} and glucagon I^{125} binding to isolated remaining liver cells of prenatal, postnatal and adult rats.

and the abundance of hepatic glycogen during this period; c) the modified hormonal binding by fetal hepatocytes could represent an example of so-called "down-regulation" of hormone receptors by relatively high circulating levels of the corresponding hormones, as proposed by Gavin et al (15). Alternatively, it could reflect a different rate of synthesis of these receptors in the fetus. In any case, the sequential development of receptors for insulin and glucagon could have an important role in controlling the metabolic activities of the fetal rat liver.

REFERENCES

1. Asplund, K. (1973): Effects of glucose on insulin biosynthesis in fetal and newborn rats. Hor. Metab. Res. 5, 410-415.

2. Baird, J. D. and Farquhar, J. W. (1962): Insulin-secreting capacity in newborn infants of normal and diabetic women. Lancet 1, 71-74.

3. Ballard, F. J. and Oliver, I. T. (1963): Glycogen metabolism in embryonic chick and neonatal rat liver. Biochim. Biophys. Acta 71, 578-588.

4. Ballard, F. J. (1971): Gluconeogenesis and the regulation of blood glucose in the neonate. In: Diabetes, R. R. Rodriguez and J. Vallance-Owen, Eds. Excerpta Medica Foundation, Amsterdam, p. 592-600.

5. Blázquez, E., Lipshaw, L. A., Blázquez, M. and Foa, P. P. (1975): The synthesis and release of insulin in fetal, nursing, and young adult rats: studies in vivo and in vitro. Pediat. Res. 9, 17-25.

6. Blázquez, E., Montoya, E. and Lopez Quijada, C. (1970): Interrelationship of the insulin concentrations on plasma and pancreas between foetal and weaning rats. J. Endocr. 48, 553-561.

7. Blázquez, E., Rubalcava, B., Montesano, R., Orci, L. and Unger, R. (1976): Development of insulin and glucagon binding and the adenylate cyclase response in liver membranes of the prenatal, postnatal and adult rat. Evidence of glucagon "resistance". Endocrinology 98, 1014-1023.

8. Blázquez, E., Sugase, T., Blázquez, M. and Foà, P. P. (1972): The ontogeny of metabolic regulation in the rat, with special reference to the development of insulin function. Acta Diabetol. Latina (Suppl. 1) 9, 13-35.

9. Blázquez, E., Sugase, T., Blázquez, M. and Foà, P. P. (1974): Neonatal changes in the concentrations of rat liver cyclic AMP and of serum glucose, free fatty acids, insulin, pancreatic and total glucagon in man and in the rat. J. Lab. Clin. Med. 83, 957-967.

10. Burch, H. D., Lowry, O. R., Kuhlman, A. M., Skerjaner, J., Diamant, E. J., Lowry, S. J. and Von Dippe, P. J. (1963): Changes in patterns of enzymes of carbohydrate metabolism in the developing rat liver. J. Biol. Chem. 238, 2267-2273.

11. Cross, K. W., Tizard, J. P. M. and Trythall, D. A. H. (1957): The gaseous metabolism of the newborn infant. Acta Paediat. Scand. 46, 265-285.

12. Dymsza, H. A., Czajka, D. M. and Miller, S. A. (1964): Influence of artificial diet on weight gain and body composition of the neonatal rat. J. Nutr. 84, 100-106.

13. Edwards, J. C., Asplund, K. and Lundquist, G. J. (1972): Glucagon release from the pancreas of the newborn rat. Endocrinology 54, 493-504.

14. Felix, J. M., Jacquot, R. and Sutter, B. C. J. (1969): Insulinémies maternelles et foetales chez le rat. Horm. Metab. Res. 1, 41-42.

15. Gavin, J. R. III, Roth, J., Neville, D. M. Jr., DeMetys, P. and Buell, D. N. (1974): Insulin-dependent regulation of insulin receptor concentrations: a direct demonstration in cell culture. Proc. Natl. Acad. Sci (USA) 71, 84-88.

16. Girard, J., Ball, D. and Assan, R. (1972): Glucagon secretion during the early postnatal period in the rat. Horm. Metab. Res. 4, 168-170.

17. Girard, J. R., Cuendet, G. S., Marliss, E. B., Kervran, A., Rieutort, M. and Assan, R. (1973): Fuels, hormones and liver metabolism at term and during the early postnatal period in the rat. J. Clin. Invest. 52, 3190-3200.

18. Girard, J. R., Kervran, A., Soufflet, M. S. and Assan, R. (1974): Factors affecting the secretion of insulin and glucagon by the rat fetus. Diabetes 23, 310-317.

19. Goodner, C. J. and Freinkel, N. (1961): Carbohydrate metabolism in pregnancy. IV. Studies on permeability of the rat placenta to I^{131}insulin. Diabetes 10, 383-392.

20. Johnston, D. I. and Bloom, S. R. (1973): Plasma glucagon levels in the term human infant and effect of hypoxia. Arch. Dis. Child. 48, 451-454.

21. Kaye, R. and Kumagal, M. (1958): Studies of unesterified fatty acid metabolism in infants. Am. J. Dis. Child. 96, 527-529.

22. Luyckx, A. S., Massi-Benedetti, F., Falorni, A. and Lefèbvre, P. J. (1972): Presence of pancreatic glucagon in the portal plasma of human neonates. Differences in the insulin and glucagon responses to glucose between normal infants and infants from diabetic mothers. Diabetologia 8, 296-300.

23. Milner, R. D. G., Chouksey, S. K., Mickelson, K. N. P. and Assan, R. (1973): Plasma pancreatic glucagon and insulin:glucagon ratio at birth. Arch. Dis. Child. 48, 241-242.

24. Neville, D. M. Jr. (1968): Isolation of an organ specific protein antigen from cell-surface membrane of rat liver. Biochim. Biophys. Acta 154, 540-552.

25. Persson, B. and Gentz, J. (1966): The pattern of blood lipids, glycerol and ketone bodies during the neonatal period, infancy and childhood. Acta Paediat. Scand. 55, 353-362.

26. Pictet, R. and Rutter, W. J. (1972): Development of the embryonic endocrine pancreas. In: Endocrine Pancreas, D. F. Steiner and N. Freinkel, eds. American Physiological Society, Washington, p. 25-66.

27. Salomon, Y., Londos, C. and Rodbell, M. (1974): A highly sensitive adenylate cyclase assay. Anal. Biochem. 58, 541-548.

28. Sodoyez-Goffaux, F., Sodoyez, J. C. and Foà, P. P. (1971): Effects of gestational age, birth and feeding on the insulinogenic response to glucose and tolbutamide by fetal and newborn rat pancreas. Diabetes 20, 586-591.

29. Taylor, C. B., Bailey, E. and Bartley, W. (1967): Changes in hepatic lipogenesis during development of the rat. Biochem. J. 105, 717-722.

30. Vernon, R. G. and Walker, D. G. (1968): Changes in activity of some enzymes involved in glucose utilization and formation in developing rat liver. Biochem. J. 106, 321-329.

ROLE OF THE INSULIN/GLUCAGON RATIO IN THE CHANGES OF HEPATIC METABOLISM DURING DEVELOPMENT OF THE RAT

J. R. Girard, P. Ferre, A. Kervran, J. P. Pegorier and R. Assan

Laboratoire de Physiologie du Developpement, College de France
Paris, France

The insulin/glucagon ratio was measured during the embryologic development of the rat and its changes were correlated with the variation in the diet and in hepatic metabolism. During late fetal life, the high carbohydrate diet and the high insulin/glucagon ratio are appropriate for an organism whose metabolism is set at a maximally anabolic mode: growth, protein synthesis, liver glycogen storage. After birth and during suckling, rapid growth can coexist with active gluconeogenesis and ketogenesis in the liver and with a low insulin/glucagon ratio, since the mother's milk is a high fat diet. The weaning period is characterized by a replacement of this high fat by a high carbohydrate diet. Suppression of gluconeogenesis and ketogenesis and appearance of an active lipogenesis in the liver occur, with a rise in insulin/glucagon ratio. These data suggest that glucagon plays a more important role during early life than during adult life.

In adult mammals, insulin and glucagon exert diametrically opposite actions upon hepatic metabolism. Glucagon, through its powerful glycogenolytic and glucogenic effects, promotes hepatic glucose production, while insulin opposes both these actions and, under appropriate conditions, can cause glucose storage as hepatic glycogen (17, 46, 50, 52). Glucagon promotes triglyceride breakdown and stimulates ketone body production by the liver (37) while insulin has been reported to have the reverse effects (66). In 1971, Unger proposed the concept that the islets of Langerhans, by varying the relative concentrations of these two hormones in the portal blood, possess the biological capacity of controlling the disposition of key nutrients and endogenous substrates in a manner appropriate to the prevailing exogenous fuel supply and energy requirements (67). The idea that the A and B cells of the islets of Langerhans act as a functional unit has been strengthened by the finding of a structural coupling between these different endocrine cells (51). The validity of this concept has now been verified in a variety of physiologic and pathologic conditions including fasting, exercise, meals, stress and diabetes (68).

In the present paper we consider whether the changes in bihormonal pattern occurring during development are appropriate to the prevailing energy needs. We will examine 3 different periods of the rat's development: late fetal life, the neonatal and suckling period and the period of weaning. These periods were chosen because they coincide with marked changes in the diet and because important changes in the levels of plasma insulin and glucagon and in hepatic metabolism may be expected.

Late Fetal Life

It is widely accepted that fetal transplacental nutrition in the rat and other non ruminant mammals is generally classified as a "high carbohydrate diet." The main oxidative substrate is glucose, of which part is used for energy metabolism and growth and part is stored as glycogen in many tissues and particularly in the liver (41). Amino acids are transferred rapidly to the fetus, are oxidized only in part by fetal tissues and largely conserved and utilized for protein synthesis which proceeds at a very rapid rate. Since, when the pregnant mother is well fed, all the glucose needs for growth and oxidative metabolism can be provided to the fetus via the placenta, it is not surprising that hepatic gluconeogenesis does not occur in fetal liver (6, 54, 69, 76). However, recent observations suggest that this conclusion may not be valid for substrates which can enter the pathway above the step catalyzed by phosphoenolpyruvate carboxykinase (PEPCK), e. g. for glycerol and serine (13, 22). The rate of lipogenesis from glucose or acetate also is very high in fetal rat liver (4, 65), while the placental transfer of free fatty acids (FFA) is very limited and the FFA are slowly degraded by fetal tissue (73) and do not contribute to ketogenesis which does not occur in fetal rat liver (1, 2, 43). As can be expected for an organism consuming a diet high in carbohydrate and whose metabolism is set in a maximally anabolic mode, the plasma insulin level (Fig. 1) and the insulin/glucagon ratio (Fig. 2) are very high. Indeed, insulin has been shown to stimulate glycogen synthesis in differentiated fetal liver (14, 15, 47, 58, 63) and to increase overall glucose and amino acid utilization by the fetus (56, 61).

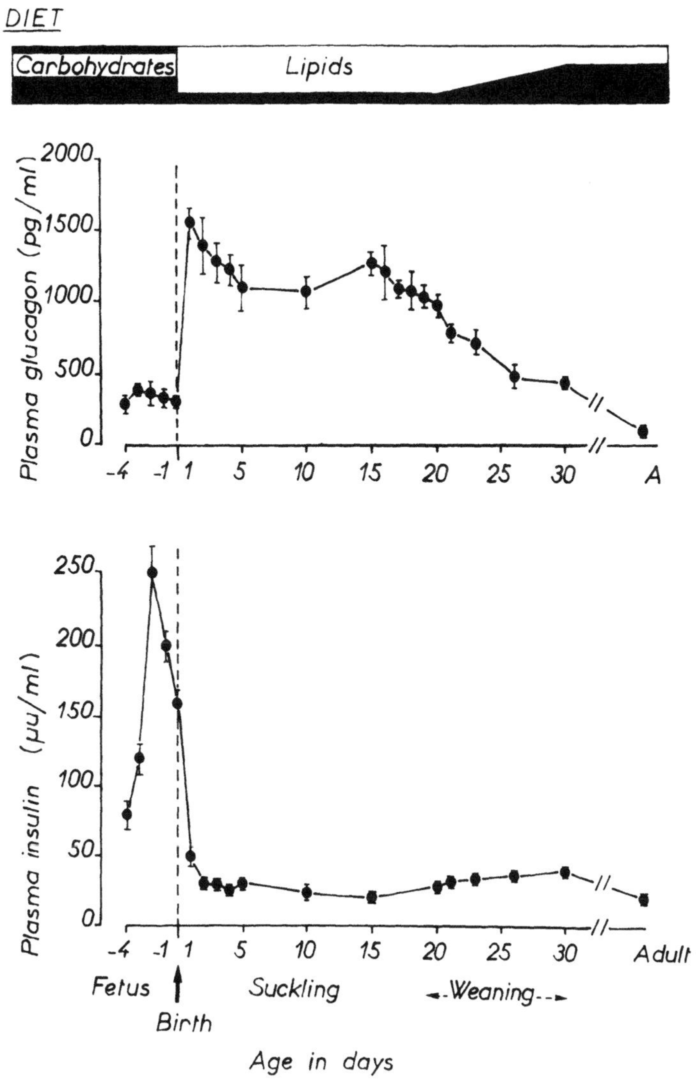

Fig. 1. Plasma insulin and glucagon during development of the rat. The values at 0 time represent the levels of circulating hormones just after delivery by cesarian section. Since caloric percentage from proteins in the diet remains constant during suckling and after weaning, only caloric percentage from fat and carbohydrate has been drawn.

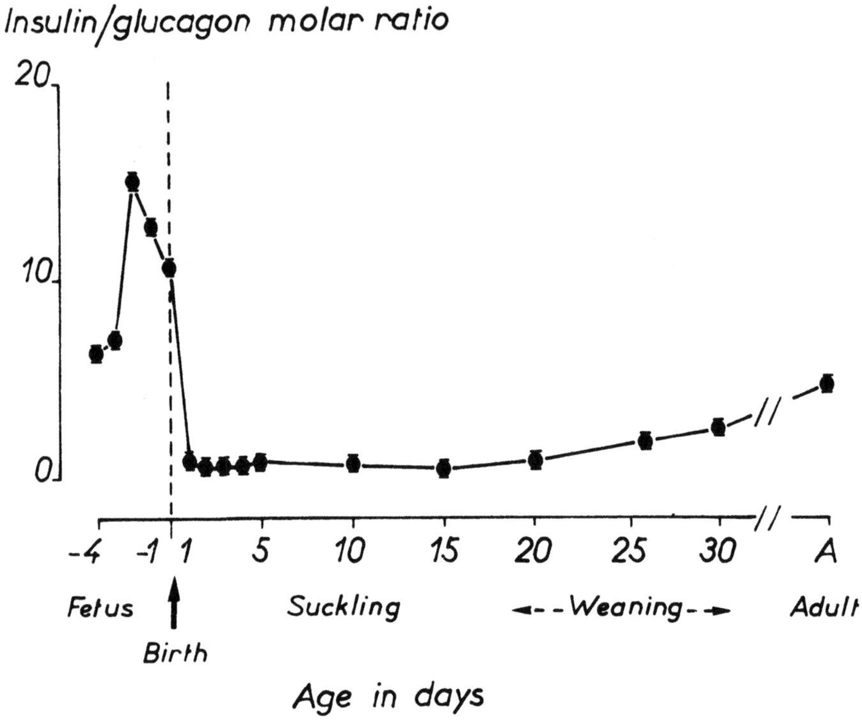

Fig. 2. Insulin/glucagon molar ratio during development of the rat

Surprisingly, plasma glucagon levels are higher in the fetus than in the fed adult rat (Fig. 1) and are similar to those observed in starved adult rats. Recent experiments have suggested that this phenomenon is due to glucagon unresponsiveness of the fetal rat liver, secondary to a limited number of glucagon receptor sites (8, 72). This glucagon "resistance" occurring at a time when substantial insulin sensitivity already exists (8) allows the accumulation of hepatic glycogen in the fetus, in spite of plasma glucagon levels comparable to or slightly higher than those of the adult animal (Fig. 1; 9). However, a glycogenolytic response to glucagon has been obtained in cultured fetal hepatocytes with glucagon concentrations in the range of the physiologic blood levels found in adult or newborn rats (57). In utero, the absence of an effect of circulating glucagon on fetal liver could result from the very high plasma insulin levels, rather than from glucagon resistance. It has been reported that glucagon injection into the fetus can induce liver glycogen degradation (24, 32, 38) and the premature appearance of liver PEPCK, the rate limiting enzyme of gluconeogenesis (36, 77). However, the increase of activity of this enzyme induced by maximal doses of glucagon in the fetus in utero is not comparable to the spontaneous rise which occurs immediately after birth (Fig. 3). An important difference between these two situations is that the level of plasma insulin decreases dramatically in the newborn, while it is increased to very high levels by the injection of glucagon into the fetus (29). Neutralization of endogenous insulin by means of anti-insulin serum markedly potentiates the effect of glucagon on liver PEPCK in the fetus (Fig. 3). This suggests that it is the insulin/glucagon ratio rather than the absolute level of glucagon which is important in the control of PEPCK activity in the liver of the neonatal rat. Recent experiments strongly support this view. Thus, it has been shown that prolonged fasting of pregnant rats during late gestation (23, 26) or prolongation of pregnancy by progesterone injection (21, 53, 59) induces the premature appearance of liver PEPCK in the rat fetus. In these two situations, the change in plasma insulin and glucagon (Fig. 4, Table 1; 21, 60) are similar to that occurring in the rat pup immediately after birth (10, 25). Furthermore, prolonged gestation produces hepatic glycogen breakdown in the fetus (60). The fall in the insulin/glucagon ratio observed in such situations is appropriate for the premature induction of liver PEPCK and glycogenolysis.

The Neonatal and Suckling Periods

With birth there is an abrupt interruption of the continuous fuel supply from the mother to the fetus; the newborn rat must rely on its own energy stores and on mother's milk. This is a high fat- low carbohydrate diet in which 69% of the calories are provided by fat; 23% by protein and 8% by carbohydrate (45). Thus,

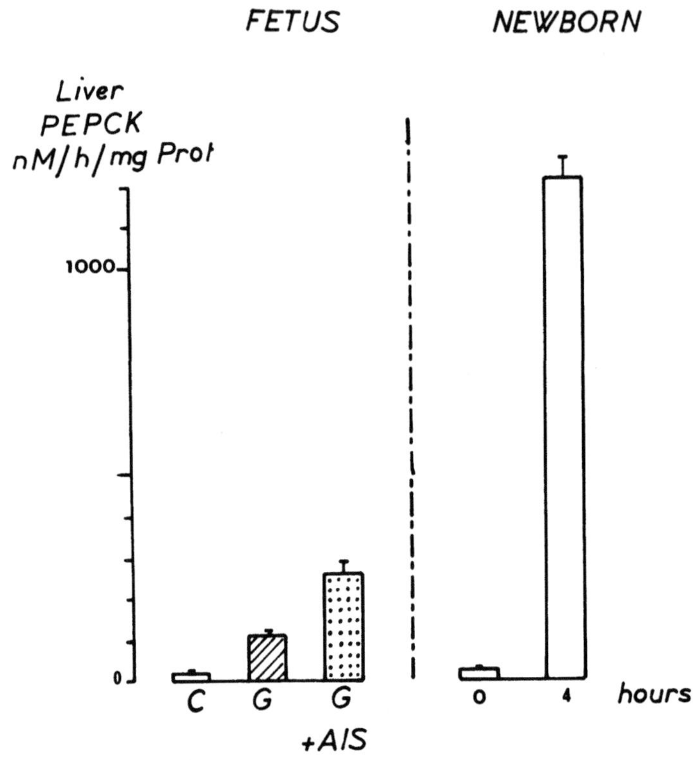

Fig. 3. Effects of injection of glucagon (G) or glucagon plus anti-insulin serum (G + AIS) on liver PEPCK activity in term rat fetus. C represents control fetuses (left panel). The spontaneous change of liver PEPCK after birth is given for comparison on the right panel (Girard, unpublished data).

TABLE 1

CIRCULATING HORMONES AND LIVER PEPCK ACTIVITY IN FETUSES FROM FED OR 96 HR FASTED PREGNANT RATS AT TERM

Means ± SEM. N = 8 to 18
(26, with permission)

	Fed	Fasted 96 hr
Insulin μU/ml	199 ± 8	120 ± 17*
Glucagon pg/ml	318 ± 25	863 ± 89*
Insulin/glucagon molar ratio	14.6	3.24
Liver PEPCK nMole/hr/mg Prot.	24 ± 3	103 ± 11*

*P <0.01 vs. fetuses of fed pregnant rats. Unpaired Student t test.

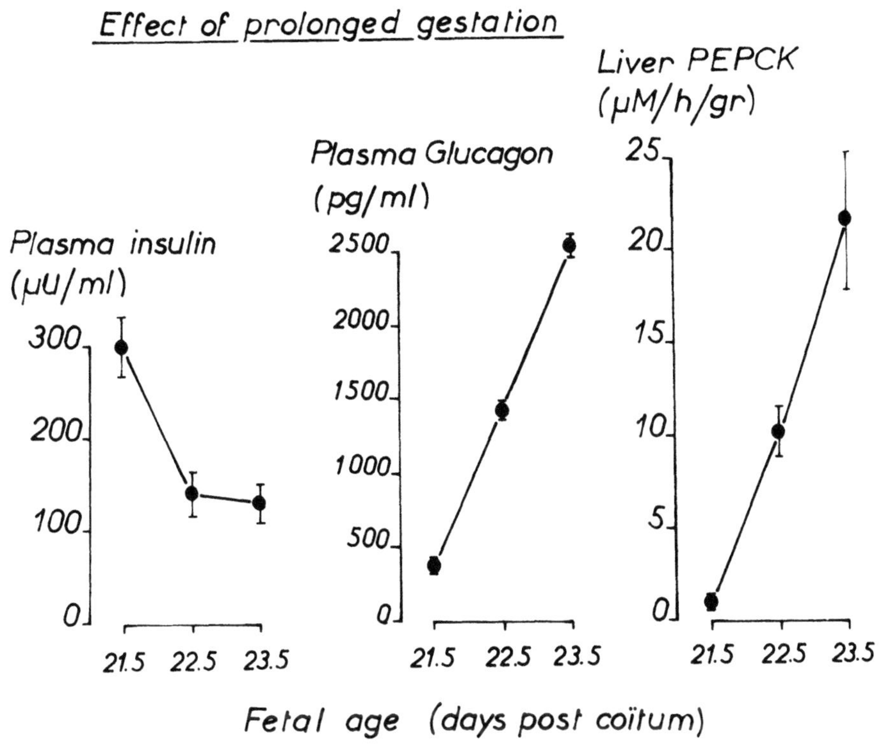

Fig. 4. Effects of the prolongation of gestation, by progesterone injection to the mother (2.5 mg on day 20.5, 21.5 and 22.5 of gestation), on circulating hormones and liver PEPCK activity in the fetus. Data on liver PEPCK activity are redrawn from Pearce et al. (53). Data on circulating hormones are from Girard and Kervran (unpublished data).

the suckling newborn must produce its own glucose to meet the energy needs of its tissues. Liver glycogen stores are mobilized rapidly after birth and are exhausted within 12 hours whether the newborn is fasting or suckling (Fig. 5). The animal is then dependent upon gluconeogenesis to maintain its blood glucose. Thus, fasting newborns develop profound hypoglycemia (25, 28), while suckling newborns maintain normal blood glucose levels in spite of the low carbohydrate diet (Fig. 5). The hypoglycemia of fasting newborn rats does not result from a delay in the development of hepatic gluconeogenesis, but rather from a reduced activity of this pathway, secondary to the lack of free fatty acid (FFA) and of gluconeogenic substrates (19). Newborn rats which receive fat and gluconeogenic substrates in their milk maintain adequate blood glucose levels by active gluconeogenesis and not as a result of glucose sparing by the high FFA and ketone body levels (18).

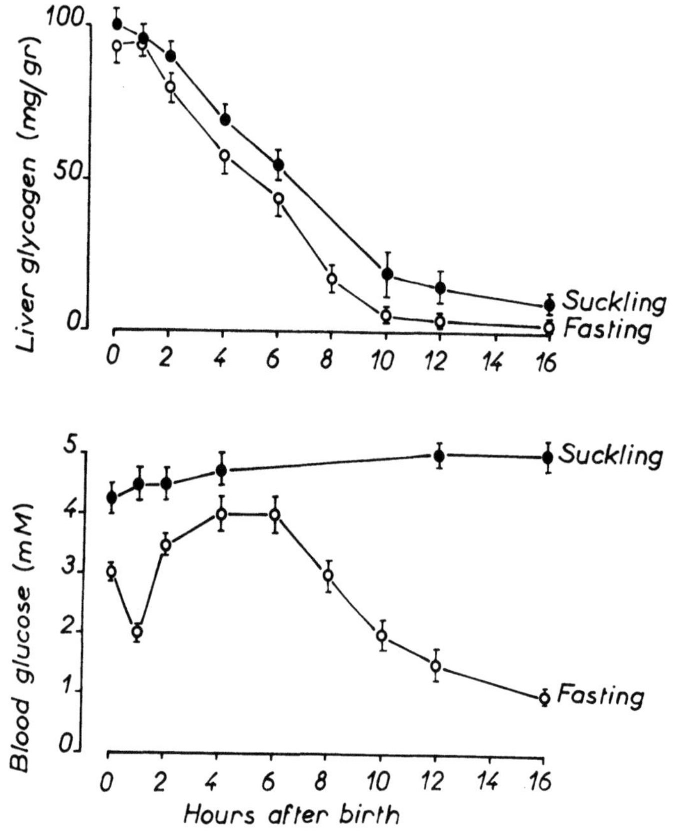

Fig. 5. Blood glucose and hepatic glycogen in fasting or suckling newborn rats (Girard, unpublished data). Suckling begins within 2 hr after birth.

The postnatal development of hepatic gluconeogenesis is dependent upon the de novo synthesis of liver PEPCK (55). The induction of this enzyme occurs whether the newborn rats are fasting or suckling (3, 5, 25, 53) and, most likely, is dependent upon the activity of the pancreatic hormones, since their changes at birth are appropriate both in time and direction for this purpose (25, 36). The transient neonatal hypoglycemia is not the factor triggering the secretion of insulin and glucagon at birth since the pancreatic A cells of neonatal rats are insensitive to variations in blood glucose levels (30, 31). Furthermore, the rise in plasma glucagon and the fall in plasma insulin occur normally in newborn rats delivered by caesarian section and maintained at 24°C (cold exposure), even though under these conditions there is no change in blood glucose level (42). The stress of removal of the fetus from its privileged intrauterine environment at birth, undoubtedly augments the activity of the sympathetic nervous system, causes epinephrine secretion or the local release of norepinephrine and, thus, may account for the changes in the levels of insulin and glucagon occurring at birth (30).

Immediately after birth, ketogenesis develops in the liver when lipogenesis is suppressed (2). This has been attributed to the change in the diet since a high fat and low carbohydrate diet in adult rats produces ketonemia and suppresses lipogenesis (33, 39). Although hepatic ketogenesis and lipogenesis are controlled by insulin and glucagon in adult rats (20, 49), the possible hormonal control of these pathways has not been studied sufficiently in newborn rats.

It has been shown that during the transition from intrauterine to extra-uterine life the surge of ketone body production is accompanied by a concomitant elevation in liver carnitine content (62) and that the changes in liver carnitine content coincide with the pattern of hepatic carnitine transferase activities (1, 43). In suckling newborn rats, the primary source of carnitine was shown to be milk (62). Recently, we have shown that fasting newborn rats which are fed a triglyceride emulsion (Intralipid, carnitine-free) 13 hours after birth are capable of increasing their blood ketone body levels 15-fold within 3 hours (19, 27). This suggests that the carnitine from milk is not absolutely necessary to the liver of newborn rats for normal ketogenesis. However, this does not negate the thesis advanced by McGarry _et al_. (48) that an increase in carnitine concentration is a prerequisite for the development of high ketogenic capacity in the liver, since carnitine accumulation may occur in the liver of fasting newborn rats via de novo synthesis, under the influence of the rise of plasma glucagon (25). Glucagon injection into a rat fetus at term has been shown to increase the fetal liver carnitine acetyl transferase and carnitine palmitoyl transferase activities within 4 hours (Hahn, Ferre and Girard, unpublished data). Thus, the rise of the plasma glucagon level which occurs at birth could explain the increase of these two carnitine transferase activities in the livers of newborn rats (1, 43). The capacity for ketone bodies production in liver homogenates of suckling rats is inversely related to plasma insulin levels and the injection of insulin suppresses plasma ketone body levels in suckling newborn rats (75). This suggests that the low level of plasma insulin in newborn and suckling rats (Fig. 1; 10, 75) could also contribute to the development of the high rate of ketogenesis in the liver of these animals.

At the present time, no studies have been carried out to learn whether the suppression of hepatic lipogenesis at birth is under hormonal control or whether it is the result of dietary changes.

During the suckling period, hepatic gluconeogenesis and ketogenesis are at a high level, while lipogenesis is very low (2, 64), two phenomena which are strongly favored by the high blood glucagon and the low insulin levels prevailing at the time (Fig. 1; 9). However, a low insulin/glucagon molar ratio (Fig. 2) in an organism whose growth rate is very fast does not fit with the concept that maximal anabolism requires a high insulin/glucagon ratio (67). This apparent discrepancy can perhaps be explained by recent observations in patients receiving total parenteral nutrition, in whom a eucaloric lipid infusion promoted positive nitrogen balance in the presence of a low plasma insulin/glucagon ratio, while a glucose infusion produced a similar positive nitrogen balance, in the presence of a high insulin/glucagon ratio (40). High plasma insulin levels seem required for maximal anabolism only when carbohydrates provide the calories necessary to meet the energy needs of the body. By contrast, anabolism can occur at normal or low plasma insulin levels when lipids are the calorie source. In suckling rats, it has been reported that dietary amino acids are used not only for protein synthesis, but that they are also utilized to provide 25% of the daily caloric needs (74). If the observations made in man can be applied to suckling rats, one can understand why rapid growth can coexist with active gluconeogenesis for, in animals fed a high fat diet, growth would not be impeded by the low plasma glucagon levels.

The factors which stimulate the release of glucagon during suckling in the rat have not yet been established. In adult dogs, the ingestion of a fat meal is accompanied by a rise in pancreatic glucagon which is mediated by a gut signal (11). Unfortunately, adult rats fed a high fat, carbohydrate-free diet for 8 to 10 days show normal plasma glucagon levels (16) and fat-induced hyperglucagonemia occurs only during the absorption of the meal. However, the feeding habits of adult rats are quite different from those of newborn rats which suckle continuously and may release the gut signal more frequently and with more prolonged effects. One such signal could be the gastric inhibitory peptide (GIP) which is released by a fat meal in the adult (12) and has been shown to stimulate the secretion of glucagon by the pancreas of the newborn rat (7). Further research is needed to clarify this point.

The Weaning Period

When rat pups remain with their mother, weaning occurs progressively between the 20th and 30th day of life. It consists primarily of a decrease in the fat and an increase in the carbohydrate content of the diet (Fig. 1). The rat pups gradually begin to nibble on standard laboratory chow (Caloric percentage: carbohydrate 55, protein 24, fat 21), and to suckle less frequently. Weaning is complete at 30 days of age.

As the milk diet is replaced by laboratory chow, a progressive rise in lipogenesis and a progressive decrease in gluconeogenesis and ketogenesis occur (2, 71). During this period, a decrease in the level of plasma glucagon and a small, but significant, increase in the plasma insulin level are observed (Fig. 1). The rise in the insulin/glucagon ratio during weaning (Fig. 2) can explain the shift in hepatic metabolism occurring at this time.

The rat can be weaned prematurely to laboratory chow or to a high carbohydrate diet after 18 days of age. This is accompanied by a lowering of the activity of the gluconeogenic enzymes: glucose-6-phosphatase, fructose-diphosphatase and PEPCK (34, 35, 70). By contrast, when rats are weaned to a high fat diet, the activities of the 3 gluconeogenic enzymes remain as high as they were during suckling (34, 35, 70). Similarly, when rats are weaned to a high fat, low carbohydrate diet, the rise in hepatic lipogenic enzymes is prevented (34, 35, 44) and the enzymes of hepatic ketogenesis are maintained at a high level of activity (34). Recently, we have studied the effects of premature weaning to several diets on plasma insulin and glucagon in the rat (Fig. 6). Weaning to a high carbohydrate diet was followed by a marked fall in plasma glucagon while weaning to a high fat diet led to the maintenance of the high level of plasma glucagon seen during suckling. In either case there were no changes in the level of plasma insulin. These data suggest that the fall in plasma glucagon level in association with the increased carbohydrate content of the diet, can play a role in the changes in hepatic metabolism at the time of weaning. This assumption is supported by the observation that, in rats weaned to a high carbohydrate diet, glucagon injections help maintain a high level of hepatic PEPCK activity and repress the rise of hepatic lipogenic enzymes (35).

CONCLUSIONS

We have reviewed the evidence that variations in the insulin/glucagon ratio, associated with changes in the diet, probably play an important role, in the adaptations of hepatic metabolism occurring at birth and at the time of weaning in the rat. In the adult animal or in man, recent experiments have suggested that the fine tuning of hepatic metabolism is primarily a function of insulin

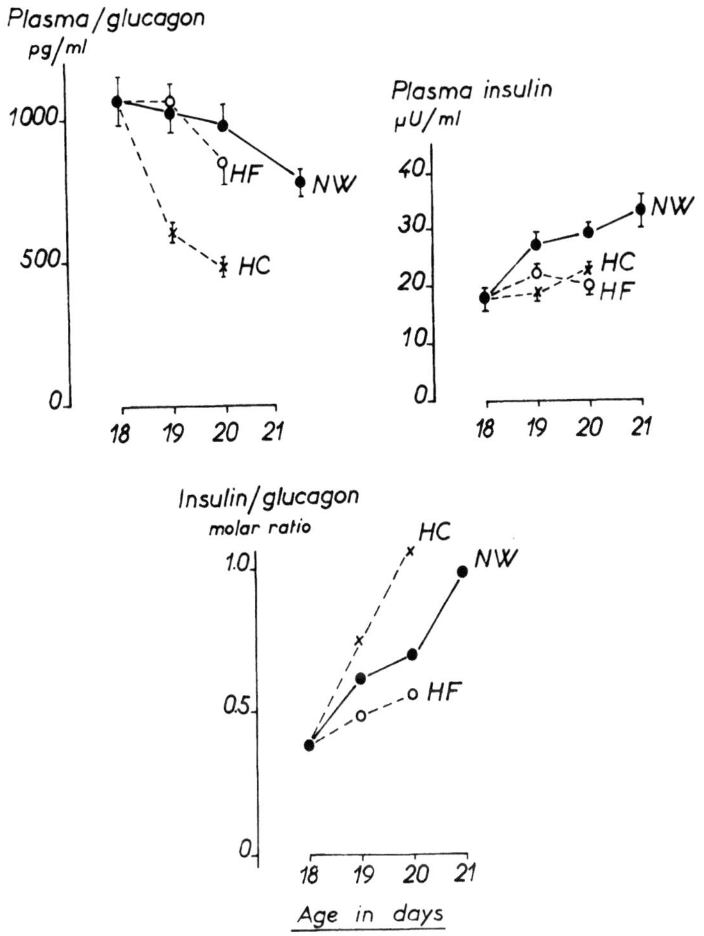

Fig. 6. Circulating hormones in suckling newborn rats which remain with their mother (NW) and in newborn rats weaned prematurely at 18 days on high fat (HF) or on high carbohydrate (HC) diets (Hahn and Girard, unpublished data). Compositions of diets are given by Hahn and Kirby (34, 35).

and substrate availability and that the presence of glucagon is necessary but not regulatory. From the present data, it appears that glucagon could play a more important role during early life (birth, weaning) than during adult life.

REFERENCES

1. Augenfeld, J. and Fritz, I. B. (1970): Carnitine palmitoyl transferase activity and fatty acid oxidation by livers from fetal and neonatal rats. Canadian J. Biochem. 48, 288-294.

2. Bailey, E. and Lockwood, E. A. (1973): Some aspects of fatty acid oxidation and ketone body formation and utilization during development of the rat. Enzyme 15, 239-253.

3. Ballard, F. J. (1970): Gluconeogenesis and the regulation of blood glucose in the neonate. In: Diabetes, R. R. Rodriguez and J. Vallance-Owen, eds. Excerpta Medica, Amsterdam, p. 592-600.

4. Ballard, F. J. and Hanson, R. W. (1967): Changes in lipid synthesis in rat liver during development. Biochem. J. 102, 952-958.

5. Ballard, F. J., Hopgood, M. F., Reshef, L. and Hanson, R. W. (1975): Changes in protein synthesis and degradation involved in enzyme accumulation in differentiating liver. In: The Biochemistry of Gene Expression in Higher Organisms, J. K. Pollak and J. W. Lee, eds. D. Reidel, Boston, p. 274.

6. Ballard, F. J. and Oliver, I. T. (1963): Glycogen metabolism in embryonic chick and neonatal rat liver. Biochim. Biophys. Acta 71, 578-588.

7. Bataille, D., Jarousse, C., Vauclin, N., Gestach, C. and Rosselin, G. (1977): Effect of vasoactive intestinal peptide (VIP) and gastric inhibitory peptide (GIP) on insulin and glucagon release by perifused newborn rat pancreas. This volume, p. 255-270.

8. Blázquez, E., Rubalcava, B., Montesano, R., Orci, L. and Unger, R. H. (1976): Development of insulin and glucagon binding and the adenylate cyclase response in liver membranes of the prenatal, postnatal and adult rat: evidence of glucagon "resistance". Endocrinology 98, 1014-1023.

9. Blázquez, E., Sugase, T., Blázquez, M. and Foà, P. P. (1972): The ontogeny of metabolic regulation in the rat, with special reference to the development of insular function. Acta Diabetol. Latina 9 suppl. 1, 13-35.

10. Blázquez, E., Sugase, T., Blázquez, M. and Foà, P. P. (1974): Neonatal changes in the concentration of rat liver cyclic AMP and of serum glucose, free fatty acids, insulin, pancreatic and total glucagon in man and in the rat. J. Lab. Clin. Med. 83, 957-967.

11. Böttger, I., Dobbs, R., Faloona, G. R. and Unger, R. H. (1973): The effects of triglyceride absorption upon glucagon, insulin and gut glucagon-like immunoreactivity. J. Clin. Invest. 52, 2532-2541.

12. Brown, J. C. (1974): Gastric inhibitory polypeptide (GIP). In: Endocrinology 1973, S. Taylor, ed. W. Heinemann, London, p. 276-284.

13. Coufalik, A. and Monder, C. (1976): Spontaneous development of gluconeogenesis in fetal rat livers during incubation in organ culture. Proc. Soc. Exptl. Biol. Med. 152, 603-605.

14. Eisen, H. J., Glinsmann, W. H. and Sherline, P. (1973): Effect of insulin on glycogen synthesis in fetal rat liver in organ culture. Endocrinology 92, 584-588.

15. Eisen, H. J., Goldfine, I. D. and Glinsmann, W. H. (1973): Regulation of hepatic glycogen synthesis during fetal development: role of hydrocortisone, insulin and insulin receptors. Proc. Natl. Acad. Sci. (USA) 70, 3454-3457.

16. Eisenstein, A. B., Strack, I. and Steiner, A. (1974): Increased hepatic gluconeogenesis without a rise of glucagon secretion in rats fed a high fat diet. Diabetes 23, 869-875.

17. Exton, J. H. and Park, C. R. (1972): Interaction of insulin and glucagon in the control of liver metabolism. In: Handbook of Physiology, Endocrine Pancreas. D. F. Steiner and N. Freinkel, Eds. Amer. Physiol. Soc. Washington, pp. 437-455.

18. Ferre, P., Pegorier, J. P. and Girard, J. R. (1977): The effects of inhibition of gluconeogenesis in suckling newborn rats. Biochem. J. 162, 209-212.

19. Ferre, P., Pegorier, J. P., Marliss, E. B. and Girard, J. R. (1977): Influence of exogenous fat of gluconeogenic substrates on glucose homeostasis in the newborn rat. Am. J. Physiol. (accepted for publication).

20. Geelen, M. J. H. and Gibson, D. M. (1975): Lipogenesis in maintenance cultures of rat hepatocytes. FEBS Letters 58, 334-339.

21. Ghisalberti, A. V. (1976): Enzyme induction in perinatal rat liver. Ph.D. Thesis, University of Western Australia.

22. Gilbert, M. (1977): Origin and metabolic fate of plasma glycerol in the rat and rabbit fetus. Pediat. Res. 11, 95-99.

23. Girard, J. R. (1975): Metabolic fuels of the fetus. Israel J. Med. Sci. 11, 591-600.

24. Girard, J. and Bal, D. (1970): Effets du glucagon-zinc sur la glycémie et la teneur en glycogène du foie foetal du rat en fin de gestation. C. R. Acad. Sci. (Paris) 271, 777-779.

25. Girard, J. R., Cuendet, G. S., Marliss, E. B., Kervran, A., Rieutort, M. and Assan, R. (1973): Fuels, hormones and liver metabolism at term and during the early postnatal period in the rat. J. Clin. Invest. 52, 3190-3200.

26. Girard, J. R., Ferre, P., Gilbert, M., Kervran, A., Assan, R. and Marliss, E. B. (1977): Fetal metabolic response to maternal fasting in the rat. Am. J. Physiol., 232, E456-463.

27. Girard, J. R., Ferre, P. and Pegorier, J. P. (1976): Glucose homeostasis in the fasted newborn rats: role of lipids and gluconeogenic substrates. Diabetologia 12, 393 (abst.).

28. Girard, J. R., Guillet, I., Marty, J. and Marliss, E. B. (1975): Plasma amino acid levels and development of hepatic gluconeogenesis in the newborn rat. Am. J. Physiol. 229, 466-473.

29. Girard, J. R., Guillet, I., Marty, J., Assan, R. and Marliss, E. B. (1976): Effects of exogenous hormones and glucose on plasma levels and hepatic metabolism of amino acids in the fetus and in the newborn rat. Diabetologia 12, 327-337.

30. Girard, J. R., Kervran, A. and Assan, R. (1975): Functional maturation of the A cell in the rat. In: Early Diabetes in Early Life, R. A. Camerini-Davalos and H. S. Cole, Eds. Academic Press, New York, pp. 57-71.

31. Girard, J. R., Kervran, A., Soufflet, E. and Assan, R. (1974): Factors affecting the secretion of insulin and glucagon by the rat fetus. Diabetes 23, 310-317.

32. Greengard, O. and Dewey, H. K. (1970): The premature deposition or lysis of glycogen in livers of fetal rats injected with hydrocortisone or glucagon. Develop. Biol. 21, 452-461.

33. Griglio, S. and Malewiak, M. I. (1975): Triglyceride storage, ketonemia, and fat diets. Nutr. Metab. 19, 131-144.

34. Hahn, P. and Kirby, L. (1973): Immediate and late effects of premature weaning and of feeding a high fat or high carbohydrate diet to weanling rats. J. Nutr. 103, 690-696.

35. Hahn, P. and Kirby, L. T. (1974): The effects of catecholamines, glucagon and diet on enzyme activities in brown fat and liver of the rat. Canadian J. Biochem. 52, 739-743.

36. Hanson, R. W., Reshef, L. and Ballard, F. J. (1975): Hormonal regulation of hepatic P-enolpyruvate carboxykinase (GTP) during development. Fed. Proc. 34, 166-171.

37. Heimberg, M., Weinstein, I. and Kohout, M. (1969): The effects of glucagon, dibutyryl cyclic adenosine 3'5' monophosphate and concentration of free fatty acid on hepatic lipid metabolism. J. Biol. Chem. 244, 5131-5139.

38. Hunter, D. J. S. (1969): Changes in blood glucose and liver carbohydrate after intrauterine injection of glucagon into foetal rats. J. Endocrinol. 45, 367-374.

39. Jansen, G. R., Hutchison, C. F. and Zanetti, M. E. (1966): Studies on lipogenesis in vivo. Effect of dietary fat or starvation on conversion of ^{14}C-glucose into fat and turnover of newly synthetized fat. Biochem. J. 99, 323-332.

40. Jeejeebhoy, K. N., Anderson, G. H., Nakhooda, A. F., Greenberg, G. R., Sanderson, I. and Marliss, E. B. (1976): Metabolic studies in total parenteral nutrition with lipid in man: comparison with glucose. J. Clin. Invest. 57, 125-136.

41. Jost, A. and Picon, L. (1970): Hormonal control of fetal development and metabolism. Adv. Metab. Dis. 4, 123-184.

42. Kervran, A., Gilbert, M., Girard, J. R., Assan, R. and Jost, A. (1976): Effect of environmental temperature on glucose-induced insulin response in the newborn rat. Diabetes 25, 1026-1030.

43. Lee, L. P. K. and Fritz, I. B. (1971): Hepatic ketogenesis during development. Canadian J. Biochem. 49, 599-605.

44. Lockwood, E. A., Bailey, E. and Taylor, C. B. (1970): Factors involved in changes in hepatic lipogenesis during development of the rat. Biochem. J. 118, 155-162.

45. Luckey, T. D., Mende, T. J. and Pleasants, J. (1954): The physical and chemical characterization of rat's milk. J. Nutr. 54, 345-359.

46. Mackrell, D. J. and Sokal, J. E. (1969): Antagonism between the effects of insulin and glucagon on the isolated liver. Diabetes 18, 724-732.

47. Manns, J. G. and Brockman, R. P. (1969): The role of insulin in the synthesis of fetal glycogen. Canadian J. Physiol. Pharmacol. 47, 917-921.

48. McGarry, J. D., Robles-Valdes, C. and Foster, D. W. (1975): The role of carnitine in hepatic ketogenesis. Proc. Natl. Acad. Sci. (USA) 72, 4385-4388.

49. McGarry, J. D., Wright, P. H. and Foster, D. W. (1975): Hormonal control of ketogenesis. Rapid activation of hepatic ketogenic capacity in fed rats by anti-insulin serum and glucagon. J. Clin. Invest. 55, 1202-1209.

50. Menahan, L. A. and Wieland, O. (1969): Interactions of glucagon and insulin on the metabolism of perfused livers from fasted rats. Europ. J. Biochem. 9, 55-62.

51. Orci, L., Malaisse-Lagae, F., Ravazzola, M., Rouiller, D., Renold, A. E., Perrelet, A. and Unger, R. H. (1975): A morphological basis for intercellular communication between A and B cells in the endocrine pancreas. J. Clin. Invest. 56, 1066-1070.

52. Parrilla, R., Goodman, M. N. and Toews, C. J. (1974): Effect of glucagon: insulin ratios on hepatic metabolism. Diabetes 23, 725-731.

53. Pearce, P. H., Buirchell, B. J., Weaver, P. K. and Oliver, I. T. (1974): The development of phosphopyruvate carboxylase and gluconeogenesis in neonatal rats. Biol. Neonate 24, 320-329.

54. Philippidis, H. and Ballard, F. J. (1969): The development of gluconeogenesis in rat liver. Experiments in vivo. Biochem. J. 113, 651-657.

55. Philippidis, H., Hanson, R. W., Reshef, L., Hopgood, M. F. and Ballard, F. J. (1972): The initial synthesis of proteins during development. Phosphoenolpyruvate carboxylase in rat liver at birth. Biochem. J. 126, 1127-1134.

56. Picon, L. (1967): Effect of insulin on growth and biochemical composition of the rat fetus. Endocrinology 81, 1419-1421.

57. Plas C. and Nunez, J. (1975): Glycogenolytic response to glucagon of cultured fetal hepatocytes. Refractoriness following prior exposure to glucagon. J. Biol. Chem. 250, 5304-5311.

58. Plas, C. and Nunez, J. (1976): Role of cortisol on the glycogenolytic effect of glucagon and on the glycogenic response to insulin in fetal hepatocyte culture. J. Biol. Chem. 251, 1431-1437.

59. Portha, B., Leprovost, E., Picon, L. and Rosselin, G. (1977): Postmaturity in the rat: phosphorylase, glucose-6-phosphatase and phosphoenolpyruvate carboxykinase activities in the fetal liver. Horm. Metab. Res. (submitted for publication).

60. Portha, B., Rosselin, G. and Picon, L. (1976): Postmaturity in the rat: impairment of insulin, glucagon and glycogen stores. Diabetologia 12, 429-436.

61. Rabain, F. and Picon, L. (1974): Effect of insulin on the maternofetal transfer of glucose in the rat. Horm. Metab. Res. 6, 376-380.

62. Robles-Valdes, C., McGarry, J. D. and Foster, D. W. (1976): Maternal-fetal carnitine relationships and neonatal ketosis in the rat. J. Biol. Chem. 251, 6007-6012.

63. Schwartz, A. L. and Rall, T. W. (1973): Hormonal regulation of glycogen metabolism in neonatal rat liver. Biochem. J. 134, 985-993.

64. Snell, K. and Walker, D. G. (1973): Gluconeogenesis in the newborn rat: the substrates and their quantitative significance. Enzyme 15, 40-81.

65. Taylor, C. B., Bailey, E. and Bartley, W. (1967): Changes in hepatic lipogenesis during development of the rat. Biochem. J. 105, 717-722.

66. Topping, D. L. and Mayes, P. A. (1972): The immediate effects of insulin and fructose on the metabolism of the perfused liver. Changes in lipoprotein secretion, fatty acid oxidation and esterification, lipogenesis and carbohydrate metabolism. Biochem. J. 126, 295-311.

67. Unger, R. H. (1974): Alpha- and beta-cell interrelationships in health and disease. Metabolism 23, 581-593.

68. Unger, R. H. and Orci, L. (1976): Physiology and pathophysiology of glucagon. Physiol. Rev. 56, 778-826.

69. Vernon, R. G., Eaton, S. W. and Walker, D. G. (1968): Carbohydrate formation from various precursors in neonatal rat liver. Biochem. J. 110, 725-731.

70. Vernon, R. G. and Walker, D. G. (1968): Changes in activity of some enzymes involved in glucose utilization and formation in developing rat liver. Biochem. J. 106, 321-329.

71. Vernon, R. G. and Walker, D. G. (1968): Adaptative behaviour of some enzymes involved in glucose utilization and formation in rat liver during the weaning period. Biochem. J. 106, 331-338.

72. Vinicor, F., Higdon, G., Clark, J. F. and Clark, C. M, Jr. (1976): Development of glucagon sensitivity in neonatal rat liver. J. Clin. Invest. 58, 571-578.

73. Warshaw, J. B. (1974): Fatty acid oxidation during development. In: Parenteral Nutrition in Infancy and Childhood, H. H. Bode and J. B. Warshaw, eds. Plenum, New York, p. 88-97.

74. White, P. K. and Miller, S. A. (1976): Utilization of dietary amino acids for energy production in neonatal rat liver. Pediat. Res. 10, 158-164.

75. Yeh, Y. Y. and Zee, P. (1976): Insulin, a possible regulator of ketosis in newborn and suckling rats. Pediat. Res. 10, 192-197.

76. Yeung, D. and Oliver, I. T. (1967): Gluconeogenesis from amino acids in neonatal rat liver. Biochem. J. 103, 744-748.

77. Yeung, D. and Oliver, I. T. (1968): Factors affecting the premature induction of phosphopyruvate carboxylase in neonatal rat liver. Biochem. J. 108, 325-331.

GLUCOSE HOMEOSTASIS IN THE NEWBORN

J. S. Bajaj and K. D. Buchanan

All-India Institute of Medical Sciences, New Delhi, India and
Queen's University of Belfast, Northern Ireland

Measurements of insulin and glucagon secretion by pancreatic islets isolated from rat pups at various times after birth (0 hours to 15 days) indicate a progressive decline in basal insulin release and a progressive increase in basal glucagon release with age. Thus, a decrease in the insulin:glucagon ratio after birth may provide the endocrine environment required for glucose homeostasis in the newborn consuming a low carbohydrate diet.

The ontogenetic differentiation of the endocrine pancreas has received considerable attention in recent years (15, 25). A cells and glucagon have apparently been demonstrated in the pancreas of the fetal rat on and after the 11th day of gestation, suggesting thereby that glucagon might play a role in the regulation of growth and differentiation of either all or certain categories of embryonic cells (26). Insulin is the anabolic hormone in the fetus. With an abundant supply of glucose to the fetus from the mother, the hormonal profile in the fetus during intrauterine life reflects a high insulin:glucagon ratio. In the chick and in the rat the release of pancreatic glucagon into the circulation seems to be correlated with the concentration of hepatic phosphorylase: both reach a maximum at or near birth when the glycogen synthetase content decreases and the activity of glucose 6-phosphatase begins to show an increase (11). These hormonal and enzymatic changes favour glycogenolysis and gluconeogenesis which are necessary for the prevention of neonatal hypoglycemia.

Blázquez, Sugase, Blázquez and Foà (4) describe several phases involving the hormonal and metabolic changes which take place in the rat during gestation, at birth and in the few weeks that follow. Beginning about 3 hours after birth and continuing during day one, a sharp rise in plasma glucagon and fall in plasma insulin has been demonstrated. The increase in plasma glucagon was both in the pancreatic IRG as well as in the gut GLI fractions; a sharp increase in the gut

GLI was demonstrable after the first act of suckling. Up to about day 15-20 (suckling period) the plasma glucagon levels remained elevated while plasma insulin showed a progressively downward trend. At the time of weaning there was a demonstrable, although transient increase in plasma insulin with a decrease in plasma glucagon. It has been suggested (4, 5) that these hormonal changes, especially affecting the insulin:glucagon molar ratio, are essential for the glucose homeostasis of the newborn.

Lernmark and Wenngren (21) studied insulin and glucagon release from the pancreas of newborn mice. Although glucose was shown to stimulate insulin release, it was without effect on glucagon release unless arginine was also present. These studies employed whole pancreas and their attempt to study the glucagon behaviour in older animals was foiled by the degradation of glucagon in the medium. This present study utilizes isolated islets from rats of 0-15 days old, and attempts to define the control by glucose of insulin and glucagon during this unusual metabolic period in the rat's life.

METHODS

Rats from the Wister strain were used and were divided into 4 age groups: 0-24 hrs., 1-4 days, 5-10 days, and 11-15 days. The pups were removed from their mother just prior to the experiment and killed by decapitation. Isolated islets were prepared from the pancreas according to the method of Buchanan and Mawhinney (9) with the modifications that less collagenase (Koch,light) was used (2 mg per 2 pancreata) than in the adult. Islets were readily identified and transferred by a very fine capillary pipette into small glass incubation flasks (10 islets/flask) into 1 ml of Krebs-Ringer-bicarbonate (KRB) buffer supplemented by 1% human albumin (Blood Products Laboratory) and 1000 units of Trasylol (FBA Pharmaceuticals). After a 30 minute preincubation with shaking at 37°C in 60 mg/dl glucose, the islets were washed and incubated for a further 30 minutes in either 30 or 300 mg/dl glucose. The flasks were gassed with 95% O_2 and CO_2 throughout the experiment.

Following the incubation, samples were removed for radioimmunoassay of insulin and glucagon. Insulin was assayed by the method of Buchanan and McCarroll (10) using human insulin as standard and antibody raised against pork insulin. Rat insulin did not crossreact fully with this antiserum and for this reason the insulin levels are lower than the actual values. Glucagon was assayed by the method of Buchanan (8). The antiserum (YY57) used was raised against pork glucagon and did not discriminate between pork and rat glucagon. It crossreacted with intestinal extracts but, as isolated islet preparations were used in these experiments, this non-specificity is not pertinent.

RESULTS

A. Basal insulin and glucagon release (Fig. 1). There was a progressive decline in basal insulin release with age ($p < .005$ between 0-24 hrs and 1-4 days; $p < .0005$ between 1-4 days and 5-10 days and $p < .005$ between 5-10 days and 11-15 days). In contrast there was a progressive increase in basal glucagon release with age ($p < .05$ between 0-24 hrs and 5-10 days; $p < .05$ between 5-10 days and 11-15 days).

B. Effect of glucose on insulin and glucagon release. As shown in Fig. 2, an increase in the concentration of glucose in the incubation medium resulted in a significant increase in insulin release ($p < .0005$ in all age groups). Glucagon release on the other hand was not altered by an increase in the glucose concentration of the medium in any of the age groups (Fig. 3).

DISCUSSION

The role of insulin and glucagon in the maintenance of glucose homeostasis in the newborn is now well recognized. Information is available in various species regarding both the presence of A cells in the fetus as well as the presence of glucagon and insulin in fetal circulation. Using electron microscopic

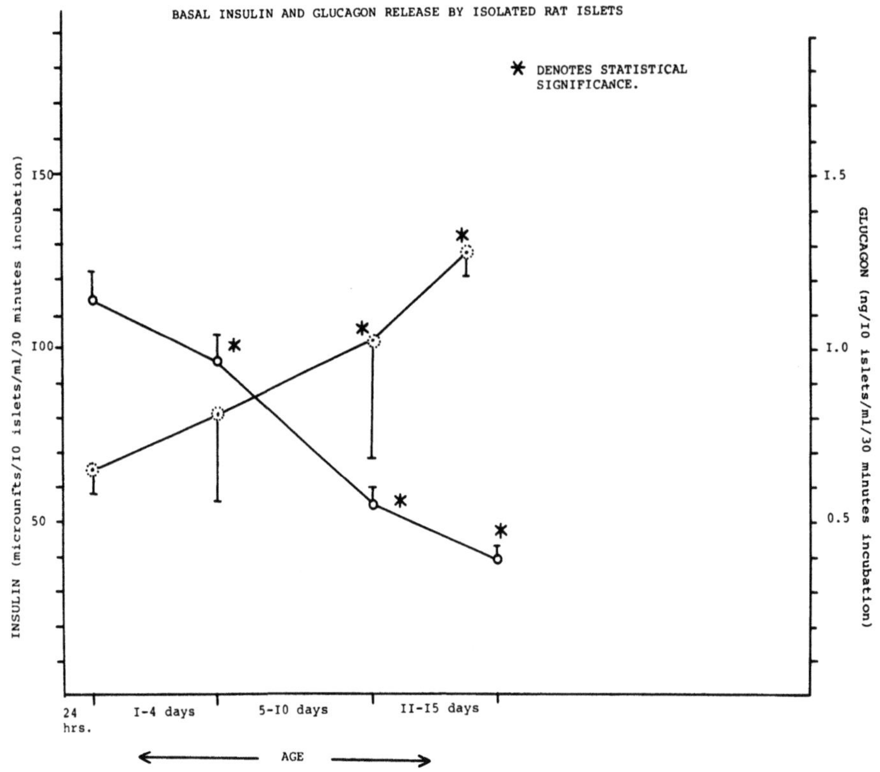

Fig. 1. Basal insulin and glucagon release by isolated rat pancreatic islets in different postnatal age groups. Mean ± S.E.

technique, typical A cells have been identified in fetal rat pancreas at gestational ages varying from 18 days (24) to 11 days (25). Some authors (25) were also able to demonstrate the presence of immunoreactive glucagon in fetal rat pancreas on the 11th day of gestation. Pancreatic glucagon has been shown in plasma of rat fetus aged 18.5 to 21.5 days (4, 16). As there is no evidence for a transplacental movement of maternal glucagon to the fetus, this strongly suggests the fetal origin of pancreatic glucagon circulating in fetal plasma. These observations in rat regarding the presence of both A cells in the early gestational age as well as the presence of circulating glucagon in fetal plasma before term have been confirmed in other species including mouse (21), rabbit (3),

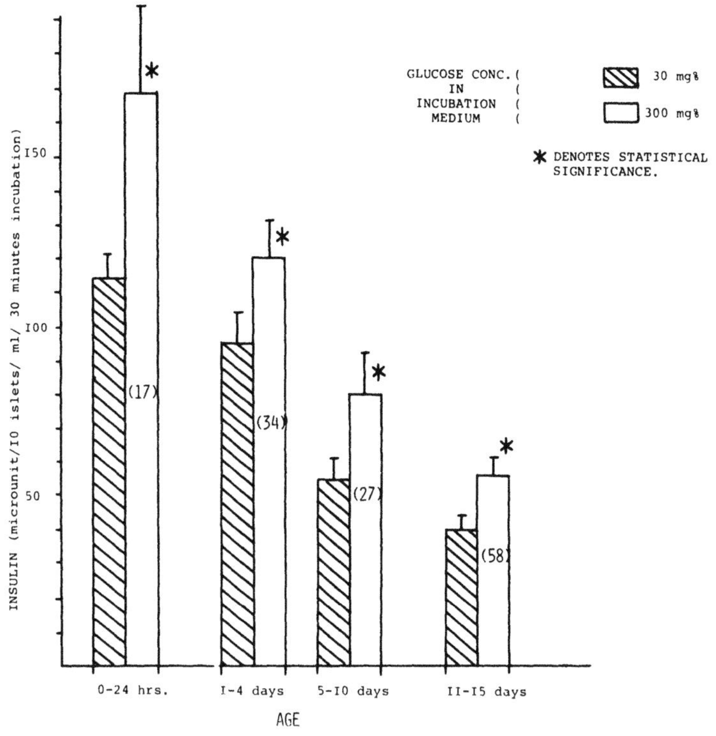

Fig. 2. Release of insulin from isolated rat pancreatic islets during incubation in varying concentration of glucose. Figures in parenthesis in the bar indicate the number of incubation vessels. Mean ± S.E.

sheep (1) and the human (22) at correspondingly appropriate gestational ages.

There is a high insulin:glucagon molar ratio in the fetal circulation both before and at birth. This ratio favours energy storage. However, soon after birth there is a rise of plasma glucagon and a fall in plasma insulin. These hormonal changes facilitate the expression of hepatic effects of glucagon namely glycogenolysis and gluconeogenesis. Indeed administration of glucagon in the full term fetal rat has been shown to produce a rise in glucose and a fall in plasma amino acid levels, with a concomitant increase in the hepatic uptake of amino acids and the conversion of lactate and amino acids into glucose (18).

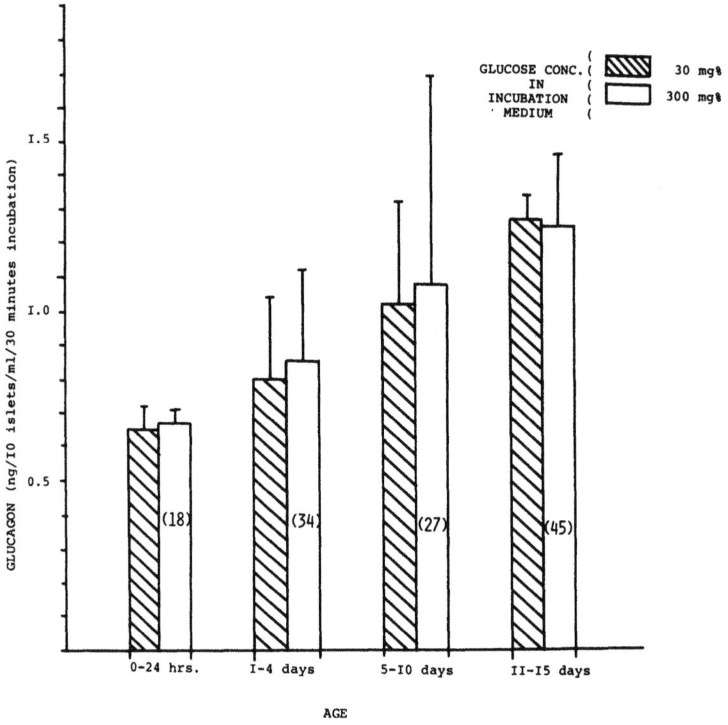

Fig. 3. Release of glucagon from isolated rat pancreatic islets during incubation in varying concentration of glucose. Figures in parenthesis in the bar indicate the number of incubation vessels. Mean ± S.E.

Most of the studies on the subject have been carried out in vivo and therefore indicate the changes in the plasma levels of circulating hormones, thereby reflecting the net effects of hormone secretion as well as hormone degradation. There is a paucity of data regarding insulin and glucagon kinetics in the postnatal period. Studies using in vitro secretion of insulin and glucagon by incubating pancreas, pancreatic pieces, or slices may not yield precise information in view of the possibility of hormone degradation in such systems. The present study therefore employed isolated islets.

It is difficult to quantitatively compare glucagon and insulin release in this study with that of adult rats also performed in this laboratory (9). Although the same number of islets were used in each study the size of the islet is much greater in the adult than in the newborn rat. Despite this, however, the newborn rat released as much insulin in the basal state (113 μu/ml/10 islets) as the adult rats (106 μu/ml/10 islets) and this therefore may represent a relative state of insulin excess in the newborn rats. Increased glucose concentration although causing insulin release had less effect than in the adult (150% of basal in the newborn compared with 250% of basal in the adult). Rat fetal pancreas has been previously reported to be relatively insensitive in its insulin response to glucose, but this is restored as soon as the rat is born (4, 21). During the first 15 days of life there was a striking reduction in basal insulin release although the magnitude of response to glucose remained approximately the same over the 15 day period. These observations are entirely consistent with those of Blázquez et al (4, 5) who determined plasma insulin concentration in the rat at varying time intervals following birth and showed progressive decline after birth.

Glucagon release in the rat has been shown to occur after 30 minutes following birth by caesarian section (17) and vaginal delivery (4). Similar observations have been reported in the normal human infant (6). Hypoxia at or during birth seem to accentuate the changes in circulating glucagon levels. However, in the newborn calf significant increases in plasma glucagon could not be demonstrated up to 24 hrs after birth (7). These variations in the magnitude and chronology of changes in plasma glucagon in the newborn period may be related to species differences in substrate availability and utilization as well as in the metabolic pathways for the maintenance of energy balance. The progressive increase in basal glucagon release of isolated islets in the present study is consistent with the observations of Blázquez et al (6) and Girard et al (19) who demonstrated persistently elevated levels of plasma glucagon until the 15th postnatal day after which there was a progressive decline until adult levels were reached.

The precise mechanism(s) which triggers the acute increase in glucagon in the immediate postnatal period are not entirely known; hypoglycemia, hypoxia and other stressful situations occurring at birth may well provide the trigger mechanism. This in turn may "switch on" glucagon release through neurogenic pathways; catecholamines and acetylcholine have been shown to increase plasma glucagon in the rat fetus while atropin and alpha adrenergic blocking drugs inhibit plasma glucagon increase in the newborn pups (19).

The relative insensitivity of pancreatic A cells to glucose changes in the newborn period has been previously documented. Acute hypoglycemia in the 21.5 day old rat fetus does not lead to a significant increase in glucagon release (2). Likewise plasma glucagon levels were unchanged after glucose infusion in sheep fetus (14). Similar observations have been reported in normal newborn infants. Glucagon release from newborn rat pancreas in vitro was unaltered by modifying the glucose concentrations (13). In the 3-day-old rat pups, glucagon secretion from the pancreas incubated in vitro was not affected by changes in glucose concentration, but was increased by the addition of a mixture of amino acids in the incubation medium (20). In the present study, using isolated rat islets, we have not been able to show any effect on glucagon release as a result of modifying the glucose concentration in the medium. These results are therefore consistent with the observations discussed above, although somewhat contradictory to the information available from the studies on monolayer cultures of newborn rat pancreas wherein high glucose concentrations were shown to decrease glucagon release (23). There is no readily available explanation for these discrepancies except that different in vitro systems have been used.

The low insulin:glucagon molar ratio during the suckling period (0-15 days) has been commented upon by other investigators. Rat milk is a high fat, low carbohydrate diet (12). The energy requirements during this period therefore depend upon oxidation of lipids and gluconeogenesis. A low insulin:glucagon molar ratio may therefore contribute to the maintenance of blood glucose in an animal consuming a high fat, high protein diet. This hormonal profile with the resultant metabolic pattern seems to continue until the rat is about 15 days old; at this time weaning starts with a cessation of milk intake and an increase in the consumption of carbohydrates (20).

In conclusion, the change from high to low insulin:glucagon molar ratio after birth and until weaning may be the key hormonal adaptation pattern required for glucose homeostasis in the newborn.

REFERENCES

1. Alexander, D. P., Assan, R., Britton, H. G. and Nixon, D. A. (1971): Glucagon in the foetal sheep. J. Endocr. 51, 597-598.

2. Assan, R., Attali, J. R., Ballerio, G., Girard, J. R., Hautecouverture, M., Kervran, A., Plouin, P. F., Slama, G., Soufflet, E., Tchobroutsky, G. and Tiengo, A. (1974): Some aspects of the physiology of glucagon. In: Diabetes. W. J. Malaisse, J. Pirart and J. Vallance-Owen, Eds. Excerpta Medica, Amsterdam, pp. 144-179.

3. Bencosme, S. A., Wilson, M. B., Aleyassine, H., De Bold, A. J. and De Bold, M. L. (1970): Rabbit pancreatic B cell. Morphological and functional studies during embryonal and post-natal development. Diabetologia 6, 399-411.

4. Blázquez, E., Sugase, T., Blázquez, M. and Foà, P. P. (1972): The ontogeny of metabolic regulation in the rat, with special reference to the development of insular function. Acta Diabetol. Latina 9 (Suppl. 1) 13-35.

5. Blázquez, E., Sugase, T., Blázquez, M. and Foà, P. P. (1974): Neonatal changes in the concentration of rat liver cyclic AMP and of serum glucose, free fatty acids, insulin, pancreatic and total glucagon in man and in the rat. J. Lab. Clin. Med. 83, 957-967.

6. Bloom, S. R. and Johnston, D. I. (1972): Failure of glucagon release in infants of diabetic mothers. Brit. Med. J. 4, 453-454.

7. Bloom, S. R., Vaughan, N. J. A. and Edwards, A. V. (1973): Pancreatic glucagon levels in the calf. Diabetologia 9, 61.

8. Buchanan, K. D. (1973): Studies on the pancreatic and enteric hormones. Ph.D. Thesis, Queen's University of Belfast.

9. Buchanan, K. D. and Mawhinney, W. A. A. (1973): Glucagon release from isolated pancreas in streptozotocin-treated rats. Diabetes 22, 797-801.

10. Buchanan, K. D. and McCarrol, A. M. (1971): In: Radioimmunoassay Methods. K. E. Kirkham, W. M. Hunter, E. Edinburgh and S. Livingstone, Eds. p. 136.

11. Dawkins, J. H. (1963): Glycogen synthesis and breakdown in fetal and newborn rat liver. Ann. N. Y. Acad. Sci. 111, 203-211.

12. Dymsza, H. A., Czajka, D. M. and Miller, S. A. (1964): Influence of artificial diet on weight gain and body composition of the neonatal rat. J. Nutr. 84, 100-106.

13. Edwards, J. C., Asplund, K. and Lundqvist, G. (1972): Glucagon release from the pancreas of the newborn rat. J. Endocr. 54, 493-504.

14. Fiser, R. H., Phelps, D. L., Erenberg, A., Sperling, M. A., Oh, W. and Fisher, D. A. (1973): Pancreatic alpha and beta cell responsiveness in fetal and newborn lambs. Pediat. Res. 7 311.

15. Foà, P. P. (1972): The secretion of glucagon. In: Endocrine Pancreas, Vol. 1, Sec. 7. D. F. Steiner and N. Freinkel, Eds. American Physiological Society, Washington, D. C., pp. 261-277.

16. Girard, J., Bal, D. and Assan, R. (1971): Rat plasma glucagon during the perinatal period. Diabetologia 7, 481.

17. Girard, J., Bal, D. and Assan, R. (1972): Glucagon secretion during the early postnatal period in the rat. Horm. Metab. Res. 4, 168-170.

18. Girard, J. R., Guillet, I., Marty, J., Assan, R. and Marliss, E. B. (1976): Effects of exogenous hormones and glucose on plasma levels and hepatic metabolism of amino acids in the fetus and in the newborn rat. Diabetologia 12, 327-337.

19. Girard, J. R., Kervran, A. and Assan, R. (1975): Functional maturation of the A cell in the rat. In: Early Diabetes in Early Life. R. A. Camerini-Davalos and H. S. Cole, Eds. Academic Press, Inc., New York, pp. 57-71.

20. Jarousse, C., Rançon, F. and Rosselin, G. (1973): Hormonogenèse périnatale de l'insuline et du glucagon chez le rat. Compt. Rend. Acad. Sci. (Paris) 276, 585-588.

21. Lernmark, A. and Wenngren, B. L. (1972): Insulin and glucagon release from isolated pancreas of foetal and newborn mice. J. Embryol. Exp. Morphol. 28, 607-614.

22. Like, A. A. and Orci, L. (1972): Embryogenesis of the human pancreatic islets: a light and electron microscopic study. Diabetes 21 (Suppl. 2), 511-534.

23. Marliss, E. B., Wollheim, C. B., Blondel, B., Orci, L., Lambert, A. E., Stauffacher, W., Like, A. A. and Renold, A. E. (1973): Insulin and glucagon release from monolayer cell cultures of pancreas from newborn rat. Eur. J. Clin. Invest. 3, 16-26.

24. Orci, L., Lambert, A. E., Rouiller, C., Renold, A. E. and Samols, E. (1969): Evidence for the presence of A-cells in the endocrine fetal pancreas of the rat. Horm. Metab. Res. 1, 108-110.

25. Pictet, R. and Rutter, W. J. (1972): Development of the embryonic endocrine pancreas. In: Endocrine Pancreas, Vol. 1, Sec. 7. D. F. Steiner and N. Freinkel, Eds. American Physiological Society, Washington, D. C., pp. 25-66.

26. Rall, L. B., Pictet, R. L., Williams, R. H. and Rutter, W. J. (1973): Early differentiation of glucagon-producing cells in embryonic pancreas: a possible developmental role for glucagon. Proc. Nat. Acad. Sci. USA 70, 3478-3482.

VI. ROLE OF GLUCAGON IN DIABETES MELLITUS, IN PREDIABETES AND IN OTHER DISEASE STATES

NEW CONCEPTS IN GLUCAGON PHYSIOLOGY

R. Sherwin and P. Felig

Section of Endocrinology, Department of Medicine
Yale University School of Medicine
New Haven, Connecticut USA

A decreased renal clearance of glucagon contributes to the hyperglucagonemia of uremia and of starvation. Thus, hyperglucagonemia does not necessarily reflect altered glucagon secretion. Experimental hyperglucagonemia produced in normal human subjects by means of intravenous infusions of glucagon resulted in fleeting elevations of splanchnic glucose production and blood glucose level. No changes in glucose tolerance were noted, even though there was no compensatory hyperinsulinism. The IV administration of glucagon in physiologic doses to insulin-treated diabetic patients did not alter the level of blood glucose and ketone bodies, or the urinary excretion of glucose and nitrogen, but caused marked hyperglycemia in insulin-deprived juvenile diabetics. Prolonged infusions of somatostatin cause temporary hypoglycemia followed by hyperglycemia. The latter is due to suppression of insulin secretion and occurs despite the concomitant persistent suppression of glucagon. Thus, basal glucagon secretion is not essential for the development of fasting hyperglycemia in diabetes and hyperglucagonemia worsens the diabetic state only when insulin is not available. The results underscore the primary role of insulin deficiency rather than glucagon excess in the pathogenesis of diabetes and suggest that by further suppressing insulin secretion, somatostatin may worsen metabolic control in adult-onset diabetic patients.

Recent investigative interest has increasingly focused on the role of glucagon in normal physiology and in the pathogenesis of glucose intolerance and diabetes mellitus. The development of radioimmunoassay procedures for measurement of circulating glucagon, infusion of crystalline glucagon into intact subjects in physiologic amounts, and the availability of an agent (somatostatin) which inhibits glucagon secretion, have markedly enhanced our understanding of glucagon physiology. In the current review we will discuss a variety of recent studies from our laboratory employing these techniques which have provided new data regarding the turnover and metabolic effects of glucagon.

Glucagon Catabolism

Recent studies from our laboratory indicate that in contrast to insulin, the kidney rather than the liver is the principal site of glucagon degradation. The fractional extraction ratio of glucagon across the kidney in the intact dog and the rat is 40-45% (3, 8). When glucagon was infused in physiologic doses (3 ng/kg/min) in normal anesthetized dogs, renal clearance of glucagon accounted for approximately 50% of total glucagon catabolism (8). The key role of the kidney in glucagon catabolism in man is suggested by studies in patients with chronic renal failure (17). In this disorder plasma concentrations of total and pancreatic (MW 3500) immunoreactive glucagon are increased 2-4 fold (11, 17). Hyperglucagonemia in uremia is solely a consequence of decreased hormonal catabolism rather than hypersecretion. In uremia, the metabolic clearance rate of glucagon as determined by infusions of pancreatic hormone is decreased by 60% as compared to healthy controls (17). In contrast, glucagon delivery (post-hepatic) is comparable in uremic and normal subjects (17). This reduction in glucagon clearance is likely secondary to alterations in renal removal processes. After induction of renal insufficiency in rats, a consistent renal uptake of glucagon is no longer observed (3).

The importance of altered hormonal catabolism in determining plasma glucagon concentration is further underscored by studies of glucagon turnover during starvation (7). In prolonged starvation, the rise in plasma glucagon observed after 3 days is primarily the result of a 20% reduction in the metabolic clearance rate of glucagon. No consistent change in glucagon delivery is observed at this time (Fig. 1). As starvation continues for 3-4 weeks, decreased glucagon secretion accounts for the return of plasma glucagon to baseline despite a further decline in glucagon catabolism (Fig. 1). These studies thus demonstrate that hyperglucagonemic disorders cannot be assumed to reflect altered secretion and that renal function must be taken into account when interpreting circulating glucagon levels.

Evanescent Effect of Hyperglucagonemia

While the pharmacologic effects of glucagon are well established, recent studies from our laboratory have raised questions regarding the significance of

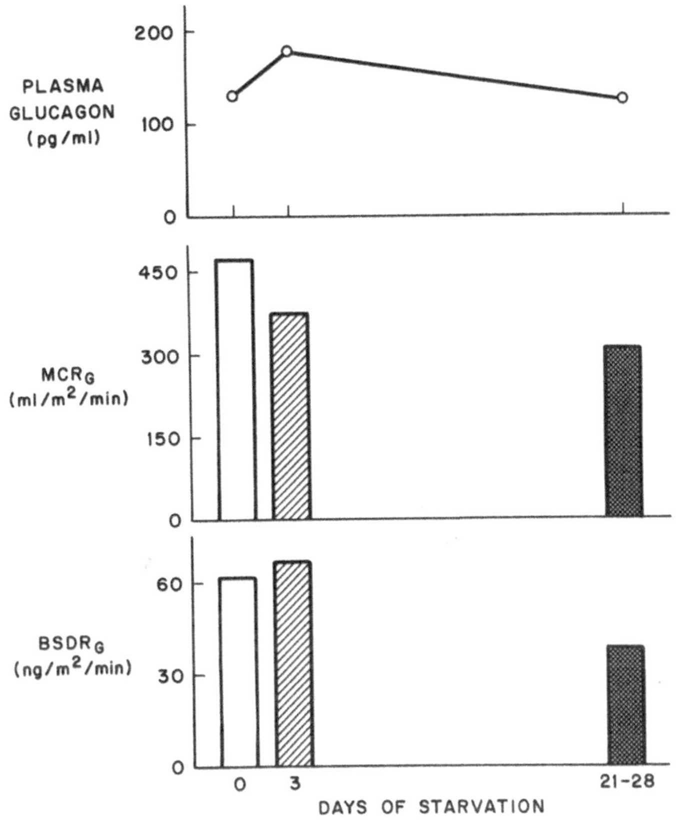

Fig. 1. Plasma glucagon concentration, metabolic clearance rate (MCR_G) and basal systemic delivery rate ($BSDR_G$) of glucagon during prolonged starvation. The initial rise in plasma glucagon after three days of starvation is due to a fall in MCR_G (catabolism) rather than an increase in secretion. The late fall in plasma glucagon is due to a decline in secretion. Based on the data of Fisher et al. (7).

these observations to normal physiology. When physiologic doses of glucagon (3 ng/kg/min) are administered to normal subjects so that plasma levels of this hormone are comparable to those observed in a variety of hyperglucagonemic states, only a small (5-10 mg/100 ml) transient increase in blood glucose is observed (18). Blood glucose concentration returns to baseline levels within 2-3 hours despite persistent hyperglucagonemia and normal peripheral levels of insulin (18). The transient rise in blood glucose after physiologic infusions of glucagon is solely a consequence of a very evanescent increase in splanchnic glucose production rather than augmented glucose disposal (Fig. 2; 6). After continuous infusion of glucagon splanchnic glucose output rapidly increases 2-3 fold, but within 30 minutes returns to baseline values (Fig. 2; 6). The transient nature of the stimulatory response suggests the rapid development of inhibition or reversal of glucagon action. It is of interest in this regard that while the liver becomes unresponsive to chronic hyperglucagonemia, it is still capable of responding to other agents known to augment hepatic glucose output. When physiologic doses of epinephrine (0.1 μg/kg/min) are added to an ongoing infusion of glucagon (3 ng/kg/min) in normal conscious dogs, there is a prompt increase in hepatic glucose output despite prior unresponsiveness to the effects of glucagon (unpublished observations). These data thus suggest that the liver adapts to chronic hyperglucagonemia so that elevations of this hormone are unable to produce sustained effects on glucose homeostasis in normal subjects and that this transitory hepatic response to glucagon is likely a selective process.

Glucagon and Glucose Disposal

The bihormonal response (hyperinsulinemia and hypoglucagonemia) to carbohydrate ingestion has prompted speculation that the combined A and B cell response (i. e., the insulin : glucagon ratio) rather than changes in insulin alone may be required for normal glucose disposal (15). To test this hypothesis we infused glucagon in physiologic amounts (3 ng/kg/min) prior to and during oral glucose tolerance testing (18). In this manner the usual decline in plasma glucagon accompanying carbohydrate feeding was prevented and circulating glucagon reached levels (300-400 pg/ml) comparable to those reported in diabetic ketoacidosis (16), uremia (17), cirrhosis (21) or trauma (14). Despite a 5 to 6-fold fall in the peripheral insulin : glucagon ratio (from 34:1 to 6:1) during the glucagon infusion, glucose tolerance was no different from that observed during

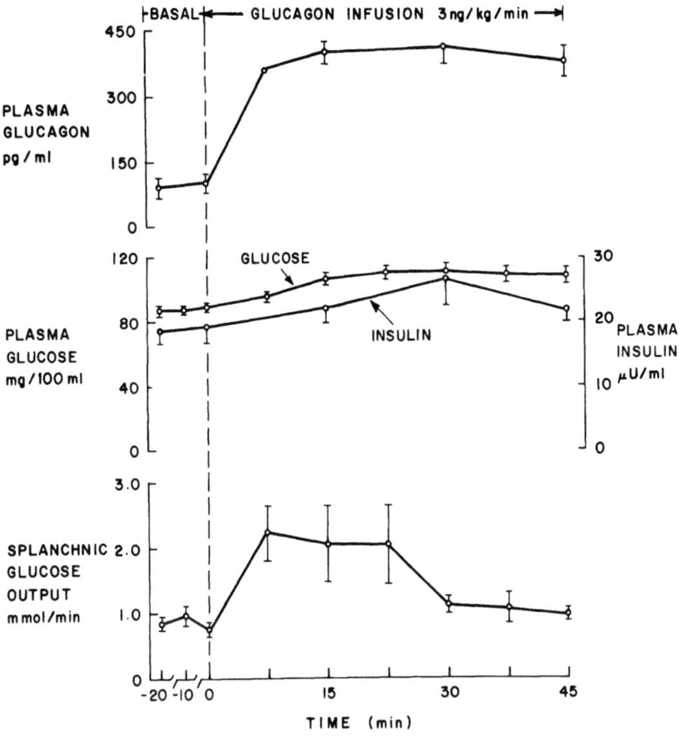

Fig. 2. The evanescent effect of physiologic hyperglucagonemia on splanchnic glucose output. Despite ongoing hyperglucagonemia, splanchnic glucose output returns to baseline within 20 minutes. Based on the data of Felig, et al. (6).

a control study in which saline was given (18). Particularly noteworthy was the fact that normal glucose tolerance was maintained in the absence of compensatory hyperinsulinemia (18). Similar results were observed in obese subjects, despite evidence of mild glucose intolerance and insulin resistance during the saline control infusion in these subjects (18). These findings thus indicate that the fall in plasma glucagon observed after carbohydrate feeding is not essential for normal glucose tolerance. Rather the primary determinant of normal glucose disposal is the insulin secretory response.

Glucagon and Diabetic Control

To determine whether hyperglucagonemia of the magnitude observed in diabetes is a sufficient hormonal alteration to cause a worsening of diabetic regulation in the absence of changes in insulin availability, we infused physiologic doses of glucagon for periods of 2-4 days in adult-onset (non-insulin dependent) and in juvenile-onset diabetics in whom insulin was administered in the usual amounts (18). The infusion of glucagon (3-9 ng/kg/min) resulted in an elevation of plasma glucagon (250-800 pg/ml) which was comparable to or in excess of that observed in most patients with diabetic ketoacidosis (16). Despite ongoing hyperglucagonemia, fasting and postprandial blood glucose levels as well as ketonemia were unchanged from those observed during a pre- and post-control period (18). Furthermore, urinary glucose and nitrogen excretion in these patients were also unchanged by glucagon administration (20). In contrast, when similar doses of glucagon were given to juvenile-onset diabetic subjects withdrawn from insulin, the glycemic response to hyperglucagonemia was 5-15 fold greater than that observed in normal controls (18).

These data thus suggest that hyperglucagonemia of itself is insufficient to bring about deterioration of diabetic control when insulin is available. On the other hand, glucagon in the insulin-deprived patient can worsen the diabetic state. These results underscore the primary role of insulin deficiency rather than glucagon excess in the pathogenesis of diabetes.

Effect of Somatostatin on Glucose Homeostasis

The hypothesis that glucagon is essential for the development of hyperglycemia in diabetes has largely derived from studies with somatostatin (23). Somatostatin is a potent inhibitor of insulin, glucagon and growth hormone secretion, while, at the same time lowers blood glucose in normal (1) and diabetic patients (10). This effect of somatostatin is a consequence of a marked reduction in hepatic glucose production (2) and appears to be mediated by suppression of endogenous glucagon secretion (9). However, such studies have generally involved short (1-2 hr) infusions of somatostatin. When we infused somatostatin (9 μg/min) for prolonged periods to normal subjects, the hypoglycemic effect of somatostatin was transitory, persisting for only 90-120 minutes (19). Beyond two hours, blood glucose rose to hyperglycemic levels (140-150 mg/100 ml), despite persistent suppression of glucagon secretion by somatostatin (Fig. 3). The increase in blood glucose during prolonged infusion of somatostatin is a consequence of an increase in hepatic glucose production to values 15-20% above basal preinfusion values, as well as a decline in glucose utilization (as determined by 3-^{3}H glucose; 19).

With respect to the mechanism of these changes in glucose homeostasis produced by prolonged somatostatin infusion, insulin deficiency appears to be the primary factor. Infusion of exogenous insulin so as to restore plasma insulin to preinfusion values or cessation of the somatostatin infusion with restoration of endogenous insulin secretion resulted in a prompt reduction in blood glucose to baseline values (Fig. 3). In contrast, it is unlikely that counter regulatory hormones are responsible, inasmuch as prevention of the initial somatostatin-induced hypoglycemia by infusion of intravenous glucose failed to prevent the delayed hyperglycemia (19).

These findings have implications regarding the potential usefulness of somatostatin in the treatment of diabetes. Inasmuch as the hyperglycemic effects of somatostatin are mediated by suppression of insulin, intensification rather than amelioration of diabetes may occur with prolonged infusion in patients with residual insulin secretion. We have in fact observed worsening of diabetic control in maturity-onset diabetes by prolonged somatostatin administration in the face of hypoglucagonemia (unpublished observations). Our data thus indicate that basal glucagon secretion is not essential for the development of fasting hyperglycemia in diabetes and support the conclusion that insulin deficiency rather than glucagon excess is the primary hormonal disturbance in the pathogenesis of diabetes.

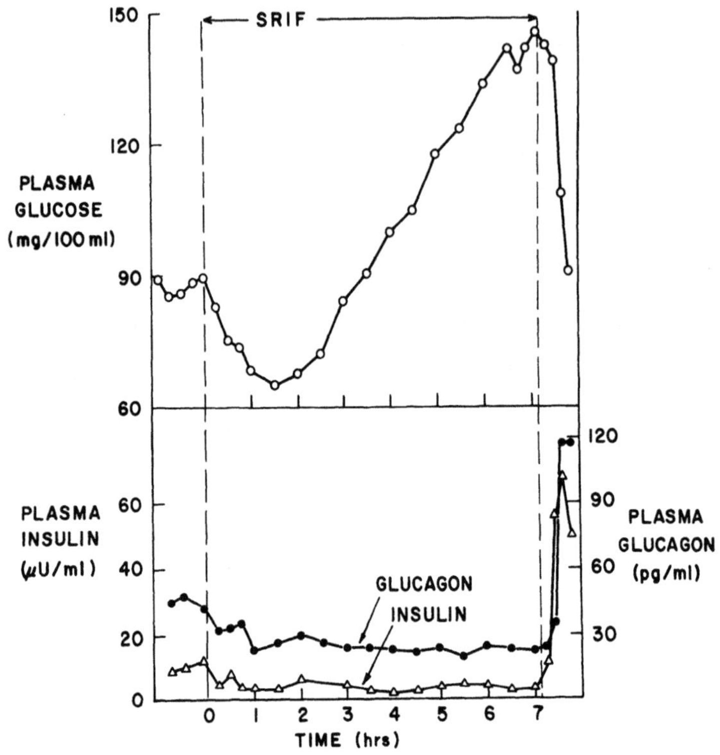

Fig. 3. The effects of prolonged administration of somatostatin on glucose homeostasis in normal subjects. Despite ongoing suppression of plasma glucagon, fasting hyperglycemia develops within 4-5 hours. Plasma glucose rapidly returns to baseline levels when the somatostatin infusion is discontinued. Based on the data of Sherwin, et al (19).

Altered Sensitivity to Glucagon

In addition to the importance of insulin deficiency, recent studies from our laboratory suggest that under certain circumstances altered tissue responsiveness to glucagon may contribute to the diabetogenic effect of this hormone. The importance of augmented tissue sensitivity to glucagon in the pathogenesis of glucose intolerance is suggested by observations in uremic patients. Chronic renal failure is characterized by an increased incidence of glucose intolerance (5), insulin resistance (12) and hyperglucagonemia (17). Following chronic dialysis, glucose tolerance and insulin sensitivity return towards normal (12) in the absence of changes in plasma glucagon (17). While these data do not support a role for glucagon in the glucose intolerance of uremia, glucagon's importance becomes evident when tissue responsiveness to this hormone is examined. When we infused physiologic doses of glucagon to undialyzed uremic patients the glycemic response was increased 3-4 fold as compared to normal controls (Fig. 4).

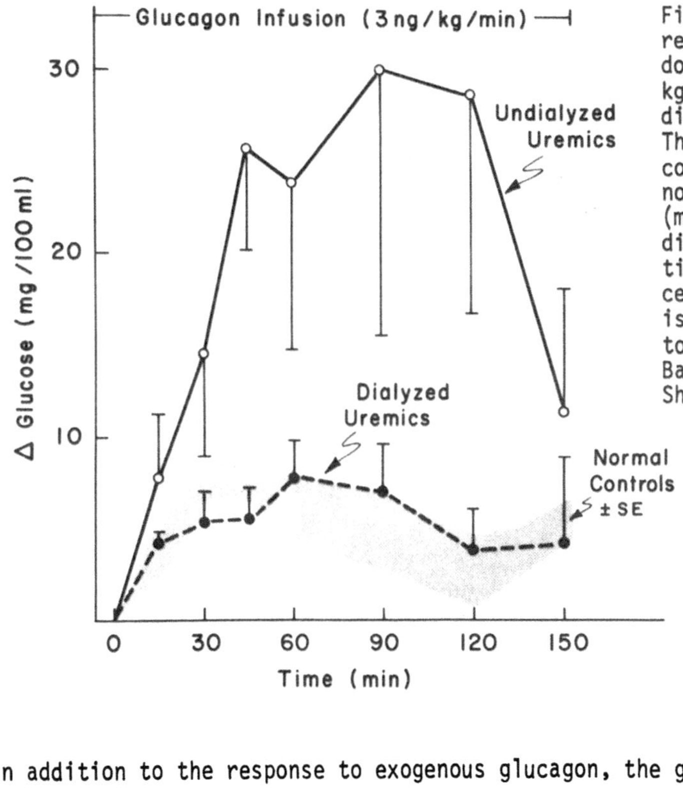

Fig. 4. The glycemic response to physiologic doses of glucagon (3 ng/kg/min) in undialyzed and dialyzed uremic patients. The response in healthy control subjects is denoted by the shaded area (mean ± SE). In the undialyzed uremics sensitivity to the hyperglycemic effects of glucagon is increased and returns to normal after dialysis. Based on the data of Sherwin et al. (17).

In addition to the response to exogenous glucagon, the glycemic response to alanine-stimulated secretion of endogenous glucagon was increased (17). This augmented sensitivity to glucagon returns to normal following dialysis, thus accounting for improved glucose tolerance despite persistence of the hyperglucagonemia (17). Recent studies using the uremic rat model provide evidence that these changes in sensitivity to glucagon are mediated by augmented binding of this hormone to target cells (22). In partially nephrectomized rats, glucagon binding to liver membranes and cyclic AMP generation in response to glucagon are increased 3-fold (22). These findings thus provide evidence of an important role of glucagon receptors in a syndrome of carbohydrate intolerance.

REFERENCES

1. Alford, F., Bloom, S., Nabarro, J., Hall, R., Besser, G., Coy, D., Kastin, A. and Schally, A. V. (1974): Glucagon control of fasting glucose in man. Lancet 2, 974-976.

2. Altszuler, N., Gottlieb, B. and Hampshire, J. (1976): Interaction of somatostatin, glucagon, and insulin on hepatic glucose output in the dog. Diabetes 25, 116-121.

3. Bastl, C., Finkelstein, F. O., Sherwin, R. S., Hendler, R., Felig, P. and Hayslett, J. R. (1977): Renal extraction of glucagon in rats with normal and reduced renal function. Am. J. Physiol. In Press.

4. Brazeau, P., Vale, W., Burgus, R., Ling, N., Butcher, M., Rivier, J. and Guillemin, R. (1973): Hypothalamic polypeptide that inhibits the secretion of immunoreactive pituitary growth hormone. Science 179, 77-79.

5. DeFronzo, R. A., Andres, R., Edgar, P. and Walker, W. G. (1973): Carbohydrate metabolism in uremia: A review. Medicine (Baltimore) 52, 469-481.

6. Felig, P., Wahren, J. and Hendler, R. (1976): Influence of physiologic hyperglucagonemia on basal and insulin-inhibited splanchnic glucose output in normal man. J. Clin. Invest. 58, 761-765.

7. Fischer, M., Sherwin, R. S., Hendler, R. and Felig, P. (1976): Kinetics of glucagon in man: Effects of starvation. Proc. Natl. Acad. Sci. USA 73, 1735-1739.

8. Forrest, J. N., Fischer, M., Hendler, R., Sherwin, R. and Felig, P. (1976): Contrasting roles of kidney in the disposal and hormonal action of physiologic concentrations of glucagon. Clin. Res. 24, 400A.

9. Gerich, J. E., Lorenzi, M., Bier, D. M., Tsalikian, E., Schneider, V., Karam, J. H. and Forsham, P. H. (1976): Effects of physiologic levels of glucagon and growth hormone on human carbohydrate and lipid metabolism: studies involving administration of exogenous hormone during suppression of endogenous hormone secretion with somatostatin. J. Clin. Invest. 57, 875-884.

10. Gerich, J. E., Lorenzi, M., Schneider, V., Karam, J. H., Rivier, J., Guillemin, R. and Forsham, P. H. (1974): Effects of somatostatin on plasma glucose and glucagon levels in human diabetes mellitus. New Engl. J. Med. 291, 544-547.

11. Kuku, S. F., Jaspan, J. B., Emmanouel, D. S., Zeidler, A., Katz, A. L. and Rubenstein, A. H. (1976): Heterogeneity of plasma glucagon. Circulating components in normal subjects and patients with chronic renal failure. J. Clin. Invest. 58, 742-750.

12. Hampers, C. L., Soeldner, J. S., Doak, P. B. and Merrill, J. P. (1966): Effect of chronic renal failure and hemodialysis on carbohydrate metabolism. J. Clin. Invest. 45, 1719-1731.

13. Leblanc, H., Anderson, J., Sigel, M. and Yen, S. (1974): Inhibitory action of somatostatin on pancreatic α and β cell function. J. Clin. Endocrinol. Metab. 40, 568-572.

14. Lindsey, A., Santeusania, F., Braaten, J., Faloona, G. R. and Unger, R. H. (1974): Pancreatic alpha-cell function in trauma. JAMA 227, 757-761.

15. Müller, W. A., Faloona, G. R., Aguilar-Parada, E. and Unger, R. H. (1970): Abnormal alpha-cell function in diabetes. Response to carbohydrate and protein ingestion. New Engl. J. Med. 283, 109-115.

16. Müller, W. A., Faloona, G. R. and Unger, R. H. (1973): Hyperglucagonemia in diabetic ketoacidosis. Its prevalence and significance. Am. J. Med. 54, 52-57.

17. Sherwin, R. S., Bastl, C., Finkelstein, F. O., Fischer, M., Black, H., Hendler, R. and Felig, P. (1976): Influence of uremia and hemodialysis in the turnover and metabolic effects of glucagon. J. Clin. Invest. 57, 722-731.

18. Sherwin, R. S., Fischer, M., Hendler, R. and Felig, P. (1976): Hyperglucagonemia and blood glucose regulation in normal, obese and diabetic subjects. New Engl. J. Med. 294, 455-461.

19. Sherwin, R. S., Hendler, R., DeFronzo, R. A., Wahren, J. and Felig, P. (1977): Glucose homeostasis during prolonged suppression of glucagon and insulin secretion by somatostatin. Proc. Natl. Acad. Sci. USA 74, 348-352.

20. Sherwin, R. S., Hendler, R. and Felig, P. (1977): Influence of physiologic hyperglucagonemia on urinary glucose, nitrogen, and electrolyte excretion in diabetes. Metab. Clin. Exp. 26, 53-58.

21. Sherwin, R., Joshi, P., Hendler, R., Felig, P. and Conn, H. O. (1974): Hyperglucagonemia in Laennec's cirrhosis: The role of portal-systemic shunting. New Engl. J. Med. 290, 239-242.

22. Soman, V. and Felig, P.: Altered glucagon and insulin receptor binding: Cellular mechanism of glucose intolerance in uremia. Clin. Res. 25, 401A.

23. Unger, R. H. and Orci, L. (1975): The essential role of glucagon in the pathogenesis of the endogenous hyperglycemia of diabetes. Lancet 1, 14-16.

THE ISSUES RELATED TO ABNORMAL A CELL FUNCTION IN DIABETES MELLITUS

K. D. Buchanan and E. R. Trimble

Department of Medicine, Queen's University of Belfast
Northern Ireland

This paper discusses the issues related to abnormal A-cell function in diabetes mellitus, under the following headings: 1) Problems of radioimmunoassay of glucagon; 2) The measurement of plasma levels of glucagon-like immunoreactivity (GLI) in diabetes mellitus; 3) Glucagon deficiency states induced by somatostatin or by pancreatectomy; 4) Glucagon excess states as in glucagonoma, glucagon infusion, renal failure, trauma, starvation etc. and 5) Histologic studies of the pancreatic islets.

The major issue under discussion is whether abnormal glucagon secretion may play a role in diabetes mellitus, the "double-trouble" hypothesis of Unger (22). Even if glucagon may not play a role in the primary genesis of the disease then disturbances of glucagon secretion due to the disturbed metabolism of diabetes mellitus may yet play an important role in some of the features of diabetes mellitus. Glucagon has the properties of a diabetogenic hormone: it can produce hyperglycaemia, lipolysis and ketoacidosis and it is catabolic. The hypothesis that glucagon plays a primary role in diabetes mellitus is attractive but the issues related to proving this are complex.

Radioimmunoassay (RIA) of Glucagon

The problems related to RIA of glucagon are discussed extensively in this volume and a further exhaustive discussion is not in place here. However, much of the data concerning glucagon in diabetes mellitus have come from the use of RIA. There remains considerable ignorance concerning the multiple species of GLI in tissues, and in the circulation. Antibodies to pancreatic glucagon which appear to discriminate between pancreatic glucagon and gut GLI have been used to specifically measure pancreatic glucagon, but this author believes that there is insufficient evidence at the present time to specifically indicate that these antibodies measure "pancreatic glucagon". The author has developed an empirical terminology when describing GLI measured by RIA, namely those antibodies which react with N-terminal sequences of glucagon are said to measure N-terminal GLI (N-GLI), while antibodies which react with the C-terminal sequence are said to measure C-terminal GLI (C-GLI). N-GLI appears to represent all species of GLI and C-GLI to represent predominantly pancreatic glucagon, but also some of the species of gut GLI.

It is important that the limitations of glucagon RIA be clearly recognized so that widespread and sometimes misleading conclusions not be made from their use.

The Measurement of GLI in Diabetes Mellitus

The issues related to measurement of plasma GLI in maturity onset diabetes (MOD, Table 1) and in juvenile onset diabetes (JOD, Table 2) will be dealt with separately as there may be important differences.

TABLE 1

PLASMA C-GLI MEASUREMENTS IN MOD

1. Relative hyperglucagonaemia
2. Poor suppression by glucose
3. Influenced by body weight
4. No change with dietary therapy
5. Respond to insulin therapy

TABLE 2

PLASMA C-GLI MEASUREMENTS IN JOD

1. Hyperglucagonaemia at diagnosis
2. Instances of hypoglucagonaemia
3. Role in ketoacidosis
4. Respond to insulin therapy

Most reports would indicate that fasting hyperglucagonaemia in MOD, if present, is only slight (4, 23). However, these levels are probably excessively high relative to the prevailing hyperglycaemia which in normal subjects would suppress the circulating glucagon.

Recently, we have noted slightly elevated plasma C-GLI levels in a large group of untreated MOD (20). We have found body weight to be the most significant factor related to C-GLI levels, thin patients having higher levels than obese patients (19), confirming the report of Wise, Hendler and Felig (26). Plasma glucagon levels in MOD do not respond to suppression by carbohydrate eating (13). We have been unable to correlate the plasma C-GLI levels in MOD with the degree of carbohydrate intolerance and following 6 months of intensive dietary management which results in improved glucose tolerance and insulin secretion, C-GLI levels remained at pre-treatment values (9, 20). It has been considered that the hyperglucagonaemia of MOD may be related to insulin deficiency and in experimental situations the hyperglucagonaemia responds to insulin (5, 14). However, in MOD the hyperglucagonaemia responds sluggishly to insulin therapy (13, 22).

Hyperglucagonaemia has also been noted in JOD (13). We have noted no absolute fasting hyperglucagonaemia in JOD, but the response to arginine infusion is excessive (19). However, some JOD may show hypoglucagonaemia, under some circumstances. Gerich, Langlois, Noacco, Karam and Forsham (6) have demonstrated a deficient glucagon response to insulin induced hypoglycaemia in JOD. It is probable that further studies of glucagon in JOD may uncover evidence of deficiency, as some JOD have autoantibodies directed towards their glucagon cells (D. Doniach, personal communication). It is conceivable that glucagon deficiency may play a part in the insulin sensitivity and "brittleness" seen in insulin requiring diabetes.

Glucagon levels are elevated 3 to 4 fold in diabetic ketoacidosis (2, 15) and with treatment of the ketoacidosis the glucagon levels fall rapidly (15). Alberti, et al. (1) have reported a correlation between plasma glucagon and acetone levels in patients with ketoacidosis.

Glucagon Deficiency States

If glucagon is implicated in the diabetic syndrome, then conditions associated with excess or deficiency of the hormone would form an experimental model to study the role and mechanism of action of glucagon in diabetes mellitus.

No chemical technique exists for destroying the glucagon cells in the body. Pancreatectomy results in glucagon deficiency, but also insulin deficiency which is of course dominant, and extrapancreatic sources of "glucagon" blur the features which develop. Vranic, Pek and Kawamori (25) observed hyperglucagonaemia in pancreatectomized dogs deprived of insulin. The dog is unusual in that it appears to have high concentrations of "glucagon" in the stomach. In 2 human pancreatectomized subjects, we have observed very low circulating levels of plasma C-GLI in one, but normal levels in another. Neither subject showed a plasma C-GLI response to arginine and when insulin was withdrawn, ketoacidosis resulted in the absence of elevations of plasma C-GLI. (K. D. Buchanan, R. W. J. Flanagan, H. Walsh, unpublished observations). In addition we have found that further characterization of circulating C-GLI in pancreatectomized human subjects shows that there is a species which cannot be differentiated from pancreatic glucagon on the basis of molecular size and RIA criteria (R. W. J. Flanagan, personal communication). This would imply that human subjects without pancreas have a circulating species of GLI similar to pancreatic glucagon, which does not respond to typical glucagon secretory stimuli, such as arginine. In addition it can be concluded that glucagon is not essential for the development of ketoacidosis.

Somatostatin has been a major tool in assessing the role of glucagon in diabetes mellitus. Somatostatin reduces both insulin (1) and glucagon secretion (12). Sakurai, Dobbs and Unger (17) found that in normal dogs somatostatin abolished insulin and glucagon secretion during amino acid infusions. Hyperglycaemia only occurred if the hypoglucagonaemia was corrected. Somatostatin in diabetic dogs suppressed plasma glucagon and lowered the plasma glucose. Gerich and collaborators (8) found that the infusion of somatostatin into 10 insulin dependent diabetic subjects resulted in a significant fall in glucagon and glucose levels. In addition, somatostatin infusion combined with insulin completely abolished post-prandial hyperglycaemia in 4 diabetic patients and was more effective than insulin alone. In diabetic ketoacidosis, there is evidence that somatostatin infusions may prevent the development of the ketoacidosis (7).

Glucagon Excess States

Table 3 lists the situations where glucagon excess states are encountered. If glucagon is implicated in diabetes mellitus then one would expect diabetic syndromes to develop whenever there is hyperglucagonaemia. Indeed, Unger and Orci (24) proposed a hypothesis that glucagon was essential for the development of diabetes mellitus. However, glucagon infusion does not worsen glucose tolerance in normal subjects (18).

TABLE 3

CONDITIONS ASSOCIATED WITH GLUCAGON EXCESS

1. Exogenous glucagon administration
2. Glucagonoma
3. Renal failure
4. Trauma
5. Starvation

In the glucagonoma syndrome where massive plasma elevations of glucagon are noted, diabetes is often mild and occasionally absent (S. R. Bloom, personal communication). Gross elevations of circulating glucagon are noted in patients with markedly impaired renal function yet glucose tolerance is usually only mildly abnormal (3). Glucagon levels are elevated following trauma (11) and this may contribute to the hyperglycaemic syndrome sometimes noted in such patients. Hyperglucagonaemia may also be implicated in "starvation diabetes" (10).

Histologic Studies

It is now possible to easily identify the glucagon cells by using specific immunohistochemical or electron microscopic criteria. It is now evident that in rat and man the outer rim of the pancreatic islet contains A and D cells whereas the B cells are in the central region (16). The possibility exists that the islet cells may function as a syncytial network (22). In our laboratory we have shown the changing distribution of islet cells during streptozotocin-induced diabetes mellitus (R. J. McFarland and K. D. Buchanan, unpublished observations). It would be exciting to apply these techniques to human diabetes mellitus, if suitable tissue could be obtained. Morphologic studies add a further dimension to the glucagon hypothesis of diabetes mellitus.

CONCLUSIONS

There is little doubt that glucagon has a role to play in diabetes mellitus. However, diabetes mellitus can develop in the absence of glucagon (pancreatectomy) and complications such as ketoacidosis can develop in the absence of glucagon. When massive hyperglucagonaemia, such as in glucagonoma and renal failure, does not always result in frank diabetes mellitus, then glucagon's role in diabetes mellitus becomes shadowy in comparison with that of insulin. Further knowledge concerning the different species of GLI in tissue is required, so that more exact and specific studies of glucagon can be made. Morphologic studies offer a new exciting frontier for hormone studies related to the pathogenesis of diabetes mellitus.

REFERENCES

1. Alberti, K. G. M. M., Christensen, N. J., Christensen, S. E., Prange-Hansen, A. A., Iversen, J., Lundbaek, K., Seyer-Hansen, K. and Ørskov, H. (1973): Inhibition of insulin secretion by somatostatin. Lancet 2, 1299-1301.

2. Assan, R., Hautecouverture, G., Guillemant, S., Dauchy, F., Protin, P. and Dérot, M. (1969): Evolution de paramètres hormonaux (glucagon, cortisol, hormone somatotrope) et énergétiques (glucose, acides gras libres, glycérol) dans dix acido-cétoses diabétiques graves traitées. Pathol. Biol. (Paris) 17, 1095-1105.

3. Bilbrey, G. L., Faloona, G. R., White, M. E. and Knockel, J. P. (1974): Hyperglucagonemia of renal failure. J. Clin. Invest. 53, 841-847.

4. Buchanan, K. D. and McCarroll, A. M. (1972): Abnormalities of glucagon metabolism in untreated diabetes mellitus. Lancet 2, 1394-1395.

5. Buchanan, K. D. and Mawhinney, W. A. A. (1973): Glucagon release from isolated pancreas in streptozotocin-treated rats. Diabetes 22, 797-801.

6. Gerich, J. E., Langlois, M., Noacco, C., Karam, J. H. and Forsham, P. H. (1973): Lack of glucagon response to hypoglycemia in diabetes: evidence for an intrinsic alpha-cell defect. Science 182, 171-173.

7. Gerich, J. E., Lorenzi, M., Bier, D. M., Schneider, V., Tsalikian, E., Karam, J. H. and Forsham, P. H. (1975): Prevention of human diabetic ketoacidosis by somatostatin: evidence for an essential role of glucagon. New Engl. J. Med. 292, 985-989.

8. Gerich, J. E., Lorenzi, M., Schneider, V., Karam, J. H., Rivier, J., Guillemin, R. and Forsham, P. H. (1974): Effects of somatostatin on plasma glucose and glucagon levels in human diabetes mellitus. Pathophysiologic and therapeutic implications. New Engl. J. Med. 291, 544-547.

9. Hadden, D. R., Montgomery, D. A. D., Skelly, R. J., Trimble, E. R., Weaver, J. A., Wilson, E. A. and Buchanan, K. D. (1975): Maturity onset diabetes mellitus: response to intensive dietary management. Brit. Med. J. 3, 276-278.

10. Henry, R. W., Ardill, J., and Buchanan, K. D. (1975): Pancreatic and gut hormones in starvation diabetes. Diabetologia 11, 349.

11. Lindsey, C. A., Santeusania, F., Braaten, J., Faloona, G. R. and Unger, R. H. (1973): Glucagon and the insulin:glucagon ratio in severe trauma. Trans. Assoc. Am. Phys. 86, 264-271.

12. Mortimer, C. H., Tunbridge, W. M. G., Carr, D., Yeomans, L., Lind, T., Coy, D. H., Bloom, S. R., Kastin, A., Mallinson, C. H., Besser, G. M., Schally, A. V. and Hall, R. (1974): Effects of growth hormone release inhibiting hormone on circulating glucagon, insulin and growth hormone in normal, acromegalic and hypopituitary patients. Lancet 1, 697-701.

13. Müller, W. A., Faloona, G. R., Aguilar-Parada, E. and Unger, R. H. (1970): Abnormal alpha-cell function in diabetes. New Engl. J. Med. 283, 109-115.

14. Müller, W. A., Faloona, G. R. and Unger, R. H. (1971): The effect of experimental insulin deficiency on glucagon secretion. J. Clin. Invest. 50, 1992-1999.

15. Müller, W. A., Faloona, G. R. and Unger, R. H. (1973): Hyperglucagonemia in diabetic ketoacidosis. Am. J. Med. 54, 52-57.

16. Orci, L., Unger, R. H. (1975): Functional subdivision of islets of Langerhans and possible role of D-cells. Lancet 2, 1243-1244.

17. Sakurai, H., Dobbs, R. and Unger, R. H. (1974): Somatostatin induced changes in insulin and glucagon secretion in normal and diabetic dogs. J. Clin. Invest. 54, 1395-1402.

18. Sherwin, R. S., Fisher, M., Hendler, R. and Felig, P. (1976): Hyperglucagonemia and blood glucose regulation in normal, obese and diabetic subjects. New Engl. J. Med. 294, 455-461.

19. Trimble, E. R., (1975): M. D. Thesis, Queen's University of Belfast, pp. 29-140.

20. Trimble, E. R., Montgomery, D. A. D. and Hadden, D. R. (1974): Change in pattern of glucagon response during O-GTT before and six months after commencement of intensive dietary management of diabetes. Diabetologia 10, 389-390.

21. Unger, R. H. (1971): Glucagon and the insulin:glucagon molar ratio in diabetes and other catabolic diseases. Diabetes 20, 834-838.

22. Unger, R. H. (1976): Diabetes and the alpha cell. Diabetes 25, 136-151.

23. Unger, R. H., Aguilar-Parada, E., Müller, W. A. and Eisentraut, A. M. (1970): Studies of pancreatic alpha cell function in normal and diabetic subjects. J. Clin. Invest. 49, 837-848.

24. Unger, R. H. and Orci, L. (1975): Hypothesis: the essential role of glucagon in the pathogenesis of diabetes mellitus. Lancet 1, 14-16.

25. Vranic, M., Pek, S. and Kawamori, R. (1974): Increased "glucagon immunoreactivity" in plasma of totally depancreatized dogs. Diabetes 23, 905-912.

26. Wise, J. K., Hendler, R. and Felig, P. (1973): Evaluation of alpha-cell function by infusion of alanine in normal, diabetic, and obese subjects. New Engl. J. Med. 288, 487-490.

ON THE CAUSES AND CONSEQUENCES OF ABNORMAL GLUCAGON SECRETION IN HUMAN DIABETES MELLITUS

J. E. Gerich

Clinical Investigation Center, Naval Regional Medical Center, Oakland;
Department of Medicine and Metabolic Research Unit, University of California
San Francisco, California USA

Abnormal glucagon secretion in human diabetes may result in part from a defect of the intrinsic A cell glucoreceptor and in part from insulin lack and may exacerbate the metabolic consequences of insulin lack in terms of fasting hyperglycemia, postprandial hyperglycemia, and the development of diabetic ketoacidosis. Insulin alone cannot completely normalize A cell function in human diabetes nor restore normal glucose homeostasis. Suppression of glucagon secretion by an agent such as somatostatin may be useful as an adjunct to insulin in the management of diabetes mellitus.

The A and B cells of the pancreas originate from the same embryologic anlage and are situated in intimate contact with one another; their secretory products (glucagon and insulin) not only act antagonistically on common target organs (liver and adipose tissue) but also influence each other's secretion. It is not surprising, therefore, that in a disease such as diabetes mellitus where B cell function is abnormal, derangements in A cell function also exist. Currently major emphasis is being placed on the development of improved methods for insulin delivery in the treatment of diabetes, i. e. artificial pancreas and transplantation of pancreatic tissue. It therefore seems of more than theoretical interest to inquire into the cause(s) and consequences of abnormal glucagon secretion in this disorder to determine whether it might also be necessary to develop specific therapy for alpha cell dysfunction.

Glucagon Secretion in Human Diabetes Mellitus

Numerous abnormalities of glucagon secretion have been described in both juvenile-onset and adult-onset varieties of human diabetes mellitus (3, 18, 73). In ketoacidosis (58), hyperosmolar nonketotic coma (48), and poorly controlled diabetes (52, 74), fasting plasma glucagon levels are usually markedly elevated.

In moderately hyperglycemic patients, glucagon levels are often within the normal range (8, 38, 74) (as insulin levels frequently are) but these values are higher than those found in nondiabetic individuals made comparably hyperglycemic (Fig. 1). This has been termed "relative hyperglucagonemia". Administration of oral glucose solutions (36), intravenous infusion of glucose, even when accompanied by insulin (21, 24, 67, 75) and ingestion of predominantly carbohydrate meals (56) fails to lower plasma glucagon levels and sometimes results in paradoxical increases in plasma glucagon levels (8).

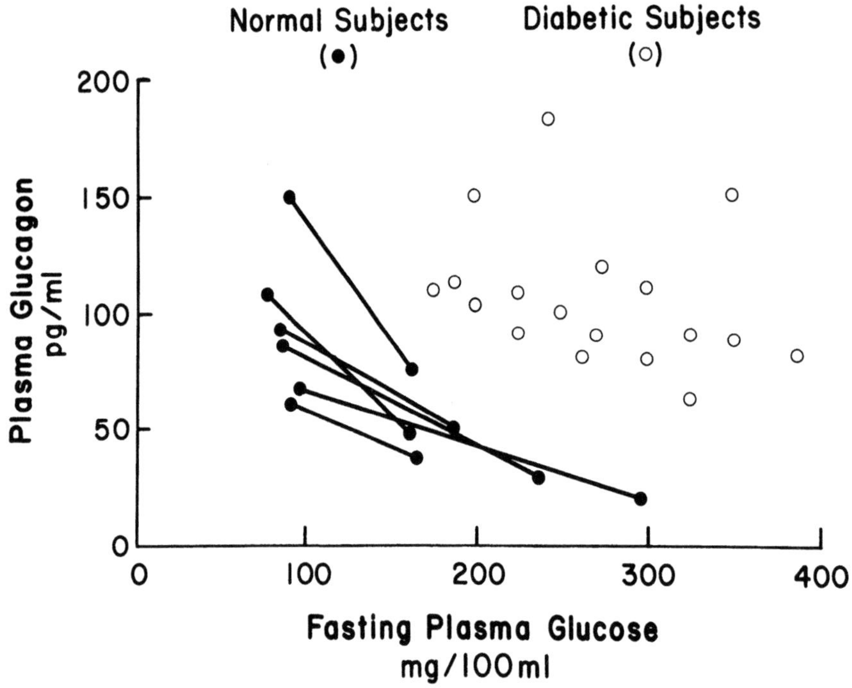

Fig. 1. Comparison of fasting plasma glucagon levels in insulin-requiring juvenile-onset diabetic patients with values found in normal subjects before and after induction of hyperglycemia by 60-minute glucose infusion.

Intravenous infusion of various amino acids (25, 35, 36, 41, 60, 66, 74, 83) or ingestion of protein containing meals (27, 56, 81) results in excessive rises in plasma glucagon levels. Characteristically, in juvenile-onset and even mild adult-onset diabetes (Fig. 2) immediately following ingestion of a balanced meal, there is an abrupt excessive rise in plasma glucagon not seen in normal individuals. It has been suggested that this exaggerates postprandial hyperglycemia in diabetic patients who are unable to compensate for it with an appropriate release of insulin (27). Generally other secretory stimuli for the A cells, such as epinephrine (33) also result in excessive glucagon responses in diabetes. One prominent exception is insulin-induced hypoglycemia (23, 61, 68) which evokes normal cortisol and growth hormone responses but little or no glucagon response. This phenomenon will be discussed in detail below.

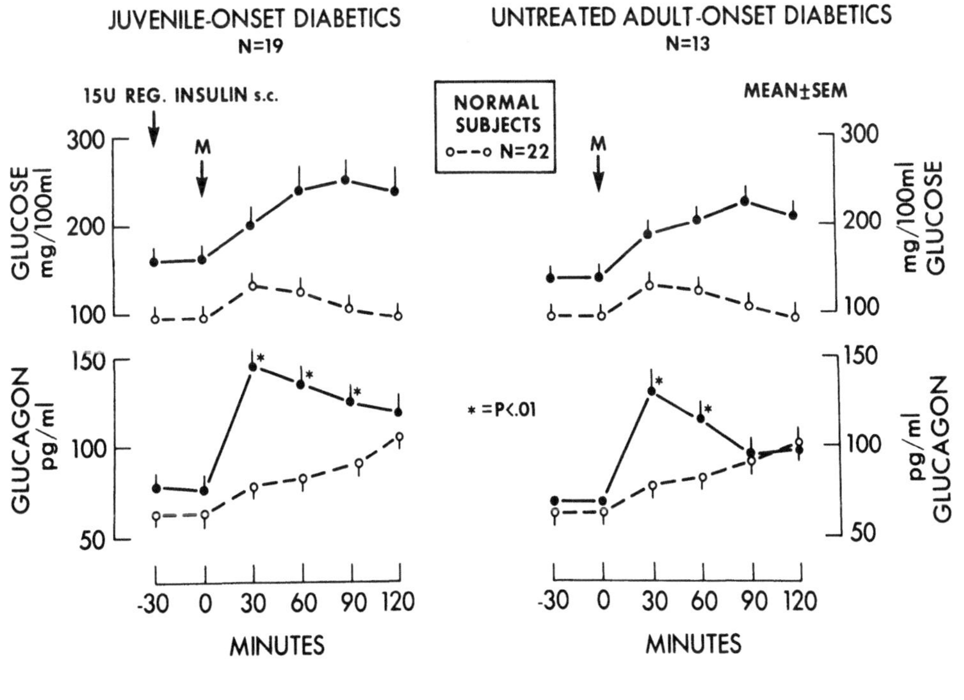

Fig. 2. Plasma glucose and glucagon levels after ingestion of a standard meal containing carbohydrate, protein, and fat in normal subjects, insulin-treated juvenile-onset diabetics and mild untreated adult-onset diabetics.

A Cell Dysfunction in Human Diabetes Mellitus

Whether the above abnormalities of glucagon secretion represent an intrinsic A cell defect or are merely secondary to insulin lack is currently under intensive investigation. Investigators who consider abnormal glucagon secretion in human diabetes to be a secondary phenomenon point out that similar abnormalities are found in animals with experimental forms of insulin deficiency (antiinsulin serum), with alloxan, and streptozotocin-induced diabetes (6, 57) and in man when diabetes occurs as a result of pancreatitis (7, 42, 65), that A cell dysfunction generally parallels the severity of insulin lack (7, 42), and finally that many of the abnormalities of glucagon secretion in human diabetes revert toward normal or disappear altogether with improvement in endogenous B cell function or with intensive exogenous insulin therapy (35, 41, 52, 60, 66, 67).

Several investigators, on the other hand, have proposed that both A and B cell dysfunction in human diabetes may result from an inborn error or environmental insult which manifests itself in functional terms as an abnormal glucoreceptor (23, 59, 73, 78). According to this concept, both A and B cells possess primary lesions induced by the pathologic process. Thus, it might be expected that defects in glucagon and insulin secretion might parallel one another.

This hypothesis is not inconsistent with what is currently known or postulated regarding the interaction of genetic and environmental factors in the etiology of human diabetes (84). For example, as shown in Figure 3, a protein within the islet cell membrane could contain a genetically-determined amino acid substitution analogous to hemoglobin in sickle cell anemia. This could alter the structure of the islet cell membrane in such a way as to render the glucoreceptor portion of the membrane less effective. Alternatively, such a membrane alteration may not directly involve the glucoreceptor but render the islet cell more susceptible to an environmental insult or endogenous injury (autoimmune) which could then induce a defective glucoreceptor. As a result of the abnormal glucoreceptor, A and B cells would not be able to respond to glucose appropriately; later or

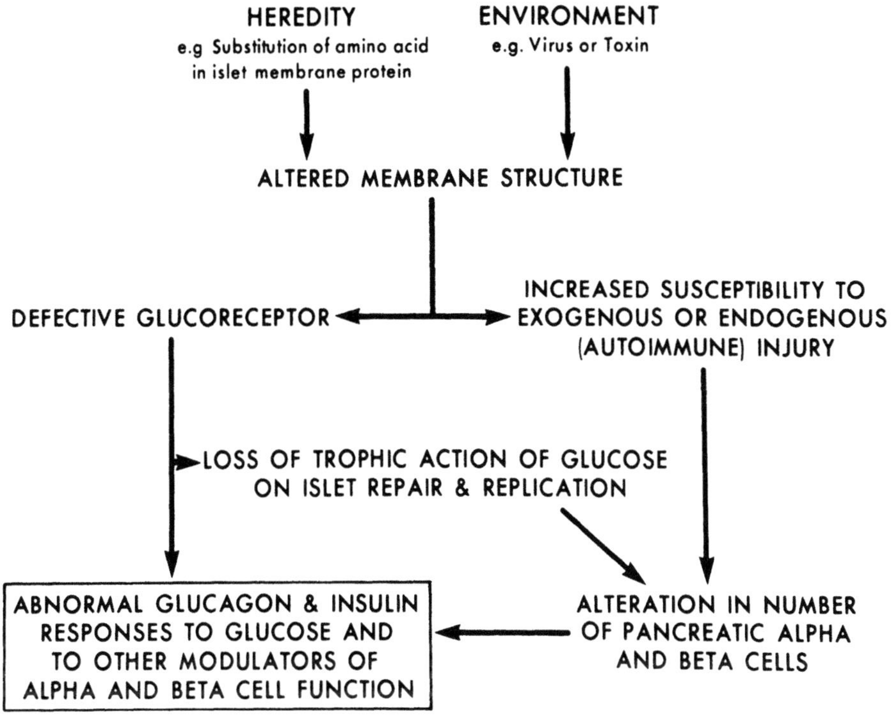

Fig. 3. Hypothesis to explain induction of glucoreceptor defect in pancreatic A and B cells.

perhaps concurrently, a decreased trophic effect of glucose on islet cell replication and repair could progressively lead to an alteration in pancreatic A (increase) and B (decrease) cell mass. The above schema is consistent with the heterogeneous pathologic changes in pancreatic islets found in both juvenile-onset or adult-onset varieties of diabetes mellitus (16, 82) and could ultimately cause abnormal glucagon and insulin responses to agents other than glucose.

Evidence for Defective Islet Cell Glucoreceptor in Diabetes

Data have recently accumulated favoring interaction of glucose with sterospecific islet cell glucoreceptors as an early step in the initiation of insulin release and modulation of glucagon secretion (2). Furthermore, there is evidence that these glucoreceptors may be defective in human diabetes rendering both the A and B cells insensitive to glucose. As for the B cell, its frequent, though transient recovery after the explosive onset of ketoacidosis indicates the presence of a potentially reversible functional defect (5); that this abnormality may be related to a glucoreceptor is suggested by the observation of frequently normal insulin responses to nonglucose stimuli concomitant with diminished insulin responses to glucose (13, 69, 77). Furthermore, a shift to the right of insulin-glucose dose response curves in adult-onset diabetics (10) indicates an altered K_m for glucose-induced insulin release rather than mere diminution in the number of cells secreting insulin.

Table 1 summarizes A cell function in human diabetes. Glucagon responses to both hyper- and hypoglycemia are abnormal. Hyperglycemia fails to suppress glucagon secretion normally even when accompanied by hyperinsulinemia (24, 32, 67, 73) although insulinization improves suppressibility (Fig. 4). This impaired suppressibility appears to be specific for glucose since elevation of plasma free fatty acids after heparin administration can suppress glucagon release normally (32; Fig. 5). Hypoglycemia induced by insulin fails to stimulate glucagon release in diabetes and is not altered by prolonged insulin administration (32; Fig. 6). This again provides support for a specific glucoreceptor defect since the diabetic A cell hyperresponds to other stimuli such as arginine or epinephrine (33).

Are these abnormalities merely secondary to insulin lack? There is certainly no question that insulin plays a major role in regulating A cell function in man, and its lack may be responsible for some of the abnormalities of glucagon secretion found in diabetes mellitus. For example, acute withdrawal of insulin (termination of prolonged insulin infusions maintaining juvenile-onset diabetics normoglycemic) rapidly results in a rise in plasma glucagon levels (35) which is equally rapidly reversed by reinfusion of physiologic amounts of insulin (15-30 μU/ml). Excessive glucagon responses to arginine can be totally corrected during

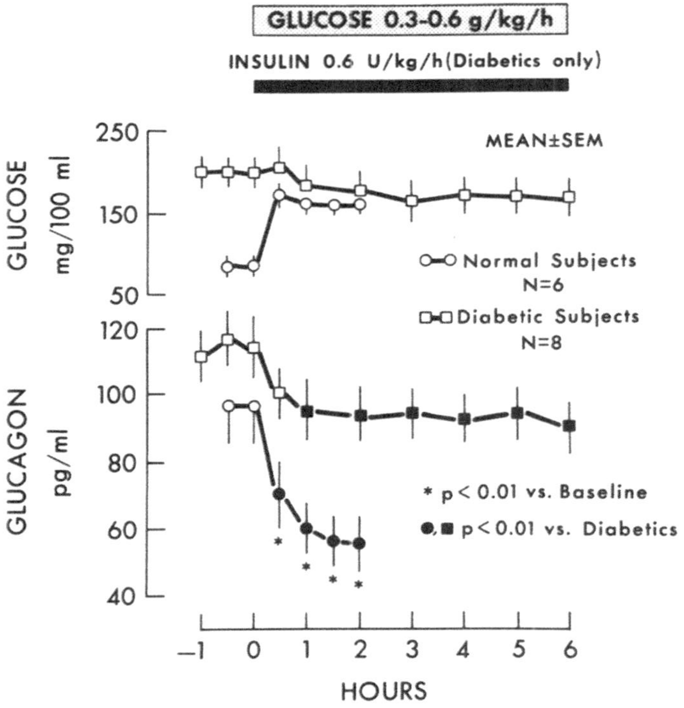

Fig. 4. Comparison of suppressibility of fasting plasma glucagon levels by glucose in normal subjects and in insulin-dependent juvenile-onset diabetic subjects infused with pharmacologic amounts of insulin (24, with permission).

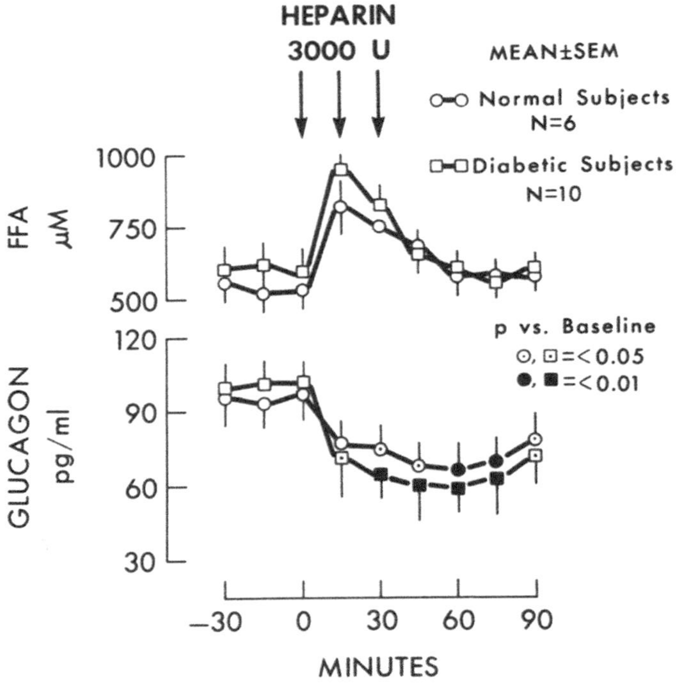

Fig. 5. Suppression of plasma glucagon by heparin-induced hyperlipidacidemia in normal subjects and in insulin-dependent diabetic subjects (24, with permission).

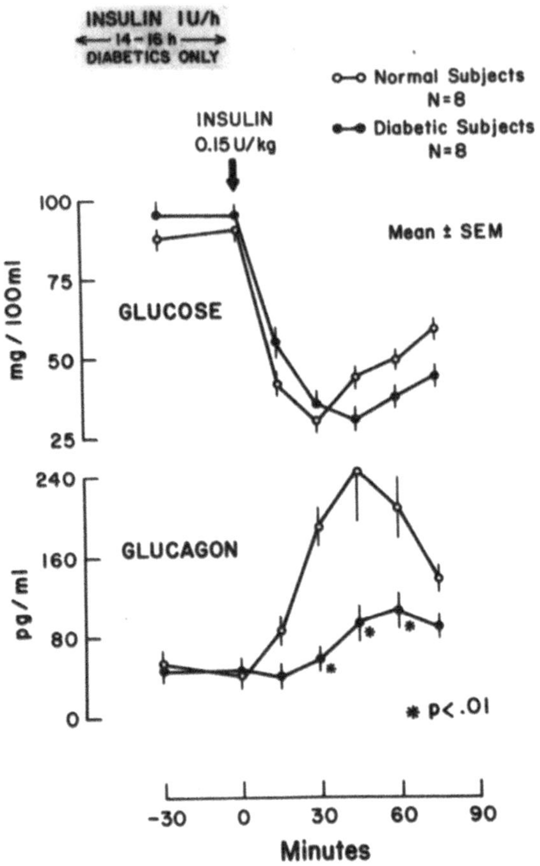

Fig. 6. Stimulation of glucagon secretion by insulin-induced hypoglycemia in normal subjects and in insulin-dependent diabetic subjects who had been maintained normoglycemic by infusion of insulin for 14-16 hours.

the infusion of small amounts of insulin, and rapidly reinstated upon discontinuation of the insulin infusion (31; Fig. 7). Excessive glucagon responses seen after ingestion of protein containing meals are also diminished by insulin. In these aspects diabetic A cell function in man resembles that seen in experimental diabetes in animals (6, 57). However, since abnormal A cell responses to glucose in human diabetes are not normalized by insulin (32, 67, 75), it would appear that not all A cell dysfunction in this disorder may result from insulin lack. Indeed, it has been shown recently that glucose-induced suppression of glucagon secretion

TABLE 1

SUMMARY OF A CELL FUNCTION IN DIABETES MELLITUS

Situation	Plasma Glucagon Response	Effect of Insulin
hyperglycemia	diminished (suppression)	partial correction
hypoglycemia	diminished (rise)	none
hyperlipidacidemia	normal suppression	-
hyperaminoacidemia	excessive	normalization
epinephrine infusion	excessive	normalization
protein-containing meals	excessive	partial correction
ketoacidosis (stress)	excessive	normalization

in rats made diabetic by alloxan can occur normally in the virtual complete absence of insulin (62). One may then ask why are glucagon responses to glucose persistently abnormal in human diabetes even when large amounts of insulin are given?

Dual Cause of Abnormal Glucagon Secretion in Human Diabetes

As illustrated in Table 1, abnormal A cell function in human diabetes can be divided into two broad categories: diminished responses to glucose (lack of suppression by hyperglycemia and lack of stimulation by hypoglycemia) and excessive responses to various stimuli. It is proposed that the latter are due predominantly to insulin lack while the former are the result of an abnormal glucoreceptor. As shown in Figure 8, intracellular c-AMP and free Ca^{++} both influence the magnitude of glucagon responses to a given stimulus, increased levels of these factors promote greater glucagon secretion (28). In tissues other than A cells, it has been shown that insulin diminishes intracellular c-AMP (39) and promotes

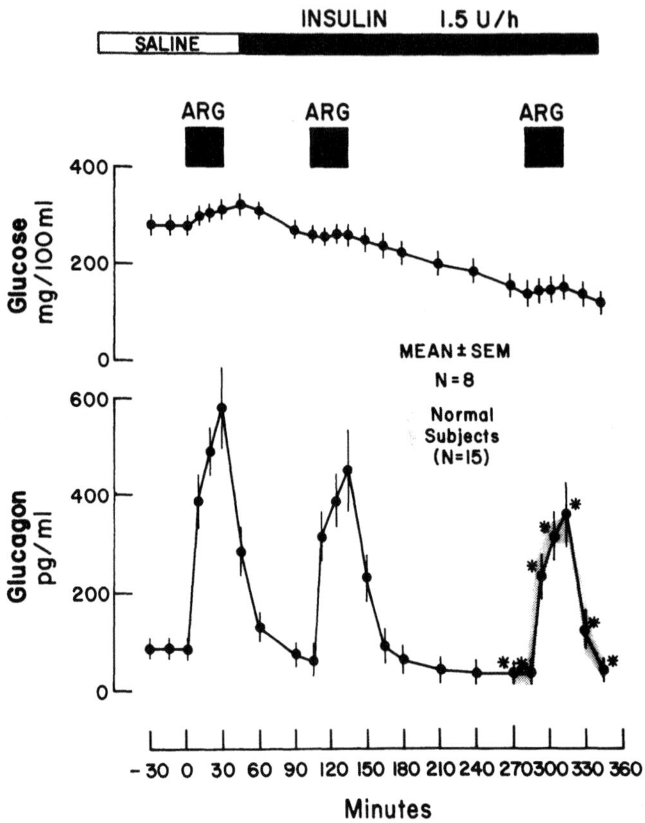

Fig. 7. Effects of infusion of insulin on glucagon responses to arginine in insulin-dependent diabetic subjects (31, with permission).

binding of free Ca^{++} (53). If this action of insulin on A cell c-AMP and Ca^{++} were deficient, one might then expect that a given stimulus would result in greater glucagon release. Thus, glucagon levels might rise to higher levels after ingestion of a meal containing protein and carbohydrate partly because the concomitant hyperglycemia did not restrain A cell secretion due to an abnormal glucoreceptor and partly because insulin lack did not restrain aminogenic stimulation of glucagon secretion.

The hypothesis proposed above can explain the ability of insulin to normalize excessive glucagon responses to arginine, to partially correct excessive glucagon responses seen after ingestion of protein containing meals as well as the in-

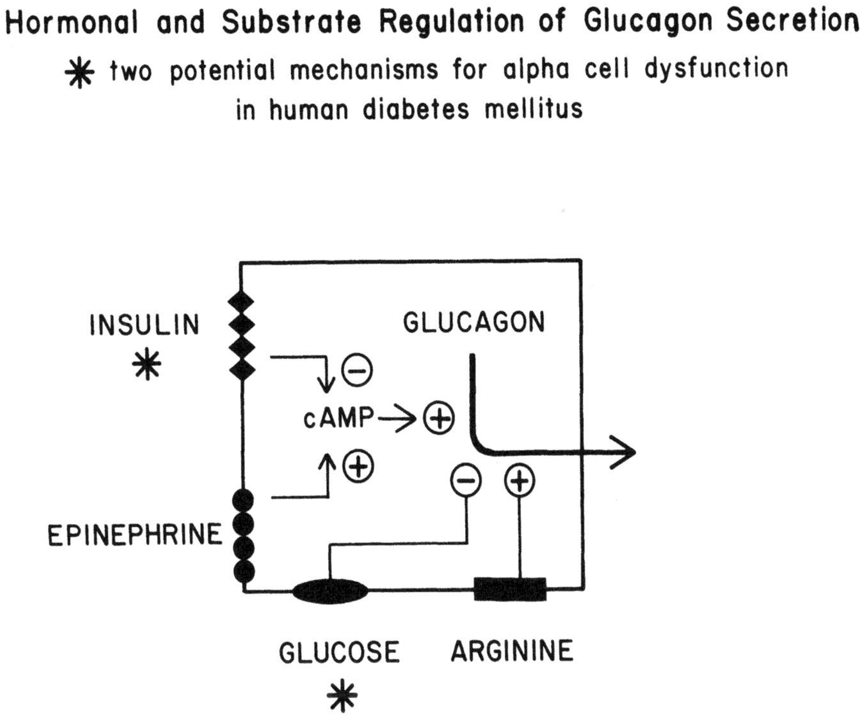

Fig. 8. Hypothesis depicting interaction of abnormal glucoreceptor and insulin deficiency on A cell function in human diabetes mellitus.

ability of insulin to normalize totally inappropriate glucagon responses to glucose which would be the result of an abnormal glucoreceptor and not insulin lack. The hypothesis thus implies that even optimal insulinization (such as with an artificial pancreas or successful pancreatic transplant) would not normalize A cell function in human diabetes.

Metabolic Consequences of A Cell Dysfunction in Diabetes Mellitus

Of course, whether or not normalization of A cell function in diabetes is important enough to require development of specific therapy depends on the extent to which abnormal glucagon secretion contributes to the metabolic consequences of insulin lack and the success of current treatment modalities in counteracting its effects. Until recently most, if not all, biochemical abnormalities in human diabetes were attributed solely to insulin lack; the role of glucagon excess was considered negligible. This was based on the fact that infusions of exogenous glucagon did not cause diabetes in normal man (72) and that pancreatectomy in man (4) and in experimental animals, causing what was thought to be total removal of insulin and glucagon still resulted in diabetes, though a somewhat milder form (4). It is now known, however, that exogenously infused glucagon can immediately stimulate sufficient release of insulin (70) to nullify its own effects (1, 47) and that the liver becomes refractory to glucagon during its prolonged administration at high constant levels (15). Furthermore, removal of the pancreas may not remove all sources of biologically active A cell glucagon both in man (63) and other species (50, 51, 71, 79).

The major advance which has permitted more accurate assessment of the role of glucagon in human diabetes was the observation that somatostatin, the growth hormone release inhibiting peptide purified originally from sheep hypothalami (76), blocked the secretion of glucagon (45). Although this peptide has numerous other actions including inhibition of insulin release and various gastrointestinal functions (37, 46), it has proved extremely useful as a research tool in evaluating the contribution of glucagon to the metabolic consequences of insulin lack in human diabetes (19).

Glucagon and Fasting Hyperglycemia

Several groups of investigators (14, 28-30, 49) have reported that infusion of somatostatin into insulin-dependent diabetic subjects who had fasted overnight and who had not received insulin for 24 hours (Fig. 9) causes a fall in plasma glucose toward normal levels. This diminution in fasting hyperglycemia is directly attributable to the suppression of glucagon secretion by somatostatin for the following reasons: (a) somatostatin itself has no direct effect on glucose (or ketone body) metabolism (12, 20); (b) somatostatin itself does not directly antagonize the action of glucagon or enhance that of insulin (28); (c) somatostatin has no effect in pancreatectomized individuals lacking measurable glucagon (22); but (d) is fully active in hypophysectomized diabetics lacking growth hormone and other pituitary hormones; and (e) its effect can be completely reversed by infusing physiologic amounts of exogenous glucagon (26). Thus, glucagon secretion definitely plays a role in the maintenance and severity of fasting hyperglycemia in human diabetics.

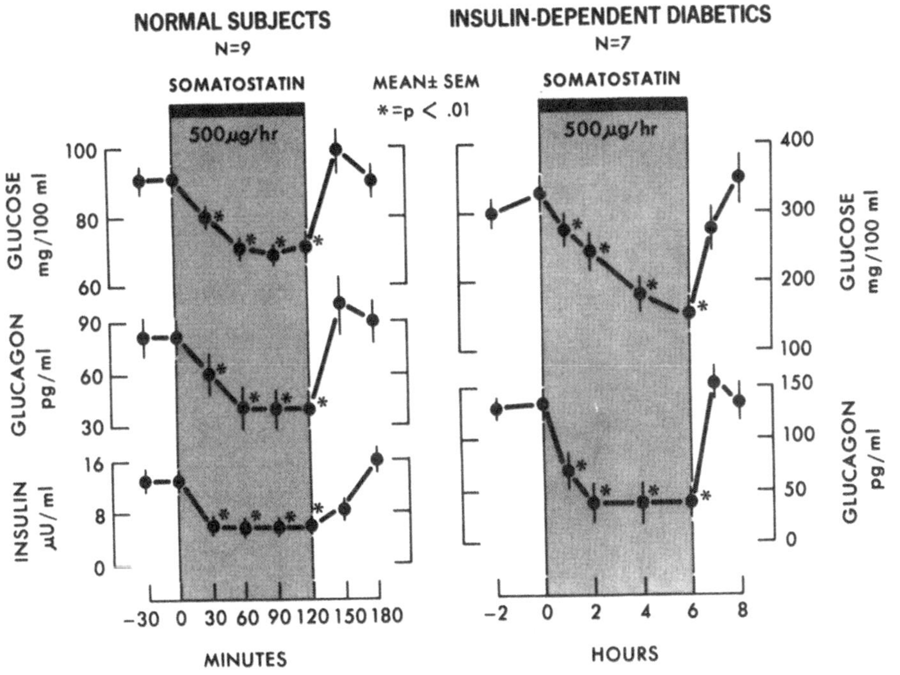

Fig. 9. Effects of somatostatin infusion on plasma glucose levels in overnight fasted insulin-dependent diabetic subjects (right) and in normal subjects (left).

Glucagon and Ketoacidosis

Glucagon can stimulate lipolysis and conversion of free fatty acids into ketone bodies in man (47). If one causes acute insulin deficiency by terminating an infusion of insulin which had maintained insulin-dependent ketosis-prone diabetic patients normoglycemic, within hours hyperglycemia and hyperketonemia ensue which is paralleled by a rise in circulating glucagon levels (25; Fig. 10). Infusion of somatostatin under identical conditions in the same patients upon stopping insulin aborts the development of ketoacidosis for as long as 18 hours. Plasma glucose levels rise to only 150-160 mg/100 ml compared to 300 mg/100 ml when glucagon secretion is not inhibited, and plasma ketone bodies rise only 0.3 mM compared to rises of over 2.0 mM without somatostatin. Since the effects observed during administration of somatostatin could be completely reversed by the simultaneous infusion of physiologic amounts of glucagon (26), one need not invoke actions of somatostatin other than suppression of glucagon secretion (such as inhibition of growth hormone release) to explain these results. Accordingly, these studies indicate that glucagon can exacerbate the metabolic consequences of insulin lack in man by promoting hyperketonemia as well as hyperglycemia and support the concept that glucagon acts to convert the liver into a ketogenic mode (54).

The above human studies recently have been supported by studies conducted in dogs employing suppression of endogenous-glucagon and insulin secretion with somatostatin during intraportal replacement of one or both of the hormones exogenously (11, 43). These studies demonstrated that hepatic production of glucose and ketone bodies was not appreciably augmented during insulin deprivation unless glucagon was present. In the longer term human studies definite increases in glucose and ketone body levels eventually did occur despite suppression of plasma glucagon below basal levels (25). Although glucagon levels were not undetectable (glucagon release was not totally inhibited), these observations would suggest that hypersecretion of glucagon is not essential to the development

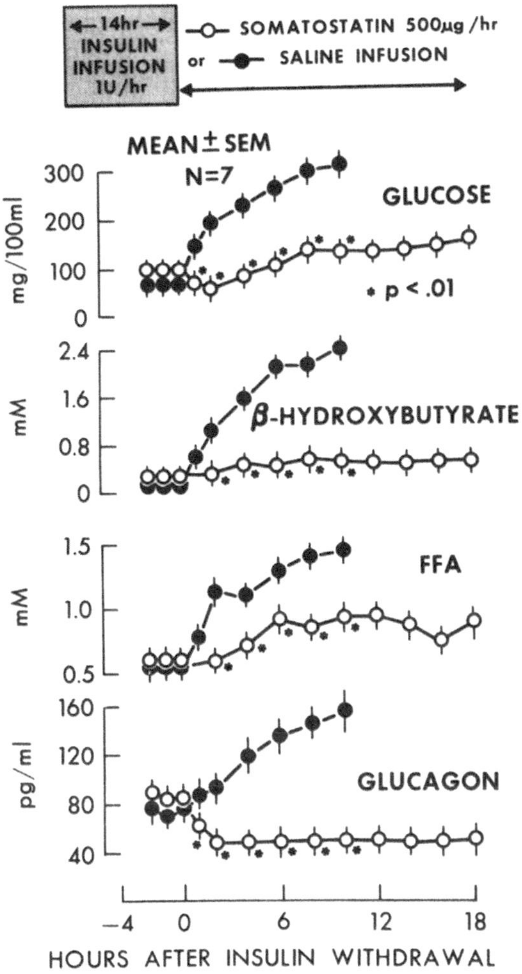

Fig. 10. Effect of somatostatin on metabolic consequences of acute insulin withdrawal in insulin-dependent juvenile-onset diabetic subjects (25, with permission).

of hyperglycemia and hyperketonemia but that the magnitude of these consequences of insulin lack is markedly diminished in absence of hyperglucagonemia.

Glucagon and Postprandial Hyperglycemia

The liver is the major organ responsible for assimilation of glucose taken orally (17, 40) and is the major target organ for glucagon (64). Thus, excessive glucagon secretion after ingestion of a meal containing carbohydrate and protein

(27, 81), as is illustrated for insulin-treated juvenile-onset and mild adult-onset diabetes in Figure 2, might exacerbate postprandial hyperglycemia. That this indeed is the case is supported by findings of several groups (27, 55, 80) that infusion of somatostatin after the ingestion of glucose or of a meal diminishes postprandial hyperglycemia in insulin-requiring diabetics. Figure 11 compares the effects of a given dose of insulin alone and that same dose of insulin given during a 24-hr infusion of somatostatin on serum glucose levels in seven insulin-requiring diabetic patients studied on two consecutive days. Somatostatin plus insulin reduced postprandial hyperglycemia and diminished fluctuations in glucose values compared to the same amount of insulin alone. Caution should be exercised, however, in attributing all of this effect to suppression of glucagon secretion since the various other actions of somatostatin (37) especially its effect on gastrointestinal function (46) may also have been partly responsible.

Practical Implications in the Management of Diabetes

The data derived from studies employing somatostatin have provided strong evidence that abnormal A cell function may contribute significantly to both fasting and postprandial hyperglycemia in human diabetes as well as to the development and severity of diabetic ketoacidosis. We must now ask how well does insulin as currently used counteract the deleterious effects of glucagon and could the addition of an agent which suppresses glucagon secretion provide better management of diabetes than optimal use of insulin alone? Figure 12 compares the results of intensive insulin (twice a day mixtures of NPH and Regular) and diet (5 meals a day) therapy on diabetic control in insulin-requiring diabetic patients with values for normal subjects studied on a metabolic ward. While insulin was clearly able to render patients asymptomatic, abolish glycosuria, and provide fasting glucose levels in the normal range, it did not prevent excessive postprandial hyperglycemia or wide fluctuation in glucose levels. Moreover, despite higher peripheral insulin levels (but not presumably portal vein insulin

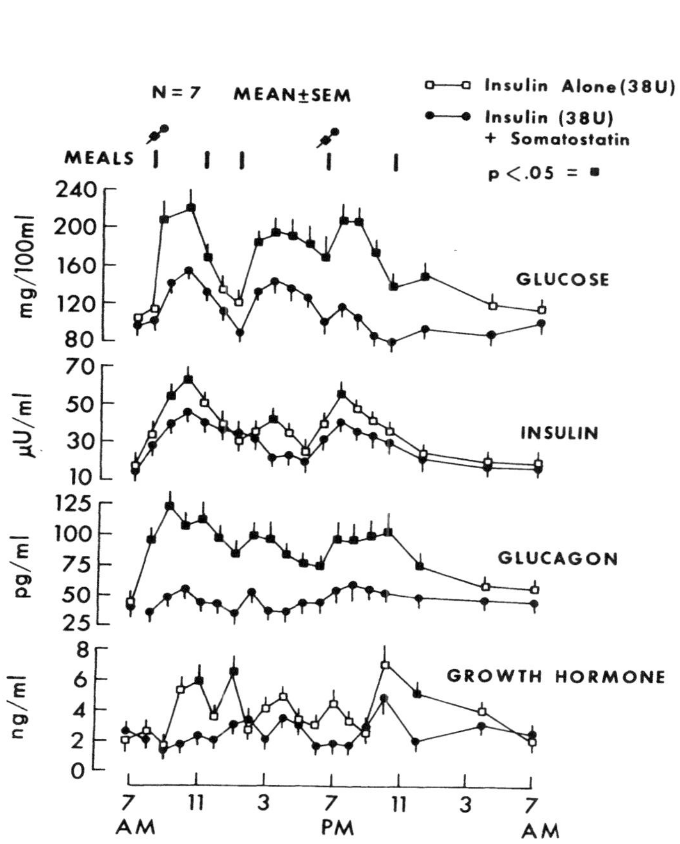

Fig. 11. Comparison of the effects of somatostatin plus insulin on diabetic control with the same dose of insulin given alone.

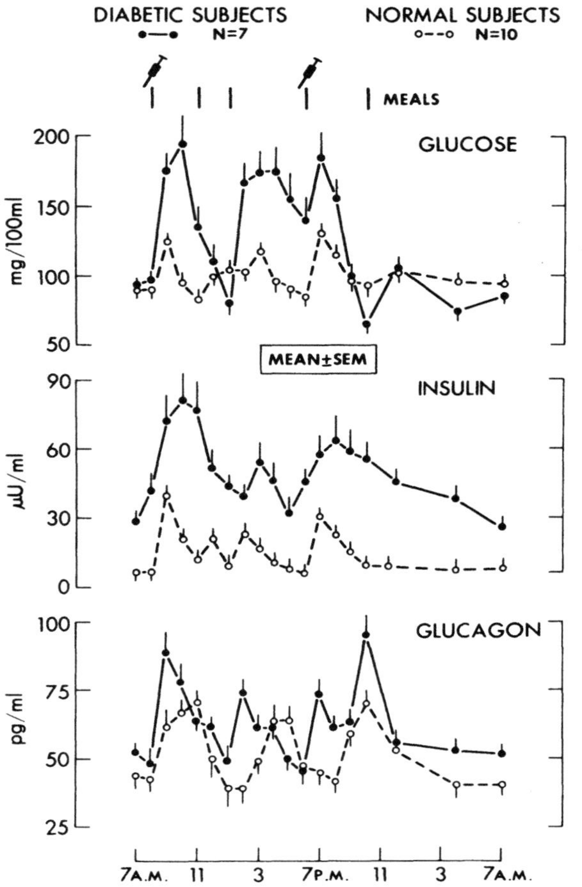

Fig. 12. Comparison of glucose, insulin and glucagon levels after meals in normal subjects with values in insulin-dependent diabetics after intensive treatment.

levels), abnormal glucagon secretion persisted. These observations suggest that conventional insulin therapy cannot reproduce the pattern of normal glucose homeostasis characterized by minor and transient elevations of glucose levels above fasting values. Unfortunately there are not data to assure one that the degree of diabetic control achievable with insulin alone as presently available is sufficient to prevent the long-term complications of diabetes. Indeed the opposite appears to be the case (44) and has led to the search for improved methods of insulin delivery such as the artificial pancreas and transplantation of pancreatic tissue.

Recently experiments were undertaken in our laboratory to determine whether an agent such as somatostatin used as an adjunct to insulin might provide better control of diabetic hyperglycemia than optimal use of insulin alone as currently employed. The effects of a three-day somatostatin infusion on diabetic control was studied in seven insulin-requiring diabetic patients whose control had been optimized on a metabolic ward for 5-7 days by diet manipulation and intensive insulin therapy. Figure 13 shows the results of this study. During infusion of somatostatin both postprandial hyperglycemia and wide fluctuations in glucose levels were diminished. This occurred despite a concomitant 50% reduction in insulin doses and appears to have been due directly to the somatostatin infusion since diabetic control rapidly deteriorated upon stopping the infusion while maintaining insulin doses constant. These results indicate that in insulin-requiring diabetic patients somatostatin plus insulin can provide better glucose homeostasis than insulin alone. Should these results be confirmed elsewhere and a longer-acting and more specific preparation be developed, somatostatin may prove useful as an adjunct to insulin in the management of diabetic hyperglycemia.

Acknowledgements

This study was supported in part by funds provided by the Bureau of Medicine and Surgery, Navy Department, for CIP 6-48-859 and in part by grants from the Susan Greenwall Foundation of New York City, the Levi J. and Mary Skaggs Founda-

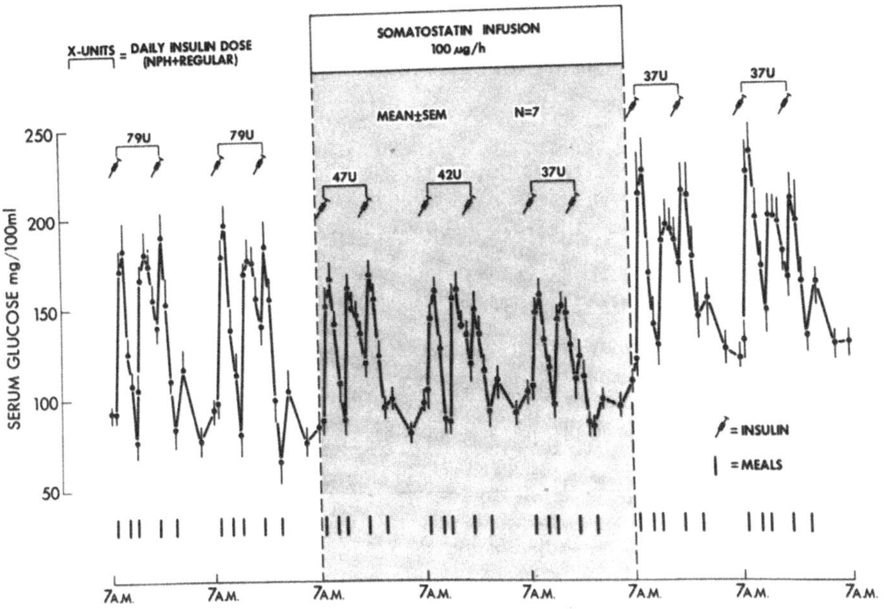

Fig. 13. Effect of 3-day somatostatin infusion on diabetic control in seven insulin-requiring diabetic subjects.

tion, Oakland, California, the Ellis L. Phillips Foundation of Jericho, New York, and by USPHS grant AM 1276 (05).

Current address for Dr. Gerich: Diabetes Research Laboratory and Endocrine Research Unit, Departments of Medicine and Physiology, Mayo Clinic Foundation and Medical School, Rochester, Minnesota, 55901, USA.

The opinions or assertions contained in this manuscript are the private ones of the author and are not to be construed as official or as reflecting the views of the Navy Department or the naval service at large.

I wish to thank my patients for their cooperation in these studies; our metabolic ward corpsmen, dietitians, and nursing staff for their dedicated work and HM3 L. Adam, Mr. G. Schmidt, and Ms. G. Gustafson for excellent technical help; I have been particularly fortunate in having M. Lorenzi, E. Tsalikian, J. Karam, S. Lewis, T. Schultz, C. Noacco, M. Langlois, W. Schneider, N. Bohannon, D. Dier, J. Penhos, G. Grodsky, and P. Forsham as collaborators over the past several years.

REFERENCES

1. Alford, F., Bloom, S., Nabarro, J., Hall, R. Besser, G., Coy, D., Kastin, A. and Schally, A. (1974): Glucagon control of fasting glucose in man. Lancet

2, 974-977.

2. Anonymous (1975): Editorial: Glucoreceptors, insulin release and diabetes. Lancet 2, 646-647.

3. Assan, R., Attale, J., Ballerio, G., Girard, J., Hautecouverture, M., Kervran, A., Plouin, P., Slama, G., Soufflet, E., Tchobroutsky, G. and Tiengo, A. (1973): Some aspects of the physiology of glucagon. In: Diabetes. Proceedings of the VIIIth Congress of the International Diabetes Federation, Brussels, July 15-20, 1973. W. Malaisse, J. Pirart and J. Vallance-Owen, Eds. International Congress Series ICS, Vol. 312, Excerpta Medica, Amsterdam, pp. 144-178.

4. Barnes, A. and Bloom, S. (1976): Pancreatectomized man: a model for diabetes without glucagon. Lancet 1, 219-221.

5. Block, M., Mako, M., Steiner, D. and Rubenstein, A. (1972): Circulating C-peptide immunoreactivity: studies in normal and diabetic patients. Diabetes 21, 1013-1026.

6. Braaten, J., Faloona, G. and Unger, R. (1974): The effect of insulin on the alpha-cell response to hyperglycemia in long-standing alloxan diabetes. J. Clin. Invest. 53, 1017-1021.

7. Buchanan, K. (1973): Gut and islet hormones in diabetes. Postgrad. Med. J. 49 (Suppl. 1), 117-121.

8. Buchanan, K. and McCarroll, A. (1972): Abnormalities of glucagon metabolism in untreated diabetes mellitus. Lancet 2, 1394-1395.

9. Carnegie, P. and MacKay, I. (1975): Vulnerability of cell-surface receptors to autoimmune reactions. Lancet 2, 684-687.

10. Cerasi, E., Luft, R. and Efendic, S. (1972): Decreased sensitivity of the pancreatic beta cells to glucose in prediabetic and diabetic subjects: a glucose dose-response study. Diabetes 21, 224-234.

11. Chiasson, J., Jennings, A., Cherrington, A., Liljenquist, J. and Lacy, W. (1975): The fine regulation of basal gluconeogenesis in vivo. Diabetes 24, 407 (abst.).

12. Chideckel, E., Palmer, J., Koerker, D., Ensinck, J., Davidson, M. and Goodner, C. (1975): Somatostatin blockade of acute and chronic stimuli of the endocrine pancreas and the consequences of this blockade on glucose homeostasis. J. Clin. Invest. 55, 754-762.

13. Deckert, T., Lauridsen, U., Madsen, S. and Morgensen, P. (1972): Insulin response to glucose, tolbutamide, secretin, and isoprenaline in maturity onset diabetes mellitus. Dan. Med. Bull. 19, 222-226.

14. Del Guercio, M., de Natale, B., Gargantini, L., Garlaschi, C. and Chiumello, G. (1976): Effect of somatostatin on blood sugar, plasma growth hormone, and glucagon levels in diabetic children. Diabetes 25, 550-553.

15. DeRubertis, F. and Craven, P. (1976): Reduced sensitivity of hepatic adenylate cyclase-cyclic AMP system to glucagon during sustained hormonal stimulation. J. Clin. Invest. 57, 435-443.

16. Doniach, I. and Morgan, A. (1973): Islets of Langerhans in juvenile diabetes mellitus. Clin. Endocrinol. 2, 233-248.

17. Felig, P., Wahren, J. and Hendler, R. (1975): Influence of oral glucose ingestion on splanchnic glucose and gluconeogenic substrate metabolism in man. Diabetes 24, 468-475.

18. Foà, P. (1972): The secretion of glucagon. In: Handbook of Physiology, Endocrinology, Sec. 7, Vol. 1. R. Greep and E. Astwood, Eds. American Physiological Society, Washington, D. C., pp. 261-278.

19. Gerich, J. (1977): Somatostatin: its possible role in carbohydrate homeostasis and the treatment of diabetes mellitus. Arch. Int. Med. 137, 656-666.

20. Gerich, J., Bier, D., Haas, R., Wood, C., Byrne, R. and Penhos, J. (1975): In vitro and in vivo effects of somatostatin on glucose, alanine, and ketone body metabolism in the rat. Progr. 57th Ann. Meet. Endocrine Soc., New York, p. 128.

21. Gerich, J., Charles, M. and Grodsky, G. (1976): Regulation of pancreatic insulin and glucagon secretion. Ann. Rev. Physiol. 38, 353-388.

22. Gerich, J., Karam, J. and Lorenzi, M. (1976): Diabetes without glucagon. Lancet 1, 855-856.

23. Gerich, J., Langlois, M., Noacco, C., Karam, J. and Forsham, P. (1973): Lack of glucagon response to hypoglycemia in diabetes: evidence for an intrinsic pancreatic alpha-cell defect. Science 182, 171-173.

24. Gerich, J., Langlois, M., Noacco, C., Lorenzi, M., Karam, J. and Forsham, P. (1976): Comparison of the suppressive effects of elevated plasma glucose and free fatty acid levels on glucagon secretion in normal and insulin-dependent diabetic subjects; evidence for selective alpha-cell insensitivity to glucose in diabetes mellitus. J. Clin. Invest. 58, 320-325.

25. Gerich, J., Lorenzi, M., Bier, D., Schneider, V., Tsalikian, E., Karam, J. and Forsham, P. (1975): Prevention of human diabetic ketoacidosis by somatostatin: evidence for an essential role of glucagon. N. Engl. J. Med. 292, 985-989.

26. Gerich, J., Lorenzi, M., Bier, D., Tsalikian, E., Schneider, V., Karam, J. and Forsham, P. (1976): Effects of physiologic levels of glucagon and growth hormone on human carbohydrate and lipid metabolism: studies involving administration of exogenous hormone during suppression of endogenous hormone secretion with somatostatin. J. Clin. Invest. 57, 875-884.

27. Gerich, J., Lorenzi, M., Karam, J., Schneider, V. and Forsham, P. (1975): Abnormal pancreatic glucagon secretion and postprandial hyperglycemia in diabetes mellitus. JAMA 234, 159-165.

28. Gerich, J., Lorenzi, M., Schneider, V. and Forsham, P. (1974): Effect of somatostatin on plasma glucose and insulin responses to glucagon and tolbutamide in man. J. Clin. Endocrinol. Metab. 39, 1057-1060.

29. Gerich, J., Lorenzi, M., Schneider, V., Karam, J., Rivier, J., Guillemin, R. and Forsham, P. (1974): Effects of somatostatin on plasma glucose and

glucagon levels in human diabetes mellitus. Pathophysiologic and therapeutic implications. N. Engl. J. Med. 291, 544-547.

30. Gerich, J., Lorenzi, M., Tsalikian, E., Bohannon, N., Brooks, R., Meyer, J., Schultz, T., Spencer, M., Karam, J. and Forsham, P. (1976): Prolonged intravenous and subcutaneous somatostatin in the treatment of human diabetes. Diabetes 25 (Suppl. 1), 340 (abst.).

31. Gerich, J., Lorenzi, M., Tsalikian, E., Bohannon, N., Schneider, V., Karam, J. and Forsham, P. (1976): Effects of acute insulin withdrawal and administration on plasma glucagon responses to intravenous arginine in insulin-dependent diabetic subjects. Diabetes 25, 955-960.

32. Gerich, J., Lorenzi, M., Tsalikian, E., Bohannon, N., Schneider, V., Karam, J., Forsham, P. and Lewis, S. (1976): Selective failure of prolonged infusion of insulin to normalize glucagon responses to glucose in human diabetes mellitus: evidence for a defective α-cell glucoreceptor. Clin. Res. 24, 361A.

33. Gerich, J., Lorenzi, M., Tsalikian, E. and Karam, J. (1976): Studies on the mechanism of epinephrine-induced hyperglycemia in man: evidence for participation of pancreatic glucagon secretion. Diabetes 25, 65-71.

34. Gerich, J., Schneider, J., Lorenzi, M., Tsalikian, E., Karam, J., Bier, D. and Forsham, P. (1976): Role of glucagon in human diabetic ketoacidosis: studies using somatostatin. Clin. Endocrinol. 5, 299s-306s.

35. Gerich, J., Tsalikian, E., Lorenzi, M., Schneider, V., Bohannon, N., Gustafson, G. and Karam, J. (1975): Normalization of fasting hyperglucagonemia and excessive glucagon responses to intravenous arginine in human diabetes mellitus by prolonged infusion of insulin. J. Clin. Endocrinol. Metab. 41, 1178-1180.

36. Gossain, V., Matute, M. and Kalkhoff, R. (1974): Relative influence of obesity and diabetes on plasma alpha-cell glucagon. J. Clin. Endocrinol. Metab. 38, 238-243.

37. Guillemin, R. and Gerich, J. (1976): Somatostatin: physiologic and clinical significance. Ann. Rev. Med. 27, 379-388.

38. Heding, L. and Rasmussen, S. (1972): Determination of pancreatic and gut glucagon-like immunoreactivity (GLI) in normal and diabetic subjects. Diabetologia 8, 408-411.

39. Illiano, G. and Cuatrecasas, P. (1972): Modulation of adenylate cyclase activity in liver and fat cell membranes by insulin. Science 175, 906-908.

40. Jackson, R., Peters, N., Advani, V., Perry, G., Rogers, J., Brough, W. and Pilkington, T. (1973): Forearm glucose uptake during oral glucose tolerance test in normal subjects. Diabetes 22, 442-458.

41. Josefsberg, Z., Laron, Z., Doron, M., Keret, R., Belinski, Y. and Weismann, I. (1975): Plasma glucagon response to arginine infusion in children and adolescents with diabetes mellitus. Clin. Endocrinol. 4, 487-492.

42. Kalk, W., Vinik, A., Bank, S., Buchanan, K., Keller, P. and Jackson, W. (1974): Glucagon responses to arginine in chronic pancreatitis: possible

pathogenic significance in diabetes. Diabetes 23, 257-263.

43. Keller, U., Chiasson, J., Liljenquist, J., Cherrington, A., Jennings, A. and Crofford, O. (1976): Glucagon and ketogenesis in acute diabetes. Clin. Res. 24, 363A.

44. Knowles, H. (1970: Control of diabetes and the progression of vascular disease. In: Diabetes Mellitus: Theory and Practice. M. Ellenberg and H. Rifkin, Eds. McGraw-Hill Book Co., New York, pp. 666-673.

45. Koerker, D., Ruch, W., Chideckel, E., Palmer, J., Goodner, C., Ensinck, J. and Gale, C. (1974): Somatostatin: hypothalamic inhibitor of the endocrine pancreas. Science 184, 482-484.

46. Konturek, S. (1976): Somatostatin and the gastrointestinal secretions. Scand. J. Gastroenterol. 11, 1-4.

47. Liljenquist, J., Bomboy, J., Lewis, S., Sinclair-Smith, B., Felts, P., Lacy, W., Crofford, O. and Liddle, G. (1974): Effects of glucagon on lipolysis and ketogenesis in normal and diabetic men. J. Clin. Invest. 53, 190-197.

48. Lindsey, C., Faloona, G. and Unger, R. (1974): Plasma glucagon in non-ketotic hyperosmolar coma. JAMA 229, 1771-1773.

49. Lundbaek, K., Hansen, A., Ørskov, H., Christensen, S., Iversen, J., Seyer-Hansen, K., Alberti, G. and Whitefoot, R. (1976): Failure of somatostatin to correct manifest diabetic ketoacidosis. Lancet 1, 215-218.

50. Mashiter, K., Harding, P., Chou, M., Mashiter, G., Stout, J., Diamond, D. and Field, J. (1975): Persistent pancreatic glucagon but not insulin response to arginine in pancreatectomized dogs. Endocrinology 95, 678-693.

51. Matsuyama, T. and Foà, P. (1974): Plasma glucose, insulin, pancreatic and enteroglucagon levels in normal and depancreatectomized dogs. Proc. Soc. Exp. Biol. Med. 147, 97-102.

52. Matsuyama, T., Hoffman, W., Dunbar, J., Foà, N. and Foà, P. (1975): Glucose, insulin, pancreatic glucagon and glucagon-like immunoreactive materials in the plasma of normal and diabetic children. Effect of initial insulin treatment. Horm. Metab. Res. 7, 452-456.

53. McDonald, J., Bruns, D. and Jarett, L. (1976): Ability of insulin to increase calcium binding by adipocyte plasma membranes. Proc. Natl. Acad. Sci. USA 73, 1542-1546.

54. McGarry, J., Wright, P. and Foster, D. (1975): Hormonal control of ketogenesis. Rapid activation of hepatic ketogenic capacity in fed rats by anti-insulin serum and glucagon. J. Clin. Invest. 55, 1202-1209.

55. Meissner, C., Thum, C., Beischer, W., Winkler, G., Schröder, K. and Pfeiffer, E. (1975): Antidiabetic action of somatostatin -- assessed by the artificial pancreas. Diabetes 24, 988-996.

56. Müller, W., Faloona, G., Aguilar-Parada, E. and Unger, R. (1970): Abnormal alpha-cell function in diabetes: response to protein and carbohydrate ingestion. N. Engl. J. Med. 283, 109-115.

57. Müller, W., Faloona, G. and Unger, R. (1971): The effects of experimental insulin deficiency on glucagon secretion. J. Clin. Invest. 50, 1992-1999.

58. Müller, W., Faloona, G. and Unger, R. (1973): Hyperglucagonemia in diabetic ketoacidosis: its prevalence and significance. Am. J. Med. 54, 52-57.

59. Niki, A. and Niki, H. (1975): Is diabetes a disorder of the glucoreceptor? Lancet 2, 658.

60. Ohneda, A., Ishii, S., Horigome, K. and Yamagata, S. (1975): Glucagon response to arginine after treatment of diabetes mellitus. Diabetes 24, 811-819.

61. Ohneda, A., Sato, M., Matsuda, K., Yanbe, A., Maruhama, Y. and Yamagata, S. (1972): Plasma glucagon response to blood glucose fall, gastrointestinal hormones and arginine in man. Tohoku J. Exp. Med. 107, 241-251.

62. Pagliara, A., Stillings, S., Haymond, M., Hover, B. and Matschinsky, F. (1975): Insulin and glucose as modulators of the amino acid-induced glucagon release in the isolated pancreas of alloxan and streptozotocin diabetic rats. J. Clin. Invest. 55, 244-255.

63. Palmer, J., Werner, P., Benson, J. and Ensinck, J. (1976): Plasma-glucagon after pancreatectomy. Lancet 1, 1290-1291.

64. Park, C. and Exton, J. (1972): Glucagon and the metabolism of glucose. In: Glucagon, Molecular Physiology, Clinical and Therapeutic Implications. P. Lefèbvre and R. Unger, Eds., Pergamon Press, New York, pp. 77-108.

65. Persson, I., Gyntelberg, F., Heding, L. and Boss-Nielsen, J. (1971): Pancreatic glucagon-like immunoreactivity after intravenous insulin in normals and chronic pancreatitis patients. Acta Endocrinol. 67, 401-404.

66. Raskin, P., Aydin, I. and Unger, R. (1976): Effects of insulin on the exaggerated glucagon response to arginine stimulation in diabetes mellitus. Diabetes 25, 227-229.

67. Raskin, P., Fujita, Y. and Unger, R. (1975): Effect of insulin-glucose infusions on plasma glucagon levels in fasting diabetics and nondiabetics. J. Clin. Invest. 56, 1132-1138.

68. Reynolds, C., Horwitz, D., Molnar, G., Rubenstein, A. H. and Taylor, W. (1974): Abnormalities of endogenous insulin and glucagon in insulin-treated unstable and stable diabetics. Diabetes 23 (Suppl. 1), 343.

69. Robertson, R. and Porte, D. (1973): The glucose receptor: a defective mechanism in diabetes mellitus distinct from the beta adrenergic receptor. J. Clin. Invest. 52, 870-876.

70. Samols, E., Marri, G. and Marks, V. (1965): Promotion of insulin secretion by glucagon. Lancet 2, 415-416.

71. Sasaki, H., Rubalcava, B., Baetens, D., Blázquez, E., Srikant, C., Orci, L. and Unger, R. (1975): Identification of glucagon in the gastrointestinal tract. J. Clin. Invest. 56, 135-145.

72. Sherwin, R., Fisher, M., Hendler, R. and Felig, P. (1976): Hypergluca-gonemia and blood glucose regulation in normal, obese, and diabetic subjects. N. Engl. J. Med. 294, 455-461.

73. Unger, R. (1972): Pancreatic alpha-cell function in diabetes mellitus. In: Glucagon, Molecular Physiology, Clinical and Therapeutic Implications. P. Lefèbvre and R. Unger, Eds., Pergamon Press, New York. pp. 245-257.

74. Unger, R., Aguilar-Parada, E., Müller, W. and Eisentraut, A. (1970): Studies of alpha cell function in normal and diabetic subjects. J. Clin. Invest. 49, 837-848.

75. Unger, R., Madison, L. and Müller, W. (1972): Abnormal alpha cell function in diabetes: response to insulin. Diabetes 21, 301-307.

76. Vale, W., Brazeau, P., Rivier, C., Brown, M., Boss, B., Rivier, J., Burgus, R., Ling, N. and Guillemin, R. (1975): Somatostatin. Recent Progr. Hormone Res. 31, 365-397.

77. Varsano-Aharon, N., Echemendia, E., Yalow, R. and Berson, S. (1970): Early insulin responses to glucose and tolbutamide in maturity-onset diabetes. Metabolism. 19, 409-417.

78. Vinik, A., Kalk, W. and Jackson, W. (1974): A unifying hypothesis for hereditary and acquired diabetes. Lancet 1, 485-487.

79. Vranic, M., Pek, S. and Kawamori, R. (1974): Increased glucagon immuno-reactivity in plasma of totally depancreatectomized dogs. Diabetes 23, 905-912.

80. Wahren, J. and Felig, P. (1976): Somatostatin (SRIF) and glucagon in diabetes: failure of glucagon suppression to improve I.V. glucose tolerance and evidence of and effect of SRIF on glucose absorption. Clin. Res. 24, 461A.

81. Wahren, J., Felig, P. and Hagenfeldt, L. (1976): Effect of protein inges-tion on splanchnic and leg metabolism in normal man and in patients with diabetes mellitus. J. Clin. Invest. 57, 987-999.

82. Warren, S., LeCompte, P. and Legg, M. (1966): The Pathology of Diabetes Mellitus. 4th Ed., Henry Klimpton, London.

83. Wise, J. K., Hendler, R. and Felig, P. (1973): Evaluation of alpha-cell function by infusion of alanine in normal, diabetic, and obese subjects. N. Engl. J. Med. 288, 487-490.

84. Zonana, J. and Rimoin, D. (1976): Inheritance of diabetes mellitus. N. Engl. J. Med. 295, 603-605.

THE EFFECTS OF EXOGENOUSLY INDUCED HYPERGLUCAGONEMIA IN INSULIN-TREATED DIABETICS*

P. Raskin, C. B. Srikant, H. Nakabayashi, R. E. Dobbs and R. H. Unger

Veterans Administration Hospital, and the University of Texas Southwestern Medical School, Department of Internal Medicine Dallas, Texas USA

A recent report that the continuous intravenous infusion of glucagon failed to worsen hyperglycemia in two insulin-treated juvenile diabetics has raised doubts as to the importance of hyperglucagonemia in insulin-treated diabetics. To reexamine this, six juvenile type patients receiving a constant 28 to 50 U/day dose of regular insulin in five doses (before each of four meals and at 1:30 a. m.) were given glucagon subcutaneously for one or two days at 4-hour intervals (four patients) or 2-hour intervals (two patients), raising their mean plasma glucagon levels to an average of 590 ± 15 pg/ml. Mean plasma glucose concentration, measured at 2-hour intervals around the clock, rose from 231 ± 17 mg/dl on the control day to 285 ± 19 pg/ml ($p < 0.01$) during glucagon administration. Glucose excretion rose significantly above the control value of 42 ± 7 gms per 24 hours to 145 ± 23 gms ($p < 0.01$) and the excretion of urea nitrogen and ketones also increased significantly during glucagon administration ($p < 0.05$; $p < 0.01$, respectively). These data indicate that exogenous hyperglucagonemia can cause metabolic deterioration in insulin-treated diabetics.

INTRODUCTION

Relative hyperglucagonemia is reportedly present in all forms of spontaneous (1, 2, 6, 9) and experimentally induced (7) diabetes and has been implicated in the glucose overproduction that may occur in that disease (11). However, a recent report that infusion of glucagon for 48 hours failed to worsen the hyperglycemia of two insulin-treated juvenile diabetics has cast doubt on glucagon's importance in the presence of insulin (8). This study was designed to reexamine the effects of exogenous glucagon in six insulin-dependent diabetics.

METHODS AND MATERIALS

Six patients with typical juvenile type diabetes mellitus, ranging in age

*Presented in part at the Annual Meeting of the American Diabetes Association, San Francisco, California, June 22, 1976.

from 16 to 49 years, were studied in the Clinical Research Unit of Parkland Memorial Hospital. All patients received a constant diet consisting of 40% carbohydrate, 40% fat, and 20% protein with identical meals throughout the study. A constant dose of regular insulin ranging from 28 to 50 U/day was administered in four divided doses 30 minutes prior to meals and, in most patients, a fifth dose was given at 0130 each morning. A placebo injection of normal saline or glucagon (0.2 mg to 0.5 mg) was administered subcutaneously at 4-hour intervals in four patients and at 2-hour intervals in two patients.

Blood was drawn at 2-hour intervals through a 19-gauge butterfly needle in a forearm vein and collected in chilled tubes containing 12 mg EDTA and Trasylol ^{R} , centrifuged at 4°, and the plasma stored at -20°C until radioimmunoassay for glucagon (3) and, in one patient, for insulin (12). Glucose in plasma and urine was measured with a Beckman glucose analyzer. Measurements of urinary urea nitrogen (4) and β-hydroxybutyrate and acetoacetate excretion (5) were also made.

RESULTS

Effects of subcutaneous administration of glucagon. The mean plasma IRG and glucose concentrations and the 24-hour glucose excretion for all patients before, during, and after glucagon administration are shown in Table 1. The mean plasma glucose concentration rose from 231 ± 17 mg/dl to 285 ± 19 ($p < 0.01$) during glucagon administration. Glucose excretion rose from 42 ± 7 gms per 24 hours to 145 ± 23 gms per 24 hours ($p < 0.01$) during glucagon administration and declined to 56 ± 18 ($p < 0.01$) following its discontinuation. Urinary excretion of urea nitrogen and ketones increased significantly during the period of glucagon administration ($p < 0.05$; Table 2).

The IRG concentration of the group averaged 590 ± 135 pg/ml during the administration of glucagon. Of the 72 glucagon determinations during glucagon administration in the six patients, 50 were less than 1000 pg/ml and 38 were less than 500 pg/ml; the IRG range observed in the portal vein of insulin-deprived

TABLE 1

MEAN OF BIHOURLY PLASMA GLUCOSE AND GLUCAGON CONCENTRATION AND 24-HOUR URINARY GLUCOSE EXCRETION BEFORE, DURING AND AFTER THE SUBCUTANEOUS ADMINISTRATION OF GLUCAGON TO JUVENILE-TYPE DIABETICS

	Pre-Glucagon Control Day			Glucagon Day			Post-Glucagon Control Day		
Patient	Glucagon (pg/ml)	Glucose (mg/dl)	Glucose Excretion (g/24 hrs)	Glucagon (pg/ml)	Glucose (mg/dl)	Glucose Excretion (g/24 hrs)	Glucagon (pg/ml)	Glucose (mg/dl)	Glucose Excretion (g/24 hrs)
M. C.	109 ± 7	193 ± 19	12	537 ± 176	285 ± 11*	109	137 ± 39	217 ± 21**	17
T. R.	136 ± 3	255 ± 11	52	609 ± 178	309 ± 9*	153	134 ± 12	277 ± 18**	80
C. L.	198 ± 10	280 ± 12	46	431 ± 72	295 ± 18	109	102 ± 5	250 ± 10**	43
P. T.	42 ± 4	170 ± 8	49	299 ± 111	190 ± 15	84	53 ± 3	154 ± 8**	19
R. I.	265 ± 6	259 ± 15	64	1228 ± 76	319 ± 20*	179	241 ± 14	250 ± 17**	43
G. B.	121 ± 2	234 ± 10	31	433 ± 77	307 ± 31*	234	140 ± 10	305 ± 29*	131
Mean	145 ± 32	231 ± 17	42 ± 7	590 ± 135	285 ± 19*	145 ± 23*	135 ± 25	242 ± 21**	56 ± 18**

* p <0.01 vs. pre-glucagon control day value; ** p <0.01 vs. glucagon day value

TABLE 2

MEAN 24-HOUR URINARY UREA NITROGEN AND KETONE EXCRETION BEFORE, DURING AND AFTER THE SUBCUTANEOUS ADMINISTRATION OF GLUCAGON TO JUVENILE-TYPE DIABETICS

	Pre-Glucagon Control Day		Glucagon Day		Post-Glucagon Control Day	
	Urea Nitrogen (gm/24 hrs)	Ketones (μMoles/24 hrs)	Urea Nitrogen (gm/24 hrs)	Ketones (μMoles/24 hrs)	Urea Nitrogen (gm/24 hrs)	Ketones (μMoles/24 hrs)
1. M. C.	8.9	---	12.7	---	7.2	---
2. T. R.	14.2	110	18.5	100	13.9	136
3. C. L.	9.4	96	12.9	366	17.4	161
4. P. T.	5.2	83	5.9	423	6.2	482
5. R. I.	15.5	64	19.3	274	14.6	647
6. G. B.	2.2	29	13.8	594	8.0	193
Mean	9.2 ± 2.0	76 ± 14	13.9 ± 1.9*	351 ± 82**	11.2 ± 1.9	323 ± 102**

* $p < 0.05$ vs. pre-glucagon control day; ** $p < 0.01$ vs. pre-glucagon control day

alloxan diabetic dogs during alanine stimulation was 700 to 1900 pg/ml (unpublished observations).

Figure 1 depicts the response, typical of the group, of M. C. the only patient in whom measurements of plasma insulin were possible.

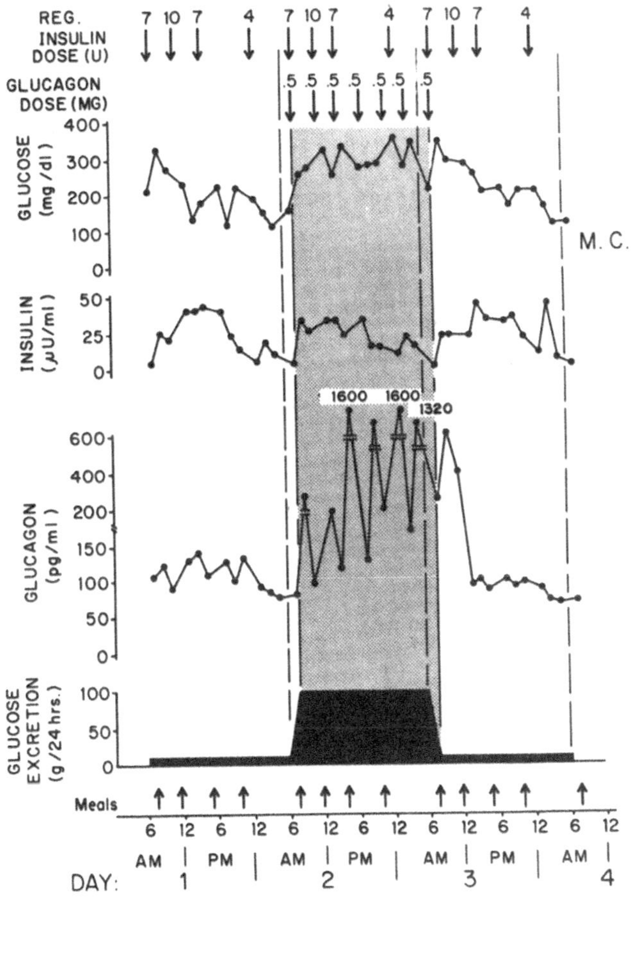

Fig. 1. The effect of subcutaneous administration of glucagon on plasma glucose, insulin, and glucagon concentration and 24-hour glucose excretion in a 26-year-old newly discovered juvenile diabetic. The arrows indicate the times at which insulin and glucagon were administered subcutaneously. The shaded area indicates the duration of glucagon administration.

DISCUSSION

The subcutaneous administration of glucagon to insulin-dependent human diabetics receiving from 28 to 50 U/day of regular insulin caused a significant increase in hyperglycemia, glycosuria, urea nitrogen excretion, and ketonuria, which receded with discontinuation of the glucagon.

The discrepancy between these findings and those of Sherwin et al. (8) are almost certainly the consequence of many differences in experimental design. In our study, blood samples were collected at 2-hour intervals, 24-hour excretion of glucose, urea nitrogen, and ketones were measured, and glucagon was administered subcutaneously. In the studies of Sherwin et al, blood specimens were obtained only at 8 a.m., 12 noon, 4 p.m. and 10 p.m., glucose excretion was not reported, and glucagon was administered by constant intravenous infusion containing 4 ml/100 ml of human blood, which reduces glycogenolytic activity within 4 hours and inactivates it completely within 8 hours. However, immunologic activity declined by only 60% in 8 hours, which could explain the continuing elevation of plasma immunoreactive glucagon levels without apparent biologic activity (12).

There were other differences in the two studies. The patients of Sherwin et al. received NPH insulin during the study and exhibited striking declines in their glucose levels, suggesting irregular surges of insulin absorption. In our studies, crystalline insulin was employed in the hope of avoiding irregular insulin absorption. In patient M. C. (Fig. 1), in whom the pattern of plasma insulin could be examined, levels ranged between 5 and 50 μU/ml on all days. However, the glucagon-induced increase in gluconeogenesis and ketogenesis suggests that the insulin levels in our patients were too low to deprive the liver of sufficient amino acids and free fatty acids to allow for glucagon-mediated acceleration of these processes.

The plasma IRG levels induced in our patients were intermittently higher than those maintained in the two patients of Sherwin et al. Although portal vein

IRG in poorly controlled human diabetic patients has never been measured, in insulin-deprived alloxan diabetic dogs, portal IRG levels above 1000 pg/ml during alanine infusion are common. It seems likely, therefore, that the hyperglucagonemia in these studies did not exceed the peak values that may occur in the portal vein plasma of poorly controlled diabetic patients and that, for most of the study, were probably below such levels.

The glucagon-induced hyperglycemia was not impressive in these patients despite their massive glycosuria. When ability to excrete glucose is unimpaired, as in this group, a glucagon-induced increase in hepatic glucose production is presumably reflected by an increase in glucose excretion rather than by striking venous hyperglycemia. In contrast, in symptomatic diabetic patients hospitalized because of poor control, volume contraction has already reduced glucose excretion and more marked venous hyperglycemia is present. A direct renal effect of glucagon on glucose excretion is not known to occur.

The study indicates that the full syndrome of metabolic deterioration in diabetes can be induced despite administration of up to 50 U/day of regular insulin by raising glucagon levels intermittently into the 500 to 1500 pg/ml range.

Acknowledgements

This work was supported by VA Institutional Research Support Grant 549-800-01; NIH Grants AM02700-16, I-R01-AM18179-02, and I-M01-RR0633; 30K Rabbit Fund; and The American Diabetes Association, North Texas Affiliate.

The authors wish to thank Margaret Bickham, Grace Chen, Loretta Clendenen, Virginia Harris, Vicki Lupean, Kay McCorkle, Cathy Mitchell, and Daniel Sandlin for technical assistance; Billie Godfrey, Susan Freeman, and Grace Mouille for secretarial assistance; and the staff and nurses of the Clinical Research Center.

REFERENCES

1. Aguilar-Parada, E., Eisentraut, A. M. and Unger, R. H. (1969): Pancreatic glucagon secretion in normal and diabetic subjects. Am. J. Med. Sci. 257, 415-419.

2. Assan, R., Hautecouverture, G., Guillémant, S., Douchy, F., Protin, P. and Dérot, M. (1969): Evolution de paramètres hormonaux (glucagon, cortisol,

hormone somatotrope) et enérgétiques (glucose, acids gras libre, glycerol) dans dix acido-cétoses diabétiques graves traitées. Pathol. Biol. 17, 1095-1105.

3. Faloona, G. R. and Unger, R. H. (1974): Glucagon. In: Methods of Hormone Radioimmunoassay. B. M. Jaffe and H. R. Behrman, Eds. Academic Press, Inc., New York, pp. 317-330.

4. March, W. H., Fingerhut, B. and Miller, H. (1965): Automated and manual direct methods for the determination of blood urea. Clin. Chem. 11, 624-627.

5. McGarry, J. D. and Foster, D. W. (1971): The regulation of ketogenesis from oleic acid and the influence of antiketogenic agents. J. Biol. Chem. 246, 6247-6253.

6. Müller, W. A., Faloona, G. R., Aguilar-Parada, E. and Unger, R. H. (1970): Abnormal alpha cell function in diabetes: response to carbohydrate and protein ingestion. New Engl. J. Med. 283, 109-115.

7. Müller, W. A., Faloona, G. R. and Unger, R. H. (1971): The effect of experimental insulin deficiency on glucagon secretion. J. Clin. Invest. 50, 1992-1999.

8. Sherwin, R. S., Fisher, M., Hendler, R. and Felig, P. (1976): Glucagon and glucose regulation in normal, obese, and diabetic subjects. New Engl. J. Med. 294, 455-461.

9. Unger, R. H., Aguilar-Parada, E., Müller, W. A. and Eisentraut, A. M. (1970): Studies of pancreatic alpha cell function in normal and diabetic subjects. J. Clin. Invest. 49, 837-848.

10. Unger, R. H., Aydin, I., Nakabayashi, H., Srikant, C. B. and Raskin, P. (1976): The effects of glucagon administration to nondiabetics and diabetics. Metab. 25 (Suppl. 1), 1523-1526.

11. Unger, R. H. and Orci, L. (1975): The essential role of glucagon in the pathogenesis of the endogenous hyperglycemia of diabetes mellitus. Lancet 1, 14-16.

12. Yalow, R. S. and Berson, S. A. (1960): Immunoassay of endogenous plasma insulin in man. J. Clin. Invest. 39, 1157-1175.

THE NATURE OF THE A CELL ABNORMALITY IN HUMAN DIABETES MELLITUS

D. J. Chisholm and F. P. Alford

Endocrinology Unit and Department of Medicine, St. Vincent's Hospital
(University of Melbourne)
Fitzroy, Melbourne, Australia

The hyperglucagonemia of human diabetes mellitus appears to be secondary to insulin insufficiency or to its metabolic consequences and not the result of a primary abnormality of the A cells.

An abnormality of A cell function in human diabetes was first clearly demonstrated when it was shown that plasma glucagon concentrations in diabetics were inappropriately elevated in relation to the hyperglycemia (3) and were not suppressed by an oral carbohydrate load as they are in normal subjects (10). Subsequently, A cell hyperactivity was demonstrated to be a direct consequence of insulin lack in streptozotocin and alloxan-induced diabetes in animals (4, 11). In human diabetes the causes of hyperglucagonemia are less clear. A probable one may be insulin deficiency (7). However, other data suggest that A cell hyperfunction in human diabetes cannot be corrected by insulin or, at least, that it is relatively unresponsive to hyperglycemia and to physiologic amounts of insulin (15, 17). These observations led to the postulate that human diabetes is characterized by a primary defect of both the A and the B cells (18).

In an effort to elucidate the nature of the A cell abnormality in human diabetes, we have measured plasma glucagon concentrations in three groups of insulin-requiring diabetic patients using an antiglucagon serum specific for pancreatic glucagon* (2). From these studies, we conclude that hyperglucagonemia in human diabetes, as in animal models, is a result of the insulin deficient state and is thus a secondary rather than a primary abnormality.

---

*Antiserum RCS5, kindly given to us by Dr. S. R. Bloom

Intravenous Administration of Insulin to Insulin Requiring, Young, Diabetic Patients

Following the intravenous administration of insulin (neutral regular insulin, 0.1 units/kg) to normal subjects, after an overnight fast, blood glucose levels fell sharply, eliciting the release of glucagon from the A cells (Fig. 1).

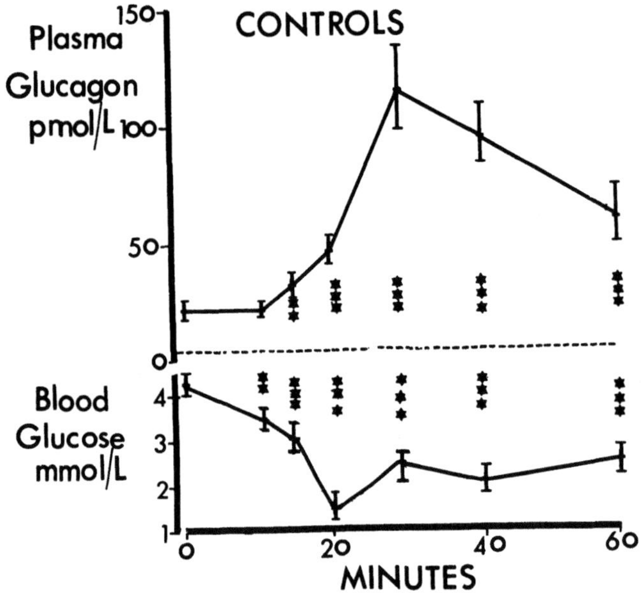

Fig. 1. The effect of rapid insulin administration (0.1 units/kg IV) on plasma glucagon and blood glucose concentrations in 8 normal subjects (mean ± SEM). The significance of the changes in blood concentrations from time zero, analyzed by paired t test are: * $p < 0.01$; ** $p < 0.002$; ***$p < 0.001$ (19, with permission).

The resulting hyperglucagonemia probably plays an important counter-regulatory role and contributes to the return of blood glucose to normal. In contrast, when insulin was administered under similar conditions, to 14 insulin-dependent juvenile diabetics, who were hyperglycemic at the time of the study, there was a profound and very consistent fall in plasma glucagon concentration. In fact, values of < 3 pmol/l (the limit of detection of the assay) were recorded in all subjects between 20 and 40 minutes after the injection of insulin (Fig. 2). The decline in blood glucose levels in these subjects was sluggish as compared to that of the controls. Only one patient became hypoglycemic and had a rebound rise in plasma glucagon (to 59 pmol/l) at 60 minutes. Other workers (5, 6) have noted that a glucagon response to hypoglycemia is often absent in insulin-treated diabetics.

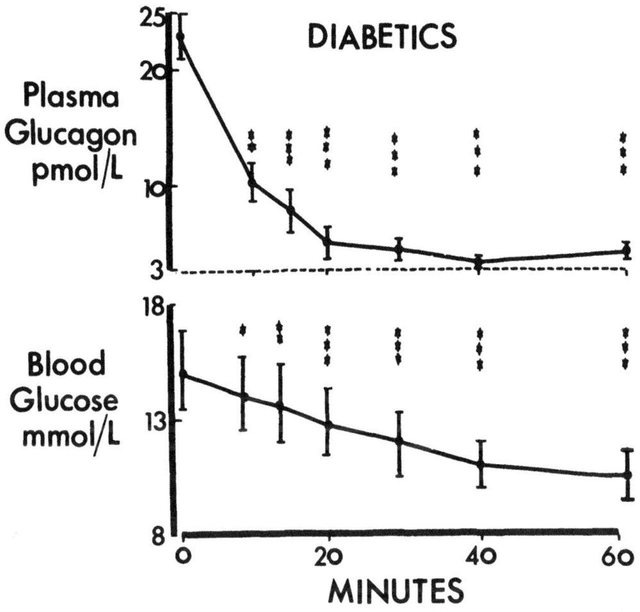

Fig. 2. The effect of rapid insulin administration (0.1 units/kg IV) on plasma glucagon and blood glucose concentration in 14 patients with juvenile-onset diabetics of 8 1/2 - 13 yrs duration (mean ± SEM). Note that the vertical scales for plasma glucagon and blood glucose are different from those in Fig. 1 (controls). The significance of the changes in blood concentrations from time zero, analyzed by paired t test are: *p <0.01; **p <0.002; ***p <0.001 (19, with permission).

It is now evident that insulin in large (pharmacologic) amounts suppresses hyperglucagonemia in human diabetes (19). However, it is not evident whether pharmacologic rather than physiologic amounts of insulin are required to overcome A cell hypersecretion and whether insulin suppresses A cell secretion in diabetes directly or by restoring the A cell responsiveness to hyperglycemia. The following study provides an answer to these questions.

A- and B-cell Function Before and After Metabolic Control in Newly Diagnosed Diabetic Patients

The responsiveness of the A and B cells was assessed in 10 newly diagnosed diabetic patients at presentation and again after therapy with insulin for an average of 3.5 months (9). The age of the patients varied between 14 and 53 years (mean 31). They were selected using the following criteria: acute onset of typical diabetic symptoms, moderate to marked ketonuria, complete absence of the rapid insulin response to I. V. glucose, absence of associated disease, therapy with insulin not urgent (i.e., treatment could safely be delayed until testing with I. V. glucose and tolbutamide had been completed).

After an overnight fast, subjects were given I. V. glucose (0.5 gm/kg in 2 min) followed 60 min later by I. V. tolbutamide (1 gm in 30 sec). The patients were then stabilized on diet and "monocomponent" pork insulin of medium duration of action*. This insulin of low antigenicity was used to minimize the production of antibodies which would interfere with the subsequent interpretation of insulin assay data. After 2 to 6 months (mean 3.5), the patients were retested following the same protocol, after an overnight fast and having had no insulin since the previous morning. Three subjects had detectable insulin antibodies at the time of the second test and their B cell responsiveness was evaluated measuring C-peptide.

B Cell Response: All diabetics failed to show an insulin response to glucose when first tested (a criterion for entering into the study), but some responded to tolbutamide. At the time of the second test, the 10 subjects clearly fell into 2 groups (Fig. 3). Group A consisted of 5 subjects in whom the glucose disappearance rate or KGTT** improved during the second test, as indicated by an insulin response to I. V. glucose greater than 10 mM/1 (or an equivalent C peptide

---

* Monotard insulin, Novo Industri A/S

**The slope of the logarithm of the glucose level between 5 and 30 minutes after intravenous glucose.

response). Group B consisted of 5 subjects in whom KGTT did not improve. In these subjects there was no restoration of insulin (or C peptide) response to I. V. glucose during the second test.

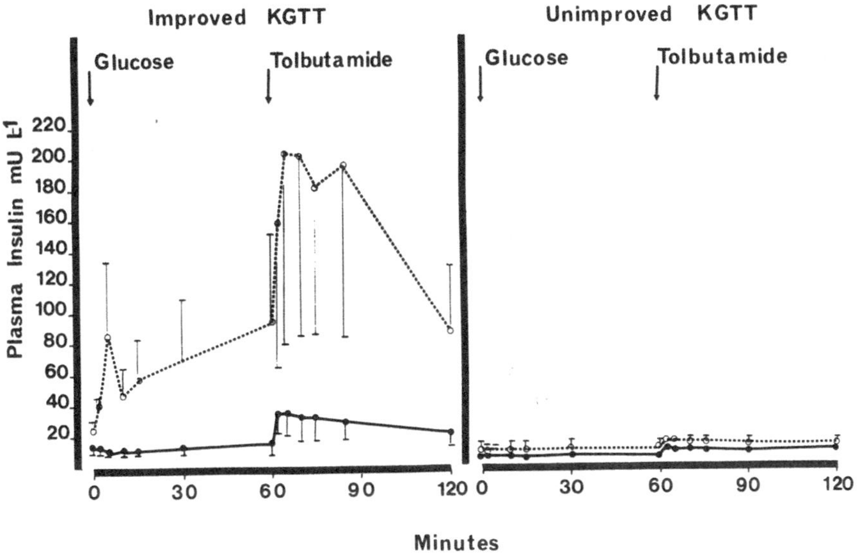

Fig. 3. Plasma insulin responses according to the change in KGTT. Left panel, patients whose KGTT improved (Group A). Right panel, patients whose KGTT did not improve (Group B). First test o——o, second test o-----o. Only data from the 7 patients whose plasma insulin could be assayed are shown. Vertical lines represent standard error of the mean (19, with permission).

A Cell Response: The basal plasma glucagon levels of our diabetic patients were grossly elevated at the time of the first test (28.8 ± 5.1 as compared to 8.3 ± 2.8 pmol/l in 5 healthy non-obese controls of similar mean age and weight, $p < .02$) and were not suppressed by I. V. glucose. However, at the time of the second test, the basal plasma glucagon concentrations had returned to the normal range and, more importantly, they were suppressed by I. V. glucose. This was true both of group A, in which plasma insulin rose and of group B in which plasma insulin did not (Fig. 4). Glucagon concentrations were reduced to <3 pmol/l (limit of detection of the assay) after I. V. glucose in 9 of the 10 subjects and, in particular, in all patients of group B. Thus, in group B, normal A cell suppression by I. V. glucose occurred in the absence of any rise in serum insulin and when the basal insulin levels were in the normal fasting range (4, 13 and 17 mU/l in the 3 subjects who had no insulin antibodies).

The two questions posed earlier can now be answered, i.e., the A cell may be suppressed in human diabetes when insulin levels are well within the physiologic range and an acute increase in insulin levels is not essential for suppression of the A cell. It would appear that insulin therapy restores the responsiveness of the A cell to the hyperglycemic "message." Recent _in vitro_ studies suggest that this effect of insulin on the A cell may be indirect (13).

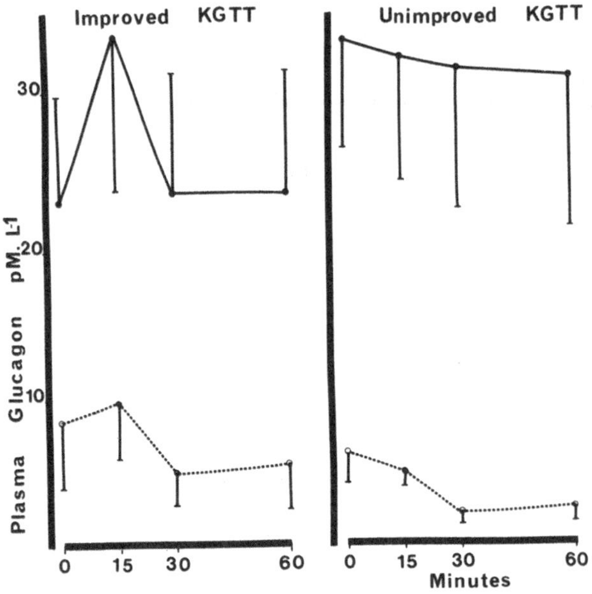

Fig. 4. Plasma glucagon responses according to the change in KGTT. Symbols are as in Fig. 3 (9, with permission).

A Cell Response During Insulin Therapy for Diabetic Ketoacidosis or Hyperosmolar Coma

A cell function was examined also in states of severe metabolic disturbance, i. e., during therapy for diabetic ketoacidosis or hyperosmolar coma (14).

Sequential venous samples were obtained for estimation of glucagon and other relevant biochemical parameters, during 8 episodes of spontaneous ketoacidosis in 7 subjects and during one episode of non-ketotic hyperosmolar coma. All patients except one (E. H.) had been on insulin therapy for diabetes of 9-22 year duration. Their age ranged between 22 and 80 years. The initial blood glucose was 12.2 - 49.0 mmol/l in the ketoacidotic subjects and 58.3 mmol/l in the patient with hyperosmolar coma. The patients were treated with one of two low-dose insulin regimens, i. e., hourly I. M. injection of 8 units neutral regular insulin similar to the regimen of Alberti *et al*. (1) or continuous I. V. infusion of neutral regular insulin (4 to 8 U/h) as suggested by Kidson *et al*. (8) and Page *et al*. (12). Bicarbonate (50-200 meq) was given only to patients with severe acidosis (pH <7.1 or serum bicarbonate <6 mmol/l). Isotonic saline was infused until the blood glucose had fallen to 10-15 mmol/l, when 0.3 M sodium chloride in 4% dextrose or 5% dextrose in H_2O was substituted. Potassium chloride replacement was commenced early, using approximately 20 meq/h and a rate of infusion determined by the plasma levels of potassium. Plasma glucagon levels fell dramatically in the first few hours of therapy in all but one subject (Fig. 5); in 5 subjects glucagon levels fell to the limits of detection of the assay, within 24 hours of initiation of insulin therapy.

Thus, in most episodes of ketoacidosis glucagon levels, which may have been markedly elevated at presentation, fell dramatically following the administration of low doses of insulin and fluid replacement. This indicates that insulin is capable of facilitating the suppression of the A cell even in the presence of a major metabolic derangement.

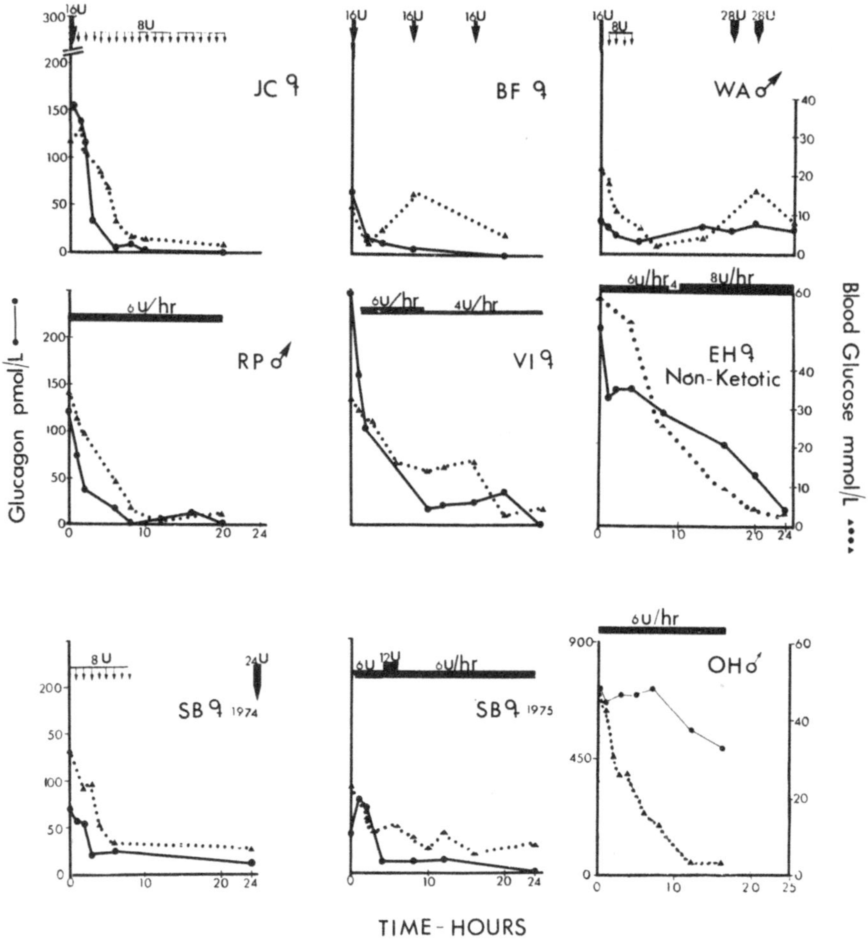

Fig. 5. Changes in blood glucose Δ.......Δ and plasma pancreatic glucagon ●———● in 7 subjects with diabetic ketoacidosis (8 episodes) and in one subject (E.H.) with non-ketotic hyperosmolar coma, during the initial 24 hours of treatment. The insulin dose administered is indicated for each subject (↓ intramuscular injection; solid histogram, intravenous infusion) (14, with permission).

It was also of interest that neither the baseline glucagon concentrations nor the rate of decline of glucagon levels during insulin administration were correlated with the rate of improvement of the hyperglycemia or of the ketoacidosis.

Thus, glucagon undoubtedly accentuates hyperglycemia and probably also aggravates ketogenesis in the presence of severe insulinopenia. However, once adequate insulin therapy is achieved, glucagon levels probably have little influence on the rate of improvement of hyperglycemia and ketosis.

SUMMARY

Not all the questions about A cell function in human diabetes have been answered (16). Our studies strongly suggest that the A cell abnormality in human diabetes is either secondary to insulin lack per se or to the metabolic consequences of insulin deficiency. However, it is possible that with long-standing disturbances of insulin and glucose levels, the A cell (or its neuro-regulatory pathways) may develop "glucose blindness." This defect, not immediately responsive to metabolic correction, might explain the observed failure of the A cell to respond to hypoglycemia in some diabetic patients (5, 6). Nevertheless, the following conclusions can be drawn about A cell behavior in human diabetics:

1. Hyperglucagonemia in uncomplicated human diabetes can be rapidly suppressed by the intravenous administration of insulin. In fact, A cell suppression can be achieved by low doses of insulin in most diabetics even when there is a severe metabolic disturbance.
2. A cell suppression by hyperglycemia is restored by insulin therapy when circulating insulin levels are in the normal fasting range, even though there is no endogenous insulin response to the hyperglycemia.

Thus, there seems little doubt that the abnormality of A cell function commonly seen in human diabetics is secondary to chronic insulinopenia and is not a primary defect of the A cell itself.

Acknowledgements

This work was supported by the National Health and Medical Research Council of Australia. The clinical studies were carried out in collaboration with Drs. R. G. Larkins, F. I. R. Martin, D. A. Perry-Keene, G. L. Warne and J. Court of the Royal Melbourne and Royal Children's Hospitals, Melbourne.

REFERENCES

1. Alberti, K. G. M., Hockaday, T. D. R. and Turner, R. C. (1973): Small doses of intramuscular insulin in the treatment of diabetic coma. Lancet 2, 515-521.

2. Alford, F. P., Bloom, S. R. and Nabarro, J. D. N. (in press): Glucagon levels in normal and diabetic subjects: Use of a specific immunoabsorbent for glucagon radioimmunoassay. Diabetologia 13, 1-6.

3. Aguilar-Parada, E., Eisentraut, A. M. and Unger, R. H. (1969): Pancreatic glucagon secretion in normal and diabetic subjects. Am. J. Med. Sci. 257, 415-419.

4. Braaten, J. T., Faloona, G. R. and Unger, R. H. (1974): The effect of insulin on the alpha cell response to hyperglycemia in long-standing alloxan diabetes. J. Clin. Invest. 53, 1017-1021.

5. Campbell, L. V., Kraegen, E. W., Lazarus, L., Meler, H., Chia, Y. O., Chisholm, D. and Alford, F. (1976): Defective hypoglycemia counter regulation during physiological insulin infusion. Current Topics in Diabetes Research. Abstr. IX Congr. Int. Diabetes Fed., Excerpta Med., I. C. S. 400, p. 35.

6. Gerich, J. E., Langlois, M., Noacco, C., Karam, J. H. and Forsham, P. H. (1973): Lack of glucagon response to hypoglycemia in diabetes: evidence for an intrinsic pancreatic alpha cell defect. Science 182, 171-173.

7. Gerich, J. E., Tsalikian, E., Lorenzi, M., Schneider, V., Bohannon, N. V., Gustafson, G. and Karam, J. H. (1975): Normalization of fasting hyperglucagonemia and excessive glucagon responses to intravenous arginine in human diabetes mellitus by prolonged infusion of insulin. J. Clin. Endocrinol. Metab. 41 1178-1180.

8. Kidson, W., Casey, J., Kraegen, E. and Lazarus, L. (1974): Treatment of severe diabetes mellitus by insulin infusion. Brit. Med. J. 2, 691-694.

9. Larkins, R. G., Martin, F. I. R., Alford, F. P. and Chisholm, D. J. (1976): Effect of prolonged metabolic control on insulin (or C-peptide), glucagon and growth hormone dynamics in ketotic diabetic subjects. Submitted for publication.

10. Müller, W. A., Faloona, G. R., Aguilar-Parada, E. and Unger, R. H. (1970): Abnormal α cell function in diabetes: response to carbohydrate and protein ingestion. N. Engl. J. Med. 283, 109-115.

11. Müller, W. A., Faloona, G. R. and Unger, R. H. (1971): The effect of experimental insulin deficiency on glucagon secretion. J. Clin. Invest. 50, 1992-1999.

12. Page, M. McB, Alberti, K. G. M. M., Greenwood, R., Grumaa, K. A., Hockaday, T. D. R., Lowy, C., Nabarro, J. D. N., Pyke, D. A., Sonksen, P. H., Watkins, P. J. and West, T. E. T. (1974): Treatment of diabetic coma with continuous low-dose infusions of insulin. Brit. Med. J. 2, 687-690.

13. Pagliara, A. S., Stillings, S. N., Haymond, M. W., Hover, B. A. and Matschinsky, F. M. (1975): Insulin and glucose as modulators of the amino acid-induced glucagon release in the isolated pancreas of alloxan and streptozotocin diabetic rats. J. Clin. Invest. 55, 244-255.

14. Perry-Keene, D. A., Alford, F, P., Chisholm, D. J., Findlay, D. M., Larkins, R. G. and Martin, F. I. R. Glucagon and diabetes. I. The failure of hyperglucagonaemia to influence the response of established diabetic ketoacidosis to therapy. Clin. Endocrinol. 6, 417-423.

15. Raskin, P., Fujita, Y. and Unger, R. H. (1975): Effect of insulin-glucose infusions on plasma glucagon levels in fasting diabetics and nondiabetics. J. Clin. Invest. 56, 1132-1138.

16. Unger, R. H. (1976): Diabetes and the alpha cell. Diabetes 25, 136-151.

17. Unger, R. H., Madison, L. L. and Müller, W. A. (1972): Abnormal alpha cell function in diabetics, response to insulin. Diabetes 21, 301-307.

18. Unger, R. H. and Orci, L. (1975): The essential role of glucagon in the pathogenesis of the endogenous hyperglycemia of diabetes mellitus. Lancet 1, 14-16.

19. Warne, G. L., Alford, F. P., Chisholm, D. J., Court, J. (in press): Glucagon and diabetes. II. Complete suppression of glucagon by insulin in human diabetes. Clin. Endocrinol.

THE NATURE OF HYPERGLUCAGONEMIA IN DIABETES MELLITUS

K. Nonaka, H. Toyoshima, T. Yoshida, T. Matsuyama, S. Tarui
and M. Nishikawa

The Second Department of Internal Medicine
Osaka University Medical School
Osaka, Japan

The nature of hyperglucagonemia in diabetes mellitus was investigated in diabetic patients and in experimental animals. The following characteristics of diabetic hyperglucagonemia were demonstrated: decreased suppressibility by hyperglycemia, decreased response to insulin induced hypoglycemia, augmented response to arginine infusion, and easy suppressibility by insulin. Extrapancreatic rather than pancreatic glucagon appeared to be mostly responsible for these features. Glucagon release in response to insulin induced hypoglycemia was found to correlate with B cell function and may be a good test of pancreatic A cell function in human diabetes.

Since the discovery of a high level of immunoreactive glucagon (IRG) in the plasma of totally pancreatectomized dog without insulin treatment (5, 6, 19), it is generally accepted that the pancreas is not the only source of plasma IRG in diabetes. Indeed, a material with hyperglycemic-glycogenolytic properties was found in the gastro-intestinal mucosa many years ago (14), while, more recently, a material indistinguishable from pancreatic glucagon has been isolated from porcine duodenum (13). Morphologic studies have revealed the presence of A cells identical to those of the pancreatic islets, in the stomach (1) and in the digestive tract (3, 9) of mammals. This paper describes our studies on the nature of diabetic hyperglucagonemia under various clinical and experimental conditions done for the purpose of investigating which glucagon, pancreatic or extrapancreatic, plays a major role in the pathogenesis of human diabetes mellitus.

IRG in Diabetes Mellitus

The mean fasting level of plasma IRG of treated diabetic patients was not different from that of normal controls (Table 1). However, diabetics with fasting blood sugar levels above 200 mg/100 ml, had higher IRG values than normal subjects or diabetics with a fasting blood sugar level lower than 200 mg/100 ml.

TABLE 1

FASTING LEVELS OF IRG IN DIABETIC PATIENTS. ALL IRG DETERMINATIONS WERE DONE USING AN ANTIGLUCAGON SERUM SPECIFIC FOR PANCREATIC GLUCAGON (30 K) ACCORDING TO THE METHOD OF YOSHIDA ET AL. (20)

	FBS mg%	Exp. No.	IRG (pg/ml) mean ± SEM	P value
Normal	92 ± 3	7	41 ± 6	N. S.
Diabetes	168 ± 13	35	57 ± 7	
Fasting Blood Glucose	> 200 mg%	15	74 ± 11	P <.05
	≤ 200 mg%	20	44 ± 8	
Duration	> 1 Year	22	62 ± 8	N. S.
	≤ 1 Year	13	48 ± 12	
Therapy	Diet and/or oral medication	19	51 ± 9	N. S.
	Insulin	16	64 ± 11	

In general, the level of IRG was proportional to the severity of the disease, especially in the presence of ketoacidosis (Fig. 1). This hyperglucagonemia was readily normalized with small doses of insulin and the normalization of IRG preceded that of blood sugar. It is possible that this easy suppressibility applies primarily to extrapancreatic IRG since pancreatic IRG in adult type diabetes has been found to be rather resistant to insulin therapy (16).

The IRG response to l-arginine was also exaggerated in diabetes (Fig. 2). In our experience, the magnitude of the response appeared related to the degree of insulin deficiency, as indicated by the degree of metabolic control of the patient. Indeed, the augmented response of IRG to l-arginine in newly diagnosed diabetic patients became less marked after treatment (Fig. 3). In this regard, there was no difference between dietary therapy and sulfonylurea therapy (Fig. 4), even though the sulfonylureas have been shown to inhibit glucagon secretion (4, 12).

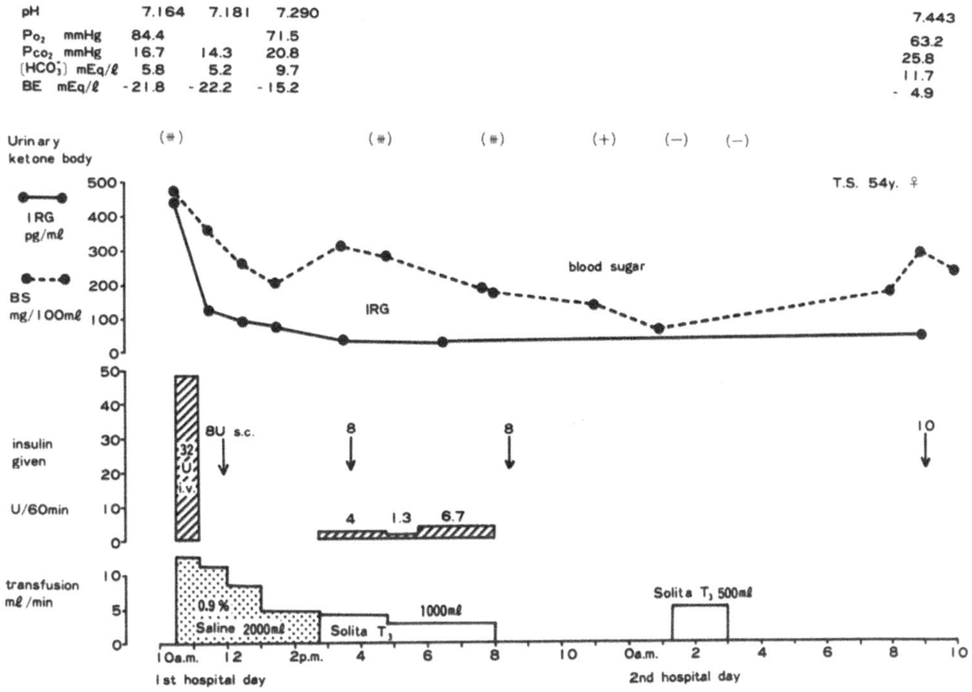

Fig. 1. A case of diabetic ketoacidosis treated with small doses of insulin.

The suppressibility of glucagon by hyperglycemia is reduced in diabetes. To elucidate this phenomenon we investigated whether amino acid-induced IRG secretion could be suppressed by the concomitant administration of glucose . Figs. 5 and 6 show that, in normal subjects, the IRG response to l-alanine infusion was reduced to about one-half by hyperglycemia and that, after cessation of the glucose infusion, an IRG rebound was observed. On the other hand, in diabetes, the response of IRG to l-alanine infusion was not altered by the concomitant hyperglycemia. Nor was a rebound observed. This defect in IRG suppressibility in

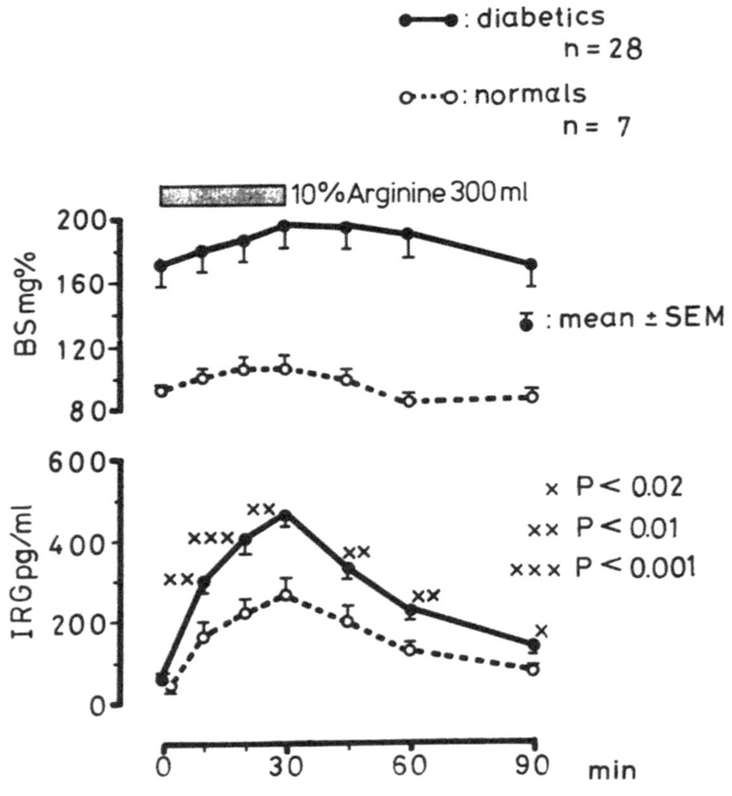

Fig. 2. Plasma IRG and blood glucose responses to arginine infusion in diabetes.

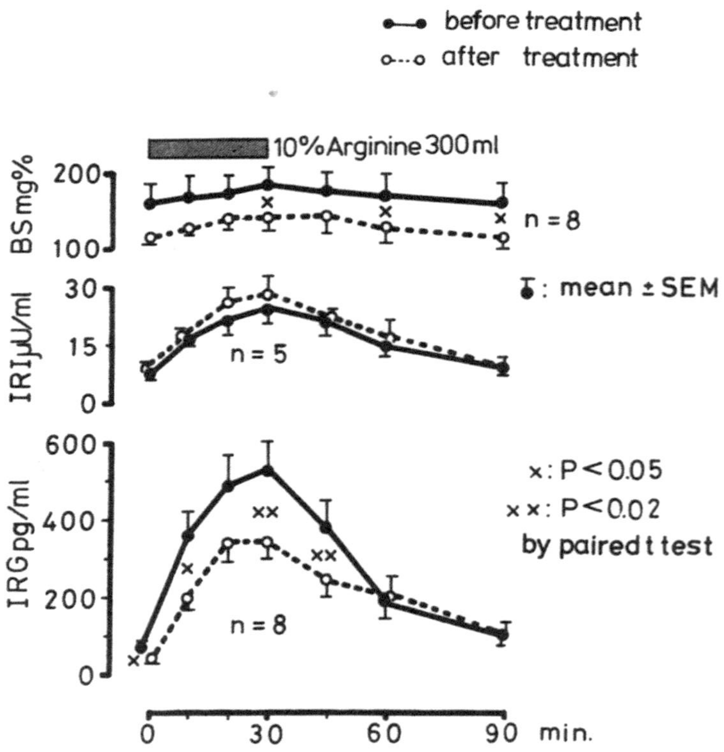

Fig. 3. Plasma IRG and blood glucose responses to arginine infusion in diabetic patients before and after treatment.

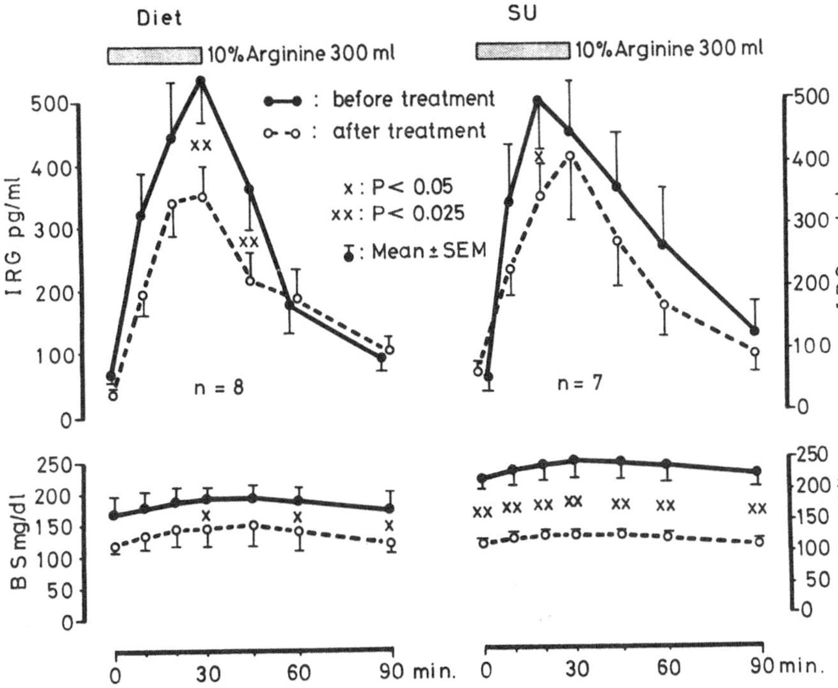

Fig. 4. Improvement of the plasma IRG response to arginine following therapy with diet and a hypoglycemic sulfonylurea.

diabetes could not be correlated with the family history of the patient, suggesting that it was a characteristic of the diabetic state rather than the result of a genetic determinant.

Extrapancreatic IRG

It has been reported that the material crossreacting with antiserum 30 K in a totally pancreatectomized dog is a mixture of at least four fractions (17, 18). Among them is a substance whose physicochemical, immunologic and biologic properties are indistinguishable from those of pancreatic glucagon (13) and which may tentatively be called extrapancreatic glucagon. Though pancreatic and extrapancreatic glucagon may be identical, their releasing mechanisms may differ (2, 7). For example, extrapancreatic glucagon does not appear to be suppressed by hyperglycemia, as indicated by the concomitant presence of hyperglycemia and high

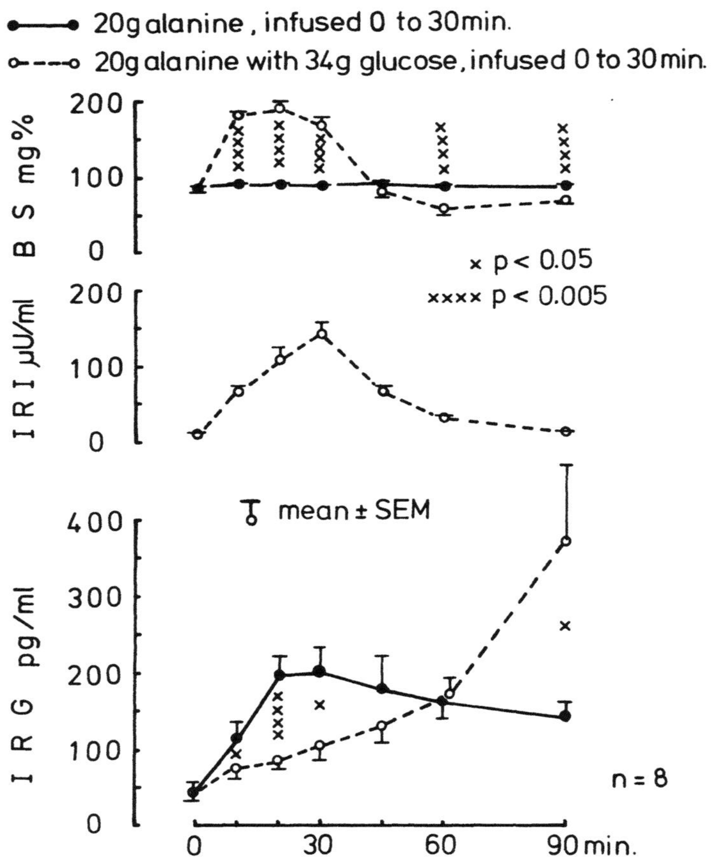

Fig. 5. Plasma IRG, IRI and blood glucose responses to alanine and to alanine with glucose in normal subjects.

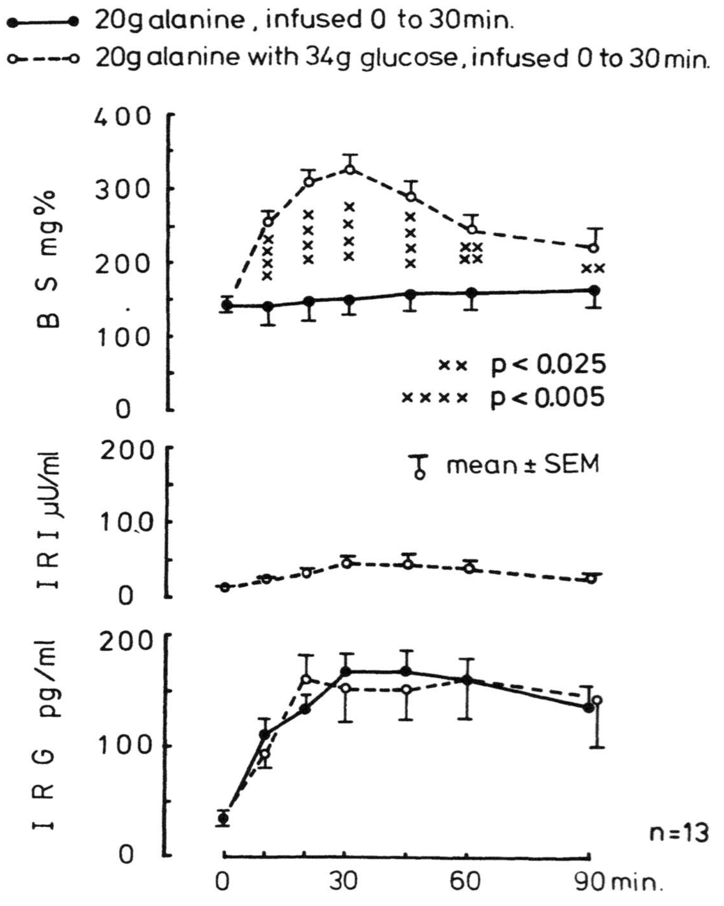

Fig. 6. Plasma IRG, IRI and blood glucose responses to alanine and to alanine with glucose in diabetic subjects.

levels of plasma IRG in totally pancreatectomized animals. In addition, extrapancreatic glucagon does not seem to be released in response to hypoglycemia. Fig. 7 shows that the IRG fails to respond to insulin induced hypoglycemia in rats

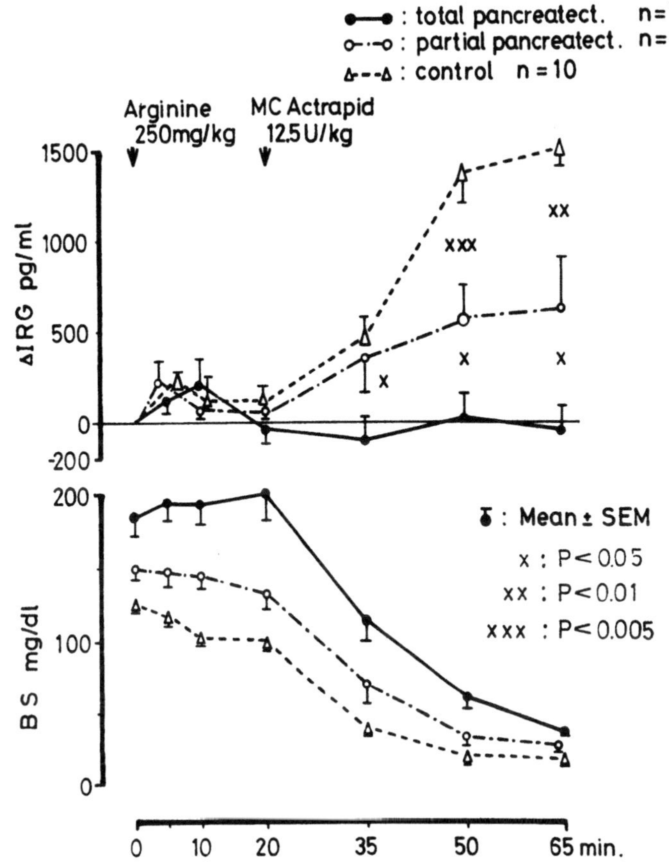

Fig. 7. Plasma IRG and blood glucose responses to intravenously administered arginine and monocomponent insulin in totally and partially pancreatectomized rats and in control animals.

60 min after total removal of the pancreas. Thus, extrapancreatic glucagon does not appear to respond to changes in plasma glucose concentration. On the other hand, it has been reported that extrapancreatic glucagon is suppressed easily by exogenous insulin (6). This observation was confirmed by the data shown in Fig. 1

and by a communication (11) reporting that, in diabetes, IRG responds to insulin administration. This phenomenon may reflect the behavior of extrapancreatic glucagon. In addition, extrapancreatic IRG is released in response to l-arginine infusion, as demonstrated by Matsuyama et al. in totally pancreatectomized dogs (6). A similar phenomenon was observed also in some, but not all, totally pancreatectomized rats (Fig. 8), and in totally pancreatectomized man (10). We have obtained some clinical evidence that the IRG released in response to arginine infusion may derive from extrapancreatic sources. In a case of acute diabetes, in whom Coxsackie virus, group B_4 infection was highly probable, the plasma IRG value was as high as 440 pg/ml and the IRI level was only 12 μu/ml despite a blood sugar

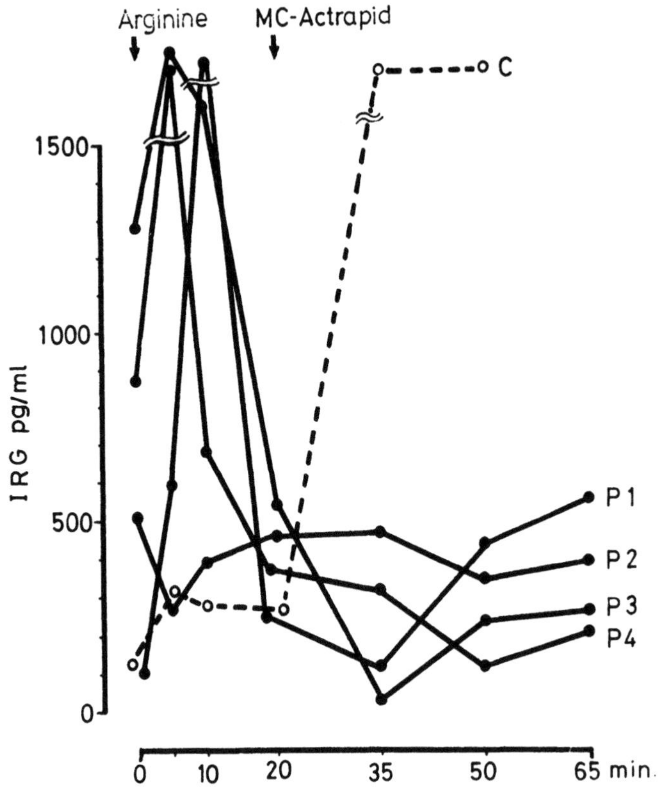

Fig. 8. Plasma IRG responses to intravenous arginine and monocomponent insulin in totally pancreatectomized and control rats.

level of 588 mg/100 ml (15). Considering that in this case the cause of diabetes was probably viral-induced insulitis, it is possible that the circulating IRG may have been of extrapancreatic origin. Moreover, in another patient with chronic calcific pancreatitis and severe brittle diabetes, the IRG response to an arginine load was almost normal. At necropsy, the pancreas weighed only 10 grams and the islets of Langerhans had almost disappeared, again suggesting that the arginine response may have been due to extrapancreatic glucagon. These facts suggest that the IRG released in response to an arginine load in severe insulin dependent diabetes is derived from extrapancreatic sources and that this response is not specific for pancreatic glucagon.

If this assumption is correct, it follows that the characteristics of circulating IRG in diabetes mellitus represent the characteristics of extrapancreatic glucagon rather than those of pancreatic glucagon. What is then the behavior of pancreatic glucagon in diabetes mellitus? In order to answer this question, one must devise a method to detect only pancreatic glucagon.

Pancreatic Glucagon in Diabetes

Since there is no extrapancreatic glucagon response to insulin-induced hypoglycemia, this response may be considered specific for pancreatic A cell function in diabetes. The test was performed by injecting monocomponent insulin (Novo) in doses sufficient to lower blood glucose to less than 40 mg/100 ml. Figs. 9-14 show the results (8). In diabetic patients, after an initial small decrease (Fig. 9), the IRG response was weaker than in normal subjects. The average of maximal increment of IRG (ΔIRG) irrespective of the time at which it occurred, was also significantly smaller in diabetes. This impairment of the IRG response to insulin-induced hypoglycemia was greater in juvenile onset than in adult onset diabetes (Fig. 10) and in diabetics with a fasting blood sugar level higher than 200 mg/100 ml than in those with a level lower than 200 mg/100 ml (Fig. 11).

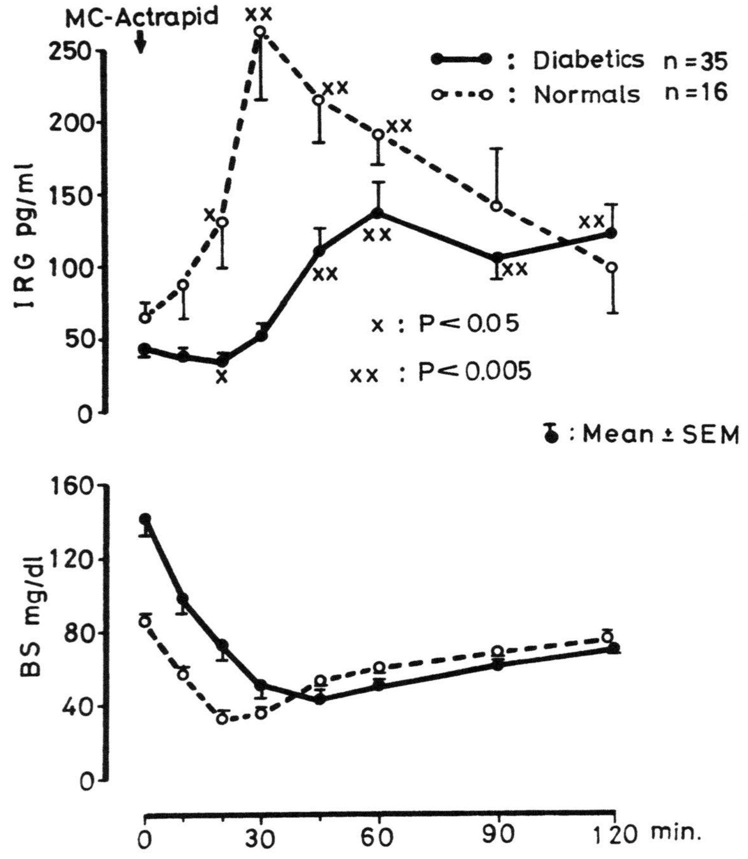

Fig. 9. Plasma IRG response to intravenous monocomponent insulin in diabetic patients.

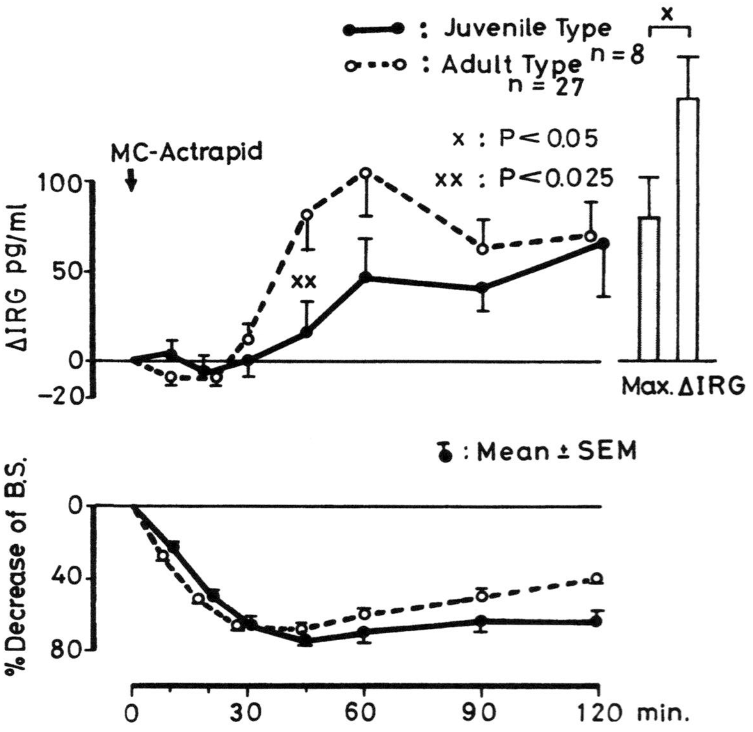

Fig. 10. Plasma IRG increment after the intravenous injection of monocomponent insulin in juvenile and adult onset diabetes. The bars at the right represent the mean maximal increment of IRG (Max. ΔIRG).

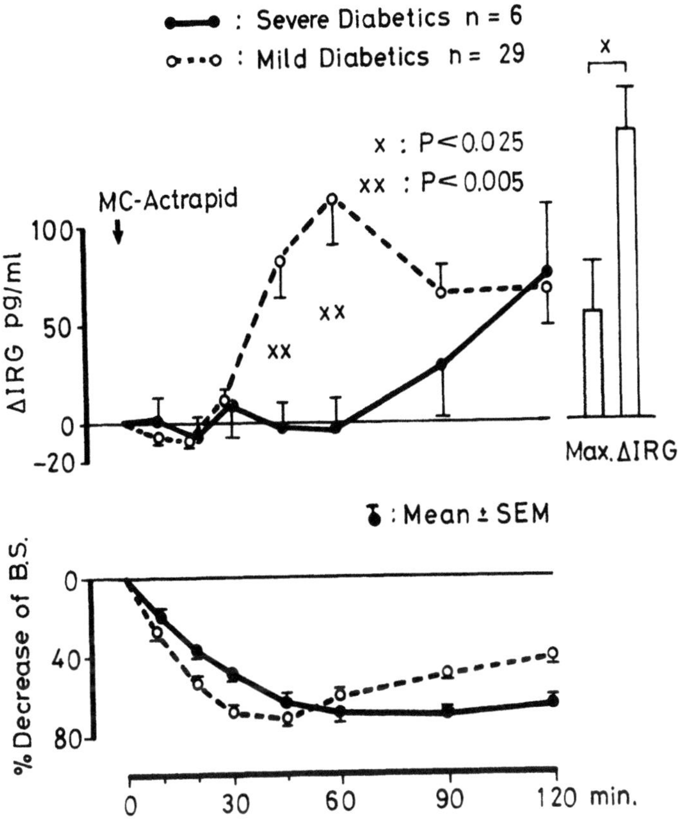

Fig. 11. Plasma IRG increment after the intravenous injection of monocomponent insulin in severe (FBS ≧ 200 mg%) and mild (FBS <200 mg%) diabetic patients. The bars at the right represent the mean maximal increment of IRG (Max. ΔIRG).

These results suggest that the severity of diabetes may correlate with the IRG response to insulin induced hypoglycemia. Indeed, a statistically significant negative correlation was found between the areas under the 90 minute IRG response curve (ΣΔIRG) and the fasting blood sugar level (Fig. 12). Since a major factor determining the severity of diabetes is the magnitude of insulin lack, we correlated A and B cell function by studying the relationship between the area under the IRI curve during a 180 min oral glucose tolerance test (ΣΔIRI) with the ΣΔIRG during the 90 minute insulin-induced hypoglycemia. As shown in

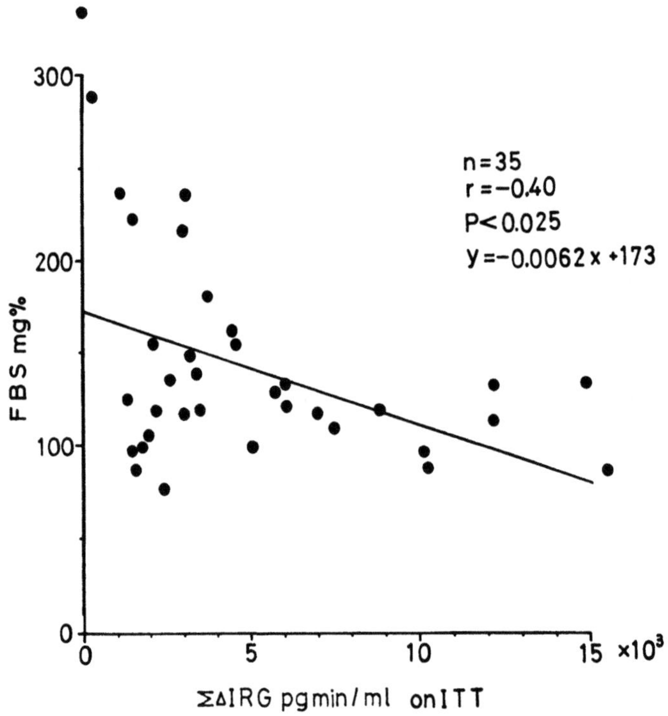

Fig. 12. Correlation between fasting blood sugar and ΣΔIRG following the intravenous injection of monocomponent insulin in diabetic patients.

Fig. 13, there was a significant positive correlation between these two parameters. Similarly, a significant correlation was found using the ratio of the insulin to the sugar increment during the first 30 minutes of the oral glucose tolerance test (ΔIRI/ΔBS), as a measure of B cell function (Fig. 14). Thus, the responsiveness of the cell and that of the B cells to changes in blood glucose concentration are intimately related to each other and the function of both cells is impaired to the same extent in diabetes mellitus.

CONCLUSIONS

The characteristics of immunoreactive glucagon (IRG) in diabetes mellitus are: 1) relative or absolute hyperglucagonemia in the presence of hyperglycemia; namely,

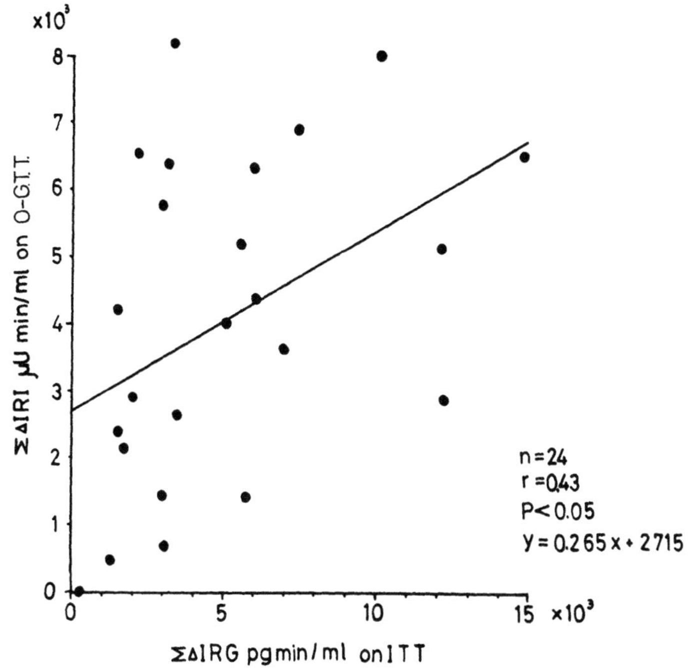

Fig. 13. Correlation between ΣΔIRI during an oral glucose tolerance test and ΣΔIRG following the intravenous injection of monocomponent insulin in diabetic patients.

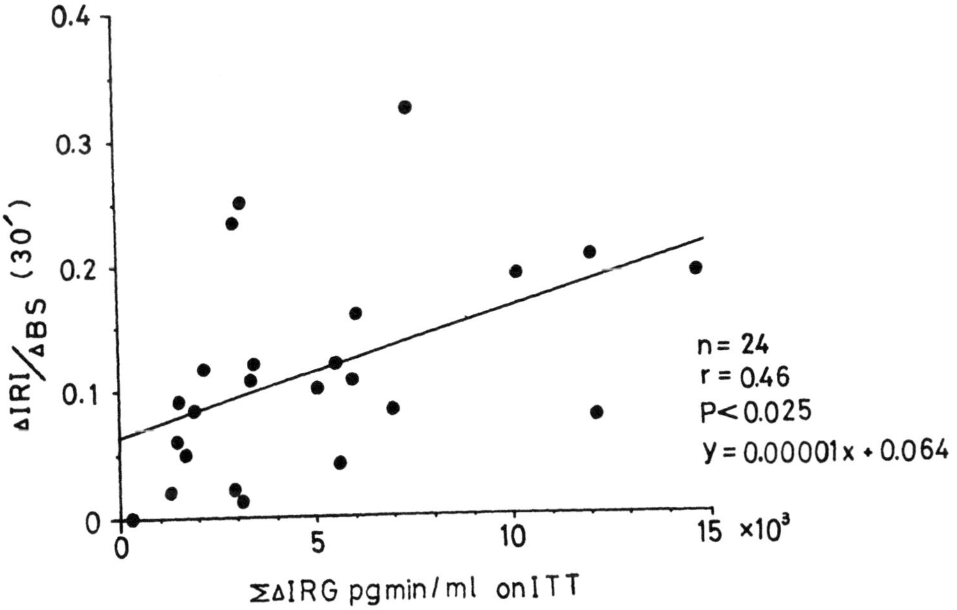

Fig. 14. Correlation between ΔIRI/ΔBS 30 minutes after the injection of 100 g of glucose and ΣΔIRG following the intravenous injection of monocomponent insulin in diabetic patients.

diminished suppressibility of glucagon secretion by glucose; 2) diminished IRG response to insulin-induced hypoglycemia; 3) augmented IRG response to arginine infusion; 4) easy suppressibility with exogenous insulin.

In totally pancreatectomized animals, these characteristics are shared by extrapancreatic IRG. A positive correlation between the degree of A cell impairment and B cell response to changes in blood glucose concentration, may indicate that, 1) insulin deficiency leads to secondary hypersecretion of extrapancreatic glucagon; 2) diabetic hyperglucagonemia involves mainly extrapancreatic glucagon and, to a lesser degree, pancreatic glucagon; 3) IRG released in response to insulin-induced hypoglycemia is specific for pancreatic glucagon and may be useful to study pancreatic A cell function in diabetes mellitus.

REFERENCES

1. Baetens, D., Rufener, C., Srikant, B. C., Dobbs, R., Unger, R. H. and Orci, L. (1976): Identification of glucagon producing cells (A-cells) in dog gastric mucosa. J. Cell. Biol. 69, 455-464.

2. Blázquez, E., Munoz-Barragan, L., Patton, G. S., Orci, L., Dobbs, R. S. and Unger, R. H. (1976): Gastric A-cell function in insulin-deprived depancreatized dogs. Endocrinology 99, 1182-1188.

3. Larsson, L.-I., Holst, J., Hakanson, R. and Sundler, F. (1975): Distribution and properties of glucagon immunoreactivity in the digestive tract of various mammals. Histochemistry 44, 281-290.

4. Loreti, L., Sugase, T. and Foà, P. P. (1974): Diurnal variations of serum insulin, total glucagon, cortisol, glucose and free fatty acids in normal and diabetic subjects before and after treatment with chlorpropamide. Horm. Res. 5, 278-292.

5. Mashiter, K., Harding, P. E., Chou, M., Mashiter, G. D., Stout, J., Diamond, D. and Field, J. B. (1975): Persistent pancreatic glucagon but not insulin response to arginine in pancreatectomized dogs. Endocrinology 96, 678-693.

6. Matsuyama, T. and Foà, P. P. (1974): Plasma glucose, insulin, pancreatic and enteroglucagon levels in normal and depancreatized dogs. Proc. Soc. Exp. Biol. Med. 147, 97-102.

7. Munoz-Barragan, L., Blázquez, E., Patton, G. S., Dobbs, R. E. and Unger, R. H. (1976): Gastric A cell function in normal dogs. Am. J. Physiol. 231, 1057-1061.

8. Nonaka, K., Toyoshima, H., Yoshida, T., Matsuyama, T., Tarui, S. and Nishikawa, M. Glucagon response to insulin induced hypoglycemia in diabetes mellitus. In preparation.

9. Orci, L., Pictet, R., Forssman, W. G., Renold, A. E. and Rouiller, C. (1968): Structural evidence for glucagon producing cells in the intestinal mucosa of the rat. Diabetologia 4, 56-67.

10. Palmer, J. P., Werner, P. L., Benson, J. W. and Ensinck, J. W. (1976): Plasma glucagon after pancreatectomy. Lancet 1, 1290.

11. Raskin, P., Fujita, Y. and Unger, R. H. (1975): Effect of insulin-glucose infusions on plasma glucagon levels in fasting diabetics and nondiabetics. J. Clin. Invest. 56, 1132-1138.

12. Samols, E., Tyler, J. and Mialhe, P. (1969): Suppression of pancreatic glucagon release by the hypoglycemic sulfonylureas. Lancet 2, 174-176.

13. Sasaki, H., Rubalcava, B., Baetens, D., Blázquez, E., Srikant, C. B., Orci, L. and Unger, R. H. (1975): Identification of glucagon in the gastrointestinal tract. J. Clin. Invest. 56, 135-145.

14. Sutherland, E. W. and de Duve, C. (1948): Origin and distribution of the hyperglycemic glycogenolytic factor of the pancreas. J. Biol. Chem. 175, 663-674.

15. Toyoshima, H., Nonaka, K., Kuroda, K., Kurimura, T., Otsu, K. and Tarui, S. (1976): Coxsackie B_4 virus neutralizing antibody titers in sera of diabetic subjects, with special reference to the relation between titers and clinical courses in two cases with acute insulin-dependent diabetes followed by the immediate complete remission. J. Japan Diab. Soc. 19, 203-211.

16. Unger, R. H., Madison, L. L. and Müller, W. A. (1972): Abnormal alpha cell function in diabetics: response to insulin. Diabetes 21, 301-307.

17. Valverde, I., Dobbs, R. and Unger, R. H. (1975): Heterogeneity of plasma glucagon immunoreactivity in normal, depancreatized and alloxan-diabetic dogs. Metabolism 24, 1021-1028.

18. Vinik, A. I. and Jackson, W. P. U. (1975): Glucagon in pathogenesis of diabetes. Lancet 1, 694-695.

19. Vranic, M., Pek, S. and Kawamori, R. (1974): Increased "glucagon immunoreactivity" in plasma of totally depancreatized dogs. Diabetes 23, 905-912.

20. Yoshida, T., Toyoshima, H., Nonaka, K. and Tarui, S. (1975): Application of a correction factor to the radioimmunoassay for plasma pancreatic glucagon. J. Japan Diab. Soc. 18, 156-163.

GLUCAGON IN DIABETIC COMA

L. Lazarus, E. W. Kraegen, H. Meler, M. Zacharin, L. Campbell and J. H. Casey

Garvan Institute of Medical Research, St. Vincent's Hospital
Sydney, Australia

Measurements of glucagon were performed in 31 patients with diabetic ketoacidosis before and during insulin therapy. Marked hyperglucagonemia was recorded in all patients. A significant correlation was found between the basal plasma levels of glucagon and those of glucose ($r = 0.51$, $p < .05$) and ketone bodies ($r = 0.75$, $p < .01$). The highest levels of glucagon were noted in the patients with non-ketotic hyperglycemia. Insulin therapy resulted in a significant suppression of glucagon levels which preceded any significant change in plasma glucose. Nevertheless, the correction of hyperglycemia occurred whilst absolute hyperglucagonemia still persisted. Although these results suggest that insulin has a direct inhibitory effect on glucagon secretion, they raise the question of the role of glucagon in the development of ketosis and hyperglycemia.

Although hyperglucagonaemia has been recorded in patients with diabetic ketosis and non-ketotic hyperglycemic states (1, 8, 11), its role in the production and treatment of the metabolic disturbances of these syndromes is obscure and the subject of debate (2, 5, 9, 13).

In order to study the role of glucagon in diabetic "coma" we have performed measurements of glucagon in a series of 31 patients before and during I.V. insulin therapy at various dose levels and also following insulin given either as an I.V. bolus or as a low-dose infusion.

PATIENTS AND METHODS

There were two groups of patients. One group consisted of 8 insulin requiring diabetic volunteers who were admitted to the hospital and then deprived of insulin for periods of 36 to 72 hours until hyperglycemia and ketosis had appeared. On the morning of the test procedure they were given an I.V. bolus of insulin at a dose of 0.1 U/kg and serial blood samples for assay were taken for 180 minutes, at which time an infusion of insulin was commenced at a rate of 4 U/hour and continued for an additional 180 minutes (7).

The other group consisted of 23 patients admitted to our unit for the treatment of diabetic "coma". There were 18 patients with diabetic ketoacidosis, four with non-ketotic hyperosmolar coma and one with phenformin-induced lactic acidosis. Five of the patients were given an initial I.V. bolus dose of 10 U of insulin followed, at an interval of one hour, by a low dose infusion and the remainder were treated by low dose infusions at rates which ranged from 2.4 U/h to to 9.6 U/h. All of the infusions were administered in polygeline to protect the insulin from adsorption to the I.V. giving set (6). Venous blood samples were handled as follows: 4 ml were collected into ice cold perchloric acid (5% w/v) for measurement of metabolites, 10 ml were collected in a chilled tube containing 5,000 KIU of aprotinin (Trasylol) and 150 U of heparin for measurement of glucagon, insulin, growth hormone and cortisol, 2 ml were collected in a tube with sodium fluoride for glucose determination. The tubes for metabolic and hormone determination were centrifuged immediately in a refrigerated centrifuge at $3^{o}C$ and the separated plasma were frozen until the assays were performed.

Glucose was measured using the AutoAnalyzer (Technicon Corp.) using the ferricyanide method. Cortisol was measured by a competitive protein binding method. Insulin and growth hormone were measured by radioimmunoassay using human standards and the growth hormone (GH) levels are reported in units of the WHO reference preparation. Glucagon was measured by radioimmunoassay using antiserum 30K supplied by Dr. R. Unger. Lactate, pyruvate, acetoacetate and hydroxybutyrate were measured by enzymatic methods (3).

Results are recorded as mean ± SEM. Statistical significance was assessed using the Wilcoxon test for pair differences (14).

RESULTS

Hyperglucagonaemia was recorded in all the patients studied (Fig. 1). The mean basal level in the 8 insulin deprived volunteers was 125 ± 21 pg/ml which is significantly greater than the normal basal level of 45.5 ± 8.5 pg/ml recorded

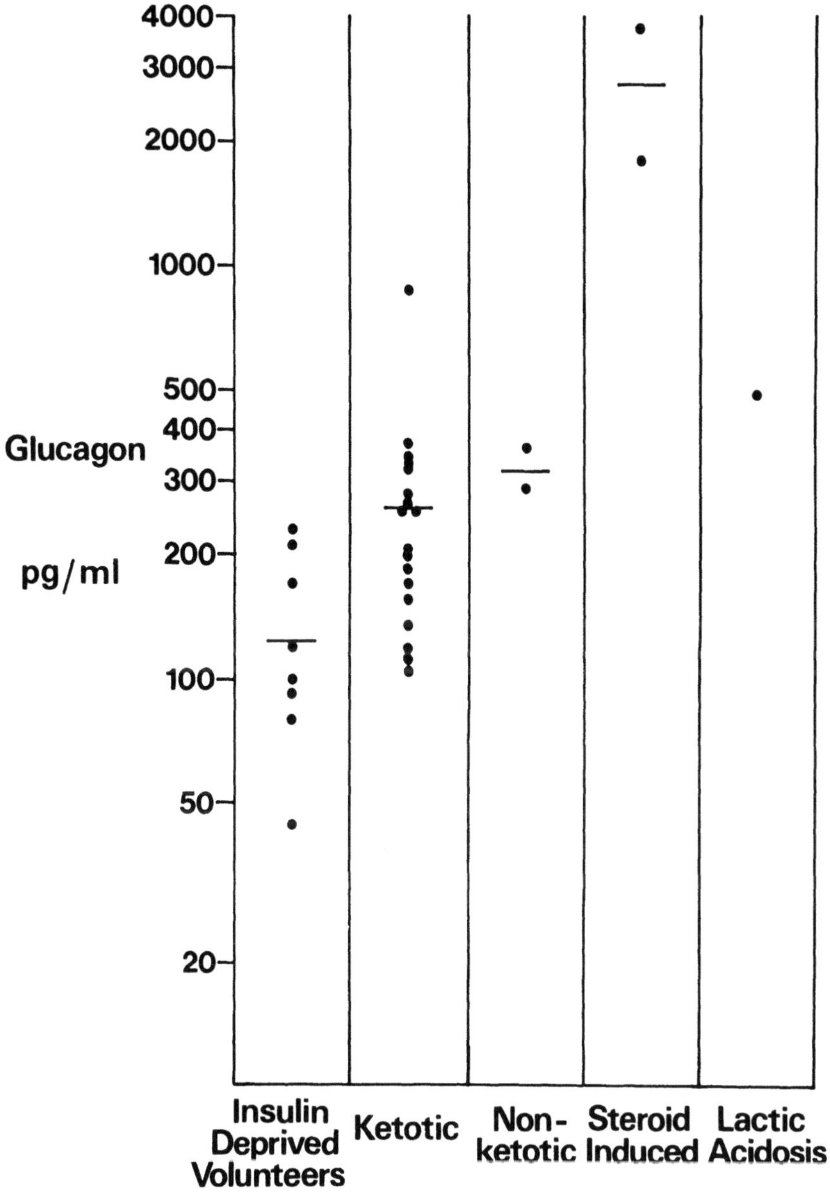

Fig. 1. Plasma glucagon levels in 31 patients with severe diabetic hyperglycaemia.

in our laboratory. The 18 patients with diabetic ketoacidosis had a mean initial glucagon level of 260 ± 41 pg/ml, with a range of 105 to 880 pg/ml. The initial glucagon levels in the four non-ketotic patients ranged from 290 to 3,700 pg/ml, with a mean of 1,540 pg/ml. The highest levels were found in two patients whose hyperglycaemic episode followed the administration of steroids in large doses. One received dexamethasone therapy for possible cerebrovascular thrombosis and the other was receiving prednisone therapy for leukaemia.

Comparison of the basal glucagon levels with the basal glucose levels revealed a significant correlation, $r = 0.51$, $p < .05$. There was also a significant correlation between the basal glucagon level and the plasma ketone bodies in the ketotic patients, $r = 0.73$, $p < .01$.

Administration of insulin to the volunteer group resulted in a significant suppression of glucagon and glucose levels, following the bolus and during the infusion (Fig. 2). In each case the glucagon suppression preceded a significant change in glucose levels. Following the bolus dose of insulin there was a rapid rebound of glucagon to a level not significantly different from the basal by 120 minutes post injection (Fig. 2).

In each of the patients with diabetic coma there was a suppression of glucagon during the infusion of insulin (Fig. 3). In the five given an initial bolus dose of 10 U of insulin there was a rapid drop in glucagon levels with a subsequent rebound as observed in the volunteer group (Fig. 4).

DISCUSSION

These studies revealed hyperglucagonaemia in all patients with diabetic "coma" regardless of the cause. The highest levels of glucagon were noted in the patients with non-ketotic hyperglycaemia and this observation raises questions regarding the role of glucagon in the development of ketosis. The liver of the non-ketotic patients appears to respond quite differently to the hyperglucagonaemia and the recent studies of Clark et al. (4) showing that the hepatic

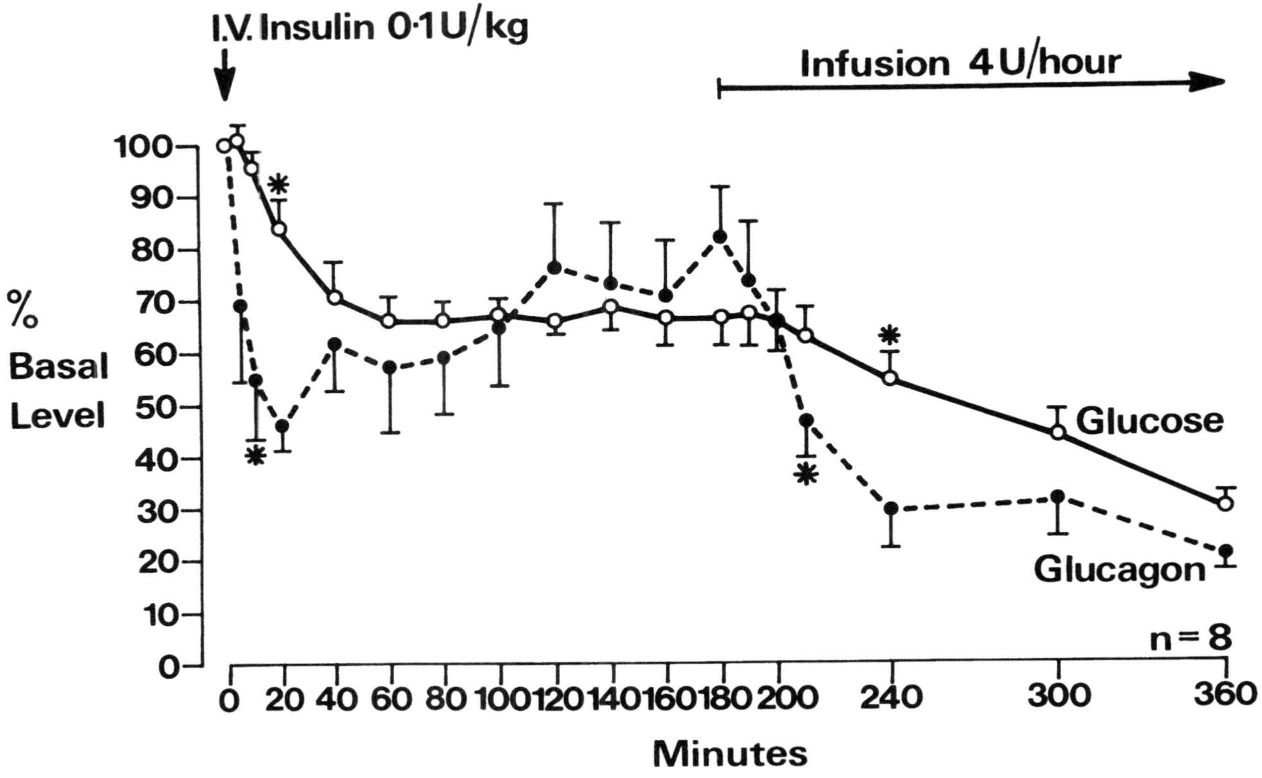

Fig. 2. Plasma glucose and glucagon levels during the bolus dose and infusion studies in 8 insulin-deprived volunteers.

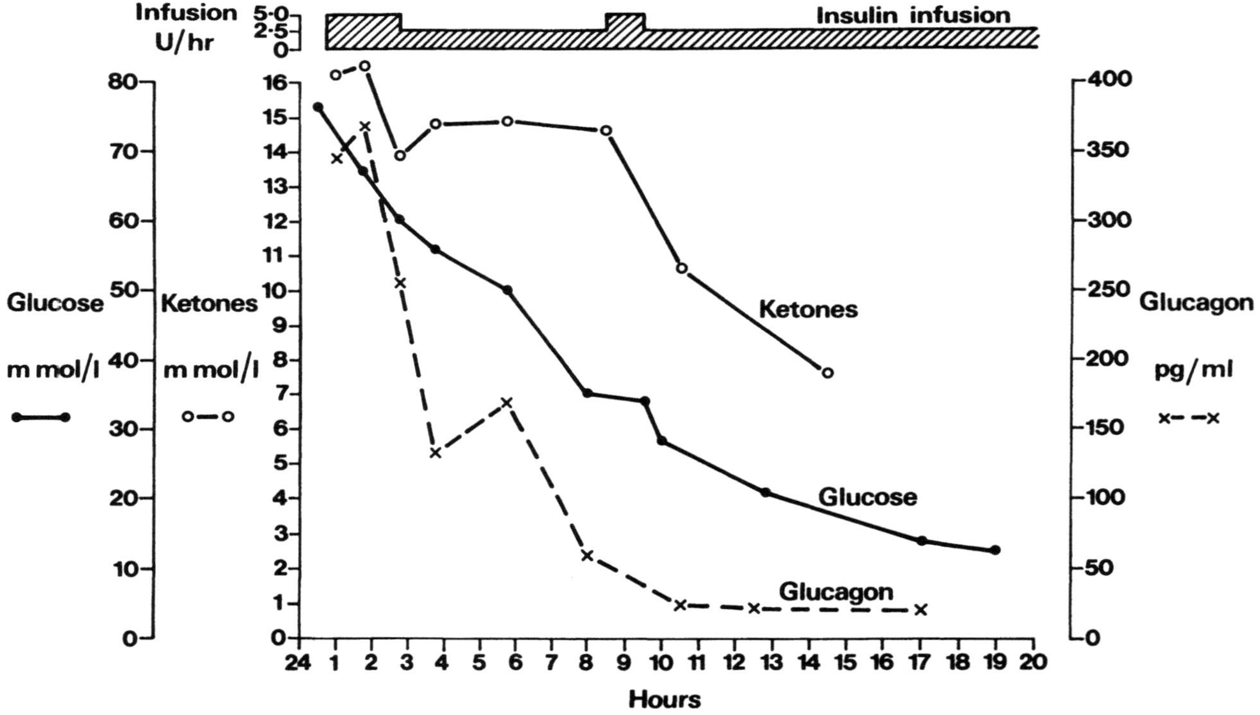

Fig. 3. Plasma glucose, glucagon and ketone body responses to insulin infusion in a patient with diabetic ketoacidosis.

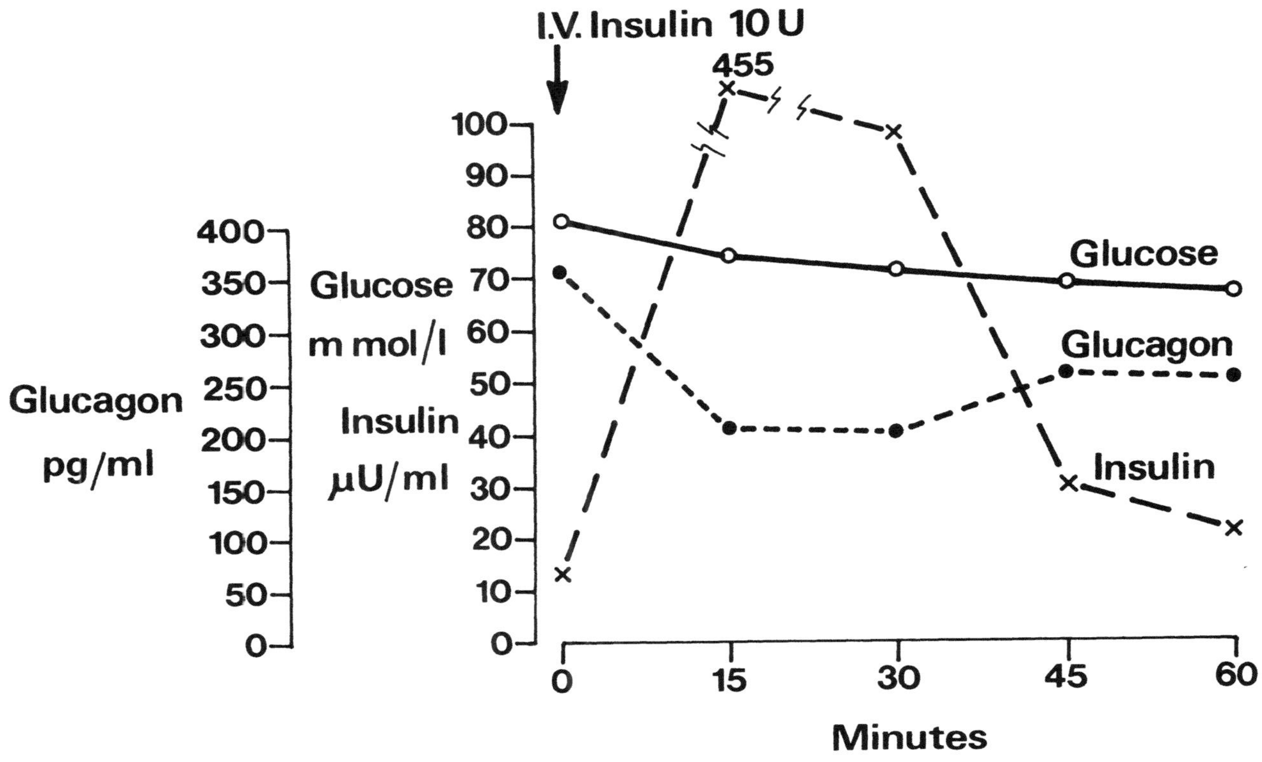

Fig. 4. Plasma glucagon, glucose and growth hormone responses to an insulin bolus in a patient with non-ketotic hyperglycaemia, showing an initial suppression and subsequent recovery.

response to glucagon depends upon the redox state may explain these observations.

In each of the patients there was a suppression of glucagon following the administration of insulin and in both the volunteer group and the coma patients there was evidence of glucagon suppression prior to any glucose change, confirming the earlier work of Samols et al. (12) that there is a direct suppressive effect of insulin on glucagon secretion. The suppressive effect of insulin on glucagon secretion also was dose related, as can be seen in Figure 5, where a decrease in the insulin infusion rate from 4.8 U/h to 2.4 U/h resulted in a significantly decreased rate of response of glucagon in each of three patients.

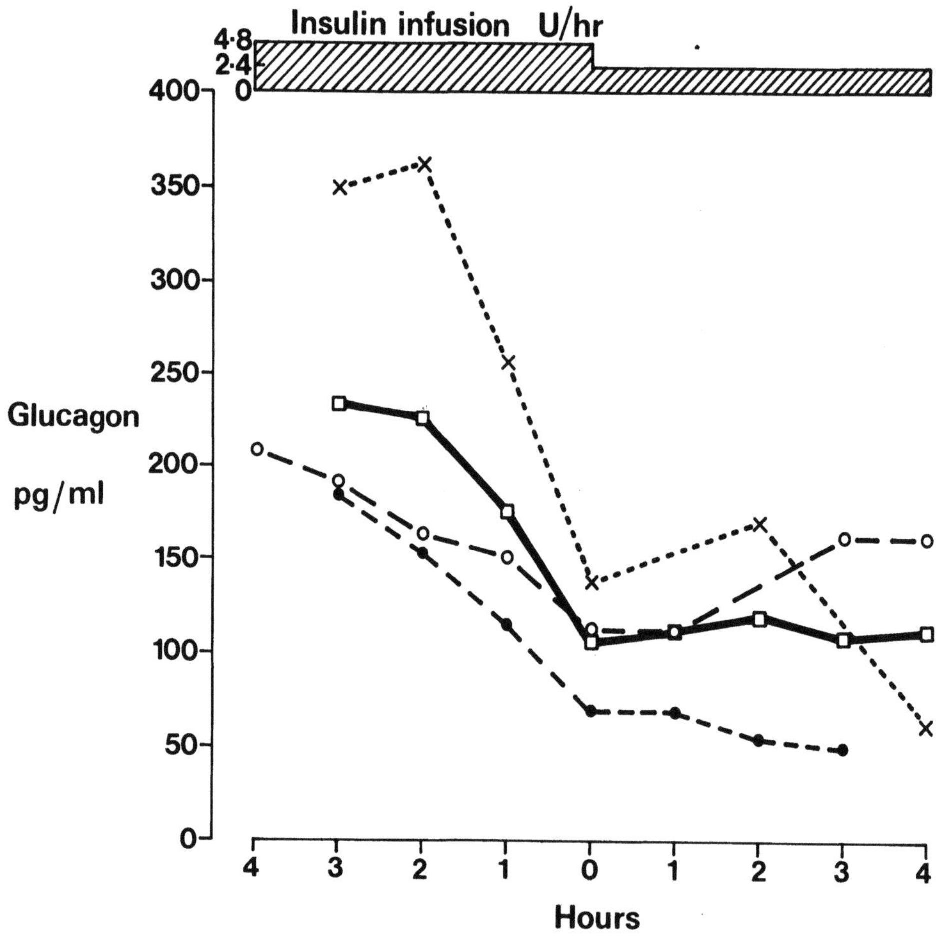

Fig. 5. Plasma glucagon responses in three patients to a decrease in insulin infusion rate from 4.8 U/h to 2.4 U/h. The thick line represents the mean.

Although the hyperglucagonaemia was suppressed in all patients by the insulin therapy it was interesting to note the correction of hyperglycaemia occurring whilst absolute hyperglucagonaemia still persisted. Such a result is shown in Figure 6 where it can be seen that the glucose level dropped from 81 to 16 mmol/l whilst the glucagon had dropped from 3,700 pg/ml to only 900 pg/ml. Although hyperglucagonaemia was still present, one assumes that the high levels of insulin (100 μU/ml) were sufficient to overcome the gluconeogenetic action of the glucagon.

These studies confirm the presence of hyperglucagonaemia in all patients with diabetic hyperglycaemia regardless of cause. They also confirm earlier studies suggessive effect of insulin on glucagon secretion. The role of glucagon in the production of the hyperglycaemia and ketonaemia remains obscure, but it is suggested that the altered responses may depend upon the hepatic redox state.

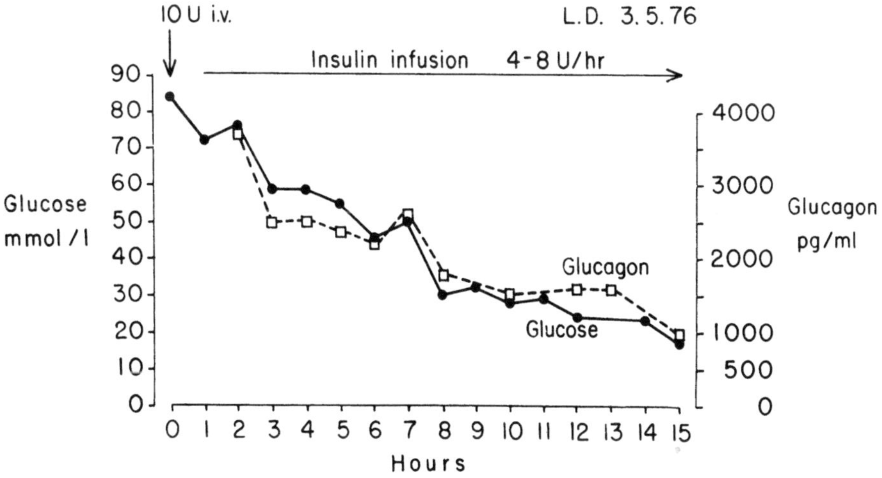

Fig. 6. Plasma glucagon and glucose levels in a patient with non-ketotic hyperosmolar coma. Note the correction of the hyperglycaemia when the plasma glucagon level was still 900 pg/ml.

Acknowledgement

This work was supported by a grant from the National Health and Medical Research Council of Australia.

REFERENCES

1. Assan, R., Hautecouverture, G., Guillemant, S., Dauchy, F., Protin, P. and Dérot, M. (1969): Evolution de paramètres hormonaux (glucagon, cortisol, hormone somatotrope) et énergétiques (glucose, acides gras, glycérol libre) dans dix acido-cétoses diabétiques graves traitées. Path. Biol. 17, 1095-1105.

2. Barnes, A. J. and Bloom, S. R. (1976): Pancreatectomized man: a model for diabetes without glucagon. Lancet 1, 219-221.

3. Bergmeyer, H. V. (1973): Methods of Enzymatic Analysis, 2nd English Edition, Academic Press, New York.

4. Clark, M. G., Filsell, O. H. and Jarrett, I. G. (1976): Gluconeogenesis in isolated intact lamb liver cells. Effects of glucagon and butyrate. Biochem. J. 156, 671-680.

5. Gerich, J. E., Lorenzi, M., Bier, D. M., Schneider, V., Tsalikian, E., Karam, J. H. and Forsham, P. H. (1975): Prevention of human diabetic ketoacidosis by somatostatin: evidence for an essential role of glucagon. New Engl. J. Med. 292, 985-989.

6. Kraegen, E. W., Lazarus, L., Meler, H., Campbell, L. and Chia, Y. O. (1975): Carrier solutions for low level intravenous insulin infusion. Brit. Med. J. 3, 464-466.

7. Lazarus, L., Kraegen, E. W., Campbell, L., Casey, J. H., Zacharin, M. and Wienholt, J. (1976): Submitted for publication.

8. Lindsey, C. A., Faloona, G. R. and Unger, R. H. (1974): Plasma glucagon in non-ketotic hyperosmolar coma. JAMA 229, 1771-1773.

9. Lundbaek, K., Christensen, S. E., Hansen, A. P., Iversen, J., Ørskov, H., Seyer-Hansen, K., Alberti, K. G. M. M. and Whitefoot, R. (1976): Failure of somatostatin to correct manifest diabetic ketoacidosis. Lancet 1, 215-218.

10. Marco, J., Calli, C., Roman, D., Diaz-Fierros, M., Villanueva, M. L. and Valverde, I. (1973): Hyperglucagonism induced by glucocorticoid treatment in man. New Engl. J. Med. 288, 128-132.

11. Müller, W. A., Faloona, G. R. and Unger, R. H. (1973): Hyperglucagonemia in diabetic ketoacidosis. Am. J. Med. 54, 52-57.

12. Samols, E., Tyler, J. M. and Marks, V. (1972): Glucagon-insulin interrelationships. In: Glucagon. Molecular Physiology, Clinical and Therapeutic Implications. P. J. Lefèbvre and R. H. Unger, Eds. Pergamon Press, New York, pp. 151-173.

13. Unger, R. H. (1976): Diabetes and the alpha cell. Diabetes 25, 136-151.

14. Wilcoxon, F. (1947): Probability tables for individual comparisons by ranking methods. Biometrics 3, 119-122.

GLUCAGON METABOLISM IN POTENTIAL DIABETES MELLITUS. STUDIES OF DISCORDANT TWINS AND PATIENTS AFTER MYOCARDIAL INFARCTION

J. L. Day

The Ipswich Hospital
Ipswich, Great Britain

Plasma glucagon responses to glucose have been restudied in 12 unaffected twins of diabetics after an interval of 5 years and contrasted with 12 normal (N) and 12 diabetic subjects (D). Once more their glucagon responses were intermediate between N and D and showed inadequate suppression by oral glucose. No significant differences were observed between those twins discordant for more or less than 10 years. Twenty-two survivors of myocardial infarction were studied over a 6 months period on 3 occasions in order to identify glucagon relationships during variable glucose tolerance. In all 3 tests those whose glucose tolerance was normal throughout ("Normal") showed suppression of plasma glucagon-like activity by oral glucose. Those with impaired glucose tolerance ("Diabetic") showed stimulation and those whose glucose tolerance improved from abnormal to normal ("Potential diabetics") showed impaired suppression similar to that observed in the twins. No significant differences between the three tests were observed in any of the 3 groups. It would appear that those subjects showing potential carbohydrate intolerance after myocardial infarction and discordant twins share a defect in A cell function which by itself does not produce carbohydrate intolerance but may contribute to its development in the presence of other aetiological stresses.

The well known genetic component in subjects with diabetes mellitus has prompted numerous attempts to identify biochemical markers in subjects genetically predisposed to develop the disease. Although twins and offspring of diabetics, as a whole, may show different glucose profiles from normal subjects, these differences are inconsistent. Likewise, convincing evidence of abnormalities of insulin secretion prior to the development of carbohydrate intolerance is unavailable (7).

Higher plasma growth hormone concentrations have been detected in subjects with diabetic potential but the pathologic significance of these are unclear (1, 8). In view of the hypersecretion of glucagon in diabetes mellitus (2, 3, 9), considerable interest has been evoked by the possibility of abnormalities of glucagon metabolism in potential diabetics. Earlier observations of absent suppressibility of plasma glucagon by oral glucose in a group of non-diabetic, monozygotic twins of diabetics (4) have been supported by studies showing enhanced glucagon response to arginine (5) and failure of suppression by intravenous glucose (6) in offspring and first degree relatives of diabetics, respectively.

Two studies have been undertaken to clarify the role of glucagon in subjects with potential diabetes. The first was a re-examination of the monozygotic twins (4) after a five-year period to determine not only whether the previously observed glucagon abnormalities persisted or worsened, but also the relationship of any changes to changes in carbohydrate tolerance. The second was a study of a group of patients following myocardial infarction, a significant proportion of whom have a temporary carbohydrate intolerance which returns to normal without specific antidiabetic therapy. Specific attention was paid to differences in glucagon metabolism in patients with persistently normal, persistently abnormal, or temporarily abnormal glucose tolerance.

METHODS

First Study. Twelve of the 16 twins originally studied in 1971 (4) were restudied in 1976 with a 50 g oral glucose tolerance test (OGTT), administered under identical conditions.

Second Study. Twenty-two male survivors of their first myocardial infarct, aged less than 65 and all within 10% of ideal body weight, were given a 100 g OGTT two weeks, 3 months and 6 months after the infarct. All were ambulant, requiring no drugs and suffered no complications. The subjects were divided into 3 groups according to their blood glucose response: group 1 included those who had a normal glucose tolerance throughout ("Normal" n = 6); group 2 included those whose glucose tolerance remained abnormal throughout ("Diabetic" n = 8) and group 3 included those whose first glucose tolerance was abnormal, but reverted to normal in the 2nd and 3rd tests ("Potential Diabetic" n = 8).

All blood samples were analyzed for glucose, insulin and glucagon concentrations. In the second study the basal concentrations of triglyceride and cholesterol were also measured. Glucagon concentrations (PG) were measured by radioimmunoassay using Unger's antiserum 30K, with the exception of the first 1971 examination of the twins which was performed using the method of Day (3).

RESULTS

First Study. The result of the OGTT of the original 16 twins are restated. These showed that the group as a whole had mean plasma glucagon concentrations that were intermediate between those of the diabetic and the normal subjects (Fig. 1). On closer examination the individual responses were shown to be heterogeneous with some showing stimulation and some showing suppression of plasma glucagon. When the twins were divided into two groups according to possible risk of developing diabetes, Group I who had remained discordant for more than 10 years and

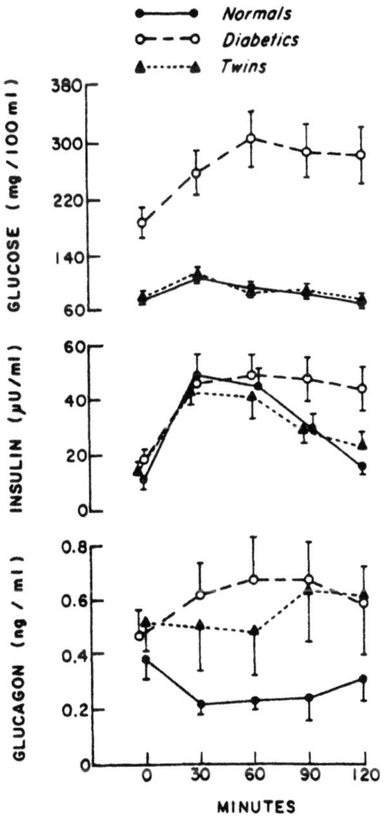

Fig. 1

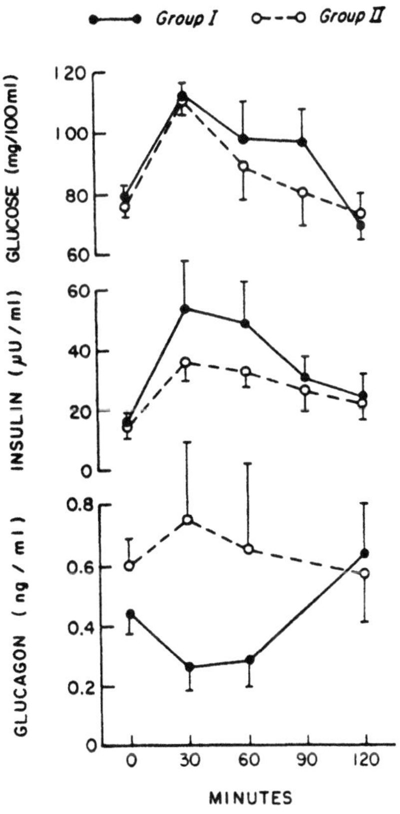

Fig. 2

Group II who had remained discordant for less than 10 years, a significant suppression of PG concentration was demonstrated in the former and a stimulation in the latter (Fig. 2). Re-examination of the 12 subjects in 1976 showed no significant differences from the parent group in relationship to age, sex or length of discordance. Six subjects originally fell into Group I and 6 in Group II. Their overall mean plasma glucagon responses (Fig. 3) revealed no significant suppression or stimulation and, even when divided into original groups, neither group showed a significant change from baseline (Fig. 3a). Insulin and glucose concentrations showed no significant differences when 1971 and 1976 tests were compared.

Second Study. The 6 normal subjects showed completely normal mean plasma glucose concentrations in all 3 tests. The 8 "diabetics" showed abnormally raised 60, 90, 120 and 180 minute glucose concentrations in all 3 tests, with some re-

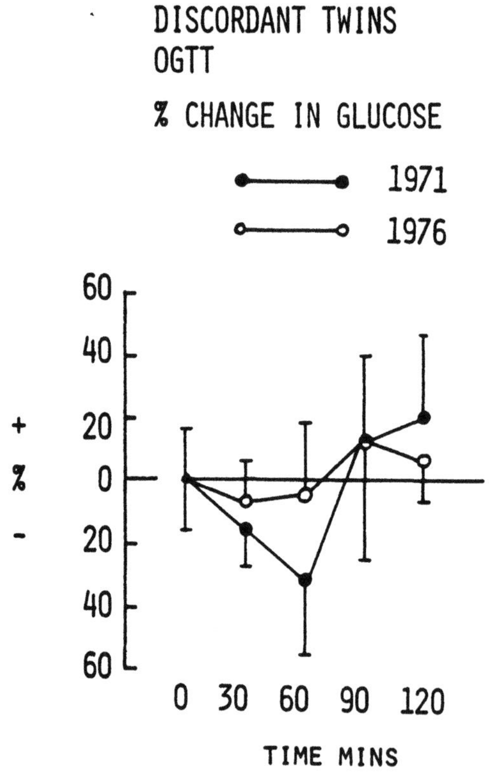

Fig. 3

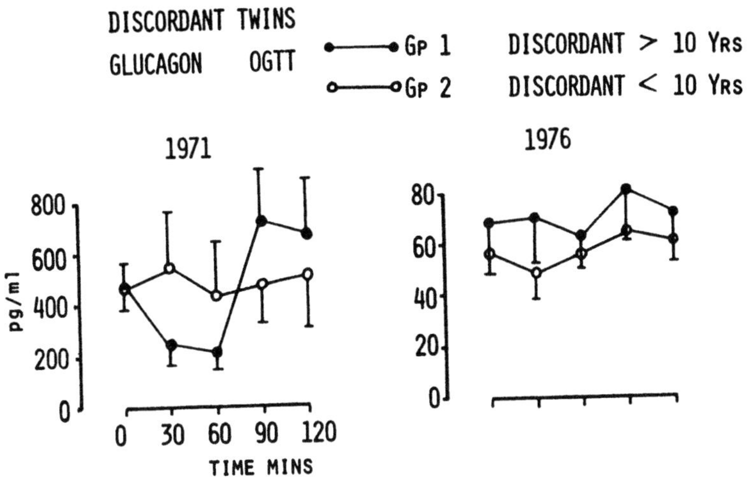

Fig. 3a

duction between the 1st and the 2nd and between the 2nd and 3rd tests. The 8 "potential diabetics" showed abnormal 60, 90 and 120 minute glucose concentrations in the 1st test with return to normal by the 3rd test (Fig. 4). Mean values in tests 1, 2 and 3 were similar in "normals" and "diabetics". The "potential diabetics" showed a significant fall in mean insulin concentrations when tests 2 and 3 were compared with test 1 (Fig. 5). Mean PG concentrations of the whole group showed no significant differences between tests 1, 2 and 3 and no significant differences when post and pre-glucose levels were compared (Fig. 6). However, when sub-divided into groups, significant changes from baseline were observed.

In the normals, significant suppression of PG from baseline was observed in all three tests. In all three tests the "diabetics" showed a significant increase and the "potential diabetics" no change. However, in the "diabetic" group, absolute values were lower in the third test (Fig. 7). The normals were younger and heavier than the "diabetics" and "potential diabetics". Mean basal triglyceride concentrations fell significantly in "potential diabetics". All groups showed a slight rise in mean cholesterol concentrations over the six month period.

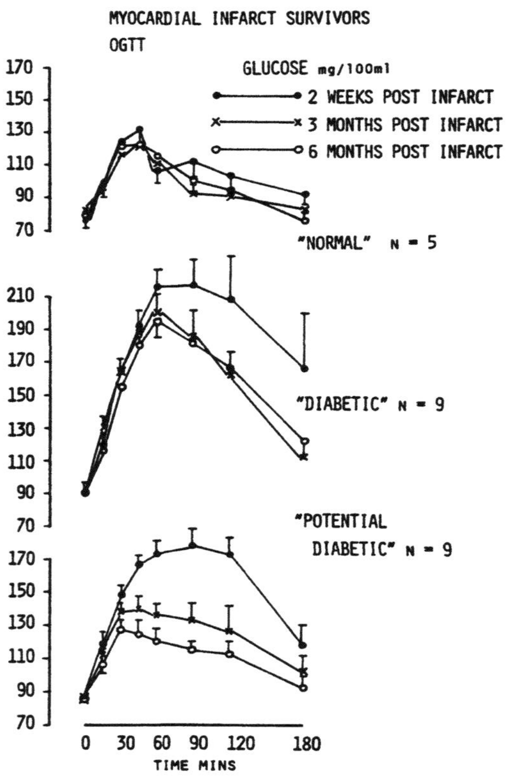

Fig. 4

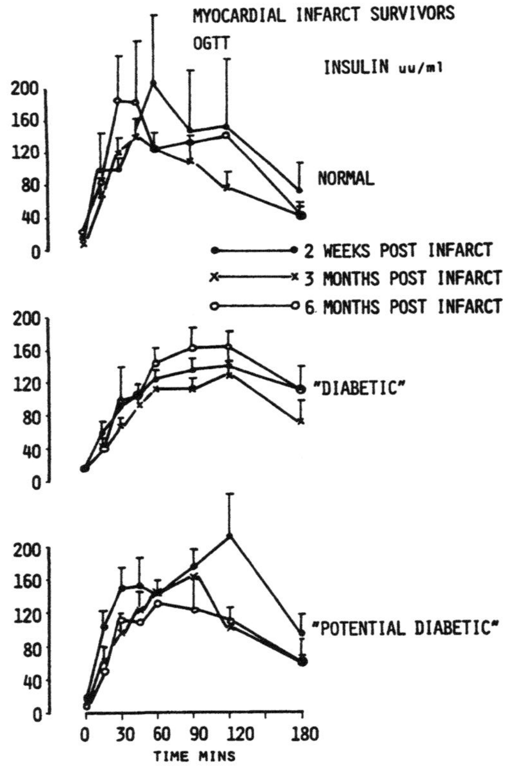

Fig. 5

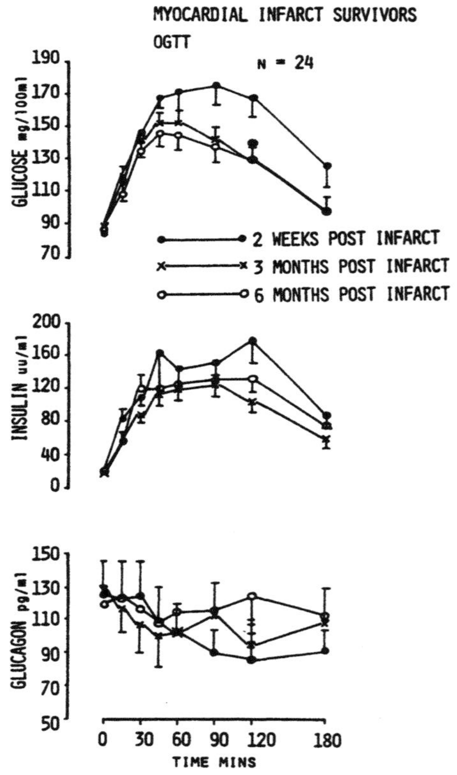

Fig. 6

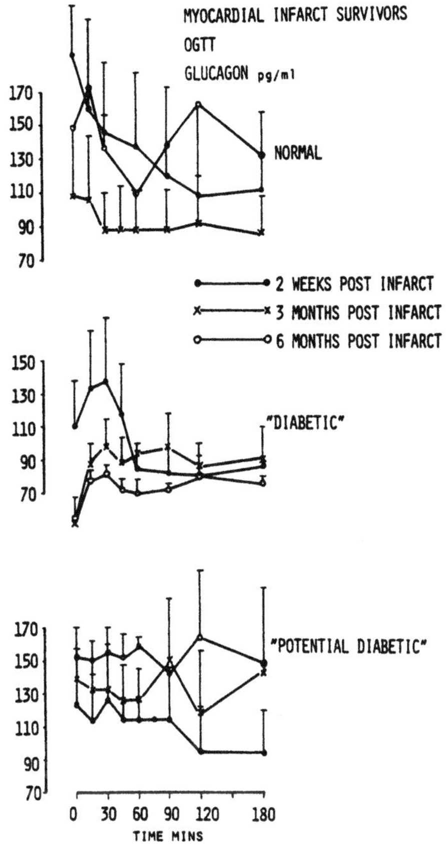

Fig. 7

CONCLUSIONS

The mean PG concentration of discordant twins of diabetics did not decrease after the ingestion of glucose, when tested on two occasions, after a 5-year interval, with two different assay systems. Some individuals, who initially showed suppression, failed to suppress five years later. With the passage of time no significant differences between subjects initially discordant for less than 10 years and between those discordant for a longer period could be detected. In patients known to have the potential for carbohydrate intolerance after myocardial infarct, glucagon concentrations did not suppress, even after their carbohydrate intolerance has resolved. It would appear that both groups share a defect in A cell function which, by itself does not produce carbohydrate intolerance, but may contribute to its development should the patient encounter other aetiological stresses.

REFERENCES

1. Boden, G., Soeldner, J. S., Gleason, R. F. and Marble, A. (1968): Elevated serum growth hormone and decreased serum insulin in prediabetic males after intravenous tolbutamide and glucose. J. Clin. Invest. 47, 729-739.

2. Buchanan, K. D. and McCarroll, A. M. (1972): Abnormalities of glucagon metabolism in untreated diabetes mellitus. Lancet 2, 1394-1395.

3. Day, J. L. and Anderson, J. A. (1973): Abnormalities of glucagon metabolism in diabetes mellitus. Clin. Endocrinol. 2, 211-217.

4. Day, J. L. and Tattersall, R. B. (1975): Glucagon secretion in unaffected monozygotic twins of juvenile diabetics. Metabolism 24, 145-151.

5. Fallucca, F., Maldonato, A., Ganbardella, S., Russo, S. and Menzinger, G. (1976): Glucagon in genetic potential diabetes. Diabetologia 12, 389 (abst).

6. Kirk, R. D., Dunn, P. J., Smith, J. R., Beaven, D. W. and Donald, R. A. (1975): Abnormal pancreatic alpha cell function in first degree relatives of known diabetics. J. Clin. Endocrinol. Metab. 40, 913-916.

7. Tattersall, R. B. and Pyke, D. A. (1972): Diabetes in identical twins. Lancet 2, 1120-1125.

8. Unger, R. H., Siperstein, M. D., Madison, L. L., Eisentraut, A. M. and Whissen, N. (1964): Apparent growth hormone hyperresponsiveness in prediabetes. J. Lab. Clin. Med. 64, 1013 (abst.).

9. Unger, R. H., Arquilla-Parada, E., Müller, W. A. and Eisentraut, A. M. (1970). Studies of pancreatic alpha cell function in normal and diabetic subjects. J. Clin. Invest. 49, 837-848.

TOLBUTAMIDE: STIMULATOR AND SUPPRESSOR OF GLUCAGON SECRETION

E. Samols and J. Harrison

Veterans Administration Hospital and
Department of Medicine, University of Louisville School of Medicine
Louisville, Kentucky USA

Tolbutamide may appear to stimulate, suppress or have no effect on glucagon secretion in vivo. We have established in vitro that the drug normally stimulates glucagon secretion. However, this stimulus may be obscured by the effect of insulin released following tolbutamide stimulation. This insulin may act locally within the islets on glucopenic or insulinopenic A cells to abolish the glucopenic stimulus to glucagon secretion. The true stimulatory effect is revealed in the glucopenic insulin deficient diabetic pancreas. Stimulation by a drug of the potent hyperglycemic hormone glucagon, when not offset by insulin release, is unlikely to benefit diabetes.

INTRODUCTION

It is well established that the anti-diabetic sulfonylureas acutely stimulate insulin secretion;: however, their effect on glucagon release, recently suggested to be as important as insulin in regulating glucose homeostasis (2), has been reported as inhibitory (5, 13, 14), stimulatory (6) or non-evident (8). It is important to solve the problem and explain the controversy, not only for academic reasons, but also because the sulfonylureas continue to be widely prescribed without adequate knowledge of their mode of action.

We propose that tolbutamide, a representative sulfonylurea, basically stimulates the release of glucagon from the A cell as well as insulin from the B cell. Under certain circumstances, provided there is some degree of glucose lack and provided there is sufficient stimulus of insulin secretion, the stimulus to glucagon secretion is masked by the immediate local action of insulin (15) on the glucose-starved (glucopenic) A cells. Thus, insulin which has not yet left the pancreas suppresses that element of glucagon secretion caused by glucopenia. This concept of concealed antagonism may explain how different investigators, depending on the circumstances of their studies, could obtain such disparate results. The corollary to this concept is that, although the effect of sulfonylureas on glucagon secretion in the normal pancreas may be concealed by intra-islet insulin secretion, the true stimulatory action would be unmasked by the diabetic, insulin-deficient pancreas. Such a glucagonotropic effect is probably not a desirable quality for a pharmaceutical agent used to treat diabetes mellitus.

Our results indicate that tolbutamide does indeed acutely stimulate glucagon release, at least simultaneously with, and sometimes preceding, its stimulation of insulin release. An increase in glucagon levels is seen as the initial effect of tolbutamide. At normal glucose concentrations, this increase continues for the duration of the infusion. In contrast, at low glucose levels which are associated with a higher level of basal glucagon secretion than occurs at normal glucose concentrations, the initial stimulation of glucagon is "converted" to a net suppression. The net suppression is probably caused by tolbutamide-induced insulin release ameliorating the effects of glucopenia (low glucose) on the A cell. If the stimulus of glucopenia (Sg) is greater than the stimulus of tolbutamide (St) with respect to glucagon secretion, then removal of Sg by insulin results in a net suppression, i. e., if $Sg \geq St$, then $St - Sg \leq 0$; conversely, if $Sg \leq St$, then $St - Sg \geq 0$.

Methodologic Considerations

We studied glucagon secretion by a truly isolated in vitro dog pancreas, perfused without recirculation, during tolbutamide infusions. The original perfusion technique (3) was modified to exclude the duodenum from the circulation to ensure that the effects we were observing were not caused by duodenal "pancreatic" glucagon. The perfusate contained either normal or low levels of glucose. In some studies a mixture of 19 amino acids in physiologic concentration (1 mM) was added to the perfusate (7).

Pancreatic glucagon was measured by the "pancreatic specific" antiserum 30K (2), insulin was measured by the Yalow and Berson technique (19), and glucose by the glucose oxidase method with the Technicon AutoAnalyzer R.

Actions of Tolbutamide on Truly Isolated Pancreas

We infused tolbutamide at a concentration known to cause suppression of glucagon levels in vivo in ducks (13), maintaining the glucose concentrations in the perfusate at a normal level of either 88 or 110 mg/100 ml. In this way we avoided the effects of hypoglycemia which would normally be produced in vivo. To our surprise, glucagon release was clearly stimulated for the duration of the infusion (Fig. 1). The same stimulatory effect was seen when amino acids were added to the perfusate (Fig. 2). Although typical single studies are shown, the means of several perfusions showed significant stimulation (p <.001) in the absence or presence of amino acids (9).

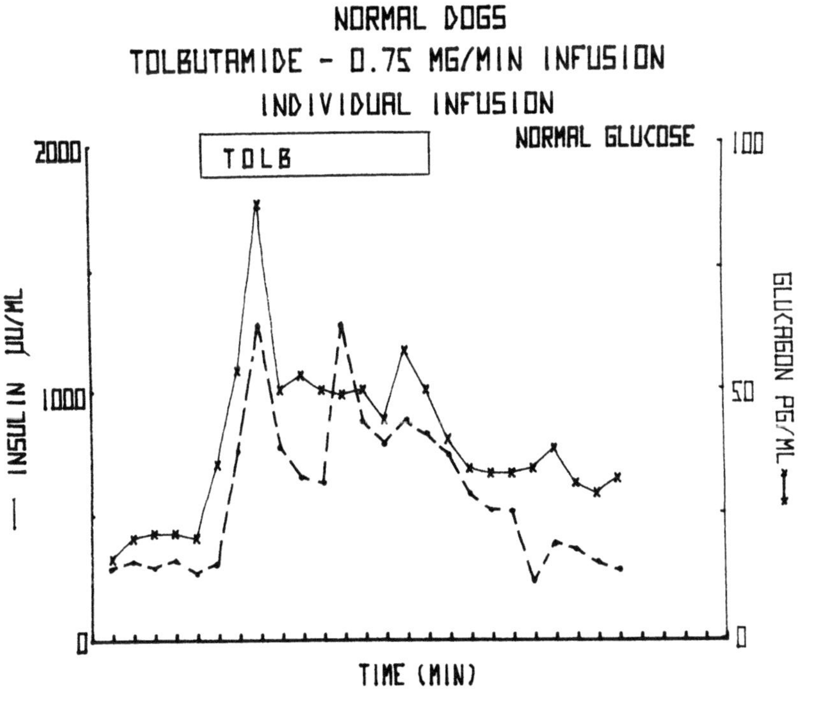

Fig. 1. Effect of tolbutamide on glucagon and insulin secretion in a representative truly isolated perfused canine normal pancreas. Perfusate contained a normal glucose concentration without the addition of amino acids (10, with permission).

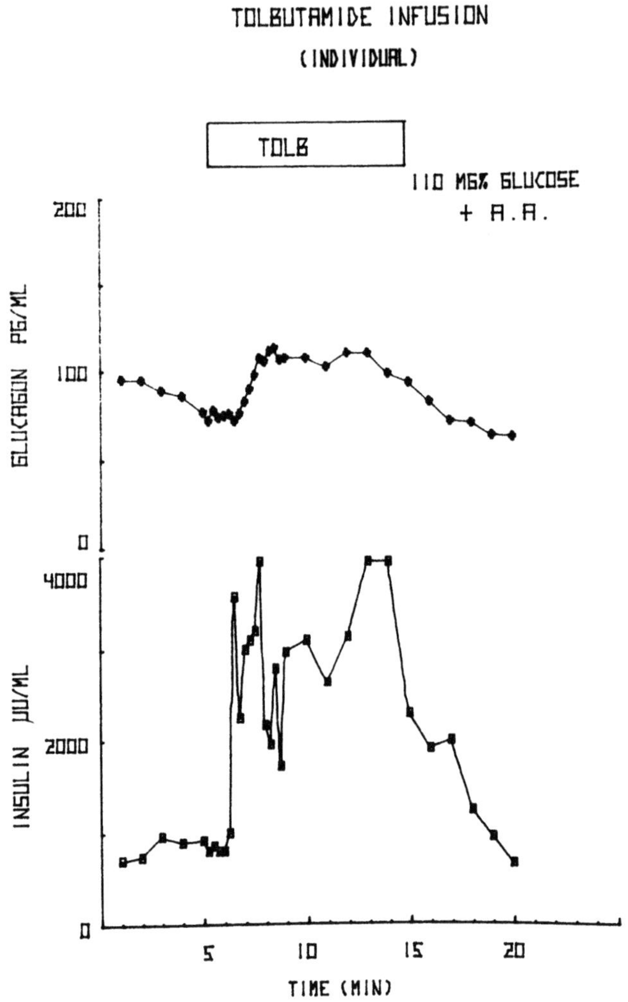

Fig. 2. Effect of tolbutamide on a representative normal pancreas when perfusate contained a normal glucose concentration with the addition of a physiologic concentration of amino acids. The scale for glucagon and insulin differs from that shown in Fig. 1.

In order to see if the insulin released by tolbutamide could influence the glucopenic A cell, the same studies were repeated using a perfusate containing glucose at a concentration of 15 mg/100 ml (Fig. 3). In the absence of the amino acid mixture, the initial stimulation was converted within 4 minutes to a net suppression, presumably because the glucopenic stimulus to glucagon secretion was removed by the action of insulin on A cells within the islets. Although the basal secretion of insulin during low glucose perfusion was much less than that for normal glucose perfusion, tolbutamide-stimulated insulin secretion was not significantly inhibited by glucopenia. Mean glucagon suppression without amino acids was -20 ± 3% ($p < .001$, N = 6) and with amino acids was -51 ± 3% ($p < .001$, N = 6). The timing, magnitude and duration of glucagon suppression in individual and in mean comparisons was directly related to the rapidity, magnitude and duration of the insulin response to tolbutamide. By studying individual infusions it appeared that a rise of at least 0.25 mU/ml of insulin in the efflux was necessary before the suppressive effect of insulin on glucagon secretion was clearly evident. This minimal "critical" efflux change may represent an intra-islet intravascular increase of 25 mU/ml, based on an estimated 100-fold dilution of islet insulin, as islets receive about 1% of the total vascular flow through the pancreas.

In order to further test our hypothesis we perfused glucopenic diabetic pancreata made insulin deficient by pretreatment with streptozotocin (50 mg/kg body weight), 1-4 days before surgery (17). In striking contrast to the suppression seen in glucopenic normal pancreata, there was a clear cut stimulation of glucagon secretion by tolbutamide for the duration of the infusion (Fig. 4). Tolbutamide-induced insulin secretion in diabetic pancreata was reduced to approximately 6% of that observed with normal pancreas and was presumably now insufficient to influence the A cell. The clear and sustained stimulation of glucagon secretion by tolbutamide in the glucopenic diabetic pancreata strongly suggests that endogenous intra-islet insulin was the cause of the net suppression of glucagon observed in the glucopenic normal pancreata. The biphasic glucagon secretion

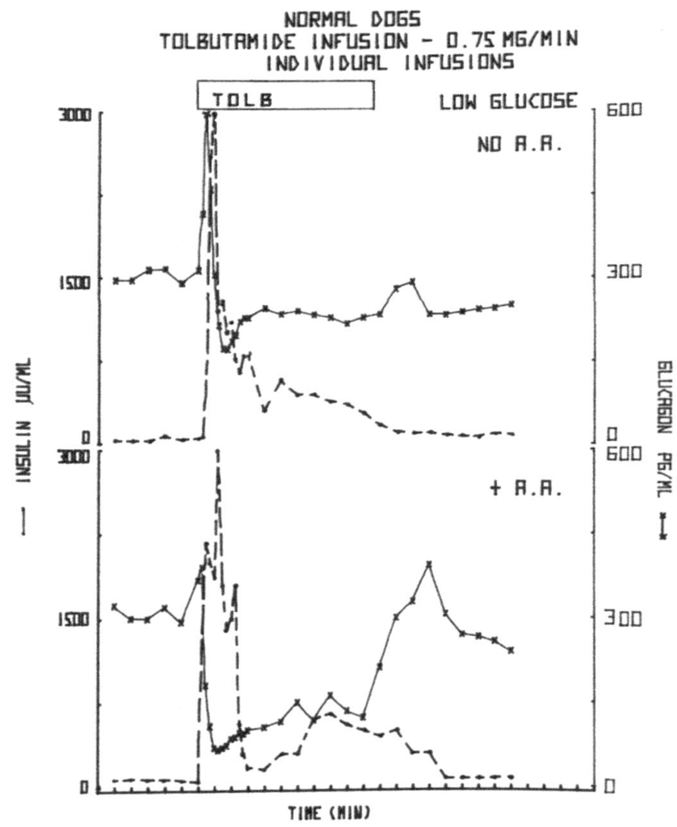

Fig. 3. Effect of tolbutamide on representative normal pancreata perfused with a low glucose concentration without (above) and with (below) amino acids. x——x glucagon, ——— insulin (10, with permission).

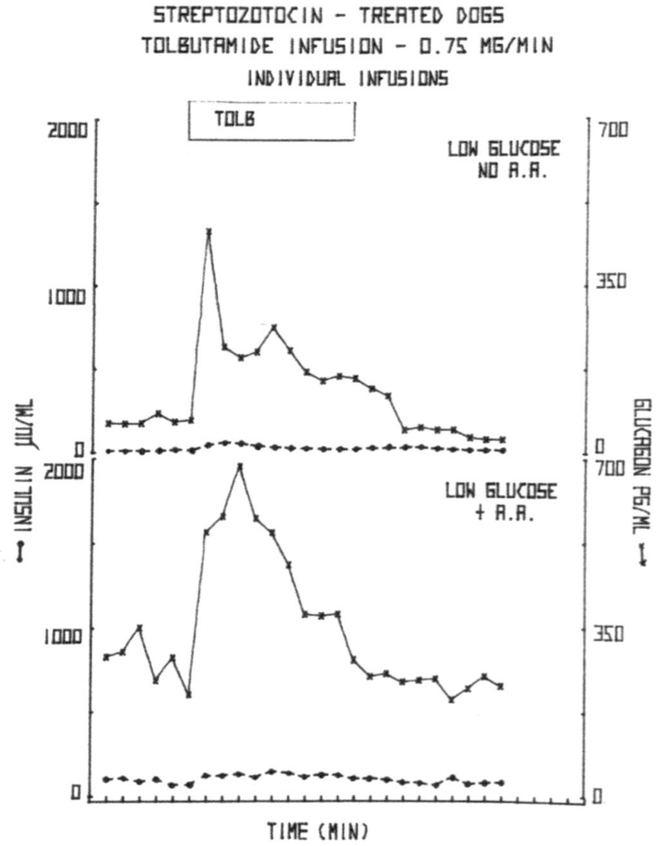

Fig. 4. Effect of tolbutamide on representative pancreata from streptozotocin treated dogs perfused with a low glucose concentration without (above) and with (below) amino acids. x——x glucagon, •——• insulin (10, with permission).

curve demonstrated in the diabetic pancreas (Fig. 4) probably reveals the true effect of tolbutamide infusion on A cells, with little or no modification by the poor insulin response.

Glucopenia and insulinopenia (insulin deficiency), separately or in combination, may modify the response of the A cell to an increase in insulin secretion. During glucopenia, insulin either facilitates the transport of the small amount of glucose into the A cell or increases its intracellular metabolism to switch off the glucopenic stimulus. In the non-glucopenic, normal glucose studies, it seems likely that the stimulation of glucagon secretion by tolbutamide was readily observed because there was very little, if any, glucopenic contribution to glucagon secretion (Sg) to be overcome by insulin. In the low glucose studies it was apparent that a minimal critical increase (approximately 0.25 mU/ml) in insulin was required to reverse Sg. In other studies using the same pancreas preparation (10), we found that an exogenous insulin concentration of 25 mU/ml or more in the influx were required to suppress glucagon secretion, presumably removing Sg. This need for a relatively high influx insulin concentration matches the requisite rise in endogenous insulin (for reversal of Sg), assuming a 100-fold dilution of endogenous intravascular islet insulin by the time it reaches the efflux. Weir et al. (18) observed that similar high concentrations of exogenous insulin (20 mU/ml) were required to suppress glucagon secretion in vitro by the rat pancreas rendered insulinopenic by epinephrine or streptozotocin, but that the same dose of insulin did not suppress glucagon secretion if endogenous insulin secretion was plentiful. It has been proposed that hormone concentration can lead to reciprocal changes in receptor number and cellular responsiveness (4). High insulin concentrations bathing A cells would tend to decrease the number of receptors for insulin and thus A cell responsiveness to an increase in insulin secretion. The converse would obtain during insulinopenia, and it is quite possible that in our glucopenic pancreas glucopenic-induced insulinopenia may have enhanced the susceptibility of the A cells to endogenous insulin action.

It seems unlikely that iatrogenic promotion of so potent a hyperglycemic hormone as glucagon is desirable in diabetes mellitus. There is evidence that diabetes may be caused by absolute or relative glucagon excess (2) as well as by absolute or relative insulin deficiency. The clinical inference from our work is that these drugs would only be advisable therapeutically if their stimulation of insulin secretion were sufficient to overcome their stimulation of glucagon secretion. Sulfonylureas could be harmful if their stimulation of glucagon secretion predominated, a possibility that could explain the behavior of many cases of "primary" and "secondary" failures in diabetic patients when administration of oral sulfonylureas is followed by an actual deterioration in diabetic control manifested by an increase in blood glucose levels. We would predict that, in some diabetic patients with poor B cell function but retaining A cell function, an infusion of tolbutamide would cause an acute increase in glucagon and perhaps even of glucose levels. In fact, we have observed an increase in circulatory immunoreactive glucagon levels following intravenous tolbutamide (1 gm) in three insulin-deficient diabetic patients, illustrated by one example (Fig. 5). Also, it has recently been noted that patients with glucagonomas, who usually masquerade as mild noninsulin-dependent diabetics, may demonstrate a large release of glucagon in response to tolbutamide treatment (1). Our studies indicate that such a response represents the summation of an essentially normal response to tolbutamide by the excessive number of tumorous A cells in the glucagonoma.

Intra-islet Interrelationships of Insulin, Glucagon and Somatostatin

Our evidence strongly suggests that insulin secretion may suppress glucopenic glucagon secretion within the islets. We have not excluded the possibility that somatostatin secretion, at an intra-islet level, may also contribute to the demonstrated difference in the effect of tolbutamide on glucagon secretion in normoglucosic islets (stimulation) and in glucopenic islets (secondary suppression). Preliminary studies suggest a greater stimulation of somatostatin by tolbutamide

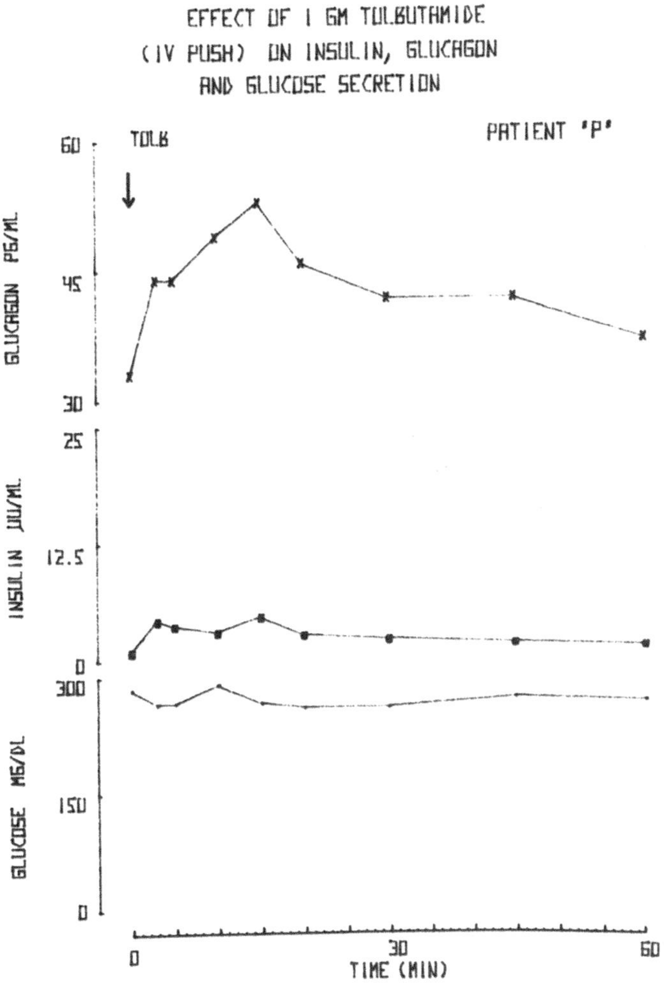

Fig. 5. Effect of 1 g tolbutamide (I.V. push over 2 min) on peripheral plasma immunoreactive glucagon and insulin levels in an insulin-deficient diabetic patient.

during glucopenia than during normal glucose perfusion (10). Our pancreas preparation is exquisitely sensitive in terms of suppression of its glucagon secretion by exogenous somatostatin (11). Therefore, the detection of increases in tolbutamide-induced somatostatin levels, presumably considerably diluted in the pancreatic efflux compared with intra-islet concentration, is likely to be of physiologic significance in the dynamic control of insular secretions.

CONCLUSIONS

We conclude that tolbutamide is a moderately potent stimulus of glucagon secretion. The decrease in glucagon observed during tolbutamide infusion given under glucopenic conditions in our in vitro experiments can be attributed to an inhibitory effect on the glucopenia-stimulated A cells by the marked increase in insulin released by tolbutamide, thereby masking the primary stimulus. This phenomenon also provides convincing support for the hypothesis that an intra-islet negative insulin-glucagon feedback mechanism (15) does exist and is best demonstrated during glucopenia. These findings also explain previous contradictory reports on the effect of the sulfonylureas on glucagon secretion. Any permutation: stimulation, suppression or no effect, is possible, depending on the balance between stimulation of glucagon release by tolbutamide (St) and suppression of glucopenic (and perhaps insulinopenic) glucagon release (Sg) by drug-stimulated insulin. It also seems likely that there may be other substances which basically stimulate glucagon secretion, yet their stimulus may be masked by an intra-islet negative insulin-glucagon effect.

Acknowledgement

This work was supported by the Medical Research Service of the Veterans Administration.

REFERENCES

1. Bloom, S. R. (1975): Hormonal interrelations and their clinical significance. Gastrointestinal hormones. Proc. Roy. Soc. Med. 68, 710-711.

2. Dobbs, R., Sakurai, H., Sasaki, H., Faloona, G., Valverde, I., Baetens, D., Orci, L. and Unger, R. (1975): Glucagon: role in the hyperglycemia of diabetes mellitus. Science 187, 544-547.

3. Iversen, J. and Miles, D. W. (1971): Evidence for a feedback inhibition of insulin on insulin secretion in the isolated, perfused canine pancreas. Diabetes 20, 1-9.

4. Kahn, C. R., Neville, D. M., Jr. and Roth, J. (1973): Insulin-receptor interaction in the obese-hyperglycemic mouse. A model of insulin resistance. J. Biol. Chem. 248, 244-250.

5. Loreti, L., Sugase, T. and Foà, P. P. (1974): Diurnal variations of serum insulin, total glucagon, cortisol, glucose and free fatty acids in normal and diabetic subjects before and after treatment with chlorpropamide. Hormone Res. 5, 278-292.

6. Loubatières, A. L., Loubatières, M. M., Alric, R. and Ribes, G. (1974): Tolbutamide and glucagon secretion. Diabetologia 10, 271-276.

7. Pagliara, A. S., Stillings, S. and Matschinsky, F. M. (1973): Interactions between glucose and amino acids in α and β cells of the isolated perfused rat pancreas. Progr. 55th Ann. Meet. Endocr. Soc., Chicago, A169.

8. Pek, S., Fajans, S. S., Floyd, J. C., Jr., Knopf, R. F. and Conn, J. W. (1972): Failure of sulfonylureas to support plasma glucagon in man. Diabetes 21, 216-223.

9. Samols, E. and Harrison, J. (1976): Intraislet negative insulin-glucagon feedback. Metabolism 25, Suppl. 1, 1443-1447.

10. Samols, E. and Harrison, J. (1976): Glucagon secretion in the absence of glucose: stimulation and suppression by tolbutamide: suppression by exogenous insulin. Clin. Res. 24, 52A.

11. Samols, E. and Harrison, J. (1976): Remarkable potency of somatostatin as a glucagon suppressant. Metabolism 25, Suppl. 1, 1495-1497.

12. Samols, E., Marri, G. and Marks, V. (1965): Promotion of insulin secretion by glucagon. Lancet 2, 415.

13. Samols, E., Tyler, J. M. and Kajinuma, H. (1971): On the mechanism of action of the hypo and hyperglycemic sulfonamides. Proc. 7th Congr. Intern. Diabetes Fed., Buenos Aires, Exc. Med. Found. ICS 231, 636-655.

14. Samols, E., Tyler, J. M. and Mailhe, P. (1969): Suppression of pancreatic glucagon release by the hypoglycemic sulfonylureas. Lancet 1, 174-176.

15. Samols, E., Tyler, J. M. and Marks, V. (1972): Glucagon-insulin interrelationships. In: Glucagon, Molecular Physiology, Clinical and Therapeutic Implications. P. J. Lefèbvre and R. H. Unger, Eds. Pergamon Press Ltd., Oxford, pp. 151-173.

16. Samols, E., Weir, G. C., Day, J. A., Harrison, J. and Patel, Y. C. Unpublished observations.

17. Schein, P. S., Rakieten, N., Cooney, D. A., Davis, R. and Vernon, M. L. (1973): Streptozotocin diabetes in monkeys and dogs and its prevention by nicotinamide. Proc. Soc. Exp. Biol. Med. 143, 514-518.

18. Weir, G. C., Knowlton, S. D., Atkins, R. F., McKennan, K. X. and Martin, D. B. (1976): Glucagon secretion from the perfused pancreas of streptozotocin-treated rats. Diabetes 25, 275-282.

19. Yalow, R. S. and Berson, S. A. (1960): Immunoassay of endogenous plasma insulin in man. J. Clin. Invest. 39, 1157-1175.

EFFECT OF LONG TERM SULFONYLUREA THERAPY ON GLUCAGON LEVELS IN MATURITY ONSET DIABETES MELLITUS

S. Podolsky and A. M. Lawrence

Medical Service, Boston VA Outpatient Clinic, Department of Medicine,
Boston University School of Medicine, Boston, Massachusetts and
Departments of Medicine and Biochemistry
Loyola Stritch School of Medicine and Hines VA Hospital
Maywood, Illinois USA

Glucagon secretion was studied before and after the control of hyperglycemia in 15 maturity onset male diabetics (mean age 56.1 yrs; mean duration of diabetes 7.3 yrs). Optimum control of hyperglycemia was achieved by addition of the sulfonylurea chlorpropamide to dietary treatment. A standardized 3 hour oral glucose tolerance test was performed before and after approximately eight months of individualized therapy with the sulfonylurea. The glucose tolerance test was repeated with each patient taking his usual dose of chlorpropamide 90 minutes prior to the administration of the glucose load. Basal hyperglucagonemia occurred both before and after reduction of fasting hyperglycemia, and glucagon levels for the group rose, rather than falling, following glucose administration. Peak glucagon level occurred at 60 min in each test, and fell but not significantly from 182.3 ± 29.4 pg/ml to 163.5 ± 26.4 pg/ml. While there were wide variations in individual responses, these data indicate that the hypoglycemic action of long term sulfonylurea treatment cannot be ascribed to lowering of circulating glucagon levels.

The hypoglycemic effect of sulfonylurea drugs has generally been presumed to be secondary to the release of endogenous insulin from the pancreatic B cells (21). Yalow et al (42) provided evidence for this when they demonstrated an acute rise in circulating plasma insulin levels after administration of tolbutamide to normal subjects. However, a number of studies have questioned whether chronic administration of sulfonylureas causes a sustained increase in insulin secretion (1, 4, 12, 28, 34, 37). There is also continuing controversy over the concept that basal glucagon levels are always elevated or inappropriate to the prevailing glucose concentration in human diabetes (5, 8, 39), or whether the relative or absolute glucagon excess in diabetics is simply secondary to insulin deficiency (18, 38). The effect of sulfonylureas on glucagon release has been reported to be inhibitory (20, 35, 36), stimulatory (22), or non-existent (27).

The present study was undertaken to investigate the effects of chronic administration of a sulfonylurea agent on glucagon secretion in maturity onset diabetes mellitus.

MATERIALS AND METHODS

Fifteen male patients with maturity onset diabetes mellitus were studied. The age range was 45-67 yrs (mean = 56.1 yrs) and the estimated duration of diabetes ranged from 2 to 22 yrs (mean = 7.3 yrs). No patient had ever received insulin therapy. All but two had previously received sulfonylurea agents. Only two of the patients had evidence of a vascular complication of diabetes such as diabetic retinopathy (15). None of the patients had a history of acute metabolic disturbances, such as hyperosmolar nonketotic coma (30), nor did any of the patients suffer from a reversible form of abnormal carbohydrate metabolism, such as occur in states of potassium depletion (29, 32).

Although all patients had been on an individualized American Diabetes Association diet calculated to attain or maintain optimal weight, each received a minimum of two individual diet instruction sessions at the beginning of the study. Fasting blood for serum glucose was taken weekly for up to three weeks with the patients receiving only diet therapy. A three hour baseline oral glucose tolerance test was then performed on each patient using 100 gm of lemon flavored glucose. After completion of control studies, chlorpropamide was begun with a daily morning dose of 250 mg because of fasting hyperglycemia on diet-alone therapy. Fasting glucose levels were taken weekly in all subjects for the first month, then biweekly the second month, then monthly. In patients whose fasting serum glucose levels continued to exceed 140 mg/dl after several months of chlorpropamide therapy and additional dietary instruction, the dosage of the sulfonylurea was gradually increased to a maximum of 500 mg each morning. At the end of an eight month interval, the glucose tolerance test was repeated, with the patients taking their usual dose of chlorpropamide 90 minutes prior to the administration of the glucose load. Each subject was studied in identical fashion so that each served as his own control.

The glucose tolerance tests were performed with the patients supine, comfortable and free from stress. After a subject had been lying down for approximately 30 minutes, a thin walled 19 gauge "butterfly" needle was inserted into an antecubital vein, and remained in place throughout the test (31). Beginning 30 minutes after the insertion of the needle, bloods were withdrawn according to the following time schedule: -20, -10, 0, 30, 60, 120 and 180 minutes. Ten ml blood samples for hormone assay were collected into chilled tubes containing 0.5 ml trasylol (6). The tubes were inverted quickly, kept chilled, then promptly spun down in a refrigerated centrifuge and the plasma separated and stored at -20°C until analyses. All determinations were performed in duplicate and all samples from each subject were assayed on the same day. Serum glucose determinations were performed on a Technicon Autoanalyzer using a modification of the neocuproine method (3). Serum glucagon levels were measured by a double antibody radioimmunoassay, using antiserum specific for pancreatic glucagon (17).

RESULTS

Dietary adherence was good, and adherence to the prescribed sulfonylurea regimen was excellent in all subjects. Mean body weight for the 15 patients was 193.6 lb and 191.7 lb at the beginning and at the end of the drug treatment interval, respectively. There was gradual lowering of fasting serum glucose levels during the 32 weeks of sulfonylurea therapy. Normal fasting glucose levels were achieved for the group as a whole and for 10 of the 15 individual patients.

Glucagon responses during the pre- and post-therapy glucose tolerance tests are shown in Fig. 1. Basal hyperglucagonemia occurred both before and after reduction of fasting hyperglycemia with chlorpropamide therapy. There was a paradoxical glucagon rise associated with the hyperglycemic phase of both glucose tolerance tests. Peak glucagon level occurred at 60 minutes in each test, and fell slightly but not significantly with therapy, from 182.3 ± 29.4 pg/ml to 163.5 ± 26.4 pg/ml (mean ± SEM).

Figure 2 shows marked lowering of elevated glucagon levels after treatment in a 59 yr old diabetic male. Diabetic control, as indicated by multiple glucose determinations, was excellent.

Figure 3 shows normalization of the glucagon curve in a 47 yr old patient. In this patient the paradoxical glucagon rise which initially occurred with hyperglycemia, was abolished. In the 45 yr old patient, whose data are plotted in Figure 4, the elevated basal glucagon values were reduced to normal, but the paradoxical glucagon rise persisted.

Figure 5 demonstrates both an increase in fasting glucagon levels and a worsening of the overall curve in the post-treatment test. The change to abnormally elevated glucagon levels after sulfonylurea treatment occurred in this 48 yr old male despite excellent control of hyperglycemia.

DISCUSSION

Chlorpropamide plus diet therapy was effective in reducing fasting hyperglycemia in the majority of our patients with maturity onset diabetes.

Although glucagon levels did not change significantly for the group, there were wide variations in individual responses. Neither the age, the duration of diabetes nor the state of control of hyperglycemia appeared to influence the pattern of the glucagon curve after chronic sulfonylurea administration. The relationship of glucagon response to intactness of endogenous insulin reserve is presently under investigation.

Since there is still considerable uncertainty concerning the actual pathogenesis of diabetes in man, it is only natural that any abnormality of A cell function in diabetes will be quickly seized as a potential explanation. Whether elevated glucagon levels causally contribute to the development of diabetes melli-

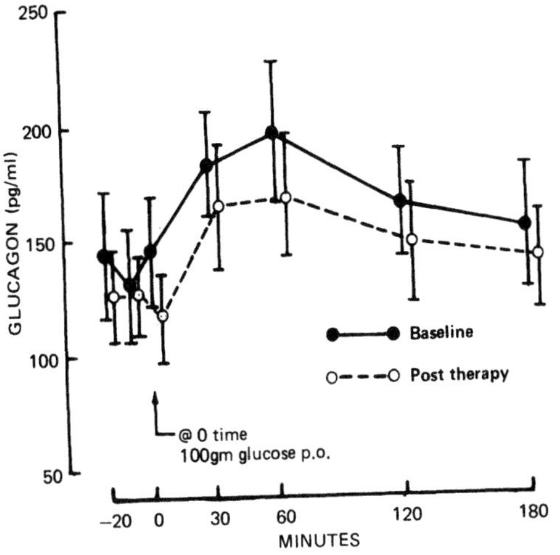

Fig. 1

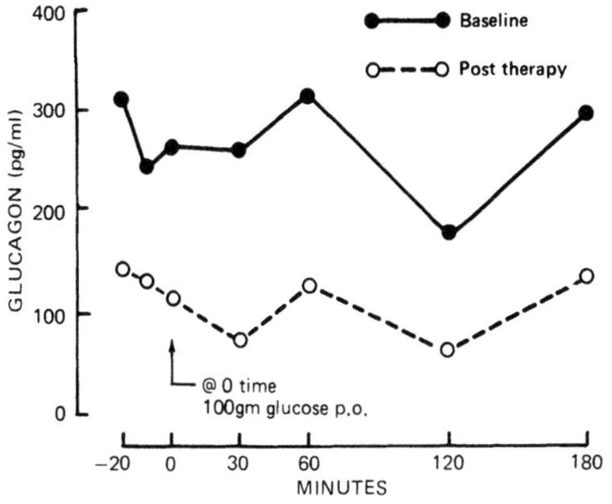

Fig. 2

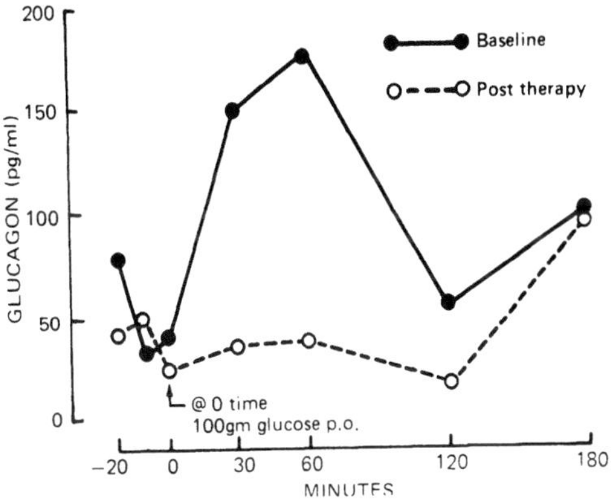

Fig. 3

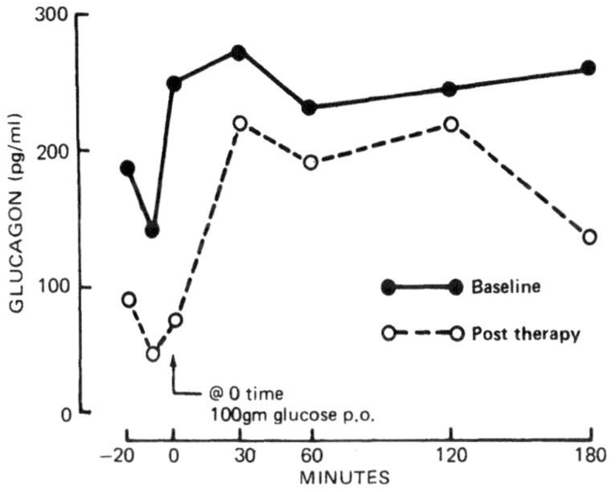

Fig. 4

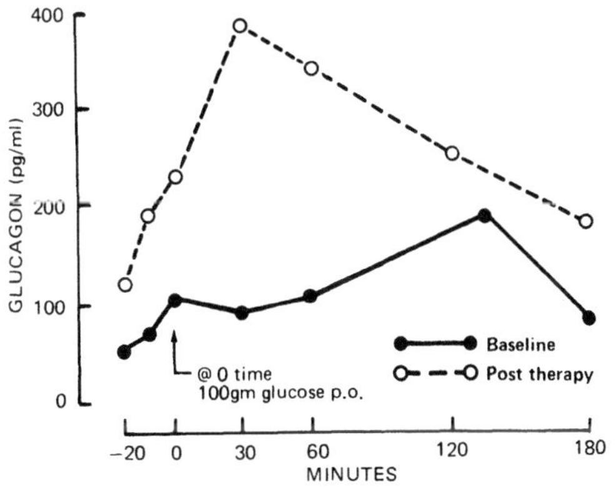

Fig. 5

tus or strongly influence metabolic abnormalities in this disease remains controversial. Nevertheless, observations have been made which can implicate glucagon in diabetes mellitus. Unger and associates pioneered glucagon research in this area almost from the time they introduced the radioimmunoassay for glucagon in the early 1960's. His group has been extraordinarily interested in the issue of glucagon in diabetes mellitus (39).

Immunoreactive glucagon levels are markedly elevated in all patients with diabetic ketoacidosis (26). As therapy corrects the gross metabolic derangements, glucagon levels decline but often remain elevated above normal basal levels. It now appears that stress, and in particular stress which theoreticaly calls for rapid mobilization of endogenous fuels, is associated with a remarkable elevation in circulating glucagon levels. It has been proposed that the insulin-resistance and impaired glucose tolerance seen in such patients is related to excessive glucagon secretion. Often there is impaired insulin secretion in these situations as well.

To date it has been demonstrated that hyperglucagonemia, insulin resistance, and impaired glucose tolerance are characteristically seen with severe burns and blunt trauma (19, 25), violent exercise (2), pyrogenic infection (24), and, in animals, psychic stress (24). It has been suggested that epinephrine may stimulate glucagon release (9).

It is not clear whether a general autonomic discharge causes hyperglucagonemia in these stressful circumstances. Thus, it is possible that diabetic ketoacidosis also reflects another in the list of "stress conditions" associated with hyperglucagonemia in man and animals. The hypothalamo-pituitary-adrenal axis may also play a role. Glucagon secretion is enhanced in Cushing's syndrome and by the administration of glucocorticoids. Thus, cortisol secreted in response to stress may be partially responsible for the elevated glucagon titers in many of these situations.

Apart from the situation of stress hyperglucagonemia, other peculiarities of glucagon secretion may be seen in patients with diabetes mellitus. As was observed in the present series, many, but not all, diabetics have truly elevated basal glucagon levels even when hyperglycemia control appears adequate (39).

Some juvenile diabetics, particularly those prone to rapid and recurring ketosis, demonstrate continuously elevated basal glucagon levels and markedly exaggerated glucagon response to arginine infusion (11). Another unusual feature is that hyperglycemic suppression of circulating glucagon levels is often absent in diabetics (26, 40). Normal subjects show a significant suppression of basal levels of glucagon following intravenous glucose. Diabetics generally fail to suppress glucagon (13, 26) and in many, a definite early rise is seen during post-absorptive hyperglycemia.

It has been argued that relative or absolute glucagon excess in diabetics more accurately reflects intracellular difficulty in metabolism of glucose secondary to insulin deficiency with elevated glucagon levels akin to what obtains in starvation (7). Indeed, the administration of insulin to animals with experimental diabetes mellitus results in correction of their hyperglucagonemia (41). Infusions of insulin for up to 72 hours in human diabetics failed to alter raised basal glucagon levels nor did it lead to hyperglycemic suppression. Perhaps 72 hours may be insufficient for restoration of normal A cell metabolism after years of relative insulin lack.

More recently, it has been shown that intravenous administration of physiologic amounts of insulin plus glucose for 2 hours to fasting diabetics results in a substantial decline in plasma glucagon (33). Extension of such studies should help clarify whether abnormalities of glucagon secretion in diabetics are causal or secondary to the diabetic state. Another feature of insulin-dependent diabetics is their apparent inability to raise their glucagon level in response to insulin-induced hypoglycemia (10). Thus, it may very well be that in diabetes mellitus, with its attendant and complex metabolic alterations, the A cell is relatively insensitive to major changes in circulating glucose levels.

The studies reported here confirm previous observations that, for whatever reason, maturity onset diabetics often demonstrate relatively or absolutely increased fasting glucagon levels, and fail to suppress glucagon (or actually show a paradoxical rise) following hyperglycemia. These findings occurred both in the presence of fasting hyperglycemia and after control of hyperglycemia by the sulfonylurea. It would appear that the hypoglycemic effect of chronic administration of sulfonylureas cannot be ascribed to any tendency of these agents to inhibit glucagon secretion.

REFERENCES

1. Barnes, A. J., Garbien, K. J. T., Crowley, M. F. and Bloom, A. (1974): Effect of short and long-term chlorpropamide treatment on insulin release and blood glucose. Lancet 2, 69-72.

2. Boettger, I., Schlein, E. M., Faloona, G. R., Knochel, J. P. and Unger, R. H. (1972): The effect of exercise on glucagon secretion. J. Clin. Endocrinol. Metab. 35, 117-125.

3. Brown, M. E. (1961): Ultra-micro sugar determinations using 2,9-dimethyl-1, 10-phenanthroline hydrochloride (neocuproine). Diabetes 10, 60-62.

4. Chu, P., Conway, M. J., Krouse, H. A. and Goodner, C. J. (1968): The pattern of response of plasma insulin and glucose to meals and fasting during chlorpropamide therapy. Ann. Int. Med. 68, 757-769.

5. Dobbs, R., Sakurai, H., Sasaki, H., Faloona, G., Valverde, I., Baetens, D., Orci, L. and Unger, R. H. (1975): Glucagon: role in the hyperglycemia of diabetes mellitus. Science 187, 544-547.

6. Eisentraut, A. M.., Whissen, N. and Unger, R. H. (1968): Incubation damage in the radioimmunoassay for human plasma glucagon and its prevention with "Trasylol." Am. J. Med. Sci. 255, 137-142.

7. Felig, P. (1970): Glucagon: physiologic and diabetogenic role. New Engl. J. Med. 283, 149-150.

8. Foà, P. P. (1975): The metabolic role of pancreatic glucagon (IRG_p), of enteroglucagon (IRG_e) and of other glucagon-like immunoreactive materials (GLI) in health and disease. In: Diabetes Mellitus, 4th Ed., K. E. Sussman and R. J. S. Metz, Eds. American Diabetes Association, New York, pp. 23-28.

9. Gerich, J. E., Karam, J. H. and Forsham, P. H. (1973): Stimulation of glucagon secretion by epinephrine in man. J. Clin. Endocrinol. Metab. 37, 479-481.

10. Gerich, J. E., Langlois, M., Noacco, C., Karam, J. H., Tantillo, J. and Forsham, P. H. (1973): Primary alpha-cell-defect in juvenile diabetes

mellitus: lack of glucagon response to insulin induced hypoglycemia. Clin. Res. 21, 624.

11. Gerich, J. E., Lorenzi, M., Tsalikian, E., Bohannon, N., Schneider, V., Karam, J. and Forsham, P. (1976): Effects of acute insulin withdrawal and administration on plasma glucagon responses to I. V. arginine in insulin dependent diabetic subjects. Diabetes 25, 955-960.

12. Hecht, A., Gershberg, H. and Hulse, M. (1973): Effect of chlorpropamide treatment on insulin secretion in diabetics: Its relationship to the hypoglycemic effect. Metabolism 22, 723-733.

13. Heding, L. G. (1971): Radioimmunological determination of pancreatic and gut glucagon in plasma. Diabetologia 7, 9-10.

14. Josefsberg, Z., Laron, Z., Doron, M., Keret, R., Belinski, Y. and Weisman, I. (1975): Plasma glucagon response to arginine infusion in children and adolescents with diabetes mellitus. Clin. Endocrinol. 4, 487-492.

15. Krall, L. P. and Podolsky, S. (1971): Long-term clinical observations of diabetic retinopathy. In: Diabetes. R. R. Rodriguez and J. Vallence-Owen, Eds. Excerpta Med., Amsterdam, pp. 268-275.

16. Lawrence, A. M. (1969): Glucagon. Ann. Rev. Med. 20, 207-222.

17. Lawrence, A. M. (1974): Glucagon. In: Nuclear Medicine In Vitro. B. Rothfield, Ed. J. B. Lippincott Co., Philadelphia, pp. 249-266.

18. Lawrence, A. M. and Abraira, C. (1977): Glucagon and its clinical significance. Lab. Med. 8, 21-26.

19. Lindsey, C. A., Wilmore, D. W., Moylan, J. A., Faloona, G. R. and Unger, R. H. (1972): Glucagon and the insulin:glucagon ratio in burns and trauma. Clin. Res. 20, 802.

20. Loreti, L., Sugase, T. and Foà, P. P. (1974): Diurnal variations of serum insulin, total glucagon, cortisol, glucose and free fatty acids in normal and diabetic subjects before and after treatment with chlorpropamide. Horm. Res. 5, 278-292.

21. Loubatières, A. (1957): The hypoglycemic sulfonamides: history and development of the problem from 1942 to 1945. Ann. N. Y. Acad. Sci. 71, 4-11.

22. Loubatières, A. L., Loubatières, M. M., Alric, R. and Ribes, G. (1974): Tolbutamide and glucagon secretion. Diabetologia 10, 271-276.

23. Marco, J., Calle, C., Roman, D., Diaz-Fierros, M., Villanueva, M. C. and Valverde, I. (1973): Hyperglucagonism induced by glucocorticoid treatment in man. New Engl. J. Med. 288, 128-131.

24. Marreiro Rocha, D., Santeusanio, F., Faloona, G. R. and Unger, R. H. (1973): Abnormal pancreatic alpha-cell function in bacterial infections. N. Engl. J. Med. 288, 700-703.

25. Meguid, M. M., Brennan, M. F., Müller, W. A. and Aoki, T. T. (1972): Glucagon and trauma. Lancet 2, 1145.

26. Müller, W. A., Faloona, G. R. and Unger, R. H. (1973): Hyperglucagonemia in diabetic ketoacidosis. Its prevalance and significance. Am. J. Med. 54, 52-57.

27. Pek, S., Fajans, S. S., Floyd, J. C., Jr., Knopf, R. F. and Conn, J. W. (1972): Failure of sulfonylureas to support plasma glucagon in man. Diabetes 21, 216-223.

28. Podolsky, S. and Lawrence, A. M. (1975): Effect of chronic sulfonylurea therapy on insulin reserve in diabetes mellitus. Ann. Meet. South. Med. Assoc., Miami, November 15.

29. Podolsky, S. and Melby, J. (1976): Improvement of impaired growth hormone response to stimulation in primary aldosteronism with correction of potassium depletion. Metabolism 25, 1027-1032.

30. Podolsky, S. and Pattavina, C. G. (1973): Hyperosmolar nonketotic diabetic coma. A complication of propranolol therapy. Metabolism 22, 685-693.

31. Podolsky, S. and Sivaprasad, R. (1972): Assessment of growth hormone reserve: comparison of intravenous arginine and subcutaneous glucagon stimulation tests. J. Clin. Endocrinol. Metab. 35, 580-584.

32. Podolsky, S., Zimmerman, H. J., Burrows, B. A., Cardarelli, J. A. and Pattavina, C. G. (1973): Potassium depletion in hepatic cirrhosis: a reversible cause of impaired growth hormone and insulin response to stimulation. New Engl. J. Med. 288, 644-648.

33. Raskin, P., Fujita, Y. and Unger, R. H. (1975): Effect of insulin-glucose infusions on plasma glucagon levels in fasting diabetics and nondiabetics. J. Clin. Invest. 56, 1132-1138.

34. Reaven, G. and Dray, J. (1967): Effect of chlorpropamide on serum glucose and immunoreactive insulin concentrations in patients with maturity onset diabetes. Diabetes 16, 487-492.

35. Samols, E., Tyler, J. M. and Kajinuma, H. (1971): Influence of the sulfonamides on pancreatic humoral secretion and evidence for an insulin-glucagon feedback system. In: Diabetes. R. R. Rodriguez and J. Vallance-Owen, Eds. Excerpta Med., Amsterdam, pp. 636-655.

36. Samols, E., Tyler, J. M. and Miahle, P. (1969): Suppression of pancreatic glucagon release by the hypoglycemic sulfonylureas. Lancet 1, 174-176.

37. Sheldon, J., Taylor, K. W. and Anderson, J. (1966): The effects of long-term acetohexamide treatment on pancreatic islet cell function in maturity onset diabetes. Metabolism 15, 874-883.

38. Sherwin, R. S., Fisher, M., Hendler, R. and Felig, P. (1976): Hyperglucagonemia and blood glucose regulation in normal, obese and diabetic subjects. New Engl. J. Med. 294, 455-461.

39. Unger, R. H. (1972): Pancreatic alpha cell function in diabetes mellitus. In: Glucagon. Molecular Physiology, Clinical and Therapeutic Implications. P. J. Lefèbvre and R. H. Unger, Eds. Pergamon Press, Oxford. pp. 245-257.

40. Unger, R. H., Aguilar-Parada, E., Müller, W. A. and Eisentraut, A. M. (1970): Studies of pancreatic alpha-cell function in normal and diabetic subjects. J. Clin. Invest. 49, 837-848.

41. Unger, R. H., Madison, L. L. and Müller, W. A. (1972): Abnormal alpha cell function in diabetes: response to insulin. Diabetes 21, 301-307.

42. Yalow, R. S. and Berson, S. A. (1960): Plasma insulin concentrations in non-diabetic and early diabetic subjects. Determinations by a new sensitive immuno-assay technic. Diabetes 9, 254-260.

EFFECT OF CHLORPROPAMIDE ON BASAL AND ARGININE-STIMULATED PLASMA LEVELS OF GLUCAGON AND INSULIN IN DIABETIC PATIENTS

A. H. Telner, C. I. Taylor, S. Pek, R. L. Crowther, J. D. Floyd, Jr. and S. S. Fajans

Department of Internal Medicine, Division of Endocrinology and Metabolism and The Metabolism Research Unit, The University of Michigan
Ann Arbor, Michigan USA

To ascertain whether chronic administration of a sulfonylurea compound affects plasma levels of glucagon, the effect of chlorpropamide (CP) on plasma glucagon was determined in 17 patients with diabetes mellitus in the basal state and in response to infused arginine. Plasma levels of insulin were measured also. In these patients the increases in plasma levels of glucagon in response to arginine were significantly reduced by prior chlorpropamide administration. Arginine-stimulated levels of insulin were reduced significantly in 10 lean patients.

A significant reduction in arginine-stimulated increases in plasma glucagon (and insulin) occurred also in patients in whom basal and arginine-stimulated levels of glucose were not altered significantly by administration of CP. Conversely, CP lowered mean basal levels of plasma glucose although basal levels of glucagon were not changed significantly. This suggests that CP has independent effects upon plasma glucagon and plasma glucose.

The mechanisms by which chronic administration of sulfonylureas lower plasma glucose in patients with diabetes mellitus are poorly understood. Observations of changes in plasma levels of insulin in diabetic patients after chronic administration of sulfonylureas differ. Evidence has suggested that prolonged treatment with sulfonylureas of patients with diabetes mellitus is associated with an increase (11, 24), a decrease (21, 26) or no change (2, 4, 18) in their plasma levels of insulin during glucose tolerance tests. In order to explain the blood glucose-lowering effect of chronic sulfonylurea therapy, investigators have looked for extrapancreatic actions of sulfonylureas (7) or for changes in secretion of other glucoregulatory hormones associated with sulfonylurea therapy.

In patients with diabetes mellitus plasma levels of glucagon are frequently elevated or inappropriately increased for the prevailing plasma concentrations of glucose (3, 17, 27). The question of whether the hyperglucagonemia is primary or secondary to insulin insufficiency is unresolved, although recent evidence has strengthened the latter interpretation (20). Since the initiation of this study, reports have indicated that the oral administration of sulfonylurea drugs is followed by a decrease of plasma levels of glucagon in patients with diabetes mellitus (6, 12, 18). Such an effect of sulfonylureas has not been observed in healthy subjects (12, 14, 19). Several authors have shown that in ducks (9, 23) and in the perfused rat pancreas (10) the acute administration of sulfonylureas suppresses glucagon secretion, but in normal dogs there is no convincing evidence that sulfonylureas suppress plasma levels of glucagon in vivo (1, 9, 13) or in vitro (8).

We have studied a heterogenous group of patients with diabetes mellitus to determine whether prolonged administration of a sulfonylurea compound affects plasma levels of glucagon in the basal state and after stimulation with arginine. Plasma levels of insulin were also measured. The results indicate that treatment with chlorpropamide (CP) is associated with a reduced response to intravenous arginine in plasma levels of glucagon and insulin.

PATIENTS, MATERIALS AND METHODS

Seventeen patients with diabetes mellitus who were not treated with insulin were studied. Six of these patients did not have fasting hyperglycemia (blood glucose greater than 120 mg/dl) but have had on numerous occasions over many years glucose tolerance tests which were diagnostic of diabetes mellitus according to the criteria of Fajans and Conn (5) and those of the U. S. Public Health Service (22, Table 1). All patients were following a diet and remained on the same diet during the entire study period. The nature of the study was explained to each patient, and informed consent was obtained.

Two experiments were performed on each patient. The control or "No CP" experiments were done either before CP was administered to the patients or two to three weeks after CP treatment had been discontinued (9 patients). For the "On CP" experiments the patients took CP orally, 100-500 mg daily, for 2 to 260 weeks. The last dose of CP was administered 24 hours prior to the beginning of the experiment.

The experiments were done after the patients had fasted a minimum of eight hours. A butterfly needle connected to a three-way stopcock was placed in a forearm vein and kept patent with heparin; blood samples in the basal state were obtained at -30, -15, and 0 minutes; a solution of 0.574 M l-arginine monohydrochloride (2.35 mmol/kg of body weight, maximal dose 172 mmol) was infused at a constant rate over a period of 30 minutes (0 to 30 minute time points). Blood samples were obtained 3, 5, 7, 10, 20 30, 40, 50, 60, 75, 90, 105, and 120 minutes after the start of the infusion. Blood samples were collected in heparinized tubes, chilled immediately and centrifuged at 4°C within 2 hours. The plasma was separated and stored at -20°C until assay.

TABLE 1

GLUCOSE TOLERANCE TESTS* OF PATIENTS NOT HAVING FASTING HYPERGLYCEMIA

	Time (minutes)						
	0	30	60	90	120	150	180
Patient	Plasma glucose (mg/dl)						
1	109	185	252	224	204	202	178
2	89	205	292	352	348	308	220
3	96	172	222	210	196	177	148
4	103	180	205	272	298	340	330
5	100	135	209	217	175	164	131
6	88	190	267	266	256	192	128

*Glucose 1.75 g/kg ideal body weight, orally

Analytical Methods. Plasma levels of glucagon and insulin were measured by radioimmunoassay, using the double antibody separation technique as modified from that described by Morgan and Lazarow (16). In one patient who had received insulin therapy in the past, plasma insulin could not be measured due to the presence of endogenous insulin antibodies.

In the glucagon assay the rabbit anti-bovine and porcine glucagon serum G9-I was employed. The immunoglobulins of this antiserum react with the carboxyl-, but not with the amino-terminal residues of the glucagon molecule; they do not react with glucagon-related compounds, such as gastric inhibitory polypeptide, secretin, somatostatin and the threonine-phenylalanine-threonine-serine tetrapeptide, nor with gastrin, insulin and proinsulin. The affinity of the immunoglobulins for glucagon-like compounds present in a purified extract of porcine intestinal mucosa ("MUC-101", Novo Research Institute, Copenhagen, Denmark) is 2 to 5% of their affinity for pancreatic glucagon. The sensitivity of the assay is 3.6 ± 1.7 (mean ± SD, N = 72) pg/ml. Plasma glucose was determined by an automated method using glucose oxidase (15). To avoid the effects of interassay variability plasma samples from paired tests were run in the same assay. Basal values were derived by averaging the 3 values (-30, -15 and 0 minutes) obtained before starting the arginine infusion. The cumulative hormonal responses to arginine were calculated as the sum of the partial areas for each time interval under the 0 to 120 minute response curve (total area). The formula used was $A_p = \frac{V_{t-1} + V_t}{2} \cdot [t-(t-1)]$ where A_p is the partial area, V_{t-1} the individual value at time t-1 and V_t the value at time t. The incremental area was computed by subtracting an area which corresponded to the product of the basal value by 120 minutes (basal area) from the total area.

The data were analyzed for the whole group of 17 patients as well as for those patients whose basal levels of plasma glucose while receiving CP were within 10% of their basal glucose without CP. These patients were assumed to have "similar glucose levels" on paired tests. The results of tests derived from the two groups are reported as means ± standard errors. The statistical significance of the differences between groups of data was assessed by the two-tailed Student's t test for paired observations (25).

RESULTS

The results are given in Table 2. In the 17 patients, mean basal levels of glucagon were not changed by CP although mean basal levels of glucose were significantly reduced. However, plasma levels of glucagon in response to arginine were significantly lower with CP treatment than those without CP (Fig. 1, Table 2).

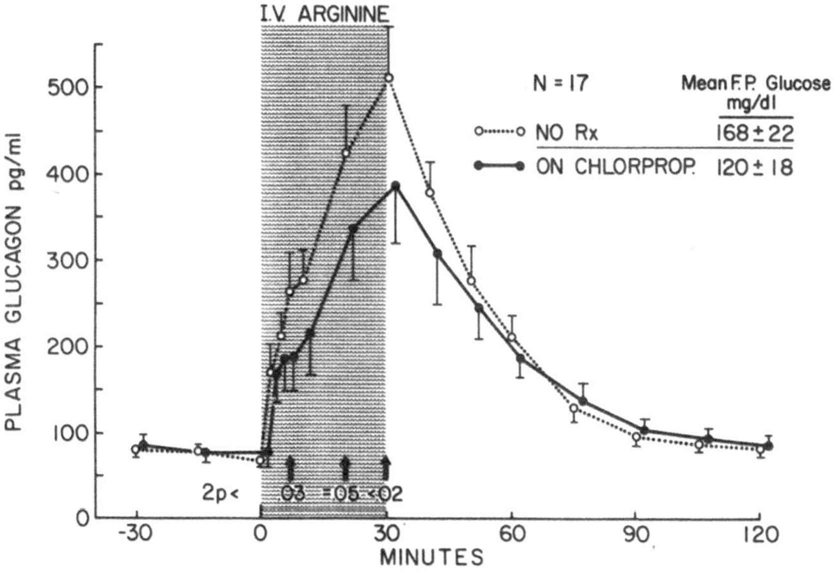

Fig. 1. Plasma glucagon responses to infused arginine (mean ± SE) without and with CP treatment in 17 patients not receiving insulin. Arrows signify time points at which corresponding values are significantly different.

With CP, there was no correlation between the degree of lowering of arginine-induced increases in glucagon and the degree of lowering of basal levels of glucose (r=0.01). In response to infused arginine the mean of maximal increments in

TABLE 2

EFFECT OF CHLORPROPAMIDE ON BASAL AND ARGININE-STIMULATED PLASMA LEVELS OF GLUCAGON AND INSULIN

(Mean ± SEM)

A - 10 males, 7 females; age 42.0 (20-58) years; body weight 109.9 (88-146) % of ideal
B - 5 males, 1 female; age 40.5 (24-52) years; body weight, 106.1 (88-127) % of ideal

	Plasma Glucagon				Plasma Insulin				Plasma Glucagon	
	Basal pg/ml		Arginine Response* ng.min/ml		Basal U/ml		Arginine Response* mU·min/ml		Basal mg/dl	
	No CP	On CP	No CP	On CP	No CP	On CP	No CP	On CP	No CP	On CP
A** All Patients	74± 9	76±13	18.4±2.5***	14.1±2.4	13±4	17±7	2.84±0.73	2.51±0.83	168±22****	120±18
B Patients having glucose "similar on paired tests"	59±14	47± 8	16.6±3.2***	10.4±1.7	12±4	12±5	2.45±0.65***	1.43±0	99± 8	98± 9

* 0-120 minute incremental areas

** As one patient had endogenous insulin antibodies, the data on insulin values are for 16 patients

*** 2p <.05

**** 2p <.01

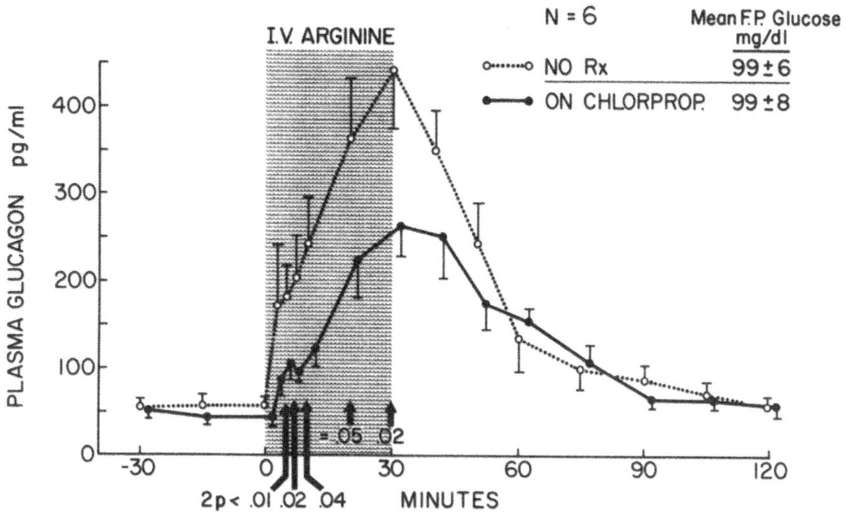

Fig. 2. Plasma glucagon responses to infused arginine in 6 diabetic patients who manifested no change in plasma glucose with CP.

plasma levels of glucagon was 442 ± 52 pg/ml while the patients were not receiving CP and 341 ± 53 pg/ml after CP treatment (2 p <.02). Maximal increments of plasma glucose were 24 ± 3 and 22 ± 3 mg/dl (NS), respectively. The change in the mean maximal increments of plasma levels of glucagon associated with CP administration was not correlated with the change in the mean maximal increments of plasma levels of glucose (v = -.22). The changes in levels of plasma glucagon in response to arginine expressed as incremental areas did not correlate with the dose of CP administered (r = .06).

Plasma levels of insulin in the basal state or in response to arginine were not altered significantly by CP (Table 2).

Patients with "similar basal glucose" on paired tests. In four of the 17 patients basal levels of plasma glucose with CP remained within 10% of the corresponding values observed without CP. In another two patients, data on additional arginine infusions were available from which to choose tests in which the basal plasma levels of glucose were within 10% of each other. In these six patients the mean paired fasting concentrations of glucose were identical (Table 2). In addition, the paired plasma levels of glucose during and following the infusion of arginine were similar (2 p <.05 at only one of 13 time points). In this group mean arginine-stimulated plasma levels of glucagon were significantly lower with CP (Fig. 2, Table 2). The mean plasma insulin response to arginine was also reduced (Table 2).

DISCUSSION

In the 17 diabetic patients the mean basal concentrations of plasma glucagon and insulin were unaffected by administration of chlorpropamide (CP). The fact that CP caused a significant lowering in mean basal levels of plasma glucose in these patients suggests that the mechanism of glucose lowering by the sulfonylurea need not involve changes in plasma levels of these hormones. The lack of suppression by CP of basal levels of plasma glucagon in diabetic patients is in agreement with previous reports (12, 18).

In an effort to uncover a possible effect of CP upon plasma levels of glucagon, arginine was infused so as to magnify changes in plasma glucagon. In the 17 patients, mean increases in plasma levels of glucagon in response to infused arginine were reduced significantly by CP administration. However, there was no correlation between the change in maximal increases in plasma glucagon and the change in maximal increases in plasma glucose in response to arginine which occurred with CP. This observation suggests that factors other than, or in addition to, the glucagon response to arginine are responsible for concomitant changes in plasma glucose. The effect of CP on plasma concentrations of glucagon was evaluated in patients in whom CP did not alter plasma levels of glucose. The mean response of plasma glucagon to arginine in these patients was significantly lower after they had received CP treatment. Thus, in the absence of a glucose-lowering effect, the sulfonylurea can result in a lowering of glucagon.

In a similar fashion it was found that in 6 patients CP lowered the mean plasma insulin response to arginine without having an effect on plasma glucose. This suggests that CP administration may increase the sensitivity of the liver and/or peripheral tissues to endogenous insulin.

Chronic administration of chlorpropamide has effects upon plasma levels of glucagon and insulin independent of changes in plasma glucose. This conclusion is borne out by (a) the reduction in arginine-stimulated plasma levels of glucagon and insulin in groups of patients who manifested no mean change in levels of plasma glucose on paired tests and (b) the lowering of mean basal glucose in the 17 patients who demonstrated no change in the basal levels of glucagon and insulin.

Acknowledgements

This study was supported in part by USPHS Grants AM-00888, and TI-AM-05001 from the National Institute of Arthritis, Metabolism and Digestive Diseases; USPHS Grant 5P11-GM-15559, National Institute of General Medical Sciences; and by grants from the Pfizer Co., New York, and the State of Michigan Department of Public Health.

REFERENCES

1. Aguilar-Parada, E., Eisentraut, A. M. and Unger, R. H. (1969): Effect of HB419 (glibenclamide) induced hypoglycemia on pancreatic glucagon secretion. Horm. Metab. Res. 1, 48-50.

2. Barnes, A. J., Crowley, M. F., Gabrien, K. J. T. and Bloom, A. (1974): Effect of short and long term chlorpropamide treatment on insulin release and blood glucose. Lancet 2, 69-72.

3. Buchanan, K. D. and McCarroll, A. M. (1972): Abnormalities of glucagon metabolism in untreated diabetes mellitus. Lancet 2, 1394-1395.

4. Chu, P., Conway, M. J., Krouse, H. A. and Goodner, C. J. (1968): The pattern of response of plasma insulin and glucose to meals and fasting during chlorpropamide therapy. Ann. Int. Med. 68, 757-769.

5. Fajans, S. S. and Conn, J. W. (1959): The early recognition of diabetes mellitus. Ann. N. Y. Acad. Sci. 82, 208-218.

6. Fallucca, F., Tinelli, F. P., Mirabella, C., Menzinger, G. and Andreani, D. (1975): Effects of glibenclamide on glucagon and growth hormone secretion in insulin treated diabetic patients. Diabetologia 11, 340-341.

7. Feldman, J. M. and Lebovitz, H. E. (1969): Appraisal of the extrapancreatic action of sulfonylureas. Arch. Int. Med. 123, 314-322.

8. Iversen, J. (1970): Secretion of immunoreactive insulin and glucagon from the isolated perfused canine pancreas following stimulation with adenosine 3',5'-monophosphate (cyclic-AMP), glucagon and theophylline. Proc. 7th Congr. Intern. Diabetes Feder. Excerpta Med. Found. I. C. S. 209, 46 (abst.).

9. Kajinuma, H., Kuzuya, T. and Ide, T. (1974): Effects of hypoglycemic sulfonamides on glucagon and insulin secretion in ducks and dogs. Diabetes 23, 412-417.

10. Laude, H., Fussgänger, R., Goberna, R., Schroeder, K., Straube, K., Sussman, K. and Pfeiffer, E. F. (1971): Effects of tolbutamide on insulin and glucagon secretion of the isolated perfused rat pancreas. Horm. Metab. Res. 3, 238-242.

11. Lauvaux, J. P., Mandart, G., Heymans, G. and Ooms, H. A. (1972): Effect of long term tolbutamide treatment on glucose tolerance and insulin secretion in maturity-onset diabetes without obesity. Horm. Metab. Res. 4, 58-62.

12. Loreti, L., Sugase, T. and Foà, P. P. (1974): Diurnal variations of serum insulin, total glucagon, cortisol, glucose and free fatty acids in normal and diabetic subjects before and after treatment with chlorpropamide. Horm. Res. 5, 278-292.

13. Loubatières, A. L., Loubatières-Mariani, M. M., Alric, R. and Ribes, G. (1974): Tolbutamide and glucagon secretion. Diabetologia 10, 271-276.

14. Marco, J. and Valverde, I. (1973): Unaltered glucagon secretion after 7 days of sulfonylurea administration in normal subjects. Diabetologia 9, 317-319.

15. Medzihradsky, F. and Dahlstrom, P. J. (1971): A specific, low cost procedure for the automated determination of blood glucose. Univ. of Mich. Med. Center J. 37, 188-190.

16. Morgan, C. R. and Lazarow, A. (1963): Immunoassay of insulin: two antibody system. Plasma insulin level of normal, subdiabetic and diabetic rats. Diabetes 12, 115-126.

17. Müller, W. A., Faloona, G. R., Aguilar-Parada, E. and Unger, R. H. (1970): Abnormal alpha-cell function in diabetes: response to protein and carbohydrate ingestion. N. Engl. J. Med. 283, 109-115.

18. Ohneda, A., Ishii, S., Horigome, K. and Yamagata, S. (1975): Glucagon response to arginine after treatment of diabetes mellitus. Diabetes 24, 811-819.

19. Pek, S., Fajans, S. S., Floyd, J. C., Jr., Knopf, R. F. and Conn, J. W. (1972): Failure of sulfonylureas to suppress plasma glucagon in man. Diabetes 21, 216-222.

20. Raskin, P., Fujita, Y. and Unger, R. H. (1975): Effect of insulin-glucose infusions on plasma glucagon levels in fasting diabetics and nondiabetics. J. Clin. Invest. 56, 1132-1138.

21. Reaven, G. and Dray, J. (1967): Effect of chlorpropamide on serum glucose and immunoreactive insulin concentrations in patients with maturity onset diabetes mellitus. Diabetes 16, 487-492.

22. Remein, O. R. and Wilkerson, H. L. C. (1961): The efficiency of screening tests for diabetes. J. Chr. Dis. 13, 6-21.

23. Samols, E., Tyler, J. M. and Mialhe, P. (1969): Suppression of pancreatic glucagon release by the hypoglycemic sulfonylureas. Lancet 1, 174-176.

24. Sheldon, J., Taylor, K. W. and Anderson, J. (1966): The effects of long term acetohexamide treatment on pancreatic islet cell function in maturity onset diabetes. Metabolism 15, 874-883.

25. Snedecor, G. W. and Cochran, W. G. (1967): Statistical Methods, 6th Ed., The Iowa State University Press, Ames, Iowa.

26. Turtle, J. R. (1970): Glucose and insulin secretory response patterns following diet and tolazamide therapy in diabetes. Brit. Med. J. 3, 606-610.

27. Unger, R. H., Aguilar-Parada, E., Müller, W. A. and Eisentraut, A. M. (1970): Studies of pancreatic alpha cell function in normal and diabetic subjects. J. Clin. Invest. 49, 837-848.

GLUCAGON AND HYPERTRIGLYCERIDEMIA

A. Tiengo, R. Nosadini, M. C. Garotti, D. Fedele, M. Muggeo and G. Crepaldi

Division of Gerontology and Metabolic Diseases
Department of Internal Medicine, University of Padova
Padova, Italy

The authors report the results of their studies on glucagon secretion in human endogenous hypertriglyceridemia and on its effects in hyperlipemic patients and in isolated rat liver.

In human hyperlipemia, increased insulin secretion is associated with fasting hyperglucagonemia and increased reactivity of the A-cell to arginine. In non-diabetic hyperlipemics, normal sensitivity of A-cells to glucose plus insulin is preserved.

In human hyperlipemia paraphysiologic doses of glucagon (1 μg/kg) exert a hypotriglyceridemic effect similar to that observed in normals. The lipolytic action is lacking, the ketogenic and hyperglycemic effects are increased while the insulinotropic effect of glucagon is normal.

During perfusion of rat liver with oleate, pharmacologic doses of glucagon induced a significant decrease in triglyceride and an increase in the β-hydroxybutyrate output.

Our results in man do not support the theory of peripheral resistance to glucagon in sustaining endogenous hyperlipemia.

Alterations in glucagon and insulin concentrations and in their molar ratio in the portal vein seem to determine the catabolic or anabolic state of the organism (34). It is well known that glucagon and insulin levels in the portal venous and peripheral blood are altered in some pathologic conditions and in changing nutritional states. These modifications are accompanied by alterations in various hepatic functions, such as the production of protein, lipid and glucose. Glucagon and insulin have direct and antagonistic effects upon hepatic metabolism, not only with regard to gluconeogenesis and glycogenolysis but also to lipid and ketone body production.

Endogenous triglyceride synthesis is influenced and conditioned by the amount of energy substrates reaching the liver as well as by the prehepatic insulin and glucagon levels. Variations in insulin and glucagon concentrations and in their metabolic activity may play an important role in determining abnormal triglyceride production by the liver.

Glucagon and Hepatic Lipid Metabolism

Insulin and glucagon have direct and antagonistic effects upon hepatic lipoprotein production. In fact, while glucagon decreases the triglyceride lipoprotein production rate, insulin has the opposite effect. It has been observed that insulin stimulates the incorporation of ^{14}C-glucose into hepatic triglycerides in vitro (24) and the production of VLDL (very low density lipoproteins) triglycerides in the perfused rat liver (33).

Glucagon could promote hypotriglyceridemia by inhibiting the esterification of FFA (free fatty acids) to triglycerides and by shifting the FFA towards intramitochondrial β-oxidation and ketone body synthesis (16).

According to Eaton (6), the hypotriglyceridemic effect of glucagon could also take place through a decrease in the synthesis of the VLDL's protein carriers, that is, of the apoproteins which limit hepatic lipoprotein production. Glucagon might activate the carnitine acyltransferase enzyme, which catalyzes the transfer of long-chain fatty acids across the mitochondrial membrane (19), constituting the first step in fatty acid oxidation. The two pathways of intrahepatic FFA utilization, esterification and β-oxidation, are antagonistic and could be regulated by this enzyme. The possible effect of glucagon on acyl carnitine transferase activity has been demonstrated indirectly by Williamson et al. (38) who observed that decanoylcarnitine, a potent inhibitor of acylcarnitine transferase, blocks glucagon's effect on the perfused liver.

According to Weinstein et al (37), the output of triglyceride and ketone bodies by the isolated perfused rat liver depends upon the concentration of glucagon in the perfusate. Glucagon has less effect on triglyceride output than on glucose, urea, and ketone body production.

Our recent experiments using perfused rat liver confirm that glucagon, at a concentration of 10^{-5}M, is capable of completely inhibiting triglyceride production, even in the presence of sodium oleate (Fig. 1). A FFA infusion, in the absence of glucagon, provoked a large output of triglycerides by the liver. The

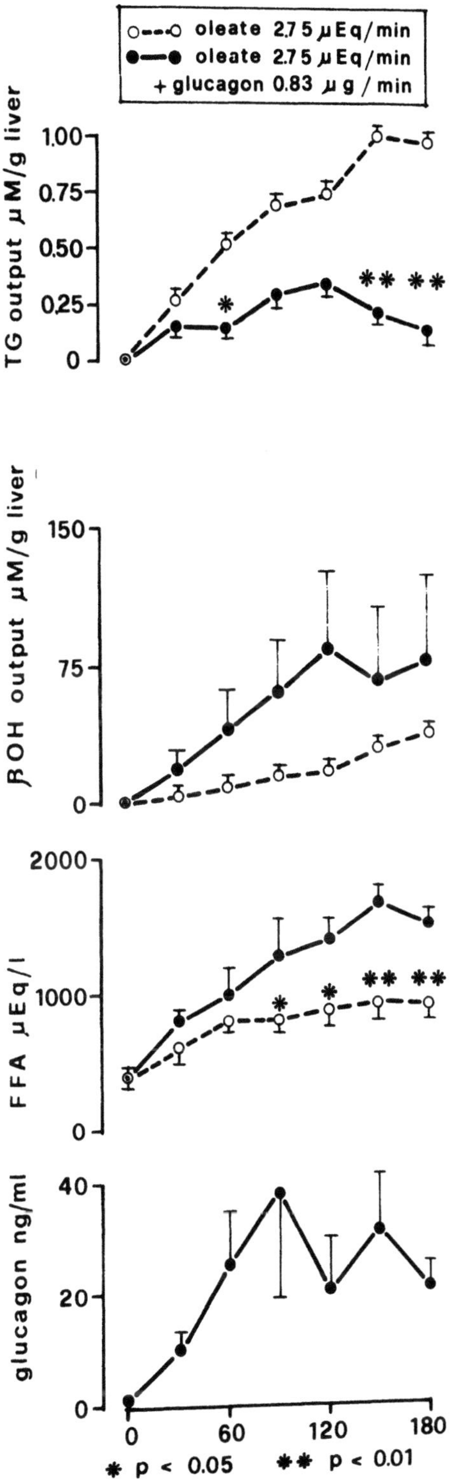

Fig. 1. Effect of oleate with or without glucagon on triglyceride and β -hydroxybutyrate output and on free fatty acid and glucagon levels (Mean ± S.E.; n = 8). Oleate (496 μEq in 60 ml of perfusion medium) was infused over a 3 hour period with or without 150 μg of glucagon.

block in triglyceride formation is associated with greater β-hydroxybutyrate production.

During oleate perfusion FFA levels in the perfusate were higher when glucagon was added to the medium. This would seem to suggest that glucagon induces a decrease in FFA liver uptake. Heimberg (16), on the contrary, noted no alteration in free fatty acid liver uptake following the addition of glucagon.

Our experiments and those of Heimberg and Weinstein seem to suggest that glucagon suppresses triglyceride and enhances ketone body production in the liver directly. As demonstrated by Bates et al. (3), insulin does not seem to interfere with the ketogenic effect of glucagon. Glucagon's autonomous ketogenic effect has been demonstrated by McGarry et al. (20), who obtained a ketogenic liver from a nonketotic animal treated with glucagon.

Glucagon might influence not only triglyceride synthesis but also its removal, by stimulating the PHLA (post heparin lipolytic activity; 5).

These data seem to confirm glucagon's hypotriglyceridemic effect and suggest the possibility that hypertriglyceridemic states may be due to a defect in the hypolipidemic action of glucagon.

Endogenous hypertriglyceridemia, or carbohydrate-induced hyperlipemia, is characterized by an increased output of VLDL triglycerides. The perfusion of oleate induced a significant increase in serum triglyceride production in the hyperlipemic livers of fructose-fed rats (28). Carbohydrate feeding did not result in an alteration in FFA uptake or in the production of triglycerides containing fatty acids synthesized by the liver. On the contrary, the increase in VLDL synthesis is derived from exogenous FFA (28). In endogenous hypertriglyceridemia, the liver extracts an abnormal amount of FFA from the plasma, esterifies the FFA and converts them into VLDL triglycerides.

In the "fed state", the FFA are utilized by the hepatocyte for triglyceride synthesis. In the "fasting-state", on the contrary, FFA undergo β-oxidation for ketone body synthesis. In diabetes, both metabolic pathways of FFA utilization are activated simultaneously, inducing a maximal intrahepatic triglyceride synthesis and the development of diabetic ketoacidosis. In endogenous hypertriglyceridemia, the esterification pathway and triglyceride synthesis are particularly enhanced.

As previously demonstrated, glucagon interferes with the metabolic steps of FFA utilization and with lipoprotein production. An alteration in glucagon secretion or a defect in glucagon's metabolic action might contribute to the genesis or maintenance of the hyperlipoproteinemic states. According to Eaton et al. (12), this hormone's activity is reduced in various hyperlipemic states due to diet, to the use of drugs, or genetic abnormalities.

Glucagon Secretion in Endogenous Hypertriglyceridemia

In endogenous hypertriglyceridemia, the use of hypercaloric, carbohydrate-rich diet, the availability of fuel and the frequent association with obesity should result in a persistent suppression of circulating glucagon levels (13, 39) and a consequent increase in the insulin/glucagon molar ratio favoring the synthesis of endogenous triglycerides.

In contrast with this hypothesis, Eaton (6) demonstrated that hypertriglyceridemia is associated with hyperfunction of the A-cell in rats with experimental hyperlipemia. An increase of GLI (glucagon-like-immunoreactivity) levels has been reported in fasting man by Marks (18) and by our group (Fig. 2). An arginine-induced hyperglucagonemia was observed in some hyperlipoproteinemic patients (Fredrickson's Types IV and V) by Eaton (8). This arginine induced glucagon hypersecretion was confirmed in a group of 13 patients with hyperprebetalipoproteinemia using an antiglucagon serum either specific or non-specific for pancreatic glucagon (30, 31). Glucagon's hyperresponse to arginine was present in patients with hypertriglyceridemia associated with normal or impaired glucose tolerance (Fig. 3). No correlation was found between fasting glucagon and triglyceride levels while a positive correlation existed between the insulin/glucagon

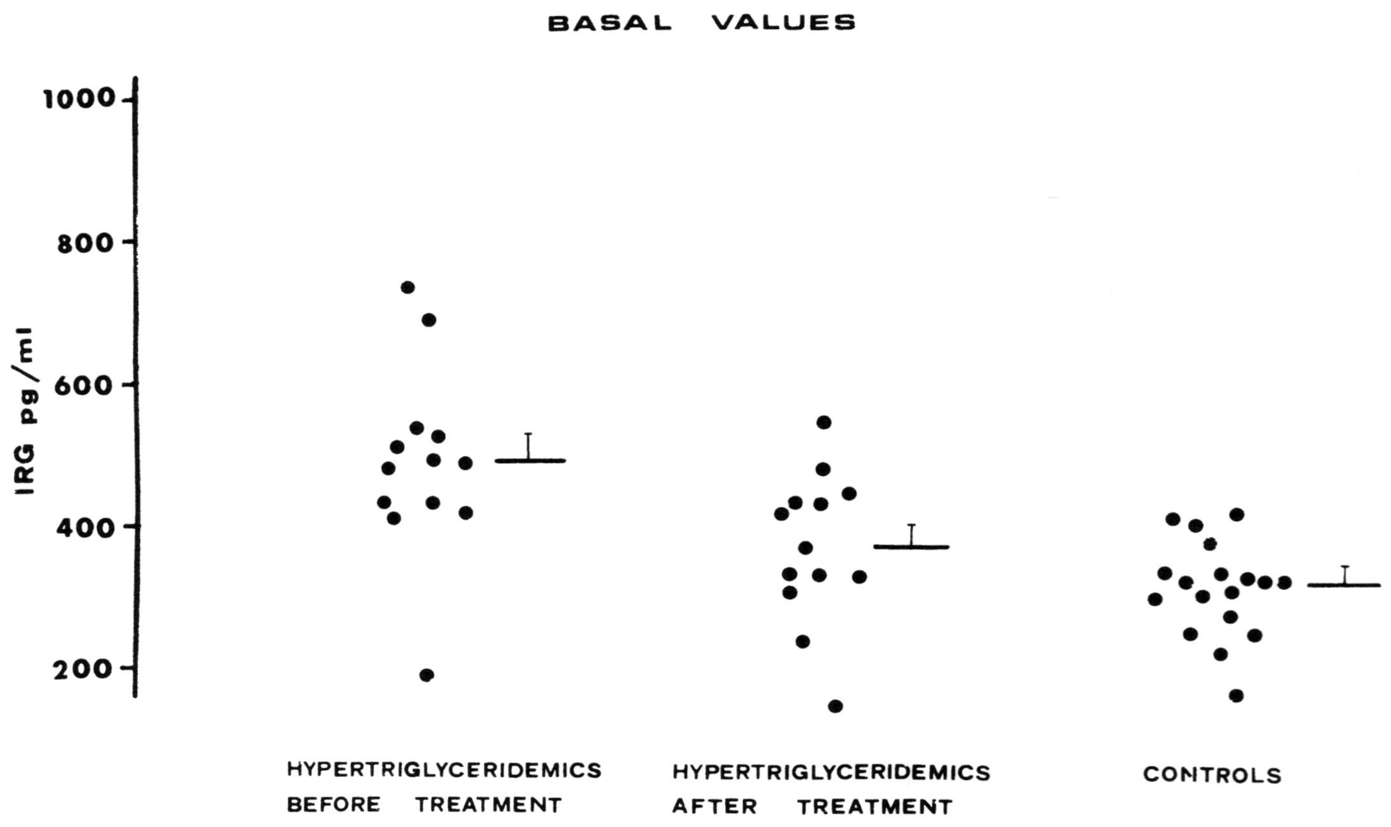

Fig. 2. Fasting plasma GLI (glucagon-like-immunoreactivity) in hypertriglyceridemic patients before and after clofibrate treatment and in a control group.

molar ratio and the triglyceride values.

After two months of clofibrate treatment (Fig. 4) and with reduction in plasma triglyceride levels, the production of glucagon in response to arginine was even more elevated, while that of insulin was reduced. The reduction in triglyceride levels induced by clofibrate was correlated to the reduction in the insulin/glucagon molar ratio (Fig. 5). These results are in agreement with those reported by Eaton and Schade in hyperlipemic rats (10) and in normal humans (11). Clofibrate could exert its hypolipidemic effect by producing a new hormonal adjustment. As affirmed by Eaton, relative insulin deficiency and glucagon excess could contribute to a reduction in the concentration of plasma triglyceride by inhibiting the hepatic synthesis of prebetalipoproteins. In genetic hyperlipemia, the lipid lowering action of other hypolipidemic drugs, such as halofenate (9) may reflect the altered bihormonal equilibrium induced by the drug.

Triglycerides do not stimulate glucagon secretion. In fact, the intravenous infusion of long chain triglycerides does not modify the concentration of glucagon under fasting conditions (4) or the glucagon response to arginine (3; Fig. 6).

In hyperlipemia, hyperglucagonemia may manifest itself even before reduced glucose tolerance and overt diabetes appear. On the other hand, hyperlipemic patients frequently have a strong family history of diabetes and a high frequency of impaired glucose tolerance and microangiopathy. Therefore, hyperglucagonemia could be due to a diabetic-like genetic alteration in the A cell function (35), even in the absence of overt diabetes. Moreover, the hyperproduction of glucagon in endogenous hypertriglyceridemia, as in maturity onset diabetes, cannot be explained by the lack of available insulin.

In human maturity onset diabetes it has been reported that glucagon levels are not suppressed by carbohydrate ingestion (21) or by exogenous insulin infused in supraphysiologic amounts during a large carbohydrate meal or during glucose infusion (35, 36). It has been suggested that the A-cell abnormality in human diabetes is not caused by insulin deficiency, but rather represents a pri-

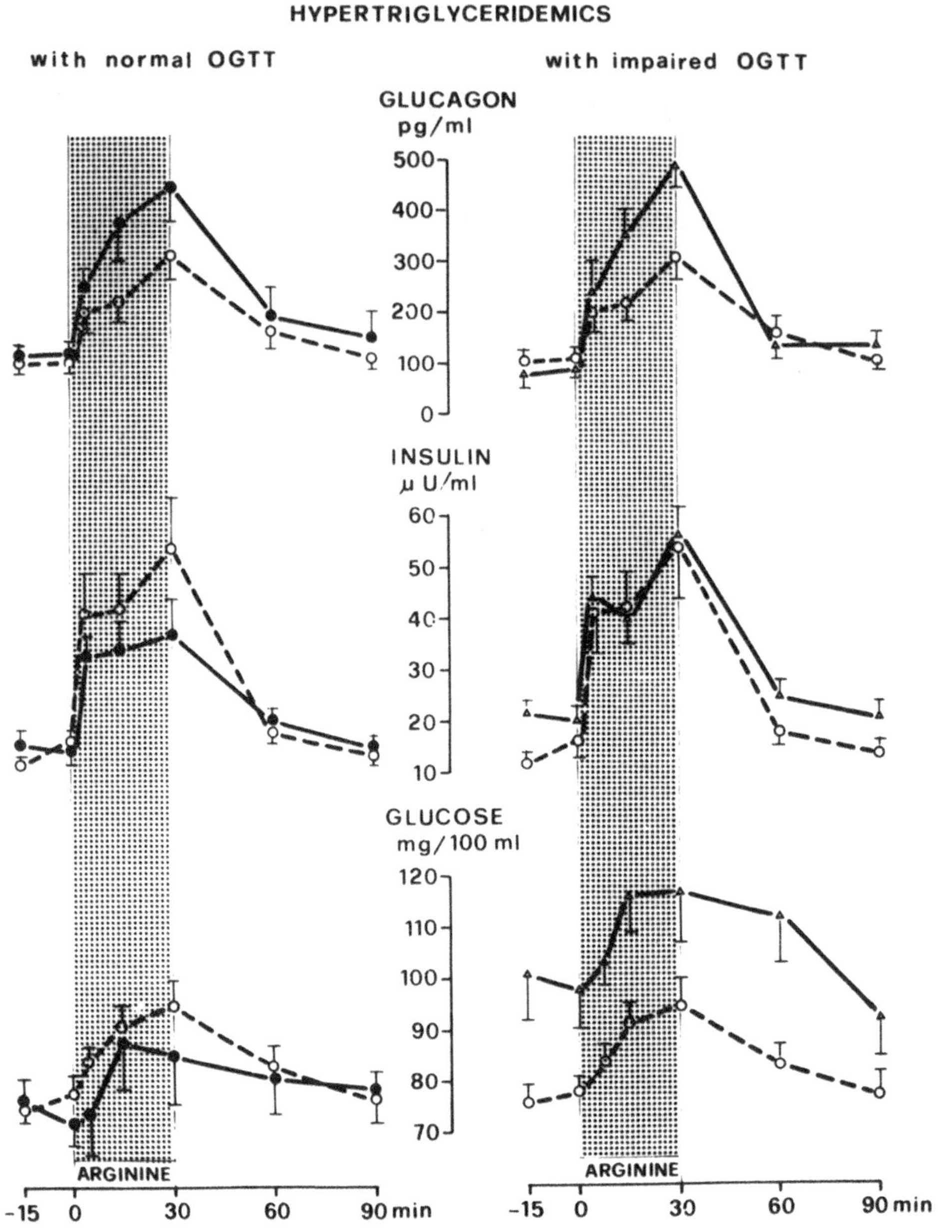

Fig. 3. Glucagon, insulin and glucose response to arginine (25 g in 30 min) in two groups of hypertriglyceridemic patients with normal and impaired glucose tolerance and in a control group (---).

mary inability of the A-cell to respond to glucose and to insulin (35). In order to investigate whether this disorder of the A-cell exists in endogenous hypertriglyceridemia as well, glucose and insulin infusions were carried out in two groups of hypertriglyceridemic patients with normal or impaired glucose tolerance. In non-diabetic hypertriglyceridemia, normal glucagon suppression was observed during glucose plus insulin infusion (Fig. 7). On the contrary, in diabetic hypertriglyceridemia, despite the presence of supraphysiologic insulin levels, little or no decrease in glucagon was observed. The same trend was observed during oral glucose loading (Fig. 8). Normal A-cell sensitivity to glucose and insulin was, therefore, demonstrated in hyperlipemia in the absence of diabetes. It is possible, on the other hand, that A-cell suppression by glucose and insulin involves mechanisms other than those implicated in the aminogenic response.

Metabolic Effects of Glucagon in Patients with Endogenous Hypertriglyceridemia

In man, glucagon stimulates hepatic glucose production (29), ketogenesis (17) and lipolysis (25). It has been demonstrated that pharmacologic doses of insulin reduce triglyceride levels even in endogenous hypertriglyceridemia (1,2). Paloyan and Harper (23) had previously demonstrated that the experimental destruction of the pancreatic A-cells induces a hyperlipemia that is reversible only following large doses of glucagon. It is still unclear if physiologic or paraphysiologic doses of glucagon, capable of producing circulating levels similar to those present in the portal vein, have the same hypotriglyceridemic effect as that observed in normal subjects (25). Glucagon was administered in doses of 1 or 2 μg/kg in hyperlipemic patients with normal or impaired glucose tolerance (22). Its effects on the endogenous production of glucose and triglycerides and on lipolysis and ketogenesis were evaluated. A significant decrease in triglyceride levels was observed in hyperlipemic and in control subjects. The maximum percent and the absolute triglyceride decrease were not significantly different in the two groups (Fig. 9). A defect in glucagon's hypotriglyceridemic effect, therefore,

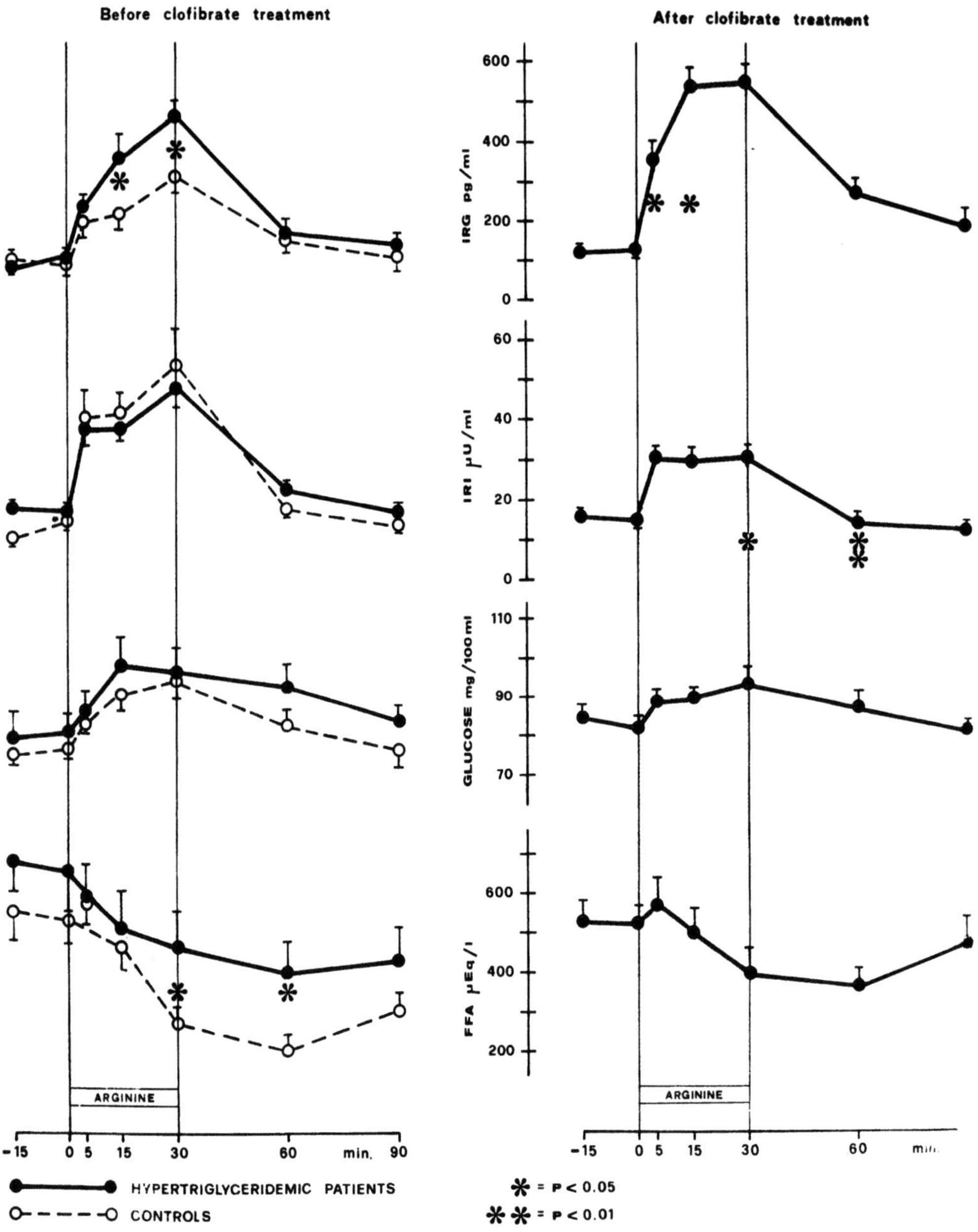

Fig. 4. Glucagon (IRG), insulin (IRI), glucose and free fatty acid (FFA) response to arginine (25 g in 30 min) in hypertriglyceridemic patients and in control subjects before and after clofibrate treatment (31, with permission).

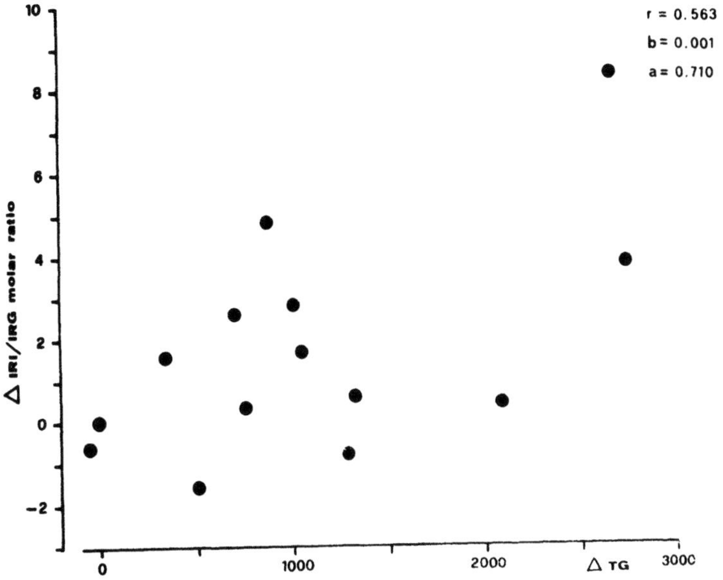

Fig. 5. Correlation between the reduction in triglyceride levels (Δ TG) and that in the IRI/IRG molar ratio (Δ IRI/IRG), induced by clofibrate treatment in hypertriglyceridemic patients (31, with permission).

does not seem to be implicated in endogenous hypertriglyceridemia. In patients and control subjects either dose of glucagon induced a significant glucose increase which was higher in the hyperlipemic patients. The insulinotropic effect of glucagon in hyperlipemic subjects was not significantly different from that of the controls (Fig. 10). In both groups, glucagon induced a decrease in circulating alanine, reflecting its utilization in enhanced gluconeogenesis. Glucagon's lipolytic effect, which was similar after both doses, was evaluated by means of the variations in the plasma concentrations of FFA and of glycerol. The significant FFA increase, also observed by Schade and Eaton (25) in control groups, was not noted in our hyperlipemic patients with normal or impaired glucose tolerance (Fig. 11). The area representing the increase in FFA within 20 minutes was significantly higher in the control than in the hyperlipemic group. The succes-

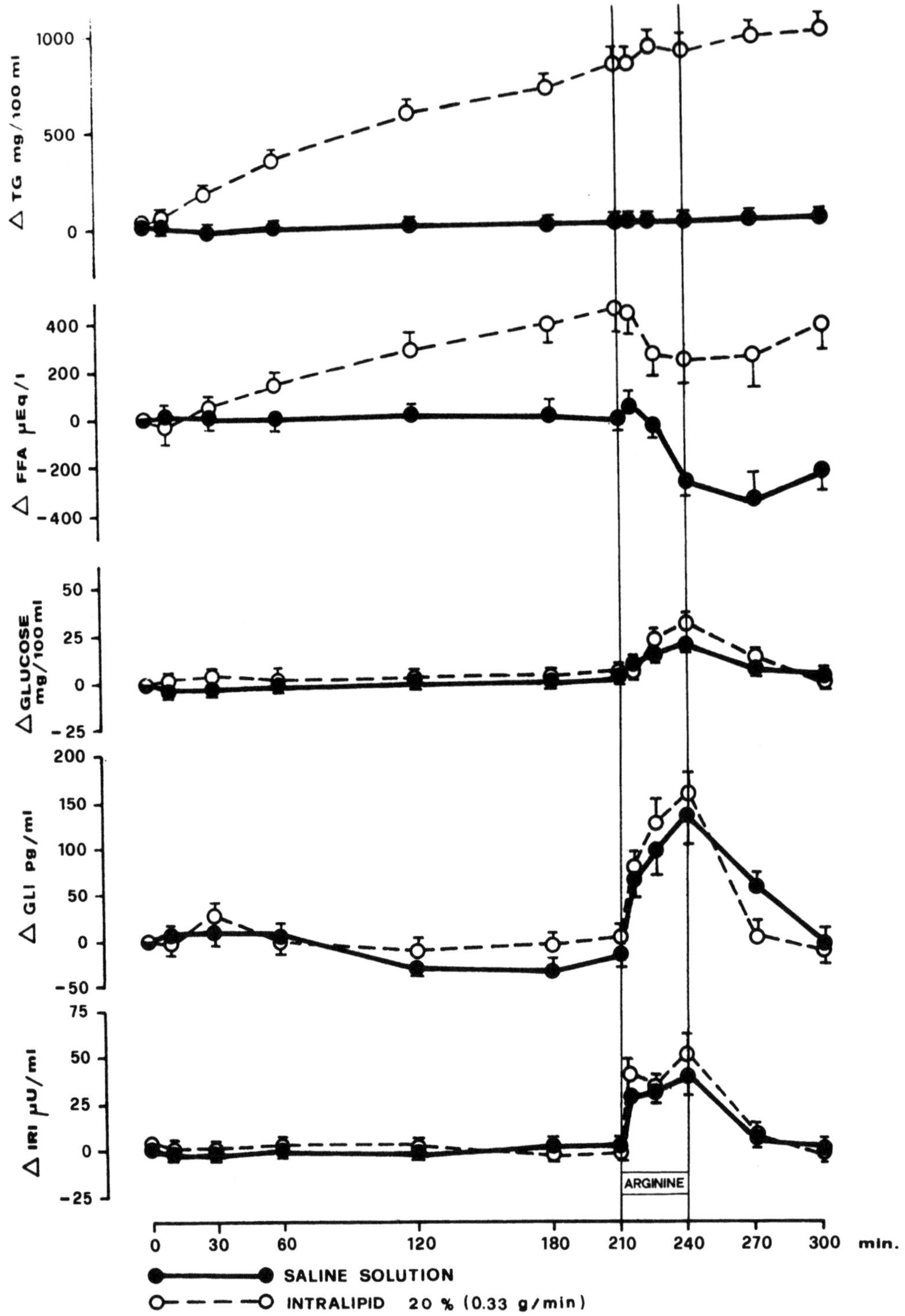

Fig. 6. Effect of continuous Intralipid infusion (0.33 g/min) on fasting triglycerides, FFA, glucose, GLI, IRI, and their respective responses to arginine in normal subjects as compared to the effects of saline infusion used as a control (31, with permission).

sive decrease in FFA after 20 minutes was more marked in hyperlipemic patients even though no significant intergroup variations were observed. Insignificant increases in glycerol levels were observed throughout the test. As already demonstrated in obese patients (9), the lipolytic effect of glucagon is not evident in those who have hyperlipemia since insulin's antilipolytic effect prevails. In metabolic conditions characterized by low insulin levels, Schade and Eaton (26) and Liljenquist *et al* (17) demonstrated a marked lipolysis by glucagon.

Glucagon's ketogenic effect was evaluated by measuring the increase in β-hydroxybutyrate (Fig. 12). This increase was greater in non-diabetic and diabetic hyperlipemic patients than in normal subjects. The total β-hydroxybutyrate increase in the first 20 minutes of the test, evaluated by the area under the curve, was significantly higher only in the diabetic, hyperlipemic groups. Despite elevated levels of circulating insulin, the ketogenic effect of glucagon is greater in the hyperlipemic than in the normal groups. This confirms the direct and autonomous effect of glucagon on ketogenesis. Moreover, we have not observed any correlation between insulin levels and ketone body production. Glucagon's ketogenic effect cannot be attributed to the hormone's lipolytic action. Glucagon stimulation of ketone body production by the liver has already been discussed. Further evidence of its physiologic role in ketone regulation has been provided by *in vivo* observations that the acute suppression of endogenous glucagon secretion by somatostatin is accompanied by a parallel reduction in the rise in plasma ketone body concentrations following insulin withdrawal (14). Physiologic levels of glucagon can stimulate hyperketonemia in man (15). Other studies have demonstrated a close correlation between glucagon values and the degree of ketoacidosis in human diabetes. Glucagon-induced hyperketonemia, associated with hyperglycemia and relative hyperinsulinemia, increases the similarity between hyperlipemia and diabetes.

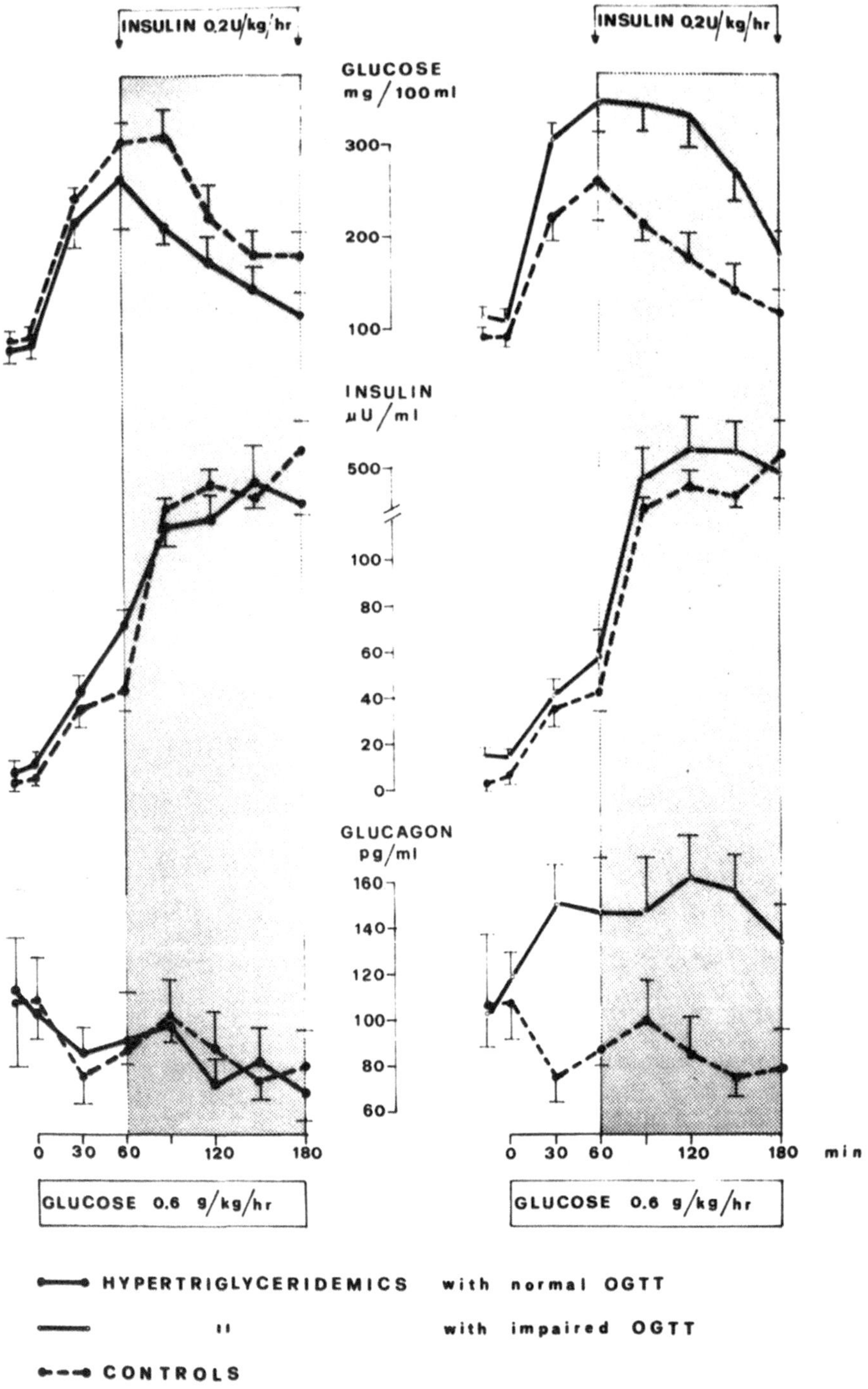

Fig. 7. Effect of glucose and insulin infusion on glucose, IRI, and IRG levels in two groups of hypertriglyceridemic patients with normal or impaired glucose tolerance and in a control group.

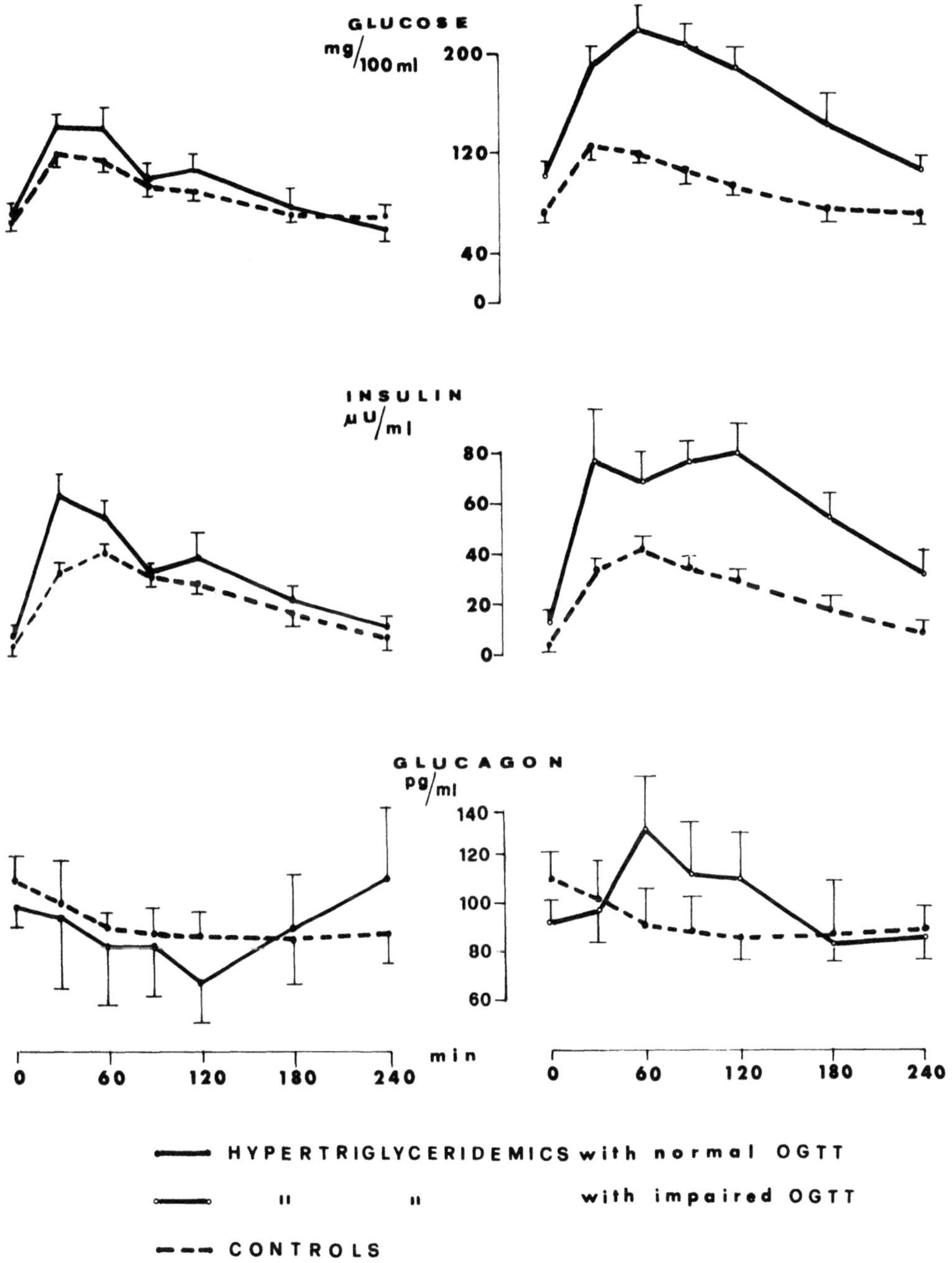

Fig. 8. Effect of oral glucose loading (100 g) on glucose, IRI, and IRG levels in two groups of hypertriglyceridemic patients with normal and impaired glucose tolerance and in a control group.

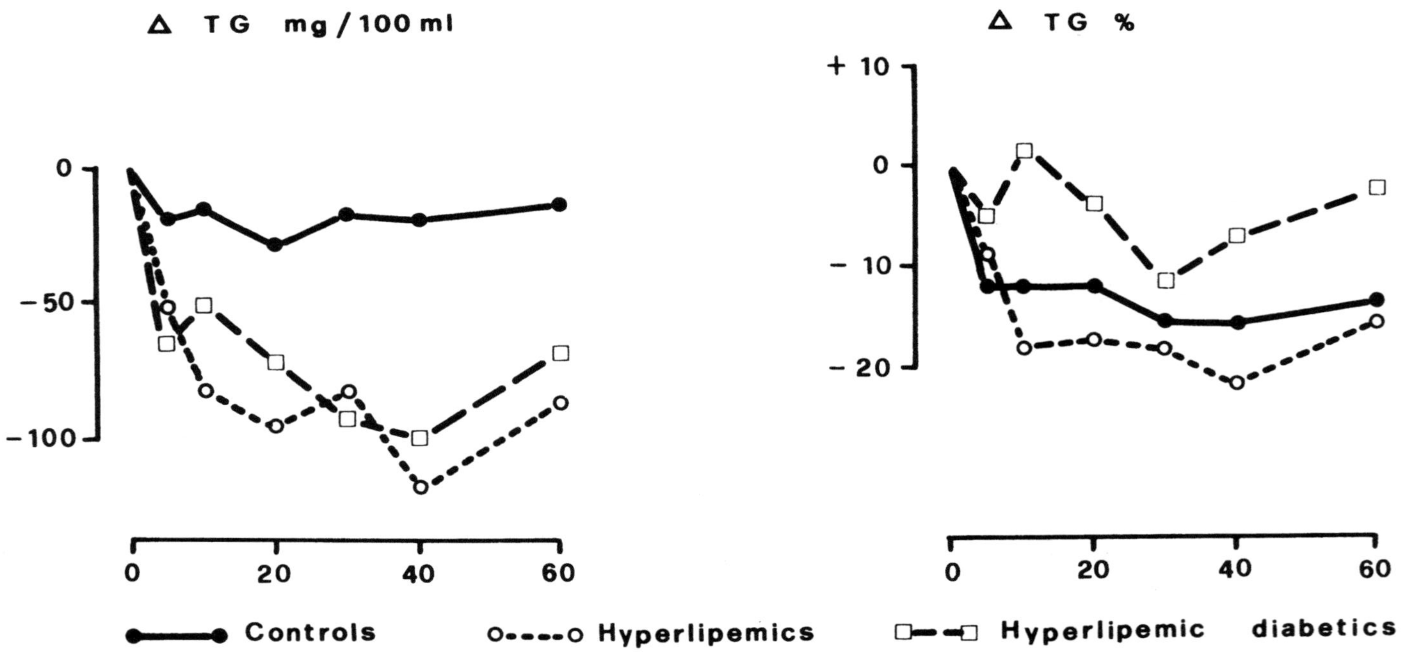

Fig. 9. Absolute and percent decrements in triglyceride after an intravenous injection of glucagon (1 μg/kg) in hyperlipemic patients, with or without chemical diabetes and in control subjects (22).

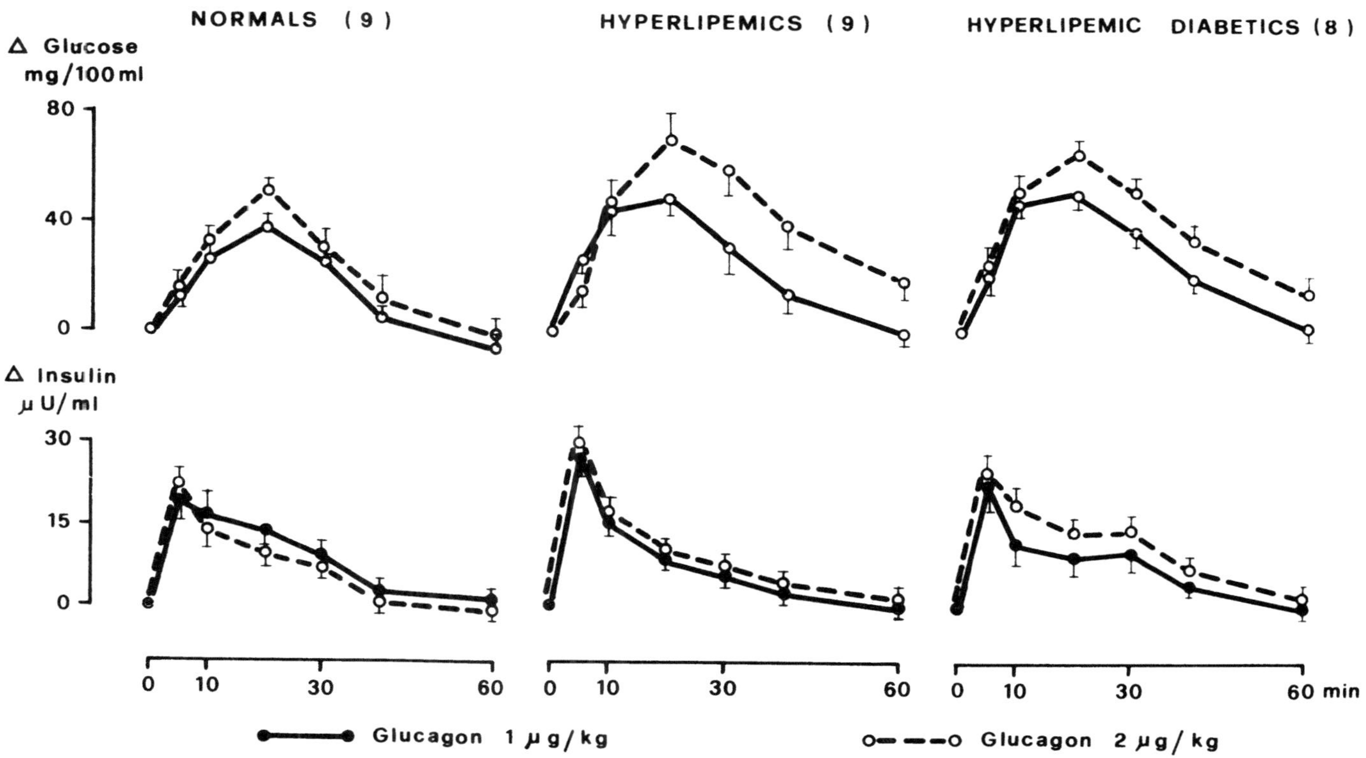

Fig. 10. Increment in glucose and insulin levels after an intravenous injection of glucagon (1 μg/kg) in hyperlipemic patients, with or without chemical diabetes and in control subjects (22).

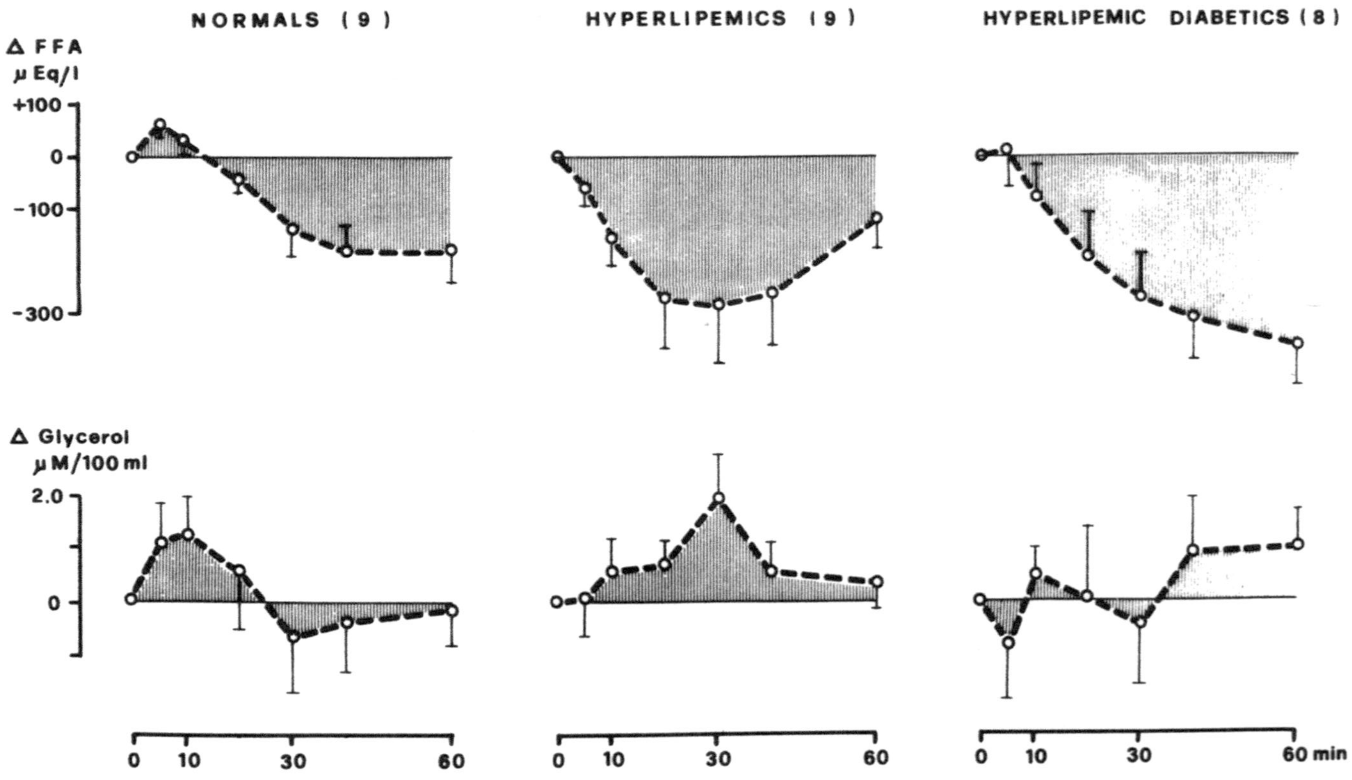

Fig. 11. Variations in free fatty acids (FFA) and glycerol levels, with respect to basal values, following the intravenous injection of glucagon (1 μg/kg) in hyperlipemic patients, with or without chemical diabetes and in control subjects (22).

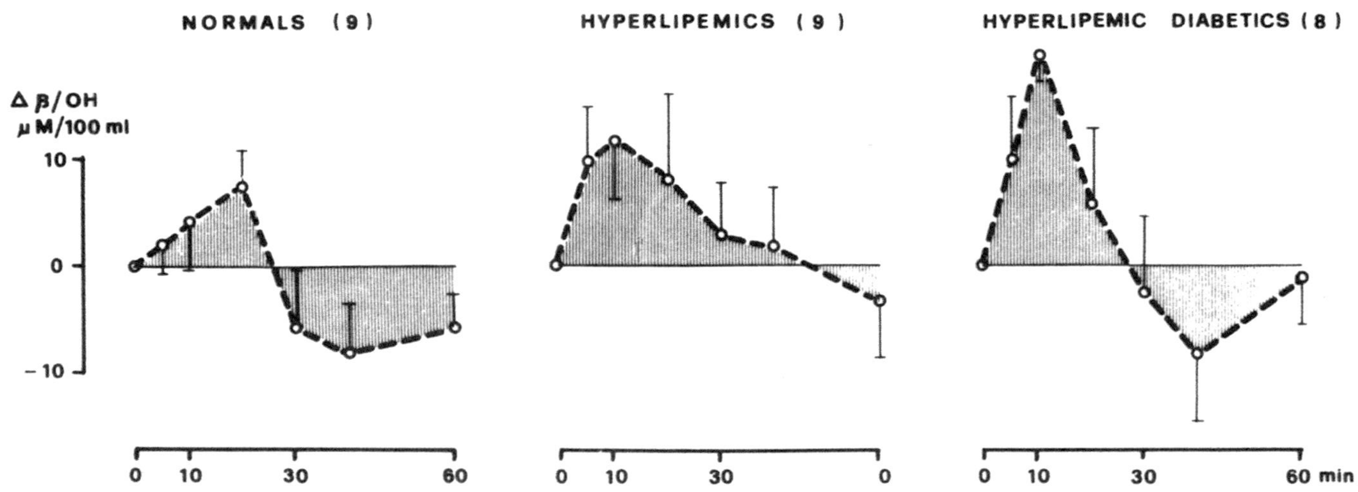

Fig. 12. Variations in β-hydroxybutyrate (β-OH) with respect to basal values, following an intravenous injection of glucagon (1 μg/kg) in hyperlipemic patients, with or without chemical diabetes and in control subjects (22).

In endogenous hypertriglyceridemia, some of glucagon's effects seem enhanced while others are unvaried or depressed. In most cases, however, they can be demonstrated suggesting that human hyperlipemia is not associated with a possible defect in glucagon action, such as that reported in experimental rat hyperlipemia by Eaton (16).

CONCLUSIONS

The bihormonal control of hepatic lipoprotein production and of ketogenesis by insulin and glucagon has been confirmed experimentally in the perfused liver and in vivo. Glucagon seems to play a determining role in stimulating FFA β-oxidation and, therefore, ketone body formation.

Arginine-induced hyperglucagonemia, present in endogenous hypertriglyceridemia, suggests the presence of a state of glucagon resistance. Moreover, glucagon's hypotriglyceridemic effect has been demonstrated in vivo in normal and in hypertriglyceridemic subjects. The hyperglycemic, insulinopoietic, and ketogenic effects of glucagon seem to be enhanced in endogenous hypertriglyceridemia, as they are in maturity onset diabetes. From these results, it is not possible to attribute the development of human hyperlipemia to a reduced metabolic effect of glucagon. Similarly, the arginine-induced hyperglucagonemia may be not as much an expression of glucagon resistance as of a primitive hyperactivity of the A-cell such as that observed in the diabetic disease. The hyperresponse of the A-cell to arginine in hyperlipemia without diabetes is not accompanied by an abnormal A-cell sensitivity to glucose and insulin, as it is in hyperlipemia associated with chemical diabetes. The meaning and metabolic consequences of hyperglucagonemia in endogenous hypertriglyceridemia is still unclear. Even if it is not possible to attribute this metabolic disorder to a reduced metabolic activity of the hormone in the organism as a whole, an endogenous glucagon resistance at the hepatic level cannot be excluded. Research dealing with glucagon's effects on perfused liver or on isolated hepatocytes in genetic or acquired hyperlipemia could help clarify this problem.

Acknowledgements

We wish to thank Drs. F. De Biasi, E. Marchiori, and G. Soldà for their helpful technical assistance and L. Inverso for the translation and editing of the manuscript.

We are indebted also to Dr. Bartosek, Istituto Mario Negri, Milano, for his valuable assistance in setting up the liver perfusion technique employed in our laboratory.

This work was supported in part by grant CT 75.00636.04 from the Consiglio Nazionale delle Richerche, Roma.

REFERENCES

1. Amatuzio, D. S., Grande, F., and Wada, S. (1962): Effect of glucagon on the serum lipids in essential hyperlipemia and hypercholesterolemia. Metabolism 11, 1240-1249.

2. Aubry, F., Marcel, L. Y. and Davignon, J. (1974): Effects of glucagon on plasma lipids in different types of primary hyperlipoproteinemia. Metabolism 23, 225-238.

3. Bates, M., Linn, L. and Huen, A. (1976): Effects of oleic acid infusion on plasma free fatty acids and blood ketone bodies in the fasting rat. Metabolism 25, 361-374.

4. Böttger, I., Dobbs, R, Faloona, G. and Unger, R. H. (1973): The effects of triglyceride absorption upon glucagon, insulin and gut glucagon-like immunoreactivity. J. Clin. Invest. 52, 2532-2541.

5. Caren, R. and Corbo, L. (1974): Glucagon and plasma lipoprotein lipase. Proc. Soc. Exp. Biol. Med. 146, 1106-1110.

6. Eaton, R. (1973): Hypolipemic action of glucagon in experimental endogenous lipemia in the rat. J. Lipid Res. 14 312-318.

7. Eaton, R. P. (1973): Glucagon secretion and activity in the cobalt chloride-treated rat. Am. J. Physiol. 225, 67-72.

8. Eaton, R. (1973): Effect of clofibrate on arginine-induced insulin and glucagon secretion. Metabolism 22, 763-767.

9. Eaton, R. P., Oase, R. and Schade, D. S. (1976): Altered insulin and glucagon secretion in treated genetic hyperlipemia: a mechanism of therapy. Metabolism 25, 245-249.

10. Eaton, R. and Schade, D. (1973): Glucagon resistance as a hormonal basis for endogenous hyperlipemia. Lancet 1, 973-974.

11. Eaton, R. P. and Schade, D. S. (1974): Effect of clofibrate on arginine-stimulated glucagon and insulin secretion in man. Metabolism 23, 445-454.

12. Eaton, R. P., Schade, D. S. and Conway, M. (1974): Decreased glucagon activity: a mechanism for genetic and acquired endogenous hyperlipemia. Lancet 2, 1545-1547.

13. Fujita, Y., Gotto, A., Unger, R. H. (1974): Insulin (I), glucagon (G) and I/G ratio in carbohydrate (CHO)-induced hyperlipemia. Diabetes 23 (Suppl. 1), 372.

14. Gerich, J. E., Lorenzi, M., Bier, D. M., Schneider, V., Tsalikian, E., Karam, J. H. and Forsham, P. H. (1975): Prevention of human diabetic ketoacidosis by somatostatin. Evidence for an essential role of glucagon. New Engl. J. Med. 292, 985-989.

15. Gerich, J. E., Lorenzi, M., Bier, D. M., Tsalikian, E., Schneider, V., Karam, J. H. and Forsham, P. H. (1976): Effects of physiologic levels of glucagon and growth hormone on human carbohydrate and lipid metabolism. Studies involving administration of exogenous hormone during suppression of endogenous hormone secretion with somatostatin. J. Clin. Invest. 57, 875-884.

16. Heimberg, M., Weinstein, I., Kohout, M. (1969): The effects of glucagon, dibutyryl cyclic adenosine 3',5' monophosphate, and concentration of free fatty acid on hepatic lipid metabolism. J. Biol. Chem. 244, 5131-5139.

17. Liljenquist, J. E., Bomboy, J. D., Lewis, S. D., Sinclair-Smith, B. C., Felts, P. W., Lacy, W. W., Crofford, O. B. and Liddle, G. W. (1974): Effects of glucagon on lipolysis and ketogenesis in normal and diabetic man. J. Clin. Invest. 53, 190-197.

18. Marks, V. (1976): Glucagon and lipid metabolism in man. Postgrad. Med. J. 49, 615-619.

19. McGarry, J. D. and Foster, D. W. (1974): The metabolism of (-)-octanoylcarnitine in perfused livers from fed and fasted rats. Evidence for a possible regulatory role of carnitine acyltransferase in the control of ketogenesis. J. Biol. Chem. 249, 7984-7996.

20. McGarry, J. D., Wright, P. H. and Foster, D. W. (1975): Hormonal control of ketogenesis. Rapid activation of hepatic ketogenic capacity in fed rats by anti-insulin serum and glucagon. J. Clin. Invest. 55, 1202-1209.

21. Müller, W. A., Faloona, G. R., Unger, R. H. and Aguilar-Parada, E. (1970): Abnormal alpha cell function in diabetes: response to carbohydrate and protein ingestion. New Engl. J. Med. 283, 109-115.

22. Nosadini, R., Solda, G., DeBiasi, F. and Tiengo, A.: Metabolic effects of glucagon in endogenous hypertriglyceridemia. Unpublished.

23. Paloyan, E. and Harper, P. U. (1961): Glucagon as a regulating factor of plasma lipids. Metabolism 10, 315-323.

24. Salans, L. B., and Reaven, G. M. (1966): Effect of insulin pretreatment on glucose and lipid metabolism of liver slices from normal rats. Proc. Soc. Exp. Biol. Med. 122, 1208-1213.

25. Schade, D. S. and Eaton, R. P. (1975): Modulation of fatty acid metabolism by glucagon in man. I. Effects of normal subjects. Diabetes 24, 502-509.

26. Schade, D. S. and Eaton, R. P. (1975): Modulation of fatty acid metabolism by glucagon in man. II. Effects in insulin-deficient diabetes. Diabetes 24, 510-515.

27. Schade, D. S. and Eaton, R. P. (1975): The contribution of endogenous insulin secretion to the ketogenic response to glucagon in man. Diabetologia 11, 555-559.

28. Schonfeld, G. and Pfleger, B. (1971): Utilization of exogenous free fatty acids for the production of very low density lipoprotein triglyceride by livers of carbohydrate-fed rats. J. Lipid. Res. 12, 614-621.

29. Sokal, J. (1966): Glucagon - an essential hormone. Amer. J. Med. 41, 331-341.

30. Tiengo, A.,Fedele, D., Muggeo, M. and Crepaldi, G. (1973): Diabetic-like behaviour of glucagon secretion in primary endogenous hypertriglyceridemia. 8th Congr. Intern. Diabetes Feder., Brussels. Excerpta Med. ICS 280, 47.

31. Tiengo, A., Muggeo, M., Assan, R., Fedele, D. and Crepaldi, G. (1975): Glucagon secretion in primary endogenous hypertriglyceridemia before and after clofibrate treatment. Metabolism 24, 901-914.

32. Tiengo, A., Muggeo, M., Fedele, D., Nosadini, R. and Crepaldi, G. (1975): La sécrétion d'insuline et de glucagon dans l'hypertriglycéridemie endogène. In: Journées Annuelles de Diabetologie de l'Hotel-Dieu. Flammarion, Paris.

33. Topping, D. L. and Mayes, P. A. (1971): The immediate effects of insulin and fructose on the metabolism of the perfused liver. Changes in lipoprotein secretion, fatty acid oxidation and esterification, lipogenesis and carbohydrate metabolism. Biochem. J. 126, 295-311.

34. Unger, R. H. (1971): Glucagon and the insulin:glucagon ratio in diabetes and other catabolic illnesses. Diabetes 20, 834-838.

35. Unger, R. H. (1972): Pancreatic alpha-cell function in diabetes mellitus. In: Glucagon, Molecular Physiology, Clinical and Therapeutic Implications. P. J. Lefèbvre and R. H. Unger, Eds. Pergamon Press, Oxford, pp. 245-257.

36. Unger, R. H., Madison, L. L. and Müller, W. A. (1972): Abnormal alpha cell function in diabetes. Response to insulin. Diabetes 21, 301-307.

37. Weinstein, I., Klausner, H. A. and Heimberg, M. (1973): The effect of concentration of glucagon on output of triglyceride, ketone bodies, glucose, and urea by the liver. Biochim. Biophys. Acta 296, 300-309.

38. Williamson, J. R., Browning, E. T., Thurman, R. G. and Scholz, R. (1969): Inhibition of glucagon effects in perfused rat liver by (+)decanoylcarnitine. J. Biol. Chem. 244, 5055-5064.

39. Wise, J. K., Hendler, R. and Felig, P. (1973): Evaluation of alpha-cell function by infusion of alanine in normal, diabetic and obese subjects. New Engl. J. Med. 288, 487-490.

GLUCAGONOMAS AND SKIN DISEASE

S. R. Bloom

Department of Medicine, Hammersmith Hospital
London W12, U.K.

The glucagonoma syndrome has now become an established entity and a major feature is the presence of a typical skin rash. The physiologic basis of this feature is unknown. The only effective treatment is early diagnosis and surgical removal of the tumour. A greater awareness of the syndrome coupled with ready availability of glucagon assays, which can provide rapid and reliable diagnosis, should lead to more successful therapy.

The pancreas, in spite of its low cell turnover rate, is an organ in which tumour formation, both exocrine and endocrine is common. Insulinomas, gastrinomas and VIPomas (2) produce highly active hormones and give rise to well defined clinical syndromes. By contrast excessive glucagon production by A cell tumours is less dramatic. Indeed the clinical features of the glucagonoma syndrome were not fully delineated until 1974 (8). One of the most important features, which was not predictable from a knowledge of glucagon's physiology, was the presence of a characteristic rash. This has been of great diagnostic help and the number of cases reported has increased dramatically (3, 4, 6, 7, 11, 13, 15).

Clinical Features

Rash. The skin disease associated with a glucagonoma syndrome has been described as a necrolytic migratory erythema (16). This rash can be sufficiently characteristic for the dermatologist to be able to make the diagnosis from the patient's appearance alone. The glucagonoma syndrome can be associated with less typical rashes, however, or no rash at all. Further, the author has been sent plasma from several patients described as having "the typical rash" where no elevation of glucagon has been detected. Good examples of the glucagonoma rash can be seen in several publications (4, 5, 8, 11, 12). The presence of bullae has led to it being compared with benign familial pemphigus or pemphigus foliaceus, and the pustule formation with generalized pustular psoriasis or toxic epidermal necrolysis.

Histology shows necrolysis in the stratum granulosum of the epidermis and in some cases a pronounced infiltrate of leukocytes and histiocytes. The acantholysis typical of pemphigus vulgaris and antiepithelial antibodies are absent. Other features of the rash are its tendency to exacerbations and remissions, which may be complete, and its localization in skin folds, dependent parts and areas easily traumatized. The commonest sites are the lower abdomen, groins, perineum and the thighs and legs. Another characteristic feature of the syndrome is the presence of a red raw painful glossitis. Less commonly an angular stomatitis with circumoral crusting is seen.

Diabetes. While an abnormal glucose tolerance test is invariable in the glucagonoma syndrome, overt diabetes is seen in only a proportion of the patients. It is clear this is not just due to destruction of the pancreas by a tumour, as a successful tumour excision can be curative. A patient requiring 68 units of insulin a day has been reported in whom removal of a pancreatic A cell adenoma gave complete remission both of rash and diabetes with a subsequent completely normal glucose tolerance curve (7). This provided the first proof that excessive glucagon alone could provoke severe diabetes in man. Of more relevance, however, in view of the present day emphasis on the glucagon abnormalities of diabetes, is the fact that many of these patients had only the mildest abnormalities of glucose tolerance in spite of massive elevations of glucagon, often a hundred-fold higher than those seen in diabetes mellitus. Thus, although hyperglucagonaemia can cause diabetes, its potential in this respect would appear to be no different from the elevations of cortisol or growth hormone seen in Cushing's disease and acromegaly where the incidence of diabetes is also increased.

Other features. Weight loss, although a rather non-specific symptom has been recorded in all patients described. This is perhaps not surprising as glucagon is a powerfully catabolic hormone. Anaemia is a common but not an invariant feature and is usually of the normochromic normocytic variety and does not appear to be associated with iron deficiency. Some observers have commented on an increased tendency to vascular thrombosis which may prove a terminal event (5).

Glucagon Measurement

Because glucagonomas present fairly late in their clinical course, the plasma glucagon elevations are usually enormous. Only a few cases have been described where the initial value was below a 1000 pg/ml while fasting plasma glucagon levels in healthy subjects are usually below 100 pg/ml. Thus, once suspected the diagnosis is easily confirmed by the measurement of a single fasting plasma sample.

The nature of the circulating glucagon is of some interest. Normal plasma when analyzed by gel chromatography shows the majority of the glucagon immunoreactivity to elute in the expected 3500 molecular weight position. There is some controversy concerning the presence of some very high molecular weight glucagon-like immunoreactivity (14) which is variably detected depending on the assay system employed. Patients with glucagonomas may show a normal pattern or may demonstrate the presence of a considerable quantity of intermediate molecular weight glucagon immunoreactivity (3, 13, 15). There has been speculation whether this might represent "proglucagon" and whether its greater amounts are a feature of abnormal tumourous production. The quantity of this big glucagon is often much larger in plasma than in extracts of the causative tumor and this suggests that it may possess a rather longer half life in the circulation than normal glucagon.

The response to stimuli such as glucose or arginine is variable. In some patients it is entirely normal and plasma glucagon release is fully suppressible by either oral or intravenous glucose while in others the response is paradoxical. Arginine always appears to be stimulatory but the degree of rise is again variable. It is interesting to note the finding in one patient of a marked stimulation by sulphonylureas (11). It was apparent on review of his history that the skin rash had coincided with treatment by these agents and it is at least possible that their use was counter-productive. Glucagon release can be suppressed by somatostatin and in one patient this resulted in rapid onset of profound hypoglycaemia (9).

Multiple Hormone Producing Tumours

It is now apparent that a high percentage of glucagonomas also contain a large number of pancreatic peptide (PP) producing cells, both in the primary tumour and in the metastases (10). The clinical effects of the high circulating PP levels are not immediately apparent. Other hormones which are produced by tumours containing A cells include gastrin, insulin and ACTH (1). In these cases, however, the clinical picture is dominated by the other more potent hormones.

Treatment

The only effective treatment is surgical removal. This implies the need for early diagnosis, as many cases have metastasised by the time they had come to light. The extent to which the glucagonoma syndrome is underdiagnosed is not known. It may well be that a number of patients attending diabetic clinics or skin clinics have occult disease and a wide scale screening programme by means of fasting glucagon measurements would seem to be desirable to sort out this question. Two major agents have been tried in metastatic glucagonomas, streptozotocin and 5-flurouracil. Neither is dramatically effective. Symptomatic treatment of the skin rash is a little more effective and while a number of drugs have been tried none has been universally successful (8).

REFERENCES

1. Belchetz, A. E., Brown, C. L. Makin, H. L., Trafford, D. F., Mason, A. S., Bloom, S. R. and Ratcliffe, J. (1973): ACTH, glucagon and gastrin production by a pancreatic islet cell carcinoma and its treatment. Clin. Endocr. 2,

307-316.

2. Bloom, S. R., Polak, J. M. and Pearse, A. G. E. (1973): Vasoactive intestinal peptide and watery-diarrhea syndrome. Lancet 2, 14-16.

3. Boden, G., Owen, O. E. and Quickel, K. E. (1976): Effect of chronic glucagon excess on glucose homeostasis and insulin secretion in man. Diabetes 25, 341.

4. Holst, J. J., Jonsson, L., Pedersen, N. B. and Thomsen, K. (1975): Glucagonomsyndromet. Ugeskr. Laeg. 137, 2631-2633.

5. Jonsson, L., Pedersen, N. B. and Holst, J. J. (1976): Necrolytic migratory erythema and glucagon cell tumor of the pancreas: glucagonoma syndrome. Acta Dermatovener 56, 391-395.

6. Kramer, S., Machina, T. and Marcus, R. (1976): Metabolic studies in the malignant glucagonoma syndrome. Diabetes 25, 370.

7. Lightman, S. L. and Bloom, S. R.(1974); Cure of insulin dependent diabetes mellitus by removal of a glucagonoma. Brit. Med. J. 1, 367-368.

8. Mallinson, C. N., Bloom, S. R., Warin, A. P., Salmon, P. R. and Cox, B. (1974): A glucagonoma syndrome. Lancet 2, 1-5.

9. Mortimer, C. H., Tunbridge, W. M., Carr, D., Lind, T., Bloom, S. R., Mallinson, C. N., Schally, A. V., Yeomans, L., Coy, D. H., Kastin, A. and Besser, G. M. (1974): Effects of growth hormone release-inhibiting hormone on circulating insulin, glucagon and growth hormone in normal, diabetic, acromegalic and hypopituitary patients. Lancet 1, 697-701.

10. Polak, J. M., Bloom, S. R., Adrian, T. E., Heights, P., Bryant, M. G. and Pearse, A. G. E. (1976): Pancreatic polypeptide in insulinomas, gastrinomas, VIPomas and glucagonomas. Lancet 1, 328-330.

11. Soler, N. G., Oates, G. D., Malins, J. M., Cassar, J. and Bloom, S. R. (1976): Glucagonoma syndrome in a young man. Proc. Roy. Soc. Med. 69, 429-431.

12. Sweet, D. (1974): A dermatosis specifically associated with a tumor of pancreatic alpha cells. Brit. J. Derm. 90, 301-308.

13. Valverde, I., Lemon, H. M., Kessinger, A. and Unger, R. H. (1976): Distribution of plasma glucagon immunoreactivity in a patient with suspected glucagonoma. J. Clin. Endocr. Metab. 42, 792-795.

14. Valverde, I., Villanueva, M. L., Lozano, I. and Marco, J. (1974): Presence of glucagon immunoreactivity in the globulin fraction of human plasma ("big plasma glucagon"). J. Clin. Endocr. 39, 1090-1098.

15. Weir, G. C., Horton, E. S., Aoki, T. T. and Slovik, D. M. (1976): Increased circulating large glucagon immunoreactivity in the glucagonoma syndrome. Diabetes 25, 326.

16. Wilkinson, D. S. (1973): Necrolytic migratory erythema with carcinoma of the pancreas. Trans. St. Johns Hosp. Dermatol. Soc. 59, 244-250.

PLASMA GLI IN GASTRECTOMIZED PATIENTS*

K. Shima, T. Matsuyama and R. Tanaka

Department of Medicine and Geriatrics and
Second Department of Internal Medicine,
Osaka University Medical School
Osaka, Japan

The plasma glucagon-like immunoreactivity (GLI) and immunoreactive insulin responses to oral glucose were significantly greater in gastrectomized patients than in normal subjects. Galactose, fructose and xylitol were less effective than glucose. In contrast, the intravenous administration of glucose did not cause an elevation of plasma GLI. Decrease in glucose absorption following treatment with phenformin did not alter basal GLI or its response to oral glucose. Thus the GLI response may depend upon contact of the sugar with the intestinal mucosa, rather than upon its absorption. GLI fractions, isolated from porcine intestine by affinity chromatography had no insulinogenic activity in the perfused rat pancreas.

The physiologic role of gut glucagon-like immunoreactivity (GLI) has not been fully established yet, primarily because the pure material is difficult to obtain and because data on secretory mechanisms are scarce (1, 9, 17, 19). Consequently, the study of the biologic activities of gut GLI gave conflicting results (13). The aim of the present study was to reinvestigate these problems and to elucidate the physiologic role of gut GLI. The secretion of gut GLI was studied in gastrectomized subjects whose GLI response to ingested sugars was exaggerated. In other experiments we explored the insulinogenic activity of gut GLI using material extracted from dog intestinal mucosa and purified by affinity chromatography on antiglucagon serum bound to sepharose (14).

MATERIALS AND METHODS

1. GLI secretion in vivo. The observations were carried out in 9 male and 2 female healthy, nonobese, 32 to 56 year-old volunteers, and in 47 patients (22 to

---

*Portions of this paper were published in Proc. Soc. Exper. Biol. Med. 139, 1042, 1972 and 150, 232, 1975 and were read at the IX Congress of the Intern. Diabetes Feder., New Delhi, November 1976.

70 years old) who had been gastrectomized for various benign gastric or duodenal diseases 8 months to 19 years previously, or, in three cases, for gastric carcinoma.

All subjects were given 100 g of glucose orally after an overnight fast. In addition, several days later, some of them received oral loads of galactose (100 g), fructose (100 g), xylitol (100 g), glucose (50 g) or an intravenous injection of glucose (25 g). In another series of experiments, the oral or intravenous glucose tolerance test was repeated within 2 months of the first one and 60 min after the ingestion of 150 mg of phenformin hydrocholoride.

2. GLI secretion in vitro. Fifty ml of a 5% solution of α-D-glucose or β-D-glucose, prepared immediately before use, were introduced into a 30 cm loop placed at the end portion of the ileum of mongrel dogs, under pentobarbital anesthesia. Blood samples for measurements of GLI and glucose concentrations were obtained from a mesenteric vein draining the loop, before and for 15 min after the introduction of the experimental solution.

3. Insulinogenic activity of gut GLI. Gut GLI was extracted from the mucosa of the porcine small intestine and of the gastric fundus with acid-ethanol according to Kenny's method (7). When applied to a Bio-Gel P-10 column and eluted with 0.1 N acetate solution, the crude extracts of the small intestine yielded two peaks, Peak I and Peak II. These peaks and the extract of the gastric fundus were purified further by affinity chromatography, using antiglucagon rabbit serum bound to sepharose 4B and were termed EG-S_1, EG-S_2 and EG-F, respectively. The perfused rat pancreas (5) was used for evaluating the insulinogenic activity of the materials thus obtained. After perfusing the pancreas with a control fluid containing glucose (60 mg/100 ml) for 10 min, the glucose concentration of the perfusate was raised to 100 mg/100 ml and the test agents were added simultaneously at the rate of 100 ng/min for 5 min. The IRI concentration in the perfusate was measured using rat insulin standards. GLI was determined using a double antibody (17, 18) or a polyethylene glycol method (Shima

et al., in press). Antiglucagon sera employed were "specific" for pancreatic glucagon (K47) or "non-specific" (AGS K52, 159 or 10). The latter had different cross reactivities for GLI. Glucose was measured with the method of Somogyi-Nelson or with the Hoffman ferricyanide technique adapted to an AutoAnalyzer. Immunoreactive insulin (IRI) was determined using an insulin assay kit.

RESULTS AND COMMENTS

Figure 1 shows the plasma GLI responses to oral glucose in normal and gastrectomized subjects. There was no difference between the mean fasting concentrations of plasma GLI in the two groups. However, after the oral administration of glucose in the gastrectomized subjects, the plasma GLI rose to values which were significantly higher than those observed in the normal subjects. The peak values of plasma IRI and glucose were also higher in the gastrectomized subjects than in the normal volunteers. In contrast to oral administration, no elevation of plasma GLI was observed after i. v. glucose in five gastrectomized subjects (Fig. 2). Figure 3 shows that, in 4 gastrectomized subjects, 100 g of glucose caused a more marked increase in plasma GLI than 50 g of glucose.

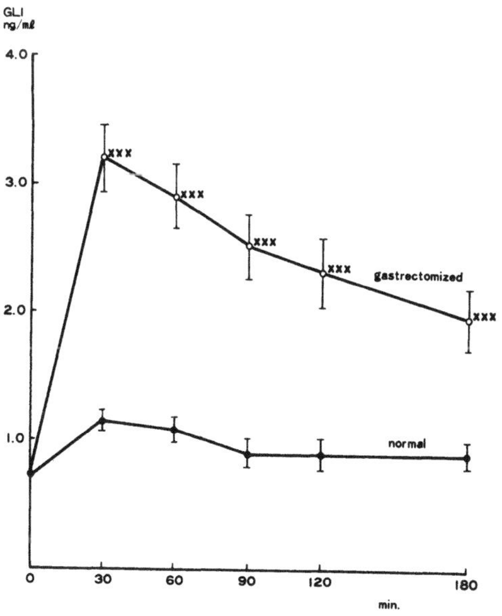

Fig. 1. Plasma GLI response to oral glucose in 21 gastrectomized and in 11 normal subjects. Mean ± S.E., xxx: p <.005

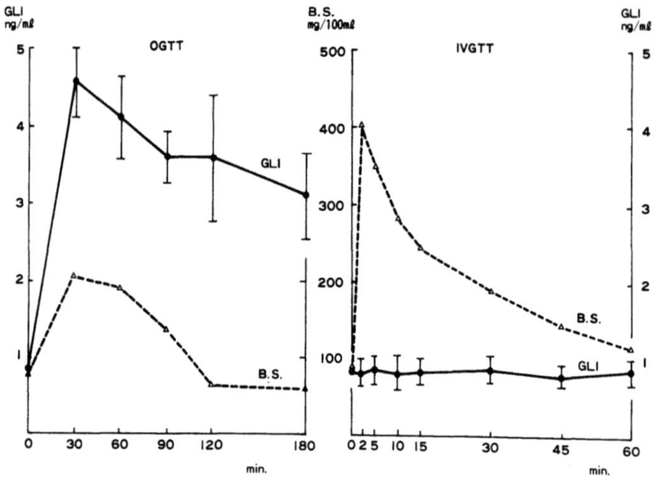

Fig. 2. Plasma GLI and glucose responses to iv glucose in 5 gastrectomized subjects. Mean ± S.E.

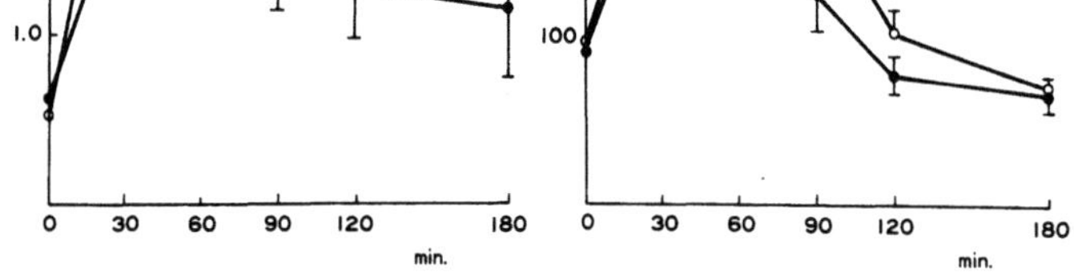

Fig. 3. Plasma GLI and glucose responses to 50 and 100 g of oral glucose in 4 gastrectomized subjects. Mean ± S.E., x: $p < .05$, xx: $p < .025$.

Thus, gastrectomized subjects whose intestine received a bolus of undiluted glucose solution demonstrated an exaggerated plasma GLI response compared to the normal subjects. This fact suggests that mechanical or osmotic stimuli may have contributed to the release of enteric GLI. Another explanation for the exaggerated plasma GLI response to oral glucose in the gastrectomized subjects is that a larger amount of ingested glucose solution may reach the ileum where enteric GLI appears to be secreted, at least in the dog (12).

To determine if chemical stimuli influence the secretion of gut GLI, we studied the effect of different sugars. No direct comparison between the effect of the four sugars was possible because not all patients received all 4 of them. Figure 4 shows the concentrations of plasma GLI after galactose, fructose or xylitol calculated in relation to those seen after glucose. In gastrectomized subjects, glucose was the most potent stimulus followed by galactose, fructose and xylitol, in that order. The data were very similar to those reported by Marco et al (10) in dogs. These results suggest that the chemical characteristics of the stimulus may be discerned by the gut. However, a possible role of the intestinal absorption of sugar upon GLI release cannot be excluded on the basis of these results because the hyperGLIemia observed in this study was always associated with hyperglycemia. In order to clarify this point, we have studied the plasma GLI response to the oral or intravenous administration of glucose in gastrectomized subjects before and after treatment with phenformin, a drug believed to impair the intestinal absorption of glucose (2, 6, 8).

Figure 5 shows the effect of phenformin on the mean blood sugar responses to oral or intravenous glucose loading. Sixty minutes after the oral administration of phenformin the mean fasting blood sugar level of 22 gastrectomized subjects remained unchanged. However, the drug supressed the plasma glucose response to an oral load. Thus, the glucose tolerance curve reached a peak of 124.5 ± 4.3 mg/100 ml in the drug-treated patients, but rose to 154.0 ± 7.4 mg/100 ml in the untreated patients ($p < .005$). A significant difference was noted also at the 60

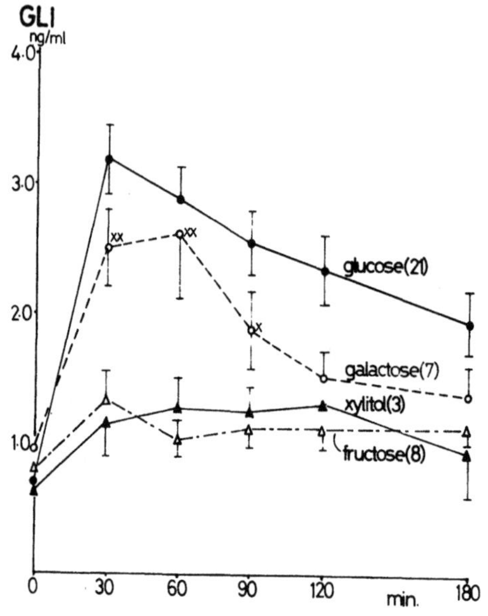

Fig. 4. Relative concentrations of plasma GLI after galactose, fructose and xylitol in gastrectomized subjects, calculated as ratios to the values obtained after oral glucose. x: $p < .05$, xx: $p < .025$, galactose vs fructose.

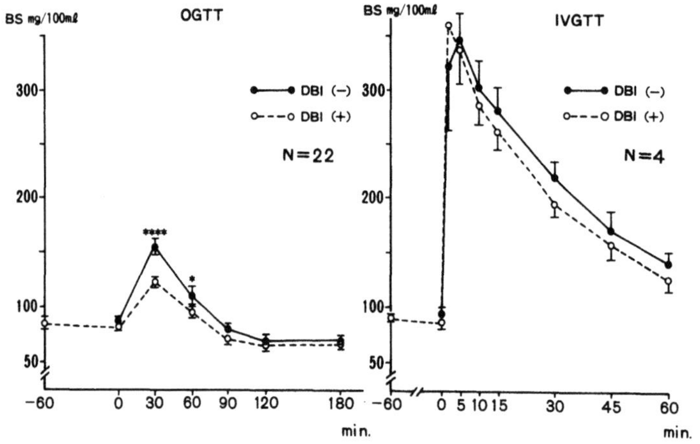

Fig. 5. Effect of phenformin (DBI) on the blood sugar response to oral and intravenous glucose loading in gastrectomized subjects. Mean ± S.E., xxxx: $p < .005$, x: $p < .05$.

min point ($p < .05$). In contrast, the rate of glucose disappearance (K value) after intravenous loading was not affected by the previous administration of phenformin. These results show that, in our subjects, the intestinal absorption of glucose was impaired to some extent by phenformin. Figure 6 shows the changes in plasma GLI levels under these conditions. The drug did not alter either the basal or the peak level of plasma GLI attained 30 min following the ingestion of glucose*. However, 2 and/or 3 hr thereafter, all but one of the subjects exhibited higher GLI concentrations after phenformin treatment than before. On the other hand, the plasma GLI response to the intravenous injection of glucose was not modified by pretreatment with the drug. In 7 of these 22 phenformin pretreated subjects, elevation of blood glucose level after an oral glucose loading was markedly reduced (more than 50 mg/100 ml difference, group I). On the other hand, 5 subjects showed no significant difference in blood glucose response to oral glucose with and without phenformin (group III). The remaining 10 subjects showed a moderate decrease in blood glucose response to an oral load after phenformin pretreatment (group II). These three groups showed no significantly different patterns of plasma GLI in response to oral glucose, despite the variable suppression of their intestinal glucose absorption (Fig. 7). These results, together with the findings of Matsuyama and Foà (12), lend support to the notion that the release of gut GLI depends more upon contact of sugar with the intestinal mucosa than upon its absorption.

Furthermore, it might be presumed that the GLI-secreting cell has the means of recognizing or discriminating between stimuli on its surface. As shown in Fig. 8, the GLI response to α-D-glucose was significantly greater than to β-D-glucose. These findings suggest that gut GLI-secreting cells may have a

---

*Plasma GLI concentrations were lower in the subjects treated with phenformin than in those in whom we studied the effect of various sugars on plasma GLI level. This is probably due to the difference in antiglucagon sera employed for the assay.

glucose sensor similar to that of the A and B cells of the pancreas (3, 4, 11, 15).

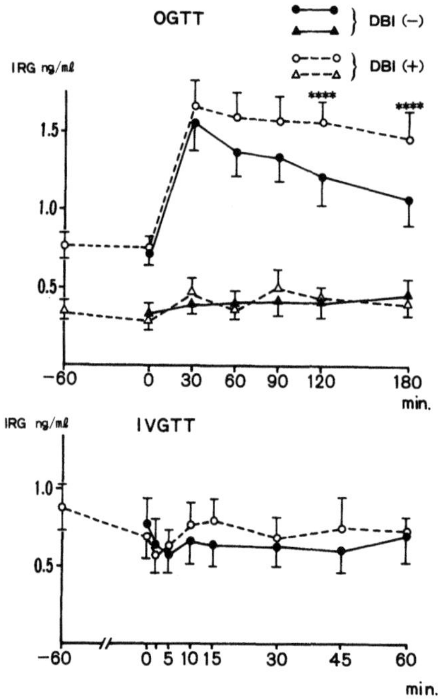

Fig. 6. Effect of phenformin (DBI) on the plasma IRG response to oral and intravenous glucose loading in gastrectomized subjects. Circles and triangles represent values measured using "non-specific" and "specific" anti-glucagon sera, respectively. Mean ± S.E.; xxxx: $p < .005$.

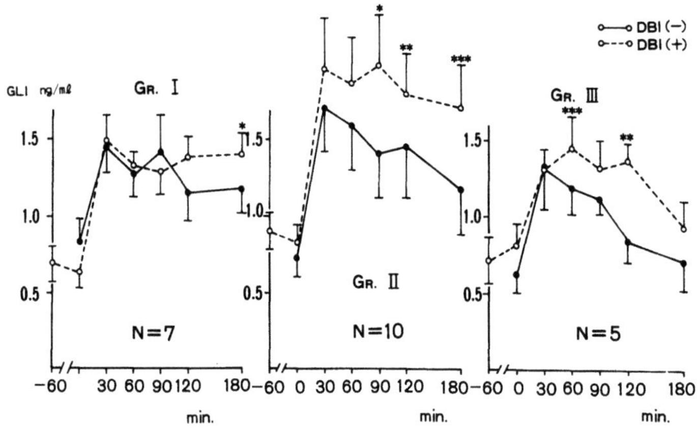

Fig. 7. Effect of phenformin (DBI) on the plasma GLI response to oral glucose loading in various groups of gastrectomized subjects. See text for details. Mean ± S.E.

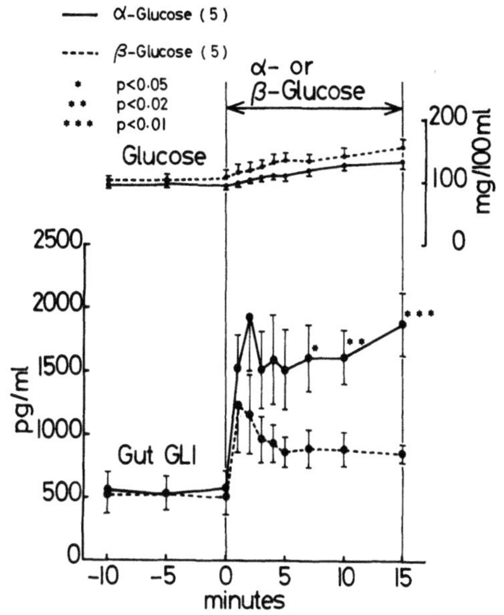

Fig. 8. Plasma GLI and glucose in the regional mesenteric blood following injection of 50 ml of 5% α- or β-D-glucose into a loop of ileum of normal dogs. Mean ± S.E.

The facts that GLI is released by the oral secretagogues which normally release insulin (notably glucose) and that the hyperinsulinemia observed in gastrectomized subjects is always associated with hyperGLIemia are compatible with the idea that GLI might be one of the enteric factors affecting insulin release. In order to clarify this point, the insulinogenic activity of various preparations of porcine gut GLI purified by affinity chromatography was explored.

Figure 9 shows the insulinogenic activity of the purified peak I GLI ($EG\text{-}S_1$). $EG\text{-}S_1$ in the presence of glucose, at a concentration of 100 mg/100 ml yielded no significant increase in insulin output over control values. Figure 10 shows the insulin output from the perfused rat pancreas following stimulation with various preparations of purified GLI, during the first 10 min interval. Pancreatic glucagon and EG-F had a stimulatory effect, but neither $EG\text{-}S_1$ nor $EG\text{-}S_2$ showed significant insulinogenic activity under these conditions. These data suggest that enteric GLI may not play a direct role in the stimulation of insulin secretion. In contrast to our data and to those of Murphy et al. (14), Ohneda et al (16) found that canine gut GLI had insulinogenic activity. Accordingly, a final conclusion as to whether or not gut GLI plays an important role in the metabolic processing of nutrients through its insulinogenic activity cannot be reached until more data on this matter will have been accumulated.

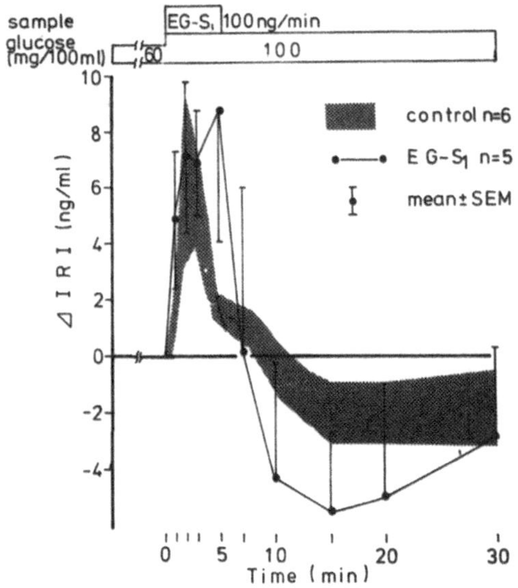

Fig. 9. Insulin releasing activity of purified peak I GLI (EG-S_1). The isolated rat pancreas was infused with EG-S_1 (100 ng/min), in the presence of glucose (100 mg/100 ml). ΔIRI represents the insulin output above baseline (60 mg/100 ml glucose alone). Mean ± S.E.

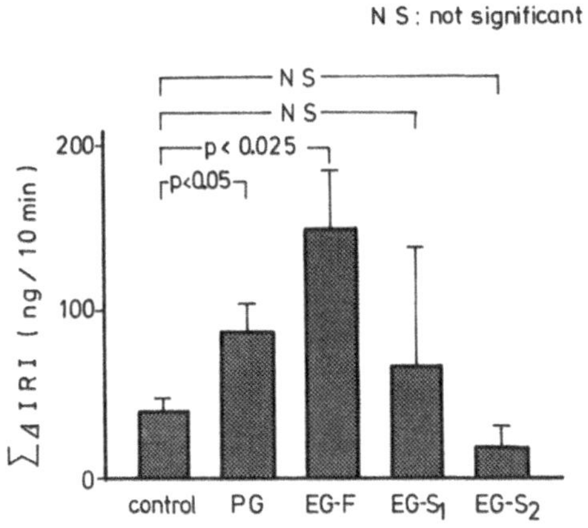

Fig. 10. Insulin releasing activity of pancreatic glucagon and gut GLI. Σ ΔIRI represents the insulin output above baseline during the first 10 min interval. PG: Pancreatic glucagon; EG-S_1: purified Peak I; EG-S_2: purified Peak II; EG-F: purified gastric GLI. Mean ± S.E.

CONCLUSIONS

We studied the secretory mechanism of GLI in normal and gastrectomized subjects and the insulinogenic activity of partially purified gut GLI extracts in the perfused rat pancreas. The following results were obtained.

1. Gastrectomized subjects showed greater GLI response to oral glucose than normal subjects.
2. Glucose was the most potent stimulus to GLI release in gastrectomized subjects, followed by galactose, fructose and xylitol, in that order.
3. Phenformin did not alter either the basal or the peak level of plasma GLI, while it reduced the elevation of blood glucose following an oral glucose load.
4. The GLI response to α-D-glucose was significantly greater than that to β-D-glucose placed in a loop of dog ileum. These data suggest that the release of intestinal GLI is not related to the intestinal absorption of glucose, but to the recognition of glucose by GLI secreting cells.
5. No insulinogenic activity of GLI was found in an in vitro system.

REFERENCES

1. Böttger, I., Dobbs, R., Faloona, G. R. and Unger, R. H. (1972): The effect of calcium and other salts upon the release of glucagon-like immunoreactivity from the gut. J. Clin. Invest. 51, 831-836.

2. Czyzyk, A., Tawecki, J., Sadowski, J., Ponikowska, I. and Szczepanik, Z. (1968): Effect of biguanides on intestinal absorption of glucose. Diabetes 17, 492-498.

3. Grill, V. and Cerasi, E. (1975): Glucose-induced cyclic AMP accumulation in rat islets of Langerhans. Preferential effect of the alpha anomer. FEBS Letters 54, 80-83.

4. Grodsky, G. M., Bennett, L. L., Smith, D. F. and Schmid, F. G. (1967): Effect of pulse administration of glucose and of glucagon on insulin secretion in vitro. Metabolism 16, 222-233.

5. Grodsky, G. M., Fanska, R., West, L. and Manning, M. (1974): Anomeric specificity of glucose-stimulated insulin release: Evidence for a glucoreceptor. Science 186, 536-538.

6. Hollobaugh, S. L., Rao, M. B. and Kruger, F. A. (1970): Studies on the site and mechanism of action of phenformin. I. Evidence for significant "nonperipheral" effects of phenformin on glucose metabolism in normal subjects. Diabetes 19, 45-49.

7. Kenny, A. J. (1955): Extractable glucagon of the human pancreas. J. Clin. Endocrinol. Metab. 15, 1089-1105.

8. Kruger, F. A., Altschuld, R. A., Hollobaugh, S. L. and Jewett, B. (1970): Studies on the site and mechanism of action of phenformin. II. Phenformin inhibition of glucose transport by rat intestine. Diabetes 19, 50-52.

9. Marco, J., Baroja, I. M., Diaz-Fierros, M., Villanueva, M. L. and Valverde, I. (1972): Relationship between insulin and gut glucagon-like immunoreactivity (GLI) secretion in normal and gastrectomized subjects. J. Clin. Endocrinol Metab. 34, 188-191.

10. Marco, J., Valverde, I., Faloona, G. R. and Unger, R. H. (1970): Endogenous gut glucagon-like immunoreactivity (GLI) and insulin secretion. Diabetologia 6, 639.

11. Matschinsky, F. M., Pagliara, A. S., Hover, B. A., Haymond, M. W. and Stillings, S. N. (1975): Differential effects of alpha- and beta-D-glucose on insulin and glucagon secretion from the isolated perfused rat pancreas. Diabetes 24, 369-372.

12. Matsuyama, T. and Foà, P. P. (1974): Plasma glucose, insulin, pancreatic and entero-glucagon levels in normal and depancreatized dogs. Proc. Soc. Exp. Biol. Med. 147, 97-102.

13. Moody, A. J. (1972): Gastrointestinal glucagon-like immunoreactivity. In: Glucagon, Molecular Physiology, Clinical and Therapeutic Implications. P. J. Lefèbvre and R. H. Unger, Eds., Pergamon Press, Oxford, pp. 319-341.

14. Murphy, R. F., Buchanan, K. D. and Elmore, D. T. (1973): Isolation of glucagon-like immunoreactivity of gut by affinity chromatography on anti-glucagon antibodies coupled to sepharose 4B. Biochem. Biophys. Acta 303, 118-127.

15. Niki, A., Niki, H., Miwa, I. and Okuda, J. (1974): Insulin secretion by anomers of D-glucose. Science 186, 150-151.

16. Ohneda, A., Horigome, K., Kai, Y., Itabashi, H., Ishii, S. and Yamagata, S. (1976): Purification of canine gut glucagon-like immunoreactivity (GLI) and its insulin releasing activity. Horm. Metab. Res. 8, 170-174.

17. Shima, K. and Foà, P. P. (1968): A double antibody assay for glucagon. Clin. Chim. Acta 22, 511-520.

18. Shima, K., Sawazaki, N., Tanaka, R., Tarui, S. and Nishikawa, M. (1975): Effect of an exposure to chloramine-T on the immunoreactivity of glucagon. Endocrinology 96, 1254-1260.

19. Unger, R. H., Ohneda, A., Valverde, I., Eisentraut, A. M. and Exton, J. (1968): Characterization of the response of circulating glucagon-like immunoreactivity to intraduodenal and intravenous administration of glucose. J. Clin. Invest. 47, 48-65.

PANCREATIC A CELL FUNCTION IN A PATIENT WITH ADDISON'S DISEASE

G. Okuno and H. Takenaka

Department of Medicine, Itami City Hospital
Kasugaoka, Itami, Hyogo 664 Japan

Pancreatic A cell function was studied in a patient with primary adrenocortical insufficiency. Plasma glucagon levels in the fasted patient were found to be 65 and 78 pg/ml, close to the upper limit of our normal range and did not change after the daily administration of 10 mg of hydrocortisone for 4 weeks. The glucagon response to insulin-induced hypoglycemia was normal. However, the glucagon response to arginine infusion was about one half that observed in normal subjects before treatment. This response became supernormal even if delayed after hydrocortisone treatment.

Relative or absolute glucagon hypersecretion has been reported in every form of endogenous hyperglycemia so far investigated (6). However, reports on pancreatic A cell function in endocrine disorders with a tendency to hypoglycemia are limited to that of Lawrence (3) who described a case of hypopituitarism with decreased basal and arginine-stimulated glucagon release.

We measured plasma pancreatic glucagon in a patient with Addison's disease and found a normal fasting glucagon concentration and a decreased response to arginine, that became supernormal after treatment with glucocorticoids.

Case Report. T. U., a 47-year -old male, was admitted to Gifu University Hospital complaining of general fatigue. A diagnosis of Addison's disease was made and he was treated successfully with hydrocortisone. In January of 1975, at the age of 52, he came to our clinic for reexamination. His general condition was good. The skin and buccal mucosa were slightly pigmented. His urine, blood count and blood chemistry, including serum glucose (98 mg/dl), electrolytes, cholesterol and urea nitrogen were normal and so was his liver function, except for a serum γ-globulin of 21%.

Endocrine Function Tests. Endocrine function was studied 4 weeks after interruption of hydrocortisone therapy. At that time, thyroid function was normal. A 50 g oral glucose tolerance test gave a relatively flat curve, with a fasting value of 93 mg/dl, a peak value of 119 mg/dl and a 3 hour value of 83 mg/dl. The serum 11-OHCS concentration was below normal (4.0 μg/dl), showed almost no increase 45 min after the intramuscular injection of 0.25 mg of ACTH, then gradually decreased for the next 75 min. The urinary excretion of 17-OHCS was zero or nearly zero on repeated occasions. As shown in Table 1, before, during and after 3 daily injections of 0.5 mg of long-acting ACTH there was very little increase in urinary 17-OHCS excretion, as compared with that observed in normal subjects or in subjects with secondary adrenal cortical insufficiency.

TABLE 1

URINARY 17-OHCS RESPONSE TO DAILY INJECTIONS OF DEPOT-ACTH FOR 3 DAYS

		17-OHCS, mg/day			
Day		Patient T. U.	Normal Subjects*.		Patients with Secondary Adrenal Cortical Insufficiency
1	control	0.3			0.9**
2	control	0.5	3.5 -	6.0	1.5
3	ACTH	2.0	13.0	5.3	4.0
4	ACTH	2.0	21.1	10.0	9.6
5	ACTH	2.0	24.2	7.4	13.4
6	control	0.5			8.3
7	control	0.1			

* Values obtained in healthy Japanese subjects.

**Value obtained in a typical case.

The fasting plasma concentration of pancreatic glucagon was measured with AGS 30K and the technique of Nonaka and Foà (5) and corrected as suggested by Weir et al (7) and by us (9). Levels of our fasted patient were found to be 78 and 65 pg/ml before treatment and 60 pg/ml after treatment. These are close to the upper limit of our normal values (41 ± 20 pg/ml, Ave. ± SD). Although the meaning of this finding is difficult to assess, one might have expected an increased secretion of glucagon in a pathologic state characterized by a tendency to hypoglycemia and undernutrition, such as Addison's disease. To study the glucagon response to insulin-induced hypoglycemia, the patient received a single injection of monocomponent rapid acting insulin (0.05 U/kg) after the infusion of a 5% solution of glucose. Forty-five min after the injection, a hypoglycemic episode occurred and the experiment was discontinued. As indicated in Fig. 1, glucose infusion caused only small changes in plasma glucose and glucagon. The subsequent insulin injection caused a decrease of plasma glucose (69 mg/dl) and an increase of 80 pg/ml in glucagon concentration. The data indicate that plasma glucagon can respond to hypoglycemia in Addison's disease, although the calculated ratio of change (Δ glucagon/Δ glucose) appears to be decreased as compared to that observed in normal subjects by Gerich et al. (2). Recently, Alford et al. (1) demonstrated directly that glucagon plays an important role in the maintenance of plasma glucose in normal fasting subjects. Thus, our data suggest that hypofunction of the pancreatic A cells exists in adrenocortical insufficiency and that glucagon hyporesponsiveness may contribute to the hypoglycemia of Addison's disease.

As shown in Fig. 2, a test dose of 10 g of arginine was infused before and after the daily administration of 10 mg of hydrocortisone for 4 weeks. Before treatment the glucagon response decreased to about one half of that seen in normal controls. After treatment, the response appeared augmented even if delayed, reaching a peak value of 432 pg/ml at 60 min. The depressed glucagon response to arginine before hydrocortisone treatment and its reversal to supernormal values

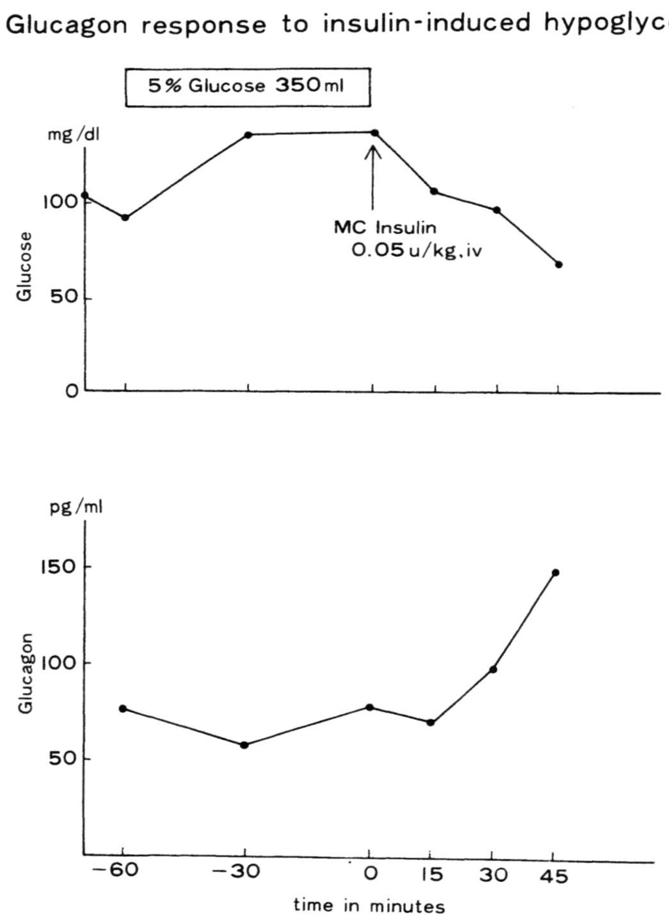
Glucagon response to insulin-induced hypoglycemia
5% Glucose 350 ml
mg/dl
Glucose
100
50
0
MC Insulin
0.05 u/kg, iv
pg/ml
Glucagon
150
100
50
-60
-30
0
15
30
45
time in minutes

Fig. 1

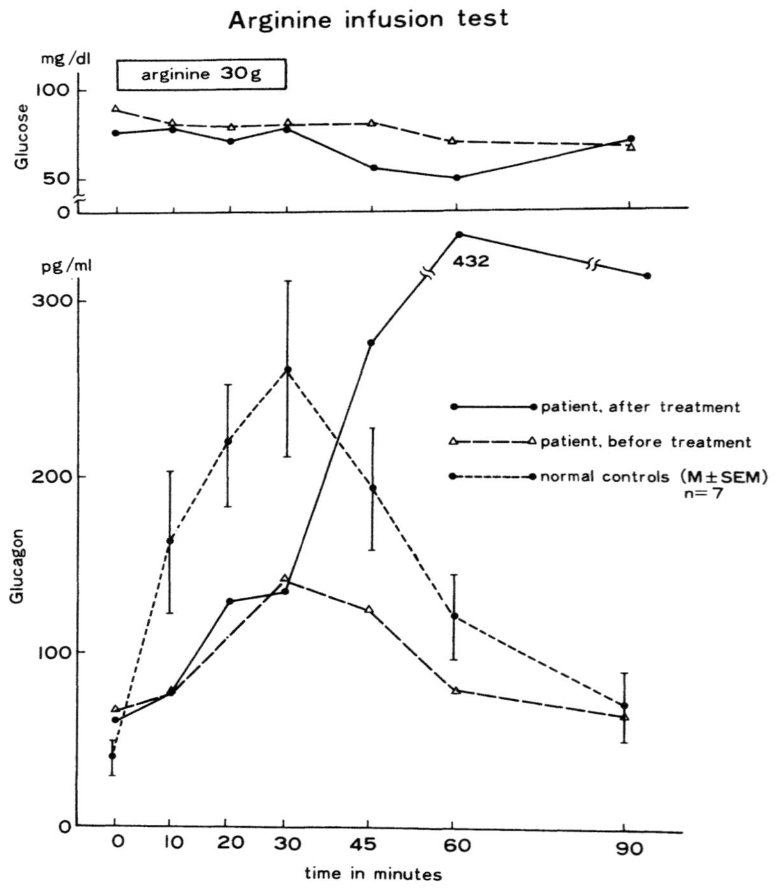
Arginine infusion test
arginine 30g
mg/dl
Glucose
100
50
0
pg/ml
Glucagon
300
200
100
0
432
patient, after treatment
patient, before treatment
normal controls (M±SEM)
n=7
0
10
20
30
45
60
90
time in minutes

Fig. 2

after treatment suggest that this steroid plays a role in the regulation of A cell function. Indeed, two recent reports (4, 8) indicate increased glucagon secretion following excessive glucocorticoid administration in normal man. In one of these studies, an exaggerated response to arginine infusion was also noted.

REFERENCES

1. Alford, F. P., Bloom, S. R., Nabarro, J. D. N., Hall, R., Besser, G. M., Coy, D. H., Kastin, A. J. and Schally, A. V. (1974): Glucagon control of fasting glucose in man. Lancet 2, 974-976.

2. Gerich, J. E., Schneider, V., Dippe, S. E., Langlois, M., Noaco, C., Karam, J. H. and Forsham, P. H. (1974): Characterization of the glucagon response to hypoglycemia in man. J. Clin. Endocrinol. Metab. 38, 77-82.

3. Lawrence, A. M. (1972): Pancreatic alpha-cell function in miscellaneous clinical disorders. In: Glucagon: Molecular Physiology, Clinical and Therapeutic Implications. P. J. Lefèbvre and R. H. Unger, Eds. Pergamon Press, Oxford and New York, pp. 259-274.

4. Marco, J., Calle, C., Roman, D., Diaz-Fierros, M., Villaneuva, M. L. and Valverde, I. (1973): Hyperglucagonism induced by glucocorticoid treatment in man. N. Engl. J. Med. 288, 128-131.

5. Nonaka, K. and Foà, P. P. (1969): A simplified glucagon immunoassay and its use in a study of incubated pancreatic islets. Proc. Soc. Exp. Biol. Med. 130, 330-336.

6. Unger, R. H. and Orci, L. (1975): The essential role of glucagon in the pathogenesis of diabetes mellitus. Lancet 1, 14-16.

7. Weir, G. C., Turner, R. C. and Martin, D. B. (1973): Glucagon radioimmunoassay using antiserum 30K: interference by plasma. Horm. Metab. Res. 5, 241-244.

8. Wise, J. K., Hendler, R. and Felig, P. (1973): Influence of glucocorticoids on glucagon secretion and plasma amino acid concentrations in man. J. Clin. Invest. 52, 2774-2782.

9. Yoshida, T., Toyoshima, H., Nonaka, K. and Tarui, T. (1975): Application of a correction factor to the radioimmunoassay for plasma pancreatic glucagon. Japan Diab. Soc. 18, 156-162.

SUBJECT INDEX

Numbers in italics refer to figures and tables in text.

A

G

I

GPSR Compliance

The European Union's (EU) General Product Safety Regulation (GPSR) is a set of rules that requires consumer products to be safe and our obligations to ensure this.

If you have any concerns about our products, you can contact us on ProductSafety@springernature.com

In case Publisher is established outside the EU, the EU authorized representative is:

Springer Nature Customer Service Center GmbH
Europaplatz 3
69115 Heidelberg, Germany

Batch number: 10692442

Printed by Printforce, the Netherlands